Intelligent Biomaterials

Zhe Liu
Editor

Intelligent Biomaterials

Fundamentals, Principles and Applications

Springer

Editor
Zhe Liu
Medical School
Tianjin University
Tianjin, China

ISBN 978-981-96-8645-2 ISBN 978-981-96-8646-9 (eBook)
https://doi.org/10.1007/978-981-96-8646-9

This Springer imprint is published by the registered company Springer Nature Singapore Pte Ltd.
The registered company address is: 152 Beach Road, #21-01/04 Gateway East, Singapore 189721, Singapore

Two Gateways to validate truth:
time and practice.

—Zhe Liu

Preface

Biomaterials have been playing an indispensable role in our daily life from everyday consumables to medical bandages, glues, gels, stents, and implants. With the emerging development of artificial intelligence (AI), telemedicine, mega-data analysis as well as their close incorporation with intelligent devices and AI-based algorithms, biomaterials nowadays have been demonstrating features of intelligence, and thus they are named as intelligent biomaterials.

Different from traditional biomaterials, intelligent biomaterials show apparent characteristics as follows:

1. Smartness or responsiveness. M. Aizawa, a Japanese scientist in Tokyo Institute of Technology, firstly proposed the term of intelligent biomaterials in 1993 describing the nature of targeting ligands specifically binding receptors or biomolecules on the cell membrane for molecular recognition, signal transduction, and consequent bio-effects. Since then, the term of smart biomaterials refers to their capability to passively or actively respond to ambient circumstances including physical, chemical, or pathological changes and give corresponding feedbacks. Presently, more and more smart biomaterials have been reported and widely applied to tremendous biomedical applications due to the feature of smartness or responsiveness.
2. Bioinformation-based health analysis and management. Intelligent biomaterials are capable of capturing bio-signals via bio-sensing, utilizing bioinformatics for data analysis, diagnosis and health management, and outputting interfering feedbacks for therapeutic treatments. Among those processes, intelligent algorithms, analysis, and decision play a vitally important role in conducting and modulating these manipulations.
3. Human-machine interactions and communications. With machine learning and deep learning techniques, intelligent biomaterials or built-in devices are endowed to simulate human brain working mechanism, so that they will display higher computation efficiency and better judgment capability, which ultimately reduces labors of clinical physicians but enhances the treatment precision with optimized outcomes.

In order to elucidate the above-mentioned features of intelligent biomaterials and systemize latest advancements, we composed this book with our colleagues and organized the contents from fundamentals to applications. Chapter 1 describes the concept of intelligent biomaterials and discusses the comparison between traditional and intelligent biomaterials. Chapters 2, 3, 4, 5, 6, and 7 depict various types of biomaterials, and mainly focus on their nature of intelligence. Chapters 8, 9, 10, and 11 introduce the cutting-edge biomedical applications of intelligent biomaterials in the fields of bio-sensing, in vitro diagnosis (IVD), microfluids, and micro/nanorobots. Chapter 12 provides a closing conclusion on current challenges and future perspectives of intelligent biomaterials.

This contribution was cordially supported by the National Natural Science Foundation of China (No. 82072057; 82311540023), and we are also grateful for all contributors' arduous efforts in composing the book chapters and sharing their valuable viewpoints and visions with readers in this area. In December of 2025, we will be honored to host a symposium titled *Intelligent Biomaterials* on the upcoming International Chemical Congress of Pacific Basin Societies 2025 (https://pacifichem.org/), and this may be the first professional symposium which is titled with the term of intelligent biomaterials. Along with this book, we look forward to sharing the ever-developing proceedings in this arena with broad readership worldwide, and kindly anticipate that this contribution will help us renew our understanding and rejuvenate our innovations in the frontier of intelligent biomaterials in the coming future.

Tianjin, China Zhe Liu

Contents

About the Editor

Zhe Liu[1] is Vice-Dean of Department of Medical Sciences and Engineering; Director of Division of Intelligent Biomaterials; and Professor of the Academy of Medical Engineering and Translational Medicine, Medical School, Tianjin University, China. He acquired his PhD from the Institute of Chemistry, Chinese Academy of Sciences in 2007, and then joined the Molecular Imaging Program at Stanford (MIPS, Stanford University, USA) and Helmholtz Institute of Biomedical Engineering (RWTH Aachen, Germany) as a postdoctoral scholar. His current research interest is the development of intelligent biomaterials for drug delivery, visualized medicine, and cutting-edge disease treatment strategies. He is the Fellow of the Royal Society of Chemistry (FRSC), the Fellow of Royal Society of Biology (FRSB) and the Fellow of Royal Society of Medicine (FRSM), and holds the (senior) membership of Radiology Society of North America (RSNA), American Society of Chemistry (ACS), Chinese Chemical Society (CCS), Chinese Anti-Cancer Association (CACA), Chinese Medical Doctor Association (CMDA), and Chinese Association of Ultrasound in Medicine and Engineering (CAUME).

[1] Medical School, Tianjin University, Tianjin, China

Chapter 1
Introduction of Intelligent Biomaterials

Zhe Liu and Yu Shi

Abstract Traditional biomaterials interact with living tissues and organs, and perform specific functions in a physiological environment with specific physical and biochemical properties. They have a wide scope of applications in areas such as tissue engineering, drug delivery, and disease treatments. In contrast, intelligent biomaterials are a kind of advanced biomaterials with self-adjustment and auto-feedback capability that communicate with nearby circumstances. Generally, intelligent biomaterials combine functions of biosensing, feedback responses, information analysis, and mega-data processing as well as artificial intelligence computation. By virtue of the characteristics of bioinformation management, two-way interactions, and integration of biomaterials-device-information, they have shown broad biomedical applications in health management, sensing and imaging, theranostics, and medical assistances. The continuing development of intelligent biomaterials offers sustainable solutions with cutting-edge medical paradigms, and possesses a promising prospect in future healthcare, precision medicine, and upcoming era of intelligent medicine.

Keywords Intelligent biomaterials · Mega-data analysis · Biosensing · Artificial intelligence · Intelligent medicine

---

Z. Liu (✉)
Division of Intelligent Biomaterials, Department of Medical Sciences and Engineering, Medical School, Tianjin University, Tianjin, China

Academy of Medical Engineering and Translational Medicine, Tianjin University, Tianjin, China

Tianjin Key Laboratory of Brain Science and Neural Engineering, Tianjin University, Tianjin, China
e-mail: zheliu@tju.edu.cn

Y. Shi
Academy of Medical Engineering and Translational Medicine, Tianjin University, Tianjin, China

Tianjin Key Laboratory of Brain Science and Neural Engineering, Tianjin University, Tianjin, China

Z. Liu (ed.), *Intelligent Biomaterials*, https://doi.org/10.1007/978-981-96-8646-9_1

1.1 Biomaterials

1.1.1 History and Definition of Biomaterials

Biomaterials are defined as a class of functional materials which interact with biological systems (tissues or organs) for diagnostic or therapeutic purposes (Park and Lakes 2007). Representative examples include therapeutic agents, imaging probes, hydrogels, catheters, and sutures that originate from natural organisms such as bone and skin, or synthesized substances including metals, ceramics, polymers, and composites with satisfactory biocompatibility.

The development and utilization of biomaterials have a long history (Fig. 1.1) (Ratner and Zhang 2020; Ratner 1996). Since ancient times, nacre teeth, wrought iron teeth, and gold sutures were applied in human bodies. In the eighteenth century, biomaterials were widely tested for surgery. Till 1950, the increasing understanding of biocompatibility and sterilization promoted the development of various sorts of biomaterials (Hench and Thompson 2010). In modern times, molecular biology in the 1970s together with the advances of genomics and proteomics in the 1990s exerted a significant influence on the design of biomaterials, and hence advanced biomaterials with bioactive components and specific properties put an emphasis on biological responses via molecular pathways (Huebsch and Mooney 2009). Nowadays, biomaterials involving material sciences, medicine, chemistry, and biology experience the flourished development with a history of around 60 years, which have attracted considerable attention by companies and research institutes for new commercial products. Dr. Larry Hench, who discovered bioglasses in 1969, initiated a new field of man-made biomaterials interacting with live tissues and proposed new-generation biomaterials. The first generation of biomaterials were exploited as prostheses specially used inside the human body during the 1960s, most of which were inert single-phase materials for commercial uses. The second generation of biomaterials were emphasized on bio-activity to live tissues and biodegradation during 1980-2010, and the third generation of biomaterials are cell- and gene-activating biomaterials from 2010 till now (Hench and Thompson 2010). Initial biomaterials were defined as "a systemically and pharmacologically inert substance designed for implantation within or incorporation with living systems" by the Clemson University Advisory Board (Black 1982), so the main feature of

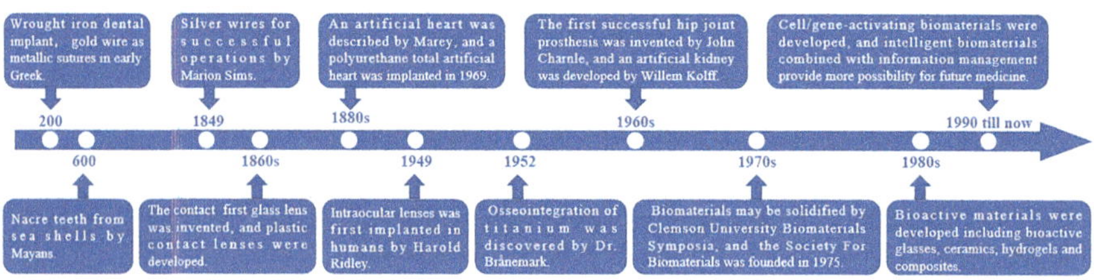

Fig. 1.1 A brief history of biomaterial developments

biomaterials was their inertness to immunological rejection as much as possible. The development of bioglasses with silicate-rich composition and hydroxyapatite as successful bioactive materials for bone grafting is of epoch-making significance in the history of biomaterials (Montazerian and Zanotto 2017; Rea and Bonfield 2004). These bioactive biomaterials elicited physiological reactions by interaction with live tissues and cells including crystallization of hydroxyl carbonate apatite, adsorption of biological moieties followed by cell attachment, differentiation, and proliferation (Hench 1998). Additionally, the conception of resorbable biomaterials was also proposed to manipulate the biomaterial-tissue interfaces (Hench 1980). These biomaterials with good biodegradability have been replaced for tissue regeneration. For example, copolymer of poly (lactic) and poly (glycolic) (PGA) acids as sutures were applied in clinics in 1984 (Hastings and Ducheyne 1984). Limited to the lifetime of artificial tissues with increasingly expensive costs, bioactive biomaterials with biochemical stimuli were explored as advanced twenty-first-century biomaterials. Based on the bioactive materials with biodegradability, chemical modifications render biomaterials with more possibility to stimulate specific cellular responses and direct cellular behaviors at the molecular level (Hench and Polak 2002).

In practice, biomaterials perform, augment, and even replace functions of organs and tissues. Given the target of biomaterials to the improvement of human health by interaction with live tissues, it is necessary to understand their fundamental properties of biomaterials applied in living systems.

(i) Non-biotoxicity

Biotoxicity contains carcinogenicity, pyrogenicity, allergenicity, and inflammatory (Thomas et al. 2015). Even though destructive drug delivery systems, such as cancerous nanoplatforms, they are also required of targeting capability to specific cells with less systematically side effects. Similarly, desirable biocompatibility means appropriate host responses facilitating with beneficial inflammatory and healing responses rather than obviously foreign body reactions, while resorbable biomaterials more often from natural products are preferable to other biomaterials with less risks of harmful effect and improved biocompatibility (Song et al. 2018).

(ii) Customized and Tunable Physical Performance

Some implanted biomaterials should possess equivalent physical properties to the replaced organs/tissues. These physical properties include mechanical performances, durability, and other surface and bulk properties according to different substitutes (Raghavendra et al. 2015). Specific surface properties for interaction with blood/tissues and bulk properties for special functions are also necessary for general biomaterials to meet diverse requirements in different scenarios, such as electricity, photothermal property, surface energy, surface textures, and so on.

1.1.2 Classification of Biomaterials

All kinds of biomaterials are often engineered models, films, foams, and coatings for biomedical use, and derive from a variety of sources including metals, ceramics, polymers, and their composites (Raghavendra et al. 2015). These biomaterials are constructed to obtain suitable combination of physical, chemical, and biological properties. They interact with living tissues and cells, and even perform within physiological media with unique customized properties (Kara 2012).

(i) Metals

Some metals and metal alloys feature superior mechanical properties, desirable corrosion resistance, and excellent biocompatibility, among which titanium and alloys are examples for medical uses (Niinomi 2002). With high tensile strength and fatigue limit, they have been harnessed as dental implants, artificial hearts, bone fixation devices, and others (Kara 2012). Especially, the metal-based biomaterials release metal ions in some physiological environment or biologic medium to interact with cell or tissues for disease treatments and tissue regeneration. Thus, assorted metal-based systems have been developed as disease theranostic platforms.

(ii) Ceramics

Ceramics are majorly exerted as hard tissue substitutes given inertness, deformability, and water resistance in the body. Remarkably, bio-ceramics of phosphates similar to bone mineral components are favorable bone substitutes owing to the high biocompatibility and bone integration, such as dental implants, bone defects, orbital and middle ear implants, cardiac valves, and coatings for fixation in bone (Vallet-Regí 2001).

(iii) Polymers

With the advantages of simple production processes, tunable properties, and low cost, polymers are applied as diverse surgical materials and implants of soft tissues. They are divided into synthetic polymers, natural polymers, and their combination copolymers. The common polymers contain acrylics, polyesters, polyethylene, chitosan, alginate, and so on, which are widely applied in contact and intraocular lenses, tissue adhesives, etc. (Anselme 2000; Mehdizadeh and Yang 2013).

(iv) Composite Biomaterials

Composites are comprised of two or more materials with complementary strengths of different materials. These constituent materials more often possess similar linear expansion constants. Composite scaffolds can be exploited as hard tissue replacements. For instance, bioglass/polymers are used for cancellous bone scaffolds, and fiber-reinforced composite biomaterials serve for skull reconstruction and bone fracture repairing (Dorozhkin 2011).

1.1.3 Development of Biomaterials

The worldwide demand for improved life quality is underscored by the imperative to the advancement of biomaterials, involving joint and limb scaffolds, artificial arteries, skin substitutes, dentures, and contact lenses (Fenton et al. 2018). In past decades, there has been notable progress in the development of biomaterials for diverse implants, medical devices as the replacement of irreversibly damaged tissues, and other medical uses (Hench and Polak 2002; Holzapfel et al. 2013). As mentioned above, biomaterials have experienced the evolution from inert materials such as stainless steel, cobalt-chromium alloys, and some polymers such as poly methyl methacrylate (PMMA) and polyethylene without tissue responses, to bioactive materials such as titanium alloys and bioceramics that interact with host tissues, to cell- and gene-activating biomaterials or responsive biomaterials with porous coatings or 3D structures supporting cell and tissue growth. Currently, there is a strong interest in the R&D of biomaterials for medical and healthcare applications. This surge is propelled by the necessity for drug delivery, tissue engineering, medical devices as well as both permanent and temporary implants. With aged populations and heightened expectations for better life quality, there is an increasing demand for precise and personalized treatments.

The focus of research is to enhance traditional biomaterial uses while also exploring novel bioactive materials for advanced purposes such as regenerative medicine and therapeutic biomaterials (Chelu and Musuc 2023; Insuasti-Cruz et al. 2022). These biomaterials can be categorized into four primary groups: proteins-based materials, polysaccharides-based materials, glycosaminoglycans-based materials, and extracellular matrix (Chelu and Musuc 2023). Protein-based materials include collagen derived from skin, and cartilage for hemostatic dressings (Yang et al. 2022c), and silk from silkworm cocoons employed as hydrogels in tissue engineering (Johari et al. 2022). Polysaccharides involve cellulose from plants, algae, bacteria used for wound dressings and drug delivery (Abdelhamid and Mathew 2022), chitin/chitosan obtained from the shells or bones of marine and non-marine organisms for drug delivery systems or hydrogels (Baharlouei and Rahman 2022), gum arabic served as a nanoscale scaffold (Alghuthaymi 2022), and alginates applied for controlled drug release (Vijian et al. 2022). Glycosaminoglycans cover hyaluronic acid originated from connective tissue for drug delivery (Kim et al. 2019), and heparin extracted from animal tissues like pig intestines utilized for neurodegenerative diseases, antimicrobial agents, and pancreatitis (Lima et al. 2017). Lastly, the extracellular matrix is harnessed to regulate angiogenesis and tissue engineering (Wu et al. 2021). Synthetic materials include metals for medical implants, ceramics and glass for bone fillers, tissue engineering scaffolds, and artificial polymers. Artificial polymers are classified into non-biodegradable polymers like polymethylmethacrylate, polyethylene, and polypropylene for eyeglass lenses and bone cement as well as biodegradable polymers such as polylactic acid (PLA), poly (ε-caprolactone) (PCL), polyurethane for drug delivery or scaffolding materials (Chelu and Musuc 2023).

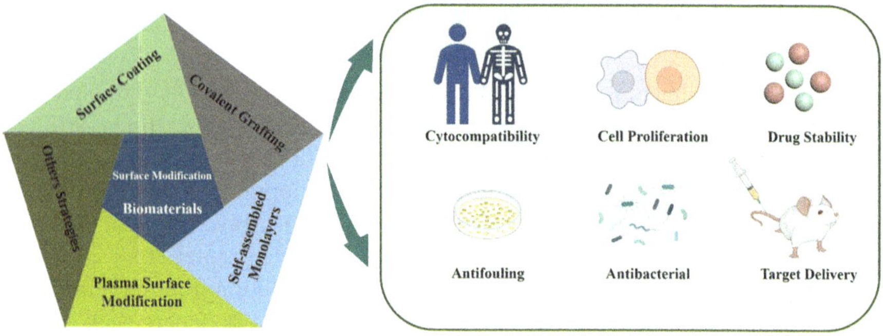

Fig. 1.2 Different surface modifications of biomaterials and their functions. (Reprinted with permission from Hu et al. 2023)

Focusing on the enhancement of the availability of biomaterial implants as well as potential risks such as immune rejection and infection, surface modification techniques can be utilized to alter the inherent physical, chemical, and biological properties of biomaterials to optimize their functionality (Hu et al. 2023). These techniques include coating, covalent grafting, self-assembled monolayers (SAMs), and plasma surface modification (Fig. 1.2). Coatings applied to biomaterials enhance properties like hemocompatibility, cellular behavior, and hydrophilicity, but they are unstable and toxic. Covalent grafting, using chemical interactions, improves stability in physiological environments and prevents contamination. However, traditional grafting methods are complex. SAMs are nanometer-thick coatings formed through the adsorption of molecules onto solid surfaces. SAMs improve biocompatibility by controlling cell adhesion without altering mechanical properties. They are flexible and have thermal and hydrolytic stability. However, their antimicrobial properties and long-term stability need further verification. Plasma surface modification changes surface properties without affecting bulk properties. It enhances biocompatibility and can be used for tissue engineering. Films created through plasma modification are thermally stable, chemically inert, and tough. Plasma modification improves surface wettability, cell adhesion, osteogenic properties, and antimicrobial properties. However, the choice of precursors can affect the desired properties of biomaterials. Furthermore, some researchers focus on developing biomaterials that possess inherent capabilities to enhance the natural healing and self-repairing abilities of the body by surface modification, rather than increasing the complexity to *ex vivo* tissue-engineered structures (Place et al. 2009) in that the surface properties such as texture, shape, and chemical composition, exert a significant influence on the attachment, migration, proliferation, and differentiation of cells. Advanced biomaterials with modifying three-dimensional properties (including porosity, surface area, and pore interconnectivity) or nature-inspired structure may enhance tissue growth, blood vessel formation, and nutrient delivery (Karageorgiou and Kaplan 2005; Hutmacher et al. 2007). The introduction of hierarchical modularity in terms of mechanical properties or degradation mechanisms also can fulfill customized function and biological requirements to improve interactions between tissues and biomaterials (Dormer et al. 2010).

As a typical application, biomaterials exert a critical role in enhancing the delivery and efficacy of various pharmaceutical compounds, including drugs, antibodies, vaccines, peptides, and enzymes, given that the therapeutic efficacy of agents is affected by their formulation and administration route (Tibbitt et al. 2016; Allen and Cullis 2004). Delivery methods impact various factors on therapeutic effectiveness, including pharmacokinetics, distribution, metabolism, clearance, toxicity, and even biological barrier penetration. Furthermore, environmental changes such as temperature, moisture, and pH levels in the body can result in loss of pharmacological activity for drugs either during storage or in vivo. As the biotechnology industry continues to advance novel biopharmaceuticals, there is a necessity to enhance our understanding of how drug delivery impacts safety and efficacy while also developing new delivery technologies. Biomaterials, particularly polymers and lipid-based substances utilized for drug delivery, are influenced by advancements in organic and synthetic chemistry, genetic engineering, materials sciences, and biotechnology (Langer 1990, 1998). They are engineered to release therapeutic drugs over extended periods and can be customized to target specific areas in the body to minimize drug dosage while mitigating harm to patients. The physicochemical properties of biomaterials and their targeted incisions of administration can be systematically adjusted to optimize therapeutic benefits. Furthermore, advancements in genetic engineering and biotechnology have led to the emergence of innovative nucleic acid, antibody, and protein therapies (Hubbell et al. 2009). These developments will require a new generation of high-performance biomaterials capable of providing therapeutic protection, specificity, and controlled release, such as implantable devices and multi-functional bio-chips. Notably, whether implanted devices or polymer particles and nanoparticles for drug delivery are inevitable to trigger strong or weak inflammatory responses at the injection site and in the target organ, contributing to immunogenicity issues and even tissue damage (Andorko and Jewell 2017). To address this challenge, endeavors have been directed toward enhancing biocompatibility through the development of novel polymers, surface chemistry modifications, and the incorporation of immunomodulators. Moreover, the body's natural defense mechanisms, encompassing both innate and adaptive immune responses, can be harnessed to bolster the eradication of potential threats to the host. By leveraging the body's own immune system against biomaterials and antigens, it is feasible to program the immune response for targeting and eliminating tumors and infections. This gives rise to two distinct approaches: utilizing biomaterials to minimize or suppress the immune response, and harnessing biomaterials to enhance the immune response against disease-causing agents (Hotaling et al. 2015).

Biomaterials is dedicated to advancing global health and well-being. In recent decades, our evolving biotechnology platforms have had a significant impact on the lives of numerous individuals worldwide by enabling therapeutic approaches that were previously considered unattainable. The advancement of biomaterials will involve progress in functional medical biomaterials and the increased utilization of biomaterials in new application areas. The utilization of natural mechanisms and the elaborated design of complex dynamic performance would render materials with more novelty properties that are currently unattainable (Chelu and Musuc 2023). Future research will focus on next-generation bioinspired "intelligent" multifunctional biomaterials for health monitoring and prevention of biological crises that are discussed in the next section. A critical rational phase in designing and synthesizing biomaterials will be to consider them as models or templates for multifunctional, active devices that imitate biological polymers and organisms, resulting in a reduction in costs associated with current chemical and drug manufacturing processes, thereby reducing economic and social impacts. This requires interdisciplinary efforts and collaborations from the structural design of biomaterials to bio-evaluations before clinical trials (Fenton et al. 2018).

1.2 Intelligent Biomaterials

Intelligent biomaterials are regarded as a class of advanced biomaterials that perceive external stimuli, respond accordingly, and output feedbacks (Liu et al. 2020). They possess the ability to set and achieve specific goals and tasks responding to different situations, and thus they can actively behave at both nano- and macro-levels under different stimuli or microenvironments such as light, sound, radiation, electricity, stress, and various physiological changes (Wang et al. 2020). Noticeably, different from traditional biomaterials, intelligent biomaterials integrate sensing, feedback response, information analysis and processing, and performance optimization together to adapt to varying situations. And compared to simple responsive biomaterials, intelligent biomaterials are able to store and memorize data, analyze external signals, transform into internal information, and adapt to optimize performance in addition to perception and response (Liu et al. 2020; Zhang et al. 2019; Amukarimi et al. 2021). Intelligent biomaterials are capable to serve synergistic diagnosis and treatment, self-repair, self-regulation, and other biomedical applications from biosensing, health monitoring to drug delivery, and human health management.

1.2.1 Characteristics of Intelligent Biomaterials

Mushrooming intelligent biomaterials have contributed to increasingly sustainable solutions on disease monitoring and treatment with advanced technology. They are adaptable to communications with surroundings, and continuously give out optimized responses. However, most traditional biomaterials are limited to the functionalities such as adaptability, self-optimization, internal analysis, and learning, even though smart biomaterials only give feedback after perceiving specific stimuli without any optimized responses (Walther 2020; Zhang et al. 2018; de León et al. 2023). Intelligent biomaterials serve as customized building blocks and sustainable

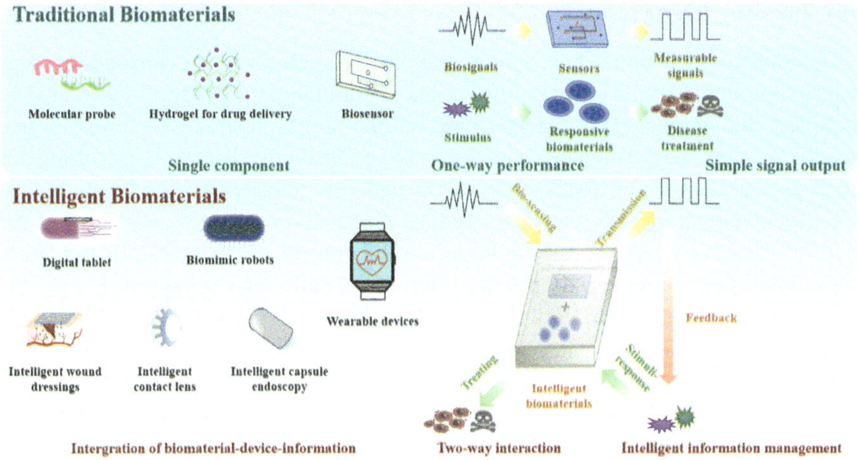

Fig. 1.3 Comparison between traditional biomaterials and intelligent biomaterials.

regenerative materials that can be triggered by user-defined or cell/tissue-dictated stimulus, and then tempo-spatially adjust their behaviors for personalized medicine. Generally, they possess the following three characteristics of intelligent information management, two-way interactions and integration of biomaterials, devices, and information (Fig. 1.3).

(i) *Integration of Biomaterials, Devices, and Information*

Different from traditional biomaterials, intelligent biomaterials possess built-in sensors, electronic components, and communication devices, which are capable of timely monitoring of the patient's physiological indicators (e.g., heart rate, body temperature, blood pressure, blood glucose, etc.) and uploading the monitored data to mobile phones or cloud servers via wireless communication such as Bluetooth (Wang et al. 2022; Xu et al. 2022). The intelligent biomaterials with functionalized composite structures consist of biosensors, control, and response modules, and are competent to long-term medical monitoring with less inferences (Lin et al. 2023). The durable materials are favorable for intelligent biomaterials such as ionic polymer membranes and carbon fibers as wearable biomaterials providing more flexible and convenient options for health management (Liu et al. 2016; Yu et al. 2022). Moreover, patients' situations also can be examined by doctors and paramedics by registration into clouds for remote guidance and management, which is especially necessary for patients with long-term chronic diseases such as diabetes, hypertension, heart disease, and Parkinson's disease (Voelker 2018; Li et al. 2023; Kwon and Dong 2022; Powers et al. 2021). Remote services (or telemedicine) supported by the biomaterial-device-information integration facilitate patient's healthcare and disease management with improved patients' life quality, availability of medical resources, and less occurrences of emergencies (Jin et al. 2020).

(ii) *Two-Way Interactions*

Compared with traditional biomaterials that perceive varying surrounding signals to output information or simply feedback, intelligent biomaterials can real-time monitor and process data analysis, and autonomously respond for programmed feedback and performances via internal control mechanisms, resulting in "perception-analysis-feedback-output" of two-way closed-loop interactions for intelligent adjustments as well as immediate diagnosis and therapy. By virtue of the ability to analysis and processing based on the control and response of surroundings, intelligent biomaterials provide innovative paths and opportunities for the development and application of digital drugs, intelligent tissue engineering materials, intelligent artificial organs, and non-invasive or minimally invasive treatment of tumors (Ouyang et al. 2023). For example, intelligent drug carriers can release a certain dosage of drugs according to patients' real-time situation to improve the effectiveness of treatment (Armenia et al. 2022). Intelligent wearable devices can implement precise treatment based on the feedback system to ensure that the therapeutic scheme meets the personalized needs of patients (Song et al. 2019).

(iii) *Intelligent Information Management*

Traditional biomaterials are often adopting universal design for specific populations of patients limited to personalized and real-time health management. Intelligent biomaterials with adaptive performance can perceive, analyze, and manage physiological variations and pathological situations of patients for monitoring and treatment of diseases (Hoare et al. 2022; Song et al. 2021). The information includes but is not limited to biological cells, tissues, or physiological parameters such as body temperature, blood pressure, pH, glucose, biomarkers, etc. (Sang et al. 2022; Min et al. 2023; Corsi et al. 2022). Real-time monitoring, recording, and feedback of patients' data not only improve accuracy in disease screening and early diagnosis, but also provide superior treatment modality and disease management for patients. In addition, intelligent biomaterials can regulate their morphology and function for personalized controlled drug release, targeted therapy and even generate biological effects through communications with exogenous or endogenous signals (Liang et al. 2020; Shodeinde et al. 2020). For example, intelligent biomaterials have been developed to release anti-inflammatory drugs or antibiotics in situ to accelerate wound repairing and healing when inflammation or infection occurs (Mostafalu et al. 2018). Resorted to intelligent information management, intelligent biomaterials render more precise treatments with greater therapeutic efficacy, less side effects, and better patient compliance, thus making a fantastic step forward to health and life quality of patients.

Above all, obviously different from the single structure and function or one-way perception and response of traditional biomaterials, intelligent biomaterials possess these characteristics including integrational structure (biomaterial-device-information), two-way interactions, and intelligent information management. The integrational components of biomaterials, devices, and information pave the foundation of intelligence capability allowing timely monitoring of patients' physiological parameters or situations, adaptive performance for more precise and personalized diagnosis and treatment. With perception-analysis-feedback-output two-way interactions, intelligent biomaterials not only sense environmental signals and changes, but also adapt to the environment and make corresponding functional adjustments and changes, i.e., presenting similar body tissue self-healing (such as self-healing polymers), environmental self-adaptation and self-repair (such as shape memory materials), self-decomposition (such as biodegradable polymers), promoting proliferation and differentiation (such as hydrogel scaffolds), responding to external stimuli (such as stimulus-responsive drug delivery systems). Furthermore, accompanied by bioinformation acquisition and information technology, intelligent biomaterials can remotely and timely monitor patients' physiological parameters or environmental conditions for personalized therapeutics.

Intelligent biomaterials involve material sciences, medical engineering, biology, electronic engineering, computer science, and other fields. Among them, material sciences focus on designing and preparation of intelligent biomaterials with high biocompatibility and availability; medical engineering is devoted to the designing and development of medical devices and instruments; electrical engineering is responsible for designing and integrating sensors and electronic components; and computer sciences and data analysis transform biological data into valuable medical information. This interdisciplinary collaboration fosters the development and innovations of intelligent biomaterials, and substantially advances the reformation of medicine and healthcare.

1.2.2 Classification of Intelligent Biomaterials

Intelligent biomaterials can be classified in different components as metal-based, inorganic non-metallic, and polymer-based materials, and they also can be assorted by varied functions as shape memory materials, stimulus-responsive materials, biomimetic materials, optoelectronic materials, piezoelectric materials, and micro- and nanorobotics. By different biomedical applications, intelligent biomaterials involve the fields of health monitoring, sensing and imaging, diagnostics and therapeutics, and medical assistances (Kowalski et al. 2018; Amukarimi and Mozafari 2021; Zheng et al. 2021a; Rajabi et al. 2015; Li et al. 2021). Over the last decades, the advent of intelligent biomaterials has renovated modern medicine including healthcare, diagnosis, drug delivery, and adjuvant medical devices. Especially with the introduction of molecular self-assembly, protein and peptide engineering, and microfabrication technologies, intelligent biomaterials with unique physicochemical properties, physiological functions, and intelligent characteristics have attracted much attentions of scientists and researchers (Kowalski et al. 2018).

(i) *Health Monitoring*

Wearable sensors can timely track physiological signs (like heart rate, blood pressure, and body temperature), analyze data, and transmit information to patients and their guardians, which gives access to monitoring of the progression of chronic diseases and early intervention. Typically, intelligently wearable textiles with embedded sensors and electronics can monitor physiological parameters and environmental information in real time by converting physical stimuli from physiological activities into electrical signals for personalized therapeutic support. Besides that, these intelligent textiles also analyze body fluid metabolites and electrolytes for disease diagnosis (Fang et al. 2021; Wicaksono et al. 2020).

(ii) *Sensing and Imaging*

Sensing-based intelligent biomaterials utilize specific biorecognition elements including enzymes, antibodies, or DNA probes to interact with target molecules. These interactions are then transformed into measurable signals by electrochemical, optical, or other means for the highly sensitive detection of various physiological and pathological parameters in living organisms such as the concentration of biomolecules (glucose, proteins, DNA, etc.). Followed by digital signal analysis and processing, early screening, diagnosis, and therapy of diseases can be achieved, which is crucial for the management of chronic diseases such as diabetes and cancers. The imaging-based intelligent biomaterials are often a class of implantable microdevices. The multifunctional capsule endoscope is a representative example. The capsule endoscopy equipped with a microcamera and other photo-/acoustical sensors is a non-invasive technique for examination of the gastrointestinal tract. The intelligent capsule acquires high-resolution images for 3D image reconstruction of suspected lesions in the digestive tract after swallowed, and then transmitted to an external terminal for analysis and diagnosis of disease (Iddan et al. 2000). Additionally, more advanced multifunctional capsule endoscopes have been developed that can be controlled by an external magnetic field to actively pass through the intestinal tract for real-time imaging, biopsies, or therapeutic operations (Zheng et al. 2022; Yim and Sitti 2011).

(iii) *Diagnostic and Therapeutic Intelligent Biomaterials*

Diagnostic and therapeutic intelligent biomaterials are a class of materials that can simultaneously perform early diagnosis of diseases and intervention therapy based on a feedback system. Among them, the diagnostic module is used to monitor and measure the physiological parameters, characteristic biomarkers, or disease indicators of living organisms (like blood glucose level, wound status, intraocular pressure, etc.), while the therapeutic module implements specific intervention or treatments based on the monitoring results (Luo et al. 2023; Guo et al. 2021; Yang et al. 2022b). It may be a drug delivery system that releases drugs based on monitored biological parameters to achieve targeted therapy (Xu et al. 2021; Kim et al. 2022). Alternatively, it also can be an electrical stimulation module capable of restoring the normal function of the organism through electrical stimulation (Yang et al. 2022a). For instance, intelligent wireless contact lenses can monitor blood glucose levels and release stored insulin to maintain blood glucose stability when necessary (Keum et al. 2020). The multifunctional microneedle patches not only monitor the inflammatory state of wounds, but also provide targeted electrical stimulation and release growth factors to accelerate wound healing (Jiang et al. 2023)

(iv) *Medical Assistance*

Medical-assisting intelligent biomaterials are commonly used for tissue repairing and regeneration with monitoring functions. They can release drugs or bioactive substances along with physical stimuli to accelerate tissue regeneration based on the monitored physiological parameters (Huang et al. 2023). For example, hydrogel microspheres for cartilage regeneration release pro-regenerative growth factors and visualize the repair process of the cartilage layer by fluorescence imaging (Liang et al. 2020). In addition, a biodegradable implantable piezoelectric nanogenerator can produce electrical stimulations driven by ultrasound to promote peripheral nerve tissue regeneration, and the nerve-repairing process is recorded by the electrophysiological response of the muscle, which opens up a new avenue for enhancing and monitoring tissue repairing (Wu et al. 2022).

1.3 Development and Trends of Intelligent Biomaterials

1.3.1 Development of Intelligent Biomaterials

Based on the research paradigm characterized by the main features of intelligent biomaterials and their clinical needs, two major development trends of "biology-biomaterial-environment" and "biology-biomaterial-information" have been presented. The former focuses on the visual-auditory-tactile feedbacks of intelligent biomaterials under sound, light, electric, magnetic, and other physical fields, and integrates the drug-device-field synergy strategy to realize efficient and precise theranostics of diseases. The latter lays stress on the high-precision, multi-channel bio-information acquisition and transmission of intelligent biomaterials, and achieves the synergistic management of chronic diseases. In general, the development of intelligent medical materials presents the following features:

(i) *Personalized and Multifunctional Development*

Intelligent biomaterials equipped with advanced devices and technologies carry out precise treatment on lesions under the stimulation, regulation, or intervention of external physical fields. Taking intelligently stimulus-responsive drug delivery systems as an example, their application greatly improves the treatment efficiency and reduces medication side effects, which ultimately enhances patients' life quality. A lipid nanocarrier composed of Brazilian carnauba wax and loaded with magnetic iron oxide nanoparticles, the anti-tumor drug doxorubicin (DOX), and a cell membrane fluorescent dye (DIO), can be utilized for magnetic field-mediated magnetothermal therapy against melanoma (Fig. 1.4) (García-Hevia et al. 2022). The applied alternating magnetic field was used to remotely control the DOX release at the

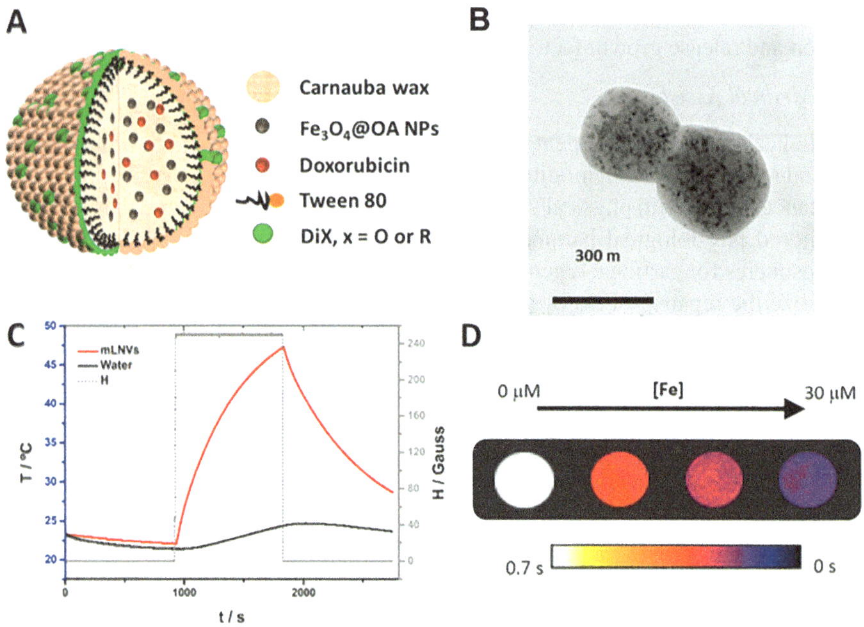

Fig. 1.4 Magnetic lipid nanocarriers for applied magnetic field-mediated chemo- and magnetothermal synergistic therapy. (Reprinted with permission from García-Hevia et al. 2022)

tumor lesion, and induce thermoablative therapy in situ. The results showed that the combination of chemo- and magnetothermal therapy enhanced its anti-tumor efficacy with reduced side effects of chemotherapy, while the loaded magnetic iron oxide nanoparticles allowed for non-invasive monitoring of melanoma treatments through magnetic resonance imaging.

(ii) *Pluralistically Application Scenarios*

Nowadays, intelligent biomaterials are not limited to single treatment modality, but rather are expanded for disease screening, health management, and personalized medicine. The novel combination of drugs and devices provides real-time and comprehensive monitoring of patients' situation including physiological and biochemical information, while the progression of diseases may be analyzed by algorithms providing early-stage warnings. The diagram of drug-device combination prompts personalized management and therapy of disease, especially for elderly or chronic illness. For instance, considering that diabetic patients must strictly control their blood glucose levels at all times, an intelligent contact lens embedded with biosensors was exerted as a real-time monitor of tear glucose levels. The tear glucose content could be converted into blood glucose to calculate and deliver appropriate doses of hypoglycemic medication by avoiding injected pain or invasive drug injection (Fig. 1.5) (Keum et al. 2020).

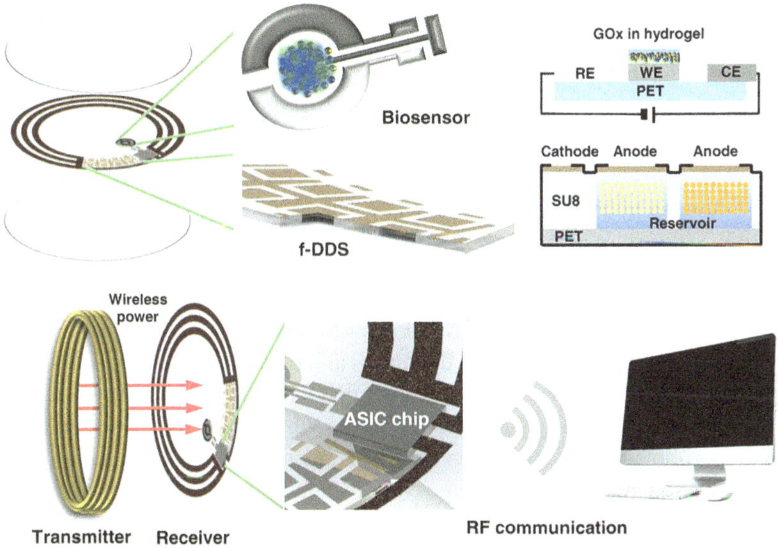

Fig. 1.5 An intelligent contact lenses for diagnosis and treatment of diabetes. (Reused with permission from Keum et al. 2020)

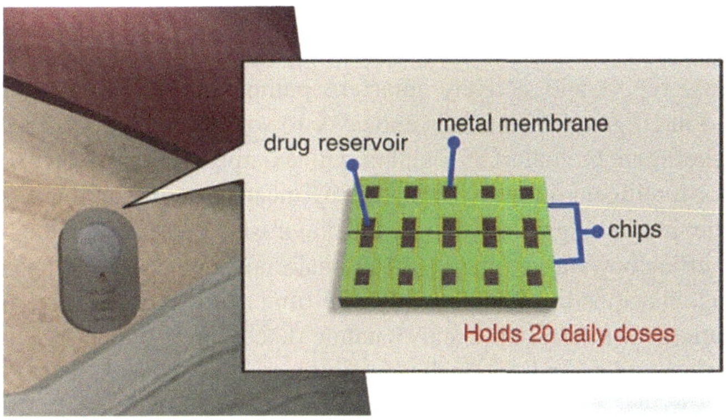

Fig. 1.6 An implantable microchip for osteoporosis in the elderly (https://www.implantable-device.com/2012/01/22/microchips-implantable-drug-delivery-device-for-the-treatment-of-osteoporosis/)

(iii) *Renovation in Both Personalized and Public Medicine*

The development, translation, and clinical applications of intelligent biomaterials provide new strategies and pathways for the whole chain of healthcare, prevention, diagnosis, treatment, and rehabilitation of human diseases, and will promote revolution in human lifestyles, public health, and modern medicine (Zhao et al. 2019). In 2015, MicroCHIPS Biotech, Inc. developed an implantable microchip for the treatment of osteoporosis, a common disease in the elderly (Fig. 1.6) (https://www.implantable-device.com/2012/01/22/microchips-implantable-drug-delivery-device-for-the-treatment-of-osteoporosis/). The microchip with independently sealed drug reservoir is capable to release the predetermined dose of the drug at the defined time via wireless electric current after subcutaneous implantation in the abdomens of patients. This microchip allows for long-term implantation without disrupting the patient's daily life or activities avoiding pain and inconvenience of long-term and repeated injections, which provides a new solution for personalized treatment of chronic diseases in the elderly, particularly those who suffer from osteoporosis. In addition, nanorobots have made significant contributions to timely prediction, effective protection, and accurate diagnosis of coronavirus (COVID-19) (Khamis et al. 2021). Maaza and his team prepared composite gold-platinum-silver (Au-Pt-Ag) nanoparticles coated with hydroxychloroquine for the prevention and treatment of COVID-19. They verified the ability to efficient drug loading and COVID-19 targeting via molecular dynamics and quantum mechanics simulations, and the nanoparticles showed good anti-COVID-19 activity (Morad et al. 2021). The researchers therefore developed erythrocyte membrane magnesium-based bionic nanorobots with active antigen delivery as nano-vaccines, which have efficiently activated the human immune system and established an effective nano-vaccine barrier for viral infestation and disease prevention (Wei et al. 2019).

1.3.2 Trends of Intelligent Biomaterials

In the future, roaring kinds of intelligent biomaterials will be developed to monitor health status, report and actively interfere pathological conditions of targeted regions, and intelligent biomaterials endeavor to solve more clinical problems. On one hand, intelligent biomaterials together with existing clinical medical devices are exploited for intelligent molecular imaging by means of artificial intelligence techniques, achieving intelligent prediction and diagnosis of disease. On the other hand, synergistic efficiency among intelligent biomaterials, devices, and bio-information performs signal acquisition, intelligently real-time analysis and output of biomedical information for microscopic and dynamic detection of living organisms, more comprehensive understanding of the occurrence, development, and evolution of diseases.

The integration of drug-device-physical fields (magnetic field, electric field, acoustic field, light field, radiation, microwave, etc.) can intervene in the process of disease occurrence and development, and achieve the purpose of effective treatment. As a new strategy and paradigm, the combination of drugs and devices, under the action of external physical fields, will provide a new type of intervention means for disease diagnosis and treatment, and develop minimally invasive or non-invasive treatment options with high efficiency and minimized side effects so as to overcome the shortcomings of the existing diagnosis and treatment such as large trauma area and long recovery period (Shupak et al. 2003; Bassett 1989). Liu's group utilized ultrasound field as a medium to study the non-invasive/minimally invasive treatment of various tumors by combining ultrasound with micro- and nanomaterials, which not only provided an effective means for the development of oncological diagnosis and treatment strategies, but also offered a new way of thinking for exploring the occurrence, development and intervention of major human diseases (Chen et al. 2017). They constructed hard-shell gas-carrying microbubbles and nanoparticles made of poly butyl cyanoacrylate (PBCA), encapsulating the chemotherapeutic drug doxorubicin (DOX) and crosslinked by genipin, which was used as multifunctional ultrasound molecular probes and drug delivery carriers for the treatment of superficial malignant tumors (SMTs). Compared with common drug loading methods, these size-adjustable poly ester micro- and nano-drug carriers increased the encapsulation and loading rate of drugs in vivo. Due to the enhanced retention effect (EPR) at the tumor site, the drug molecules were directed and controlled to be released in the tumor tissues under external ultrasound, which dramatically improved the concentration and bioavailability of the drugs, and verified the feasibility and validity of the combination of ultrasound field with the micro/nano-sized drug carriers for the controlled delivery in the tumor incisions (Zhang et al. 2021).

Synergistic efficiency among intelligent biomaterials, devices, and bio-information technology initiates new tendency of digital healthcare in disease management. Digital healthcare is a class of technology that collects, transmits, and analyzes medical data to monitor and manage the physiological and pathological status of patients. Based on gene editing, biosensing, cloud computation, artificial intelligence, and other information technology, it is expected to promote the close integration of intelligent medical materials and bioinformation technology for prevention and control of diseases. The US Food and Drug Administration (FDA) approved the world's first digital drug (Abilify MyCite) which was developed by Japan's Otsuka Pharmaceutical Company for clinical uses in 2017. With inserted biosensors, an intelligent patch, an external device application, and a web-based data panel were integrated for the treatment of schizophrenia, bipolar disorder, and other psychiatric disorders to improve medication adherence (https://www.Abilifymycite.com/). The chips are embedded with copper- and magnesium-coated sensors that emit a signal when they dissolve in the stomach, allowing real-time recording of the time of drug intake and the state of digestion. A smart patch attached to the patient's skin records information from the sensors along with the patient's heartbeat, breathing, temperature, activity, and other physical parameters. These data can be transmitted wireless via Bluetooth to mobile terminals such as smartphones and tablet PCs, which can help physicians monitor, compute, and analyze the patients' medication so as to guide the patients to adjust the dosage and time for the next medication. Zheng et al (Zheng et al. 2021b) developed an intelligent hydrogel epidermal patch constructed by flexible two-color polyacrylamide and chitosan, which were used to sterilize, hemostasis, and promote healing of injured wounds while its loaded wound indicator (carbon quantum dots and phenol red) monitored the dynamic pH, reactive oxygen species and other indicators at the wounded site, and plotted a curve of each indicator via artificial intelligence by a smartphone. The study showed that the smart epidermal patch had good antimicrobial activity, effectively monitored wound healing, shortened the wound recovery cycle, and presented great market potentials for clinical applications. The integration of biotechnology and information technology has revolutionized the diagnosis and treatment of traditional diseases. Through precise targeted delivery and in situ release of drugs, personalized management of patients' diseases has been achieved, and real-time health monitoring also facilitated the transformation from disease treatment to disease prevention strategy, and fundamentally realized early-stage prevention, detection, treatment, and recovery.

References

Abdelhamid HN, Mathew AP (2022) Cellulose-based nanomaterials advance biomedicine: a review. Int J Mol Sci 23(10):5405

Alghuthaymi MA (2022) Antibacterial action of insect chitosan/gum Arabic nanocomposites encapsulating eugenol and selenium nanoparticles. J King Saud Univ Sci 34(7):102219

Allen TM, Cullis PR (2004) Drug delivery systems: entering the mainstream. Science 303(5665):1818–1822

Amukarimi S, Mozafari M (2021) 4D bioprinting of tissues and organs. Bioprinting 23:e00161

Amukarimi S, Ramakrishna S, Mozafari M (2021) Smart biomaterials—A proposed definition and overview of the field. Curr Opin Biomed Eng 19:100311

Andorko J, Jewell C (2017) Designing biomaterials with immunomodulatory properties for tissue engineering and regenerative medicine. Bioeng Transl Med 2(2):139–155

Anselme K (2000) Osteoblast adhesion on biomaterials. Biomaterials 21(7):667–681

Armenia I, Ayllón CC, Herrero BT, Bussolari F, Alfranca G, Grazú V, de la Fuente JM (2022) Photonic and magnetic materials for on-demand local drug delivery. Adv Drug Deliv Rev 191:114584

Baharlouei P, Rahman A (2022) Chitin and chitosan: prospective biomedical applications in drug delivery, cancer treatment, and wound healing. Marine Drugs 20(7):460

Bassett C (1989) Fundamental and practical aspects of therapeutic uses of pulsed electromagnetic fields (PEMFs). Crit Rev Biomed Eng 17(5):451–529

Black J (1982) The education of the biomaterialist: report of a survey, 1980–81. J Biomed Mater Res 16(2):159–167

Chelu M, Musuc AM (2023) Advanced biomedical applications of multifunctional natural and synthetic biomaterials. Processes [Online]:2696

Chen W, Yang Y, Shangguan D, Wu Y, Liu Z (2017) Multifunctional hard-shelled microbubbles for differentiating imaging, cavitation and drug release by ultrasound. RSC Adv 7(42):25892–25896

Corsi M, Paghi A, Mariani S, Golinelli G, Debrassi A, Egri G, Leo G, Vandini E, Vilella A, Dähne L (2022) Bioresorbable nanostructured chemical sensor for monitoring of pH level in vivo. Adv Sci 9(22):2202062

de León EH-P, Valle-Pérez AU, Khan ZN, Hauser CA (2023) Intelligent and smart biomaterials for sustainable 3D printing applications. Curr Opin Biomed Eng 26:100450

Dormer NH, Berkland CJ, Detamore MS (2010) Emerging techniques in stratified designs and continuous gradients for tissue engineering of interfaces. Ann Biomed Eng 38:2121–2141

Dorozhkin SV (2011) Biocomposites and hybrid biomaterials based on calcium orthophosphates. Biomatter 1(1):3–56

Fang Y, Chen G, Bick M, Chen J (2021) Smart textiles for personalized thermoregulation. Chem Soc Rev 50(17):9357–9374

Fenton OS, Olafson KN, Pillai PS, Mitchell MJ, Langer R (2018) Advances in Biomaterials for Drug Delivery. Adv Mater 30(29):1705328

García-Hevia L, Casafont Í, Oliveira J, Terán N, Fanarraga ML, Gallo J, Bañobre-López M (2022) Magnetic lipid nanovehicles synergize the controlled thermal release of chemotherapeutics with magnetic ablation while enabling non-invasive monitoring by MRI for melanoma theranostics. Bioact Mater 8:153–164

Guo M, Wang Y, Gao B, He B (2021) Shark tooth-inspired microneedle dressing for intelligent wound management. ACS Nano 15(9):15316–15327
Hastings, G. W.; Ducheyne, P., Macromolecular biomaterials. 1984.
Hench LL (1980) Biomaterials. Science 208(4446):826–831
Hench LL (1998) Bioceramics. J Am Ceram Soc 81(7):1705–1728
Hench LL, Polak JM (2002) Third-generation biomedical materials. Science 295(5557):1014–1017
Hench LL, Thompson I (2010) Twenty-first century challenges for biomaterials. J R Soc Interface 7(suppl_4):S379–S391
Hoare D, Tsiamis A, Marland JR, Czyzewski J, Kirimi MT, Holsgrove M, Russell E, Neale SL, Mirzai N, Mitra S (2022) Predicting cardiovascular stent complications using self-reporting biosensors for noninvasive detection of disease. Adv Sci 9(15):2105285
Holzapfel BM, Reichert JC, Schantz J-T, Gbureck U, Rackwitz L, Nöth U, Jakob F, Rudert M, Groll J, Hutmacher DW (2013) How smart do biomaterials need to be? A translational science and clinical point of view. Adv Drug Deliv Rev 65(4):581–603
Hotaling NA, Tang L, Irvine DJ, Babensee JE (2015) Biomaterial strategies for immunomodulation. Annu Rev Biomed Eng 17:317–349
Hu X, Wang T, Li F, Mao X (2023) Surface modifications of biomaterials in different applied fields. RSC Adv 13(30):20495–20511
Huang Y, Zhang L, Ji Y, Deng H, Long M, Ge S, Su Y, Chan SY, Loh XJ, Zhuang A (2023) A non-invasive smart scaffold for bone repair and monitoring. Bioact Mater 19:499–510
Hubbell JA, Thomas SN, Swartz MA (2009) Materials engineering for immunomodulation. Nature 462(7272):449–460
Huebsch N, Mooney DJ (2009) Inspiration and application in the evolution of biomaterials. Nature 462(7272):426–432
Hutmacher DW, Schantz JT, Lam CXF, Tan KC, Lim TC (2007) State of the art and future directions of scaffold-based bone engineering from a biomaterials perspective. J Tissue Eng Regen Med 1(4):245–260
Iddan G, Meron G, Glukhovsky A, Swain P (2000) Wireless capsule endoscopy. Nature 405(6785):417–417
Insuasti-Cruz E, Suarez-Jaramillo V, Urresta KM, Pila-Varela K, Fiallos-Ayala X, Dahoumane S, Alexis F (2022) Natural biomaterials from biodiversity for healthcare applications. Adv Healthcare Mater 11(1)
Jiang Y, Trotsyuk AA, Niu S, Henn D, Chen K, Shih C-C, Larson MR, Mermin-Bunnell AM, Mittal S, Lai J-C (2023) Wireless, closed-loop, smart bandage with integrated sensors and stimulators for advanced wound care and accelerated healing. Nat Biotechnol 41(5):652–662
Jin H, Yu J, Lin S, Gao S, Yang H, Haick H, Hua C, Deng S, Yang T, Liu Y (2020) Nanosensor-based flexible electronic assisted with light fidelity communicating technology for volatolomics-based telemedicine. ACS Nano 14(11):15517–15532
Johari N, Khodaei A, Samadikuchaksaraei A, Reis RL, Kundu SC, Moroni L (2022) Ancient fibrous biomaterials from silkworm protein fibroin and spider silk blends: Biomechanical patterns. Acta Biomater 153:38–67
Kara S (2012) A roadmap of biomedical engineers and milestones. BoD Books Demand
Karageorgiou V, Kaplan D (2005) Porosity of 3D biomaterial scaffolds and osteogenesis. Biomaterials 26(27):5474–5491
Keum DH, Kim S-K, Koo J, Lee G-H, Jeon C, Mok JW, Mun BH, Lee KJ, Kamrani E, Joo C-K (2020) Wireless smart contact lens for diabetic diagnosis and therapy. Sci Adv 6(17):eaba3252
Khamis A, Meng J, Wang J, Azar AT, Prestes E, Takács Á, Rudas IJ, Haidegger T (2021) Robotics and intelligent systems against a pandemic. Acta Polytech Hungarica 18(5):13–35
Kim K, Choi H, Choi ES, Park M-H, Ryu J-H (2019) Hyaluronic acid-coated nanomedicine for targeted cancer therapy. Pharmaceutics 11(7):301
Kim TY, Mok JW, Hong SH, Jeong SH, Choi H, Shin S, Joo C-K, Hahn SK (2022) Wireless theranostic smart contact lens for monitoring and control of intraocular pressure in glaucoma. Nat Commun 13(1):6801

Kowalski PS, Bhattacharya C, Afewerki S, Langer R (2018) Smart biomaterials: recent advances and future directions. ACS Biomater Sci Eng 4(11):3809–3817

Kwon SH, Dong L (2022) Flexible sensors and machine learning for heart monitoring. Nano Energy 102:107632

Langer R (1990) New methods of drug delivery. Science 249(4976):1527–1533

Langer R (1998) Drug delivery and targeting. Nature 392(6679 Suppl):5–10

Li J, Xin M, Ma Z, Shi Y, Pan L (2021) Nanomaterials and their applications on bio-inspired wearable electronics. Nanotechnology 32(47):472002

Li L, Sheng S, Liu Y, Wen J, Song C, Chen Z, Xu W, Zhang Z, Fan W, Chen C (2023) Automatic and continuous blood pressure monitoring via an optical-fiber-sensor-assisted smartwatch. PhotoniX 4(1):21

Liang Y, Zhu H, Wang L, He H, Wang S (2020) Biocompatible smart cellulose nanofibres for sustained drug release via pH and temperature dual-responsive mechanism. Carbohydr Polym 249:116876

Lima M, Rudd T, Yates E (2017) New applications of heparin and other glycosaminoglycans. Molecules 22(5):749

Lin R, Fan Y, Xie Y, Ge D, Liang S, Guan H, Chen M, Zhang Y, Xing L, Xue X (2023) A self-powered wearable seizure-monitoring/brain-stimulating system for potential epilepsy treatment. Nano Energy 107:108121

Liu Y, Hu Y, Zhao J, Wu G, Tao X, Chen W (2016) Self-powered piezoionic strain sensor toward the monitoring of human activities. Small 12(36):5074–5080

Liu K, Tebyetekerwa M, Ji D, Ramakrishna S (2020) Intelligent materials. Matter 3(3):590–593

Luo X, Yu Q, Yang L, Cui Y (2023) Wearable, sensing-controlled, ultrasound-based microneedle smart system for diabetes management. ACS Sensors 8(4):1710–1722

Mehdizadeh M, Yang J (2013) Design strategies and applications of tissue bioadhesives. Macromol Biosci 13(3):271–288

Min S, Kim DH, Joe DJ, Kim BW, Jung YH, Lee JH, Lee BY, Doh I, An J, Youn YN (2023) Clinical validation of a wearable piezoelectric blood-pressure sensor for continuous health monitoring. Adv Mater 35(26):2301627

Montazerian M, Zanotto ED (2017) A guided walk through Larry Hench's monumental discoveries. J Mater Sci 52(15):8695–8732

Morad R, Akbari M, Rezaee P, Koochaki A, Maaza M, Jamshidi Z (2021) First principle simulation of coated hydroxychloroquine on Ag, Au and Pt nanoparticles. Sci Rep 11(1):2131

Mostafalu P, Tamayol A, Rahimi R, Ochoa M, Khalilpour A, Kiaee G, Yazdi IK, Bagherifard S, Dokmeci MR, Ziaie B (2018) Smart bandage for monitoring and treatment of chronic wounds. Small 14(33):1703509

Niinomi M (2002) Recent metallic materials for biomedical applications. Metall Mater Trans A 33:477–486

Ouyang, W.; Lu, W.; Zhang, Y.; Liu, Y.; Kim, J. U.; Shen, H.; Wu, Y.; Luan, H.; Kilner, K.; Lee, S. P.; Lu, Y.; Yang, Y.; Wang, J.; Yu, Y.; Wegener, A. J.; Moreno, J. A.; Xie, Z.; Wu, Y.; Won, S. M.; Kwon, K.; Wu, C.; Bai, W.; Guo, H.; Liu, T.-l.; Bai, H.; Monti, G.; Zhu, J.; Madhvapathy, S. R.; Trueb, J.; Stanslaski, M.; Higbee-Dempsey, E. M.; Stepien, I.; Ghoreishi-Haack, N.; Haney, C. R.; Kim, T.-i.; Huang, Y.; Ghaffari, R.; Banks, A. R.; Jhou, T. C.; Good, C. H.; Rogers, J. A., A wireless and battery-less implant for multimodal closed-loop neuromodulation in small animals. Nat Biomed Eng 2023, 7 (10), 1252-1269.

Park J, Lakes RS (2007) Biomaterials: an introduction. Springer

Place ES, Evans ND, Stevens MM (2009) Complexity in biomaterials for tissue engineering. Nat Mater 8(6):457–470

Powers R, Etezadi-Amoli M, Arnold EM, Kianian S, Mance I, Gibiansky M, Trietsch D, Alvarado AS, Kretlow JD, Herrington TM (2021) Smartwatch inertial sensors continuously monitor real-world motor fluctuations in Parkinson's disease. Sci Transl Med 13(579):eabd7865

Raghavendra GM, Varaprasad K, Jayaramudu T (2015) Biomaterials: design, development and biomedical applications. In: Nanotechnology applications for tissue engineering. Elsevier, pp 21–44
Rajabi AH, Jaffe M, Arinzeh TL (2015) Piezoelectric materials for tissue regeneration: A review. Acta Biomater 24:12–23
Ratner BD (1996) Biomaterials science: an interdisciplinary endeavor. In: Biomaterials science. Elsevier, pp 1–8
Ratner BD, Zhang G (2020) A history of biomaterials. In: Biomaterials science. Elsevier, pp 21–34
Rea S, Bonfield W (2004) Biocomposites for medical applications. J Australas Ceram Soc 40(1):43–57
Sang M, Kang K, Zhang Y, Zhang H, Kim K, Cho M, Shin J, Hong JH, Kim T, Lee SK (2022) Ultrahigh sensitive Au-doped silicon nanomembrane based wearable sensor arrays for continuous skin temperature monitoring with high precision. Adv Mater 34(4):2105865
Shodeinde AB, Murphy AC, Oldenkamp HF, Potdar AS, Ludolph CM, Peppas NA (2020) Recent advances in smart biomaterials for the detection and treatment of autoimmune diseases. Adv Funct Mater 30(37):1909556
Shupak NM, Prato FS, Thomas AW (2003) Therapeutic uses of pulsed magnetic-field exposure: a review. URSI Radio Sci Bull 2003(307):9–32
Song R, Murphy M, Li C, Ting K, Soo C, Zheng Z (2018) Current development of biodegradable polymeric materials for biomedical applications. Drug Des Develop Therapy:3117–3145
Song Y, Min J, Gao W (2019) Wearable and implantable electronics: moving toward precision therapy. ACS Nano 13(11):12280–12286
Song Y, Mukasa D, Zhang H, Gao W (2021) Self-powered wearable biosensors. Acc Mater Res 2(3):184–197
Thomas S, Grohens Y, Ninan N (2015) Nanotechnology applications for tissue engineering, William Andrew
Tibbitt MW, Dahlman JE, Langer R (2016) Emerging frontiers in drug delivery. J Am Chem Soc 138(3):704–717
Vallet-Regí M (2001) Ceramics for medical applications. J Chem Soc Dalton Trans 2:97–108
Vijian RS, Yusefi M, Shameli K (2022) Plant extract loaded sodium alginate nanocomposites for biomedical applications: A review. J Res Nanosci Nanotechnol 6(1):14–30
Voelker R (2018) Blood glucose readings made possible via smartphone. JAMA 319(17):1758–1758
Walther A (2020) From responsive to adaptive and interactive materials and materials systems: A roadmap. Adv Mater 32(20):1905111
Wang Z, Wu J, Wei X, Saleemi S, Liu W, Li W, Marriam I, Xu F (2020) Bioinspired microstructure-reorganized behavior of carbon nanotube yarn induced by cyclic stretching training. J Mater Chem C 8(1):117–123
Wang M, Yang Y, Min J, Song Y, Tu J, Mukasa D, Ye C, Xu C, Heflin N, McCune JS (2022) A wearable electrochemical biosensor for the monitoring of metabolites and nutrients. Nat Biomed Eng 6(11):1225–1235
Wei X, Beltrán-Gastélum M, Karshalev E, Esteban-Fernández de Ávila B, Zhou J, Ran D, Angsantikul P, Fang RH, Wang J, Zhang L (2019) Biomimetic micromotor enables active delivery of antigens for oral vaccination. Nano Lett 19(3):1914–1921
Wicaksono I, Tucker CI, Sun T, Guerrero CA, Liu C, Woo WM, Pence EJ, Dagdeviren C (2020) A tailored, electronic textile conformable suit for large-scale spatiotemporal physiological sensing in vivo. npj Flexible. Electronics 4(1):1–13
Wu S-C, Hsu H-C, Liu M-Y, Ho W-F (2021) Characterization of nanosized hydroxyapatite prepared by an aqueous precipitation method using eggshells and mulberry leaf extract. J Korean Ceramic Soc 58:116–122
Wu P, Chen P, Xu C, Wang Q, Zhang F, Yang K, Jiang W, Feng J, Luo Z (2022) Ultrasound-driven in vivo electrical stimulation based on biodegradable piezoelectric nanogenerators for enhancing and monitoring the nerve tissue repair. Nano Energy 102:107707

Xu G, Lu Y, Cheng C, Li X, Xu J, Liu Z, Liu J, Liu G, Shi Z, Chen Z (2021) Battery-free and wireless smart wound dressing for wound infection monitoring and electrically controlled on-demand drug delivery. Adv Funct Mater 31(26):2100852

Xu J, Tao X, Liu X, Yang L (2022) Wearable eye patch biosensor for noninvasive and simultaneous detection of multiple biomarkers in human tears. Anal Chem 94(24):8659–8667

Yang SM, Kim H, Ko G-J, Choe JC, Lee JH, Rajaram K, An B, Han WB, Kim D-J, Shin J-W (2022a) Soft, wireless electronic dressing system for wound analysis and biophysical therapy. Nano Today 47:101685

Yang C, Wu Q, Liu J, Mo J, Li X, Yang C, Liu Z, Yang J, Jiang L, Chen W (2022b) Intelligent wireless theranostic contact lens for electrical sensing and regulation of intraocular pressure. Nat Commun 13(1):2556

Yang Y, Zhang Y, Min Y, Chen J (2022c) Preparation of methacrylated hyaluronate/methacrylated collagen sponges with rapid shape recovery and orderly channel for fast blood absorption as hemostatic dressing. Int J Biol Macromol 222:30–40

Yim S, Sitti M (2011) Design and rolling locomotion of a magnetically actuated soft capsule endoscope. IEEE Trans Robot 28(1):183–194

Yu Y, Jiang C, Zheng XT, Liu Y, Goh WP, Lim R, Tan S, Yang L (2022) Three-dimensional highway-like graphite flakes/carbon fiber hybrid electrode for electrochemical biosensor. Mater Today Adv 14:100238

Zhang K, Wang S, Zhou C, Cheng L, Gao X, Xie X, Sun J, Wang H, Weir MD, Reynolds MA (2018) Advanced smart biomaterials and constructs for hard tissue engineering and regeneration. Bone Res 6(1):31

Zhang X, Chen L, Lim KH, Gonuguntla S, Lim KW, Pranantyo D, Yong WP, Yam WJT, Low Z, Teo WJ (2019) The pathway to intelligence: using stimuli-responsive materials as building blocks for constructing smart and functional systems. Adv Mater 31(11):1804540

Zhang C, Liu C, He W, Jiao D, Liu Z (2021) Cross-linking-enhanced and ultrasound-mediated drug delivery: From fabrication, mechanisms to translations. Appl Mater Today 22:100897

Zhao Z, Ukidve A, Krishnan V, Mitragotri S (2019) Effect of physicochemical and surface properties on in vivo fate of drug nanocarriers. Adv Drug Deliv Rev 143:3–21

Zheng Y, Liu J, Lu X, Li Y (2021a) Shape Memory Biomaterials and Their Clinical Applications. In: Narayan R (ed) Biomedical Materials. Springer International Publishing, Cham, pp 195–255

Zheng K, Tong Y, Zhang S, He R, Xiao L, Iqbal Z, Zhang Y, Gao J, Zhang L, Jiang L (2021b) Flexible bicolorimetric polyacrylamide/chitosan hydrogels for smart real-time monitoring and promotion of wound healing. Adv Funct Mater 31(34):2102599

Zheng L, Guo S, Kawanishi M (2022) Magnetically controlled multifunctional capsule robot for dual-drug delivery. IEEE Syst J 16(4):6413–6424

Chapter 2
Shape Memory Biomaterials

Yu Shi and Zhe Liu

Abstract Shape memory materials (SMMs) are defined as materials that are capable of recovering permanent shapes from temporary deformations when subjected to specific stimuli given the shape memory effect (SME). Shape memory effect is the phenomenon whereby a material recovers its phase transition temperature through residual strain during annealing after unloading an applied stress. Shape memory materials, including shape memory alloys (SMAs), shape memory ceramics (SMCs), and shape memory polymers (SMPs), are considered to be ideal materials for reconfiguring structures in response to external stimuli. Their performance can be enhanced through modification and processing. Emerging technologies such as additive manufacturing and computer-aided designing have the potential to improve cost-effectiveness, adaptability, and versatility, while also enabling the replacement of thermal actuation with fields such as light, electricity, and magnetism-providing opportunities for biomedical devices. Shape memory materials have been extensively applied in tissue engineering including wound healing and self-adjusting implants as well as disease theranostics including medical smart actuator, drug delivery, and biosensors. It is anticipated that shape memory materials will become more sophisticated—offering customized services for personalized medicine—and will exhibit considerable promise for biomedical applications.

Keywords Shape memory alloy · Shape memory polymer · Shape memory ceramic · Intelligent biomaterials

Y. Shi · Z. Liu (✉)
Academy of Medical Engineering and Translational Medicine, Tianjin University, Tianjin, China

Tianjin Key Laboratory of Brain Science and Neural Engineering, Tianjin University, Tianjin, China
e-mail: zheliu@tju.edu.cn

Z. Liu (ed.), *Intelligent Biomaterials*, https://doi.org/10.1007/978-981-96-8646-9_2

2.1 Overview of Shape Memory Biomaterials

2.1.1 Concept of Shape Memory

Shape memory materials (SMMs) refer to a kind of materials that can memorize a permanent shape and recover the original shape from temporary deformations responding to specific stimuli, e.g., thermomechanical and magnetic variations. Generally, the shape memory effect (SME) refers to a phenomenon that a kind of material recovers its phase-transition temperature thanks to residual strain during subsequent annealing after unloading an applied stress. Typically, a folded flower (temporary shape) made by SMMs at low temperature will transform to blooming flower (original shape) with temperature increasing (Hu and Chen 2010). In past decades, SMMs have been widely applied in aerospace technologies, textile engineering, robotics, biomedical industry, and others (Ma et al. 2010). Owing to aptitude to assorted shape changeability, shape memory materials are deemed as fantastic aspirants to reconfigure structures for offering unique opportunities for biomedical devices, especially adaptable medical implants and smart biomedical garment-integrated devices (Tzou et al. 2004). Shape memory biomaterials are classified into three major groups: shape memory alloys (SMAs), shape memory ceramics (SMCs), and shape memory polymers (SMPs). Conventional SMAs, developed SMCs, and novel SMPs have their own intrinsic merits as shown in Table 2.1 (Hornbogen 2006; Korde and Kandasubramanian 2020).

2.1.2 History of Shape Memory Materials

The shape memory effect was first discovered by the Swedish metallurgist Olander at the meeting of the Swedish Metallurgical Society of 1932, and was described as the peculiar rubber-like properties of a gold-cadmium alloy (Ölander 1932).

Table 2.1 The characteristics of different shape memory materials

Characteristics	SMAs	SMCs	SMPs
Phase transformation	Martensitic transformation	Martensitic transformation	Glass transition
Density	High density (6 ~ 8 g/m^3)	High density (2 ~ 8 g/m^3)	Low density (0.9 ~ 1.25 g/m^3)
Transformation temperature	0 ~ 500 °C	0 ~ 1000 °C	<100 °C
Physiochemical merits	Wear resistance	Corrosion resistance	Light weight and portable
Biological properties	Partial biocompatibility	Good biocompatibility	Desired biodegradability
SME	Less strain (<8%) but fast recovery	Larger strain (< 15%) but poor recovery	High strain (<800%) but slow recovery
Consume	Expensive	Expensive	Low cost

Adapted from Korde and Kandasubramanian (2020)

Subsequently, Benedicks, Bystrom, and Almin made the research of Olander's alloy (Bystrom and Almin 1947). Chang and Read also noticed the reversibility of the Au-Cd alloy by both metallographic observations and the changes of resistivity (Chang and Read 1951). Similarly, Greninger and Mooradian also discovered the shape memory effect of Cu-Zn and Cu-Sn alloys in 1938 (Greninger and Mooradian 1938). The term shape memory was coined by Vernon in 1941 (Lester and Vernon 1941). Shape memory effect was further observed and first explained by Chang and Read in 1951 (Chang and Read 1951). They reported reverse shape memory through the tests of a deformed straight rod in cubic phase reformed into its original straight shape under heat. Shape memory effect was also presented in copper-zinc alloys in 1952 and indium-thallium alloys in 1953 (Christian 2002). The research of nickel-titanium alloys has freshen up since Buehler discovered in 1963 (Buehler et al. 1963). Until 1975, Andreasen from Iowa University made the first superelastic SMAs orthodontic device as a medical implant (Machado and Savi 2003). With regard to SMCs, the shape memory effect of partly stabilized zirconia ceramics was observed in 1986 by Swain, and he also reported zirconia ceramics limited by unique brittleness have higher transformation temperature, relatively smaller reversible deformation strains than reported for metallic alloys in the journal of Nature (Swain 1986). Afterward, Reyes-Morel and Chen discovered the pseudoelasticity shape memory effect of CeO_2-ZrO_2 tetragonal polycrystals in 1988 (Reyes-Morel et al. 1988). However, the practical application of zirconia-based SMCs was limited on grounds of poor recoverable strains of 0.5–1.0% and the premature fracture. Until 2013, Lai et al. (Lai et al. 2013; Du et al. 2015) alleviated the cracking problems of SMCs by the utilize of oligo-crystalline or single-crystalline micro/submicron-scale pillars. Nevertheless, the nano/microstructural evolution during microcracking has not been revealed (Christian 2002). Some researchers proposed that SMCs designed with preferred crystal orientations may improve the sustained shape memory effect when SMCs transformed along particular orientations, fracture, or plastic slip (Zeng et al. 2016a). As for SMPs, cross-linked polyethylene was utilized for heat-shrinkable tubes and films in the 1960s, and then SMPs became known to researchers that were exploited for textiles after two decades, making great progress in the early twenty-first century (Zhao et al. 2015; Hu et al. 2012). Since shape memory was highlighted in third IUMRS international conference on advanced materials in 1994 (Abrahamsson et al. 1994), SMMs application design and related commercial application have flourished in varieties of fields including automotive, biomedicine, aerospace, robotics, and so on (Mohd Jani et al. 2014) (Fig. 2.1).

2.1.3 Shape Memory Effect

According to observable results of SME, there are four phenomena: one-way SME, two-way/reversible SME, superelastic effect, and rubber-like elasticity. Notably, the first two effects can be only deformed during the stress at given temperature without

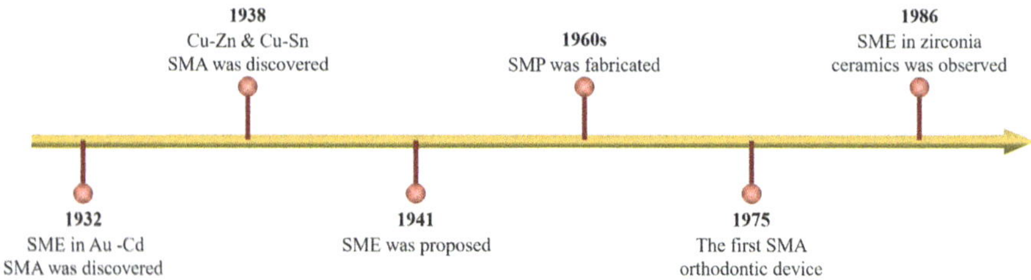

Fig. 2.1 History and development of shape memory materials

variances in chemical driving force compared with the last two effects (Christian 2002). The behaviors of one-way or two-way/reversible shape memory effect are distinctly performed in SMAs such as Ti-Ni alloys, which the former can only remember one shape when heated, while the later can also remember another shape when cooled down. In SMAs, the shape memory effect is a solid phase transition in martensite and austenite, or other martensite with rubber-like elasticity, rhombohedral, and bainite (Mohd Jani et al. 2014; Tadaki et al. 1988). As metallic alloys, SMAs can conduct reversible deformations of crystal structure responding to temperature or stress, which is interpreted by the alteration of the plastic martensite (M), a disordered crystal in product phase, and the austenite (A), a well-organized crystal in parent phase by heating (Huebsch et al. 2012). Specifically, as shown in Fig. 2.2, stable austenite at high temperature ($T > A_f$, austenite finish temperature) will transform into twinned martensite upon cooling ($T < M_S$, martensite start temperature), and then below the temperature M_f (martensite finish temperature) the twinned martensite deformed into detwinned martensite with residual strain upon loading. Afterward, the detwinned martensite can easily recover its original austenite without exceptional stress by heating within the limited yield stress for dislocation motion. Reversely, pseudo-elasticity such as superelastic effect and rubber-like elasticity is a transformation of the specimen, e.g. gold-cadmium and indium-thallium alloys, from the austenite into the martensite state. In addition, superelasticity is a reversible transformation from austenite to twinned martensite upon mechanical load at $T > A_{f,}$ and original austenite together with fully or partially strain can be recovered after unloading without any temperature regulation different from SMAs (Duerig et al. 1990). Such a capability of strain recovery depends on the ambient temperature, composition, and microstructure texture of crystalline.

Shape memory polymers (SMPs) have the same property as SMAs that can transform one programmed or temporary conformation to another permanent conformation responding to external excitation such as light and temperature. The SME of SMPs is an entropy-driven phenomenon that a lower-entropy state at freeze recovers to its original high-entropy state upon reheating by means of initial programming step (Sabahi et al. 2020). To a certain extent, most polymers exhibit shape memory behavior to some extent. In practical programming steps (Fig. 2.3), SMPs firstly soften to trigger high chain mobility of SMPs by heating above a transformation temperature, T_{Trans}; secondly, the external load is applied to rearrange the SMPs chains into high degree of order at the low entropy state with high

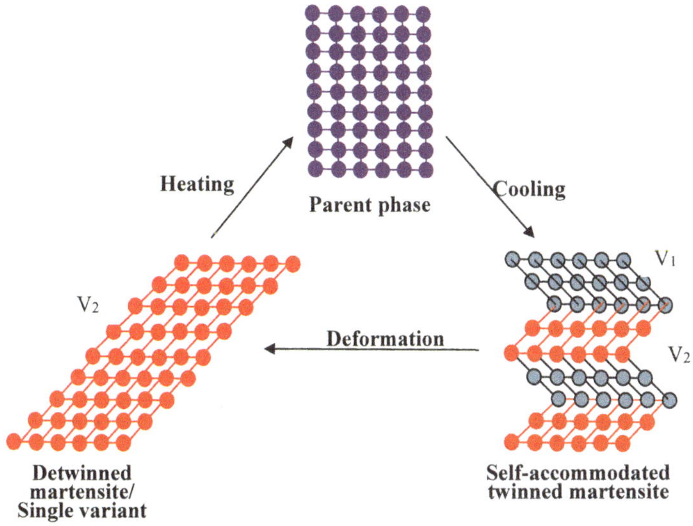

Fig. 2.2 Crystallographic mechanism of shape memory effect in SMAs (V_1 for one martensite variant and V_2 represents another variant). (Reprinted with permission from Yuan et al. 2018)

Fig. 2.3 Typical shape memory effect of thermally actuated SMPs. (Reprinted with permission from Lei et al. 2019)

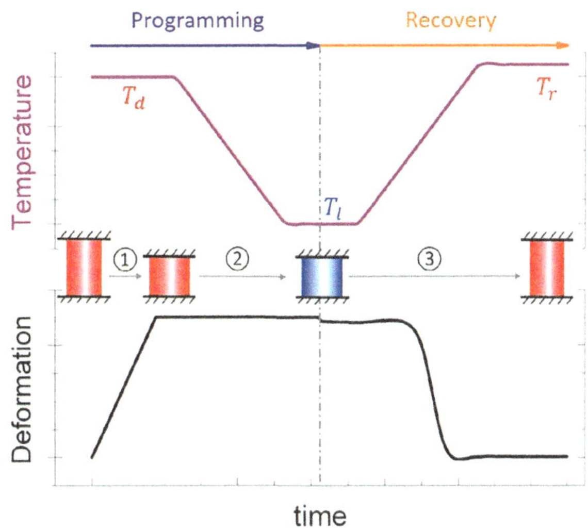

temperature above T_{Trans} maintained; next, the chain segments of SMPs freeze for trap of the lower-entropy state by cooling SMPs with the applied load maintaining; finally, SMPs can memorize the new shape after unloading and original shape at higher-entropy state when heating up above T_{Trans} as elastic energy stored in the programmed SMPs offers a driving force of shape memory. Different from SMAs, the mechanism of shape memory effect in SMPs includes melting/glass transitions, bondings by specific ions or photo, reversible covalent/supramolecular interactions, change in hydrophilicity, physical cross-links, and others (Zhang et al. 2018). The shape (strain) fixity (R_f) and shape (strain) recovery ratio (R_r) are two critical factors to quantify the SME of SMPs. By cyclic thermomechanical testing, these two parameters are defined as following (Li and Xu 2011):

$$R_f = \frac{\varepsilon_{temp}}{\varepsilon_{load}} \times 100\%$$

$$R_r = \frac{\varepsilon_{load} - \varepsilon_{rec}(N)}{\varepsilon_{load} - \varepsilon_{rec}(N-1)} \times 100\%$$

Here, in the programming process, ε_{temp} refers to the final strain of the temporary shape, when ε_{load} for the maximum strain, ε_{rec} for recovered strain after one full programming of heating cycle, and ε_{rec} for the difference of the current (N) and previous ($N - 1$) thermomechanical cycles. Opposing to the equation of hot programmed indicated above, R_f and R_r for cold-programmed SMPs can be defined as follows:

$$R_f = \frac{\varepsilon_{res}}{\varepsilon_{pr}} \times 100\%$$

$$R_r = \frac{\varepsilon_{rec}}{\varepsilon_{res}} \times 100\%$$

Here, ε_{rec} is the residual viscoplastic damage strain, that is the deduction of the rebound strain from the total programming strain ε_{pr}, and ε_{rec} refers to the recovered strain. Noticeably, the SME of SMPs depends on molecular architecture and composition rather than chemical structure in repeating units. SMPs are programmed by the regulation of hard and soft segments in polymer network. Hard segments as net points remain permanent networks during the programming process, while soft segments as switchable line forms a temporary shape during the fixing stage responding to external stimuli. The driving force of recovery is elastic potential energy stored by the hard segments like a spring once the soft segments are stimulated by a suitable stimulus (Xia et al. 2021; Kumar Patel and Purohit 2018). Compared with SMAs stimulated by stress or temperature, shape memory polymers can be also excited by UV and infrared light as well as moisture, and the process of reformation can be achieved many times.

2.1.4 *Modification of Shape Memory Materials for Biomedical Application*

Currently, additive manufacturing (AM) as a pioneering manufacturing technique facilitates the development of highly customized medical devices of shape memory materials to meet personalized demands. In the manufacturing technique of AM, solid model of the computer-aided design (CAD) is first produced through three-dimensional (3D) modeling software, or medical tomographic imaging data; next, solid model is recorded by the AM equipment to manufacture three-dimensional (3D) or four-dimensional (4D) objects by layer-by-layer arrangement or successive addition of various materials (Huang et al. 2016; Gao et al. 2015). Customized implantable devices comprised of shape memory materials would be more promising by AM according to detailed CT and/or MRI data (Sabahi et al. 2020).

AM techniques for SMAs can be classified into powder bed fusion-based technologies and directed energy deposition methods. In powder bed fusion techniques, fuse powder particles are fused in selected areas of the powder bed by a laser or electron beam for the fabrication of highly accurate, minimal, and exquisite parts, especially biomedical devices with controllable mechanical properties and special function (Elahinia et al. 2012). Powder bed fusion-based techniques contain selective laser melting (SLM), electron beam melting (EBM), and selective laser sintering (SLS). Among of these techniques, SLM and EBM require high laser or electric energy density to fully melt metal powder for high porosity 201,193, while the SLS process is that a scanning laser beam makes singer-layer powder particles selectively sinter together (Santos et al. 2006). Laser-engineered net shaping, as a kind of directed energy deposition technique, utilizes a laser beam to melt and solidify metal specimen injected into a melt pool on a substrate for layer-by-layer materials strongly attached to the substrate with bonded particles (Elahinia et al. 2012).

The AM techniques of SMP are commonly stereolithography (SLA) and fused deposition modeling (FDM). Photopolymerization-based SLA of a liquid thermoset resin can scan predefined paths by UV laser beam under the control of computer, and harden plane graph layer by layer on the surface of the resin (Stansbury and Idacavage 2016; Melchels et al. 2010). FDM, as a user-friendly, simple, and inexpensive measure of 3D printing adaptable for thermoplastic SMPs, can print deposited layers through polymeric filament extruded by a heated nozzle for 3D object (Ning et al. 2015).

2.2 Classification of Shape Memory Materials

2.2.1 Shape Memory Alloys (SMAs)

SMAs as a kind of structure-function-integrated material are usually made up of Fe, Cu, or NiTi alloys. SMAs have unique and intrinsic nature of SME and superelastic effects on account of a reversible martensitic phase transformation in two distinct crystal (Chluba et al. 2015). Actually, SMAs possess two phases with three crystallographic structures for six possible transformations among twinned martensite, detwinned martensite, and austenite (Sun and Huang 2009; Mihálcz 2001). In the process of shape memory, heated SMAs will transform from martensite that is stable in low temperatures into the austenite that is stable at high temperature. The start and finish temperatures both of martensitic and austenite are the characteristic parameters of phase transition. The phase transition of SMAs will occur when the temperature is above A_s, the start temperatures of austenite, even with high applied loads, and then SMAs recover the original shape of martensite as a consequence of high actuation energy densities (Lagoudas 2010). Similarly, the revert phase transition starts at M_s, the start temperature of martensite and end up at M_f, the finish temperature of martensite (Buehler et al. 2004). Nevertheless, the SME of SMAs will permanently disappear when the martensite of SMAs cannot be induced by stress at the highest temperature (Duerig and Pelton 1994). These SMEs are described by pseudoelasticity (or superelasticity) mentioned above.

Hysteresis, which determines the applicably operating range, is generally defined as the temperature difference of phase transition (i.e., $\Delta T = A_f - M_s$), and the two critical temperatures (A_f and M_s) can make the material in 50% transform to another phase upon heating or cooling (Buehler and Wang 1968). These transition temperatures and the hysteresis loop behavior are concerned with compositions, thermomechanical processing, and applied stress, and are commonly measured by differential scanning calorimetry (DSC), dilatometry, constant stress thermal cycling experiments, and others (Stöckel 1995). Mechanical properties also play important roles in SMAs including Young's modulus, electrical resistivity, thermal conductivity, and thermal expansion coefficient (Sreekumar et al. 2009; Mertmann and Vergani 2008). The hard austenite has a higher Young's modulus, while the soft martensite is more malleable, stretchable, and deformable.

Owing to the phase transformation between martensite and austenite, SMAs show outstanding mechanical behavior including shape memory effect (SME) and superelasticity (Otsuka and Ren 2005). There are three major groups within the systems of alloys with SME: NiTi alloy, Cu-based SMAs, and Fe-based SMAs, which cover around 20 metal elements including Mn, Zn, Co, Au, and so on for the modification of SMAs (Dasgupta 2014). In engineering applications, NiTi alloy has been exploited for commercial biomedical applications with distinctive biological compatibility and toughness, while low-cost Cu-based SMAs are more attractive owing to their great thermal/electrical conductivity and thermal stability in the promise of research (Sun et al. 2012; Zuperl and Cus 2004). Advanced SMAs integrate temperature sensing as well as intelligent driving into one group applied for actuators, healthcare, and theranostics (Sun et al. 2023). So far, SMAs have been greatly developed with desired shape memory effect and mechanical properties for multitudinous fields (Qiu et al. 2015) including microelectromechanical systems (Kahn et al. 1998), medicine (Prokoshkin et al. 2016) as well as automatics (Stoeckel 1990).

2.2.1.1 NiTi-Based SMAs

NiTi-based SMAs as the studied hotspot of SMAs have preponderance of expected SME, ductility, stability, and high work output (Karaca et al. 2013). The crystal structure of NiTi is a body-centered cube with a B2 structure for austenite, monocline with a B19′ structure for martensite, or possible intermediate rhombohedral phase with a trigonal (rhombohedral) structure for transitional state (Wang et al. 2015; Polatidis et al. 2015). It is worth noting that the composition of Ni in NiTi-based alloy is responsible for the shape memory characteristics, thoroughness, sensitivity to post-processing heat treatments, and the hysteresis and damping capacity (Saedi et al. 2023). Ni-rich NiTi SMAs perform greater superelasticity and shape memory effect with mechanical hysteresis of 185 MPa in contrast to equiatomic NiTi with less 50.6% of Ni with higher hysteresis of up to 300 MPa (Saedi et al. 2018). Not only that, other alloying elements, for instance, Cr, Fe, and Cu, are doped in NiTi-based SMAs to design ternary and quaternary SMAs for the command of transformation hysteresis properties and the properties of shape memory effect. Alloying elements dramatically decrease the mechanical hysteresis of down to 70 MPa in most ternary and quaternary NiTi-based SMAs such as Ni49.0Ti49.6Cu5.4 (Gil and Planell 1999). The incorporation of Ag can render antibacterial properties to SMAs, while the addition of Nb can enhance the elastic properties of SMAs. Surface modification and coatings can reduce the release of harmful ions (Zwawi 2022). Thus Ni-Ti based-SMAs are widely exploited to biomedical use including orthopedics, orthodontics, implantable stent, and even surgical instruments (Wen et al. 2010; Hao et al. 2013).

Ni-Ti alloys as representative biomedical SMAs originated from the birth of nitinol, a titanium-based intermetallic material. Early in 1963, nitinol deprived of nickel (Ni) and titanium (Ti) alloys was discovered by William J. Buehler and Frederick Wang in Naval Ordnance Laboratory (NOL), USA, and "niti" and "nol" in the term of "nitinol" are on behalf of nickel-titanium and NOL, respectively (Otsuka and Ren 2005; Kauffman and Mayo 1997). In physical properties, near-equiatomic nitinol exhibits distinguishing shape memory effect and superelasticity given thermoelastic martensitic transformation (Saigal and Fonte 2011). In practical application, nitinol has been utilized for various fields ranging from manufacturing industry to daily merchandise, and even for high-temperature applications since the mid-1970s (Shabalovskaya et al. 2009; Morgan 2004; Venugopalan and Trépanier 2000). Biomedical nitinol shows great advantages with suitable phase transformation temperature close to human body, good mechanical properties, distinctive biocompatibility, and corrosion resistance (Li et al. 2022). In comparison with traditional medical devices constituted by ordinary alloys like stainless steel, nitinol possesses the capability of real-time adjusted correction and support to achieve one-off installation and implantation. Moreover, the shape memory effect facilitates the tissue growth along or into the porous of implant, which strikingly improves the therapeutic efficiency (Zwawi 2022; Petrini and Migliavacca 2011).

2.2.1.2 Cu-Based SMAs

With cost-effective raw materials and production, Cu-based SMAs, as a class of emerging materials with desired transition temperatures and mechanical properties, could be a candidate for other commercial SMAs applied at high temperature (Mohd

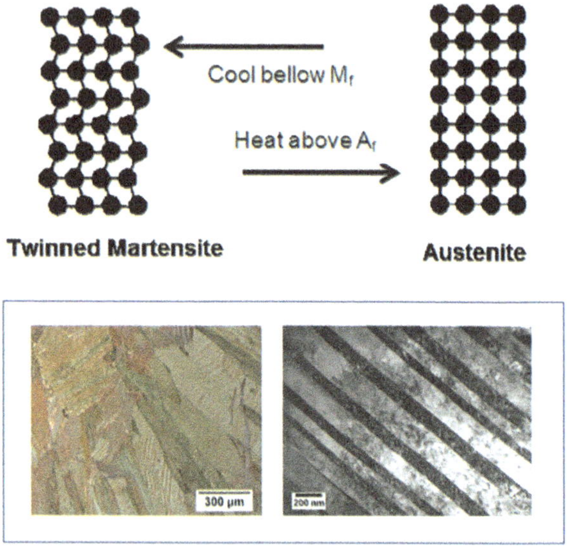

Fig. 2.4 A schematic illustration of the reverse transformation in Cu-based SMAs and β′ martensitic microstructure of a Cu-Al-Ni alloy. (Reprinted with permission from Mazzer et al. 2022)

Jani et al. 2014). They are categorized into two major groups, i.e., Cu-Al and Cu-Zn systems. The shape memory effect of Cu-based SMAs is connected with thermoelastic transformation from B2 (or DO3) phase to 18 R, 3 R, and 2 H martensite according to their composition (Sun 2002). In Cu-based SMAs, the austenite of the high-temperature phase is normally cube called β, while martensite has distinctive types named as γ′, β″, and β′ thinking to the constitution and applied stress, for instance, the β′ martensite in Cu-Al–Ni as shown in Fig. 2.4 (Mazzer et al. 2022). Traditional fabrications of Cu-based SMAs are casting or powder metallurgy with subsequent heat treatments (e.g., solution treatments) followed by quenching (Mazzer et al. 2022). Nevertheless, poly-crystal Cu-based SMAs fabricated by traditional processing have highly elastic anisotropy leading to brittle intergranular fracture given the concentrated stress of the grain boundaries induced by the incoordinate of plastic and elastic deformations among the grains (Chen and Schuh 2011; Ueland and Schuh 2012). Hence grain refinement or precipitate formation is implemented by processing and modification with the addition of alloying elements (Vajpai et al. 2013). Oligocrystal Cu-based SMAs possess larger free surface area than the grain boundary areas for preferable shape memory effect but are limited to SMA wires or ribbons (Ueland et al. 2012; Ueland and Schuh 2013). Bulk single crystal Cu-Al-Mn with superior superelastic properties is synthesized by the innovative processing routes of new cyclic heat treatment, in contrast to hot/cold-working by forging, rolling, and extrusion for sheets or wires of Cu-based SMAss (Mazzer et al. 2022; Omori et al. 2013a). Besides, additive manufacturing (AM) along with high cooling rates during processing are available for complicated Cu-based SMAs with higher strength and ductility (Mazzer et al. 2014). Cu-based shape memory alloys have been developed as following several alloy systems: Cu-Al-Ni for mechanical improvement, Cu-Al-Mn for high ductility and resistance to low-cycle fatigue, Cu-Zn-Al and Cu-Sn for high transformation temperatures (Dasgupta 2014). The addition of a few alloying elements could modify the shape memory effect and mechanical properties of SMAs (Dasgupta et al. 2014), for instance, Cu-Al-Fe-Mn with instantaneous shape memory effect that martensite is uncommonly retained in the microstructure different from traditional shape memory effect (Yang et al. 2016), Al, Mn, and Ti for damping capability, ductility, transformation temperature, and the control of grain size (Sutou et al. 2004; Mallik and Sampath 2008; Chen et al. 2009), Ti, Zr, and Si for the reduction of grain size (Waitz et al. 2008).

2.2.1.3 Fe-Based SMAs and Other Magnetic SMAs (MSMAs)

Fe-Pd and Fe-Pt were the first designed Fe-based SMAs (Kakeshita and Ullakko 2002), the representative examples of phase transition from face-centered cubic austenite into face-centered tetragonal martensite. However, Pd and Pt in the SMA are high cost, and the required pseudoelasticity is difficult to be exhibited at room temperature (Cladera et al. 2014). The utilization of inexpensive Mn element designs the second-generation Fe-based SMAs, in addition to Si for the suppression of the magnetic phase transition (Saedi et al. 2023). Fe-Mn-Si SMAs cannot achieve full strain recovery in the phase transformation from face-centered cubic γ-austenite to hexagonal close-packed ε-martensite, so are considered as partial or semi-superelastic alloys (Sawaguchi et al. 2016). In order to improve the properties of Fe-Mn-Si, other elements are doped or as substitution, for example, Cr and Ni for the improvement of shape memory effect and corrosion resistance, Al and Ni as the replacement of Si for greater superelasticity and more strain recovery cycles (Tseng et al. 2015, 2016; Vollmer et al. 2016).

The Fe-Ni-Co-Al SMAs have gained fantastic attraction as well given suitable pseudoelasticity at room temperature, and usually transform from face-centered cubic austenite into body-centered tetragonal martensite in the process of shape memory. The Ta, Ti elements could be availed to improve their superelasticity (Tanaka et al. 2010; Omori et al. 2013b; Chumlyakov et al. 2020). All the same, the main shortcomings of most Fe-based SMAs are rapid functional fatigue and poor superelasticity following cyclic loading (Abuzaid and Sehitoglu 2018; Krooß et al. 2014).

Some Fe-based SMAs, for example, Fe-Ni alloys discovered by Ullakko in 1996, have also been found shape memory effect induced by magnetic fields, termed as magnetic SMAs (MSMAs) (Ullakko 1996). Subsequently, more MSMAs have been explored including Ni-based (and some Co-based) Heusler alloys such as Ni-Mn-Ga, Ni-Mn-Al, Fe-Pd, and Fe-Pt alloy (Planes et al. 2009; Acet et al. 2011). The martensite in Fe-based alloys is face-centered tetragon (Cui et al. 2004), while the martensite in Heusler alloys can be tetragon or modulated monocline in dependence on their composition (Acet et al. 2011). The shape memory effect of MSMAs depends on material composition and crystallization with limited working temperature below 100 °C (Pérez-Checa et al. 2018; Pagounis et al. 2014). The actuation strain of MSMAs is defined as magnetic-field-induced strain (Kim et al. 2023). Single crystal MSMAs exhibit preferable MFIS of 1–10% modulated by the microstructures, in contrast by polycrystalline of <0.01% due to the suppression of the grain boundary to twin boundary motion (Dunand and Müllner 2011). Rapid actuation response of 1–10 ms in MSMAs triggers immediate phase transition and faster motion with less energy consumption and comparable work output compared to traditional thermal-induced SMAs (Likhachev et al. 2006). MSMAs are commonly divided into two categories according to the magnetic order in martensite: MSMAs with ferromagnetical martensite and metamagnetic MSMAs with weakly magnetism (Mañosa and Planes 2016). It is noteworthy that metamagnetic MSMAs possess intrinsic magnetic superelasticity and inverse magnetocaloric effects in that great alteration of magnetization during the martensitic transition brings about transformation temperature susceptible to magnetic field (Krenke et al. 2005, 2007).

2.2.1.4 Optimization of SMAs

Latterly, the performance of SMAs has been enhanced by taking various approaches by means of improving composition, optimizing fabrication processes, changing forms or shapes, and fabricating hybrid or composite SMM (Mohd Jani et al. 2014). Generally, two major measures are implemented to improve the shape memory effect of SMAs: small size for relaxing free surface, and a few grains like oligocrystal or single crystal to reduce resistance from the grain boundaries. Small-scale samples with high surface-to-volume ratio, mismatch stress, and lattice dilation during transformation could possess more relaxing free surface. According to the study of NiTi-based SMAs, their transformation and shape memory effect are susceptible to the nano-scale grain size, for instance, excellent shape memory effect is acquired when the grain size d is comparable to the sample size D, and martensitic transformation is constrained when d is smaller than a marginal value (Waitz et al. 2009). The related thermodynamic equation during the transformation is (Waitz et al. 2008):

$$\Delta G^{\mathrm{A-M}} = E_{\mathrm{b}},$$

$$\text{while}\, \Delta G^{\mathrm{A-M}} = G^{\mathrm{A}} - G^{\mathrm{M}}, E_{\mathrm{b}} = E_{\mathrm{s}} + E_{\mathrm{i}} + E_{\mathrm{f}}$$

Here, $\Delta G^{\mathrm{A-M}}$ refers to the chemical free energies difference per unit volume of the austenite G^{A} and martensite G^{M}, while E_{b} means the energy barrier per unit of transformed volume. E_{s} for the total elastic strain energy and E_{i} interface energy are attributed to the martensitic formation, and E_{f} for the total irreversible frictional energy of the transformation. According to the equation above, the E_{b} of the transformation to martensite is inversely proportional to the grain size, and the marginal grain size of the constrained martensitic phase transformation is 50 nm. Otherwise, austenite and martensite in micro/nano-scale pillars will be more stable than in bulk given that higher figure of merit for mechanical damping during the martensitic transformation (Juan et al. 2009).

In another measure, oligocrystal or single crystal SMAs can decrease the mismatch stress and transformation from grain boundary area (Chen and Schuh 2011; Ueland et al. 2012). Fracture in bulk SMAs often occurs at minimal strain; on the contrary, oligocrystal or single crystal possesses more extensive surface area than grain boundary area leading to less elastic mismatch stress during the phase transformation. In the thermodynamic equation, the elastic mismatch stresses contribute to energy barrier E_s, and therefore oligocrystal or single crystal SMAs are not easily fractured. Especially for microscale pillar or oligocrystal SMAs, relaxing mismatch stress at large free surface can heavily weaken E_s and reduce transformation constraint from the grain boundaries.

In the fabrication methods for SMAs, powder metallurgy technique for porous and lightweight structures is the preference of biomedical applications, while the thermomechanical techniques are adaptable for SMAs wire. Additionally, additive manufacturing (AM) and three-dimension (3D) printing is state of the art. In terms of AM, powder or wire as raw material resorts to heating sources like a laser, electron beam, or electric arc, and is deposited layer by layer to fabricate finished product with less wastage and post-processing requirement in comparison to the powder metallurgy (Singh Rajput et al. 2022). With the assistance of 3D CAD model software, additive manufacturing also can fabricate sophisticated geometry SMAs with accurate dimension. 3D printing has access to precise geometry of SMAs with minimal wastes, and selective laser melting (SLM) has become popular 3D printing technique for shape memory materials fabrication.

2.2.2 *Shape Memory Ceramics (SMCs)*

2.2.2.1 Performances of SMCs

Ceramics refer to a class of hardened non-metallic and inorganic solids with metallic components by heating process as the forms of oxides, nitrides, and carbides, for instance, inorganic semiconductors, diamond, and graphite (Mirzaali et al. 2022). Ceramics with the merits of hard, corrosion-resistant, and brittle materials have been applied in various industrial and medical applications. Conventional SMAs, especially in nanoscaled dimension, have the problems of reduced shape memory effect and poor mechanical performance due to oxidation and instability, for example, deformation of the surface and interfacial layers in Ti-SMAs because of oxidation (Waitz et al. 2009; Waitz et al. 2004; König et al. 2015). Owing to higher operating temperature, SMCs as a unique class of shape memory materials possess better thermal stability, higher actuation stress, and superior oxidation/corrosion resistance compared with SMAs (Fig. 2.5). However, the mechanism of SMCs includes reversible martensitic phase transformation, such as zirconia (ZrO_2)-based ceramics, and other ferroelectrics (e.g., bismuth ferrite – $BiFeO_3$) (Asle Zaeem et al. 2019). The shape memory effect of SMCs relates to paraelectric-ferroelectric and antiferroelectric-ferroelectric phase transformations (Kimura et al. 1981). The

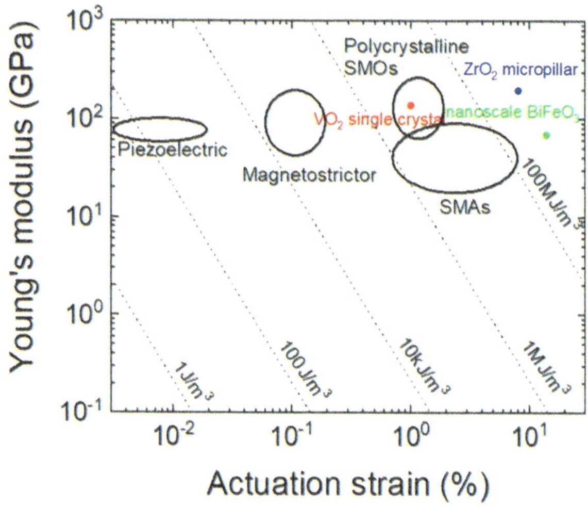

Fig. 2.5 Young's modulus and actuation strain of shape memory effect for different shape memory materials. Dashed lines: contours of constant specific work density. (Reprinted with permission from Wang and Ludwig 2020)

former refers to the thermally induced shape memory behavior of zirconia ceramics, while the latter indicates the electric field-induced phase transition of ferroelectrics. Even in the stage of development and optimization, pure SMCs still have possible potential in the biomedical field (Nayak et al. 2023). In practical application, SMCs with higher transformation stress and temperature as well as larger hysteresis in a load-unload cycle have an extensive variety of applications such as latching relays, high-temperature actuator, and mechanical clampers.

2.2.2.2 Types of SMCs

Zirconia

Zirconia, as the largest family of SMCs, provides advantages including great wear resistance, low coefficient of friction, ionic conductivity, low thermal conductivity, high strength, and fracture toughness (Asle Zaeem et al. 2019), along with shape memory effect similar to SMAs (Hannink et al. 2000), and broadly utilized for fuel cells (Qian et al. 2013), thin-film coatings (Hannink et al. 2000), and sensors (Satsuma et al. 2007). Zirconia has four times stronger strength and reversibly damp more mechanical energy than typical SMAs, in addition to higher tunable transformation temperature of 0 ~ 1200 °C than that of SMAs of 0 ~ 500 °C (Asle Zaeem et al. 2019). There are three polymorphs in zirconia: monoclinic (m), tetragonal (t), and cubic (c), and the high-temperature tetragonal→low-temperature monoclinic athermal martensitic transformation is accompanied by shape memory effect with around 5% transformation strain and around 15% shear strain at ~950 °C (Wang and Ludwig 2020). The shape memory effect of zirconia was first discovered in magnesia-partially-stabilized-zirconia (MgO-PSZ) ceramics with reversible deformation strain of only 0.5% upon heating above 800 °C (Swain 1986). Generally, the monoclinic phase exists room temperature, the tetragonal phase exists between ~1170 °C and ~ 2370 °C, and the cubic phase as the high-temperature phase exists 2680 °C. Undoped zirconia shows phase transitions under ambient during thermal cycling as follow (Kisi and Howard 1998):

$$m-ZrO_2 \underset{1170\ ^\circ C}{\overset{950\ ^\circ C}{\rightleftarrows}} t-ZrO_2 \overset{2370\ ^\circ C}{\rightleftarrows} c-ZrO_2 \overset{2680\ ^\circ C}{\rightleftarrows} \text{liquid}$$

Although the pure tetragonal zirconia can achieve displacive tetragonal to monoclinic (t → m) phase transitions by cooling down to 950 °C, crack and porosity will occur due to residual stresses causing terrible structure integrity and durability of the zirconia substrate (Zhang and Asle Zaeem 2019). Whereas, tetragonal zirconia modified by other oxides (ceria and yttria) can be stable at room temperature for lower transformation temperatures given that the plastic deformation mechanisms of zirconia with yttria and titania stabilized has a strong crystal orientation dependence (Reyes-Morel et al. 1988; Zeng et al. 2016a, b; Zhang and Asle Zaeem 2016). Similar to SMAs, small-scale zirconia also perform excellent shape memory effect including micro-particles (Du et al. 2017a), submicrometer pillars (Lai et al. 2013), granular packing (Yu et al. 2017), and foam (Zhao et al. 2017) because of smaller mismatch stress from grain boundaries and larger relaxing free surface during the phase transformation. During the martensitic transformation, the transformation stress and strain of highly anisotropic zirconia are affected by the crystallographic orientation, hence orientation or texture of zirconia is of importance for desired shape memory effect. Not only that, eliminating the grain boundaries in the SMO is another approach to obtain superior shape memory effect performance at both micro-scale and mm-scale (Crystal et al. 2020).

Ferroelectrics

Multiferroic perovskite with piezoelectric properties exists in different symmetric crystalline structures covering cubic, tetragonal and rhombohedral, and its shape memory effect is induced by the reorientation of polar domains upon electric field. Driven by electric field, ferroelectrics possess lower recoverable strain of <1% and faster response speed compared to SMAs (Zhuo et al. 2018). Pb-free ferroelectric systems $BaTiO_3$ was also reported to exhibit shape memory effect with an electro-induced nonlinear strain of 0.75% at a field of 200 V/mm (Ren 2004). Multiferroic material $BiFeO_3$ with tetragon structure has ferroelectricity and ferromagnetism (Silva et al. 2011; Čebela et al. 2017), and can martensite-transform from rhombohedral (R) phase to tetragonal (T) phase with recoverable strain of up to 14% induced by heating/electricity (Rojac et al. 2011). In other words, the shape memory effect behavior of $BiFeO_3$ induced by structural phase transformation is the consequence of the interactions of stress, electrical fields, and temperature.

As shown in Fig. 2.6a, traditional SMAs are deformed in the martensitic state and recovered in the reverse martensitic state by heating or stress, and re-deformed

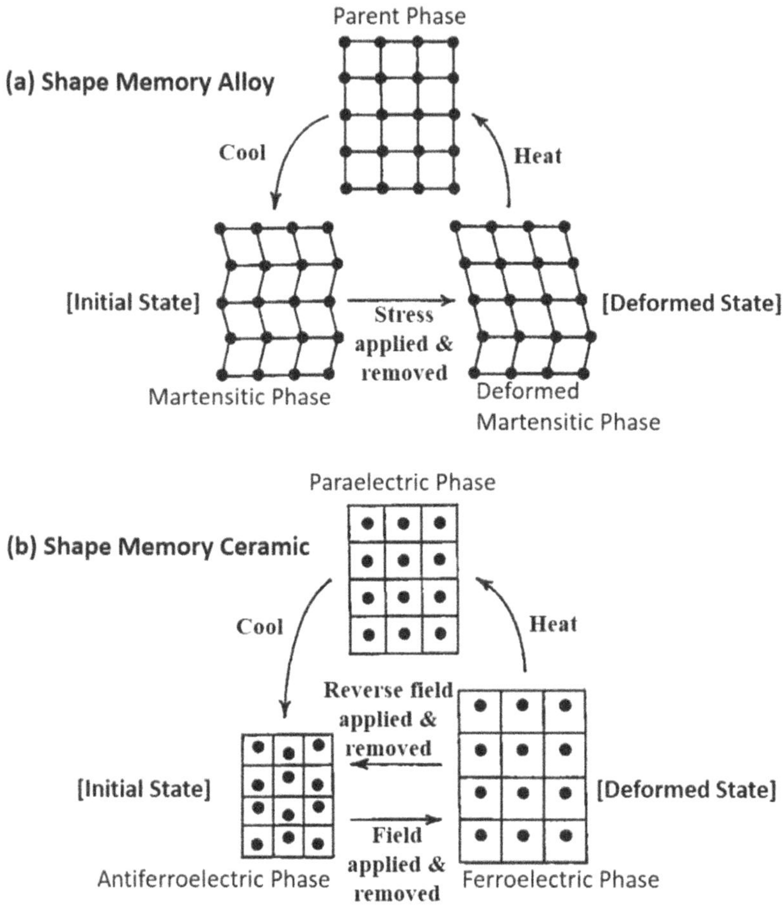

Fig. 2.6 Comparison of shape memory effect mechanisms between SMAs (**a**) and ferroelectric ceramics (**b**). (Reprinted with permission from Uchino 2016)

in the original martensitic state when cooled. In regard to the transitions of antiferroelectric-ferroelectric of ferroelectrics, ferroelectrics presents large displacement of 0.4% with a shape memory function under the control of electric field opposite to the analog nature of conventional piezoelectric with electrostrictive strains of 0.1%, of which shape memory effect mechanism is illustrated in Fig. 2.6b (Uchino 2016). Interestingly, antiferroelectrics in the electric dipole are antiparallelly arranged in each crystal unit cell without polarization instead of parallel arrangement as ferroelectric with polarization can also exhibit shape memory effect under electric field, with 0.26% volumetric expansion and 2.5 ms response speed in the antiferroelectric→ferroelectric transitions, for example, $Pb_{0.99}Nb_{0.02}[(Zr_{0.6}Sn_{0.4})_{1-y}Ti_y]_{0.98}O_3$ (PNZST) system (Uchino 2016). The study on SME of antiferroelectrics originated in the 1980s, and antiferroelectrics could be exploited as ceramic actuators or piezoelectric/electrostrictive materials for micro-mechatronic or electromechanical actuator devices.

2.2.2.3 Optimization of SMCs

Different from other shape memory materials, SMCs with smaller recoverable residual strains and good environmental inertness are more suitable for high temperature, high strength, or corrosive environments (Reyes-Morel et al. 1988; Lai et al. 2013). However, large and polycrystalline brittle ceramics have difficulty in keep integrity during the phase transition in that the transformation strain is above the elastic limit of the lattice, thus constraint and mismatch stress in the neighboring grains of the bulk SMCs during the martensitic transformation leads to early cracking and lower recoverable residual deformation (Du et al. 2015; Kelly and Francis Rose 2002). Some SMCs have explored recently that exist the strain glass state rather than macroscopic martensitic transition system not leading to microcracking, for example, Ni-rich or doped Ti-Ni alloys and Cr-doped Ti-Pd alloys (Wang et al. 2010; Hao et al. 2019; Zhou et al. 2009). Moreover, SMCs at nano/micro-scales or in the form of particles with few grains also can perform excellent shape memory effect with superelasticity by means of the reduction of the amounts of grain boundaries, strain-engineering, or the formation of nano-films (Wang and Ludwig 2020).

By the mesostructure and microstructure engineering of granular packings, metal matrix composites, and cellular architectures, local and sequential transformation can occur in the meso−/micro-structured SMCs rather than simultaneous transformation as bulk SMCs on grounds of heterogeneous driving force, weak mechanical constraint, and low nucleation barrier in martensitic transformation (Erb et al. 2023). In the investigation of granular SMCs packings, taken $(ZrO_2)_{1-x}$-$(CeO_2)_x$ as an example, in the shape memory regime ($T < M_f$, the martensite finish temperature), the peak intensity of preferred crystallographic re-orientation of martensite increases with external loading under the confined uniaxial compression; In the intermediate transition regime ($M_f < T < A_s$), the transformed volume of 25–40% in irreversible martensitic transformation increases with external loading under confined uniaxial compression; in the superelastic regime ($T < A_f$, the austenite finish temperature), forward and reverse martensitic transformation occurs with large hysteresis loop of high energy dissipation density in each load-unload cycle under confined uniaxial compression (Yu et al. 2017). Regarding oligocrystals or single crystals, Lai et al. (2013) firstly showed small zirconia micropillars allowed recoverable strains of up to 7%. Du et al. (2017b) also discovered superelastic microscale SMCs with oligocrystal and single crystal particles exhibit reproducibly cyclic superelasticity of over one hundred and particle compressions of 3.8% compared to fractured polycrystalline during loading by micro-compression experiments.

2.2.3 *Shape Memory Polymers (SMPs)*

SMPs cover poly blends and polymer networks such as hydrogels besides pure polymers (Behl and Lendlein 2007; Chatterjee et al. 2015; Hu et al. 2015). Compared to complex SMAs with high melting temperature, exquisite craft, and stiff texture, SMPs show great advantages with lightweight, low-priced, easily adjustable, and especially higher compatibility, resulting in tiny and functional devices for personal medical use (Lee and Kang 2021). On the basis of shape recovery ability, SMPs with desirable mechanical properties (softness and flexibility) could be as actuating material applied in biomedicine fields, like implantable devices (Ansari et al. 2019), wound dressing (Vakil et al. 2022), and drug release systems (Kashif et al. 2016).

2.2.3.1 Conventional Shape Memory Polymers

Although early in 1941 SMPs were birthed by Vernon and his coworkers in their US patent related to dental material with elastic memory (Vernon 1941), limited technique influence was found until numerous patents associated with polyethylene as heat-shrinkable tubing were registered in the USA in the 1960s (PE W 1967; RJ P 1967). Polynorbornene or polyurethane-based SMP are landmark products with ambient transition temperature of about 40 °C (Santiago et al. 2014; Leng et al. 2009; Gall et al. 2000). According to chemical structure, SMPs can be categorized into three main groups: block co-polymers, supramolecular polymers, and polymer blends (Melly et al. 2020). Block copolymers are constructed by blocks of different polymerized monomers that are composed of various chemical components by covalent bonds in sequence. In block copolymers, the cross-linking netpoints can be physically cross-linked for thermoplastics, or chemically for thermosets (Suchao-in and Chirachanchai 2013), and the thermosets are more stable and irreversible owing to the covalent binding of polymer chains compared to the thermoplastics formed by crystalline or glassy phases (Leng et al. 2009). Supramolecular polymers pioneered by Lehn refer to polymers with non-covalent intermolecular interactions, more often hydrogen bonding, which with supramolecular units as switches provide opportunities for SMPs leveraging on shape memory effect, self-healing, and other mechanical properties (Hu et al. 2012; Chen et al. 2019). Polymer blends are the miscible, immiscible, or compatible combination of two or more polymer matrices for enhanced composite with great practicability (Chatterjee et al. 2015; Arnebold and Hartwig 2016). Among these polymer blends, immiscible blends in separate phases allow individual transition temperature and other properties observed; miscible blends exist in a single phase structure only with one transition temperature (Jing et al. 2016); whereas compatible polymer blends indicate the immiscible blends with strong interaction that presents similar properties at macroscopic level. For the optimization of SMPs, thermomechanical properties are adjusted by combining different polymers, and the property of shape recovery is modulated by the crystallinity of polymers and adding nanoparticles. SMPs are universally reinforced with nanomaterials like carbon nanotubes for expected mechanical properties (Memarian et al. 2016), graphene particles for heat conductivity (Khademeh Molavi et al. 2017; Babaahmadi et al. 2017), nanosized hydroxyapatite for thermal activation, and biodegradability in biomedical application (Peponi et al. 2018).

Besides the one-way shape memory effect, the two-way shape memory effect is the same as SMAs. SMAs also can remember one or more number of temporary shapes when stimulus is removed, thus classified as dual-shape, triple-shape, and multi-shape memory, resulting in a total of six categories of shape memory effect in SMPs (Karger-Kocsis and Kéki 2018). Most of the available SMPs are one-way dual-shape memory because of flexibility, functionality, and applicability (Athimoolam and Moorthy 2012). To achieve temperature-dependent triple or multi-shape memory, more than two transition temperatures in SMPs are required for shape fixation at each transition temperature, and these SMPs can recover the corresponding shape at different temperatures to accommodate a variety of usage scenarios. Multi-shape memory polymers, for example artificial muscles, are promising in medical industry (Wang and Zhu 2018).

The common materials of SMPs contain poly (lactic acid) (PLA), polyurethane (PU) as well as co-polymers due to superior biodegradability and biocompatibility (Ning et al. 2015). Due to the merits of shape memory and biodegradability, SMPs have access to minimally invasive surgery along with automatic actuation by stimuli-response to activate memorize shapes for precise medicine (Lee and Kang 2021). Concretely, SMPs-based drug delivery system with various targets could replace implanted materials for tissue repair and regeneration (Montgomery et al. 2017; Bao et al. 2014), while SMPs-based stents could avoid damages of secondary surgery, restenosis, and potential inflammatory response with great feasibility and safety (Yang et al. 2013). Moreover, SMPs show the promise to be as an actuating biomedical device with electronic systems for soft robotics.

2.2.3.2 Advanced Shape Memory Hydrogels

Although hydrophobic SMPs possess the capability of multiple or reversible shape-switching with varieties of functions, some challenges persist in their use, for instance, possible dissolve in water when SMPs fixed temporary shape, and actuated by some molecular switches responding to multi stimuli (Löwenberg et al. 2017). Even worse, some SMPs possess highly dense netpoints together with less polymer-chain mobility resulting in poor strain capacity, that is loss of recovery capacity when strains over 1000% (Ding et al. 2022). Contrarily, SMHs, like tough organohydrogel possess desired strain capacity given their superior stretchability (Wang et al. 2021). To further improve the properties of shape memory effect, SMHs with unique space structure are brought into being, such as dual-shape hydrogels, triple-shape hydrogels, and organohydrogel with a binary cooperative phase. Similar to stimuli-sensitive hydrogels, elastic SMHs can vary their shape, size, and other physical properties with external stimuli including mechanics, wettability, and light. However, they also can be fixed at temporal permanent structures or programmed conformation beyond or below critical temperature (T_c) through cross-linked network and retrieve their perpetual structure under stimuli on reheating above $T_{c\ (Kabir\ et\ al.}$ $2014_{)}$. Additionally, sanctions constituents of SMHs could regulate preceding structure responding to extraneous stimuli by sophisticated chemical approaches, for example, sol-gel transition (Korde and Kandasubramanian 2020). As favored candidate for biomedical materials, intelligent and soft SMHs analogous to human skin tissues possess exceptional water content and self-adaptability in addition to prospective biocompatibility and biodegradability, and they are rapidly employed in biomedical fields, such as soft actuators, tissue engineering, and drug delivery (Korde and Kandasubramanian 2020). For further meet actual demands, SMHs are fabricated in complex shape along with suitable stereostructure or sophisticated microstructured surfaces tightly bonding to their functions by the aid of special processing, for instance, electrospinning, four-dimension printing, origami-inspired manufacturing, and photo-induced shape transformation (Zhao et al. 2015).

2.2.3.3 Novel Shape Memory Polymer Composites

Despite remarkable advances of SMPs in multi-industries, intrinsic shortcomings of SMPs such as high transition temperature, low thermal conductivity, low recovery stress, and low stiffness limit their applications (Lei et al. 2019). Especially, SMPs with weak mechanical properties are hard to be as structural materials, and low heat conductivity results in delayed thermal response as well as slower recovery than SMA (Kim et al. 2006). Hence composite materials that comprised of two or more components can retain excellent physical properties of individual components and has been proposed to overcome these drawbacks. Shape memory polymer composites (SMPCs) are developed for enhanced recovery stresses, tunable transition temperature, desired mechanical properties, and even more function compared to those of pure SMPs (Ohki et al. 2004). Based on the disordered or ordered distribution of fillers, SMPCs are classified into hybrid composites and patterned composites.

The hybrid composites, also called nanocomposites, are filled with randomly distributed composites including inorganic materials and polymer matrix in the forms of particles or films. Inorganic materials cover silicon, carbons, metals, and metallic oxides, which possess higher modulus than polymer matrix, for example, celluloses, elastomers, and crystalline polymers with versatility. In the practical application, SiC and SiO_2 as traditional fillers can enhance the micro-hardness, elastic modulus, and the free recovery speed (Luo and Daniel 2003; Gall et al. 2002); carbon fillers, metal, and metal oxide are fabricated for remote-stimulated SMPCs with superb electrical and thermal conductivity (Zhang and Zhao 2013; Schmidt 2006; Fei et al. 2012); cellulose with good biodegradability facilitates with the mechanical properties and shape memory effect responding to water (Auad et al. 2008; Luo et al. 2012; Zhu et al. 2012); elastomers as an amorphous polymer are leveraged to accelerate the response of shape recovery (Shirole et al. 2016);

crystalline polymers are utilized for the healing of the crack at high temperature by the introduce of the self-healing properties into SMPCs (Li and Uppu 2010).

The patterned composites with ordered filling are also applied to provide more designability of SMPCs. The primary patterned fillers align carbon nanotube fiber under the applied field for better conductivity, and fracture toughness along the fiber orientation (Khan et al. 2013). Afterward, nanopapers made by carbon nanotubes/carbon nanofibers as thin membranes with specific geometry are fabricated for multiple-layer composites with high thermal and electrical conductivity (Wang et al. 2004; Lu et al. 2010, 2011). Additionally, the surface pattern as an another pattered filling technique exhibits excellent designability, and easy manipulation employed for 3D printing (Liu et al. 2012). For example, Liu et al. (2017) designed the temperature gradient-dependence foldable three-layer lotus by the use of color ink patterns that selectively absorb light. Oriented fibers and laminas, in addition to multi-blocks, are also applied for complex 3D structure and sequentially folding (Ge et al. 2013; Yu et al. 2015).

2.3 Biomedical Applications

2.3.1 Tissue Engineering

2.3.1.1 Wound Healing

Mechanical regulation for the avoidance of keloid along with electric stimulation for wound healing metabolism could be a fashionable strategy in skin repair, and SMAs can be responsible for the mechanical force of wound contraction (PE H 2017; Harn et al. 2019; Boerckel et al. 2011). Yao et al. (2022) designed electromechanical synergistic dressings (EMSDs) activated by body temperature for wound healing (Fig. 2.7). The wound dressing was comprised of programmed SMAs and electret thin film that can generate electric field for infection. The electromechanical synergistic stimulation could modulate the wound microenvironment, and increase the secretion of epidermal growth factor, transforming growth factor–β, and vascular endothelial growth factor, so as to improve the wound healing rate of over 50% versus the blank control group in rat model with both linear and circular wounds.

Materials used for wound healing are usually desired adaptable wound geometries with antibacterial and moist properties in the form of wound dressing, suture, foams, and so on. Simply, the deformable shape memory materials can temporally change the bulky and complex structure into small simple pieces for the convenience of transportation or implantation (Pilate et al. 2016). Early in 2002, Lendlein and Langer (2002) developed thermoplastic biodegradable SMPs as the surgical suture with suitable stress to avoid tissue necrosis or scar. Perfect wound dressing compactly covers wound bed to prevent infection for great prognosis surgical incisions or chronic wounds. Porous SMPs foaming SMAs can be easily deformed into temporary shape by heating, compressing, and cooling for storage and wound

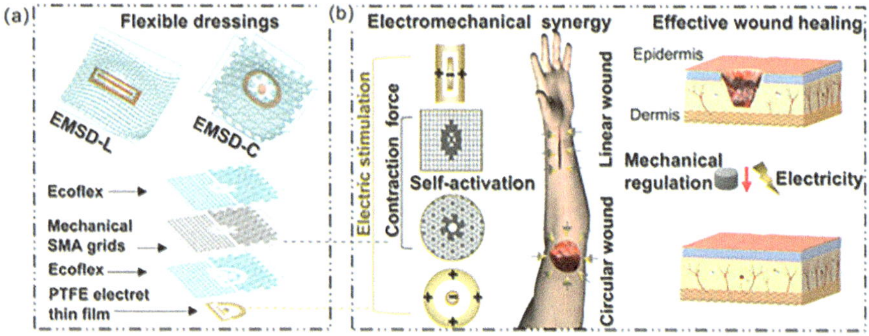

Fig. 2.7 Electromechanical synergistic dressing for wound healing. (**a**) The design principle of the flexible dressing. (**b**) Working principle of EMSDs activated by body temperature for wound healing. (Reused with permission from Yao et al. 2022)

fillings. In particular, polyurethane SMPs foams with shape memory effect could be as hemostatic material for trauma-related wounds. SMPs polyurethane foams with antimicrobial phenolic acids (p-coumaric, vanillic, and ferulic acids) were also prepared by Du et al. for traumatic wound healing, with great strengths of hemocompatibility and antioxidant properties (Du et al. 2022). Similarly, biodegradable SMPs foams modified with ether groups and ester groups were designed as shape-filling hemostatic dressing, which could be degraded completely within 30–60 days with the clinically relevant and tunable degradation rates (Vakil et al. 2021). Besides foam, SMPs-based polyvinyl alcohol (PVA)/marjoram/kaolin composite sponges with significant thrombogenicity were also developed as homeostasis and antibacterial wound dressing (Tamer et al. 2021). The response of bacterial protease for SMPs could significantly promote infected wound healing. Ramezani and Monroe fabricated a bacterial protease-responsive SMPs polyurethane with poly(glutamic acid) peptide (PU-Pep) as infected wound dressing (Fig. 2.8) (Ramezani and Monroe 2023). Poly (glutamic acid) will trigger the shape memory effect of PU-Pep films by degradation responding to bacterial proteases, at the same time, shape recovery of strained PU-Pep will also hinder the formation of biofilm on the sample surfaces with incorporated antimicrobials killing isolated bacteria. Admirably, bacterial protease-responsive wound dressing could remind clinicians of biofilm-associated infections by the specific deformation during bacterial colonization for the prevention of antibiotic abuse.

SMHs could absorb wound exudate and fill the wound completely by thermal-deformation. For example, cytocompatible polyurethane-based poly (ethylene glycol) (PEG) SMHs-incorporated plant-based phenolic acids were developed for antimicrobial wound dressing (Vakil et al. 2022). Phenolic acids with antimicrobial properties can adjust intermolecular bonding in the hydrogels to improve mechanical properties and shape memory effect, and retain antimicrobial efficacy over 20 days owing to its structures and interactions with the polymer backbone. Hydrophilic PEG with a polyurethane network as porous foams possesses shape memory effect making wound sealing, and could enhance exudate absorption to maintain a moist environment for analgesic effect.

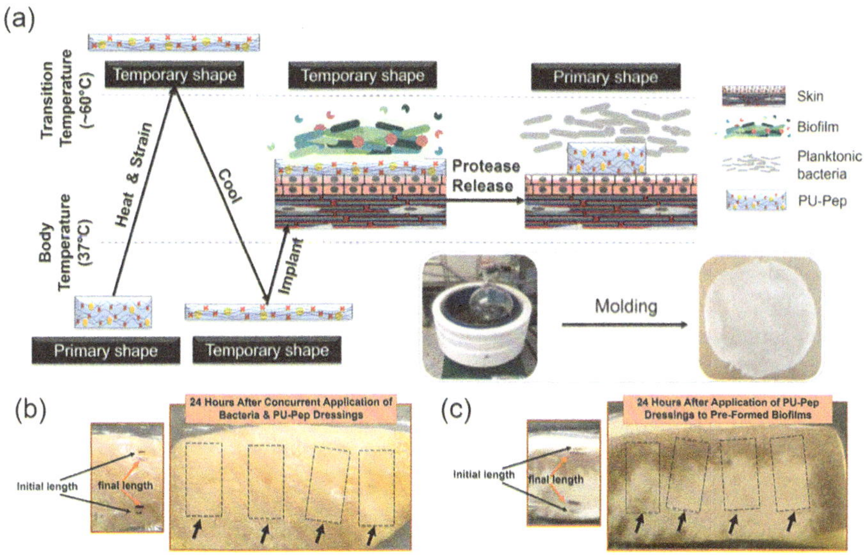

Fig. 2.8 The bacterial protease-responsive SMPs polyurethane (PU-Pep) for infected wound healing. (**a**) The PU-Pep work mechanism of biofilm suppression in an infected wound and the photography of PU-Pep. (**b**, **c**) The application of PU-Pep to wounded pig skin with bacterium (**b**) and with pre-formed biofilms (**c**). Arrows mean the applied PU-Pep wound dressings. (Reused with permission from Ramezani and Monroe 2023)

2.3.1.2 Self-Adjusting Implants

Self-adjusting implantable devices with shape memory effect are sought after in personalized medicine, as surgical implants, orthopedic implants, and cardiovascular devices are usually expected to be actuated to narrow location as a small shape in the body for minimally invasive surgeries, the reduce of surgical complications and rapid recovery (Bajpai et al. 2020). Pseudo-elastic nitinol SMAs, as common biomedical, with approximately equiatomic lattice structures hold advantages of combination of thermal and mechanical shape memory effect, desired corrosion resistance, and biocompatibility properties, and could recover rather large strains of 8% at a constant temperature, thus become the attractive material of implants (Mirzaali et al. 2022; Haberland et al. 2014). For instance, a contactless thermo-responsive SMAs nitinol implant with adjustable stiffness was developed for potential orthopedic fixation (Pfeifer et al. 2013). So far, nitinol has been extensively serviced as orthopedic implants, cardiovascular stents, and others, and the further optimization of strength, corrosion resistance, and biocompatibility of nitinol are still urged in clinical practice (Shrivastava 2004). Due to the poor osteogenesis and osseointegration, distinctive process was implemented to SMAs for the improvement of bioavailability. In one case, nanocrystalline $Ti_{49.2}Ni_{50.8}$ SMAs by the process of equal channel angular pressing, was developed that possess high bioactivity including cell attachment, bone mineralization, and bone regeneration (Li et al. 2019). In another case, additive manufacturing-assisted investment casting was manufactured in nitinol SMAs for bone plates, which allows simple or complex shapes, and even exquisite customization (Da Cruz Gomes et al. 2021). Additionally, surface modification of SMAs is necessary to avoid possible bio-inadaptability. Electrophoretic deposited nitinol SMAs with hydroxyapatite coating were fabricated to attain high hydrophilicity and biocompatibility (Fig. 2.9) (Dudek et al. 2023).

Anything else, pseudoelastic nitinol SMAs with unique nature are favorable of fracture healing and joint fusion: matching osteal mechanobiological factors during the compression of shape recovery (Safranski et al. 2020). They can be exploited as expandable pedicle screw for the enhance fixation of osteoporotic bone (Tabesh et al. 2012), minimally invasive nitinol hinge-based cage for spinal fusion surgery (Taheri Andani et al. 2014), and adaptive pedicle screws for spine surgery (Werner et al. 2020). Not only that, SMAs are also worded as artificial anal sphincter (Wang and Yu 2019) and ventricular assistive devices of heart failure patients (Hosseinipour and Elahinia 2013).

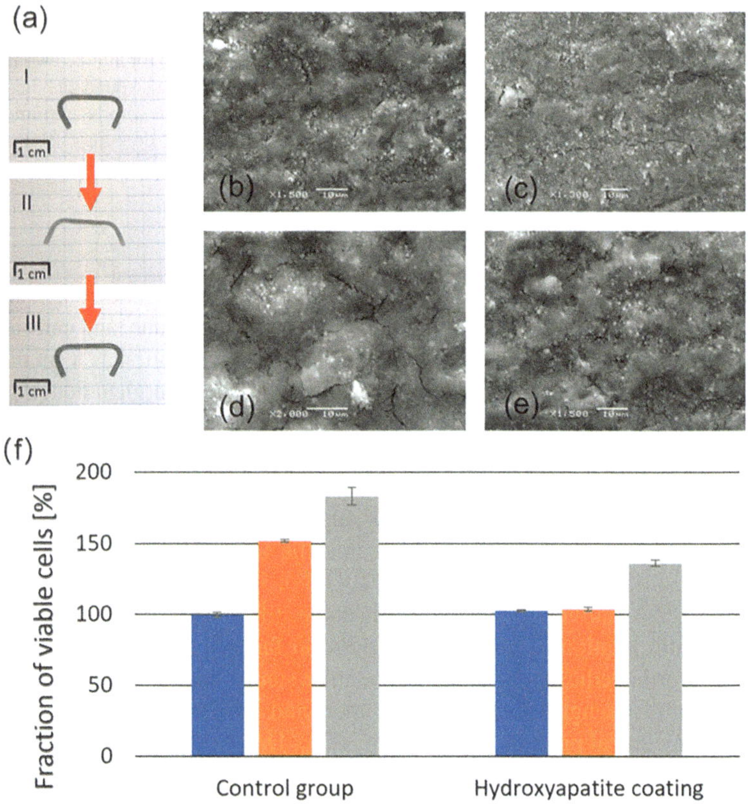

Fig. 2.9 Electrophoretic deposited nitinol SMAs. The deformation of SMAs is shown in photos (**a**) and SEM images (**b**–**e**) including the first circle (**b**-tension, **c**-compression) and the second circle (**d**-tension, **e**-compression). (**f**) The viability of normal human dermal fibroblast cells on the hydroxyapatite coating. (Reprinted with permission from Dudek et al. 2023)

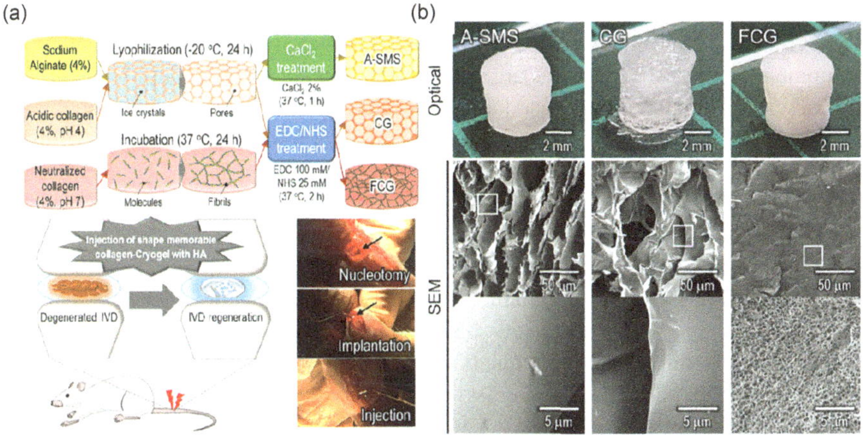

Fig. 2.10 SMPs-based scaffold for the repair of IVD. (**a**) Schematic illustrations of fabrication and implantation of SMAs-based scaffold. (**b**) Optical and scanning electron microscopy (SEM) images of the scaffold. (Reprinted with permission from Koo et al. 2023)

SMPs could transform into temporary shapes with inelastically responding to external excitation by the programs of shape memory effect, which become favorable candidates of intravascular stents, and aneurysm occluders (Senatov et al. 2017; Bodaghi et al. 2018). At the cellular level, since cell can respond to soft materials with spatial structure and load carrying, SMHs are able to support cell adhesion, proliferation, and differentiation (Nan et al. 2013; Xu et al. 2011; Han et al. 2014). Not only that, porous SMHs with interconnected structure can support the cell migration with the diffuse of oxygen, nutrients, and then can be tailored and biodegraded to satisfy the demands of tissue regeneration at different stages in tissue engineering (Annabi et al. 2010; Neffe et al. 2015). As for living bodies, Koo et al. (2023) designed SMPs-based scaffold constituted of collagen-cryogel and hyaluronic acid for the repair of Intervertebral disc degeneration (IVD) (Fig. 2.10). In the model of rat tail nucleotomy, the implant of SMPs-based scaffold with prominent chondrogenic activities can not only alleviate mechanical allodynia, but also preserve the matrix proteins matrix with sufficient water content to repair IVD.

2.3.2 *Disease Theranostics*

2.3.2.1 Medical Smart Actuator

Some special intelligent materials as self-actuators have the behavior of complex and accurate shape transformations akin to biological organs. These actuators constituted by shape memory materials have the benefits of easy controllability, high robustness after frequent deformation to achieve continuous changes and rapid actuation, which has a bright future in the fashionable applications including wearable electronic devices and bionic soft robots (Liu et al. 2023). Diversified designs and strategies of human-machine interaction have been explored though rehabilitation equipment and assistive devices are still in their infancy. SMAs could be a class of soft active materials with precise nature of rigid applied in the fields of rehabilitation and assistance with virtues of convenient interaction, successively complex motions, and resilience to a variety of environments (Pan et al. 2022). The mechanism of SMAs-based actuators is based on the thermo-mechanical deformation (from martensite to austenite) when heated. A SMAs-based three-segmented finger inspired by human musculoskeletal system was developed with compact and compliant actuation. Individual segment of the actuator can achieve bending, extend, and even full bending motion to grasp something by coordinated control of fingers combined with developed mathematical model of the SMAs (Ali et al. 2021). Furthermore, Xie et al. (2023) designed a wearable SMAs-based rehabilitation gloves with the weight of only 120 g by finger force analysis driven by distinct modes. The wearable glove driven by only energizing (2–6 V) was integrated by the modules of actuation, sensing, and execution, and has potential for bidirectional movements of fingers with cycles below 20 s and the appropriate temperature of 25 to 67 °C, which provides possible path for artificial soft exoskeletons. Additionally, SMAs-based artificial muscle was also developed to avail disabled people, especially patients with stroke and spinal cord injuries, of elbow flexion as well as forearm supination/pronation (Jeong et al. 2022). Not only that, removable SMAs clipper was used for hemostasis during laparoscopic operations, and was also applied for tubular structures, blood vessels, and soft tissues (Valiev et al. 2020). Moreover, SMAs-based actuators also could be applied for surgical robot. Magnetic resonance imaging (MRI) surgical robots could offer distinct soft-tissue contrast during therapies and surgery, and robots with gripper are required to grasp and hold different objects for doctors. Hence, a tendon-driven gripper with SMAs spring actuators and stiffness joints was designed (Fig. 2.11) (Do et al. 2023). SMAs spring actuators with water-cooling system work on the contact force of the gripper link, while stiffness of joints is affected by applied current to the SMAs coil. The capturing ability of the SMAs was also verified by a series of experiments including SMAs spring actuators, joint stiffness, and a whole gripper, that is this SMAs gripper could be applied for various surgical robots required highly precise and efficient capturing ability.

Soft actuator created by soft polymers with favorable sensitivity and agility can accommodate changeable conditions preventing mechanical damage contrasted by rigid-bodied robots with less compliance (Löwenberg et al. 2017). SMHs-based soft actuators are activated by SME under wider stimuli including external physical signals and molecular-mediated shape switches (e.g., ions, salts, and redox potential) (Beck et al. 2023). Smart actuators generally are exploited for soft robotics, artificial muscles, or microswimmers (Elango et al. 2023). Multifunctional and self-healing soft robots accommodating to varying environment have been applied in postoperative rehabilitation. Inspired by natural cephalopods, Liu and her

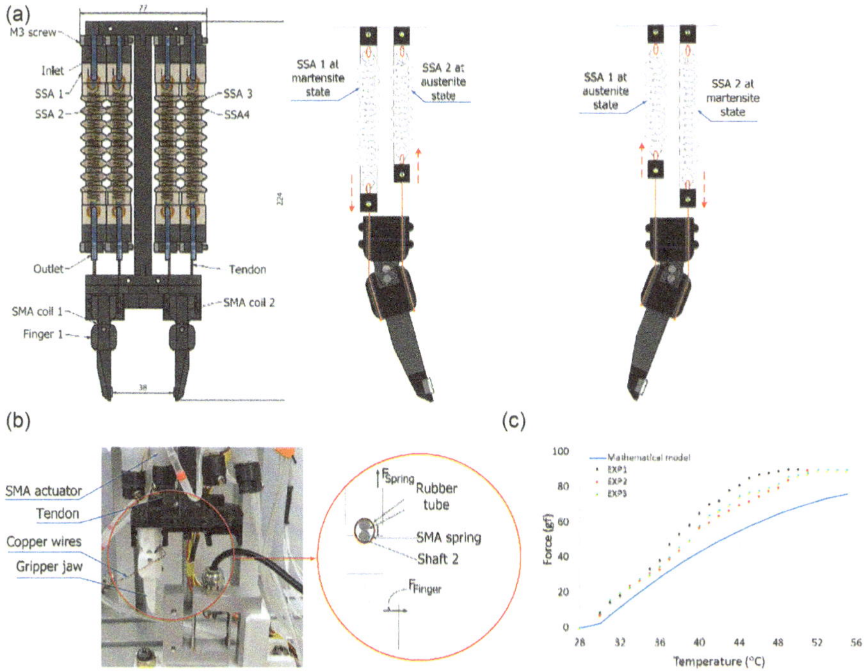

Fig. 2.11 Tendon-driven gripper with stiffness joint and SMAs springs actuators for surgical robot. (**a**) 3D design schematic of SMAs gripper (front view) and its SMAs spring actuators (SSA). The finger force measurement (**b**) and responding to the results of contact force (**c**). (Reprinted with permission from Do et al. 2023)

coworkers developed a self-healing light-driven SMAs actuator for multifunctional soft robot by photo-welding (Liu et al. 2021). Semicrystalline poly (ethylene-co-vinyl acetate) (EVA) containing silver nanowires (AgNWs) performs reversible light-trigged morphing behavior in both air and underwater and self-healing effect under photothermal effect triggering the quick assembly of the soft robotics with desired 3D structures. Significantly, the unit of these robots can be removed and exchanged for re-programmed unit on the body by photo-welding if environment alters, not to mention the self-healing function of damaged robots. In addition to SMAs gripper, a magneto-active SMPs claw made from polycaprolactone and thermoplastic polyurethane was fabricated to improve the manufacturability, and actuation modes (Yao et al. 2023). Aided by high-precision 4D printing, the claw with superior shape memory effect, magnetorheological effects, and variable stiffness shows great grasping and release capability under magnetic field, which holds huge potential to biomedicine and intelligent flexible robotics.

2.3.2.2 Drug Delivery

The dosing frequency plays an important role in medication compliance of patients with chronic disease, thus retentive drug delivery systems are proposed to prolong residence and release of drug inside the body during the last 50 years (Ingersoll and Cohen 2008). Shape memory materials-based injectable or implantable systems are not limited to gastrointestinal residence time, making possible successive doses with less side effect and discomfort (Maroni et al. 2020). For instance, Kotb et al. (2021) developed SMAs capsule micropump with a built-in replaceable drug reservoir actuated by joule heating of NiTi-alloy SMAs wires. The micropump acted as a spring made from elastic rubber material to recover the contract shape following the actuation, which structure is suitable for drug delivery owing to insulation, flexibility, as well as compatible and reusable for different drug. Further, SMAs are also made into hierarchical actuator with high actuation forces of 150 N and light weight inspired by pennate muscle, and have fantastic prospect in precise drug delivery (Chaurasiya et al. 2022).

SMHs containing bioactive molecules are capable of deliver drug to aqueous environment of body cavities (e.g., abdomen and urogenital,) by the specific anchor of medical devices. SMHs severed as a seal through spatial movements and shaped recovery before implantation (Kozlovskaya et al. 2006). In the process of drug delivery, porous morphology and switchable network structures may accelerate drug loading into the bulk, while repeated stimuli can enhance drug release during cross-link cleavage for shape recovery (Xu et al. 2015). Besides, ions also can be as temporary cross-links and released from SMHs for special medical use. For example, Xu et al. (2015) designed zinc ion-triggered multiwalled hydrogel tubes (PVV) with high strength for antibacterial and anti-inflammatory implants. By the binding of zinc with imidazole, PVV SMHs possess enhanced mechanical properties along with shape memory effect, and zinc ions as temporary cross-links can be continuously released for anti-bacterium, and holds potential as a tissue engineering scaffold. The release of ions from SMPs can also facilitate endometrial repair and regeneration. Dong et al. (2022) designed a NIR-responsive SMPs-based composite encapsulating with cuprorivaite nanosheets as photothermal conversion agents for the prevention of intrauterine adhesions (IUAs) and endometrial regeneration (Fig. 2.12). Under NIR, the shape memory composite possesses excellent shape memory effect for manipulation as an antiadhesion barrier for IUA therapy. Meanwhile, ions (Cu, Si) from cuprorivaite nanosheets were released so as to stimulate endometrial regeneration. The proliferation of human endometrial epithelial cells (HEECs) demonstrated the shape memory effect and bioactivities of the composites, while the therapeutic efficiency was verified by the recovery of the VEGF expression as well as the enhancement of phosphorylated ERK1/2 (P-ERK1/2) connected with angiogenesis and endometrial regeneration activity respectively in an IUA-induced rat model, which offer a novelty scheme to prepare artificial intrauterine devices for IUA prevention and endometrium repair.

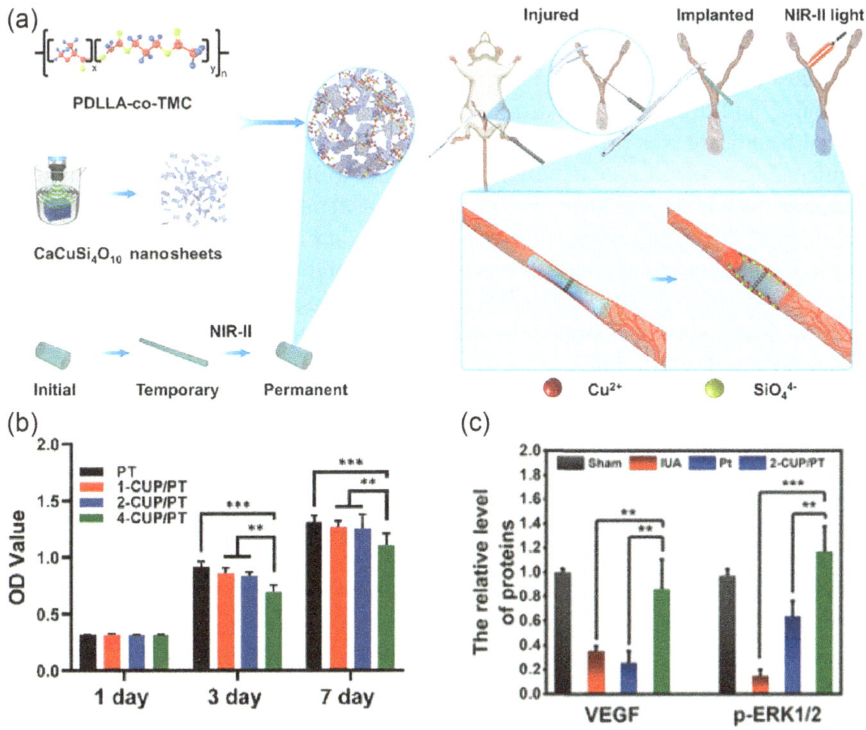

Fig. 2.12 NIR-responsive SMPs-based composite for IUA prevention and endometrial regeneration. (**a**) A schematic illustration of the composites and therapeutic plan. (**b**) Proliferation of HEECs in cubulation with composites for 1, 3, and 7 days. (**c**) Optical density analysis of the protein expressions of VEGF and P-ERK1/2 by western blotting assay. (Reprinted with permission from Dong et al. 2022)

In regard to drug delivery, the SMPs-based latanoprost solid delivery was developed by Borecka et al. for glaucoma treatment (Borecka et al. 2023). Biodegradable poly (l-lactide-co-glycolide-co-trimethylene carbonate) could be a universal drug carrier of intravitreal, subconjunctival, and subcutaneous administration apart from general subcutaneous administration with the outstanding mechanical properties of low stiffness and ductile deformation, and applied electron beam irradiation can regulate the rate of degradation and release. SMPs as drug carriers are also made into different configurations. For example, adjustable SMPs polyurethane microcapsules with morphological core-shell structure will rupture stimulated by organic solvents at the body temperature for potential drug release (Zhang et al. 2023). In another example, SMPs balloon was comprised of poly (ε-caprolactone), doxorubicin as an anticancer drug, and heat-generating magnetic nanoparticles for simultaneous thermo/chemotherapies of osteosarcoma, which allows inflation at the damaged areas with a expansion rate of 400% by simple blow molding without leakage (Ouchi et al. 2021).

2.3.2.3 Biosensors

Low-cost and portable lab-on-a-chip devices for point-of-care detection of health screening and disease diagnostics have attracted wide interests, since sustainable global healthcare was proposed (Chin et al. 2007). Some unique metal oxides or semiconductor-based nanomaterials have access to the construction of individual nanodevices with hierarchical mechanical nano-assembling for biosensors of single nano-objects that can register infinitesimal traces of biological information including individual bacteria or large molecules due to high specific surface area (Wang 2003; Janissen et al. 2017). Koledov and co-workers (Koledov et al. 2019) have made a SMAs-based nanogripper by nano-assembling to manipulation of nanowires and nanoparticles for biological nanosensors. The nanogripper made by Ti_2NiCu/Pt by focused ion beam chemical vapor deposition to generate shape memory effect active layer with the thickness of 70 nm to 1000 nm, and it can stretch up to 1% driven by laser radiation or Joule heating to manipulate (e.g., select, remove, and assemble) nanowires or nanomaterials. Similar nanosensors based on lab-on-a-chip have potential to revolutionize the medical diagnostics for precise medicine. In another case, SMAs actuator was made into a wearable wristwatch as a health monitoring system (HMS) to monitor blood glucose level of diabetic patient (Kazuyoshi et al. 2005). In this system, SMAs-based microneedles are responsible for skin penetration, while piezoelectric actuator as a pumping unit extracts the blood coupled with gold electrode as a biosensor for the detection of glucose. Interestingly, the blood collection in the device is similar to the way of female mosquito that with muscle motion to relax.

Lab-on-a-chip devices often require microfluidic electrochemical system from expensive and cumbersome semiconductor industries in the past. SMPs are engineered 3D micro/nanostructured surfaces on miscellaneous thin films with increased electroactive, which provides a cost-effective bench-top alternative for the fabrication of flexible and transparent electronic biosensors (Zhu and Moran-Mirabal 2016; Sharma et al. 2011). Poly (dimethylsiloxane) (PDMS) with shape memory effect shows great advantages with fully transparent, sufficient bond strength, and flexible mechanism, and was used for the fabrication of microfluidic electrochemical cell sensor by encapsulating gold and platinum electrodes for the detection of the adhesion of murine 3 T3 fibroblasts in connection with wound healing (Fig. 2.13) (Saem et al. 2017). For the cell sensor, the sensing ability was verified by cyclic voltammetry, and then sensed fibroblasts incubated onto the structured electrode by a redox reporter given that cell adhesion induces the decrease of charge transfer across the working electrode. The cell sensor presented the decreased signal of 61% after the adhesion of about 24 cells, showing great promise for portable point-of-care flexible devices.

Conducting hydrogels with high stretchability and flexibility are also utilized for biosensors. Poly (N-isopropylacrylamide) (PNiPAAm) SMPs with thermal response were prepared by incorporation of conducting graphite oxide/polyaniline (GO/PANI) nanocomposites for electrochemical imprinted biosensor (Wei et al. 2019).

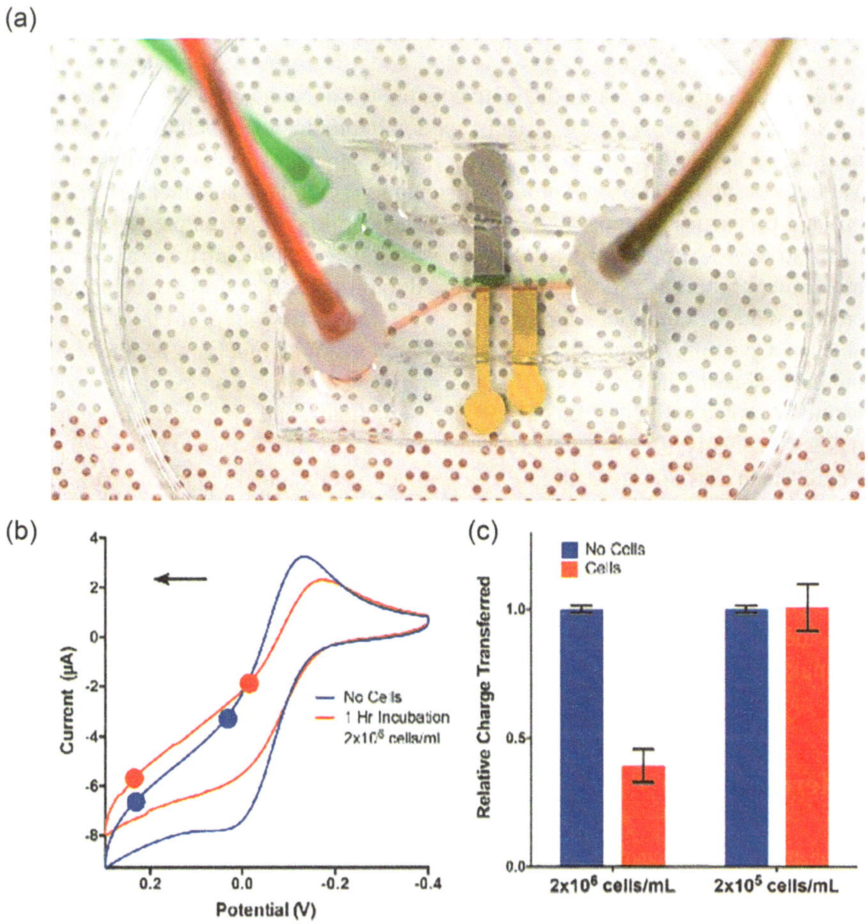

Fig. 2.13 Microfluidic electrochemical cell sensor fabricated by poly (dimethylsiloxane) (PDMS) with shape memory effect. (**a**) The structure of all-PDMS on-chip microfluidic sensor integrating gold and platinum electrodes. (**b**) The detection of murine 3 T3 fibroblast cells by cyclic voltammogram. (**c**) Relative charge transferred across the working electrode with presence of murine 3 T3 fibroblast cells. (Reprinted with permission from Saem et al. 2017)

Under near-infrared (NIR), GO/PANI converted the light energy into heat leading to reversible conformation conversion transition of PNiPAAm containing bovine serum albumin (BSA) template protein, and then the templates from BSA binding with hydrogel removed to attain sensing nanocavities with specific affinity to the BSA. Thus, the NIR-responsive electrochemical protein imprinting biosensor could absorb and desorb protein with favorable stability and sensitivity by the switch of NIR, in addition to NIR-based cleaning toward BSA without elution solution, which opens up a route for remote extraction protein smart devices.

2.4 Summary

SMAs and SMPs possess the distinctive nature of shape memory effect under external stimuli, and have been extensively applied in biomedicine including tissue engineering, drug delivery, general surgery, and other more disease theranostics. Varieties of modification and processing techniques render shape memory materials optimal properties to improve biological practicability. For example, surface treatment strategies are utilized to minimize the release of metal ions, and enhance mechanical properties such as fatigue and durability. More novelty techniques will create favorable conditions for shape memory materials with cost-effective, adaptable, and customized mechanical properties and multi-function. Moreover, with more physical fields developed, the thermal-driven force of shape memory effect would be replaced by light, electricity, magnetic fields, and even their superimposed fields. Absolutely, the bio-safety of shape memory materials should be initially considered such as tissue compatibility, biological toxicity, potential inflammation, and other body reactions. Different biomedical application has different requirements. Medical devices of tissue engineering are required to accommodate mechanical strains and transitional conditions for cell attachment, proliferation, migration, and differentiation, while implantable scaffold and wound dressing are expected to be degraded after performing their functions. In the future, shape memory materials as biomaterials would be smarter that can be customized for personalized medicine.

References

Abrahamsson P, Bjämemo R (1994) The need for product design tools in shape memory technology. In: Yamamoto R, Furubayashi E, Doi Y, Fang R, Liu B, Otsuka K, Liu CT, Shimizu K, Suzuki Y, Van Humbeeck J, Fukai Y, Ono S, Suda S (eds) Ecomaterials. Elsevier, pp 1171–1174

Abuzaid W, Sehitoglu H (2018) Superelasticity and functional fatigue of single crystalline FeNiCoAlTi iron-based shape memory alloy. Mater Des 160:642–651

Acet M, Mañosa L, Planes A (2011) Chapter four - magnetic-field-induced effects in martensitic Heusler-based magnetic shape memory alloys. In: Buschow KHJ (ed) Handbook of magnetic materials, vol 19. Elsevier, pp 231–289

Ali HFM, Khan AM, Baek H, Shin B, Kim Y (2021) Modeling and control of a finger-like mechanism using bending shape memory alloys. Microsyst Technol 27(6):2481–2492

Annabi N, Nichol JW, Zhong X, Ji C, Koshy S, Khademhosseini A, Dehghani F (2010) Controlling the porosity and microarchitecture of hydrogels for tissue engineering. Tissue Eng Part B Rev 16(4):371–383

Ansari M, Golzar M, Baghani M, Abbasishirsavar M, Taghavimehr M (2019) Force recovery evaluation of thermo-induced shape-memory polymer stent: material, process and thermo-viscoelastic characterization. Smart Mater Struct 28(9):095022

Arnebold A, Hartwig A (2016) Fast switchable, epoxy based shape-memory polymers with high strength and toughness. Polymer 83:40–49

Asle Zaeem M, Zhang N, Mamivand M (2019) A review of computational modeling techniques in study and design of shape memory ceramics. Comput Mater Sci 160:120–136

Athimoolam M, Moorthy TV (2012) Polymer nanocomposite materials and shape memory applications-a review. Procedia Eng 38:3399–3408
Auad ML, Contos VS, Nutt S, Aranguren MI, Marcovich NE (2008) Characterization of nanocellulose- reinforced shape memory polyurethanes. Polym Int 57(4):651–659
Babaahmadi M, Sabzi M, Mahdavinia GR, Keramati M (2017) Preparation of amorphous nanocomposites with quick heat triggered shape memory behavior. Polymer 112:26–34
Bajpai A, Baigent A, Raghav S, Brádaigh CÓ, Koutsos V, Radacsi N (2020) 4D printing: materials, technologies, and future applications in the biomedical field. Sustain For 12(24):10628
Bao M, Lou X, Zhou Q, Dong W, Yuan H, Zhang Y (2014) Electrospun biomimetic fibrous scaffold from shape memory polymer of PDLLA-co-TMC for bone tissue engineering. ACS Appl Mater Interfaces 6(4):2611–2621
Beck A, Obst F, Gruner D, Voigt A, Mehner PJ, Gruenzner S, Koerbitz R, Shahadha MH, Kutscher A, Paschew G, Marschner U, Richter A (2023) Fundamentals of hydrogel-based valves and chemofluidic transistors for lab-on-a-chip technology: a tutorial review. Adv Mater Technol 8(3):2200417
Behl M, Lendlein A (2007) Shape-memory polymers. Mater Today 10(4):20–28
Bodaghi M, Damanpack AR, Liao WH (2018) Triple shape memory polymers by 4D printing. Smart Mater Struct 27(6):065010
Boerckel JD, Uhrig BA, Willett NJ, Huebsch N, Guldberg RE (2011) Mechanical regulation of vascular growth and tissue regeneration in vivo. Proc Natl Acad Sci 108(37):E674–E680
Borecka A, Rech J, Janeczek H, Wilińska J, Kasperczyk J, Kobielarz M, Grieb P, Turek A (2023) Development of the Latanoprost solid delivery system based on poly(l-lactide-co-glycolide-co-trimethylene carbonate) with shape memory for glaucoma treatment. Appl Sci [Online] 13
Buehler WJ, Wang FE (1968) A summary of recent research on the nitinol alloys and their potential application in ocean engineering. Ocean Eng 1(1):105–120
Buehler WJ, Gilfrich JV, Wiley RC (1963) Effect of low-temperature phase changes on the mechanical properties of alloys near composition TiNi. J Appl Phys 34(5):1475–1477
Buehler WJ, Gilfrich JV, Wiley RC (2004) Effect of low-temperature phase changes on the mechanical properties of alloys near composition TiNi. J Appl Phys 34(5):1475–1477
Bystrom A, Almin KE (1947) X-ray investigation of gold-cadmium alloys rich in gold. Acta Chem Scand 1(1):76–89
Čebela M, Zagorac D, Batalović K, Radaković J, Stojadinović B, Spasojević V, Hercigonja R (2017) $BiFeO_3$ perovskites: a multidisciplinary approach to multiferroics. Ceram Int 43(1, Part B):1256–1264
Chang LC, Read TA (1951) Plastic deformation and Diffusionless phase changes in metals — the gold-cadmium Beta phase. JOM 3(1):47–52
Chatterjee T, Dey P, Nando GB, Naskar K (2015) Thermo-responsive shape memory polymer blends based on alpha olefin and ethylene propylene diene rubber. Polymer 78:180–192
Chaurasiya KL, Harsha AS, Sinha Y, Bhattacharya B (2022) Design and development of non-magnetic hierarchical actuator powered by shape memory alloy based bipennate muscle. Sci Rep 12(1):10758
Chen Y, Schuh CA (2011) Size effects in shape memory alloy microwires. Acta Mater 59(2):537–553
Chen Y, Zhang X, Dunand DC, Schuh CA (2009) Shape memory and superelasticity in polycrystalline Cu–Al–Ni microwires. Appl Phys Lett 95(17):171906
Chen Y, Sun S, Lu D, Shi Y, Yao Y (2019) Water-soluble supramolecular polymers constructed by macrocycle-based host-guest interactions. Chin Chem Lett 30(1):37–43
Chin CD, Linder V, Sia SK (2007) Lab-on-a-chip devices for global health: past studies and future opportunities. Lab Chip 7(1):41–57
Chluba C, Ge W, Lima de Miranda R, Strobel J, Kienle L, Quandt E, Wuttig M (2015) Ultralow-fatigue shape memory alloy films. Science 348(6238):1004–1007
Christian JW (2002) CHAPTER 26 – shape memory alloys. In: Christian JW (ed) The theory of transformations in metals and alloys. Pergamon, Oxford, pp 1102–1113

Chumlyakov YI, Kireeva IV, Kuksgauzen IV, Poklonov VV, Pobedennaya ZV, Bessonova IG, Kirillov VA, Lauhoff C, Niendorf T, Krooß P (2020) Superelasticity and shape memory effect under tension and compression in the [001]-oriented single crystals of non-equiatomic FeNiCoAlTi high-entropy alloy. Russ Phys J 62(12):2296–2305

Cladera A, Weber B, Leinenbach C, Czaderski C, Shahverdi M, Motavalli M (2014) Iron-based shape memory alloys for civil engineering structures: an overview. Constr Build Mater 63:281–293

Crystal IR, Lai A, Schuh CA (2020) Cyclic martensitic transformations and damage evolution in shape memory zirconia: single crystals vs polycrystals. J Am Ceram Soc 103(8):4678–4690

Cui J, Shield TW, James RD (2004) Phase transformation and magnetic anisotropy of an iron–palladium ferromagnetic shape-memory alloy. Acta Mater 52(1):35–47

Da Cruz Gomes AA, Grassi END, da Silva PCS, de Araújo CJ (2021) Mechanical behavior of a NiTi superelastic bone plate obtained by investment casting assisted by additive manufacturing. Smart Mater Struct 30(2):025009

Dasgupta R (2014) A look into cu-based shape memory alloys: present scenario and future prospects. J Mater Res 29(16):1681–1698

Dasgupta R, Jain AK, Kumar P, Hussein S, Pandey A (2014) Effect of alloying constituents on the martensitic phase formation in some Cu-based SMAs. J Mater Res Technol 3(3):264–273

Ding Q, Wu Z, Tao K, Wei Y, Wang W, Yang B-R, Xie X, Wu J (2022) Environment tolerant, adaptable and stretchable organohydrogels: preparation, optimization, and applications. Mater Horizons 9(5):1356–1386

Do PT, Le QN, Luong QV, Kim H-H, Park H-M, Kim Y-J (2023) Tendon-driven gripper with variable stiffness joint and water-cooled SMA springs. Actuators [Online] 12

Dong C, Yang C, Younis MR, Zhang J, He G, Qiu X, Fu L-H, Zhang D-Y, Wang H, Hong W, Lin J, Wu X, Huang P (2022) Bioactive NIR-II light-responsive shape memory composite based on Cuprorivaite nanosheets for endometrial regeneration. Adv Sci 9(12):2102220

Du Z, Zeng XM, Liu Q, Lai A, Amini S, Miserez A, Schuh CA, Gan CL (2015) Size effects and shape memory properties in ZrO2 ceramic micro- and nano-pillars. Scr Mater 101:40–43

Du Z, Ye P, Zeng XM, Schuh CA, Tamura N, Zhou X, Gan CL (2017a) Synthesis of monodisperse CeO_2–ZrO_2 particles exhibiting cyclic superelasticity over hundreds of cycles. J Am Ceram Soc 100(9):4199–4208

Du Z, Zeng XM, Liu Q, Schuh CA, Gan CL (2017b) Superelasticity in micro-scale shape memory ceramic particles. Acta Mater 123:255–263

Du C, Liu J, Fikhman DA, Dong KS, Monroe MBB (2022) Shape memory polymer foams with phenolic acid-based antioxidant and antimicrobial properties for traumatic wound healing. Front Bioeng Biotechnol 10:10

Dudek K, Goryczka T, Dulski M, Psiuk B, Szurko A, Lekston Z (2023) Functionalization of the implant surface made of NiTi shape memory alloy. Materials 16(4):1609

Duerig TW, Pelton AR (1994) TiNi shape memory alloys. Asvanced materials. Materials Park, pp 1035–1048

Duerig TW, Melton KN, Stockel D, Wayman CM (1990) Engineering aspects of shape memory alloys. Butterworth Heinemann Publishing, London

Dunand DC, Müllner P (2011) Size effects on magnetic actuation in Ni-Mn-Ga shape-memory alloys. Adv Mater 23(2):216–232

Elahinia MH, Hashemi M, Tabesh M, Bhaduri SB (2012) Manufacturing and processing of NiTi implants: a review. Prog Mater Sci 57(5):911–946

Elango J, Lijnev A, Zamora-Ledezma C, Alexis F, Wu W, Marín JMG, Sanchez de Val JEM (2023) The relationship of rheological properties and the performance of silk fibroin hydrogels in tissue engineering application. Process Biochem 125:198–211

Erb DJ, Rauch HA, Knight KP, Yu HZ (2023) Viewpoint: tuning the martensitic transformation mode in shape memory ceramics via Mesostructure and microstructure design. Shape Memory Superelast 9(1):116–126

Fei G, Li G, Wu L, Xia H (2012) A spatially and temporally controlled shape memory process for electrically conductive polymer–carbon nanotube composites. Soft Matter 8(19):5123–5126
Gall K, Mikulas M, Munshi NA, Beavers F, Tupper M (2000) Carbon fiber reinforced shape memory polymer composites. J Intell Mater Syst Struct 11(11):877–886
Gall K, Dunn ML, Liu Y, Finch D, Lake M, Munshi NA (2002) Shape memory polymer nanocomposites. Acta Mater 50(20):5115–5126
Gao W, Zhang Y, Ramanujan D, Ramani K, Chen Y, Williams CB, Wang CCL, Shin YC, Zhang S, Zavattieri PD (2015) The status, challenges, and future of additive manufacturing in engineering. Comput Aided Des 69:65–89
Ge Q, Qi HJ, Dunn ML (2013) Active materials by four-dimension printing. Appl Phys Lett 103(13):131901
Gil FJ, Planell JA (1999) Effect of copper addition on the superelastic behavior of Ni-Ti shape memory alloys for orthodontic applications. J Biomed Mater Res 48(5):682–688
Greninger AB, Mooradian VG (1938) Strain transformation in metastable beta copper-zinc and beta copper-Ti alloys. Aime Trans 128:337–369
Haberland C, Elahinia M, Walker JM, Meier H, Frenzel J (2014) On the development of high quality NiTi shape memory and pseudoelastic parts by additive manufacturing. Smart Mater Struct 23(10):104002
Han Y, Bai T, Liu W (2014) Controlled heterogeneous stem cell differentiation on a shape memory hydrogel surface. Sci Rep 4(1):5815
Hannink RHJ, Kelly PM, Muddle BC (2000) Transformation toughening in zirconia-containing ceramics. J Am Ceram Soc 83(3):461–487
Hao S, Cui L, Jiang D, Han X, Ren Y, Jiang J, Liu Y, Liu Z, Mao S, Wang Y, Li Y, Ren X, Ding X, Wang S, Yu C, Shi X, Du M, Yang F, Zheng Y, Zhang Z, Li X, Brown DE, Li J (2013) A transforming metal nanocomposite with large elastic strain, low modulus, and high strength. Science 339(6124):1191–1194
Hao Y, Ji Y, Zhang Z, Yin M, Liu C, Zhao H, Otsuka K, Ren X (2019) Strain glass in Ti50–xNi35+xCu15 shape memory alloys. Scr Mater 168:71–75
Harn HIC, Ogawa R, Hsu C-K, Hughes MW, Tang M-J, Chuong C-M (2019) The tension biology of wound healing. Exp Dermatol 28(4):464–471
Hornbogen E (2006) Comparison of shape memory metals and polymers. Adv Eng Mater 8(1–2):101–106
Hosseinipour M, Elahinia M (2013) Design and simulation of an intra-ventricular assistive device for end stage congestive heart failure patients. ProcSPIE:86860M
Hu J, Chen S (2010) A review of actively moving polymers in textile applications. J Mater Chem 20(17):3346–3355
Hu J, Zhu Y, Huang H, Lu J (2012) Recent advances in shape–memory polymers: structure, mechanism, functionality, modeling and applications. Prog Polym Sci 37(12):1720–1763
Hu Y, Lu C-H, Guo W, Aleman-Garcia MA, Ren J, Willner I (2015) A shape memory acrylamide/DNA hydrogel exhibiting switchable dual pH-responsiveness. Adv Funct Mater 25(44):6867–6874
Huang R, Riddle M, Graziano D, Warren J, Das S, Nimbalkar S, Cresko J, Masanet E (2016) Energy and emissions saving potential of additive manufacturing: the case of lightweight aircraft components. J Clean Prod 135:1559–1570
Huebsch WW, Hamburg SD, Guiler RW (2012) 3 – aircraft morphing technologies. In: Young TM, Hirst M (eds) Innovation in aeronautics. Woodhead Publishing, pp 37–55
Ingersoll KS, Cohen J (2008) The impact of medication regimen factors on adherence to chronic treatment: a review of literature. J Behav Med 31(3):213–224
Janissen R, Sahoo PK, Santos CA, da Silva AM, von Zuben AAG, Souto DEP, Costa ADT, Celedon P, Zanchin NIT, Almeida DB, Oliveira DS, Kubota LT, Cesar CL, de Souza AP, Cotta MA (2017) InP nanowire biosensor with tailored biofunctionalization: ultrasensitive and highly selective disease biomarker detection. Nano Lett 17(10):5938–5949

Jeong J, Hyeon K, Jang SY, Chung C, Hussain S, Ahn SY, Bok SK, Kyung KU (2022) Soft wearable robot with shape memory alloy (SMA)-based artificial muscle for assisting with elbow flexion and forearm supination/pronation. IEEE Robot Automat Lett 7(3):6028–6035
Jing X, Mi H-Y, Huang H-X, Turng L-S (2016) Shape memory thermoplastic polyurethane (TPU)/poly(ε-caprolactone) (PCL) blends as self-knotting sutures. J Mech Behav Biomed Mater 64:94–103
Juan JS, Nó ML, Schuh CA (2009) Nanoscale shape-memory alloys for ultrahigh mechanical damping. Nat Nanotechnol 4(7):415–419
Kabir MH, Makino M, Gong J, Furukawa H (2014) Structural analysis of shape memory gel. Adv Mater Res 922:248–253
Kahn H, Huff MA, Heuer AH (1998) The TiNi shape-memory alloy and its applications for MEMS. J Micromech Microeng 8(3):213
Kakeshita T, Ullakko K (2002) Giant Magnetostriction in ferromagnetic shape-memory alloys. MRS Bull 27(2):105–109
Karaca HE, Kaya I, Tobe H, Basaran B, Nagasako M, Kainuma R, Chumlyakov Y (2013) Shape memory behavior of high strength Ni54Ti46 alloys. Mater Sci Eng A 580:66–70
Karger-Kocsis J, Kéki S (2018) Review of Progress in shape memory epoxies and their composites. Polymers [online] 10
Kashif M, Yun B-M, Lee K-S, Chang Y-W (2016) Biodegradable shape-memory poly(ε-caprolactone)/polyhedral oligomeric silsequioxane nanocomposites: sustained drug release and hydrolytic degradation. Mater Lett 166:125–128
Kauffman GB, Mayo I (1997) The story of Nitinol: the serendipitous discovery of the memory metal and its applications. Chem Educ 2(2):1–21
Kazuyoshi T, Naoyuki N, Eiji N (2005) Development of blood extraction system designed by female mosquito's blood sampling mechanism for bio-MEMS. Proc SPIE:379–388
Kelly PM, Francis Rose LR (2002) The martensitic transformation in ceramics — its role in transformation toughening. Prog Mater Sci 47(5):463–557
Khademeh Molavi F, Ghasemi I, Messori M, Esfandeh M (2017) Nanocomposites based on poly(l-lactide)/poly(ε-caprolactone) blends with triple-shape memory behavior: effect of the incorporation of graphene nanoplatelets (GNps). Compos Sci Technol 151:219–227
Khan SU, Pothnis JR, Kim J-K (2013) Effects of carbon nanotube alignment on electrical and mechanical properties of epoxy nanocomposites. Compos A: Appl Sci Manuf 49:26–34
Kim B, Lee MG, Lee YP, Kim Y, Lee G (2006) An earthworm-like micro robot using shape memory alloy actuator. Sensors Actuators A Phys 125(2):429–437
Kim M-S, Heo J-K, Rodrigue H, Lee H-T, Pané S, Han M-W, Ahn S-H (2023) Shape memory alloy (SMA) actuators: the role of material, form, and scaling effects. Adv Mater 35(33):2208517
Kimura T, Newnham RE, Cross LE (1981) Shape-memory effect in PLZT ferroelectric ceramics. Phase Transit 2(2):113–130
Kisi EH, Howard CJ (1998) Crystal structures of zirconia phases and their inter-relation. Key Eng Mater 153-154:1–36
Koledov V, von Gratowski S, Dan Nguyen H, Le Ngo Thi H, Ky V, Santos GN, Oo TZ (2019) Nano-manipulation and nano-assembling using shape memory alloy nanogripper of metal oxide and semiconductor single nanowires and nanoparticles for biological nanosensors. Adv Nat Sci Nanosci Nanotechnol 10(3):035003
König D, Naujoks D, de los Arcos T, Grosse-Kreul S, Ludwig A (2015) X-ray photoelectron spectroscopy investigations of the surface reaction layer and its effects on the transformation properties of nanoscale Ti51Ni38Cu11 shape memory thin films. Adv Eng Mater 17(5):669–673
Koo YW, Lim CS, Darai A, Lee J, Kim W, Han I, Kim GH (2023) Shape-memory collagen scaffold combined with hyaluronic acid for repairing intervertebral disc. Biomater Res 27(1):26
Korde JM, Kandasubramanian B (2020) Naturally biomimicked smart shape memory hydrogels for biomedical functions. Chem Eng J 379:122430
Kotb Y, Elgamal I, Serry M (2021) Shape memory alloy capsule micropump for drug delivery applications. Micromachines [Online]

Kozlovskaya V, Kharlampieva E, Mansfield ML, Sukhishvili SA (2006) Poly(methacrylic acid) hydrogel films and capsules: response to pH and ionic strength, and encapsulation of macromolecules. Chem Mater 18(2):328–336

Krenke T, Duman E, Acet M, Wassermann EF, Moya X, Mañosa L, Planes A (2005) Inverse magnetocaloric effect in ferromagnetic Ni–Mn–Sn alloys. Nat Mater 4(6):450–454

Krenke T, Duman E, Acet M, Wassermann EF, Moya X, Mañosa L, Planes A, Suard E, Ouladdiaf B (2007) Magnetic superelasticity and inverse magnetocaloric effect in Ni-Mn-In. Phys Rev B 75(10):104414

Krooß P, Somsen C, Niendorf T, Schaper M, Karaman I, Chumlyakov Y, Eggeler G, Maier HJ (2014) Cyclic degradation mechanisms in aged FeNiCoAlTa shape memory single crystals. Acta Mater 79:126–137

Kumar Patel K, Purohit R (2018) Future prospects of shape memory polymer nano-composite and epoxy based shape memory polymer – a review. Materials Today: Proceedings 5(9, Part 3):20193–20200

Lagoudas DC (2010) Shape memory alloys, 1st edn. Springer, New York

Lai A, Du Z, Gan CL, Schuh CA (2013) Shape memory and Superelastic ceramics at small scales. Science 341(6153):1505–1508

Lee J, Kang S-K (2021) Principles for controlling the shape recovery and degradation behavior of biodegradable shape-memory polymers in biomedical applications. Micromachines [Online] 12

Lei M, Chen Z, Lu H, Yu K (2019) Recent progress in shape memory polymer composites: methods, properties, applications and prospects. 8(1):327–351

Lendlein A, Langer R (2002) Biodegradable, elastic shape-memory polymers for potential biomedical applications. Science 296(5573):1673–1676

Leng J, Lu H, Liu Y, Huang WM, Du S (2009) Shape-memory polymers—a class of novel smart materials. MRS Bull 34(11):848–855

Lester BV, Vernon HM (1941) Process of manufacturing articles of thermoplastic synthetic resins

Li G, Uppu N (2010) Shape memory polymer based self-healing syntactic foam: 3-D confined thermomechanical characterization. Compos Sci Technol 70(9):1419–1427

Li G, Xu W (2011) Thermomechanical behavior of thermoset shape memory polymer programmed by cold-compression: testing and constitutive modeling. J Mech Phys Solids 59(6):1231–1250

Li HF, Nie FL, Zheng YF, Cheng Y, Wei SC, Valiev RZ (2019) Nanocrystalline Ti49.2Ni50.8 shape memory alloy as orthopaedic implant material with better performance. J Mater Sci Technol 35(10):2156–2162

Li H, Wang P, Wen C (2022) Recent progress on nanocrystalline metallic materials for biomedical applications. Nanomaterials [Online] 12

Likhachev AA, Sozinov A, Ullakko K (2006) Modeling the strain response, magneto-mechanical cycling under the external stress, work output and energy losses in Ni–Mn–Ga. Mech Mater 38(5):551–563

Liu Y, Boyles JK, Genzer J, Dickey MD (2012) Self-folding of polymer sheets using local light absorption. Soft Matter 8(6):1764–1769

Liu Y, Shaw B, Dickey MD, Genzer J (2017) Sequential self-folding of polymer sheets. Sci Adv 3(3):e1602417

Liu M, Zhu S, Huang Y, Lin Z, Liu W, Yang L, Ge D (2021) A self-healing composite actuator for multifunctional soft robot via photo-welding. Compos Part B 214:108748

Liu H, Liu R, Chen K, Liu Y, Zhao Y, Cui X, Tian Y (2023) Bioinspired gradient structured soft actuators: from fabrication to application. Chem Eng J 461:141966

Löwenberg C, Balk M, Wischke C, Behl M, Lendlein A (2017) Shape-memory hydrogels: evolution of structural principles to enable shape switching of hydrophilic polymer networks. Acc Chem Res 50(4):723–732

Lu H, Liu Y, Gou J, Leng J, Du S (2010) Electrical properties and shape-memory behavior of self-assembled carbon nanofiber nanopaper incorporated with shape-memory polymer. Smart Mater Struct 19(7):075021

Lu H, Liang F, Gou J (2011) Nanopaper enabled shape-memory nanocomposite with vertically aligned nickel nanostrand: controlled synthesis and electrical actuation. Soft Matter 7(16):7416–7423

Luo J-J, Daniel IM (2003) Characterization and modeling of mechanical behavior of polymer/clay nanocomposites. Compos Sci Technol 63(11):1607–1616

Luo H, Hu J, Zhu Y, Zhang S, Fan Y, Ye G (2012) Achieving shape memory: reversible behaviors of cellulose–PU blends in wet–dry cycles. J Appl Polym Sci 125(1):657–665

Ma J, Karaman I, Noebe RD (2010) High temperature shape memory alloys. Int Mater Rev 55(5):257–315

Machado LG, Savi MA (2003) Medical applications of shape memory alloys. Braz J Med Biol Res 36:683

Mallik US, Sampath V (2008) Effect of composition and ageing on damping characteristics of cu–Al–Mn shape memory alloys. Mater Sci Eng A 478(1):48–55

Mañosa L, Planes A (2016) Mechanocaloric effects in shape memory alloys. Philos Trans R Soc A Math Phys Eng Sci 374(2074):20150310

Maroni A, Melocchi A, Zema L, Foppoli A, Gazzaniga A (2020) Retentive drug delivery systems based on shape memory materials. J Appl Polym Sci 137(25):48798

Mazzer EM, Kiminami CS, Gargarella P, Cava RD, Basilio LA, Bolfarini C, Botta WJ, Eckert J, Gustmann T, Pauly S (2014) Atomization and selective laser melting of a Cu-Al-Ni-Mn shape memory alloy. Mater Sci Forum 802:343–348

Mazzer EM, da Silva MR, Gargarella P (2022) Revisiting cu-based shape memory alloys: recent developments and new perspectives. J Mater Res 37(1):162–182

Melchels FPW, Feijen J, Grijpma DW (2010) A review on stereolithography and its applications in biomedical engineering. Biomaterials 31(24):6121–6130

Melly SK, Liu L, Liu Y, Leng J (2020) Active composites based on shape memory polymers: overview, fabrication methods, applications, and future prospects. J Mater Sci 55(25):10975–11051

Memarian F, Fereidoon A, Ghorbanzadeh Ahangari M (2016) The shape memory, and the mechanical and thermal properties of TPU/ABS/CNT: a ternary polymer composite. RSC Adv 6(103):101038–101047

Mertmann M, Vergani G (2008) Design and application of shape memory actuators. Eur Phys J Spec Topics 158(1):221–230

Mihálcz I (2001) Fundamental characteristics and design method for nickel-titanium shape memory alloy. Periodica Polytech Mechan Eng 45:75–86

Mirzaali MJ, Moosabeiki V, Rajaai SM, Zhou J, Zadpoor AA (2022) Additive manufacturing of biomaterials–design principles and their implementation. Materials [online] 15

Mohd Jani J, Leary M, Subic A, Gibson MA (2014) A review of shape memory alloy research, applications and opportunities. Mater Des (1980–2015) 56:1078–1113

Montgomery M, Ahadian S, Davenport Huyer L, Lo Rito M, Civitarese RA, Vanderlaan RD, Wu J, Reis LA, Momen A, Akbari S, Pahnke A, Li R-K, Caldarone CA, Radisic M (2017) Flexible shape-memory scaffold for minimally invasive delivery of functional tissues. Nat Mater 16(10):1038–1046

Morgan NB (2004) Medical shape memory alloy applications—the market and its products. Mater Sci Eng A 378(1):16–23

Nan W, Wang W, Gao H, Liu W (2013) Fabrication of a shape memory hydrogel based on imidazole–zinc ion coordination for potential cell-encapsulating tubular scaffold application. Soft Matter 9(1):132–137

Nayak VV, Slavin BV, Bergamo ETP, Torroni A, Runyan CM, Flores RL, Kasper FK, Young S, Coelho PG, Witek L (2023) Three-dimensional printing bioceramic scaffolds using direct-ink-writing for Craniomaxillofacial bone regeneration. Tissue Eng Part C Methods 29(7):332–345

Neffe AT, Pierce BF, Tronci G, Ma N, Pittermann E, Gebauer T, Frank O, Schossig M, Xu X, Willie BM, Forner M, Ellinghaus A, Lienau J, Duda GN, Lendlein A (2015) One step creation of multifunctional 3D Architectured hydrogels inducing bone regeneration. Adv Mater 27(10):1738–1744

Ning F, Cong W, Wei J; Wang S, Zhang M (2015) ASME 2015 International Manufacturing Science and Engineering Conference MSEC 2015, p 369

Ohki T, Ni Q-Q, Ohsako N, Iwamoto M (2004) Mechanical and shape memory behavior of composites with shape memory polymer. Compos A: Appl Sci Manuf 35(9):1065–1073

Ölander A (1932) The Crystal structure of AuCd. Zeitschrift für Kristallographie Crystall Mater 83(1–6):145–148

Omori T, Kusama T, Kawata S, Ohnuma I, Sutou Y, Araki Y, Ishida K, Kainuma R (2013a) Abnormal grain growth induced by cyclic heat treatment. Science 341(6153):1500–1502

Omori T, Abe S, Tanaka Y, Lee DY, Ishida K, Kainuma R (2013b) Thermoelastic martensitic transformation and superelasticity in Fe–Ni–Co–Al–Nb–B polycrystalline alloy. Scr Mater 69(11):812–815

Otsuka K, Ren X (2005) Physical metallurgy of Ti–Ni-based shape memory alloys. Prog Mater Sci 50(5):511–678

Ouchi S, Niiyama E, Sugo K, Uto K, Takenaka S, Kikuchi A, Ebara M (2021) Shape-memory balloon offering simultaneous thermo/chemotherapies to improve anti-osteosarcoma efficacy. Biomater Sci 9(20):6957–6965

Pagounis E, Chulist R, Szczerba MJ, Laufenberg M (2014) High-temperature magnetic shape memory actuation in a Ni–Mn–Ga single crystal. Scr Mater 83:29–32

Pan M, Yuan C, Liang X, Dong T, Liu T, Zhang J, Zou J, Yang H, Bowen C (2022) Soft actuators and robotic devices for rehabilitation and assistance. Adv Intell Syst 4(4):2100140

PE H (2017) Electrical stimulation therapy to promote healing of chronic wounds: a review of reviews. Chronic Wound Care Manage Res 4:25–44

PE W (1967) Elastic memory articles

Peponi L, Sessini V, Arrieta MP, Navarro-Baena I, Sonseca A, Dominici F, Gimenez E, Torre L, Tercjak A, López D, Kenny JM (2018) Thermally-activated shape memory effect on biodegradable nanocomposites based on PLA/PCL blend reinforced with hydroxyapatite. Polym Degrad Stab 151:36–51

Pérez-Checa A, Feuchtwanger J, Barandiaran JM, Sozinov A, Ullakko K, Chernenko VA (2018) Ni-Mn-Ga-(Co, Fe, Cu) high temperature ferromagnetic shape memory alloys: effect of Mn and Ga replacement by Cu. Scr Mater 154:131–133

Petrini L, Migliavacca F (2011) Biomedical applications of shape memory alloys. J Metall 2011:501483

Pfeifer R, Hustedt M, Wesling V, Hurschler C, Olender G, Mach M, Gösling T, Müller CW (2013) Noninvasive induction implant heating: an approach for contactless altering of mechanical properties of shape memory implants. Med Eng Phys 35(1):54–62

Pilate F, Toncheva A, Dubois P, Raquez J-M (2016) Shape-memory polymers for multiple applications in the materials world. Eur Polym J 80:268–294

Planes A, Mañosa L, Acet M (2009) Magnetocaloric effect and its relation to shape-memory properties in ferromagnetic Heusler alloys. J Phys Condens Matter 21(23):233201

Polatidis E, Zotov N, Bischoff E, Mittemeijer EJ (2015) The effect of cyclic tensile loading on the stress-induced transformation mechanism in superelastic NiTi alloys: an in-situ X-ray diffraction study. Scr Mater 100:59–62

Prokoshkin S, Brailovski V, Dubinskiy S, Zhukova Y, Sheremetyev V, Konopatsky A, Inaekyan K (2016) Manufacturing, structure control, and functional testing of Ti–Nb-based SMA for medical application. Shape Memory Superelast 2(2):130–144

Qian J, Tao Z, Xiao J, Jiang G, Liu W (2013) Performance improvement of ceria-based solid oxide fuel cells with yttria-stabilized zirconia as an electronic blocking layer by pulsed laser deposition. Int J Hydrog Energy 38(5):2407–2412

Qiu Y, Young ML, Nie X (2015) Influence of dynamic compression on phase transformation of martensitic NiTi shape memory alloys. Metall Mater Trans A 46(10):4661–4668

Ramezani M, Monroe MBB (2023) Bacterial protease-responsive shape memory polymers for infection surveillance and biofilm inhibition in chronic wounds. J Biomed Mater Res A 111(7):921–937

Ren X (2004) Large electric-field-induced strain in ferroelectric crystals by point-defect-mediated reversible domain switching. Nat Mater 3(2):91–94

Reyes-Morel PE, Cherng J-S, Chen IW (1988) Transformation plasticity of CeO2-stabilized tetragonal zirconia polycrystals: II, Pseudoelasticity and shape memory effect. J Am Ceram Soc 71(8):648–657

RJ P (1967) Heat-shrinkable articles made from silicone

Rojac T, Kosec M, Damjanovic D (2011) Large electric-field induced strain in BiFeO3 ceramics. J Am Ceram Soc 94(12):4108–4111

Sabahi N, Chen W, Wang C-H, Kruzic JJ, Li X (2020) A review on additive manufacturing of shape-memory materials for biomedical applications. JOM 72(3):1229–1253

Saedi S, Shayesteh Moghaddam N, Amerinatanzi A, Elahinia M, Karaca HE (2018) On the effects of selective laser melting process parameters on microstructure and thermomechanical response of Ni-rich NiTi. Acta Mater 144:552–560

Saedi S, Acar E, Raji H, Saghaian SE, Mirsayar M (2023) Energy damping in shape memory alloys: a review. J Alloys Compd 956:170286

Saem S, Zhu Y, Luu H, Moran-Mirabal J (2017) Bench-top fabrication of an all-PDMS microfluidic electrochemical cell sensor integrating micro/nanostructured electrodes. Sensors 17(4):732

Safranski D, Dupont K, Gall K (2020) Pseudoelastic NiTiNOL in Orthopaedic applications. Shape Memory Superelast 6(3):332–341

Saigal A, Fonte M (2011) Solid, shape recovered "bulk" Nitinol: part I—tension–compression asymmetry. Mater Sci Eng A 528(16):5536–5550

Santiago D, Ferrando F, De la Flor S (2014) Influence of holding time on shape recovery in a polyurethane shape-memory polymer. J Mater Eng Perform 23(7):2567–2573

Santos EC, Shiomi M, Osakada K, Laoui T (2006) Rapid manufacturing of metal components by laser forming. Int J Mach Tools Manuf 46(12):1459–1468

Satsuma A, Shimizu K-I, Hattori T, Nishiyama H, Kakimoto S, Sugaya S, Yokoi H (2007) Polytungstate clusters on zirconia as a sensing material for a selective ammonia gas sensor. Sensors Actuators B Chem 123(2):757–762

Sawaguchi T, Maruyama T, Otsuka H, Kushibe A, Inoue Y, Tsuzaki K (2016) Design concept and applications of Fe–Mn–Si-based alloys–from shape-memory to seismic response control. Mater Trans 57(3):283–293

Schmidt AM (2006) Electromagnetic activation of shape memory polymer networks containing magnetic nanoparticles. Macromol Rapid Commun 27(14):1168–1172

Senatov FS, Zadorozhnyy MY, Niaza KV, Medvedev VV, Kaloshkin SD, Anisimova NY, Kiselevskiy MV, Yang K-C (2017) Shape memory effect in 3D-printed scaffolds for self-fitting implants. Eur Polym J 93:222–231

Shabalovskaya SA, Tian H, Anderegg JW, Schryvers DU, Carroll WU, Humbeeck JV (2009) The influence of surface oxides on the distribution and release of nickel from Nitinol wires. Biomaterials 30(4):468–477

Sharma H, Nguyen D, Chen A, Lew V, Khine M (2011) Unconventional low-cost fabrication and patterning techniques for point of care diagnostics. Ann Biomed Eng 39(4):1313–1327

Shirole A, Sapkota J, Foster EJ, Weder C (2016) Shape memory composites based on electrospun poly(vinyl alcohol) fibers and a thermoplastic polyether block amide elastomer. ACS Appl Mater Interfaces 8(10):6701–6708

Shrivastava S (2004) Medical device materials: proceedings from the materials & processes for medical devices conference 2003, 8–10 September 2003, anaheim, California. ASM International

Silva J, Reyes A, Esparza H, Camacho H, Fuentes L (2011) BiFeO3: a review on synthesis, doping and crystal structure. Integr Ferroelectr 126(1):47–59

Singh Rajput G, Vora J, Prajapati P, Chaudhari R (2022) Areas of recent developments for shape memory alloy: a review. Mater Today Proc 62:7194–7198

Sreekumar M, Nagarajan T, Singaperumal M (2009) Application of trained NiTi SMA actuators in a spatial compliant mechanism: experimental investigations. Mater Des 30(8):3020–3029

Stansbury JW, Idacavage MJ (2016) 3D printing with polymers: challenges among expanding options and opportunities. Dent Mater 32(1):54–64
Stöckel D (1995) The shape memory effect – phenomenon, alloys, applications. In: Shape memory alloys for power systems (EPRI), pp 1–12
Stoeckel D (1990) Shape memory actuators for automotive applications. Mater Des 11(6):302–307
Suchao-in K, Chirachanchai S (2013) "Grafting to" as a novel and simple approach for triple-shape memory polymers. ACS Appl Mater Interfaces 5(15):6850–6853
Sun QP (2002) IUTAM symposium on mechanics of martensitic phase transformation in solids, held in Hong Kong, China, 11–15 June 2001. Solid Mech Appl 163–170. Springer, Dordrecht. Hong Kong, China
Sun L, Huang WM (2009) Nature of the multistage transformation in shape memory alloys upon heating. Metal Sci Heat Treat 51(11):573–578
Sun L, Huang WM, Ding Z, Zhao Y, Wang CC, Purnawali H, Tang C (2012) Stimulus-responsive shape memory materials: a review. Mater Des 33:577–640
Sun F, Feng X, Wang H, Cao X, Liu X, Liu Y, Li Y, Liu Z, Yi X (2023) Brief overview of functionally graded NiTi-based shape memory alloys. Adv Eng Mater 25(9):2201434
Sutou Y, Omori T, Wang JJ, Kainuma R, Ishida K (2004) Characteristics of Cu–Al–Mn-based shape memory alloys and their applications. Mater Sci Eng A 378(1):278–282
Swain MV (1986) Shape memory behaviour in partially stabilized zirconia ceramics. Nature 322(6076):234–236
Tabesh M, Goel V, Elahinia MH (2012) Shape memory alloy expandable pedicle screw to enhance fixation in osteoporotic bone: primary design and finite element simulation. J Med Dev 6(3)
Tadaki T, Otsuka K, Shimizu K (1988) Shape memory alloys. Annu Rev Mater Sci 18(1):25–45
Taheri Andani M, Anderson W, Elahinia M (2014) Design, modeling and experimental evaluation of a minimally invasive cage for spinal fusion surgery utilizing superelastic Nitinol hinges. J Intell Mater Syst Struct 26(6):631–638
Tamer TM, Alsehli MH, Omer AM, Afifi TH, Sabet MM, Mohy-Eldin MS, Hassan MA (2021) Development of polyvinyl alcohol/kaolin sponges stimulated by marjoram as hemostatic, antibacterial, and antioxidant dressings for wound healing promotion. Intl J Mol Sci [Online] 22
Tanaka Y, Himuro Y, Kainuma R, Sutou Y, Omori T, Ishida K (2010) Ferrous polycrystalline shape-memory alloy showing huge superelasticity. Science 327(5972):1488–1490
Tseng LW, Ma J, Wang SJ, Karaman I, Kaya M, Luo ZP, Chumlyakov YI (2015) Superelastic response of a single crystalline FeMnAlNi shape memory alloy under tension and compression. Acta Mater 89:374–383
Tseng LW, Ma J, Vollmer M, Krooß P, Niendorf T, Karaman I (2016) Effect of grain size on the superelastic response of a FeMnAlNi polycrystalline shape memory alloy. Scr Mater 125:68–72
Tzou HS, Lee HJ, Arnold SM (2004) Smart materials, precision sensors/actuators, smart structures, and Structronic systems. Mech Adv Mater Struct 11(4–5):367–393
Uchino K (2016) Antiferroelectric shape memory ceramics. Actuators [Online] 5
Ueland SM, Schuh CA (2012) Superelasticity and fatigue in oligocrystalline shape memory alloy microwires. Acta Mater 60(1):282–292
Ueland SM, Schuh CA (2013) Transition from many domain to single domain martensite morphology in small-scale shape memory alloys. Acta Mater 61(15):5618–5625
Ueland SM, Chen Y, Schuh CA (2012) Oligocrystalline shape memory alloys. Adv Funct Mater 22(10):2094–2099
Ullakko K (1996) Magnetically controlled shape memory alloys: a new class of actuator materials. J Mater Eng Perform 5(3):405–409
Vajpai SK, Dube RK, Sangal S (2013) Application of rapid solidification powder metallurgy processing to prepare Cu–Al–Ni high temperature shape memory alloy strips with high strength and high ductility. Mater Sci Eng A 570:32–42
Vakil AU, Petryk NM, Shepherd E, Beaman HT, Ganesh PS, Dong KS, Monroe MBB (2021) Shape memory polymer foams with tunable degradation profiles. ACS Appl Bio Mater 4(9):6769–6779

Vakil AU, Ramezani M, Monroe MBB (2022) Antimicrobial shape memory polymer hydrogels for chronic wound dressings. ACS Appl Bio Mater 5(11):5199–5209

Valiev RZ, Prokofiev EA, Kazarinov NA, Raab GI, Minasov TB, Stráský J (2020) Developing nanostructured Ti alloys for innovative implantable medical devices. Materials [Online] 13

Venugopalan R, Trépanier C (2000) Assessing the corrosion behaviour of Nitinol for minimally-invasive device design. Minim Invasive Ther Allied Technol 9(2):67–73

Vernon LB (1941) V. H. Producing molded articles such as dentures from thermoplastic synthetic resins

Vollmer M, Krooß P, Kriegel MJ, Klemm V, Somsen C, Ozcan H, Karaman I, Weidner A, Rafaja D, Biermann H, Niendorf T (2016) Cyclic degradation in bamboo-like Fe–Mn–Al–Ni shape memory alloys—the role of grain orientation. Scr Mater 114:156–160

Waitz T, Kazykhanov V, Karnthaler HP (2004) Martensitic phase transformations in nanocrystalline NiTi studied by TEM. Acta Mater 52(1):137–147

Waitz T, Antretter T, Fischer FD, Karnthaler HP (2008) Size effects on martensitic phase transformations in nanocrystalline NiTi shape memory alloys. Mater Sci Technol 24(8):934–940

Waitz T, Tsuchiya K, Antretter T, Fischer FD (2009) Phase transformations of nanocrystalline martensitic materials. MRS Bull 34(11):814–821

Wang ZL (2003) Nanobelts, nanowires, and Nanodiskettes of semiconducting oxides—from materials to nanodevices. Adv Mater 15(5):432–436

Wang X, Ludwig A (2020) Recent developments in small-scale shape memory oxides. Shape Memory Superelast 6(3):287–300

Wang M, Yu H (2019) A novel artificial anal sphincter with constant force. Int J Appl Electromagn Mech 59:1135–1141

Wang K, Zhu XX (2018) Two-way reversible shape memory polymers containing polydopamine nanospheres: light actuation, robotic locomotion, and artificial muscles. ACS Biomater Sci Eng 4(8):3099–3106

Wang Z, Liang Z, Wang B, Zhang C, Kramer L (2004) Processing and property investigation of single-walled carbon nanotube (SWNT) buckypaper/epoxy resin matrix nanocomposites. Compos A: Appl Sci Manuf 35(10):1225–1232

Wang D, Zhang Z, Zhang J, Zhou Y, Wang Y, Ding X, Wang Y, Ren X (2010) Strain glass in Fe-doped Ti–Ni. Acta Mater 58(18):6206–6215

Wang X, Verlinden B, Van Humbeeck J (2015) Effect of post-deformation annealing on the R-phase transformation temperatures in NiTi shape memory alloys. Intermetallics 62:43–49

Wang Y-Q, Zhu Y, Wang J-H, Li X-N, Wu X-G, Qin Y-X, Chen W-Y (2021) Fe3+, NIR light and thermal responsive triple network composite hydrogel with multi-shape memory effect. Compos Sci Technol 206:108653

Wei Y, Zeng Q, Wang M, Huang J, Guo X, Wang L (2019) Near-infrared light-responsive electrochemical protein imprinting biosensor based on a shape memory conducting hydrogel. Biosens Bioelectron 131:156–162

Wen CE, Xiong JY, Li YC, Hodgson PD (2010) Porous shape memory alloy scaffolds for biomedical applications: a review. Phys Scr 2010(T139):014070

Werner M, Hammer N, Rotsch C, Berthold I, Leimert M (2020) Experimental validation of adaptive pedicle screws—a novel implant concept using shape memory alloys. Med Biol Eng Comput 58(1):55–65

Xia Y, He Y, Zhang F, Liu Y, Leng J (2021) A review of shape memory polymers and composites: mechanisms, materials, and applications. Adv Mater 33(6):2000713

Xie Q, Meng Q, Yu W, Wu Z, Xu R, Zeng Q, Zhou Z, Yang T, Yu H (2023) Design of a SMA-based soft composite structure for wearable rehabilitation gloves. Front Neurorobot 17:17

Xu X, Davis KA, Yang P, Gu X, Henderson JH, Mather PT (2011) Shape memory RGD-containing networks: synthesis, characterization, and application in cell culture. Macromol Symp 309-310(1):162–172

Xu B, Li Y, Gao F, Zhai X, Sun M, Lu W, Cao Z, Liu W (2015) High strength multifunctional multiwalled hydrogel tubes: ion-triggered shape memory, antibacterial, and anti-inflammatory efficacies. ACS Appl Mater Interfaces 7(30):16865–16872

Yang C-S, Wu H-C, Sun J-S, Hsiao H-M, Wang T-W (2013) Thermo-induced shape-memory PEG-PCL copolymer as a dual-drug-eluting biodegradable stent. ACS Appl Mater Interfaces 5(21):10985–10994

Yang S, Omori T, Wang C, Liu Y, Nagasako M, Ruan J, Kainuma R, Ishida K, Liu X (2016) A jumping shape memory alloy under heat. Sci Rep 6(1):21754

Yao G, Mo X, Yin C, Lou W, Wang Q, Huang S, Mao L, Chen S, Zhao K, Pan T, Huang L, Lin Y (2022) A programmable and skin temperature-activated electromechanical synergistic dressing for effective wound healing. Sci Adv 8(4):eabl8379

Yao H, Yu M, Fu J, Zhu M, Li Y, Li S, Gan R, Zhou H, Qi S (2023) Shape memory polymers enable versatile magneto-active structure with 4D printability, variable stiffness, shape-morphing and effective grasping. Smart Mater Struct 32(9):095005

Yu K, Ritchie A, Mao Y, Dunn ML, Qi HJ (2015) Controlled sequential shape changing components by 3D printing of shape memory polymer multimaterials. Procedia IUTAM 12:193–203

Yu HZ, Hassani-Gangaraj M, Du Z, Gan CL, Schuh CA (2017) Granular shape memory ceramic packings. Acta Mater 132:455–466

Yuan B, Zhu M, Chung CY (2018) Biomedical porous shape memory alloys for hard-tissue replacement materials. Materials [Online] 11

Zeng XM, Lai A, Gan CL, Schuh CA (2016a) Crystal orientation dependence of the stress-induced martensitic transformation in zirconia-based shape memory ceramics. Acta Mater 116:124–135

Zeng XM, Du Z, Schuh CA, Tamura N, Gan CL (2016b) Microstructure, crystallization and shape memory behavior of titania and yttria co-doped zirconia. J Eur Ceram Soc 36(5):1277–1283

Zhang N, Asle Zaeem M (2016) Competing mechanisms between dislocation and phase transformation in plastic deformation of single crystalline yttria-stabilized tetragonal zirconia nanopillars. Acta Mater 120:337–347

Zhang N, Asle Zaeem M (2019) Nanoscale self-healing mechanisms in shape memory ceramics. NPJ Computat Mater 5(1):54

Zhang H, Zhao Y (2013) Polymers with dual light-triggered functions of shape memory and healing using gold nanoparticles. ACS Appl Mater Interfaces 5(24):13069–13075

Zhang X, Tan BH, Li Z (2018) Biodegradable polyester shape memory polymers: recent advances in design, material properties and applications. Mater Sci Eng C 92:1061–1074

Zhang F, Wang L, Geng Q, Liu Y, Leng J, Smoukov SK (2023) Adjustable volume and loading release of shape memory polymer microcapsules. Intl J Smart Nano Mater 14(1):77–89

Zhao Q, Qi HJ, Xie T (2015) Recent progress in shape memory polymer: new behavior, enabling materials, and mechanistic understanding. Prog Polym Sci 49-50:79–120

Zhao X, Lai A, Schuh CA (2017) Shape memory zirconia foams through ice templating. Scr Mater 135:50–53

Zhou Y, Xue D, Ding X, Otsuka K, Sun J, Ren X (2009) High temperature strain glass in Ti50(Pd50−xCrx) alloy and the associated shape memory effect and superelasticity. Appl Phys Lett 95(15)

Zhu Y, Moran-Mirabal J (2016) Highly bendable and stretchable electrodes based on micro/nano-structured gold films for flexible sensors and electronics. Adva Electron Mater 2(3):1500345

Zhu Y, Hu J, Luo H, Young RJ, Deng L, Zhang S, Fan Y, Ye G (2012) Rapidly switchable water-sensitive shape-memory cellulose/elastomer nano-composites. Soft Matter 8(8):2509–2517

Zhuo F, Li Q, Zhou Y, Ji Y, Yan Q, Zhang Y, Xi X, Chu X, Cao W (2018) Large field-induced strain, giant strain memory effect, and high thermal stability energy storage in (Pb,La)(Zr,Sn,Ti)O3 antiferroelectric single crystal. Acta Mater 148:28–37

Zuperl U, Cus F (2004) Tool cutting force modeling in ball-end milling using multilevel perceptron. J Mater Process Technol 153-154:268–275

Zwawi M (2022) Recent advances in bio-medical implants; mechanical properties, surface modifications and applications. Eng Res Express 4(3):032003

Chapter 3
Intelligent Biomaterials for Organ Regeneration and Tissue Engineering

Ziming Liao, Jiamei Song, Ximing Zhu, Yanping Guo, Yuhong Wang, and Bin Yao

Abstract Organ regeneration and tissue engineering hold a significant position in life science research. Tissue repair and regeneration are typically accompanied by alterations in the complex microenvironment and endogenous pathways within the organism. In conjunction with engineering techniques, biomaterial-mediated reactions can be employed to expedite tissue repair processes. Among these, intelligent biomaterials responsive to stimuli can achieve the construction of specific cells and microenvironments under human guidance, providing more advantageous conditions for the regeneration of tissues and organs. These external triggers interact with intelligently responsive biomaterials, enabling the biomaterials to act as a bridge between in vivo and in vitro, bridging pathological gaps for electrical signal transmission, biochemical signal changes, drug release, cell loading, and mechanical stress regulation. Future research should focus on the development and application of stimulus-response biomaterials to further optimize biomaterial-based combination strategies. With the deepening of research, more intelligent responsive biomaterials will enter clinical applications and hold great potential in the field of biomedical research.

Keywords Organ regeneration · Tissue repair · Stimulus-responsive · Intelligent biomaterials

---

Z. Liao · J. Song · X. Zhu · Y. Guo · Y. Wang · B. Yao (✉)
Academy of Medical Engineering and Translational Medicine, Tianjin University, Tianjin, China
e-mail: hongyaobin_1212@tju.edu.cn

Z. Liu (ed.), *Intelligent Biomaterials*, https://doi.org/10.1007/978-981-96-8646-9_3

3.1 Skin Regeneration

3.1.1 The Significance of Skin Repair and Regeneration

Skin, being the largest organ of the human body, covers the entire surface and serves as the first barrier against the external environment. It is also the primary organ for perceiving the surroundings, with its softness enabling it to sense physical trauma, and its sensitivity allowing it to respond to temperature changes. Its abundant neurons transmit sensory signals to the brain, eliciting responses such as comfort, fear, and avoidance. Composed of the epidermis, dermis, subcutaneous tissue, neurons, and skin appendages (Fig. 3.1), skin possesses a remarkable regenerative capability superior to most other organs. However, injuries like physical impact, burns, chronic diabetic ulcers, and genetic blistering diseases can lead to massive skin loss, which often heals through scarring or results in difficult-to-heal wounds. Extensive skin defects can lead to fatal consequences such as dehydration shock, while scarred repair can impact a patient's appearance and, if located in joints, restrict mobility. Additionally, skin appendages like hair follicles and sweat glands are difficult to regenerate after scarring.

Sweat glands, which are widespread throughout the body, play a vital role in thermoregulation. Extensive skin damage often leaves sweat glands unregenerated, negatively impacting the patient's quality of life post-surgery and potentially posing life-threatening risks in hot climates. Hair covers over 90% of the skin, regulating body temperature, protecting the skin, and enhancing aesthetic appeal. Hair loss can significantly impact a patient's appearance, potentially leading to psychological conditions like depression. Given their morphological diversity, close association with aging, and observable regenerative process, hair follicles serve as an ideal

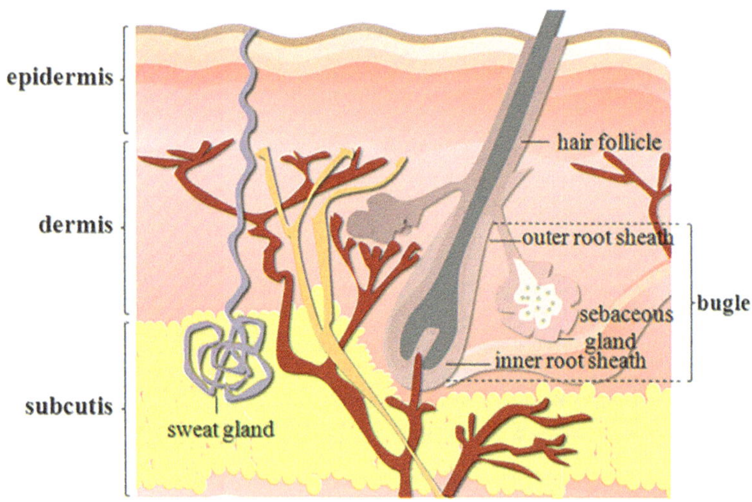

Fig. 3.1 An illustration of skin structure

model for studying organ regeneration and the molecular mechanisms underlying size, shape, and regenerative capacity changes with aging in specific organs. Therefore, the regeneration of skin appendages like sweat glands and hair follicles holds significant biological importance (Wang et al. 2017).

3.1.2 Current Research Status of Skin Repair and Regeneration

Currently, autologous skin transplantation is the primary clinical treatment for skin defects. However, its limitations in donor site availability, secondary damage, and infection risks render it unsuitable for most patients. Allogeneic skin transplantation faces challenges of immune rejection and poor survival rates. As a result, tissue-engineered skin, comprising biomaterials, cells, and cytokines, has emerged as the preferred option for wound closure and skin substitution. Tissue-engineered skin can cover wounds, accelerate healing, inhibit scarring, and promote vascularization. Various tissue-engineered skin products are widely used, including Apligraf, Integra, Dermagraft, and Alloderm, approved by the US Food and Drug Administration (FDA). Apligraf contains two layers of dermis and epidermis, with allofibroblasts and epidermis cells. Integra and Dermagraft are artificial dermis formed by inoculating fibroblasts with artificial dermal substitutes, including dermis; Alloderm is acellular dermal tissue; the recombinant human acellular dermis introduced by Shaanxi Ailskin Tissue Engineering Co., Ltd. contains only the dermis. While these products demonstrate effectiveness, they still lack normal skin structure, mechanical properties, appendages, and neuroresponsive functions, making them insufficient for functional skin restoration. Researchers have used inactivated sweat gland cells to co-culture with MSC (Li et al. 2006), sweat gland induction medium to induce umbilical mesenchymal stem cell differentiation (Xu et al. 2012; Zhang et al. 2012), overexpression of Eda in MSC (Cai et al. 2011), transfer NF-kB and Lef1 into fibroblasts (Yao et al. 2018), and transfer Irf6 into epidermal stem cells to induce sweat gland cell regeneration and participate in wound skin repair (Li et al. 2006; Zhao et al. 2015). Liang et al. induced amniotic fluid stem cells to differentiate into sweat gland cells with sweat gland medium (Liang et al. 2016). Huang et al. also used tissue engineering technology to construct skin analogs containing sweat gland tissue (Huang et al. 2010, 2016). Wu et al. used neonatal dermal cells and epidermal cells amplified in vitro and transplanted them into the back wounds of nude mice, and observed that hair and sebaceous glands could be generated on the healed wounds (Wu et al. 2014). Wang et al. transplanted mouse and human dermal stem cells (SKP) together with epidermal stem cells into the wound of nude mice to form new hair follicles (Wang et al. 2016). Takagi used iPS cells to form embryoids after 3 days of culture in vitro, which differentiated into epithelial cells and stromal cells after 7 days of culture. By using collagen to compound multiple embryoid bodies and implant them under the renal capsule of nude mice, relatively complete skin structures including epidermis, dermis, fat, nerve fibers, hair follicles, and sebaceous glands can be formed (Takagi et al. 2016).

3.1.3 Challenges of Skin Repair and Regeneration

Skin is composed of a variety of cells, with a regular three-layer structure, the perfect regeneration of skin requires the construction of a three-layer skin structure and contains a variety of orderly distribution of cells, including epidermal cells, fibroblasts, vascular endothelial cells, nerve cells, sweat gland cells, and hair follicle cells, to form functional skin. Firstly, in terms of cell sources, epidermal cells and fibroblasts are easy to obtain, while other types of cells are difficult to obtain and not easy to expand. Secondly, from the perspective of orderly structure, traditional treatment methods such as dressings and hydrogels do not have complex structures, and only promote healing through simple covering. A scaffold similar to an extracellular matrix structure can be constructed by electrospinning, but it does not contain cells and the distribution of cells in the scaffold is not uniform, nor can it form a heterogeneous structure and cell distribution, which is difficult to simulate the real situation of skin.

3D bioprinting technology is a rapidly developing technology in recent years, which integrates computer-aided design (CAD), computer-aided manufacturing (CAM), computer numerical control (CNC), precision servo drive, biomaterials, cell biology and molecular biology, and other disciplines and technologies. Cell 3D printing technology provides a new clinical medical technology for tissue and organ repair and also provides a research tool for regenerative medicine, tissue engineering, stem cells, cancer, and other research fields. Pati et al. made bioink from the acellular matrix of myocardium and cartilage tissue, constructed a 3D structure in vitro through 3D printing technology, and successfully induced adipose mesenchymal stem cells to express characteristic markers of cardiomyocytes and chondrocytes (Pati et al. 2014). Lee et al. successfully treated sheep with damaged meniscus by constructing a 3D-printed scaffold that could continuously release a variety of growth factors, inducing MSC to differentiate into meniscus in vivo (Pati et al. 2014). Kimberly et al. found that precisely controlling the pore structure of 3D scaffolds through 3D printing technology can regulate the proliferation and differentiation of mesenchymal stem cells, which indicates that 3D bio-printing technology can provide seed cells for skin regeneration by inducing stem cell differentiation (Ferlin et al. 2016).

In addition, bioprinting can quickly print artificial skin matching the wound through computer scanning imaging technology according to the shape and depth of the wound, with high throughput and high repeatability. By flexibly and accurately printing different bio-inks (including living cells, nucleic acids, growth factors pre-gummed materials, etc.) to construct tissue structures that exhibit morphology and physiology similar to normal skin, it has great advantages and application prospects

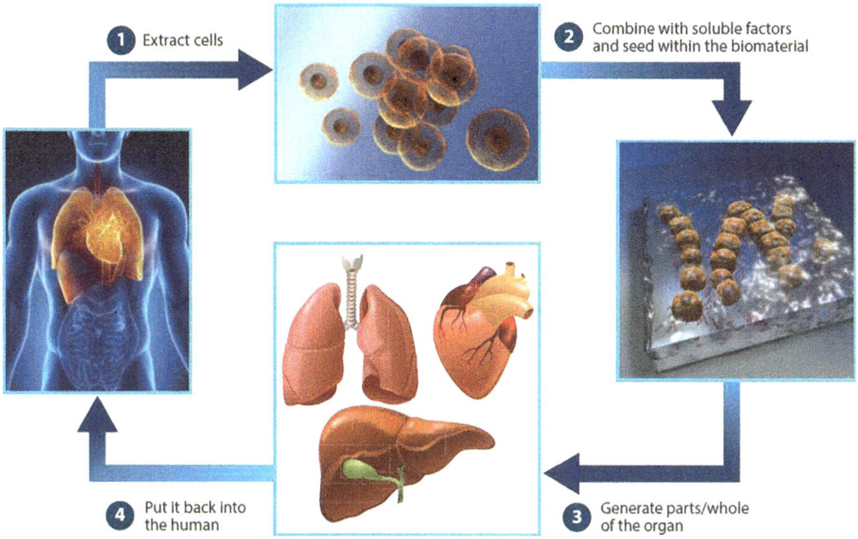

Fig. 3.2 A schematic diagram of the 3D bioprinting process

in constructing new tissue-engineered skin and promoting the perfect regeneration of wound surfaces (Fig. 3.2).

3.1.4 Strategies for Biomaterials in Skin Repair and Regeneration Research

3.1.4.1 Dressing

In the case of large skin defects, due to problems such as increased metabolism, water loss, and immune system disorders, it can cause fatal hazards such as shock and sepsis. The dressing is a kind of simulated human skin, which has certain physical and chemical properties of human skin, and can cover the wound, prevent the loss of body fluid and protein, and prevent bacterial infection. It needs to meet the following characteristics: (1) biocompatibility, no stimulation, and no immune stress response; (2) it can adhere to the wound surface and will not fall off easily; (3) it has a certain mechanical strength and can be sutured; (4) antibacterial, inhibit bacterial growth and kill bacteria; (5) moisturizing to provide a moist environment conducive to wound healing; (6) low cost. Traditional dressings are mainly linens, cotton, and cotton fabrics, which can only play a simple role in covering and stopping bleeding. In the later stage, with the improvement of wound healing theory and the development of material science, a variety of materials have been used in the research and development of dressings, and many new medical dressings have emerged. According to their role, they can be divided into the following types:

contact dressings, active dressings, passive dressings, interactive dressings, and antibacterial dressings.

Contact dressings are mainly traditional dressings, including gauze, cotton fabrics, and synthetic fiber dressings, which are generally simple and direct used for wound dressing and fixation. Active dressings refer to dressings containing active ingredients such as growth factors, active peptides, and small molecular agonists, which can promote the activation, proliferation, and secretion of wound cells. It also includes autologous or allogeneic skin, which can self-fill the wound defect; Cell-containing products such as tissue-engineered skin also belong to this category, but there are great differences in properties and preparation methods between tissue-engineered skin and cell-free dressings. Moreover, the development direction of tissue-engineered skin is to build truly fully functional skin, so this part is introduced in detail later. Passive dressings refer to the use of dressings to create an environment that is conducive to wound healing, but the dressings themselves do not interfere with the physiological state of the wound or interact with each other, including film dressings and hydrogel dressings. They have certain water retention and poor water absorption functions, so they are mostly used for relatively clean wounds with less exudation.

Interactive dressings refer to dressings that can change the physiological state of the wound by contacting with the exudate of the wound, such as forming a gel-like covering on the wound, to promote wound healing. Interactive dressings include hydrogel dressings, alginate-based dressings, and hydrophilic fiber dressings, which have certain hemostatic and antibacterial functions. Antibacterial dressing refers to the addition of certain antibacterial ingredients such as silver, honey, iodine, chlorhexidine, etc., or the preparation of the dressing body with certain antibacterial functions so that the dressing has a local antibacterial effect, such as pharmaceutical dressings.

3.1.4.2 Bionic Scaffold

An organ is not simply a cell accumulation, but a functional assembly with a specific morphological structure formed through the self-organization of a variety of cell interactions. "When the cells are eluted, what is left is a scaffold that still maintains its tissue morphology, with certain mechanical properties, protein composition, and the ability to induce cell differentiation, blood vessel ingrowths, and secondary organ formation." After skin injury, dressings can only provide simple coverage, active factor stimulation, antibacterial or inflammatory regulation, and ultimately achieve the purpose of promoting repair. In order to achieve the perfect regeneration of skin tissue, it is necessary to construct a bionic skin scaffold and provide a composite scaffold with mechanical support, topology, and chemical factors.

3.1.4.3 Electrospinning Scaffold

High voltage electrospinning (Helectrostatic spinning) is a new fiber manufacturing technology, which pulls charged molten liquid into nanofibers with a diameter of several hundred nanometers to micrometers through an electric field. Its specific surface area is large, and it can better regulate the release of drugs. In addition, due to the fineness and maneuverability of the fibers, electrospun fibers can be used to simulate the microstructure of extracellular matrix to form tissue engineering scaffolds that promote the healing of skin and other tissues. The electrospun fibers have good biocompatibility, and stable drug loading, and can better induce tissue regeneration by adjusting the topology of the scaffold. It can provide physical support for skin regeneration. Due to its applicability to a variety of materials, electrospun fibers can also be used to construct scaffolds for bone regeneration, cartilage regeneration, and angiogenesis. Electrospun fibers can also be loaded with active factors to promote the activation and proliferation of stem cells. At the same time, electrospinning can also be prepared from a mixture of a variety of natural materials, which have good biocompatibility and spatial structure and can promote the growth of cells in it.

3.1.4.4 Acellular Scaffold

For organ regeneration, the most suitable organs for transplantation are autologous organs, and the only organs that can be used for autologous transplantation are skin. However, autologous transplantation is not feasible for patients with large-area trauma, and it will cause secondary trauma. The second is allogeneic transplantation, which, however, involves immune rejection. Therefore, the cellular components of allogeneic tissue are removed by surfactant perfusion and enzymatic digestion. The residual extracellular matrix retains the tissue scaffold and appropriate extracellular matrix protein components, which is conducive to promoting cell proliferation, migration, and blood vessel ingressions. The third is xenotransplantation, xenotransplantation is an important way of organ regeneration, organ donor shortage is an important reason for the shortage of organ transplantation, and xenotransplantation is a cheap and easy source of organ. However, the immune rejection is more severe than that of allogeneic transplantation. However, the difference between the extracellular matrix is relatively small, so the xenogeneic extracellular matrix scaffold has a high value for organ transplantation. In 2021, the only allogeneic organ transplantation company in the world was listed. Its method is to construct organs for transplantation by implanting human cells into decellularized scaffolds, which shows its application prospects.

3.1.4.5 3D Printing Scaffold

3D bioprinting is an emerging comprehensive technology in recent years. Its precision, controllability, speed, and high repeatability have brought new breakthroughs in the development of tissue engineering. Traditional tissue engineering does not have the characteristics of customization, while 3D printing technology can perform CT and ultrasound detection of wounds or tissues and organs, and then integrate the relevant data for modeling, forming a graphic guide printer to print, which perfectly meets the needs of patients. Extrusion 3D printing can construct heterogeneous structures and compound the heterogeneous needs of tissues and organs. Stereolithography 3D printing can rapidly form complex organ structures with micrometer accuracy. During the printing process, the monitoring software can observe the changes in the printed scaffold in real time to adjust it at any time to make it consistent with the set parameters. By adjusting the parameters of 3D printing, tissue regeneration can be precisely induced. At the same time, 3D-printed scaffolds can also add active ingredients to the bioink for sustained release. The release can be regulated by adjusting the diameter and pore size of the printed scaffold, and the distribution of different active factors can be regulated by adjusting the heterogeneity of the printed structure to play a differential regulation role. Traditional tissue engineering is mostly for uniform release conditions.

3.1.4.6 3D Printing Cellular-Containing Skin

Extrusion 3D Bioprinting

Extrusion bioprinting is the most widely used and earliest 3D printing method, which first requires the material to form a colloid-like shape, and then the material fiber thread is extruded through continuous pressure, and the cross-linking is carried out after printing to maintain the structure of the print block. This printing method can print materials with different viscosity and different kinds of cells, and its material has a wide range of applications and can build a heterogeneous organizational structure with good strength. However, due to the stacking accumulation of materials in the printing process, the printing accuracy is slightly poor, and the printing accuracy of cells in the printing process is not high enough. According to the different driving methods of the printing system, it can be divided into pneumatic extrusion, piston extrusion, and spiral extrusion printing. The temperature, nozzle type, dynamic parameters, movement parameters, and structural parameters in the printing process all have important effects on the final shape.

Droplet Bioprinting

Droplet bioprinting, as the name suggests, the basic unit of its printing is discontinuous, independent droplets, a droplet can contain one or more cells, with higher printing accuracy. The droplets can also contain growth factors, drugs, biological materials, etc. According to the different principles of droplet formation, droplet printing can be divided into inkjet printing, electrohydrodynamic jet bioprinting, and laser-assisted bioprinting. Compared with the popularity of extrusion bioprinters, the application of commercial droplet printing is relatively small, but based on its characteristics of precise regulation of cell concentration, this printing method has a good application prospect in the construction of multi-layer cell structures and is more suitable for in situ and in vivo printing.

Photocuring Bioprinting

Photocurable bioprinting is a bioprinting method that uses a photosensitive polymer to generate light curing under precisely controlled light of a specific wavelength. Compared with other bioprinting methods, the accuracy of photocurable bioprinting can reach the nm level and the molding speed is very fast. This printing technology was first popularized to build a variety of complex models such as houses, insoles, and even missiles, and most of the applications in bioprinting are to print scaffolds and then inoculate cells on them. With the development of low and non-toxic photocuring materials and photoinitiators in recent years, more and more cell carrier printing has been done. Photocuring printing and extrusion printing, droplet printing is not completely independent, when extrusion printing or droplet printing using photopolymer materials, the final molding also belongs to the category of photocuring. According to the different ways of light scanning, light curing printing can be divided into two ways: stereolithography and digital processing bioprinting.

Microsphere Bioprinting

Strictly speaking, microsphere bioprinting belongs to extrusion bioprinting because it requires additional auxiliary facilities and its printing method is more complex, which is different from extrusion printing, and the organizational structure formed by printing is more perfect and the function is more comprehensive, so it is explained separately. The formation of microspheres can be obtained by hanging drop method or low adhesion culture, and the number of cells can be controlled, and the formed structure has preliminary histogenesis and function. The cell microspheres are collected by special equipment or pipette gun and then printed by the printer with structural guidance. After culture, the interaction between the cell microspheres occurs, and secondary tissue rearrangement and self-histogenesis occur. The time required to form functional tissues is shortened, and the advantage of extrusion printing to build heterogeneous structures is retained. Multiple functional tissues can be constructed through the combination of different cell types, which accelerates the development of artificial organs in vitro and has a good application prospect (Daly et al. 2021).

3.1.4.7 3D Bioprinted Full-Function Artificial Skin

Skin is the largest organ in the human body and also the most accessible organ and skin regeneration is currently the field where 3D printing technology is applied more. There have been many studies using fibroblasts and epidermal cells to construct 3D-printed skin containing dermis and epidermis and for wound repair. Most of them adopt the printing strategy from the bottom layer to the high layer, first printing the dermis containing fibroblasts, and then printing the epidermis layer. The two layers formed after culture are closely connected and have the formation of basement membrane-like structures. Some people also construct 3D-printed skin with the vascular network, which can promote the growth of vascular endothelial cells and the formation of blood vessels, and accelerate wound healing through more nutrient supply (Jin et al. 2021). Compared with dressings, printed skin has more cellular components and nutrient channels, but it also lacks the diversity of structure and functional appendages that skin should have. The construction of artificial skin with appendages through 3D printing is the focus and difficulty of current research in the intersection of skin regeneration and 3D printing (Zhang et al. 2021).

3.1.4.8 3D Bioprinting Artificial Skin Containing Sweat Glands and Hair Follicles

Development of Sweat Glands and Hair Follicles

The mammalian epidermis is formed by the differentiation of cells on the surface of the ectoderm, that is, a single layer of cells covering the outside of the embryo after the formation of the neural embryo (Jackson et al. 1981). Epidermal development includes epidermal layering and the generation of skin appendages. Sweat glands, hair follicles, and sebaceous glands are all differentiated from epidermal stem cells (Paus et al. 1999). Hair follicles and sweat glands have similar development patterns. Take mice as an example: at 14.5d (E14.5d) of the embryo, BMP expression in dermal stromal cells is inhibited, and Shh expression is up-regulated to induce the epidermis to form hair follicle substrate. E15.d5 induced basal plate downward growth by dermal papilla cells; After birth, 4d (P4d) forms mature hair follicles. At 17.5 days (E17.5 days), the expression of BMP and FGF in dermal stromal cells was up-regulated, and the expression of Shh from the epidermis was inhibited, which induced the formation of sweat gland substrate in the epidermis. At 0d-5d after birth (P0-P5d), the basal plate grows downward to form a single catheter and forms a coiled secretory part at the end of the catheter; At P21d, the sweat glands are fully developed, including a large coiled secretory part and a single ductal part. When mice mature, the epidermal stem cells in the toe pads differentiate into sweat gland tissue, while the epidermal stem cells in the back differentiate into hair follicles and sebaceous glands (Liu et al. 2013). The different cell fates in different sites provide animal models for studying the role of the microenvironment in the development and formation of appendages.

Microenvironment Is the Determining Factor in Inducing Hair Follicle and Sweat Gland Differentiation

The cellular microenvironment includes adjacent cells and an extracellular matrix. Adjacent cells are fibroblasts. The extracellular matrix is composed of macromolecular complexes with different biochemical properties, such as collagen, glycoprotein, proteoglycan, polysaccharide, etc. In addition, it also contains many cytokines secreted by basal cells and neighboring cells, which play an important role in regulating cell localization, cell behavior, and cell differentiation (Bissell and Labarge 2005). In order to observe the influence of characteristic microenvironment on cell fate selection, Ferraris et al. combined corneal epithelial stem cells with the foot dermis of mouse embryos at 15.5 days, the upper lip dermis of mouse embryos at 12.5 days, and the back dermis of mouse embryos at 14.5 days, and implanted them under the renal capsule of nude mice with AGAR medium. After 21 days, the complex of the dermis and cells was removed. It was detected that epithelial stem cells in the foot dermis as microenvironment showed sweat gland morphology and expressed the characteristic keratin of sweat gland; epithelial stem cells in the back dermis as microenvironment showed hair follicle morphology and expressed the characteristic keratin of hair follicle; Epithelial stem cells in the lip dermis as microenvironment differentiated into normal layered epidermis (Ferraris et al. 2000). Fuchs et al. transplanted sweat gland progenitor cells into mouse mammary fat pads. After differentiation, some sweat gland cells showed mammary duct morphology, decreased expression of sodium and potassium channel protein, and increased expression of milk protein. The progenitor cells of sweat glands were transplanted into the back skin, and the differentiated tissues showed lower expression of the characteristic keratin of sweat glands and stratified epidermis instead of curly sweat glands. After being transplanted into the shoulder fat pad, the progenitor cells of sweat glands can still form a curly shape and express the characteristic keratin of sweat glands (Lu et al. 2012).

The Role of Cytokines in Cell Differentiation

Cytokines are a kind of small molecular proteins secreted by basal cells and neighboring cells, such as epidermal growth factor, bone morphogenetic protein, etc. which have a broad spectrum of biological activities and participate in the regulation of cell growth, differentiation, and metabolism. Shikiji et al. found that epidermal cells were cultured with a collagen gel in a skin analog composed of 15 ng/ml epidermal growth factor, 10% fetal bovine serum and fibroblasts, and the epidermal cells formed a curly sweat gland shape and expressed the characteristic keratin of sweat glands after 10 days (Shikiji et al. 2003). Fushs et al. demonstrated that TGF-β2 promotes hair follicle development by competing with BMP signaling pathways in the cellular microenvironment (Oshimori and Fuchs 2012). Plikus et al. found that the BMP antagonist Noggin promoted the formation of hair follicles in the part of the foot of mice that should have formed sweat glands by stimulating Lef1 overexpression (Plikus et al. 2004). Bmpr1a knockout mice did not form sweat gland tissue (Leung et al. 2013).

Studies have shown that Cthrc1 is highly expressed in the sweat gland-developing microenvironment and lowly expressed in the non-sweat gland-developing microenvironment. Inhibition of Cthrc1 expression during the induction period of sweat gland substrate can lead to delayed development of sweat glands, weakened sweating function, and decreased expression of secretion-related genes in fetal rats. Adding Cthrc1 to 3D printing ink can induce MSC transdifferentiation into sweat gland cells and repair sweat gland injury on skin wounds. It is suggested that Cthrc1 is a key cytokine in inducing sweat gland development in microenvironment. Because of the similar development pattern and induction process of sweat glands and hair follicles, there should be key cytokines in the microenvironment to induce hair follicle development. 3D printing technology can play a huge role in simulating the microenvironment to induce cell differentiation and tissue regeneration. Researchers constructed artificial skin containing sweat glands by adding mouse sweat gland-producing microenvironment protein Cthrc1 to the biomaterial (Yao et al. 2020). The combination of screening and addition of hair follicle-inducing cytokines in the microenvironment and 3D biomimetic microenvironment can build fully functional artificial skin.

3.1.5 Biomedical Materials in the Field of Skin Repair and Regeneration Research

With the continuous development of biomedical materials, especially with the progress of life science technology and material science since the twentieth century, biomedical materials have been widely used in the field of life medicine. The development of biomaterials can be roughly divided into three stages. The first stage was the use of bone substitutes for biomimetic tissue morphology, such as titanium metal scaffolds, in the 1960s. In the second stage, bioactive bone substitutes, such as bioceramics, were widely used in the 1980s. The third stage is biomaterials with biological activity and biodegradability used in recent years, such as hydrogels and bioinks. This article mainly introduces the materials that can be used for 3D bioprinting.

3.1.5.1 Natural Polymer Biomaterials

Extracellular Matrix

An extracellular matrix (ECM) is a 3D macromolecular network of tissues after decellularization. It is composed of collagen, proteoglycan, glycosaminoglycan, elastin, fibronectin, laminin, and glycoprotein. In the process of cell removal, the genetic material and soluble proteins in the tissue are also washed away, and the remaining is mainly insoluble protein, which is lyophilized and ground into powder and dissolved with acetic acid to obtain extracellular matrix protein solution. At this time, the obtained protein solution has good biocompatibility, low immunogenicity, easy degradation, and can be formed at 37 °C, which is a good natural biological material. Moreover, extracellular matrix derived from different tissues also has certain tissue-specific mechanical properties and differentiation ability. For example, a decellularized adipose tissue matrix can induce adipogenic differentiation of adipose-derived mesenchymal stem cells, and cardiac extracellular matrix tends to induce muscle cell maturation. The chondrogenic extracellular matrix tended to induce chondrogenic differentiation of turbinate-derived MSCS, suggesting its potential application in the direction of cell induction.

The commonly used matrix glue is a kind of murine sarcoma acellular matrix, which is soft in texture and can provide a good 3D environment. It is mostly used for 3D cell experiments, but not suitable for animal experiments and human experiments.

Polysaccharide

Polysaccharides are the most abundant renewable resources on earth and are structurally and functionally diverse natural polymeric materials. Polysaccharides derived from animals include chitosan, hyaluronic acid, chondroitin sulfate, etc. The polysaccharides from sea algae included alginate and AGAR. Polysaccharides from plants include cellulose, starch, pectin, etc. Among them, chitosan, alginate, and hyaluronic acid are suitable for 3D bioprinting.

Chitosan is not easy to dissolve and is soluble in an acidic environment, which is not conducive to cell culture. Researchers have improved its solubility by chemical modification. Among them, carboxymethyl chitosan is a widely used chitosan derivative. While improving its solubility, chitosan with a high degree of carboxymethylation also has certain antibacterial properties. Chitosan and its derivatives have been used in many fields, such as drug delivery, dressings, tissue engineering products, and transplantation materials, due to their good biocompatibility, hydrophilicity, and gelation and film-forming ability.

Alginate is extracted from brown algae and consists of guluronic acid units and manuronic acid units in different ratios, which can be cross-linked by dimethyl ions to obtain a gel with good mechanical strength. However, the stability of this structure in vivo is relatively limited, and ion exchange will occur under the infiltration of ion-rich body fluids, resulting in the structure disintegration. Alginate has been widely used in biomedicine due to its good hydrophilicity, biocompatibility, and low immunogenicity.

Hyaluronic acid (HA) is widely distributed in the extracellular matrix of vertebrate tissues. In addition to its role as a structural support, it also plays an important role in morphogenesis, inflammation, and disease by regulating cell function. Hyaluronic acid forms a wide relative molecular weight range through enzymatic hydrolysis. Normally, hyaluronic acid is not cell adhesive, and its activity can be increased by adding functional groups through chemical modification. The hyaluronic acid hydrogel can be used as a cell carrier to maintain cell activity well.

Collagen

Collagen is the most widely used natural polymer and the main component of extracellular matrix in mammals. Among them, type I collagen is the most common, and its macroscopic structure is fibrous. The microstructure is a triple helix with each peptide chain left-handed and three peptide chains forming a right-handed structure. Due to the good biological characteristics of collagen and its important role in the extracellular matrix, collagen is the main material for traditional tissue-engineered skin and dressings. It has also been fully applied and developed as a 3D bioprinting ink.

Gelatin

Gelatin is a derivative of collagen that is subjected to acid treatment followed by thermal denaturation or alkali treatment to destroy the triple helix structure of collagen, resulting in a single-stranded structure. The gelatins obtained by the two treatments were of type A and B, respectively. Gelatin retains a lot of RGD sequences with adhesion, and it is soluble in water and has low immunogenicity. However, the mechanical properties of gelatin are poor, and it is mostly mixed with sodium alginate as an extruded bio-ink. Gelatin is often modified by introducing a methylacrylamide group to give GelMA, which is a commonly used photo-crosslinked bioink.

3.1.5.2 Synthetic Polymeric Biomaterials

Compared with natural polymers, synthetic polymers have the advantages of simple structure, good mechanical properties, high repeatability, and low price. Synthetic polymer materials are mainly used in daily life, such as nylon in textiles and polytetrafluoroethylene in non-stick cooking POTS. In the field of biomedicine, synthetic polymer materials are mostly used in medical devices, human scaffolds, and dressings, but are rarely used in 3D bioprinting.

3.1.6 Frontier Exploration of the Application of Biomedical Materials in Skin Repair and Regeneration Research

Bioinks mixed with gelatin and sodium alginate are widely used in skin regeneration, and based on their good biocompatibility and adjustable mechanical properties, they can be used for the printing of double-layer skin and the construction of artificial skin containing sweat glands and hair follicles. However, since sodium alginate is derived from brown algae, the human body lacks the enzyme to degrade sodium alginate, which is difficult to degrade after implantation in the body, seriously hindering its application as an organ regeneration. By adding different concentrations of sodium alginate lyase into the ink, the researchers observed the degradation of 3D-printed blocks in vivo and in vitro, as well as the stretching, living, and dying characteristics of cells in the 3D-printed blocks. It was found that the appropriate sodium alginate lyase could not only maintain its form in vitro for a certain period of time, but also promote its degradation in vivo. Reducing its retention in the body and promoting mutual communication between the cells carried by the 3D-printed block and the host have greatly improved the feasibility of bio-ink for regeneration (Yao et al. 2019).

Chitosan has good biocompatibility, but its water solubility is poor and it lacks sufficient mechanical properties. By grafting PEG to chitosan to increase its water solubility and mixing α-cyclodextrin into it, cross-linking sodium glycerophosphate-chitosan and the host-guest reaction between PEG and α-cyclodextrin was used to construct a supramolecular double network system to improve the mechanical strength of chitosan scaffold. It also regulates the differentiation direction of bone marrow mesenchymal stem cells (Hu et al. 2020).

The application of biomedical materials in bio-3D printing and its development in skin repair and regeneration suggest that a single material cannot meet the needs of bio-ink, material modification and mixing of a variety of materials is a new strategy for the development of bio-ink, and its application in the field of regenerative medicine needs further exploration.

3.1.7 Shortcomings and Challenges in the Application of Biomedical Materials to Skin Repair and Regeneration Research

The development of biomaterials is a key factor limiting bioprinting technology. Biological inks require good printability, biocompatibility, certain mechanical strength, cell adhesion, rapid prototyping after printing, moisture, and other properties; the current biological macromolecular materials cannot meet the above requirements at the same time. The combination of multiple materials, combining their respective advantages to meet the requirements of bio-ink, is a development strategy, but it will also bring new limitations. By chemically modifying one or more materials and then combining them together, multiple problems can be solved simultaneously. Proposing the direction of biomaterials evolution according to the needs of bio-ink, and then achieving the goal through chemical modification, are the shortcomings and challenges in the field of biomedical materials application and skin repair and regeneration (Liu et al. 2021a).

3.1.8 Prospects for the Application of Biomedical Materials in Skin Repair and Regeneration Research

At present, an important obstacle to skin repair and regeneration, and even organ repair and regeneration, is to establish cellular and structural heterogeneity in vitro. Extrusion 3D bioprinting and microsphere 3D bioprinting have made some progress in building heterogeneous organs that can support multi-layer skin construction with vascular networks and induced differentiation of accessory cells. With the development of biological materials, the printing accuracy is improved, the migration and self-organization of cells are promoted, and the purpose of cells leading the organogenesis and regulating the degradation of materials is achieved, so as to respond more intelligently and stimulate the signals of the body, and finally build functional artificial skin for regeneration.

3.2 Cardiac Regeneration

3.2.1 The Importance of Cardiac Repair and Regeneration

The heart, being the inaugural organ to materialize during embryonic development, serves a cardinal role in the perfusion of organs and tissues with ample blood supply. This perfusion is essential for the delivery of oxygen and a plethora of nutrients, as well as the removal of metabolic byproducts, thereby sustaining cellular metabolism and functionality (Woodcock and Matkovich 2005). The heart's functionality is intricately linked to its architectural design, with each of its components meticulously tailored to fulfill specific physiological roles. For instance, the left side of the heart is responsible for the reception of oxygen-enriched blood from the pulmonary veins into the left atrium, which subsequently propels this blood into the left ventricle. The left ventricle then ejects the oxygen-saturated blood to the systemic circulation, while the mitral valve, situated between the atrium and ventricle, ensures unidirectional blood flow and prevents regurgitation. Numerous cardiac pathologies, such as myocardial infarction and heart failure, often perturb this equilibrium, inducing alterations in the cardiac structure and function. Distinct from many other tissues, the myocardium of adult mammals exhibits a restricted capacity for regeneration. Following injury, cardiomyocytes are commonly supplanted by fibroblasts, thereby presenting formidable challenges to the realm of cardiac repair and regeneration (Wu et al. 2023).

3.2.2 The Current State and Challenges of Cardiac Repair and Regeneration Research

Myocardial infarction (MI) is one of the prevalent diseases leading to necrosis of cardiomyocytes, typically resulting from obstruction of blood flow in the coronary arteries, and is a leading cause of patient mortality and disability worldwide (Procopio et al. 2022). Current clinical treatments for cardiac infarction encompass pharmacotherapy (including antiplatelet agents, anticoagulants, beta-blockers, angiotensin-converting enzyme inhibitors (ACEI), and statins), thrombolytic therapy, percutaneous coronary intervention (PCI), and coronary artery bypass grafting (CABG) (Arslan et al. 2018). However, these therapeutic modalities are often limited to salvaging ischemic myocardium in the early stages or improving patient prognosis by inhibiting cardiac fibrosis. The repair of damaged myocardium or its regeneration remains a significant challenge that is the focus of extensive research and an urgent need for resolution.

Cardiac transplantation remains the ultimate treatment for patients with end-stage heart disease. Despite an increase in the rate of heart transplantation over the past decade with economic development, there is an observed aging of the donor population and an increase in the prevalence of diseases such as diabetes, hypertension, and obesity among donors. These donor risk factors are significantly correlated with short-term mortality, long-term mortality, and the occurrence of vascular diseases in allogeneic heart transplantation. Consequently, cardiac transplantation continues to be constrained by a scarcity of donor organs (Sun et al. 2021) (Fig. 3.3).

3.2.3 Strategies for Traditional Biomaterials in the Field of Cardiac Repair and Regeneration Research

Biomaterials are typically designed to enhance myocardial regeneration, promote neovascularization, improve cardiac contractility, limit adverse ventricular remodeling, modulate macrophage phenotype, and facilitate conductivity (Sedláková et al. 2024). Despite the preclinical testing of thousands of biomaterials, only a handful have progressed to clinical trials. Hereafter, we will focus on discussing several biomaterials that have entered clinical trials.

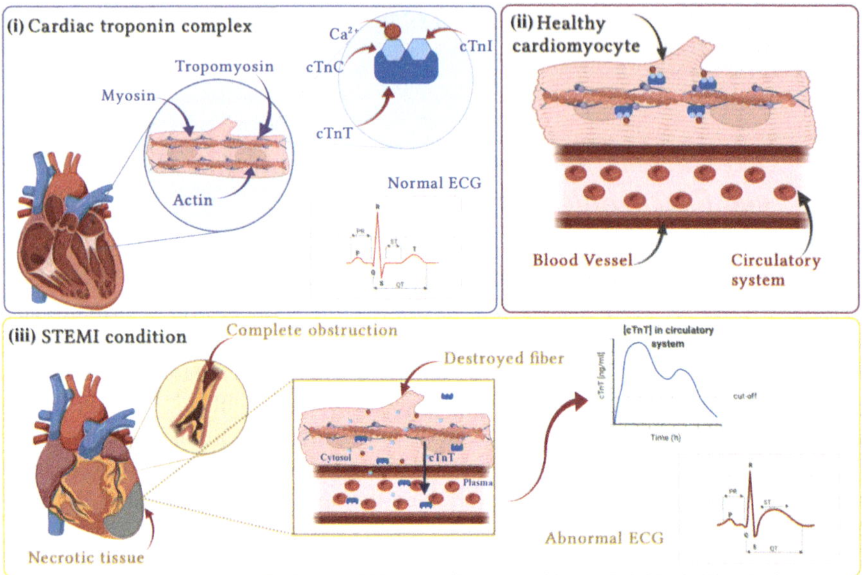

Fig. 3.3 Composition of cardiac troponin and changes during infarction. (Reprinted with permission from Procopio et al. 2022)

3.2.4 Types and Applications of Biomaterials

3.2.4.1 Decellularized Extracellular Matrix

Decellularized extracellular matrix (dECM) of the heart is an acellular scaffold derived from native cardiac tissue through a process that meticulously removes cellular components, leaving behind the intricate extracellular architecture. This procedure retains the complex structural and bioactive molecules of the cardiac ECM, including collagen, elastin, proteoglycans, and growth factors, which play crucial roles in facilitating tissue repair and regeneration (De Santis et al. 2021). VentriGel, an acellular hydrogel derived from porcine myocardial tissue, represents the first of its kind to be utilized in human cardiac repair. However, the study (NCT02305602) is designed to assess safety and feasibility rather than to demonstrate the efficacy of VentriGel in improving cardiac function (Traverse et al. 2019).

In parallel, CorMatrix ECM, a product introduced by CorMatrix Cardiovascular, has garnered FDA approval. Originating from the submucosal layer of porcine intestine, CorMatrix ECM can be fabricated into a bioprosthetic scaffold patch. Clinical trials (NCT01535807) have revealed that patients treated with CorMatrix ECM following isolated coronary artery bypass grafting (CABG) exhibited a 54% reduction in the incidence of postoperative atrial fibrillation (POAF) compared to those who did not receive the patch treatment. These advancements underscore the potential of dECM in cardiac regenerative medicine, highlighting the need for ongoing research to fully realize the therapeutic benefits of these biomaterials.

3.2.4.2 Collagen

Collagen is one of the primary constituents of the cardiac extracellular matrix, with Type I collagen being the most abundant, providing structural support and tensile strength to the myocardium (Dixon et al. 1996). Concurrently, collagen boasts excellent biocompatibility, biodegradability, and low immunogenicity, thereby finding extensive applications in the biomedical field, particularly in tissue engineering and regenerative medicine (Tani et al. 2023). In 2020, He et al. (2020) developed an injectable porous collagen scaffold hydrogel derived from bovine collagen tissue and utilized it as a delivery vehicle, co-injecting it with human umbilical cord mesenchymal stromal cells (hUC-MSCs) into the myocardium of patients with chronic ischemic heart disease (CIHD). In this randomized, single-center clinical trial, they observed that the collagen hydrogel enhanced the application efficiency and retention rate of cells. Although the size of the infarct scar did not exhibit a statistically significant difference compared to the control group, the study demonstrated the safety and feasibility of cell-laden collagen gels. Additionally, clinical trials involving gelatin, a derivative of collagen, have been reported.

3.2.4.3 Alginate

Alginate is a naturally occurring polysaccharide predominantly extracted from brown seaweed species such as kelp. Recognized for its favorable biocompatibility, degradability, gelation properties, and low toxicity, alginate has garnered considerable attention in recent years (Zoller et al. 2025). To date, three alginate-based biomaterials have proceeded to clinical trials. In 2009, Algisyl-LVR was utilized for the treatment of heart failure; this hydrogel has been demonstrated to be safe and has overall improved left ventricular function (Lee et al. 2013). Subsequently, a clinical trial involving a second alginate-based biomaterial was conducted in China (NCT04781660). Another alginate hydrogel, developed by BioLineRx and named IK-5001, is posited to aid in the treatment of myocardial infarction. Upon exposure to the excess calcium present in the infarct area, IK-5001 assembles into a hydrogel with structural properties akin to the extracellular matrix, thereby assisting in the alleviation of mechanical stress imposed on the infarct region (Frey et al. 2014). These advancements underscore the therapeutic potential of alginate-based biomaterials in cardiovascular regenerative medicine.

3.2.4.4 Fibrin

Fibrin is a protein that plays a pivotal role in the coagulation cascade of blood. Upon encountering injury, the coagulation cascade is activated, culminating in the conversion of fibrinogen to fibrin (Blanco et al. 2024). In 2018, a fibrin hydrogel was integrated with CD15 + Isl-1+ progenitor cells derived from human embryonic stem cells to form a patch. The clinical study (ESCORT) revealed that the hydrogel undergoes gelation within approximately 5–10 min and provides mechanical support. This innovative approach highlights the utility of fibrin in delivering cell-based therapies, offering a rapid-setting and supportive scaffold for cardiac repair.

3.2.5 Cutting-Edge Exploration of Intelligent-Response Biomaterials Applied to the Field of Cardiac Repair Research

The application of biomaterials in the realm of cardiac repair represents a significant branch of regenerative medicine and tissue engineering. While traditional therapeutic modalities, such as pharmacological interventions, interventional surgeries, and heart transplantation, have provided some alleviation of conditions, they do not fundamentally address the loss of cardiac tissue and function. Consequently, the development and application of biomaterials have introduced innovative approaches and methods to the field of cardiac repair (Wen et al. 2024).

The role of intelligent-response materials in cardiac repair is a cutting-edge area of interdisciplinary research involving several disciplines such as materials science, biomedical engineering, and regenerative medicine. These materials are able to sense changes in an organism's internal environment and respond accordingly, for example by altering the rate of drug release or signaling, thus playing a key role in the heart repair process. Following MI, the cardiac tissue microenvironment undergoes alterations due to excessive or persistent inflammation, such as the production of reactive oxygen species, which can lead to necrosis of cardiomyocytes. The construction of biomaterials capable of recognizing and responding to changes in the cardiac tissue microenvironment has garnered considerable interest in recent years. Zhu et al. (2021) designed a nitrate-functionalized patch that, when implanted into the hearts of rats with myocardial infarction, significantly enhanced the generation of NO upon recognizing the ischemic microenvironment, thereby exerting a mitochondria-targeting cardioprotective effect and enhancing cardiac repair. Lu et al. (2024) developed a novel multifunctional reactive oxygen species-responsive extracellular vesicle carrying rapamycin to address inflammation macrophage-mediated rejection reactions in cardiac transplantation. Similarly, biomaterials that recognize overexpressed MMP-2/9 in the infarction area, acidic microenvironments, and other cues have been reported (Chen et al. 2023).

Intelligent-response materials can also be used in drug delivery systems to achieve targeted release of drugs in damaged areas of the heart through specific stimulus-response mechanisms. This targeting not only improves the efficacy of the drug, but also reduces its side effects. Oral administration of cardiovascular drugs is the most common method of drug delivery. However, oral medications must undergo intestinal absorption and hepatic first-pass metabolism, which may affect the bioavailability and concentration of the drug reaching the heart. Intravenous injection can rapidly deliver drugs into the bloodstream, but this method has poor targeting, often affecting other tissues (Kulkarni et al. 2020). Future research needs to develop new drug delivery systems to improve the cardiac targeting, bioavailability, and therapeutic efficacy of drugs while reducing side effects. A hydrogen sulfide-releasing cardiac patch has been reported, which can alleviate inflammation after myocardial infarction and promote angiogenesis, thereby restoring cardiac function (Li et al. 2022). Jae Young Lee's research team developed a hydrogel patch composed of dextran aldehyde (Dex-ald) and gelatin, which has the properties of being paintable and adhesive. By incorporating the anti-inflammatory protein ANGPTL4 into the hydrogel, it continuously releases directly to the infarcted heart to reduce inflammation (Lee et al. 2024). At the same time, biomaterials that target the heart through the bloodstream have also attracted attention. Tannic acid-modified proteins can bind to the extracellular matrix of cardiomyocytes in a thermodynamic manner, reducing the infarct size and enhancing cardiac function in rats with myocardial ischemia-reperfusion injury (Shin et al. 2018). In addition, small nanoparticles can quickly translocate from the lungs to the blood and myocardium through inhalation. Pig large animal models have provided evidence that inhalable calcium phosphate nanoparticles loaded with peptides are an effective method for targeted drug delivery to the heart (Miragoli et al. 2018).

3D printing technology, a revolutionary advanced technology in the biomedical field in recent years, has shown great potential in the field of cardiac repair. Intelligent-response materials can be 3D-printed into scaffolds that mimic the behavior of natural heart tissue, and this method has been proven to promote the maturation of cardiomyocytes in vitro (Liu et al. 2023). However, there is still a long way to go to construct patient-structure-specific patches for cardiac repair using 3D printing technology. Due to the lack of vascular perfusion, the thickness of the cardiac patch is also limited to some extent. Therefore, developing a thick patch with stable perfusable blood vessels and composite grafting is also one of the directions that scientists are striving for (Zhang et al. 2016). Silberman (Silberman et al. 2023) used the Post-Maturation Reinforcement strategy to allow the generation of thick, fully cellularized, and vascularized cardiac tissues using 3D printing technology, and surgical operations will not affect their structure and function. In addition, by adding other components, such as primitive macrophages, the stability of blood vessels can also be maintained. These strategies have laid the foundation for the subsequent development of thick patches with blood vessels.

The development of regenerative medicine has also provided new ideas for the biomedical field, such as reprogramming with patient-derived tissues, and the obtained cells can reduce the risk of immune rejection (Lyra-Leite et al. 2022). Changing the energy metabolism of cardiomyocytes to achieve cardiac regeneration is also a milestone development in cardiac repair. How to combine biomaterials to construct a stable and effective carrier is still a challenge (Li et al. 2023).

3.2.6 Shortcomings and Challenges in the Field of Intelligent-Response Biomaterials Applied to Cardiac Repair Research

The application of intelligent-response materials in the field of cardiac repair holds vast promise, yet it is not without a series of deficiencies and challenges that must be addressed. Initially, the introduction of biomaterials into the body may elicit an immune response; thus, mitigating the rejection reaction and enhancing biocompatibility is a critical issue (Cui et al. 2024a). Secondly, biomaterials must possess mechanical properties akin to cardiac tissue to withstand the cyclical stresses and strains experienced by the heart. Thirdly, the degradation rate of biomaterials must be aligned with the pace of tissue repair; degradation that is either too rapid or too slow can compromise therapeutic outcomes Fourthly, within the realm of cardiac tissue engineering, fostering the formation of new blood vessels to support the survival and functionality of the repaired tissue presents a formidable challenge. Fifthly, given that the heart functions as a syncytium, the integration of exogenous materials without inducing complications such as arrhythmias is a significant hurdle (Cui et al. 2024b). Sixthly, the use of certain biomaterials, particularly those derived from human or animal sources, may engender ethical and legal concerns (Gilbert et al. 2018).

3.2.7 Prospects for the Application of Intelligent-Response Biomaterials in Cardiac Repair Research

Intelligent-response biomaterials hold immense research and application prospects in the field of cardiac repair. Although current biomaterials used in clinical trials for heart disease are predominantly natural macromolecules, the exceptional biocompatibility, mechanical strength, and conductivity of synthetic materials suggest that more safe and functionally compatible biomaterials for human use will emerge in future clinical trials. With the rapid development of material science, biotechnology, and regenerative medicine, a multidisciplinary approach to strategy also provides more effective, personalized, and secure treatment options for future cardiac repair.

3.3 Nerve Regeneration

3.3.1 The Critical Role of Nerve Injury Repair and Rehabilitation

The human neural architecture manifests as a dichotomous organization, with the cerebromedullary axis (CNS) governing integrative functions, while the peripheral neural network (PNS) mediates environmental interactions through sensorimotor pathways. These two remarkable entities collaborate to regulate the activities of the human body, ensuring proper functioning and facilitating interaction with the surrounding environment. However, it must be noted that dysfunction and degeneration of these systems have a profound and extensive impact, affecting over one billion people globally. Neurodegenerative pathologies exemplified by amyloidogenic Alzheimer's dementia (AD) and α-synucleinopathic Parkinsonism exhibit progressive deterioration in higher-order cognitive processing and basal ganglia-mediated motor coordination, culminating in progressive activity of daily living (ADL) impairment. Acquired neural insults including cortical contusive encephalopathy and axonal transection injuries frequently precipitate irreversible sensorimotor deficits manifesting as flaccid paralysis and somatosensory extinction syndromes. These conditions have a significant and lasting imprint on the quality of life of individuals, necessitating long-term care and imposing substantial socioeconomic burdens on both individuals and society (Houshyar et al. 2019). The CNS exhibits intrinsically constrained regenerative potential under pathological conditions. Current therapeutic modalities ranging from surgical interventions to pharmacological regimens and rehabilitative strategies have shown marginal efficacy in eliciting neurogenesis or achieving functional recovery. Surgical options are restricted to specific indications, with optimal intervention timing remaining a clinical conundrum. Pharmacological approaches, though valuable, are constrained by systemic toxicity profiles and the blood-brain barrier's (BBB) selective permeability, which collectively limit drug bioavailability and therapeutic durability in neurological disorders (La Riva et al. 2023). Conventional physical therapy methods, such as phototherapy, magnetic therapy, acupuncture, and functional training, offer only partial benefits to nerve function recovery. However, meeting the needs of patients remains a formidable challenge. The development of next-generation therapeutic interventions demands a paradigm shift toward compounds and biomaterials with optimized bioavailability and minimized off-target effects. Such agents must be engineered to precisely modulate pathological microenvironments, promote neurogenic processes, facilitate neural tissue repair, and achieve functional neurological recovery.

3.3.2 Research Status of Nerve Repair and Regeneration

The PNS comprises neurons with somata extending elongated axonal projections to innervate target tissues. Myelinated axons are ensheathed by Schwann cell-derived myelin and endoneurial layers, with nodes of Ranvier enabling saltatory conduction through localized ion exchange. These axons are hierarchically organized: individual fibers are bundled into fascicles by the perineurium, while multiple fascicles are encapsulated within the epineurium a vascularized outer layer supporting axonal metabolism through nutrient and oxygen delivery. Despite decades of neuroscientific advancements, PNS-related pathologies persist as formidable clinical challenges, underscoring the complexity of neurological disease mechanisms and therapeutic interventions (López-Cebral et al. 2017). Current standard treatments for peripheral nerve injury (PNI) continue to show limited clinical efficacy, leaving patients with enduring movement restrictions and sensory dysfunction that severely compromise their quality of life. While synthetic nerve guidance conduits (NGCs) have been developed and FDA-approved as substitutes for autografts, most designs merely bridge nerve defects without providing essential regenerative bioactivity. This fundamental limitation frequently results in irreversible neurological deficits, as illustrated in Fig. 3.1 (Zhou et al. 2024). Fundamentally, every PNI treatment strategy aligns with the intrinsic process of nerve self-repair, a principle that also forms the foundation of tissue engineering approaches (Fig. 3.4).

3.3.3 Stimulus-Responsive Intelligent Biomaterials in the Field of Nerve Injury Repair and Regeneration Research

3.3.3.1 Stimulus-Responsive Intelligent Hydrogel

Hydrogels are 3D polymer networks able of absorbing and maintain significant quantity of liquid, creation them adaptable practical materials (Ahmed 2015). Derived from abundant and cost-effective natural sources, hydrogels have seen a surge in research and applications. However, traditional hydrogel scaffolds often face challenges in precisely matching the dimensions of SCI defects, and their invasive implantation can risk secondary damage (Abdulghani and Mitchell 2019). Injectable hydrogels address these limitations by enabling minimally invasive delivery and conforming to irregular defect shapes, while also portioning as effective carriers for sustained medication delivery with high fill efficiency and controlled release (Ehsanipour et al. 2021). Furthermore, they can encapsulate stem cells to enhance nerve regeneration and microenvironment repair. Stimuli-responsive intelligent hydrogels, such as conductive (Huang et al. 2021) and thermal variants, represent an innovative advancement, offering dynamic properties that respond to external cues like temperature changes, enabling precise drug release and environmental modulation. For instance, thermosensitive phase-change gels leverage

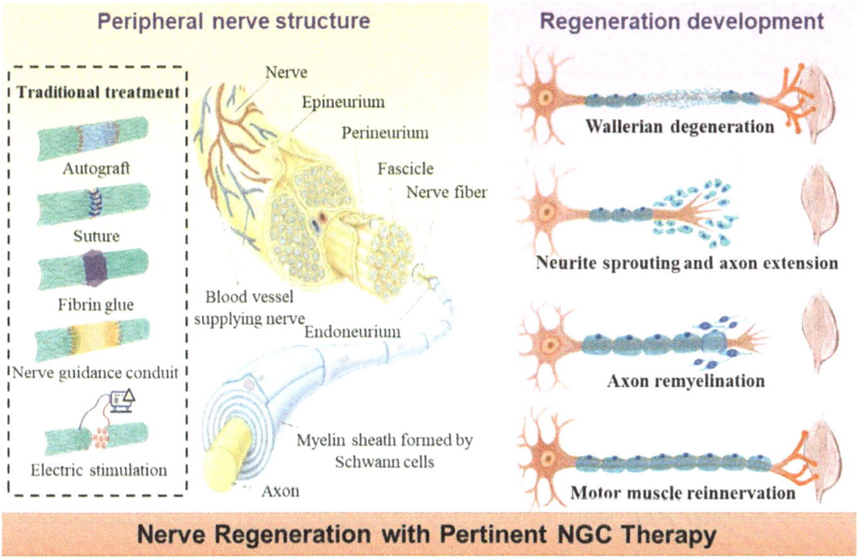

Fig. 3.4 Designing appropriate guidance conduits is crucial for effective nerve regeneration. Current clinical approaches for peripheral nerve injury encompass a range of strategies, such as autograft bridging, suturing, fibrin glue adhesion, nerve guidance conduit-assisted reconnection of severed nerve ends, and externally applied programmed electrical stimulation, all of which are employed to promote nerve repair and functional recovery. (Reprinted with permission from Zhou et al. 2024)

temperature shifts to alter their physical and chemical states, optimizing conditions for nerve repair. Despite their potential, challenges remain, including limited mechanical strength (particularly in natural polymer hydrogels) and excessive swelling behavior, which require further refinement.

3.3.3.2 Stimulus-Responsive Smart 3D-Printed Bionic Scaffolds

Given the intricate spatial organization and 3D architecture of natural spinal cord tissue, conventional 3D scaffolding methods often fall short in precisely controlling both external shapes and internal structures (Liu et al. 2021b). To direction this, 3D printing has emerged as a groundbreaking solution, enabling the creation of customized gallows tailored to the accurate dimensions of SCI. This technology not only replicates the injury site's structure but also fosters a conducive microenvironment to promote and guide axon regeneration.

3.3.3.3 Stimulus-Responsive Smart Electrospinning Scaffolds

Electrospinning is a process that creates fibers from molten or dissolved polymers using a high-voltage electrostatic field. This method, able of cause fibers ranging from nanometers to micrometers in diameter, is extensively utilized in constructing neural scaffolds due to its straightforward equipment and precise control over fiber morphology and surface porosity (Reis et al. 2018). Moreover, by integrating smart-responsive capabilities, electrospinning allows for the incorporation of diverse substances—such as proteins, lipophilic drugs, and nucleic acids—into nanofibers, enabling flexible and controlled loading and release. This advancement significantly enhances therapeutic efficiency.

3.3.4 Stimulus-Responsive Biomedical Materials Are Used as Therapeutic Strategies in the Field of Nerve Injury Repair and Regeneration Research

3.3.4.1 Reduces Inflammation

Post-SCI, infiltrating immune cells converge at the lesion epicenter, initiating a pathogenic cascade through sustained release of pro-inflammatory cytokines, chemokines, and reactive oxygen species (ROS). This persistent immune activation triggers a maladaptive inflammatory cascade, exacerbating secondary neurodegeneration through oxidative stress and bystander tissue damage, as evidenced by recent studies (Al Mamun et al. 2021). Given this, implementing anti-inflammatory treatment strategies to regulate the immune microenvironment in SCI cases can have significant therapeutic benefits. The controlled release of loaded anti-inflammatory drugs using pH and temperature-sensitive intelligent response biomaterials is a common method for inflammation treatment in neurotrauma. This is mainly accomplished by controlling the phase transition of the response biomaterials to trigger dissolution/expansion, resulting in the collapse of the system and thus controlling the anti-inflammatory capability (Karimi et al. 2016).

3.3.4.2 Reduces Oxidative Stress

Three principal ROS superoxide (O_2^-), hydrogen peroxide (H_2O_2), and hydroxyl radicals (OH^-) engage in tightly coordinated redox signaling cascades. Their biosynthesis and catabolism are governed by homeostatic regulation, which is critically dependent on spatiotemporal control of enzymatic/non-enzymatic antioxidant systems. Maintenance of this redox equilibrium is essential for cellular homeostasis, as disruption of this dynamic balance precipitates oxidative stress, macromolecular damage, and functional impairment across biological pathways. Following SCI, an excess of ROS is released into the extracellular microenvironment, thereby disrupting this equilibrium. The reduction of ROS is often a by-product of inflammation reduction; however, it is also a matter of concern as inflammation is not the sole source of ROS (Bradbury et al. 2002). Additionally, a substantial number of literatures have reported that upon external stimulation, responsive intelligent materials are prone to form various ROS substances due to the formation of holes and transition state electrons in valence and conduction bands, thus exacerbating the damage of local oxidative stress (Xie et al. 2018). While antioxidant supplementation serves as a strategic intervention against ROS, its efficacy remains constrained by inherent physicochemical limitations including susceptibility to degradation, low aqueous solubility, and suboptimal bioavailability. A critical challenge lies in employing stimuli-responsive intelligent materials to achieve spatiotemporal-controlled release while maintaining the delicate equilibrium between ROS generation and antioxidant capacity post-activation. Notably, advancements in understanding the redox-modulating properties of both natural derivatives and engineered compounds have catalyzed the emergence of novel therapeutic paradigms (Zhu et al. 2024).

3.3.4.3 Reduces Film Scarring

Following SCI, apathetic astrogliosis leads to the development of glial scars distinguished by cellular hypertrophy and multiplication. These fibrotic lesions create dual regenerative barriers a physical obstruction to axonal regrowth combined with inhibitory biochemical microenvironments rich in chondroitin sulfate proteoglycans (CSPGs). A seminal 2002 investigation demonstrated that chondroitinase ABC (ChABC)-mediated CSPG degradation attenuates scar formation and enhances functional recovery post-SCI (Bradbury et al. 2002). This breakthrough established a therapeutic paradigm shift, with subsequent research focusing on two principal strategies: modulating astrocytic activation dynamics and developing targeted CSPG neutralization approaches (Ham and Leipzig 2018). In addition to the conventional approach of controlling the release of drug loads to mitigate scarring, studies have demonstrated that the stress contraction of material responses following external stimulation increases local pressure at the wound site, which is conducive to reducing scar generation (Cao et al. 2023). Moreover, local photothermal (Yaron et al. 2024) and electroresponsive (Yang et al. 2024) biomaterials have been reported to be beneficial in reducing scar formation during spinal cord injury repair.

3.3.5 Summary and Prospect of Stimulus-Responsive Intelligent Biomedical Materials Applied to Nerve Injury Repair and Regeneration

SCI triggers a cascade of intricate pathophysiological events, creating a microenvironment that hinders nerve regeneration. The functions of various cells in this setting are still debated. For example, oligodendroglial progenitor cells, which naturally replenish spinal cord cells, might help mitigate the worsening of spinal cord injury. However, some studies suggest these progenitors could also contribute to glial scar formation. Similarly, astrocytes exhibit multifaceted roles, with recent evidence indicating that astrocyte scars might support, rather than hinder, axon regeneration in the central nervous system. Understanding the mechanisms of these cells is vital for developing new treatments for SCI. Furthermore, the cellular microenvironment operates through interconnected components, and future research should focus on clarifying these regulatory relationships to enhance nerve repair.

Intelligent responsive biomaterials for spinal cord injury treatment are advancing rapidly. After extensive research, biocompatibility, low toxicity, and degradability have become essential features of these biomaterial scaffolds. Beyond bridging pathological gaps, these scaffolds can regulate the release of bioactive molecules, cells, and drugs through stimulus-responsive feedback, aiding microenvironment reconstruction. However, challenges persist. For instance, the precise mechanisms governing cellular behavior and microenvironment regulation remain unclear. Material safety also lacks standardized international evaluation criteria, and factors such as transplantation timing, injury location, and severity require careful consideration.

Future efforts should prioritize the development and application of innovative regenerative biomaterials. Ultimately, refining biomaterial-based combination strategies will be central to advancing spinal cord injury repair. As research progresses, more intelligent responsive materials are expected to transition to clinical use, offering improved outcomes for patients (Chen et al. 2024).

3.4 Vascular Regeneration

3.4.1 The Importance of Vascular Regeneration

Blood vessels are a series of tubes supporting blood flows. Blood vessels are found all over the body except cornea, hair, nails, dentine, and epithelium. The vessel system is a network of arteries, veins, and capillaries, conducted into tree-like structures (Eelen et al. 2020). The blood vessels are among the earliest organs to form during the development of an embryo and are essential for the proper functioning of all other organs (Kabirian et al. 2018). It expands beyond the point where essential nutrients can be delivered solely by diffusion, in order to support the increasing demands of the developing tissues and organs (Potente and Mäkinen 2017). No tissue and organ of the human body can be separated from the nourishment from blood, and vessels are the channels that transport blood, and play a vital role in human health. Vessels can be said to be a large system from large blood, medium-sized, small and gradually extend to microvessels. Among them, the heart aorta is the largest, with a diameter of about 3 cm, while the microvessels which are densely covered with organs of the whole body are the smallest, with a diameter of only 10

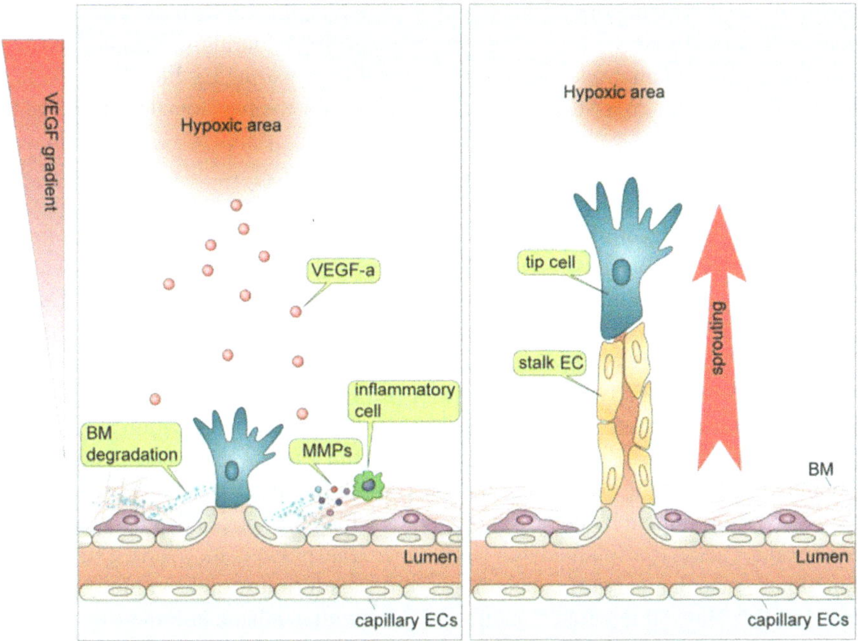

Fig. 3.5 The impact on vessel growth via growth factors and oxygen

microns, Sprouting angiogenesis, which involves the formation of new blood vessels, is triggered by gradients of proangiogenic factors (such as VEGF and FGF) released from oxygen- and nutrient-deprived areas. These factors activate quiescent endothelial cells to initiate vessel growth (Rafii et al. 2016; Rajala 2021; Trimm and RED-Horse K. 2023) (Fig. 3.5). A lot of blood vessel formation in adult diseases recapitulates many of the molecular mechanisms observed in development. Therefore, the regeneration of vessels is crucial for human health.

3.4.2 *Current Research on Vascular Regeneration*

The formation of the vascular system originates from endothelial progenitor cells that arise from the mesoderm during the embryonic phase. It constitutes an intricate and interconnected network capable of transporting oxygen, nutrients, waste products, and carbon dioxide. Over the past several decades, technological advancements have broadened our understanding of the vascular system, revealing that blood vessels are not just inert channels but also exert a dynamic regulatory influence through paracrine signaling, which is instrumental in guiding tissue regeneration and maintaining homeostasis and metabolic balance. Despite these insights, the creation of sophisticated vascular networks has consistently posed a significant hurdle in tissue engineering, primarily due to technical complexities. The ultimate goal of this field is to fabricate fully operational and enduring organs and tissues suitable for transplantation. The quest for a viable substitute for blood vessels has been a driving force in tissue engineering research for the past two decades (Zhang et al. 2024).

A variety of approaches have been developed to construct substitutes for blood vessels that possess biological functionality. These methods include the use of cell-seeded materials, decellularization scaffolds, 3D bioprinting techniques, and intelligent responsive materials. Reiner A Wimmer described a method for creating self-organizing 3D human blood vessel organoids from human pluripotent stem cells (hPSCs), which display the morphological, functional, and molecular characteristics of human microvasculature (Heydarkhan-Hagvall et al. 2008).. These organoids are developed by inducing mesoderm formation in hPSC aggregates, followed by differentiation into endothelial networks and pericytes within a 3D collagen I-Matrigel matrix. Blood vessels emerge within 2 to 3 weeks and can be further expanded in a scalable suspension culture (Mahara et al. 2015). For further study, they also report human blood vessel organoids which are surrounded by a basement membrane, can be transplanted into mice. Once transplanted, they form a stable and functional vascular network, including arteries, arterioles, and venules. Sepideh Heydarkhan-Hagvall integrated natural proteins with synthetic polymers to develop electrospun fibrous scaffolds, which exhibit desirable mechanical and biological characteristics suitable for applications in cardiovascular diseases (Wimmer et al. 2019).

3.4.3 Challenges in Vascular Regeneration

Vascular regeneration remains a challenge due to a lot of difficulties that researchers must address. One is the need for improved biocompatibility in vascular grafts, which is essential for minimizing immune response and promoting cellular integration. Current materials often cause inflammatory responses, hindering the utilization of the grafts. Furthermore, achieving adequate mechanical properties that mimic natural blood vessels is essential to deal with physiological pressures and prevent failure post-implantation. The other significant challenge lies in the complex vascular microenvironment. The dynamic interplay between endothelial cells, smooth muscle cells, and the extracellular matrix is crucial for successful vascular regeneration. Finally, while advancements in tissue engineering and 3D bioprinting offer promising avenues, the translation of these technologies into clinical practice is fraught with difficulties. Issues related to scalability, reproducibility further complicate the pathway to successful vascular graft implementation. Addressing these challenges is essential for the development of effective strategies for vascular regeneration, ultimately improving outcomes in patients with vascular diseases.

3.4.4 Methods and Strategies for Vascular Regeneration

3.4.4.1 Natural Biodegradable Polymers

Collagen and Gelatin

Collagen is recognized for its compatibility with most biological tissues and is considered non-toxic. It is also widely accessible. Predominant in the animal, collagen accounts for roughly 30% of the total protein content in vertebrates. It is ubiquitous throughout the body, with a notable presence in the skin, tendons, and bones. As a primary constituent of the extracellular matrix, collagen plays a pivotal role in the molecular framework of higher organisms, endowing various tissues with robustness and structural integrity. Two physical techniques for cross-linking collagen include short wavelength ultraviolet (UV) exposure at 254 nanometers and dehydration followed by thermal treatment under vacuum conditions at temperatures reaching up to 100 degrees Celsius for an extended period. It is important to note that these physical cross-linking methods result in a less extensive degree of cross-linking (Sorushanova et al. 2019; Du et al. 2021).

Albumin

Albumin stands as the predominant protein found in blood plasma. The compatibility of biomaterials with blood can be substantially enhanced by adhering albumin molecules to their surfaces, thereby improving their blood compatibility. Given its exceptional affinity for blood, albumin has been considered a promising candidate for a carrier matrix in drug delivery systems that operate within blood vessels. Additionally, albumin microspheres that have been labeled with radioactive isotopes are utilized in the field of diagnostic nuclear medicine for visualizing organs and tissues. The preparation of albumin microspheres typically involves the process of emulsion stabilization, which entails emulsion cross-linking. This cross-linking can be executed thermally within a temperature range of 100–180 °C, or with chemical cross-linking agents such as formaldehyde or glutaraldehyde disulfide to get bonds or form lysine-alanine cross-links. Albumin particles with a broad size range (spanning from 1 to 1000 μm in diameter) can be generated. Both hydrophobic and hydrophilic drugs, including L-epinephrine, adriamycin, prednisolone, thermally decomposed 5-fluorouracil, and insulin, have been successfully loaded into albumin particles. However, low drug loading during thermal cross-linking, the use of chemical cross-linkers, and the enzymatic degradation of the matrix (leading to bulk erosion, which is typically uncontrollable and unpredictable) are identified as the primary drawbacks of albumin carriers (Merlot et al. 2014).

Chitin

Chitin is the primary polysaccharide found in the shells of crustaceans and the exoskeletons of insects, and it is typically extracted from these natural sources. Chitosan, a deacetylated form of chitin, can be produced through alkaline treatment. Commercially, chitosan is available in various forms such as powders, fibers, particles, gels, and solutions. These products are biodegradable, possess hemostatic properties, and can promote cell growth (Kratz 2014; Sultankulov et al. 2019).

3.4.4.2 Synthetic Polymers

Poly(Glycolic) Acid (PGA)

Poly(glycolic) acid (PGA) was among the earliest synthetic biodegradable polymers utilized in biomedical applications (for instance, DEXON™ was the first biodegradable synthetic suture approved by the FDA) (Nair and Laurencin, 2007). Owing to its high crystallinity (45–55%) and a glass-transition temperature between 35–40 °C, PGA has a higher tensile modulus compared to other biodegradable polymers. However, PGA degrades rapidly in bulk, with a significant loss of mechanical properties in a short period. This degradation results in acidic by-products that can cause local inflammation in surrounding tissues, thereby limiting its use in tissue engineering (Hamed et al. 2016)..

Poly(Lactic)Acid (PLA)

Lactide is a chiral molecule that exists in two optically active forms: L-lactide and D-lactide. When these enantiomers are polymerized, they form different types of polylactic acid (PLA). Poly(L-lactide) (PLLA) is a semi-crystalline polymer with a crystallinity of about 37% and a glass-transition temperature between 60 and 65 °C. In contrast, poly(DL-lactide) (PDLLA) is amorphous due to the random distribution of L- and D-lactide units, resulting in a lower glass-transition temperature of 55 to 60 °C. Polylactic acid (PLA) degrades through bulk erosion via hydrolysis, ultimately breaking down into lactic acid, which is a natural metabolic by-product in humans. This lactic acid is then metabolized into water and carbon dioxide via the citric acid cycle. PLLA is highly resistant to hydrolysis and degrades slowly over a long period. It has excellent tensile strength, minimal deformation, and a high modulus. Specifically, the modulus of PLLA (around 5 GPa) is much higher than that of native blood vessels (approximately 15 MPa), making it ideal for stent fabrication. High molecular weight PLLA takes 2 to 5.6 years to be fully absorbed in the body and gains strength over about 6 months during hydrolysis. PLLA was approved by the FDA in 1971 for use as a superior suture material, representing a significant improvement over the DEXON suture (Nakajima et al. 2017).

Poly(Lactic-Co-Glycolic) Acid (PLGA)

The copolymerization of glycolide and lactide has been extensively studied to tailor the characteristics of polyglycolic acid (PGA) and polylactic acid (either PLLA or PDLLA). Poly(lactic-co-glycolic acid) (PLGA) has garnered significant interest, driven by its FDA clearance for clinical applications, favorable processing attributes, adjustable rate of degradation, affordability in the commercial market, and superior compatibility with biological systems as well as mechanical durability. The mechanical resilience, propensity for swelling, susceptibility to hydrolysis, and the consequent rate of biodegradation of PLGA are all influenced by the polymer's crystallinity. This crystallinity is determined by the nature and the molar proportion of the constituent monomers within the copolymer chain. The balance of these monomers is crucial for achieving the desired properties in PLGA, making it a versatile material for various biomedical applications (Nakajima et al. 2017).

3.4.4.3 Cell Combination Repair

The combination of specific cell types, such as endothelial cells and mural cells, which can self-assemble into vessel-like structures within engineered tissues, promotes graft–host vessel anastomosis and graft perfusion. Some research indicates that the mere presence of endothelial cells in the implanted engineered tissue is enough to ensure graft integration and survival, regardless of their organization. In contrast, other studies have demonstrated that the formation of vessel-like structures by these cells within the engineered tissue may improve graft performance after implantation. However, it has been observed that engineered vessels can induce thrombosis upon transplantation, resulting in decreased graft perfusion and an increased risk of graft failure. Vessel maturation, which involves the stabilization of vessels through adherence junctions, rearrangements of the vessel basement membrane, and the close interaction between mural cells and the vessel network, is a crucial final step in the development and remodeling of the vasculature (Bruzauskaite et al. 2016).

3.4.4.4 Electrospinning

The electrospinning technique, which allows for the creation of continuous polymer micro- or nanofibers with diameters ranging from 50 to 5000 nm from polymer solutions or melts under a high electric field, has emerged as a promising method for fabricating vascular scaffolds. This is due to its simple setup, cost-effective production, and the ability to easily control fiber diameter and porosity. The high surface area-to-volume ratio of electrospun nanofibers provides more surface for cell attachment compared to other structures. Using the electrospinning process, fibrous matrices at micro- or nano-scale dimensions have been developed from various natural polymers such as collagen and fibrin, as well as a wide range of absorbable synthetic polymers like poly-e-caprolactone (PCL), poly(D-lactide) (PDLA), poly(L-lactide) (PLLA), and copolymers including poly(L-lactide-co-caprolactone) (P(LLA-CL)) and poly(D,L-lactide-co-glycolide) (PLGA). Biocompatibility tests conducted both in vitro and in vivo have confirmed the potential of electrospun nanofiber scaffolds as effective substitutes for vascular grafts (Hasan et al. 2014).

3.4.4.5 Decellularization Scaffold

Decellularization of native blood vessels is an extensively studied and promising strategy in vascular engineering. The optimal decellularization process involves the complete removal of cellular components from tissues while preserving the composition, mechanical integrity, and biological activity of the native extracellular matrix (ECM). By eliminating host cellular antigens, decellularization reduces the risk of inflammatory responses and minimizes the likelihood of immune-mediated tissue rejection. Moreover, the preservation of the complex ECM structure supports the adhesion, migration, proliferation, and differentiation of recipient cells. These benefits may help address some of the current challenges in vascular graft developmen (Porzionato et al. 2018).

3.4.4.6 3D Bioprinting Blood Vessel

Recent advancements in 3D printing, also known as additive manufacturing, have unlocked vast potential for engineering tissues and organs that can replicate the native structure and function to a significant degree. The biomimetic strategy in 3D bioprinting enables the creation of blood vessels with specific shapes, while multi-material bioinks designed for vascular support can bridge the gap from laboratory research to clinical applications. Additionally, during the 3D bioprinting process, it is possible to form perfused vessels with various elastic properties, such as Young's Modulus and Poisson's ratio, that closely resemble those of natural blood vessels (Jia et al. 2016).

3.4.4.7 Intelligent Responsive Materials

When an organ does not receive normal blood supply, it can result in pathological conditions like ischemia and reperfusion injury. Therefore, restoring vascular function is a crucial initial step for both tissue regeneration and alleviating ischemic conditions. In addition to sustained release mechanisms, "intelligent" materials that can detect and respond to signals in the physiological environment are increasingly being explored for applications in drug delivery, tissue engineering, and biomedical devices (Maltepe et al. 1997). In the context of blood vessel regeneration, intelligent and responsive materials have shown great promise by reacting to various physiological signals, such as enzymes, pH levels, temperature, cell signaling, mechanical forces, and other factors.

Enzyme-Responsive Materials

Wen et al. (2011) introduced a new delivery system featuring controlled-release bovine serum albumin (BSA) and vascular endothelial growth factor (VEGF) using protein nanocapsules. This system enabled the controlled release of VEGF to promote and direct blood vessel formation. The innovative protein nanocapsules can release their contents at a controllable rate and with specificity to a particular enzyme. This research pointed to a new approach for responsive protein delivery tailored to specific cellular events or local environments, which is highly significant for therapeutic and other applications. Additionally, the unique structure of the nanocapsules provides robust protection for the encapsulated proteins against the reaction system, facilitating the creation of a variety of hydrogel scaffolds for extensive regenerative tissue applications.

pH and Temperature-Responsive Materials

From Rucha V Joshi et al. (2013), microspheres have been developed that are suitable for percutaneous tissue injection without the need for an immediate phase change during the injection process. These dual-responsive microspheres have been designed to remain in solid form when injected into slightly acidic ischemic sites and to gradually dissolve as they release their contents and the ischemic site normalizes to physiological pH. The p(NIPAAm-co-PAA-co-BA) microsphere platform described here allows for sustained, pH-dependent release of a bioactive growth factor in vitro and enhances the persistence following injectable delivery to ischemic sites in vivo. This innovative delivery system holds great promise for sustained local activation of reparative processes stimulated by FGF-2 or other therapeutic proteins in ischemic tissues damaged by peripheral artery disease (PAD), trauma, or other disease processes.

Cell Receptor-Responsive Materials

Growth factors (GFs) trigger a cellular response by binding to specific receptors on the surface of target cells. This specificity offers a unique opportunity to localize GF delivery by targeting specific cell receptors. Nori Yamaguchi et al. (2007)developed hydrogels through the noncovalent interaction between heparin-functionalized, star-PEG polymers and the therapeutically relevant, heparin-binding protein VEGF. The specificity of this interaction was demonstrated by the lack of hydrogel erosion when incubated with particles modified with a protein that does not bind to

VEGF. The VEGF released from the hydrogels remains bioactive and thus promotes blood vessel regeneration.

Mechanical-Responsive Materials

In addition to the chemical environment, the physiological environment is also mechanically dynamic. Carmeliett (2000) employed alginate hydrogels containing trypan blue as a model drug to study the effects of mechanical loading. The hydrogels were subjected to repeated compressive loading using a mechanical tester. The study revealed that mechanical compression can regulate drug release from polymeric matrices. Hydrogels loaded with VEGF and subjected to cyclic mechanical stimulation exhibited a statistically significant increase in granulation layer thickness and vascularization compared to non-stimulated VEGF-releasing gels.

3.4.5 Inadequacies in Vascular Regeneration

Despite numerous methods reported for studying vascular repair and regeneration, each approach presents its advantages and disadvantages. Transplantable vascular grafts have demonstrated excellent biocompatibility; however, their degradation performance in vivo requires significant improvement. Furthermore, vascular grafts are typically implanted only at specific necrotic sites, serving primarily to maintain vascular patency, while their biological activity still needs enhancement. In contrast, 3D printing faces challenges due to the lack of more natural biomaterials and the necessity for cellular support to provide corresponding bioactivity, which can lead to specific immune rejection responses. Thus, there remains a considerable gap before these technologies can be translated into clinical practice, necessitating more in-depth research by scientists.

3.4.6 Perspectives on Vascular Regeneration

Considering the essential role of the vascular system within the broader circulatory system, research in this field holds utmost significance. At present, the primary challenge is to achieve a balance between biocompatibility and vascular functionality. Although reconciling non-immunogenicity with functional grafts remains a complex issue, we are optimistic that progress in biomaterials and ongoing advancements in 3D printing technologies will pave the way for the development of more effective vascular grafts tailored for vascular repair.

3.5 Bone Regeneration

3.5.1 *The Importance of Bone Repair and Regeneration*

As a foundational biomechanical system, osseous tissue achieves structural maturity through dual functional imperatives: providing load-bearing capacity via mineralized matrix stiffness, and maintaining remodeling plasticity through coupled osteoblastic anabolism/osteoclastic catabolism a dynamic equilibrium fundamental for skeletal homeostasis (Karsenty and Khosla 2022). This osseous coupling mechanism in mature bones involves precisely orchestrated interactions between matrix-synthesizing osteoblasts and resorptive osteoclast populations. The tissue's remarkable regenerative potential originates from its hierarchically organized nanocomposite architecture, comprising nanoscale hydroxyapatite crystals aligned with collagen fibrils across multi-level structural dimensions (Xie et al. 2016). While minor osseous defects typically undergo spontaneous repair through phased biological processes (recruitment cascade, immunomodulation, neovascularization, and matrix mineralization), critical-sized defects surpass intrinsic regenerative thresholds (Einhorn and Gerstenfeld 2015). This clinical challenge has propelled innovations in bone tissue engineering, particularly in developing tripartite regenerative systems integrating cellular, biomolecular, and material components (Ruiz-Alonso et al. 2021). Current translational barriers stem from limitations in viable cell sources, suboptimal graft survivability, and cost-efficacy challenges associated with growth factor delivery systems. Consequently, material science advancements targeting functional biomimetic constructs are emerging as a strategic focus to circumvent these biological constraints and bridge the tissue engineering-clinical application gap.

3.5.2 *Difficulties and Challenges in Restoration and Regeneration*

The advent of nanotechnological innovations has revolutionized bone tissue engineering through bioengineered constructs combining functional nanomaterials with advanced scaffolding architectures. These developments have enabled novel intervention strategies spanning stem cell modulation to defect reconstruction. Hydroxyapatite (HA)-based systems remain pivotal in osteogenic regulation, demonstrating exceptional capacity to direct stem cell differentiation via transcriptional programming (Ramaswamy et al. 2021). Contemporary material design emphasizes hybrid systems where HA synergizes with organic matrices (e.g., collagenase-encapsulated HA-calcium sulfate-hyaluronic acid composites for alveolar regeneration) or structural modifiers (e.g., gelatin formamide/HA porous scaffolds with biomimetic topography enhancing rabbit cranial repair) (Wang et al. 2014). While growth factor supplementation (BMP-2/VEGF-functionalized chitosan-HA

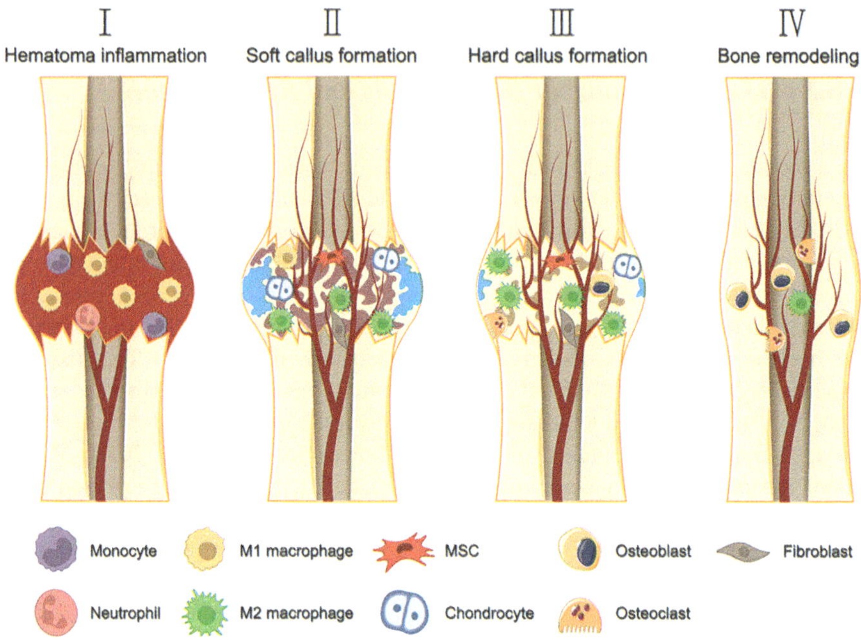

Fig. 3.6 Schematic representation of osseous regeneration dynamics with concomitant macrophage polarization across tetra-phasic progression. (Reprinted with permission from Wang et al. 2022)

systems) and hydrogel carriers (BMP-4-loaded chondroitin sulfate hydrogels (Casarrubios et al. 2020), SF/RGD-enhanced matrices (Yin et al. 2019)) have improved osteogenic outcomes, their clinical translation faces dual constraints: transient bioactivity profiles and prohibitive cost-to-efficacy ratios. This paradigm shift has intensified the pursuit of growth factor-independent strategies, focusing on optimizing material-driven osteoinduction through: (1) Biomimetic nanostructural design replicating bone's hierarchical organization. (2) Intelligent release systems for sustained ionic/cellular modulation. (3) Cost-effective fabrication of multifunctional hybrid scaffolds. The field now prioritizes creating self-sufficient regenerative microenvironments that minimize exogenous biological inputs while maximizing intrinsic healing mechanisms (Fig. 3.6).

3.5.3 Stimulus-Responsive Biomedical Materials in the Field of Bone Repair and Regeneration Research

3.5.3.1 Piezoelectric Materials

The piezoelectric phenomenon, characterized by electric field generation under mechanical deformation, has emerged as a pivotal mechanism in bone tissue engineering, mirroring native bone's inherent electromechanical properties (Hosoyama et al. 2019). This electroactive behavior enables implanted materials to transduce mechanical stimuli into bioelectric signals, triggering sequential cellular responses including protein adsorption modulation, membrane potential alteration, and calcium channel regulation. Extensive preclinical evidence substantiates the therapeutic efficacy of piezoelectric biomaterials in localized osseous regeneration (Khare et al. 2020). Current material systems exploit this mechanism through two principal categories: Piezoelectric polymers: PVDF, P(VDF-TrFE), PLLA, and PHA; Piezoceramic composites: $BaTiO_3$, ZnO, and KNN-based systems. These materials establish dynamic electromechanical microenvironments that recapitulate native bone's bioelectric signaling during remodeling.

3.5.3.2 Electrical Stimulus-Responsive Materials

Conductive biomaterials for osseous regeneration primarily comprise three classes: carbon-based systems, metallic oxides, and electroactive polymers. Recent research has particularly focused on carbon-derived architectures and polymeric conductors due to their enhanced electromechanical responsiveness (Liu et al. 2021c). These materials enable in situ generation of bioelectric stimuli under external field induction, facilitating precise modulation of cellular dynamics and osteogenic processes through endogenous electrical signaling pathways.

3.5.3.3 Magnetic Stimulus-Responsive Materials

Magnetoresponsive biomaterials incorporating ferromagnetic elements (Fe, Co, Mn, Ni) have emerged as promising platforms for regenerative medicine (Liu et al. 2021c). Iron oxide nanoparticles (IONPs) exhibit distinctive therapeutic advantages stemming from their multifunctional physicochemical profile: superparamagnetic responsiveness, tunable biocompatibility, nanoscale geometry enabling high surface reactivity, and magnetically guidable targeting precision. These intrinsic properties underpin therapeutic innovations spanning mineralized tissue regeneration, stem cell lineage specification, and smart scaffold engineering through magnetic actuation mechanisms (Xia et al. 2018). Naturally occurring IONP variants (magnetite Fe_3O_4, maghemite γ-Fe_2O_3, hematite α-Fe_2O_3) provide versatile building blocks for tissue engineering strategies. Recent advancements have explored both standalone and composite IONP systems: Monocomponent applications: Magnetic field-activated Fe_3O_4 significantly promotes MC3T3 osteoblast differentiation; Hybrid nanocomposites: Integration with structural scaffolds through advanced fabrication techniques (selective laser sintering, surface grafting, electrospinning-assisted layer-by-layer assembly, co-deposition) (Chen et al. 2018). The therapeutic mechanism involves magnetic dipole realignment under static magnetic fields (SMF), generating localized field intensification that enhances cellular responses at scaffold interfaces.

3.5.3.4 Photothermal Materials

In photothermal therapy, brightness strength is change into heat, and a photothermal agent (PTA) is usually irradiated with light from the mark tissue to raise the community temperature. Due to its selective and non-invasive benefits, photothermal therapy is becoming increasingly important in the external non-invasive thermal treatment strategy of BTE. photothermal therapy typically produces an in-situ thermal effect by applying tissue-clear NIR light in areas where PTA is present (Wu et al. 2022). Thus, mild hyperthermia therapy, which enhances bone regeneration, can be easily achieved when PTA at the site of injury has a thermal response to external near-infrared light. Recent studies have shown that mild photothermal radiation can promote MSC differentiation, osteoblast maturation, and mineralization (Wan et al. 2020). As a result, photothermal therapy is receiving increasing attention in bone tissue as a relatively mild and targeted treatment. PTAs applied to tissue engineering include carbon-based PTAs, metal-based PTAs, and other types of PTAs. Among them, graphene and GO are commonly used due to their excellent NIR brightness soaking up rate and powerful photothermal conversion effectiveness (Zhao et al. 2018). Scaffolding containing graphene and GO, including nHA/GO scaffolding (Nie et al. 2022).and HA/GO/CS scaffolding, which can not only promote the in vitro osteogenesis of BMSCs, but also exhibit excellent photothermal properties in vivo, thereby significantly supporting bone regeneration. In addition, other carbon-based difluoric acids, including carbon dots (CD), carbon-fluorofluoride, and carbon-aerogels, have also shown promising potential in bone tissue.

3.5.4 Prospects for the Application of Stimulus-Responsive Biomedical Materials in the Field of Bone Repair and Regeneration

Current biocompatibility assessments for many material classes—especially inorganic constituents and persistent polymeric matrices—remain preliminary, necessitating systematic elucidation of their interfacial interactions within physiological environments.

This characteristic often necessitates a second surgical procedure for their removal, thereby giving rise to secondary trauma. Moreover, the primary biosafety assessment of materials, which is typically conducted by observing whether the structure of major internal organs is damaged following implantation, fails to comprehensively evaluate the long-term biosafety and the potential impact on other body tissues. Consequently, there is an escalating skepticism regarding the long-term biosafety of biomaterials employed in clinical settings. Future endeavors should, therefore, be aimed toward the plan and building of biomaterials that are entirely biodegradable and show optimal biocompatibility. Furthermore, aside from osteogenesis, both the immune response and angiogenesis have demonstrated remarkable potency in promoting bone regeneration (Bezsonov et al. 2022). Current research significantly underrepresents the combinatorial effects of material-derived physical cues on the critical triad of vascularization, osteoinduction, and immune regulation. The immunomodulatory dimension is particularly crucial, as bone regeneration initiates through a precisely orchestrated inflammatory cascade: initial immune activation stimulates cellular recruitment and cytokine/chemokine secretion, establishing a regenerative microenvironment at the defect site (Rodrigues et al. 2019).

The complicated interplay between osteogenic cells particularly the link to remodeling activities of osteoblasts and osteoclasts, mediated by osteocyte-derived signaling networks constitutes a fundamental regulatory axis in bone repair and metabolic homeostasis (Robling and Bonewald 2020). This biological triad presents a compelling therapeutic target for biomaterial-mediated physical modulation to enhance site-specific regeneration. However, current research disproportionately focuses on BMSC and osteoblast responses, with insufficient attention to osteoclast dynamics, osteocyte signaling, and their intercellular crosstalk under physical stimulation. Emerging evidence highlights the therapeutic potential of biomaterial-engineered physical cues (nanotopographical features, electromechanical signals, photothermal effects, and magnetic fields) in bone tissue engineering. Future advancements require: Development of multifunctional, biodegradable platforms capable of integrating multiple physical stimuli; Systematic optimization of stimulation parameters for clinical relevance; Elucidation of immunomodulatory-angiogenic-osteogenic regulatory networks.

This paradigm shift toward physical cue-mediated regeneration strategies holds significant promise for advancing reconstructive medicine, though requires concerted efforts to bridge fundamental discoveries with clinical translation.

References

Abdulghani S, Mitchell GR (2019) Biomaterials for in situ tissue regeneration: a review. Biomol Ther 9(11)

Ahmed EM (2015) Hydrogel: preparation, characterization, and applications: a review. J Adv Res 6(2):105–121

Al Mamun A, Monalisa I, Kubra KT et al (2021) Advances in immunotherapy for the treatment of spinal cord injury. Immunobiology 226(1):152033

Arslan F, Bongartz L, Ten Berg JM et al (2018) 2017 ESC guidelines for the management of acute myocardial infarction in patients presenting with ST-segment elevation: comments from the Dutch ACS working group. Neth Hear J 26(9):417–421

Bezsonov EE, Gratchev A, Orekhov AN (2022) Macrophages in health and non-infectious disease 2.0. Biomedicines. 10(6)

Bissell MJ, Labarge MA (2005) Context, tissue plasticity, and cancer: are tumor stem cells also regulated by the microenvironment? Cancer Cell 7(1):17–23

Blanco J, Alonso AG, Hermida-Nogueira L et al (2024) How to explain the beneficial effects of leukocyte- and platelet-rich fibrin. Periodontol 2000 97:74

Bradbury EJ, Moon LDF, Popat RJ et al (2002) Chondroitinase ABC promotes functional recovery after spinal cord injury. Nature 416(6881):636–640

Bruzauskaite I, Bironaite D, Bagdonas E et al (2016) Scaffolds and cells for tissue regeneration: different scaffold pore sizes-different cell effects. Cytotechnology. 68(3):355–369

Cai S, Pan Y, Han B et al (2011) Transplantation of human bone marrow-derived mesenchymal stem cells transfected with ectodysplasin for regeneration of sweat glands. Chin Med J 124(15):2260–2268

Cao Y, Wang L, Zhang X et al (2023) Double-crosslinked PNIPAM-based hydrogel dressings with adjustable adhesion and contractility. Regen Biomater 10:rbad081

Carmeliet P (2000) Mechanisms of angiogenesis and arteriogenesis. Nature Med 6(4):389–395

Casarrubios L, Gomez-Cerezo N, Sanchez-Salcedo S et al (2020) Silicon substituted hydroxyapatite/Vegf scaffolds stimulate bone regeneration in osteoporotic sheep. Acta Biomater. 101:544–553

Chen HM, Sun JF, Wang ZB et al (2018) Magnetic cell-scaffold interface constructed by superparamagnetic IONP enhanced osteogenesis of adipose-derived stem cells. Acs Appl Mater Inter. 10(51):44279–44289

Chen W, Wang CY, Liu W et al (2023) A matrix-metalloproteinase-responsive hydrogel system for modulating the immune microenvironment in myocardial infarction. Adv Mater 35(13)

Chen KY, Yu W, Zheng GJ et al (2024) Biomaterial-based regenerative therapeutic strategies for spinal cord injury. NPG Asia Mater 16(1)

Cui JK, Xu H, Yu JZ et al (2024a) Targeted depletion of PD-1-expressing cells induces immune tolerance through peripheral clonal deletion. Sci Immunol 9(94):eadh0085

Cui MR, Wang WS, Han XY et al (2024b) Designing microneedle patch for prophylaxis of postoperative atrial fibrillation. ACS Nano 18(29):18889–18899

Daly AC, Prendergast ME, Hughes AJ et al (2021) Bioprinting for the biologist. Cell 184(1):18–32

De Santis MM, Alsafadi HN, Tas S et al (2021) Extracellular-matrix-reinforced bioinks for 3D bioprinting human tissue. Adv Mater 33(3):e2005476

Dixon IM, Ju H, Jassal DS et al (1996) Effect of ramipril and losartan on collagen expression in right and left heart after myocardial infarction. Mol Cell Biochem 165(1):31–45

Du J, Dai HJ, Wang HX et al (2021) Preparation of high thermal stability gelatin emulsion and its application in 3D printing. Food Hydrocolloid 113:113

Eelen G, Treps L, Li XR et al (2020) Basic and therapeutic aspects of angiogenesis updated. Circ Res 127(2):310–329

Ehsanipour A, Sathialingam M, Rad LM et al (2021) Injectable, macroporous scaffolds for delivery of therapeutic genes to the injured spinal cord. Apl Bioeng 5(1):016104

Einhorn TA, Gerstenfeld LC (2015) Fracture healing: mechanisms and interventions. Nat Rev Rheumatol. 11(1):45–54

Ferlin KM, Prendergast ME, Miller ML et al (2016) Influence of 3D printed porous architecture on mesenchymal stem cell enrichment and differentiation. Acta Biomater 32:161–169

Ferraris C, Chevalier G, Favier B et al (2000) Adult corneal epithelium basal cells possess the capacity to activate epidermal, pilosebaceous and sweat gland genetic programs in response to embryonic dermal stimuli. Development 127(24):5487–5495

Frey N, Linke A, Suselbeck T et al (2014) Intracoronary delivery of injectable bioabsorbable scaffold (IK-5001) to treat left ventricular remodeling after ST-elevation myocardial infarction: a first-in-man study. Circ Cardiovasc Interv 7(6):806–812

Gilbert F, O'connell CD, Mladenovska T et al (2018) Print me an organ? Ethical and regulatory issues emerging from 3D bioprinting in medicine. Sci Eng Ethics 24(1):73–91

Ham TR, Leipzig ND (2018) Biomaterial strategies for limiting the impact of secondary events following spinal cord injury. Biomed Mater 13(2):024105

Hamed I, Özogul F, Regenstein JM (2016) Industrial applications of crustacean by-products (chitin, chitosan, and chitooligosaccharides): a review. Trends Food Sci Tech 48:40–50

Hasan A, Memic A, Annabi N et al (2014) Electrospun scaffolds for tissue engineering of vascular grafts. Acta Biomater. 10(1):11–25

He X, Wang Q, Zhao Y et al (2020) Effect of Intramyocardial grafting collagen scaffold with mesenchymal stromal cells in patients with chronic ischemic heart disease: a randomized clinical trial. JAMA Netw Open 3(9):e2016236

Heydarkhan-Hagvall S, Schenke-Layland K, Dhanasopon AP et al (2008) Three-dimensional electrospun ECM-based hybrid scaffolds for cardiovascular tissue engineering. Biomaterials 29(19):2907–2914

Hosoyama K, Ahumada M, Goel K et al (2019) Electroconductive materials as biomimetic platforms for tissue regeneration. Biotechnol Adv. 37(3):444–458

Houshyar S, Bhattacharyya A, Shanks R (2019) Peripheral nerve conduit: materials and structures. ACS Chem Neurosci 10(8):3349–3365

Hu T, Cui X, Zhu M et al (2020) 3D-printable supramolecular hydrogels with shear-thinning property: fabricating strength tunable bioink via dual crosslinking. Bioact Mat 5(4):808–818

Huang S, Xu Y, Wu C et al (2010) In vitro constitution and in vivo implantation of engineered skin constructs with sweat glands. Biomaterials 31(21):5520–5525

Huang S, Yao B, Xie J et al (2016) 3D bioprinted extracellular matrix mimics facilitate directed differentiation of epithelial progenitors for sweat gland regeneration. Acta Biomater 32:170–177

Huang F, Chen TY, Chang J et al (2021) A conductive dual-network hydrogel composed of oxidized dextran and hyaluronic-hydrazide as BDNF delivery systems for potential spinal cord injury repair. Int J Biol Macromol 167:434–445

Jackson BW, Grund C, Winter S et al (1981) Formation of cytoskeletal elements during mouse embryogenesis. II. Epithelial differentiation and intermediate-sized filaments in early postimplantation embryos. Different Res Biol Divers 20(3):203–216

Jia WT, Gungor-Ozkerim PS, Zhang YS et al (2016) Direct 3D bioprinting of perfusable vascular constructs using a blend bioink. Biomaterials. 106:58–68

Jin Z, Li Y, Yu K et al (2021) 3D printing of physical organ models: recent developments and challenges. Adv Sci (Weinheim, Baden-Wurttemberg, Germany) 8(17):e2101394

Joshi RV, Nelson CE, Poole KM et al (2013) Dual pH- and temperature-responsive microparticles for protein delivery to ischemic tissues. Acta Biomater. 9(5):6526–6534

Kabirian F, Ditkowski B, Zamanian A et al (2018) An innovative approach towards 3D-printed scaffolds for the next generation of tissue-engineered vascular grafts. Mater Today-Proc 5(7):15586–15594

Karimi M, Eslami M, Sahandi-Zangabad P et al (2016) pH-sensitive stimulus-responsive nanocarriers for targeted delivery of therapeutic agents. Wires Nanomed Nanobi 8(5):696–716

Karsenty G, Khosla S (2022) The crosstalk between bone remodeling and energy metabolism: a translational perspective. Cell Metab. 34(6):805–817

Khare D, Basu B, Dubey AK (2020) Electrical stimulation and piezoelectric biomaterials for bone tissue engineering applications. Biomaterials. 258

Kratz F (2014) A clinical update of using albumin as a drug vehicle - a commentary. J Control Release 190:331–336

Kulkarni P, Rawtani D, Kumar M et al (2020) Cardiovascular drug delivery: a review on the recent advancements in nanocarrier based drug delivery with a brief emphasis on the novel use of magnetoliposomes and extracellular vesicles and ongoing clinical trial research. J Drug Deliv Sci Tec:60

La Riva A, Sayegh AS, Perez LC et al (2023) Obturator nerve injury in robotic pelvic surgery: scenarios and management strategies. Eur Urol 83(4):361–368

Lee LC, Wall ST, Klepach D et al (2013) Algisyl-LVR with coronary artery bypass grafting reduces left ventricular wall stress and improves function in the failing human heart. Int J Cardiol 168(3):2022–2028

Lee M, Kim YS, Park J et al (2024) A paintable and adhesive hydrogel cardiac patch with sustained release of ANGPTL4 for infarcted heart repair. Bioact Mater 31:395–407

Leung Y, Kandyba E, Chen YB et al (2013) Label retaining cells (LRCs) with myoepithelial characteristic from the proximal acinar region define stem cells in the sweat gland. PLoS One 8(9):e74174

Li H, Fu X, Ouyang Y et al (2006) Adult bone-marrow-derived mesenchymal stem cells contribute to wound healing of skin appendages. Cell Tissue Res 326(3):725–736
Li W, Chen P, Pan Y et al (2022) Construction of a band-aid like cardiac patch for myocardial infarction with controllable H(2) S release. Adv Sci (Weinh) 9(35):e2204509
Li X, Wu F, Günther S et al (2023) Inhibition of fatty acid oxidation enables heart regeneration in adult mice. Nature 623(7986):E7
Liang H, Sun Q, Zhen Y et al (2016) The differentiation of amniotic fluid stem cells into sweat glandlike cells is enhanced by the presence of sonic hedgehog in the conditioned medium. Exp Dermatol 25(9):714–720
Liu S, Zhang H, Duan E (2013) Epidermal development in mammals: key regulators, signals from beneath, and stem cells. Int J Mol Sci 14(6):10869–10895
Liu N, Ye X, Yao B et al (2021a) Advances in 3D bioprinting technology for cardiac tissue engineering and regeneration. Bioact Mater. 6(5):1388–1401
Liu XY, Hao MM, Chen ZJ et al (2021b) 3D bioprinted neural tissue constructs for spinal cord injury repair. Biomaterials 272:120771
Liu ZR, Wan XY, Wang ZL et al (2021c) Electroactive biomaterials and systems for cell fate determination and tissue regeneration: design and applications. Adv Mater. 33(32)
Liu ST, Wang ZH, Chen XY et al (2023) Multiscale anisotropic scaffold integrating 3D printing and electrospinning techniques as a heart-on-a-chip platform for evaluating drug-induced cardiotoxicity (Adv. Healthcare Mater. 24/2023). Adv Healthc Mater 12(24)
López-Cebral R, Silva-Correia J, Reis RL et al (2017) Peripheral nerve injury: current challenges, conventional treatment approaches, and new trends in biomaterials-based regenerative strategies. ACS Biomater Sci Eng 3(12):3098–3122
Lu CP, Polak L, Rocha AS et al (2012) Identification of stem cell populations in sweat glands and ducts reveals roles in homeostasis and wound repair. Cell 150(1):136–150
Lu XY, Xu ZX, Shu F et al (2024) Reactive oxygen species responsive multifunctional fusion extracellular nanovesicles: prospective treatments for acute heart transplant rejection. Adv Mater 36(35):e2406758
Lyra-Leite DM, Gutiérrez-Gutiérrez O, Wang MM et al (2022) A review of protocols for human IPSC culture, cardiac differentiation, subtype-specification, maturation, and direct reprogramming. Star Protoc 3(3)
Mahara A, Somekawa S, Kobayashi N et al (2015) Tissue-engineered acellular small diameter long-bypass grafts with neointima-inducing activity. Biomaterials 58:54–62
Maltepe E, Schmidt JV, Baunoch D et al (1997) Abnormal angiogenesis and responses to glucose and oxygen deprivation in mice lacking the protein ARNT. Nature 386(6623):403–407
Merlot AM, Kalinowski DS, Richardson DR (2014) Unraveling the mysteries of serum albumin-more than just a serum protein. Front Physiol 5:5
Miragoli M, Ceriotti P, Iafisco M et al (2018) Inhalation of peptide-loaded nanoparticles improves heart failure. Sci Transl Med 10(424)
Nakajima H, Dijkstra P, Loos K (2017) The recent developments in biobased polymers toward general and engineering applications: polymers that are upgraded from biodegradable polymers, analogous to petroleum-derived polymers, and newly developed. Polymers-Basel 9(10)
Nie W, Dai XY, Copus JS et al (2022) Rapid mineralization of graphene-based 3D porous scaffolds by semi-dry electrodeposition for photothermal treatment of tumor-induced bone defects. Acta Biomater. 153:573–584
Oshimori N, Fuchs E (2012) Paracrine TGF-β signaling counterbalances BMP-mediated repression in hair follicle stem cell activation. Cell Stem Cell 10(1):63–75
Pati F, Jang J, Ha DH et al (2014) Printing three-dimensional tissue analogues with decellularized extracellular matrix bioink. Nat Commun 5:3935
Paus R, Müller-Röver S, Van Der Veen C et al (1999) A comprehensive guide for the recognition and classification of distinct stages of hair follicle morphogenesis. J Invest Dermatol 113(4):523–532

Plikus M, Wang WP, Liu J et al (2004) Morpho-regulation of ectodermal organs: integument pathology and phenotypic variations in K14-Noggin engineered mice through modulation of bone morphogenic protein pathway. Am J Pathol 164(3):1099–1114
Porzionato A, Stocco E, Barbon S et al (2018) Tissue-engineered grafts from human decellularized extracellular matrices: a systematic review and future perspectives. Int J Mol Sci 19(12)
Potente M, Mäkinen T (2017) Vascular heterogeneity and specialization in development and disease. Nat Rev Mol Cell Bio 18(8):477–494
Procopio A, De Rosa S, Montefusco F et al (2022) Analysis of a cardiac-necrosis-biomarker release in patients with acute myocardial infarction via nonlinear mixed-effects models. Appl Sci-Basel 12(24)
Rafii S, Butler JM, Ding BS (2016) Angiocrine functions of organ-specific endothelial cells. Nature 529(7586):316–325
Rajala R (2021) How big is the endothelium? Comment on "spatial and temporal dynamics of the endothelium". J Thromb Haemost 19(10):2634–2635
Ramaswamy Y, Roohani I, No YJ et al (2021) Nature-inspired topographies on hydroxyapatite surfaces regulate stem cells behaviour. Bioact Mater. 6(4):1107–1117
Reis KP, Sperling LE, Teixeira C et al (2018) Application of PLGA/FGF-2 coaxial microfibers in spinal cord tissue engineering: an in vitro and in vivo investigation. Regen Med 13(7):785–801
Robling AG, Bonewald LF (2020) The osteocyte: new insights. Annu Rev Physiol. 82:485–506
Rodrigues M, Kosaric N, Bonham CA et al (2019) Wound healing: a cellular perspective. Physiol Rev. 99(1):665–706
Ruiz-Alonso S, Lafuente-Merchan M, Ciriza J et al (2021) Tendon tissue engineering: Cells, growth factors, scaffolds and production techniques. J Control Release. 333:448–486
Sedláková V, Mourcos S, Pupkait J et al (2024) Biomaterials for direct cardiac repair – a rapid scoping review 2012–2022. Acta Biomater 180:61–81
Shikiji T, Minami M, Inoue T et al (2003) Keratinocytes can differentiate into eccrine sweat ducts in vitro: involvement of epidermal growth factor and fetal bovine serum. J Dermatol Sci 33(3):141–150
Shin M, Lee HA, Lee M et al (2018) Targeting protein and peptide therapeutics to the heart via tannic acid modification. Nat Biomed Eng 2(5):304–317
Silberman E, Oved H, Namestnikov M et al (2023) Post-maturation reinforcement of 3D-printed vascularized cardiac tissues. Adv Mater 35(31)
Sorushanova A, Delgado LM, Wu ZN et al (2019) The collagen Suprafamily: from biosynthesis to advanced biomaterial development. Adv Mater 31(1):e1801651
Sultankulov B, Berillo D, Sultankulova K et al (2019) Progress in the development of chitosan-based biomaterials for tissue engineering and regenerative medicine. Biomol Ther 9(9)
Sun YF, Wang ZW, Zhang J et al (2021) Current status of and opinions on heart transplantation in China. Curr Med Sci 41(5):841–846
Takagi R, Ishimaru J, Sugawara A et al (2016) Bioengineering a 3D integumentary organ system from IPS cells using an in vivo transplantation model. Sci Adv 2(4):e1500887
Tani H, Kobayashi E, Yagi S et al (2023) Heart-derived collagen promotes maturation of engineered heart tissue. Biomaterials 299:122174
Traverse JH, Henry TD, Dib N et al (2019) First-in-man study of a cardiac extracellular matrix hydrogel in early and late myocardial infarction patients. Jacc-Basic Transl Sc 4(6):659–669
Trimm E, RED-Horse K. (2023) Vascular endothelial cell development and diversity. Nat Rev Cardiol 20(3):197–210
Wan ZQ, Zhang P, Lv LW et al (2020) NIR light-assisted phototherapies for bone-related diseases and bone tissue regeneration: a systematic review. Theranostics. 10(25):11837–11861
Wang G, Qiu J, Zheng L et al (2014) Sustained delivery of BMP-2 enhanced osteoblastic differentiation of BMSCs based on surface hydroxyapatite nanostructure in chitosan-HAp scaffold. J Biomater Sci Polym Ed. 25(16):1813–1827

Wang X, Wang X, Liu J et al (2016) Hair follicle and sebaceous gland De novo regeneration with cultured epidermal stem cells and skin-derived precursors. Stem Cells Transl Med 5(12):1695–1706

Wang X, Chen H, Tian R et al (2017) Macrophages induce AKT/β-catenin-dependent Lgr5(+) stem cell activation and hair follicle regeneration through TNF. Nat Commun 8:14091

Wang YL, Zhang H, Hu Y et al (2022) Bone repair biomaterials: a perspective from immunomodulatory. Adv Funct Mater 32(51)

Wen J, Anderson SM, Du JJ et al (2011) Controlled Protein Delivery Based on Enzyme-Responsive Nanocapsules. Adv Mater. 23(39):4549

Wen YF, Yang HX, Hong Y (2024) Transcriptomic approaches to cardiomyocyte-biomaterial interactions: a review. ACS Biomater Sci Eng 10:4175

Wimmer RA, Leopoldi A, Aichinger M et al (2019) Human blood vessel organoids as a model of diabetic vasculopathy. Nature 565(7740):50

Woodcock EA, Matkovich SJ (2005) Cardiomyocytes structure, function and associated pathologies. Int J Biochem Cell B 37(9):1746–1751

Wu X, Scott L, Washenik K et al (2014) Full-thickness skin with mature hair follicles generated from tissue culture expanded human cells. Tissue Eng Part A 20(23–24):3314–3321

Wu YT, Zhang X, Tan BW et al (2022) Near-infrared light control of GelMA/PMMA/PDA hydrogel with mild photothermal therapy for skull regeneration. Biomater Adv.:133

Wu HK, Qian XK, Liang GY (2023) The role of small extracellular vesicles derived from mesenchymal stromal cells on myocardial protection: a review of current advances and future perspectives. Cardiovasc Drug Ther 38:1111

Xia Y, Sun J, Zhao L et al (2018) Magnetic field and nano-scaffolds with stem cells to enhance bone regeneration. Biomaterials. 183:151–170

Xie J, Peng C, Zhao QH et al (2016) Osteogenic differentiation and bone regeneration of iPSC-MSCs supported by a biomimetic nanofibrous scaffold. Acta Biomater. 29:365–379

Xie ZQ, Guo WW, Guo NN et al (2018) Targeting tumor hypoxia with stimulus-responsive nanocarriers in overcoming drug resistance and monitoring anticancer efficacy. Acta Biomater 71:351–362

Xu Y, Huang S, Ma K et al (2012) Promising new potential for mesenchymal stem cells derived from human umbilical cord Wharton's jelly: sweat gland cell-like differentiative capacity. J Tissue Eng Regen Med 6(8):645–654

Yamaguchi N, Zhang L, Chae BS et al (2007) Growth factor mediated assembly of cell receptor-responsive hydrogels. J Am Chem Soc. 129(11):3040–+

Yang SB, Yuan ZD, Wang TT et al (2024) Rapid and scar free wound repair by using a biologically flexible and conductive dressing under electrical stimulation. Adv Funct Mater 34

Yao B, Song W, Li Z et al (2018) Irf6 directs glandular lineage differentiation of epidermal progenitors and promotes limited sweat gland regeneration in a mouse burn model. Stem Cell Res Ther 9(1):179

Yao B, Hu T, Cui X et al (2019) Enzymatically degradable alginate/gelatin bioink promotes cellular behavior and degradation in vitro and in vivo. Biofabrication 11(4):045020

Yao B, Wang R, Wang Y et al (2020) Biochemical and structural cues of 3D-printed matrix synergistically direct MSC differentiation for functional sweat gland regeneration. Sci Adv 6(10):eaaz1094

Yaron JR, Gosangi M, Pallod S et al (2024) In situ light-activated materials for skin wound healing and repair: a narrative review. Bioeng Transl Med 9(3):e10637

Yin HM, Liu W, Huang YF et al (2019) Surface epitaxial crystallization-directed nanotopography for accelerating preosteoblast proliferation and osteogenic differentiation. Acs Appl Mater Interfaces. 11(46):42956–42963

Zhang Y, Hao H, Liu J et al (2012) Repair and regeneration of skin injury by transplanting microparticles mixed with Wharton's jelly and MSCs from the human umbilical cord. Int J Low Extrem Wounds 11(4):264–270

Zhang YS, Arneri A, Bersini S et al (2016) Bioprinting 3D microfibrous scaffolds for engineering endothelialized myocardium and heart-on-a-chip. Biomaterials 110:45–59

Zhang Y, Enhejirigala YB et al (2021) Using bioprinting and spheroid culture to create a skin model with sweat glands and hair follicles. Burns & Trauma 9:tkab013

Zhang R, Yao YT, Gao HW et al (2024) Mechanisms of angiogenesis in tumour. Front. Oncologia 14:14

Zhao Z, Xu M, Wu M et al (2015) Direct reprogramming of human fibroblasts into sweat gland-like cells. Cell Cycle (Georgetown, Tex) 14(21):3498–3505

Zhao C, Zeng ZY, Qazvini NT et al (2018) Thermoresponsive citrate-based graphene oxide scaffold enhances bone regeneration from BMP9-stimulated adipose-derived mesenchymal stem cells. Acs Biomater Sci Eng. 4(8):2943–2955

Zhou WX, Rahman MSU, Sun CM et al (2024) Perspectives on the novel multifunctional nerve guidance conduits: from specific regenerative procedures to motor function rebuilding. Adv Mater 36(14)

Zhu DS, Hou JL, Qian M et al (2021) Nitrate-functionalized patch confers cardioprotection and improves heart repair after myocardial infarction via local nitric oxide delivery. Nat Commun 12(1):4501

Zhu YY, Wang K, Jia XY et al (2024) Antioxidant peptides, the guardian of life from oxidative stress. Med Res Rev 44(1):275–364

Zoller K, To D, Bernkop-Schnurch A (2025) Biomedical applications of functional hydrogels: innovative developments, relevant clinical trials and advanced products. Biomaterials 312:122718

Chapter 4
Self-Powered Biomaterials

Yu Shi, Yandai Lin, and Zhe Liu

Abstract Self-powered materials, including piezoelectric, pyroelectric, and photoelectric materials, have been playing a critical role due to their excellent physical features responsive to external pressure, thermal and photo-stimuli to generate an intrinsic electric field for self-powered motivation or follow-up molecular medical catalysis for an efficient cell apoptosis or lesion treatment efficacy. This chapter aims to discuss the fundamentals, categories, constructions, and mechanisms of various self-powered materials, and furthermore elucidate their versatile biomedical applications such as biocatalysis, cell elimination, tumor therapy, bioelectric devices, contaminant degradation, etc.

Keywords Self-powered biomaterials · Piezoelectric effect · Pyroelectric effect · Photoelectric effect · Redox catalysis

Y. Shi · Y. Lin
Academy of Medical Engineering and Translational Medicine, Tianjin University, Tianjin, China

Tianjin Key Laboratory of Brain Science and Neural Engineering, Tianjin University, Tianjin, China

Z. Liu (✉)
Academy of Medical Engineering and Translational Medicine, Tianjin University, Tianjin, China

Tianjin Key Laboratory of Brain Science and Neural Engineering, Tianjin University, Tianjin, China

Division of Intelligent Biomaterials, Department of Medical Sciences and Engineering, Medical School, Tianjin University, Tianjin, China
e-mail: zheliu@tju.edu.cn

Z. Liu (ed.), *Intelligent Biomaterials*, https://doi.org/10.1007/978-981-96-8646-9_4

4.1 Piezoelectric Biomaterials

4.1.1 Concept

In 1880, the piezoelectric properties of quartz and Rochelle salt crystals were discovered, and since then piezoelectric materials attracted people's attention. Piezoelectricity shows the capability of materials to convert mechanical energy into electrical energy. When a crystal is exposed with a certain external force in a particular direction, an equal amount of charges with opposite polarity is generated on the surface. After removing the external force, the surface charges also disappear. This "pressure-electricity" phenomenon was named as positive voltage effect. Generally, it is necessary to meet the requirements of non-central symmetry for satisfying piezoelectric properties. Compared with non-ferroelectric materials, ferroelectric materials usually have a better piezoelectric effect. This includes ten non-centered symmetry point groups with polar axes (self-polarizing) that are ferroelectric since ferroelectric materials belong to the scope of piezoelectric materials.

The piezoelectric materials can receive external vibration and absorb external energy converting into electrical energy. Initially, piezoelectric dynamics was applied to sensors and receivers, the decomposition of organic dyes, and surgical implanted materials. Different from traditional invasive sensors, inorganic piezoelectric materials can be used to detect the transformation of specific biomolecular changes of tissues and a variety of biological reactions, such as antigen-antibody reaction, ligand and receptor binding, gene hybridization, etc. These materials may be non-toxic and biocompatible, minimally invasive or non-invasive, so they are receiving increasing attention in tissue engineering, biosensors, drug delivery, and exogenous stimuli responsiveness to regulate cell biological behaviors and promote bone repairing.

Through external mechanical stress, the deformation processes of piezoelectric materials generate a polarized electric field and piezoelectric catalysis (Fig. 4.1). It is easy to deform and generate polarized electric fields, thus various mechanical energies can be used to induce piezoelectric catalysis. Ultrasonic vibration is the most popular method to induce material deformation, and its advantages include that high-pressure forces can be applied to the sample, and the cyclic stress induced by ultrasonic waves can be shielded by free carriers or photo-generated current carriers to avoid the voltage polarization charge. The total pressure is composed of the cyclic acoustic pressure of ultrasonic waves and the pressure generated by the rupture of the cavitation bubble. With a stress force to piezoelectric materials, an internal electric field is generated.

4.1.2 Classification

Through the deformation of piezoelectric materials, piezoelectric catalysis induces spontaneous polarization when they capture mechanical vibration energy to generate bound charges on the surface of the materials. The formation of the piezoelectric

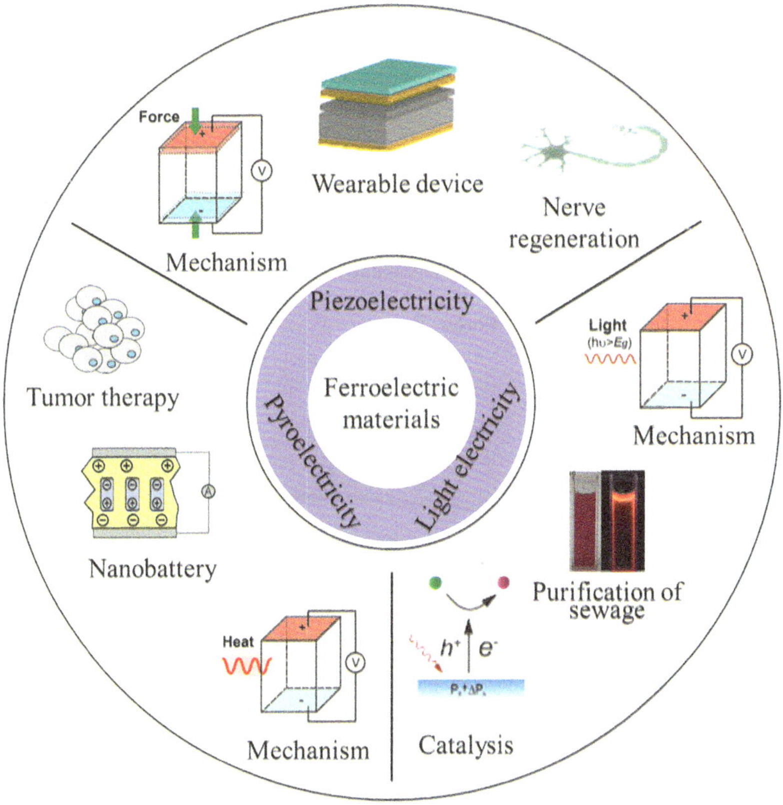

Fig. 4.1 Ferroelectric materials, such as piezoelectric, pyroelectric, and light-electric materials, attracted more and more attention in the biomedical fields, presenting a unique advantage in biosensing, acoustic tweezers, biotherapy, tissue engineering, and stimulation of biological processes

potential can occur redox reactions with the surrounding reactants. By photocatalysis and electrocatalysis, piezoelectric catalysis can not only use mechanical energy to degrade organic pollutants, but also convert mechanical energy into clean energy for hydrogen storage. At present, the research on piezoelectric catalysis mainly focuses on the following four types of piezoelectric materials. As shown in Fig. 4.2, the classifications of fundamental concepts are necessary to explore the mechanisms of their related properties. The spontaneous polarization possession is the characteristic of ferroelectrics via external electric field to induce P, -E hysteresis loop. The spontaneous polarization is an inducer to produce negative and positive bound polarization charge on the correspondingly polarized surfaces (Fig. 4.2b). In aqueous conditions, the surface charges are filled by ambient ions, including hydroxyl ions, protons, and hydrocarbons. For example, the negative polarization charges can be screened by cation and positively charged molecules.

With the exploration of advanced materials, the barium titanate ($BaTiO_3$) ceramic has superior ferroelectric and piezoelectric properties as well as a simpler

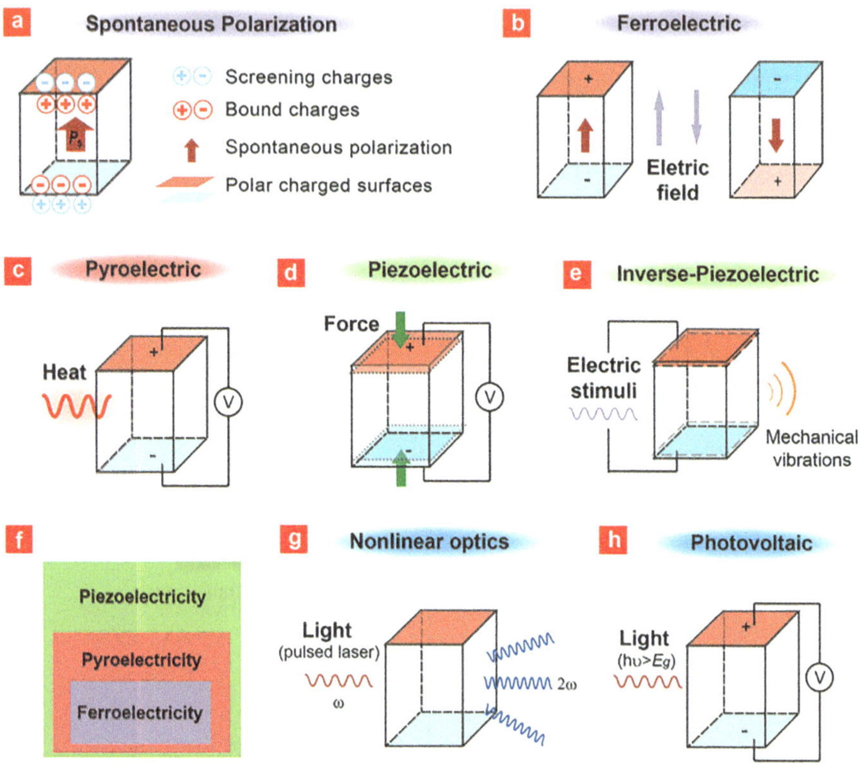

Fig. 4.2 Working mechanisms of ferroelectric materials interacting with physical stimuli. Diagrams illustrating the corresponding mechanisms in a unit cell: (**a**) spontaneous polarization (P) with induced polar charged surfaces; (**b**) ferroelectricity, inversion of polarization by applied electric field; (**c**) pyroelectricity, generation of electric signal upon temperature change; (**d**) piezoelectricity, generation of electric signal by applied force; (**e**) inverse-piezoelectricity, generation of mechanical strain/vibrations by applied electric signals; (**f**) the relationship between piezoelectricity, pyroelectricity, and ferroelectricity; (**g**) nonlinear optics, generation of harmonic photon when excited by a pulsed laser; (**h**) photovoltaic effect, generation of photovoltage when excited by illumination with photon energy (hv) higher than the band gap (Eg). (Reprinted with permission from Wang et al. 2021)

perovskite-type crystalline structure. Jaffe et al. prepared the ceramic solid solution system of lead zirconate titanate (Pb (Zr, Ti) O_3, PZ T), and found that it had better electrical properties (Shrout and Zhang 2007). Since then, researchers have discovered more and more ferroelectric materials, especially ferroelectric ceramics with perovskite structures. With the continuous depth of research, piezoelectric materials are widely used in drivers, sensors, energy exchangers, and other aspects.

1. ABO_3 perovskite structural materials are an important class of piezoelectric materials and are also the most widely used materials in piezoelectric catalysis (Yuan et al. 2019). $BaTiO_3$, represented by the most typical perovskite ferroelectric material, has the displacement of Ti atoms from the central position of the

TiO_6 octahedron and results in spontaneous polarization. When it is stressed, piezoelectric polarization changes the original polarized charge distribution, thereby enhancing or decreasing the charge flow. Ferroelectric polarization, piezoelectric response, and morphology in perovskite nanomaterials can be characterized by piezoelectric-force microscopy, phase loop, and amplitude curve.

Organic lead halide perovskite materials (denoted as ABX_3) have also been found excellent piezoelectric coefficients, which consist of coplanar metal (B) halogen (X) octahedral arrays separated by organic molecular cations. Unlike conventional pero vskites, the rotation of molecular cations in ABX_3 may amplify polarization and provide ultra-high ferroelectric and piezoelectric properties to the materials. The macroscopic polarity of ABX_3 can be well controlled by controlling the content of molecular cations. It is noteworthy that both A and B sites in perovskite crystals can affect polarization, thus enriching modification strategies.

Perovskite nanomaterials can be prepared by hydrothermal method, sol-gel method, and molten salt method. Bao et al. fabricated (Ba, Sr)TiO_3 (BST) nanomaterials method with different morphologies by adopting the template hydrothermal (Yuan et al. 2019). The prepared BST achieved catalytic degradation of methyl orange by ultrasonic vibration. Tang et al. prepared $BaTiO_3$ nanomaterials with highly exposed [001] surfaces by combining with molten salt and hydrothermal method (Tang et al. 2022). Under ultrasonic vibration, $BaTiO_3$ nanomaterials showed excellent piezoelectric catalytic degradation performance and hydrogen production performance of cracking water, which was much higher than that of $BaTiO_3$ nanoparticles. In addition, researchers prepared $CH_3NH_3Pbi_3$ nanomaterials by a co-precipitation method. Due to the large polarization displacement of Pb atom, $CH_3NH_3PBI_3$ nanomaterial has strong polarization effect, and also shows excellent performance of piezoelectric catalytic cracking water to produce hydrogen.

2. Bismuth basic-like compounds have a unique crystal structure consisting of $[Bi_2O_2]^+$ layers and interlayer ions or groups. Due to the lone pair electrons of Bi, many layered bismuth-based compounds have asymmetric crystal structures such as $BiOIO_3$ and $Bi_4Ti_3O_{12}$, and thus also have excellent piezoelectric catalytic activity (Huang et al. 2017). Huang et al. reported that $Bi_4Ti_30_{12}$ microspheres synthesized by hydrothermal method showed excellent piezoelectric catalytic activity against a variety of pollutants and antibiotics, including bisphenol A, Rhodamine B, and tetracycline hydrochloride (Tu et al. 2017). The efficiency of piezoelectric catalytic degradation was positively correlated with the ultrasonic power and the production of superoxide free radicals ($\cdot O_2$) and hydroxyl free radicals ($\cdot OH$) were the main active species of piezoelectric catalytic degradation. Under ultrasonic vibration, experimental results showed that ultra-thin $Bi_2O_2(OH)(NO_3)$ nanosheets have excellent piezoelectric catalytic activity against various pollutants. Photocatalyst BiOC1 can also generate H_2O_2 under ultrasonic vibration (Wu et al. 2021a).

3. ZnO is also one of the most studied piezoelectric catalysts, and its spontaneous polarization is distributed along the longitudinal (C-axis) (Ning et al. 2021). According to the results of first-principles calculation and finite element simulation, ZnO nanofibers are subjected to transverse stress to produce the negative pressure potential and appear on the compression side to produce positive piezoelectric potential, respectively (Pan et al. 2019). As an N-type semiconductor, ZnO has a free electron concentration and the positive potential on the tensile side can only be detected by a small positive potential because it is shielded by the free electron flow. It was also proved that the reduction reaction of oxygen and the oxidation reaction of OH occurred on the compression side and the stretching side respectively, and the degradation of dyes was realized through the generation of ·OH (Xu et al. 2018a). Xu et al. prepared ZnO nanofibers in the direction of [001](Hong et al. 2010). When ultrasonic vibration, H_2 and O_2 with a stoichiometric ratio of 2:1 can be generated and the conversion efficiency of mechanical and chemical energy in the whole reaction process is calculated to be 18%. Carbon nanotubes have high aspect ratio and excellent tensile strength, so they can be used to further improve the piezoelectric catalytic performance of ZnO (Gao et al. 2014). Dispersion of the piezoelectric catalyst in solution can be effectively solved by embedding ZnO in carbon nanotubes. In addition, carbon nanotubes with excellent electrical conductivity can be used as high-speed channels for charge transfer, thus achieving enhancement of piezoelectric catalytic activity.

Organic piezoelectric materials are also known as piezoelectric polymers, such as polyvinylidene fluoride (PVDF) and its diblock copolymers, PVDF-TRFE, PVDF-HFP, poly (tetrafluoroethylene) (PTFE), etc. PVDF is a semi-crystalline polymer arranged in different chain conformations with a total of five crystal phases a, β, y, δ, and ε. In the process of traditional melting and solution casting film forming, the ferroelectric β phase and y phase in PVDF have higher potential energy. In the study of PVDF-based PENGs with high performance, the key is to increase the ferroelectric phase content of materials (Yan et al. 2019). At present, the effective ways are to convert a phase to β phase in PVDF including in stretching, annealing, and polarization strategy. The electrospinning technology integrates the drawing and polarization process, which can effectively dipole orient the PVDF molecular segments and increase the electroactive phase content in PVDF.

The negative and positive polarities were produced on the Cu electrode and PTFE film after touching positively charged PTFE film. With moving along on the surface of Cu, the PTFE film can accumulate charges to flow Cu electrode and produce electric current. Spinning process parameters such as solution concentration, spinning speed, voltage, distance, ambient temperature, and humidity have great influence on the spinning morphology, and then affect the morphology and piezoelectric properties of PVDF spun fibers. In addition, studies have shown that nano-fillers are added to PVDF-based polymers as nucleating agents, such as Ag, Gr/GrO, GR-Ag, Pt, Yb^{3+}, TiO, $Ni(OH)_2BiVO_4$, montmorillonite, etc. It can effectively improve the formation of ferroelectric β phase (or y phase) in the crystallization process of PVDF, which is mainly through the interface between the filler and PVDF chain segment (electrostatic or hydrogen bonding). PVDF-based organic polymer piezoelectric materials are characterized by flexibility, low density, high transparency, and easy processing.

The piezoelectricity of organic materials is achieved by high-pressure stressing or stretching, such as glycine, collagen, silk and biological tissues, skeletal muscle, which are generally rich in collagen, and have a helical structure. In inorganic piezoelectric materials, the internal structure changes linearly with the outer boundary stress, which is called piezoelectric property, typical such as barium zirconate titanate, sodium bismuth titanate, gallium oxide, and so on. The composite piezoelectric material is an organic polymer substrate material embedded in sheet, rod, linear, or granular piezoelectric material, which has the advantages of flexible organic polymer and high voltage piezoelectric material. According to a certain connection mode, composite piezoelectric materials combine inorganic phase and organic phase materials, and their characteristics and performance are mainly determined by the connection mode of various materials.

4.1.3 Piezoelectric Effect and Modulation

The piezoelectric effect involves the coupling between mechanics and electricity, and the relationship between them can be described by the crystal piezoelectric constitutive equation. In the actual research of piezoelectric devices, the resistance of the measuring circuit is much smaller than the internal resistance of the piezoelectric material, so the external circuit can be considered to be in the state of electrical short circuit. Therefore, the mechanical boundary condition can be considered to be in the state of mechanical freedom. Such piezoelectric equation is listed as follows:

$$D = dT + \varepsilon TE$$

(T: stress; E: external electric field intensity; D: the dependent variable; d: the piezoelectric coefficient; ε: the dielectric constant).

When the external electric field E is zero, the piezoelectric equation can be simplified as $D = dT$, and the surface charge density of the material is proportional to the applied stress.

For enhanced piezoelectricity, a fundamental requirement is a non-centrosymmetric arrangement of ions or charged matter, resulting in the presence of permanent dipoles. When piezoelectric materials are subjected to mechanical stimulation, they become electrically polarized and vice versa. When they are subjected to an electric field, they generate strain, which are the positive and inverse piezoelectric effects, respectively. The piezoelectric effect is a molecular phenomenon that creates a macroscopic potential energy, the piezoelectric potential, through the continuous superposition of dipolar polarization. Ferroelectricity is a subset of piezoelectricity, in which ferroelectric materials exhibit built-in spontaneous electrical polarization.

Mechanical deformation, such as elongation, distortion, bending, or compression, is an indispensable requirement for enhancing a piezoelectric potential. Different methods are developed to generate piezoelectric potential, including in speed control motors, ultrasound, mechanical brush/slide, thermal stress, and internal strain of crystals caused by water flow. Ultrasound pressure cavitation effects play a major role in driving the deformation of piezoelectric materials when using ultrasound as a mechanical source (ultrasound waves with frequencies higher than 20 kHz). The sound pressure causes the tiny bubble core to expand and close rapidly, resulting in shock waves. Ultrasonic cavitation is a dynamic process including expansion, closure, and oscillation.

Through piezoelectric effect, ultrasonic waves can activate piezoelectric materials to produce local electrical stimulation. For example, Zhu et al. reported that the maximum piezoelectric potential generated by 110 nm tetravalactic barium titanate nanoparticles under 10^8 Pa sound pressure is 0.45 V (Zhu et al. 2020). In addition, regular arrangement of domain orientation can be induced to enhance piezoelectricity by applying high polarization voltage to the piezoelectric material at a specific temperature (Ramadan et al. 2014). For example, by optimizing the anisotropy characteristics of potassium sodium niobate particles and the domain structure in the textured ceramics, piezoelectric coefficients up to 700 pC/N were obtained (Li et al. 2018).

Specifically, in $BaTiO_3$, the Ba^{2+} ion occupies the apex of the cubic cell and the O^{2-} locates in the face center of the cubic cell forming an octahedron. The Ti^{4+} moves slightly upward from the center of the cubic cell relative to the O^{2-} anion, an asymmetry that leads to spontaneous electrical polarization. In contrast, ZnO crystallizes in a wurtzite structure do not show built-in electrical polarization, unless the lattice mechanically deformed. As for organic polymer, the dielectric properties of PVDF with dominant α and β phases are derived from the dipole moment generated along the F to H direction due to the different electronegativity atoms. In the β phase, the highly electronegative F atoms are arranged on the side of the carbon chain, and the parallel orientation of the dipole moment results in the highest net electric dipole moment, thus producing the highest piezoelectricity.

4.1.4 Applications

- *Tumor Therapy.*

Piezoelectric catalysis exhibits more and more important role in biological application. For example, Shi et al. designed Gel-BTO to produce ROS for piezoelectric catalytic tumor therapy with ultrasound microscopic stressor (Wu et al. 2021b). The mechanism is that a strong internal electric field generates internal $BaTiO_3$ nanoparticles under the ultrasonic vibration, and then catalyzed in situ to produce ·OH and $O_2^{\cdot-}$ (Zhao et al. 2022). Compared with typical sonodynamic therapy, the advantages of piezoelectric materials are that the catalysis process is stable and be controlled. Similarly, Wang and his colleagues found gold-coated barium titanate can promote wound repair and antibacterial under ultrasonic radiation.

Under the ultrasonic mechanical force, $BaTiO_3$ nanoparticles undergo mechanical deformation to produce electron-hole pairs and enhance the separation and migration of charge carriers through the Schottky barrier with Au particles. The further reactions with water and oxygen molecules in aqueous occur to produce OH and superoxide radical with ability of antibacterial and tumor cells. In addition, the combination of piezoelectric catalysis with oral care was developed by the toothbrush vibration to stimulate the piezoelectric response. The piezoelectric $BaTiO_3$ material continuously generates active free radicals (•OH and $•O_2^-$) for oxidizing and degrading the pigment macromolecules into colorless small molecules (Wang et al. 2020). Compared with conventional whitening techniques, teeth whitening is achieved with efficient, safe, and non-destructive.

- *Bioelectronic Devices.*

Life and health are the eternal themes of human exploration and research. Diseases caused by defects of human tissues and organs are one of the main hazards to people's health, which seriously affect people's quality of life and average life span. In recent decades, implantable biomedical devices have been developed rapidly with the continuous exploration of medical electronic technology. These devices can detect many physiological parameters (such as heart rate, blood pressure, and temperature) for diseases related to the heart, brain, and other vital organs, and support real-time treatment (such as muscle or nerve stimulation), thus becoming an indispensable medical tool for biological science research and clinical disease diagnosis, monitoring, and treatment (Choi et al. 2016). At present, implantable biomedical devices such as pacemakers, nerve and bone stimulators, and sensors have been widely used in the diagnosis and treatment of various diseases.

Although the application of implantable biomedical devices has made great progress, implantable biomedical devices still have many shortcomings due to the limitation of energy storage technology. For example, the battery life should be predetermined within 3–5 years when using deep brain stimulators and cardiac pacemakers to generate electrical pulses (Timmermann et al. 2013). After the end of battery life, these devices must be surgically removed to induce an increasing risk of adverse complications and costs. Therefore, new power supply ways should be explored for implantable biomedical devices.

Wang et al. proposed a "self-drive" conception, meaning that collects and converts energy from the environment into electricity, so as to achieve energy self-sufficiency (Wang and Song 2006). Mechanical energy is considered to be one of the most abundant energy sources in physical, owing to the movement of the body, the beating of the heart or pulse, and the contraction and relaxation of muscles. Energy harvesting technology can effectively convert these biological energies into electrical energy, which can realize the self-drive of implantable electronic devices. Due to the irregular shape of human organs, the complex morphology of tissue surfaces, and the curvature of joints, the hard texture of energy collectors makes it difficult to fit closely with human detection and treatment sites. In order to meet the development needs of micro-miniaturization and lightweight of biomedical devices, the application of flexible electronic technology in the biomedical field has been paid attention to develop. Flexible electronics technology is an emerging technology that attaches electronic devices of organic or inorganic materials to flexible and ductile plastic substrates.

- *Degradation of Organic Contaminants.*

Reactive oxygen species, such as superoxide free radicals, singlet oxygen, hydroxyl free radicals, etc., can be generated through piezoelectric catalysis in aqueous. Therefore, piezoelectric catalysis technology can also be used degradation of pollutants. Xu et al. designed piezoelectric $BaTiO_3$ micro-dendritic crystals with the collection of mechanical energy to effectively degrade the organic dye acid orange (Hong et al. 2012). Similarly, the bending of piezoelectric zinc oxide nanorods can also generate a large amount of charge to generate reactive oxygen species through the piezoelectric effect. Moreover, the piezoelectric catalytic degradation performance of $Pb(Zr_{0.32}Ti_{0.48})O_3$ (PZT) fiber is significantly higher than that of ZnO, which could be accounted for the excellent piezoelectric properties of PZT fiber to induce a large amount of charge (Lin et al. 2014).

Li et al. embedded piezoelectric ZnO nanoparticles and nanofibers into carbon nanotubes for the piezoelectric catalytic degradation of methylene blue (Li et al. 2013). The results showed that the electrons generated by the piezoelectric effect in ZnO can be transferred to carbon nanotubes, while the holes remain in ZnO to accelerate charge transport and separation, thus enhancing the piezoelectric catalytic activity. Lu and his colleague reported $BiFeO_3$ materials also attracted large attention, owing to the spontaneous polarization of more than 100 pC/cm^2 and high piezoelectric coefficient (d33) ~70 pm V^{-1} (Long et al. 2022). The $BiFeO_3$ nanofibers were prepared by electrospinning and constructed heterojunctions with Cds for efficient piezoelectric catalytic degradation of bisphenol A with a degradation rate constant up to 0.098 min^{-1}. The excellent catalytic performance can be attributed to the construction of heterojunctions and the enhancement of piezoelectric response.

In summary, piezoelectric catalysis has attracted a wide range of interest, and it has shown many interesting applications in environmental pollution remediation, water disinfection, medical health, energy conversion, and other fields. At present, there are two theories of piezoelectric potential controlling the piezoelectric catalytic process, which have obvious differences in explaining how piezoelectric potential contributes to catalysis. In band theory, the internal potential field mainly regulates semiconductor properties in an indirect way, including regulating band structure bending to provide spatial driving force for charge carrier separation and conduction. The free charge in the piezoelectric material participates in the reaction. In the mechanism of shielding charge effect, the piezoelectric potential directly determines the energy of charge carriers, and the piezoelectric potential must reach or exceed the redox potential that triggers the reaction to drive the reaction, which is very similar to the electrocatalytic process. The ability of the piezoelectric catalyst to complete the chemical reaction is determined by the piezoelectric potential. Therefore, future research will focus on the study of piezoelectric catalysis mechanism to develop and design new piezoelectric catalytic materials and explore new strategies to enhance the efficiency of piezoelectric catalysis.

4.2 Pyroelectric Biomaterials

4.2.1 Concept

Pyroelectric materials are an important branch of piezoelectric materials with spontaneous polarization changes. When the temperature vibration occurs in the surface of the pyroelectric material, it will cause the change of the polarization intensity inside the material, thereby releasing electric charge to the outside world, this phenomenon is named as pyroelectric effect. For perovskite ferroelectrics, the migration of the central atom causes a certain polarization intensity inside the material. When the temperature changes, the increase of the lattice temperature induces the energy level of the central atom to change, resulting in a certain displacement along the polar axis direction. Thus, the change of the overall polarization intensity of the materials shows a macroscopic pyroelectric effect.

The pyroelectric catalytic nanowire displays a superior pyroelectric performance with an elongated polar axis to degrade pollutant (Wu et al. 2018). By means of temperature variation, these $BaTiO_3$ nanowires may provide a promising and facile new reusable strategy for the catalytic degradation of organic dye pollutant. Pyroelectric materials are very sensitive to temperature changes and have the characteristics of fast response rate, high sensitivity, and high resolution (Stewart et al. 2020). In recent years, at the nanoscale, the applications of pollutant degradation, hydrogen preparation, and the bacteria elimination based on the pyroelectric catalytic effect have also become a hot topic (Sharma and Vaish 2021).

4.2.2 Classification

When the temperature increases, the pyroelectric material will respond to the temperature change in a very short time (about 10–12 s), showing a decrease in polarization intensity (Li et al. 2017a). The released charge undergoes a series of chemical reactions to form active free radicals, until the material returns to electrostatic equilibrium. A certain amount of free charge will remain in the environment and form free radicals to participate in the catalytic reaction (Wu et al. 2016). The human body cannot have large temperature fluctuations, and the pyroelectric catalytic effect can use small temperature changes to generate free radicals, which makes it well adapted to the human environment and the internal sterilization and treatment of the human body. Gutmann et al. used crystal powders of lithium niobate ($LiNbO_3$) and lithium tantalate ($LiTaO_3$) to co-culture Escherichia coli. The elimination of Escherichia coli was achieved through cold and hot cycling at 20–45 °C (Gutmann et al. 2012). In addition, some progress has been made in the study of pyroelectric catalytic properties of two-dimensional materials.

It can be classified into three categories about pyroelectric materials, including in single crystals, ceramics, and polymer and composites. Herein, representative structure and related performance of pyroelectric materials are systematically concluded and discussed. In single crystals, the triglycine sulfate is usually prepared by solution evaporation as the most famous single crystals. As a promising pyroelectric material, ZnO with various morphologies show huge application in piezoelectric and pyroelectric field and flexible electronics (Fig. 4.3) (Li et al. 2023). The ZnO nanowire, a single pyroelectric material, was prepared for harvesting thermal energy and their pyroelectric current/voltage coefficients were 1.2–1.5 nC cm^{-2} K^{-1}, which ascribes to the less defects of single crystalline and the same preferred orientations along the c-axis (Lang 2005).

Secondly, ceramics play an important role in piezoelectric, pyroelectric, and ferroelectric applications. Pyroelectric ceramics have great advantages such as relatively simple equipment, easy preparation, low cost, excellent mechanical and controllable electrical properties. The ceramic pyroelectric properties can be improved by adjusting the synthesis process and adding appropriate dopants or additives, to optimize the crystal phase. Especially, inorganic perovskite materials attracted great attention in superconductors, solar cells, catalysis, and piezoelectrics, due to unique chemical and physical performances. There were many classical perovskite materials applied and researched, such as $BaTiO_3$ and $BiFeO_3$ (Liu et al. 2018). $BaTiO_3$ was activated by Au-mediated 808 nm laser irradiation to produce temperature-mediated O_2-independent tumor therapy (Chang et al. 2021). The Au cores with plasmonic heating were excited to transfer a heat to pyroelectric $BaTiO_3$

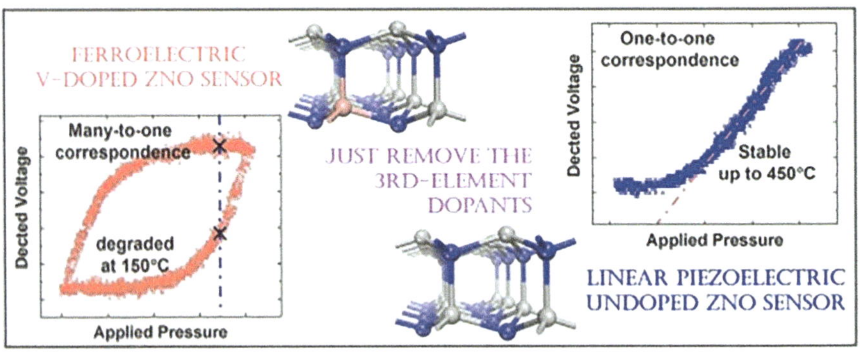

Fig. 4.3 Thermally stable sensors for quantitative pressure sensing based on ZnO thin films. (Reprinted with permission from Li et al. 2023)

shell, which the elevation of temperature could decrease spontaneous polarization and lead to abundant electron and hole released for promoting hydroxyl radical formation.

Pyroelectric ceramics have been paid more and more attention because of their advantages such as easy preparation, relatively simple equipment, low cost, controllable electrical properties, and excellent mechanical properties. The pyroelectric properties of ceramics can be improved by adding appropriate additives or dopants, optimizing the crystal phase, and improving the synthesis process. In this paper, the pyroelectric properties of several ceramics with different improvement strategies and typical crystal structures such as perovskite structures are discussed and illustrated.

You et al. reported two-dimensional black phosphorus materials to achieve hydrogen production from cracking water under the action of hot and cold cycling (Chen et al. 2016). The simulation results show that black phosphorus can not only generate 3.7 V pyroelectric potential in the pyroelectric catalysis process, but also its energy band structure meets the conditions of hydrogen preparation. SnSe-PVP nanorods were prepared with about 95% inhibition ratio for 4 T1 cells by irradiation of 1064 nm laser for 5 min with an intensity of 1.5 W cm^{-2} (Tang et al. 2018). Through utilizing near-infrared-II light as the excitation trigger, SnSe-PVP nanorods. Here we describe a specific application of a pyroelectric material, specifically SnSe-polyvinylpyrrolidone (SnSe-PVP) nanorods, for tumor photoacoustic imaging and PTT, accompanied by pyroelectric dynamic therapy (PEDT), utilizing near-infrared-II (NIR-II) light as the excitation trigger. Moreover, we used PEDT to generate reactive oxygen species (ROS) from SnSe-PVP nanorods and to have the ROS attack HSPs and cancer cells during heating and cooling.

4.2.3 Pyroelectric and Modulation

Thermoelectric materials can not only convert waste heat energy into electricity, but also achieve fluorine-free refrigeration, as a new green energy material has been paid attention. Thermoelectric power generation and refrigeration devices made of thermoelectric materials have the advantages of small size, lowest noise, no pollution, high reliability, no mechanical transmission parts, etc. However, the current conversion efficiency is still lower than traditional generators. Therefore, improving the performance of thermoelectric materials and finding high thermoelectric superior materials are the most important research topics in the field of thermoelectric. A thermoelectric material can convert heat energy into electricity by the Seebeck effect. Typically, the heterojunction is composed of P-type semiconductors and N-type semiconductors with electrically series and thermally parallel.

In order to improve the pyroelectric performance of the material, it is necessary to optimize the transport coefficients in the formula:

$$ZT = \frac{S^2 \sigma T}{\kappa}$$

where S is the Seebeck coefficient (μV/K); σ is the electrical conductivity (S/m); $S^2\sigma$ is called the power factor of the material; K is the thermal conductivity (W/mK); T is the absolute temperature. As shown in formula, relatively high Seebeck coefficient, electrical conductivity, and a relatively low thermal conductivity are beneficial for pyroelectric characteristic. From the variation trend, we can find that the Seebeck coefficient gradually decreases with the increase of carrier concentration, while the electrical conductivity and thermal conductivity both increase with the increase of carrier concentration.

It can be seen from the above description that the carrier concentration has a great influence on the properties of thermoelectric materials. The concentration of the carrier in the transport process is related to the band structure of the material. The electric power of the material system decreases significantly with the increase of the band distance. Energy band engineering, mainly the multi-energy valley effect, or energy band degeneracy, can make an otherwise isolated energy band degeneracy. In this case, the increase of state density near the Fermi level can significantly improve the effective mass and thus increase the Seebeck coefficient. At the same time, when the two bands overlap, the power factors of the two bands increase significantly.

After nanomaterials were applied to thermoelectric, researchers found that they significantly improved the performance of thermoelectric materials, which has attracted extensive attention from researchers. As we know, nanomaterials have an obvious quantum confinement effect, and it is found that they can remove the coupling relationship between the thermoelectric transport coefficients to a certain extent, thereby improving the thermoelectric performance. The quantum confinement effect gradually presents a discontinuous distribution as the Fermi energy level of the material changes with the decrease of dimension. The more significant discontinuity of Fermi level exhibits the higher density of states, and the Seebeck coefficient of the thermoelectric material increases with the large mass, thus the power increases significantly. At the same time, the generation of nanization enhances the phonon scattering inside the material, which can effectively reduce the thermal conductivity of the material and increase the Seebeck coefficient, but does not induce the decrease of the conductivity, thus significantly increasing the ZT value of the material.

Ji et al. fabricated $Bi_2Te_{2.8}Se_{0.2}$ (BTS) and $Bi_{0.5}Sb_{1.5}Te_3$ (BST) nanoplates composed of Z-scheme thermoelectric heterojunction (BST/BTS) via a two-step hydrothermal method (Fig. 4.4) (Yuan et al. 2023). Due to Fermi energy level rearrangement, the BST and BTS construct interfacial electric field, inducing electron in the conductive band of BTS movement to the holes in the valance band of BST. This characteristic can not only separate the electron and hole, but also contain stronger oxidation/reduction potentials of holes and electrons in the VB of BTS and CB of BST. At the same time, they also reported the self-triggered

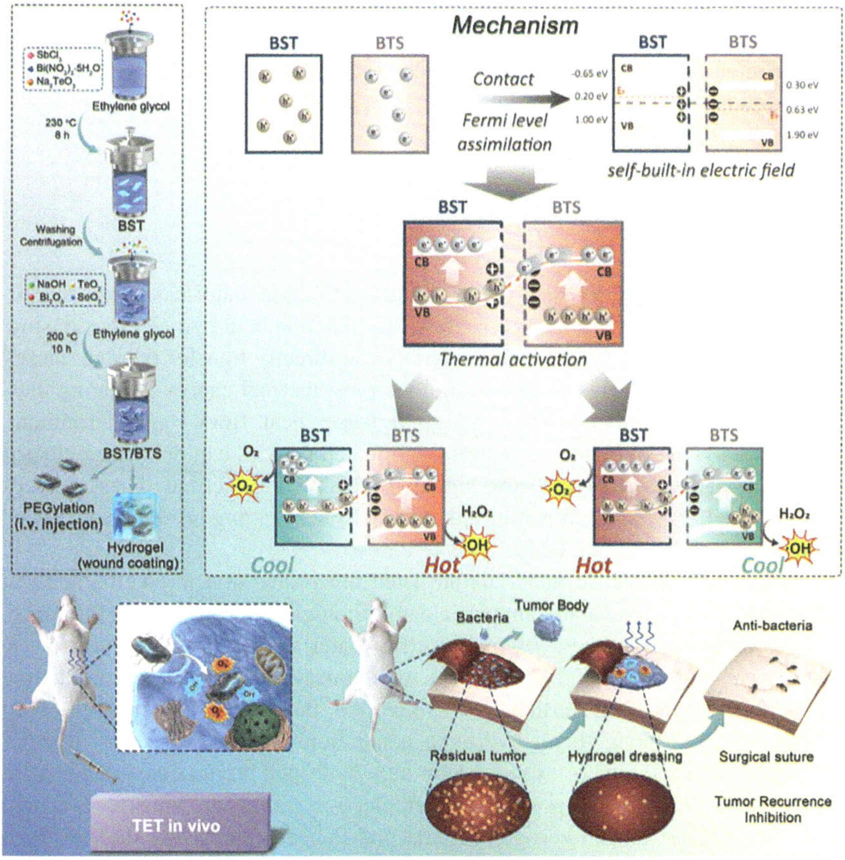

Fig. 4.4 Schematic illustration of preparation and mechanism of BST/BTS Z-scheme heterojunction-based thermoelectric therapy (TET). BST/BTS Z-scheme heterojunction was fabricated via a two-step hydrothermal approach, and used as both i.v. injected nanomedicine and postoperative adjuvant treatment. (Reprinted with permission from Yuan et al. 2019)

nanoheterojunction with thermoelectric for cancer catalytic and immunotherapy, which exhibited excellent antitumor performance (Yuan et al. 2023).

N-type and P-type doping are common methods to improve the electrical conductivity of materials. Pei et al. found that the solid solubility of Ag in PbTe can be changed by temperature, thus significantly improving its thermoelectric performance and providing a new idea for the optimization of thermoelectric performance (Pei et al. 2011).

As a parameter being relatively and independently regulated in thermoelectric systems, the inhibition of lattice thermal conductivity can directly improve material ZT value, which is also an important research direction in the field of thermoelectric in recent years (He et al. 2010). The suppression of lattice thermal conductivity can be achieved by increasing phonon scattering. He et al. reported PbTE-$PbSNs_2$

complex with 0.8 W/mK and this displacement layer structure reduced the lattice thermal conductivity (He et al. 2012). Biswas et al. introduced phonon scattering centers at atomic by doping Na and SrTe to PbTe, which greatly reduced the lattice thermal conductivity of the materials (Biswas et al. 2012).

4.2.4 Applications

The pyroelectric applications are mainly included in three major categories including thermal energy harvesting, electrochemistry reaction, and pyroelectric sensing. As the name indicates, pyroelectric materials can directly transfer thermal energy into electricity energy, thus widely using to harvest thermal energy including thermal energy from human body or external waste heat from the environment. Simultaneously, the pyroelectric material fabricated devices can deliver an electrical signal consisting of pyroelectric signals by harvesting thermal. These devices can achieve self-powered temperature detection by scavenging bio-mechanical and thermal energy from human body.

In addition, pyroelectric materials attracted much attention in electrochemical applications owing to the abundant particle morphologies, remarkable performance, and easy fabrication process of pyroelectric materials. At present, pyroelectric materials are widely applied in pyro-electrochemical including dye degradation, hydrogen generation, and sterilization disinfection, through optimization and controllable design. Among them, hydrogen generation is one of the most popular research directions for energy source. Common hydrogen production strategies are photocatalytic water decomposition, water electrolysis, etc. Owing to excellent pyroelectric performance, hydrogen generation has been extensively studied based on pyroelectric materials in the energy field (Xu et al. 2018b).

With excellent physical and chemical properties, pyroelectric materials have dramatically developed in recent years due to the growing technology demands in various areas. It is worth to note that the temperature fluctuation of spontaneous polarization can convert heat to electricity. To facilitate comparison and analysis, pyroelectric materials include ceramics, pyroelectric crystals, polymers, thin films, and their composites (Zhang et al. 2021). Through optimizing and designing the performance of device structures and pyroelectric materials, pyroelectric properties can be regulated as far as possible. In single crystals, pyroelectric performance is a crystallographic orientation-dependent difference.

Owing to the pyroelectric and ferroelectric properties, Mn-doped single crystal was researched, and it was found that the pyroelectric property of the crystal was directly influenced by the thickness of sample (Zhao et al. 2020). It was observed that the small thickness of crystal showed the enhanced pyroelectric performance obviously increasing by over 300%. On the other hand, ceramic presents a significant role in pyroelectric, piezoelectric, and ferroelectric applications. Pyroelectric property of ceramic could be improved via optimizing crystal phase, appropriate additives or dopants, and synthesis process. Typically, perovskite structure can be described as a formula of ABX_3 where B refers to the transition element ion, A is rare earth or alkaline, and X is a halide anion. Their crystal structure stays almost unchanged when both A and B were replaced partially with similar radius metal ions. The ceramic thickness decreases and their charge density and output current signal obviously increase.

In tumor therapy, pyroelectric materials can produce reactive oxygen species with high oxidative potential for high-efficiency tumor treatment in situ. $Bi_{13}S_{18}I_2$ nanorods were designed with high-performance pyroelectric nanocatalyst and photothermal conversion performance (Wang et al. 2022a). After irradiation with 808 nm laser, $Bi_{13}S_{18}I_2$ nanorods induce temperature elevation to generate reactive oxygen species through pyroelectric effect. Meanwhile, the abundant production of reactive oxygen species causes to deplete HSPs, and the reduction of thermos-resistance significantly augments the photothermal therapeutic efficacy in tumor cells. After intravenous administration, the tumor suppression rate reached 97.2% by synergizing PTT with the pyroelectric dynamic therapy (Wang et al. 2022a). Wei et al. reported Janus pyroelectric nanoparticles to overcome the delivery barriers with self-powered motion (Wei et al. 2023). The design of pyroelectric treatment has realized motion-promoted tissue diffusion, electric field-enhanced cell internalization, and photothermal-EDT-antitumor efficacy.

4.3 Photoelectric Biomaterials

4.3.1 Concept

In 1977, Japanese scientist Fujishima found the generation of hydrogen on the surface of nanoparticles after applying ultraviolet light to TiO_2, which opened the prelude to the study of photocatalysis. Photocatalysis technology refers to the catalytic technology that uses solar energy to degrade organic matter, and its main principle is shown in (Chen et al. 2021). With the development of semiconductor photocatalysis technology, new semiconductor materials have been used in photocatalysis, such as perovskite, halide, graphene, nitride, etc. It is worth noting that the photocatalytic degradation of pharmaceutical wastewater mainly relies on the surface of the photocatalyst active site via oxidation reaction of organic pollutants. Photocatalytic technology can induce hydrogen generation on the surface of the photocatalyst after excitation. By photocatalytic technology, the organic pollutants and hydrogen production can be simultaneously realized to alleviate the energy crisis.

When the semiconductor material is irradiated, the energy of the electrons in the valence band absorbing the photon transitions to the conduction band and leaves holes in the valence band to form an electron-hole pair, as long as the energy of the incident photon is greater than or equal to the bandgap width of the semiconductor. Then, the electron-hole pair separated and migrated to the surface of the photocatalyst to form the active site. In the process of migration, nearly 90% of the electron-hole pairs will recombine immediately, and only 10% of the electron-hole pairs can reach the surface of the material. These electrons and holes undergo a series of reactions to generate active free radicals with strong oxidation and achieve catalytic degradation of organic matter.

At present, semiconductor photocatalysts can produce photocatalytic electrons and holes under the condition of light. However, the amount of time that sunlight can reach the Earth is limited, which will not be able to produce hydrogen and degrade organic pollutants on the whole time. In order to solve the above problems, researchers also found that long-afterglow materials, as photoluminescent materials, store part of the energy. When the excitation stops, light emission continues, which proves their great potential to be used as a supplementary light source for photocatalytic reactions at night. In previous studies, researchers used long-afterglow materials as a supplementary light source to assist photocatalysts to perform photocatalytic reactions in the absence of external light (Li et al. 2017b). However, for the photocatalytic system, there is serious photogenerated carrier recombination inside the monomer photocatalyst and low photocatalytic activity. Therefore, it is necessary to find a more efficient all-weather photocatalytic system.

4.3.2 Classification

A large amount of photocatalysis materials have focused on the design and preparation of TiO_2 and other semiconductor materials. Semiconductor is a kind of material with electrical conductivity between conductor (such as a metal) and insulator (such as a ceramic). Unique electronic properties of semiconductors are closely related to their valence band (VB) and conduction band (CB) (Low et al. 2017). The VB of semiconductors is formed by the highest occupied molecular orbital (HOMO) interaction, while CB is formed by the lowest unoccupied molecular orbital (LUMO) interaction. There is no electronic state between the top of VB and the bottom of CB, and the energy range between CB and VB is called the bandgap (also known as the energy gap: Eg).

The band structure, including the band gap and the position of VB and CB, is one of the important properties of the semiconductor photocatalyst, because it determines the light absorption performance and redox capacity of the semiconductor. When the semiconductor photocatalyst absorbs photons, energy is equal or greater than its Eg, the electrons in VB will be excited to CB, leaving holes in VB, and the resulting photogenerated electron-hole pair may further participate in the following three possible processes: (1) Successful migration to the semiconductor surface; (2) Trapped by defect sites in the bulk phase and/or surface region of the semiconductor; (3) Recombination releases energy in the form of heat or photons. The latter two processes are generally regarded as inactivated because in both of them photogenerated electrons and holes do not contribute to the photocatalytic reaction.

In photocatalytic materials, traditional semiconductor has problems such as insufficient utilization of visible light and high electron-hole pair recombination rate, which limit the application of photocatalytic effect. Therefore, researchers adjust the band structure of the photocatalyst to help the separation of photogenerated electron-hole pairs and improve the photocatalytic performance by doping noble metal, heterojunction recombination. For example, modified Au on the surface of TiO_2 nanorods to construct an asymmetric Au-TiO_2 nanorods array, which improved the CO_2 reduction to prepare CH_4 under sunlight irradiation efficiency (Zeng et al. 2020). Chen et al. enhanced the visible light absorption of the photocatalyst by growing NiO nanowire arrays on the surface of TiO_2 nanosheets to construct heterojunctions (Chen et al. 2020). It not only achieves efficient degradation of methylene blue, but also shows excellent photocatalytic hydrogen production performance.

In conclusion, typical photocatalytic materials include sulfide-type photocatalytic materials, metal oxide photocatalytic materials, carbon nitride-based photocatalytic materials, bismuth-based photocatalytic materials, etc. The popular sulfide photocatalysts are CdS, ZnS, and $Zn_xCd_{1-x}S$ solid solutions. In addition, the semiconductor oxide catalyst materials are TiO_2, WO_3, Ga_2O_3, and CuO. The organic polymer graphite phase carbon nitride (g-C3N4) has excellent photocatalytic properties. Bismuth is a chemically stable metal with high photocatalytic activity due to its relatively large atomic mass. In addition, materials such as metal-organic framework materials (MOFs), conjugated microporous polymers (CMPs), and covalent organic frameworks (COFs) have also been used in the field of photocatalysis.

4.3.3 *Photoelectric and Modulation*

It is expected to improve the activity of photocatalyst by broadening the light absorption capacity of photocatalyst by enhancing the separation and migration of carriers and inhibiting the recombination of photogenerated carriers (Divyasri et al. 2021). This purpose can be achieved by the following methods:

1. Metal and non-metal doping through UV-driven photocatalyst.

The doping of metal and non-metal ions can effectively broaden the photocatalyst to visible light. For example, impurity levels can be formed after modifying the metal ions entering the "narrowed" TiO_2 lattice, thus widening the light absorption range. Zhao et al. reported Y^{3+}-$Bi^5Nb^3O^{15}$ doped with high photocatalytic activity under visible light irradiation (Zhao et al. 2012). Kim et al. developed Al ion doped B-Ga_20_3 nanomaterials and discovered that less than 0.6% doped Al ion with shallow energy level traps to enhance the photocatalytic activity (Kim et al. 2021). When Al ion concentration exceeds 0.6%, the crystal disorder, oxygen vacancy, and grain boundary-related defects show Al concentration-dependent increase, thus reducing the photocatalytic activity.

Therefore, metal ion dopants improve the activity of photocatalysts by acting as photoinduced charge traps to regulate the recombination rate of photogenerated carriers. However, the presence of excess electrons (or holes) in the photoinduced charge trap may reduce the amount of photogenerated charge carriers and reduce photocatalytic activity. Hence, metal ions should be doped near the surface of the photocatalyst and their concentration should be controlled accordingly to inhibit the recombination of electrons and holes. Compared with metal ions, it is difficult for non-metals to form recombination centers. Divyasri et al. reported that nitrogen (N)-doped TiO_2 nanotubes improved the photocatalytic efficiency of dye degradation and hydrogen generation under sunlight irradiation (Divyasri et al. 2021).

2. Design and construction of heterojunction photocatalytic agents.

Heterojunction photocatalytic systems, formed by two semiconductors with matching band gaps, have received more and more attention owing to their effective method for separating photogenerated charge carriers. Among them, heterojunction can effectively inhibit the recombination of electrons and holes, which has great potential to achieve the highest photocatalytic efficiency. Many scholars have enhanced the photocatalytic activity of materials by constructing type II heterojunction. Dashtian et al. reported a new nanosword n-VO_2/n-$CuWO_4$-NSs photocatalyst with a band gap of 1.75 eV, which can simultaneously degrade cationic azurin II and auroamine O dyes under blue light irradiation (Dashtian et al. 2018).

Surface engineering of photocatalyst is to select or optimize some important parameters of catalyst, including surface composition, phase, crystal surface, defect, surface state, and band bending. In order to improve the surface catalytic activity and selectivity of photocatalysis, surface engineering of catalysts usually has the following three purposes: (1) Adjust the surface parameters described above to create an active surface with high adsorption and activation capacity for reactants; (2) Control the surface parameters to make more carriers reach the activation site on the catalyst surface for reaction, which is mainly achieved by increasing the charge transport rate and reducing the charge transport distance; (3) Improve the reduction or oxidation capacity of electrons or holes on the catalyst surface.

Photocatalytic technology involves complex physical and chemical reaction processes, resulting in a lot of factors affecting the reaction rate of photocatalytic degradation of pollutants, which are mainly affected by semiconductor characteristics and reaction conditions. From the semiconductor, the band structure is the key factor affecting the photocatalytic reaction rate of semiconductor: (1) the band gap width determines the light absorption range of semiconductor materials; (2) the redox capacity of electrons and holes depends on the CB/VB band edge position. Therefore, it is a strategy to improve the semiconductor band structure to adjust the valence band, conduction band, or the edge position of conduction band and valence band simultaneously.

Secondly, crystal structure is another key factor affecting the photocatalytic reaction rate. The reason is listed as follows: (1) the adsorption of the catalytic material to the reactant molecules depends on the arrangement of the surface atoms, which affects the activity and selectivity of the catalytic reaction; (2) the crystal surface determines the electronic structure of the surface of the catalytic material, which then regulates the redox capacity of the photogenerated charge; (3) the crystal orientation determines the separation and transport efficiency of the photogenerated charge, which directly affects the carrier density. In addition, the morphology and particle size also affect the photocatalytic reaction rate of semiconductor. When the particle size reaches the nanometer level, the diffusion distance of photogenerated carrier transition to the surface of the catalytic material is greatly shortened, and the separation efficiency of photogenerated carrier is promoted.

At the same time, the electronic structure and coordination environment of the surface of the catalytic material are changed, and the quantum size effect is induced, which further widths the band gap and decreases VB. The upward shift of CB improves the redox capacity of semiconductor materials. Therefore, reasonable regulation of the morphology and size of catalytic materials can promote photogenerated carrier separation and greatly improve the performance of photocatalytic degradation of pollutants. In addition to the intrinsic properties of semiconductor materials, the parameters of the reaction system, including the pH of the reaction liquid, the amount of catalyst, the initial concentration of pollutants, the type and concentration of inorganic ions, the light intensity, the reaction temperature and the concentration of oxygen, will have a certain impact on the reaction rate of photocatalytic degradation of pollutants.

4.3.4 Applications

Traditional fossil energy is limited, and burning fossil causes greenhouse effect and environmental pollution, how to produce clean renewable energy is a research hotspot. Using photocatalysis to decompose water into H_2 and O_2 to replace fossil energy has great advantages with hydrogen energy, ecological environmental protection, and low cost (Kampouri and Stylianou 2019). However, the current hydrogen production efficiency is still relatively low, and there is still a long way to go for industrial application. On the other hand, the greenhouse effect is becoming more and more obvious to induce extreme climate frequently with the increasing concentration of CO_2 in the atmosphere. How to reduce the atmospheric CO_2 content is a major problem to be solved urgently. CO_2 can be reduced to methane, methanol, formic acid, and other organic compounds by using photocatalysis technology, which has high application value.

With the continuous development of industrialization and modernization, the problem of environmental pollution, especially water pollution, is becoming more and more serious. Compared with traditional water pollution treatment methods, photocatalysis method presents green strategy without secondary pollution. In addition to common dyes, such as methylene blue (MB), rhodamine, methyl orange (MO), etc., and colorless pollutants, such as phenol, bisphenol A (BPA), can be degraded. At the same time, photocatalysis can also reduce toxic heavy metal ions in water, such as Cr^{6+}, Pt^{4+}, and Au^{3+} to low-toxic ions, reducing their toxicity.

Since the discovery that TiO_2 can photo-catalyze the decomposition of aquatic hydrogen and oxygen production under ultraviolet irradiation, TiO_2 has been rapidly developed as a photocatalyst, and its nanoparticle P25 has been the first commercial application as a UV semiconductor photocatalyst. However, TiO_2 can only achieve photocatalytic reactions under ultraviolet light, which accounts for only ~5% of the total solar energy incident on the surface. From the perspective of energy utilization, the development of inexpensive, efficient, highly selective, and highly stable visible light semiconductor photocatalysts is the key to realize semiconductor photocatalysis technology from laboratory research to practical application. Therefore, on the one hand, researchers modified TiO_2 to enable it to be excited by visible light. For example, nitrogen-doped TiO_2 had visible light response.

Although photocatalysis technology has great application prospects in the field of clean energy and environmental protection. However, as an emerging technology, there are still some thorny problems in the current photocatalysis technology, which can be mainly divided into three points: (1) The semiconductor photocatalyst has low utilization efficiency of solar energy, especially for the visible light that accounts for the total amount of incident 45% solar energy; (2) The photoelectron-hole pair recombination rate of semiconductor photocatalyst is very high; (3) The semiconductor photocatalyst is difficult to achieve effective separation from the reactants. For problem (1), the conversion efficiency of almost all semiconductor photocatalysts to solar energy is still far below 1%. Therefore, how to improve the utilization rate of semiconductor photocatalyst to visible light is an urgent problem for semiconductor photocatalysis technology. For problem (2), the time required for photoelectron-hole pair recombination on the surface of semiconductor photocatalyst is between 0.110 ns, while the time required for interfacial charge transfer of semiconductor photocatalyst is between 10–100 ns and the time required for adsorption of H_2O, O_2 or pollutants on its surface between 10–10^6 ns. It can be seen that the time required for the photogenerated electron-hole pair to recombine on the surface of the semiconductor photocatalyst is much less than the time required for the interface charge transfer and adsorption of the reactants, that is, most of the photogenerated electrons and holes migrating to the surface of the semiconductor photocatalyst have no time to participate in the photocatalytic reaction. Therefore, inhibiting the photoelectron-hole recombination of semiconductor photocatalysts is the key to improve their photocatalytic activity. For problem (3), semiconductor photocatalysts combining nanotechnology develop in the direction of small size and ultra-thin structure. But small-size or ultra-thin structure semiconductor photocatalysts are difficult to completely separate from the reaction solution. For this reason, researchers prepare semiconductor photocatalysts into thin films, load magnetic materials, or load the photocatalysts on substrate materials (such as glass) to solve the problem of separation and recovery from the reaction solution. Through these methods, their photocatalytic activity has decreased to different degrees compared with powder semiconductor photocatalysts. Therefore, it is necessary to develop a method for separating semiconductor photocatalysts from reactants while focusing on improving the photocatalytic utilization and photocatalytic activity of semiconductor photocatalysts.

In tumor treatment, photoactivated catalysis tumor treatment also exhibited a promising field and attracted extensive attention (Fig. 4.5) (Hu et al. 2022). It was reported photoactivated Pt(IV) coordination carbon point (PT-CDS) and its bovine serum albumin (BSA) complex (Pt-CDs@BSA) with solvothermal synthesis as a

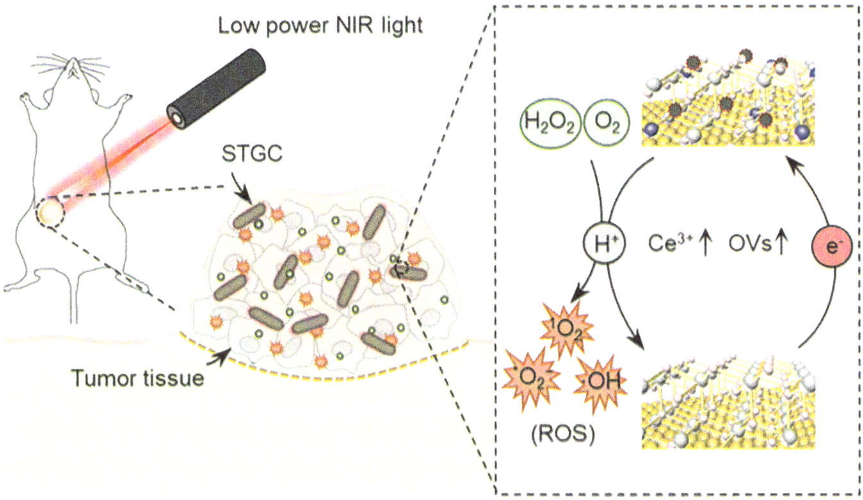

Fig. 4.5 The internal convertible gold@cerium dioxide (STGC-PEG) is triggered by the conversion between the electron-rich states of CeO_2 and CeO_{2-x} and the activation of gold to cerium dioxide by active oxygen vacancy generated by hot electron injection. (Reprinted with permission from (Hu et al. 2022))

novel orange photoinitiated antitumor therapeutic agent (Guo et al. 2022). The uniform distribution of Pt(IV) in PT-CDS (Pt: 17.2 wt%) and its carbon nuclei with significant visible light absorption show excellent photocatalytic properties, which not only effectively release cytotoxic Pt(IV) substances, but also promote the production of hydroxyl radicals in water under orange light. When triggered with a 589 nm laser, a cancer-killing behavior occurs with Pt-CDs@BSA of intracellular Pt(IV) species release, hydroxyl radical production, and acidification, inducing powerful immunogenic cell death.

In addition, a reasonable strategy based on near-infrared photo-stimulated enzymatic hydrolysis and photocatalytic decomposition of tumor interstitial solid and liquid phase is proposed to promote tumor penetration of nanomaterials. Specifically, a cuprous oxide/silver nanoparticle photocatalyst (Cu_2O/Ag) was designed and coated with tumor homologous cell membranes containing heat-sensitive papain to form nanomaterials (Cu_2O/Ag-P@M). The photocatalyzed water decomposition of Cu_2O/Ag reduces the volume of tumor interstitial fluid, thereby reducing the elevated hydrostatic pressure and promoting tumor penetration of nanomaterials.

4.4 Summary and Conclusion

Overall, piezo-, pyro-, and photoelectric materials have important applications in temperature sensing, energy harvesting, infrared imaging, and other fields. At the nanoscale, pyroelectric catalysis based on the degradation of pollutants, hydrogen production, and bacteria killing has also become a research hotspot in recent years. In terms of biomedicine, Wang Yao et al. published braces hydrogel with thermoelectric effect to teeth whitening after eating hot or cold food (Wang et al. 2022b). In addition, the pressure changes generated during brushing are used to mix piezoelectric materials in toothpaste and generate free radicals to achieve the effect of tooth whitening.

Dating back to the development of electric catalysis materials, since the piezoelectric properties of quartz crystals and Rochelle salt crystals were discovered in 1880, piezoelectric materials have entered people's vision. In 2020, Prof. Jianlin Shi reported a piezoelectric perovskite nanoparticle with piezoelectric performance for tumor treatment for the first time and achieved well therapeutic effect (Zhu et al. 2020). Since then, piezoelectric and thermoelectric materials have attracted much attention in biological applications, and have been widely used in tissue engineering, nerve regeneration, wearable devices, and so on. However, many related mechanisms still be unclear and need to be further explored due to the late relevant research in the applications of disease therapy.

References

Biswas K, He J, Blum ID, Wu C-I, Hogan TP, Seidman DN, Dravid VP, Kanatzidis MG (2012) High-performance bulk thermoelectrics with all-scale hierarchical architectures. Nature 489(7416):414

Chang Y, Cheng Y, Zheng R, Wu X, Song P, Wang Y, Yan J, Zhang H (2021) Plasmon-pyroelectric nanostructures used to produce a temperature-mediated reactive oxygen species for hypoxic tumor therapy. Nano Today 38:101110

Chen W, Ouyang J, Liu H, Chen M, Zeng K, Sheng J, Liu Z, Han Y, Wang L, Li J (2016) Black phosphorus nanosheet-based drug delivery system for synergistic photodynamic/photothermal/chemotherapy of cancer. Adv Mater (Deerfield Beach, Fla.) 29(5)

Chen J, Wang M, Han J, Guo R (2020) TiO2 nanosheet/NiO nanorod hierarchical nanostructures: p–n heterojunctions towards efficient photocatalysis. J Colloid Interface Sci 562:313

Chen F, Ma T, Zhang T, Zhang Y, Huang H (2021) Atomic-level charge separation strategies in semiconductor-based photocatalysts. Adv Mater 33(10):2005256

Choi S, Lee H, Ghaffari R, Hyeon T, Kim DH (2016) Recent advances in flexible and stretchable bio-electronic devices integrated with nanomaterials. Adv Mater 28(22):4203

Dashtian K, Ghaedi M, Shirinzadeh H, Hajati S, Shahbazi S (2018) Achieving enhanced blue-light-driven photocatalysis using nanosword-like VO2/CuWO4 type II n–n heterojunction. Chem Eng J 339:189

Divyasri YV, Reddy NL, Lee K, Sakar M, Rao VN, Venkatramu V, Shankar MV, Reddy NCG (2021) Optimization of N doping in TiO2 nanotubes for the enhanced solar light mediated photocatalytic H2 production and dye degradation. Environ Pollut 269:116170

Gao Y, Li S, Zhao B, Zhai Q, Lita A, Dalal NS, Kroto HW, Acquah SF (2014) A synergistic approach to light-free catalysis using zinc oxide embedded multi-walled carbon nanotube paper. Carbon 77:705

Guo D, Lei JH, Rong D, Zhang T, Zhang B, Tang Z, Shen HM, Deng CX, Qu S (2022) Photocatalytic Pt (IV)-coordinated carbon dots for precision tumor therapy. Adv Sci 9(36):2205106

Gutmann E, Benke A, Gerth K, Böttcher H, Mehner E, Klein C, Krause-Buchholz U, Bergmann U, Pompe W, Meyer DC (2012) Pyroelectrocatalytic disinfection using the pyroelectric effect of nano-and microcrystalline LiNbO3 and LiTaO3 particles. J Phys Chem C 116(9):5383

He J, Girard SN, Kanatzidis MG, Dravid VP (2010) Microstructure-lattice thermal conductivity correlation in nanostructured PbTe0. 7S0. 3 thermoelectric materials. Adv Funct Mater 20(5):764

He J, Girard SN, Zheng JC, Zhao L, Kanatzidis MG, Dravid VP (2012) Strong phonon scattering by layer structured PbSnS2 in PbTe based thermoelectric materials. Adv Mater 24(32):4440

Hong K-S, Xu H, Konishi H, Li X (2010) Direct water splitting through vibrating piezoelectric microfibers in water. J Phys Chem Lett 1(6):997

Hong K-S, Xu H, Konishi H, Li X (2012) Piezoelectrochemical effect: a new mechanism for azo dye decolorization in aqueous solution through vibrating piezoelectric microfibers. J Phys Chem C 116(24):13045

Hu X, Wang N, Guo X, Liang Z, Sun H, Liao H, Xia F, Guan Y, Lee J, Ling D (2022) A subnanostructural transformable nanozyme for tumor photocatalytic therapy. Nano-Micro Letters 14(1):101

Huang H, Tu S, Zeng C, Zhang T, Reshak AH, Zhang Y (2017) Macroscopic polarization enhancement promoting photo-and piezoelectric-induced charge separation and molecular oxygen activation. Angew Chem Int Ed 56(39):11860

Kampouri S, Stylianou KC (2019) Dual-functional photocatalysis for simultaneous hydrogen production and oxidation of organic substances. ACS Catal 9(5):4247

Kim S, Ryou H, Lee IG, Shin M, Cho BJ, Hwang WS (2021) Impact of Al doping on a hydrothermally synthesized β-Ga 2 O 3 nanostructure for photocatalysis applications. RSC Adv 11(13):7338

Lang SB (2005) Pyroelectricity: from ancient curiosity to modern imaging tool. Phys Today 58(8):31

Li S, Zhang M, Gao Y, Bao B, Wang S (2013) ZnO–Zn/CNT hybrid film as light-free nanocatalyst for degradation reaction. Nano Energy 2(6):1329

Li J-K, Ge C, Jin K-J, Du J-Y, Yang J-T, Lu H-B, Yang G-Z (2017a) Self-driven visible-blind photodetector based on ferroelectric perovskite oxides. Appl Phys Lett 110(14):142901

Li F, Li Z, Cai Y, Zhang M, Shen Y, Wang W (2017b) Afterglow photocatalysis of Ag3PO4 through different afterglow coatings and photocatalysis mechanism. Mater Lett 208:111

Li P, Zhai J, Shen B, Zhang S, Li X, Zhu F, Zhang X (2018) Ultrahigh piezoelectric properties in textured (K, Na) NbO3-based lead-free ceramics. Adv Mater 30(8):1705171

Li Y, Lin H, Lin J, Zhou C, Luo A, Yang J, Zhang J, Hong Z, Hou X, Xiao P (2023) Thermally stable piezoelectric sensors for quantitative pressure sensing based on linear piezoelectric zinc oxide thin films. Mater Des 235:112466

Lin H, Wu Z, Jia Y, Li W, Zheng R-K, Luo H (2014) Piezoelectrically induced mechano-catalytic effect for degradation of dye wastewater through vibrating Pb (Zr0. 52Ti0. 48) O3 fibers. Appl Phys Lett 104(16):162907

Liu Y, Wang X, Qiao Y, Min M, Wang L, Shan H, Ma Y, Hao W, Tao P, Shang W (2018) Pyroelectric synthesis of metal–BaTiO3 hybrid nanoparticles with enhanced pyrocatalytic performance. ACS Sustain Chem Eng 7(2):2602

Long J, Ren T, Han J, Li N, Chen D, Xu Q, Li H, Lu J (2022) Heterostructured BiFeO3@ CdS nanofibers with enhanced piezoelectric response for efficient piezocatalytic degradation of organic pollutants. Sep Purif Technol 290:120861

Low J, Yu J, Jaroniec M, Wageh S, Al-Ghamdi AA (2017) Heterojunction photocatalysts. Adv Mater 29(20):1601694

Ning X, Hao A, Cao Y, Jia D (2021) Boosting piezocatalytic performance of ag decorated ZnO by piezo-electrochemical synergistic coupling strategy. Appl Surf Sci 566:150730

Pan C, Zhai J, Wang ZL (2019) Piezotronics and piezo-phototronics of third generation semiconductor nanowires. Chem Rev 119(15):9303

Pei Y, May AF, Snyder GJ (2011) Self-tuning the carrier concentration of PbTe/Ag2Te composites with excess ag for high thermoelectric performance. Adv Energy Mater 1(2):291

Ramadan KS, Sameoto D, Evoy S (2014) A review of piezoelectric polymers as functional materials for electromechanical transducers. Smart Mater Struct 23(3):033001

Sharma M, Vaish R (2021) Piezo/pyro/photo-catalysis activities in Ba0. 85Ca0. 15 (Ti0. 9Zr0. 1) 1-xFexO3 ceramics. J Am Ceramic Soc *104*(1):45
Shrout TR, Zhang SJ (2007) Lead-free piezoelectric ceramics: alternatives for PZT? J Electroceram 19:113
Stewart JW, Vella JH, Li W, Fan S, Mikkelsen MH (2020) Ultrafast pyroelectric photodetection with on-chip spectral filters. Nat Mater 19(2):158
Tang Z, Zhao P, Ni D, Liu Y, Zhang M, Wang H, Zhang H, Gao H, Yao Z, Bu W (2018) Pyroelectric nanoplatform for NIR-II-triggered photothermal therapy with simultaneous pyroelectric dynamic therapy. Materials Horizons 5(5):946
Tang Q, Wu J, Kim D, Franco C, Terzopoulou A, Veciana A, Puigmartí-Luis J, Chen XZ, Nelson BJ, Pané S (2022) Enhanced piezocatalytic performance of BaTiO3 nanosheets with highly exposed {001} facets. Adv Funct Mater 32(35):2202180
Timmermann L, Schüpbach M, Hertel F, Wolf E, Eleopra R, Franzini A, Servello D, Skogseid I-M, Rumia J, Aliaga AS (2013) A new rechargeable device for deep brain stimulation: a prospective patient satisfaction survey. Eur Neurol 69(4):193
Tu S, Huang H, Zhang T, Zhang Y (2017) Controllable synthesis of multi-responsive ferroelectric layered perovskite-like Bi4Ti3O12: Photocatalysis and piezoelectric-catalysis and mechanism insight. Appl Catal B Environ 219:550
Wang ZL, Song J (2006) Piezoelectric nanogenerators based on zinc oxide nanowire arrays. Science 312(5771):242
Wang Y, Wen X, Jia Y, Huang M, Wang F, Zhang X, Bai Y, Yuan G, Wang Y (2020) Piezo-catalysis for nondestructive tooth whitening. Nat Commun 11(1):1328
Wang W, Li J, Liu H, Ge S (2021) Advancing versatile ferroelectric materials toward biomedical applications. Adv Sci 8(1):2003074
Wang Y, Dai X, Dong C, Guo W, Xu Z, Chen Y, Xiang H, Zhang R (2022a) Engineering electronic band structure of binary thermoelectric nanocatalysts for augmented pyrocatalytic tumor nanotherapy. Adv Mater 34(4):2106773
Wang Y, Wang S, Meng Y, Liu Z, Li D, Bai Y, Yuan G, Wang Y, Zhang X, Li X (2022b) Pyrocatalysis for tooth whitening via oral temperature fluctuation. Nat Commun 13(1):4419
Wei J, Liu Y, Li Y, Zhang Z, Meng J, Xie S, Li X (2023) Photothermal propelling and pyroelectric potential-promoted cell internalization of Janus nanoparticles and Pyroelectrodynamic tumor therapy. Adv Healthc Mater 12(18):2300338
Wu J, Mao W, Wu Z, Xu X, You H, Jia Y (2016) Strong pyro-catalysis of pyroelectric BiFeO 3 nanoparticles under a room-temperature cold–hot alternation. Nanoscale 8(13):7343
Wu J, Qin N, Yuan B, Lin E, Bao D (2018) Enhanced pyroelectric catalysis of BaTiO3 nanowires for utilizing waste heat in pollution treatment. ACS Appl Mater Interfaces 10(44):37963
Wu Y, Hu Y, Han M, Ouyang Y, Xia L, Huang X, Hu Z, Li C (2021a) Mechanism insights into the facet-dependent photocatalytic degradation of perfluorooctanoic acid on BiOCl nanosheets. Chem Eng J 425:130672
Wu M, Zhang Z, Liu Z, Zhang J, Zhang Y, Ding Y, Huang T, Xiang D, Wang Z, Dai Y (2021b) Piezoelectric nanocomposites for sonodynamic bacterial elimination and wound healing. Nano Today 37:101104
Xu X, Jia Y, Xiao L, Wu Z (2018a) Strong vibration-catalysis of ZnO nanorods for dye wastewater decolorization via piezo-electro-chemical coupling. Chemosphere 193:1143
Xu X, Xiao L, Jia Y, Wu Z, Wang F, Wang Y, Haugen NO, Huang H (2018b) Pyro-catalytic hydrogen evolution by Ba 0.7 Sr 0.3 TiO 3 nanoparticles: harvesting cold–hot alternation energy near room-temperature. Energy Environ Sci 11(8):2198
Yan J, Han Y, Xia S, Wang X, Zhang Y, Yu J, Ding B (2019) Polymer template synthesis of flexible BaTiO3 crystal nanofibers. Adv Funct Mater 29(51):1907919
Yuan B, Wu J, Qin N, Lin E, Kang Z, Bao D (2019) Sm-doped Pb (Mg1/3Nb2/3) O3-xPbTiO3 piezocatalyst: exploring the relationship between piezoelectric property and piezocatalytic activity. Appl Mater Today 17:183

Yuan X, Kang Y, Dong J, Li R, Ye J, Fan Y, Han J, Yu J, Ni G, Ji X (2023) Self-triggered thermoelectric nanoheterojunction for cancer catalytic and immunotherapy. Nat Commun 14(1):5140
Zeng S, Vahidzadeh E, VanEssen CG, Kar P, Kisslinger R, Goswami A, Zhang Y, Mahdi N, Riddell S, Kobryn AE (2020) Optical control of selectivity of high rate CO2 photoreduction via interband-or hot electron Z-scheme reaction pathways in Au-TiO2 plasmonic photonic crystal photocatalyst. Appl Catal B Environ 267:118644
Zhang D, Wu H, Bowen CR, Yang Y (2021) Recent advances in pyroelectric materials and applications. Small 17(51):2103960
Zhao J, Yao B, He Q, Zhang T (2012) Preparation and properties of visible light responsive Y3+ doped Bi5Nb3O15 photocatalysts for Ornidazole decomposition. J Hazard Mater 229:151
Zhao J, Zhu R, Du Q, Liu F, Wu Y, Lin D, Xu H, Di W, Chen J, Luo H (2020) Surface and thickness effect on the ferroelectric, dielectric and pyroelectric properties of Mn-doped Pb (Mg1/3Nb2/3) O3–0.28 PbTiO3 single crystals. J Alloys Compd 816:152500
Zhao Y, Wang S, Ding Y, Zhang Z, Huang T, Zhang Y, Wan X, Wang ZL, Li L (2022) Piezotronic effect-augmented Cu2–x O–BaTiO3 sonosensitizers for multifunctional cancer dynamic therapy. ACS Nano 16(6):9304
Zhu P, Chen Y, Shi J (2020) Piezocatalytic tumor therapy by ultrasound-triggered and BaTiO3-mediated piezoelectricity. Adv Mater 32(29):2001976

Chapter 5
Stimuli-Responsive Biomaterials

Yanxi Yang, Xinru Kong, Xueli Ren, and Zhe Liu

Abstract Stimuli-responsive biomaterials, as a category of intelligent biomaterials, are capable of demonstrating controllable and reversible biological responses to internal or external stimuli. These materials draw inspiration from the intricate regulatory systems found within living organisms, thereby driving innovation in various fields such as biomedical research, drug delivery, and tissue engineering. Key characteristics of stimuli-responsive biomaterials include their sensitivity to stimuli and corresponding controllability, which may be either endogenous or exogenous. The utilization of stimulus-responsive materials holds the potential to enhance personalized treatment, thereby improving treatment efficacy and safety. The advancement of such materials has propelled the progress of precision medicine, enabling tailored treatment plans based on individual patient differences while minimizing treatment-related side effects. In the realm of biomedical engineering, the controllability and adaptability of these materials play a crucial role in promoting disease diagnosis and therapy as well as tissue repair and regeneration. Furthermore, stimuli-responsive biomaterials find widespread use in biosensing technologies, medical diagnostics, and imaging applications by providing advanced technological support. Overall, research and application of stimulus-responsive biomaterials are pivotal for the advancement of personalized therapy, precision medicine, and innovative medical devices.

Keywords Stimuli-responsive materials · Intelligent biomaterials · Drug delivery · Controlled release · Biomedicine

Y. Yang · X. Kong · X. Ren · Z. Liu (✉)
Academy of Medical Engineering and Translational Medicine, Tianjin University, Tianjin, China

Tianjin Key Laboratory of Brain Science and Neural Engineering, Tianjin University, Tianjin, China
e-mail: zheliu@tju.edu.cn

Z. Liu (ed.), *Intelligent Biomaterials*, https://doi.org/10.1007/978-981-96-8646-9_5

Stimuli-responsive biomaterials are a type of intelligent biomaterials that have attracted much attention, and they show controllable and reversible biological responses when subjected to internal/external stimuli or changes in specific environmental conditions. The design of stimuli-responsive biomaterials is inspired by the complex regulatory systems in living organisms, and it brings new possibilities and innovations to biomedicine, drug delivery, tissue engineering, and other related fields.

The core features of stimuli-responsive biomaterials are their sensitivity to stimuli and corresponding controllability. This stimulus can be endogenous or exogenous. Endogenous stimuli refer to changes originating from the internal environment of organisms, such as physiological pH, ion concentration fluctuations, changes in enzyme activity, etc. External stimulation includes physical factors, such as temperature, light, ultrasound, magnetic field, etc. The potential utility of stimuli-responsive biomaterials is huge in terms of medical applications. In the field of drug delivery, these materials can be designed as carriers to release drugs under specific physiological conditions, enabling more precise treatment and reduced side effects. In tissue engineering, the controllability and adaptability of these materials can promote tissue repairing and regeneration. In addition, stimuli-responsive biomaterials are widely used in biosensing, medical diagnosis, and imaging, providing advanced tools and technologies for medical research.

With the continuous advancements, research on stimuli-responsive biomaterials is continuously expanding its application scope. Through the careful design of these materials, scientists can better understand and simulate the complex biological response processes in organisms, bringing more innovations to the intersection of medicine, biology, and materials sciences. In the future, with the in-depth understanding of stimuli-responsive biomaterials and the continuous improvement of technology, it is believed that it will continue to make important contributions to solve major medical problems and promote biomedical advances.

5.1 Typical Stimuli-Responsive Biomaterials

Stimuli-responsive biomaterials are a type of material that can respond reversibly to external stimuli. According to different components, stimuli-responsive biomaterials can be divided into the following typical categories: Stimuli-responsive biomaterials based on inorganic substances, biomacromolecules, polymers, and hydrogel composites.

5.1.1 Inorganic Stimuli-Responsive Biomaterials

5.1.1.1 Boron-Based Stimuli-Responsive Biomaterials

Boron-containing luminescent compounds are a large and versatile class of stimuli-responsive biomaterials. Boron exists in various forms such as triarylboranes, borate dyes, and boron clusters, and its excellent photophysical properties have attracted considerable attention (Mukherjee and Thilagar 2016). Since boron atoms lack electrons, doping boron atoms into π-conjugated scaffolds is a popular way to endow them with many unique properties (Mellerup and Wang 2019). When boron is incorporated into π-conjugated systems, these compounds often exhibit efficient luminescence, stimuli-responsive emission, nonlinear optical properties, and unique photophysical properties (Frath et al. 2014; Ji et al. 2017). When these boron-doped molecules are exposed to external stimuli such as light, heat, and force, their intramolecular or intermolecular interactions will change, resulting in unique electronic, photophysical, or mechanical properties, which have a wide range of applications.

Suenaga et al. (2020) designed alternating copolymers containing dense boronides as monomer units and changed the shape of polymer chains to regulate the degree of molecular interaction by changing the connection points of boron complexes. The obtained polymers exhibited strong emission in both solution and film states. One of these linear polymers exhibited thermochromic luminescence in solution, and it was speculated that assembly should play an important role in the color change of their stimulated response to luminescence. In addition, mechanical luminescence was also observed in the solid state, and the luminescence behavior may result from the change of conjugated polymer morphology after grinding. Qiu et al. (2022) proposed a bottom-up assembly method to prepare borate dynamic nanostructures (BDN) with controllable morphology and interfacial crosslinking, from which robust self-healing elastomers are prepared. The BDN was optimized to build a dense and strong interfacial boron crosslink that gave the elastomer excellent stretchability (2050%), high strength (17.9 MPa), and healing efficiency (77.1%). In addition, the elastomer also exhibited fluorescence properties in response to pH stimulation and excellent functional repair, so it has great potential application prospect in the field of smart materials such as information encoding and encryption.

5.1.1.2 Zinc Oxide-Based Stimuli-Responsive Biomaterials

Zinc oxide is a commonly used chemical additive with large energy band gap and exciton binding energy, high transparency, and excellent room temperature luminescence properties. It is also widely used in the semiconductor field (Weng et al. 2023). Zinc as an essential trace element in human tissues actively participates in the body's metabolic process, and plays an important role in maintaining body health. Thus, it is important to explore the application of zinc oxide in biomedicine (Ruszkiewicz et al. 2017). Stimuli-responsive biomaterials based on zinc oxide usually take advantage of the special properties of ZnO, such as changes that occur under external stimuli such as light, electricity, pressure, etc.

Zinc oxide-based stimuli-responsive biomaterials can be divided into exogenous stimuli (light, sound, force, etc.) and endogenous stimuli (pH, enzymes, etc.) according to the type of stimulus responses. When subjected to various stimuli, they can intelligently respond to changes in the environment. Their own properties change and perform specific biological functions. Liu et al. (2016) developed a doxorubicin delivery system based on zinc oxide nanomaterials (Fig. 5.1). Experiments have shown that zinc oxide carriers could degrade in acidic

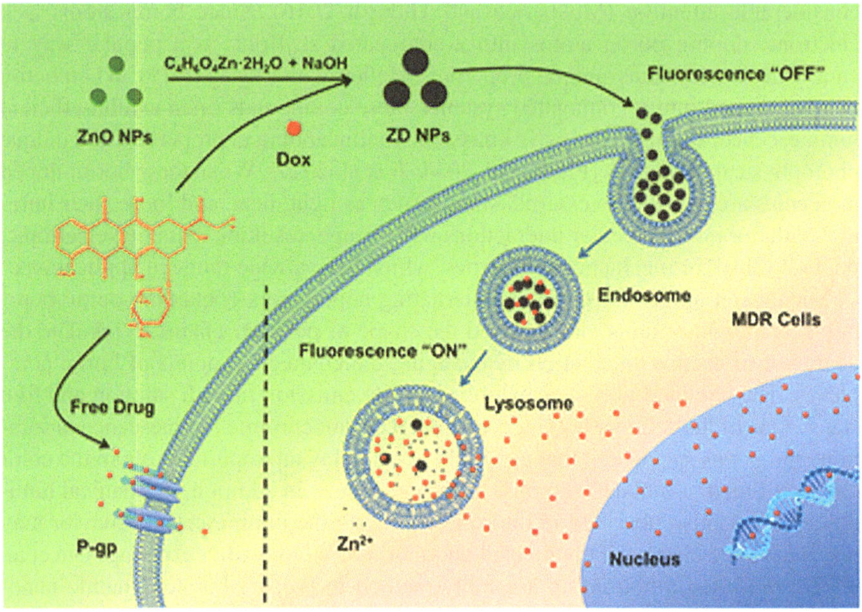

Fig. 5.1 Schematic presentation showed the formation of ZD NPs and their combined effect of intensive cellular internalization and a pH-triggered intracellular drug release for overcoming drug resistance. The drug underwent acid-triggered release from ZD NPs with fluorescence recovery (Fluorescence "ON"), and then entered into the nucleus to execute its anticancer function. (Reprinted with permission from Liu et al. 2016)

environments, triggering drug release. In addition, multiple stimuli can be combined to construct a multiple stimulus-response system. Wang et al. (2019) designed and prepared a drug-loading system with dual stimulation responses of redox and pH, using zinc oxide quantum dots as the capping agent to encapsulate the drug. Due to the acidic dissolution and disulfide bond-breaking properties of zinc oxide quantum dots, the complex had redox- and pH-responsive drug release behavior.

5.1.1.3 Silica-Based Stimuli-Responsive Biomaterials

Silica-based stimuli-responsive biomaterials are typically made by combining silica with other functional molecules or polymers, most commonly mesoporous silica nanoparticles. Mesoporous silica nanoparticles (MSNs) can produce particles with large surface area, ordered structure, and adjustable pore size and have attracted attention from the scientific community (Jafari et al. 2019). These materials can respond to stimuli inside and outside the organism and achieve specific functions such as drug delivery. Vallet-Regı et al. (2001) first proposed the use of silica-encapsulated drugs. Since then, mesoporous silica drug-loaded systems have developed rapidly, and the stimuli-responsive controlled release systems have been developed.

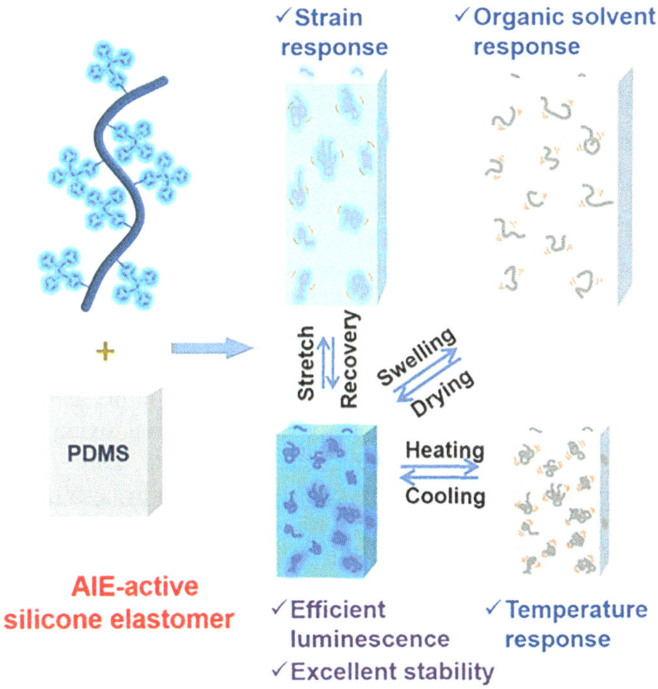

Fig. 5.2 Schematic diagram of aggregation-induced emissive silicone elastomer with multiple stimuli responsiveness. (Reprinted with permission from Chen et al 2022)

Wang et al. (2021a) reported a GSH stimuli-responsive silica nanoparticle (SNP) that integrated the disulfide cross-linker into the silica and encapsulated active ingredients such as nucleic acids and CRISPR genome editor to achieve in vivo targeted drug delivery. Chen et al (2022) have prepared a series of fluorescent silicone elastomers by combining aggregation-induced emission (AIE) active polymers with silicone matrix (Fig. 5.2). The AIE active macromolecule was synthesized by reversible addition-fragmentation chain transfer polymerization of vinyl monomer containing tetraphene side chain. These results showed that the macromolecular fluorescent compounds had good dispersibility and resistance to leaching in silicone elastomers. The fluorescence properties of silicone elastomers could be adjusted by changing the molecular weight or aggregation state of the fluorescent macromolecules. The prepared fluorescent silicone films showed multiple stimulus responses to strain, temperature, and organic solvents.

5.1.1.4 Iron Oxide-Based Stimuli-Responsive Biomaterials

Iron oxide is a transition metal oxide with many different crystal structures, often existing as α-Fe_2O_3 (hematite phase) and γ-Fe_2O_3 (maghemite phase). Among them, α-Fe_2O_3 is a stable form of iron oxide, and α-Fe_2O_3 can be converted into γ-Fe_2O_3

form under certain conditions (Grønvold and Westrum 1959). γ-Fe_2O_3 is a metastable form obtained by oxidation at a lower temperature. It will be completely converted into the α-Fe_2O_3 form when heated at a certain temperature. The magnetic characteristics of iron oxide nanoparticles introduce a magnetically responsive drug delivery method that optimizes drug distribution and improves delivery efficiency through external magnetic fields. Moreover, magnetic nanoparticles can be degraded into iron ions in the body, reducing the potential toxicity caused by the residence of nanoparticles (Pourmadadi et al. 2022).

Roy et al. (2016) prepared a stimulus-responsive polymer-modified gadolinium-doped iron oxide nanoparticles (poly@Gd-MNPs) as a cancer therapeutic diagnostic agent (Fig. 5.4a). The responsive polymer consisted of poly (N-isopropyl acrylamide)-co-tyrosine unit, which showed excellent loading for anticancer drugs (methotrexate) and stimulus-dependent release (changes in pH and temperature). Akbar et al. (2019) synthesized superparamagnetic iron oxide nanoparticles (SPIONs) by ultrasonic-based co-precipitation method. Poly (2-isopropyl-2-oxazoline) (PIPOX) and tannic acid (TA) were coated layer by layer (LbL) at pH = 4 and 25 °C. The PIPOX/TA coated SPION was then loaded with doxorubicin (DOX) at pH = 7.5 and 25 °C. LbL-coated SPIONs exhibit dual response behavior and release the most DOX at pH = 6 and 42 °C. Increasing the number of layers decreases the colloid stability and saturation magnetization. The superparamagnetism of SPIONs remains unchanged after coating.

5.1.2 Biomacromolecule-Based Stimuli-Responsive Biomaterials

5.1.2.1 Polysaccharide

Sugar is an important component of the body and the main source of energy required for all life activities. It can be divided into monosaccharides, disaccharides, and polysaccharides (Zhang et al. 2022). Polysaccharides are polymers composed of monosaccharide units connected by glycosidic bonds, and can be divided into oligosaccharides (oligosaccharides, polymerized from 2–10 sugar molecules) and polysaccharides (composed from >10 monosaccharide molecules). Polysaccharides are considered promising biomaterials due to their excellent biocompatibility, low toxicity, and degradability. The polysaccharide surface has a series of reactive functional groups, such as -OH, -COOH, -NH_2, etc., which makes the polysaccharide have excellent reactivity and modifiability (Rial-Hermida et al. 2021). Stimuli-responsive biomaterials composed of sugars have attracted much attention in recent years. Li et al. (2021a) used N-carboxyethyl chitosan and hyaluronic acid aldehyde as raw materials to prepare a pH-responsive drug-loaded hydrogel through acyl hydrazone bonds and imine bonds (Fig. 5.3). The insulin incorporated continued to respond to pH for 14 days. The results showed that this polysaccharide-based hydrogel dressing loaded with insulin could shorten the inflammatory cycle, improve peripheral neuropathy, and accelerate the healing of insulin foot ulcers.

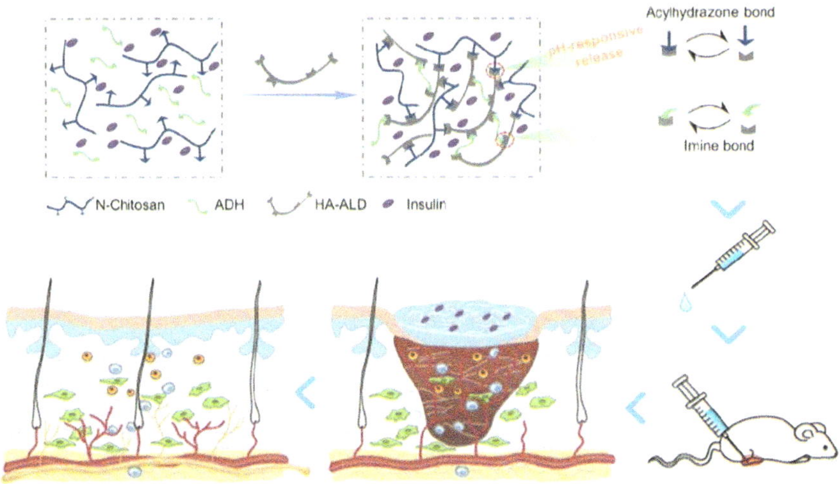

Fig. 5.3 The overall synthetic process of the insulin-loaded hydrogel which releases insulin in a pH-sensitive manner to promote wound healing and improve peripheral neuropathy. (Reused with permission from Li et al. 2021a)

Cyclodextrin (CD) is a cyclic oligosaccharide produced by hydrolysis of starch linked through 1,4-glycosidic bonds and is widely used in drug delivery (Liu et al. 2020). The structure of cyclodextrin is like a round truncated cone with a hydrophobic cavity in the middle and a hydrophilic surface (Harada 2001). Hydrophobic drugs of appropriate size can enter the cavity structure of cyclodextrin, thereby increasing its water solubility and facilitating bioavailability (Tian et al. 2021). With the deepening of research, simple drug-carrying systems can no longer meet the needs of precision medicine, and drug carriers that can achieve drug release effects according to the required environment are needed. Stimuli-responsive biomaterials based on cyclodextrins have attracted widespread attention. Bai et al. (2019) used lithocholic acid as a linker to connect two β-CD molecules to form a drug carrier. A host-guest interaction occurred between lithocholic acid and β-CD. This complex had high drug loading efficiency, and its polymer side chain had different stretching states in response to different pH values, resulting in a significantly higher drug release rate at pH = 5.0 than at pH = 7.4. In order to effectively utilize the photosensitive compound azobenzene and its derivatives, Bian et al. (2019) used aminated β-CD modified mesoporous silica nanoparticles and loaded model drugs p-Coumarin acid (CA) (Fig. 5.4). Visible light was used to control the inclusion and exclusion of β-CD, thereby reversibly controlling the release of CA.

Cellulose is a macromolecular polysaccharide composed of glucose. It is the main structural component of many organisms and has the characteristics of stable structure, renewable, and good biocompatibility. There are many potential applications for stimuli-responsive materials based on cellulose, a natural polymer that is responsive to stimuli such as changes in temperature, pH, and chemicals. Methylcellulose is a cellulose derivative that exhibits reversible swelling and gelation under temperature changes. As the temperature changes, its solubility and gel

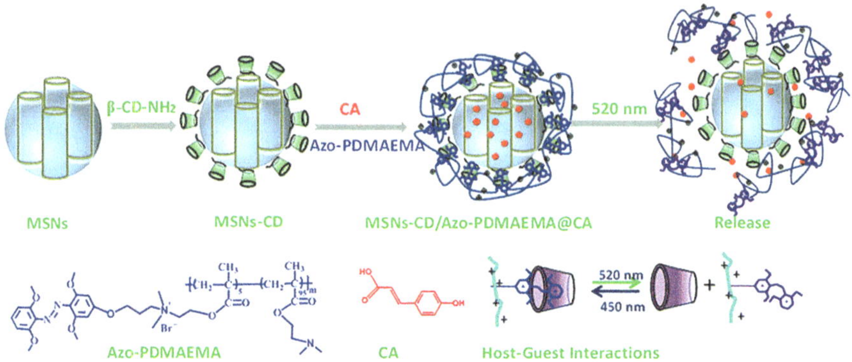

Fig. 5.4 Schematic depiction of visible-light triggered cargo release from MSNs-CD/Azo-PDMAEMA and the supramolecular interaction between Azo-PDMAEMA and β-CD. (Reused with permission from Bian et al. 2019)

state will change accordingly, and it can be used for drug delivery and tissue engineering. Shang et al. (2023) developed temperature-responsive hydrogels through carboxymethylcellulose and acrylamide and loaded urea into them as a fertilizer. This compound could respond to changes in temperature and exhibit deswelling behavior when the temperature rises. Fertilizer could be released quickly, solving the problems of soil water loss and low fertilizer efficiency. It has shown great application potential in the controlled release of water and fertilizer in agriculture and forestry. Carboxymethylcellulose is sensitive to changes in environmental pH and can undergo changes in solubility and structure under different pH conditions. This property is widely used in sustained-release systems in the pharmaceutical field. Wang et al. (2021b) prepared carboxymethylcellulose/chitosan interpenetrating network hydrogel to adsorb and desorb anionic and cationic dyes under different pH conditions. Experiments had proven that the prepared hydrogel exhibited different swelling behaviors under different pH conditions and had great application potential in controlled-release carriers.

5.1.2.2 Protein

Protein-based stimuli-responsive biomaterials are a class of materials with intelligent response properties. Their design is inspired by the structural or functional changes that occur in proteins in organisms under specific stimuli. Researchers combine knowledge from biology, chemistry, and material sciences to create new materials that can respond controllably to external stimuli. The uniqueness of these materials is that their responsive properties can be achieved by regulating the structure, function, or interaction of proteins, providing broad application prospects in fields such as biosensing, medicine, and smart materials.

Natural proteins play an important role in this field as basic ingredients. Natural proteins such as collagen, casein, and serum albumin are widely used in stimuli-responsive materials due to their natural biocompatibility and bioactivity (Du et al. 2022). These proteins can exhibit reversible changes in shape or function in response to external stimuli through changes in their native structure or interactions with other biomolecules. For example, collagen exhibits a reversible gel-sol transition under changes in environmental factors such as temperature and pH, which provides a theoretical basis for the design of thermosensitive gels. Nie et al. (2021) reported that a natural enzyme lipoic acid protein ligase A (LplA) formed stimuli-responsive and reversible hydrogels through self-assembly in response to temperature, pH, and the presence of ions.

As a further development of natural proteins, engineered proteins involve targeted changes in the structure of proteins through genetic engineering or chemical modification. Such modifications enable proteins to respond more precisely to external stimuli, for example, triggering conformational changes in the protein or releasing specific bioactive molecules under specific environmental conditions. This controllable response provides new possibilities for application fields such as drug delivery systems and biosensors. Artificially synthesized proteins represent the cutting-edge exploration in this field, designing and constructing new artificial proteins through chemical synthesis or biosynthesis technology. These proteins have greater design freedom to achieve precise structure and property control, providing a new approach to the custom design of stimuli-responsive biomaterials. Research on artificially synthesized proteins not only expands the plasticity of materials, but also lays the foundation for the construction of more complex and intelligent bionic materials. Vural et al. (2018) designed 2D titanium carbide (MXene) inks based on tandem repeats of squid teeth to synthesize proteins. These inks could be printed on various substrates as stimuli-responsive electrodes.

5.1.2.3 Nucleic Acid

Nucleic acids (including DNA and RNA) are substances that store and transmit genetic information and play an important role in many important biological functions (He et al. 2021). Nucleic acid-based stimuli-responsive materials are a type of smart materials that have attracted much research attention. They have made significant progress at the intersection of molecular biology, nanotechnology, and material sciences, and provide applications such as biosensing, drug delivery, and nanorobots. The natural sequence specificity and self-assembly ability of nucleic acid molecules such as DNA and RNA provide a solid foundation for the design of stimuli-responsive materials. By designing appropriate nucleic acid sequences, it is possible to achieve two- or three-dimensional structural changes under specific stimuli, such as the unfolding and merging of helical structures. This self-assembly property allows nucleic acid materials to achieve reversible shape or structural changes under different environmental conditions, providing the possibility of intelligent responsiveness of materials. The design of nucleic acid-based stimuli-responsive materials often involves the introduction of responsive nucleic acid elements. These elements include but are not limited to DNAzymes, aptamers, small interfering RNA, etc., which exhibit specific functions such as catalytic activity, binding affinity, or targeting under specific stimuli. By rationally designing and combining these responsive nucleic acid elements, material systems with precise responsiveness can be constructed for detecting specific molecules, regulating gene expression, or achieving directional release of drugs. Liu et al. (2022a) developed a signal conversion and amplification strategy based on stimuli-responsive nucleic acids (SR-NA). The target could recognize the responsive adapter and trigger the degradation of SR-NA, thereby converting the target into a DNA signal by generating new toehold nucleic acid strands, which was further enhanced by nucleic acid amplification. This strategy significantly improved the detection accuracy of small biomolecules (H_2O_2) and enzymes (esterase, β-Gal).

5.1.3 Polymer-Based Stimuli-Responsive Biomaterials

Stimuli-responsive biomaterials based on N-isopropyl acrylamide (NIPAAm) are a type of material that can respond reversibly to external stimuli, such as temperature, pH, ion concentration, etc. It has a low critical solution temperature and is the best known of the stimuli-responsive polymers that has become a staple in polymer science. NIPAAm is a hydrophobic acrylamide monomer that forms a hydrophobic gel at low temperatures and dissolves at high temperatures. This temperature-sensitive property enables NIPAAm and its copolymers to adjust their physical structure and properties under temperature changes. Ortega-García et al. (2022) prepared a temperature- and pH-responsive poly (N-isopropylacrylamide-co-N-isopropylmethacrylamide)/chitosan core/shell nanohydrogel. The nanohydrogel was loaded with the drug doxycycline hydrochloride, and its release kinetics showed that a higher amount of drug was released at acidic pH. Zhu et al. (2023) prepared hyperbranched polysiloxane/poly(N-isopropylacrylamide) (HBPS@PNIPAm) by copolymerizing hyperbranched polysiloxane (HBPS) with N-isopropylacrylamide (NIPAm) microgel. Using luteolin as a model drug, its drug loading and drug release behaviors were studied through photoluminescence. The results showed that as the temperature increases, the thermoresponsive active fluorescence of the microgel increased.

Among all polymers, polyacrylic acid derivatives have many unique properties, such as satisfactory biocompatibility, good stability, sensitivity to pH and temperature, and thus have been widely studied (Li et al. 2021b). Polyacrylic acid is a polymer polymerized from acrylic acid monomers. The carboxyl groups on its main chain make it highly polar. Polyacrylic acid has a high degree of solubility in water, which allows it to exhibit many interesting properties in aqueous solutions, such as pH sensitivity. Under different pH conditions, polyacrylic acid molecular chains may expand or contract. The temperature-responsive properties of materials can also be controlled by introducing temperature-responsive monomers or controlling the crystallinity of molecular chains. Polyacrylic acid and its derivatives can undergo structural or property changes under external stimuli, so they have been widely studied for the preparation of stimuli-responsive materials. These materials have potential applications in biomedicine, sensors, smart textiles, and other fields.

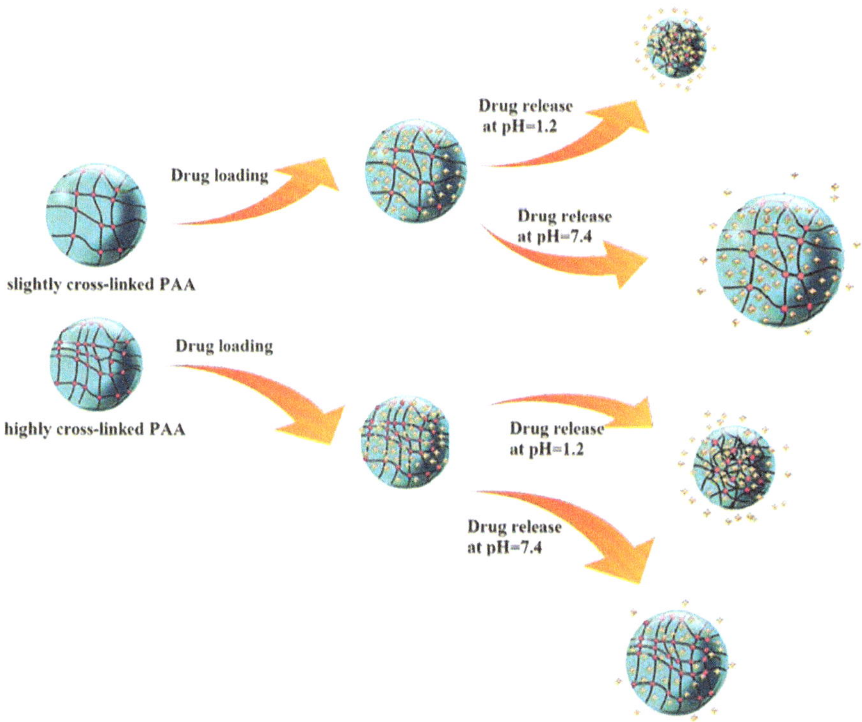

Fig. 5.5 pH-responsive behavior of poly (acrylic acid) (PAA) nanoparticles with high and low AA: EGDMA molar ratios and its effect on drug release behavior at 37 °C. (Reprinted with permission from Nikravan et al. 2019)

Nikravan et al. (2019) synthesized cross-linked poly (acrylic acid) (PAA) nanoparticles by distillation precipitation polymerization of acrylic acid and ethylene glycol dimethacrylate with different molar ratios (Fig. 5.5). The nanoparticles exhibited temperature- and pH-responsive behaviors. Yan et al. (2021) designed multi-stimuli-responsive supramolecular hydrogels based on methoxy-azobenzene-grafted poly acrylic acid (PAA-mAzo) and β-CD dimers. This complex had triple stimuli response property of thermal response, light response, and pH response to enhance drug release.

5.1.4 Hydrogel-Based Stimuli-Responsive Biomaterials

Hydrogel-based stimuli-responsive biomaterials are a class of polymer networks with special properties that form gel structures in water and can respond reversibly to external stimuli. This type of material has broad application potential in biomedicine, intelligent sensing, drug delivery, and other fields (Lavrador et al. 2021). It has the following characteristics (Serpico et al. 2023): (1) Water phase compatibility:

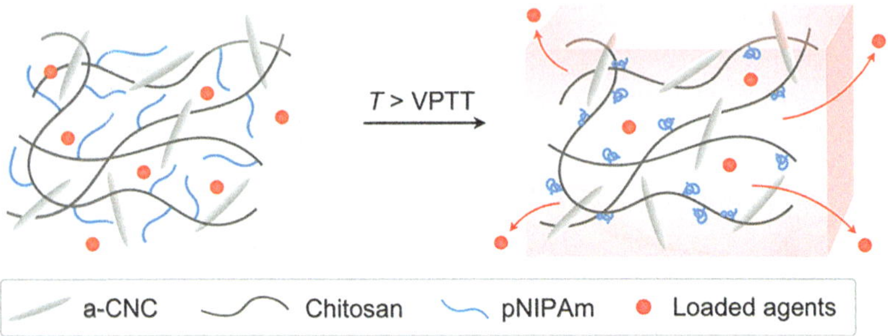

Fig. 5.6 The temperature-responsive changes in the CCN-Gel structure, resulting in enhanced diffusion-mediated mass transport of encapsulated species. (Reprinted with permission from Huang et al. 2022a)

Hydrogel is a material formed in water and has good water phase compatibility, making it widely used in organisms or other aqueous environments. (2) Controllable gel-swelling behavior: Hydrogels can achieve reversible gel-swelling transitions through external stimuli (such as temperature, pH, ion concentration), thereby regulating their volume and structure. (3) Biocompatibility: Many hydrogel materials have good biocompatibility with living organisms and are suitable for biomedical applications such as drug delivery and tissue engineering. (4) Sustainable release: Due to the reversible response of hydrogels, they can be designed to sustainably release active substances embedded in them, such as drugs, biomolecules, etc. (5) Smartness: The reversible changes of hydrogels in response to stimulation make them ideal as smart materials for making sensors and actuators that respond to environmental changes. Huang et al. (2022a) reported a nanocolloidal hydrogel that exhibited a temperature-controlled increase in pore size and therefore enhanced transport of encapsulated species in the gel (Fig. 5.6). At elevated temperatures, the size of the pores increased four-fold without significant changes in volume. Liu et al. (2022b) reported the formation of multi-stimuli-responsive supramolecular hydrogels through self-assembly of oxidized glutathione derivatives. In addition to thermosensitivity, the hydrogel also exhibited redox and photo responsiveness, which was attributed to the presence of a disulfide bond and two azobenzene end groups in the hydrogel molecular structure, respectively.

5.2 Stimulus Sources

5.2.1 Interior Stimulus Sources

Endogenous stimuli-responsive materials refer to materials that can respond specifically to internal signals or stimuli in biomedical applications. The development and application of this type of material have broad significance in the biomedical field. Endogenous stimuli-responsive materials can be used to design smart drug delivery systems. These systems can adjust the release rate or location of drugs based on changes in signals within the body to achieve precise drug delivery in the body.

Highly sensitive biosensors can be constructed using endogenous stimuli-responsive materials to monitor signal changes within the organism. This is of great significance for early-stage diagnosis and real-time monitoring of diseases. The application of endogenous stimuli-responsive materials in tissue engineering and regenerative medicine has also attracted much attention. These materials can be used to design artificial tissues and biomedical devices that respond to internal stimuli in the body, promoting tissue repair and regeneration.

Endogenous stimuli-responsive materials are widely used in the design of various biomaterials in the biomedical field, such as degradable stents, medical glues, etc. These materials can change under specific biological conditions to meet clinical needs. Nanomaterials have special applications in response to endogenous stimuli. Nanoparticles or nanomaterials can be designed to release drugs or perform imaging when specific stimuli are encountered inside the body, providing new avenues for the development of nanomedicine.

The application of endogenous stimuli-responsive materials allows for more individualized treatment. By adjusting treatment strategies based on the patient's internal signals, the effectiveness and safety of treatment can be improved. The development of such materials promotes the realization of precision medicine. The treatment plan can be adjusted according to the individual differences of patients, improving the accuracy of medicine. The application of endogenous stimuli-responsive materials can reduce side effects during treatment because the drug or treatment is only released or activated when needed. Technologies such as biosensors constructed using endogenous stimuli-responsive biomaterials facilitate early diagnosis of disease and improve the success rate of treatment. The development of endogenous stimuli-responsive biomaterials has promoted continuous innovation in the field of biomedical engineering, providing more possibilities for the design of new therapeutic and diagnostic tools. Overall, the development and application of endogenous stimuli-responsive biomaterials have far-reaching significance in the biomedical field, providing new ideas and approaches for the development of personalized treatment, precision medicine, and novel medical devices. We highlight a selection of examples of endogenous stimuli-responsive biomaterials using various biological markers such as pH, enzymes, redox, glucose, gas ATP, and nucleic acids.

5.2.1.1 pH

The pH-responsive biomaterials are a class of materials that can produce specific responses to pH changes in organisms. This responsiveness is based on the structure or chemical properties of the material. Its chemical structure contains groups that are sensitive to acid-base environments, allowing it to sense and adapt to changes in the pH of the surrounding environment. These materials are usually designed to undergo structural or property changes within a specific pH range to achieve functions such as controlled release, sensing, and imaging. Nanomaterials with pH

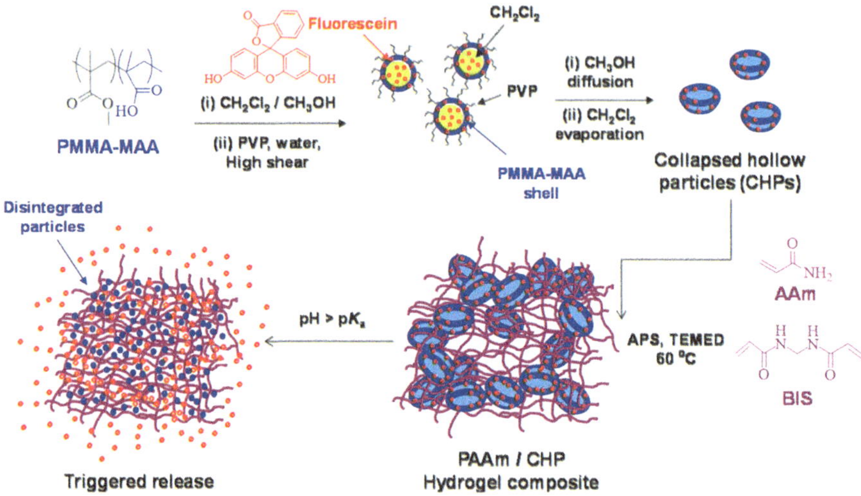

Fig. 5.7 Depiction of the preparation of PAAm/collapsed hollow particle (CHP) hydrogel composites and pH-triggered release of a model drug. (Reprinted with permission from Pafiti et al. 2016)

responsiveness, such as nanoparticles through surface modification or coating with polymers with acid-base sensitive properties, Li et al. (2023a) prepared polymer-modified Fe_3O_4@C using 4-vinylphenylboronic (VB) acid and polyethylene glycol methyl ether methacrylate (PEGMA) as polymerized monomers and confirmed that Fe_3O_4@C-VB-PEGMA exhibited effective inhibition to the growth of HepG2 cells. Pafiti et al. (2016) prepared hydrogels containing acrylic acid groups, which swelled in an alkaline environment and shrank in an acidic environment. The hydrogel composites showed dual-action pH-triggered softening with simultaneous drug release which occurred without a volume increase (Fig. 5.7).

The pH-responsive materials are used to design drug delivery systems. By selecting appropriate pH-responsive materials, drug release under specific pH conditions can be achieved, thereby improving the local efficacy of the drug and reducing adverse effects on normal tissues. Using pH-responsive materials to regulate drug release can slow down the release rate of drugs in the blood and reduce the concentration of drugs in the entire body, thereby reducing the systemic side effects of drugs. This is crucial to improve therapeutic efficacy and reduce toxicity. In tumor diagnosis, pH-responsive imaging probes can achieve highly sensitive imaging of tumor tissues by responding to the acidic microenvironment of tumors. Materials such as pH-responsive hydrogels can be used in tissue engineering to regulate the pH sensitivity of the extracellular matrix and affect cell attachment, proliferation, and differentiation. This has potential value for the construction of biocompatible and controllable artificial tissues. Overall, the application of pH-responsive materials can be adjusted according to the physiological conditions and disease states of different individuals to achieve individualized treatment. This is of significant significance for improving treatment effects, reducing adverse reactions, and promoting the development of personalized medicine.

5.2.1.2 Enzymes

There are many enzymes in living organisms, and their activities vary in different tissues and pathological states. Therefore, enzyme-responsive materials can be used to create drug delivery systems or diagnostic materials that change in the context of specific enzyme activity. The activity levels of some enzymes in tumor tissue are different from those in normal tissue. By designing materials that respond to these enzymes, drugs can be released in a targeted manner at the tumor site, improving the therapeutic effect while reducing the impact on healthy tissue. Hu et al. (2019a) reported on designed fabrication of biodegradation-medicated enzymatic activity-tunable molybdenum oxide nanourchins (MoO_3-xNUs), which first induce catalase (CAT)-like reactivity to decompose hydrogen peroxide (H_2O_2) in tumor microenvironment, producing a considerable amount of O_2 for subsequent oxidase (OXD)-like reactivity of MoO_3-xNUs; a substantial cytotoxic superoxide radical ($^{\cdot}O_2^-$) is thus generated for tumor cell apoptosis (Fig. 5.8). Real-time monitoring of enzyme activity changes in organisms is very important for early diagnosis and treatment monitoring of diseases. By

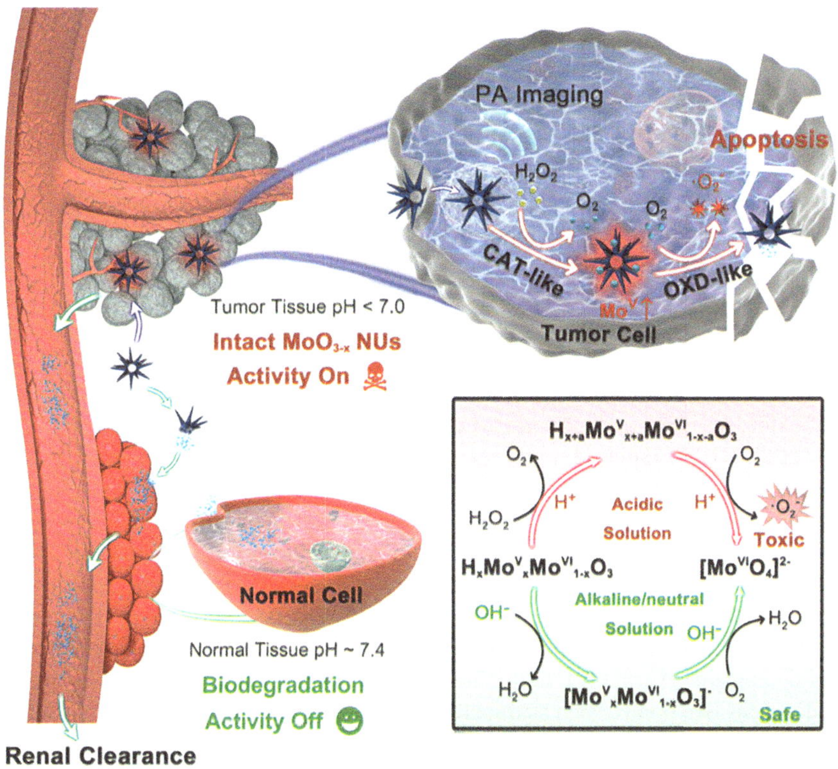

Fig. 5.8 An illustration of biodegradation-medicated enzymatic activity-tunable molybdenum oxide nanourchins (MoO3–xNUs) with the highly specific toxicity to tumor tissues via a multi-enzyme stepwise cascade catalysis in acidic tumor microenvironment, while leaving normal tissues unharmed due to their pH-responsive biodegradation and subsequent renal excretion in physiological environment. (Reprinted with permission from Hu et al. 2019a)

designing sensors that respond to target enzymes, accurate and sensitive signals can be provided to help doctors understand the physiological status of patients. Enzyme-responsive materials can be used to design scaffolds or repair materials that can change accordingly based on the presence or activity of specific enzymes in the surrounding tissue, helping to promote tissue regeneration and repair. Yu et al. (2022) reported an in situ cartilage tissue engineering porous scaffold with enzyme-responsive kartogenin (KGN). The release dose of PEGKGN is found to be accelerated by matrix metalloproteinase-2 (MMP-2), and the high-dose PEGKGN can stimulate the migration of bone marrow mesenchymal stem cells (BMSCs) obviously, and the low-dose PEGKGN can promote the chondrogenesis of BMSCs in vitro. The design principle of these materials lies in specific chemical structures in that enzymes can recognize and interact with specific groups or structures on the materials. This intelligent response to enzymes in organisms makes these materials key tools in the biomedical field, providing new possibilities for precision medicine and treatment options.

5.2.1.3 Redox (GSH and ROS)

Redox-responsive biomaterials are a sort of materials that can respond to changes in the redox environment within organisms. When these materials sense changes in the redox state within cells or in the body, they can show specific changes, such as releasing drugs, adjusting the structure or properties of the material, etc. This responsiveness makes these materials potentially useful for drug delivery, imaging, and therapy in biomedical applications. Metal oxide nanoparticles, such as Fe_3O_4, MnO_2, etc., are often used as redox-responsive materials, and their surfaces contain parts that can regulate redox reactions. These nanomaterials can release drugs or perform imaging when stimulated by a reducing environment (such as the highly reducing environment in tumor cells) to achieve targeted therapy of tumors. For example, Fe_3O_4 is used in magnetic resonance imaging (MRI) and tumor treatment. In a reducing environment, such as inside a tumor, iron oxide can be reduced to iron suboxide, resulting in the release of contrast agent and enhanced therapeutic effect. Yu et al. (2019) reported a reactive oxygen species (ROS) nanoreactor based on core–shell-structured iron carbide (Fe_5C_2@Fe_3O_4) nanoparticles (NPs) through the catalysis of the Fenton reaction. These NPs are able to release ferrous ions in acidic environments to disproportionate H_2O_2 into •OH radicals, which effectively inhibits the proliferation of tumor cells both in vitro and in vivo. MnO_2 nanoparticles are a material that can decompose in a reducing environment. In tumors, they can be reduced to soluble manganese ions, resulting in a reduction in local pH, thereby enhancing the efficacy of tumor treatment drugs. Polymer materials containing redox-sensitive bonds, such as thioether bonds and selenyl ether bonds, can degrade in a reducing environment, thereby releasing drugs or changing the properties of the material. This feature is widely used in tumor treatment. Bioactive molecules containing redox-reactive molecules, such as molecules containing disulfide bonds, can release bioactive molecules, such as antioxidants, when stimulated by a redox environment, thereby combating oxidative stress. Luo et al. (2019) developed dual pH/

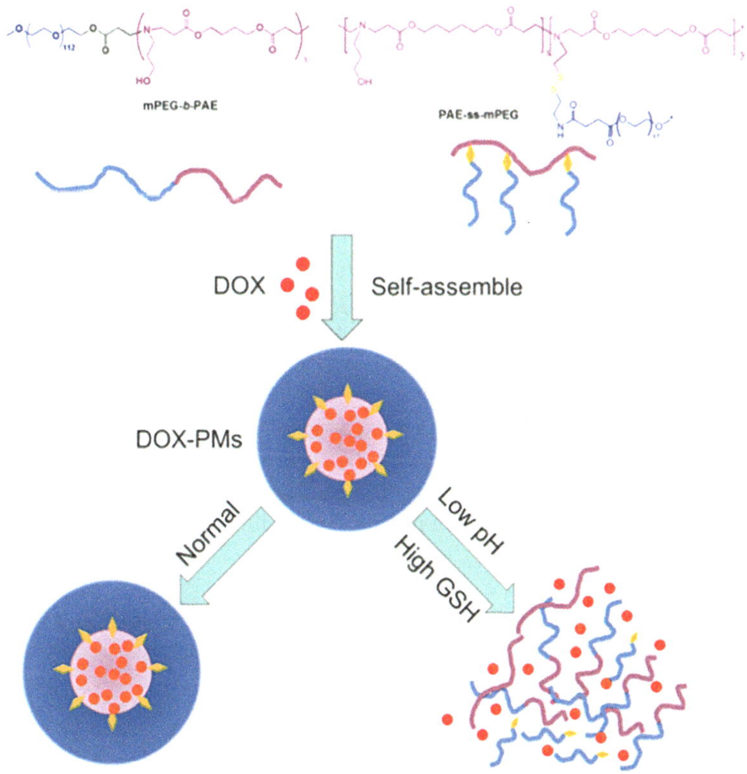

Fig. 5.9 Co-micellization of pH/redox-responsive diblock copolymers for drug delivery and controlled release triggered by pH and glutathione (GSH). (Reprinted with permission from Luo et al. 2019)

redox-responsive mixed PMs which are self-assembled from two kinds of amphiphilic diblock copolymers (poly(ethylene glycol) methyl ether-b-poly(β-amino esters) (mPEG-b-PAE) and poly(ethylene glycol) methyl ether-grafted disulfide-poly(β-amino esters) (PAE-ss-mPEG)) for anticancer drug delivery and controlled release (Fig. 5.9).

5.2.1.4 Glucose

Glucose-responsive biomaterials are a type of material that can respond specifically to changes in glucose concentration. Its design principle is based on the interaction between the material and glucose. This interaction can be physical, such as changes in structure or chemical, and sexual, such as the formation or breaking of chemical bonds. These materials can sense the presence or concentration changes of glucose, exploiting glucose's biological importance and association with disease states to enable more effective medical applications. The interaction between glucose and the probe causes the fluorescence of the glucose-sensing fluorescent probe to change with changes in the surrounding glucose concentration. It can be used for glucose monitoring, especially in patients with diabetes, and can help achieve non-invasive or low-invasive glucose monitoring. Liang et al. (2022) presented a bioorthogonal light-up fluorescent probe consisting of two reagents, Glu-HT-Me+AzGlu2, for rapidly responsive (within 25 min), highly specific and sensitive (20-fold enhancement) detection of live-cell glucose uptake based on aryl phosphine-induced a-PET effect and Staudinger ligation (Fig. 5.10). Glucose-responsive polymers or nanomaterials control the release of insulin and can be used to treat diabetes. They automatically release insulin according to the glucose level in the body to maintain blood sugar levels within the normal range. Chen et al. (2018) designed a Zeolitic Zn^{2+}-imidazolate cross-linked framework nanoparticles, ZIF-8 NMOFs, used as smart glucose-responsive carriers for the controlled release of drugs. In the presence of glucose, the nanoparticles are unlocked, resulting in the release of insulin. Nanoparticles containing components on the surface or inside that can respond to changes in glucose concentration can be used for drug delivery, controlling the release of drugs based on changes in glucose levels, thereby achieving more precise

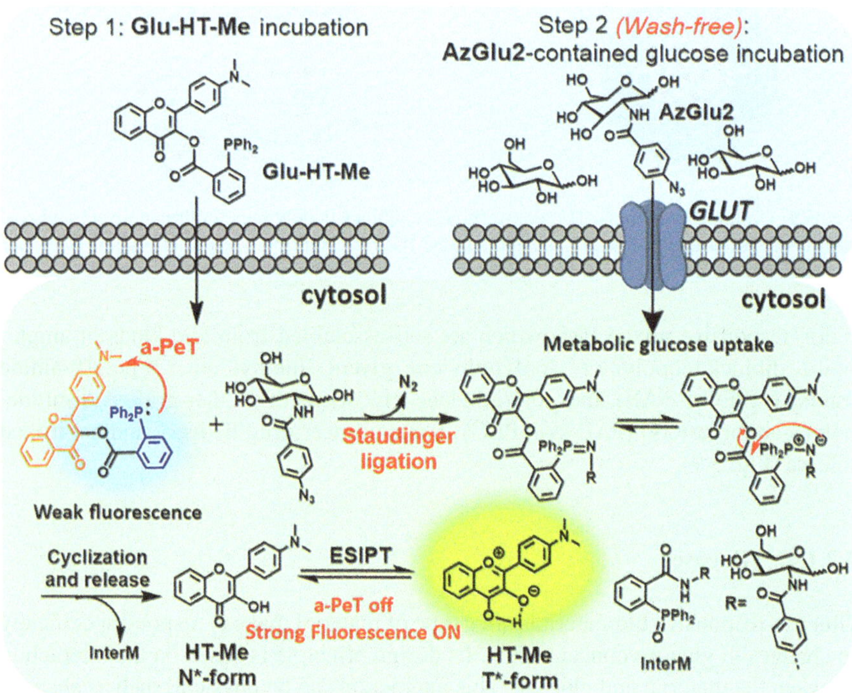

Fig. 5.10 A schematic illustration of light-up fluorescent probe presented in this work based on bioorthogonal Staudinger ligation for detection and imaging of cellular glucose uptake as well as chemical structures of Glu-HT-Me, AzGlu2, and HT-Me. (Reprinted with permission from Liang et al. 2022)

treatment. Biosensors usually contain biological recognition elements (such as enzymes or cells) that undergo specific biological reactions with glucose and can sense the presence of glucose and generate specific signal responses. They can be used to study glucose metabolism, monitor diabetes, and develop drugs.

5.2.1.5 Gas (Oxygen, CO_2, H_2S, and N_2)

Gas-responsive biomaterials sense changes in concentration of gases (including oxygen, CO_2, H_2S, N_2, etc.) in the body environment and respond accordingly. The design of these materials is often based on changes in their physical, chemical, or electrical properties upon gas exposure. Xu et al. (2022) reported hypoxia-triggered platinum supernanoparticles dissociated into ultrasmall platinum nanoclusters, which could be cleared by the kidney through urine, in response to the tumor microenvironment. After collection of the excreted nanosensors, the disease status was monitored visually and quantitatively by using a microfluidic chip. In the biomedical field, some gas-level-responsive materials may exploit the selective interaction of biomolecules with specific gases. The structure of proteins, enzymes, or other biomolecules may be affected by changes in gas concentration, enabling sensitive detection of gases.

5.2.1.6 ATP

ATP-responsive biomaterials usually refer to materials that can respond to changes in ATP concentration. ATP is an important biological molecule in cells, and changes in its concentration can reflect the physiological state, metabolic activities of cells, and the occurrence of many diseases. ATP is the main energy reserve molecule in cells, and changes in its concentration are often closely related to biological processes such as cellular metabolic activity, vitality, and cell death. ATP-responsive materials are designed to undergo measurable physical or chemical changes in the presence of ATP, such as changes in optical properties, electrical properties, magnetic properties, etc. This is a class of materials that can recognize, capture, or convert changes in ATP concentration into measurable signals. Jiang et al. (2022) developed a self-assembled three-dimensional (3D) DNA nanostructure drug delivery system with adenosine-5′-triphosphate (ATP)-triggered drug release for tumor fluorescence imaging analysis and targeted drug delivery (Fig. 5.11). These materials typically have high selectivity and sensitivity, enabling the detection of minute changes in ATP concentration in complex biological systems.

5.2.1.7 Nucleic Acids (DNA and RNA)

Nucleic acid-responsive biomaterials are a class of materials that can respond to the presence or changes in specific sequences of nucleic acid molecules (such as DNA and RNA). These materials are often designed to detect and transmit signals to

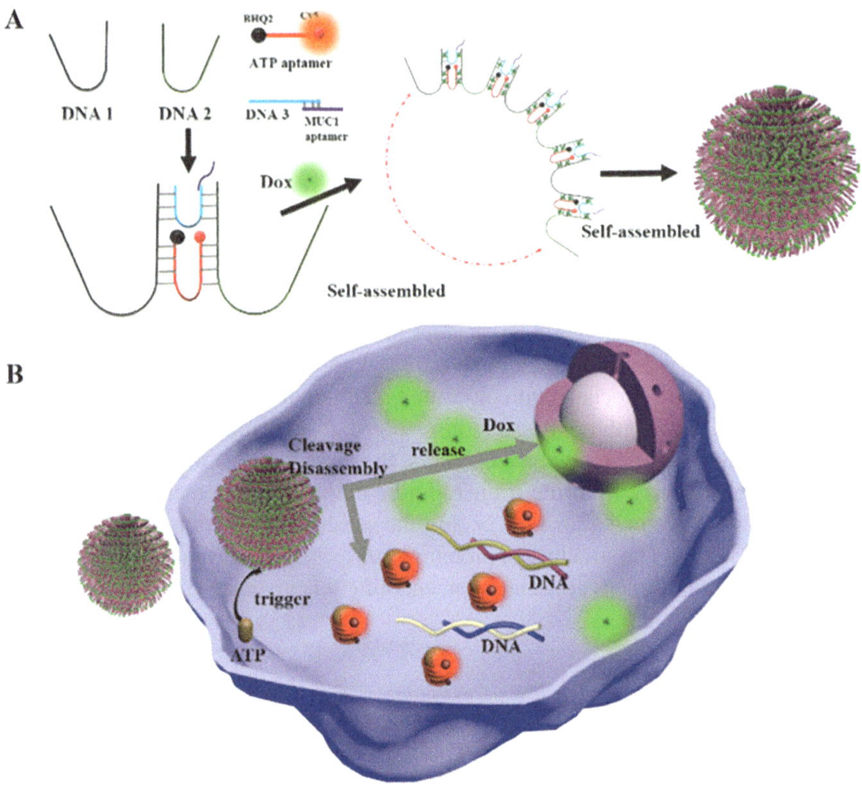

Fig. 5.11 (**a**) Design and construction of the Dox@3D DNA nanostructure drug delivery system-based aptamer and (**b**) ATP-triggered drug release for tumor fluorescence imaging analysis and targeted drug delivery. (Reprinted with permission from Jiang et al. 2022)

achieve biological or medical purposes based on changes in nucleic acids within cells. Nucleic acid-responsive biomaterials are based on the principle of specific affinity binding of nucleic acids. These materials interact with nucleic acids through specific structures, sequences, or chemical properties. The design of the material allows for structural changes upon binding to nucleic acids that may result in changes in the optical, electrical, magnetic, or fluorescent properties. Yin et al. (2022) designed chimeric DNA/RNA hairpins (displacers) that can be destabilized by activated CRISPR-Cas12a in the presence of target DNA as an ultrabright Raman reporter, and its disassembly leads to a drastic decrease of SERS intensity as signal readouts. In gene therapy, nucleic acid-responsive biomaterials can be used to regulate gene expression and achieve precise control of cell functions. Wu et al. (2021) reported a facile strategy for the construction of target mRNA (scaffold) and antisense (staple strands) co-assembled RNA/DNA hybrid "origami" elicited a noticeable inhibition of cell proliferation based on the silencing of the tumor-associated gene polo-like kinase 1 (PLK1) and provided a novel strategy for the further development of gene therapy.

5.2.2 *External Stimuli Sources*

In addition to internal stimuli sources, external stimuli sources such as light, magnetism, temperature, ultrasound, etc. also provide unique means for triggering stimuli-responsive biomaterials to produce biological effects. The advantage of these methods is that they can achieve targeted and precise control, thereby improving treatment effectiveness and reducing side effects, providing new possibilities for the development of personalized therapy and precision medicine.

5.2.2.1 Light

Light-responsive biomaterials are a class of materials that produce specific biological effects on light stimuli. These biomaterials produce measurable effects by absorbing, reflecting, conducting or converting light energy. Common light-sensitive materials include photosensitizer, photothermal materials, etc. Among them, photosensitizers are a type of substances that produce ROS when exposed to light, which can be divided into organic photosensitizers (such as porphyrins, phthalocyanines), metal complexes, metal-organic frameworks (MOFs), and so on. Nowadays, photodynamic therapies (PDT) mediated by photosensitizers have been developed and increasingly widely used in the treatment of malignant tumors (Zhang et al. 2023), bacterial infections (Zhang et al. 2021), and arteriosclerosis (Ryabova et al. 2019). Zhang et al. (2018) synthesized a metal-organic framework nanoshuttle (Zr-FeP MOF) that can generate hydroxyl radical (·OH) and singlet oxygen ($^1O_2$) under near-infrared (NIR) irradiation, while also having significant photothermal effects and showing excellent anti-tumor effects in vivo and in vitro. In addition to photosensitizers capable of producing ROS, photothermal materials with photothermal conversion capabilities are also important light-responsive materials. Photothermal materials can absorb light energy and convert it into thermal energy. In biomedicine, photothermal materials are used in photothermal therapy (PTT), which is the treatment of tumors or other diseases by local heating. Gold nanoparticles are common photothermal materials that can be excited to generate thermal effects (Kim et al. 2023). Furthermore, some light-responsive nanoparticles, such as light-sensitive polymers, hydrogels, and proteins, can undergo structural changes when stimulated by light. Smart drug delivery systems designed based on these responsive biomaterials are able to effectively avoid drug degradation and achieve spatio-temporal regulation of drugs in the body. The photo-cleavable low-molecular-weight hydrogelator (LMWG) constructed by Liu et al. can undergo a gel-sol transition when exposed to light, thereby releasing drugs (Liu et al. 2019).

5.2.2.2 Magnetism

Magnetic-responsive biomaterials are those that undergo physical or chemical changes under the influence of an applied magnetic field. In the biomedical field, magnetic-responsive biomaterials are often designed for use in diagnosis, therapy, or drug delivery. These materials are controlled by an external magnetic field to achieve target localization in vivo, control release, image recognition, and other functions. Magnetic-response biomaterials can be classified by use as contrast agents for magnetic resonance imaging (MRI), materials for magnetic hyperthermia, materials for magnetic navigation and directional delivery, and materials for tissue engineering. MRI contrast agents include superparamagnetic materials that exhibit high magnetization in the presence of an external magnetic field but lose their magnetism when the field is removed, and can be used to improve the contrast of MRI images (Hu et al. 2019b). To give MRI contrast agents specific tumor targeting ability and extended circulating time in vivo, Yin et al. (2019) prepared USPIOs@F127-WSG particles by modifying ultrasmall superparamagnetic iron oxides with peptides. USPIOs@F127-WSG particles specifically aggregate in the tumor region, resulting in T_2-weighted MRI contrast enhancement (Fig. 5.12). In magnetic hyperthermia, iron oxide nanoparticles or other magnetic materials are usually used. They generate heat under the action of magnetic fields, which can be used to mediate tumor thermal damage or treat other diseases (Perigo et al. 2015).

Magnetic-responsive biomaterials can also be aggregated to target sites under magnetic field guidance for targeted drug delivery in the body or image-guided therapy (Kim et al. 2019). Additionally, in tissue engineering, scaffold materials with magnetic properties can be used to guide cell growth and orient tissue formation. For example, ECM/RSF/MNP scaffold doped with magnetic nanoparticles effectively promotes cell migration, osteogenic differentiation, endothelial cell regeneration, and new bone formation in a critical-size femur defect rat model under the action of external static magnetic field, providing an effective strategy for bone regeneration (Liang et al. 2023).

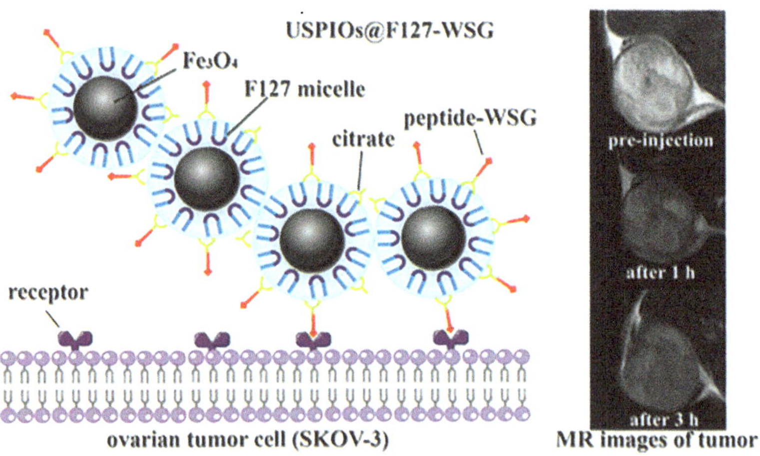

Fig. 5.12 USPIOs@F127-WSG nanoparticles for tumor imaging by MRI. (Reprinted with permission from Yin et al. 2019)

5.2.2.3 Temperature

Responsive biomaterials that are sensitive to temperature and tend to exhibit reversible or gradually changing properties, and they can be designed for specific application requirements and target temperature ranges. Such biomaterials exploit temperature changes in an organism's body or external environment for a wide range of functions, such as drug delivery, diagnostics, thermal therapy, and more. Some common examples of temperature-responsive biomaterials include thermosensitive polymers, thermosensitive liposomes, metal nanomaterials, thermosensitive hydrogels, etc., which produce a thermal response over a specific temperature range to achieve the desired function. Thermosensitive polymers and thermosensitive liposomes undergo phase transitions at specific temperatures that change their structure and properties, and such structural changes can be used to design temperature-sensitive drug delivery systems to control drug release. Regenold et al. (2023) synthesized ThermoVRL by encapsulating the commonly used chemotherapy drug vinorelbine in thermosensitive liposomes, and ThermoVRL is able to release vinorelbine via heat triggers. Compared to free drug delivery, this approach can effectively enhance the delivery of chemotherapy drugs to the tumor site and improve the targeting of chemotherapy drugs. Additionally, thermosensitive hydrogels can reversibly swell or contract in response to temperature changes, and thus can also be used for drug delivery (Fig. 5.13) (Zhou et al. 2020). The thermosensitive PDA NP-Cip/GC hydrogel can act as a drug reservoir, releasing antibiotics that effectively destroy bacteria due to local high temperatures induced by NIR light (Gao et al. 2019). In addition to targeted drug delivery, metal nanoparticles, such as gold and silver, are able to generate thermal effects when excited by an external heat source for thermal therapy (Yougbaré et al. 2021; Nakal-Chidiac et al. 2020).

5.2.2.4 Ultrasound

Ultrasound-responsive biomaterials are capable to provide a controllable response and are frequently used to enhance ultrasound imaging, enabling precise drug delivery and therapy. Ultrasound-responsive biomaterials typically include gas-filled microbubbles, ultrasound-sensitive nanoparticles and polymers, etc. Microbubbles are commonly used ultrasound-response biomaterials. The gas core of the microbubble is enclosed in a shell. When excited by ultrasound waves, the vibrations of the gas in the microbubble can enhance the contrast of the ultrasound imaging (Wang et al. 2023a). These ultrasound-sensitive microbubbles are widely used in ultrasound molecular imaging to improve the imaging quality of vascular systems and tissues. In addition to medical diagnosis, ultrasound-responsive biomaterials also play an important role in the treatment of diseases. Some ultrasound-sensitive nanoparticles can be used as sonosensitizers to produce ROS under ultrasound for disease treatment. Wu et al. (2022) prepared ZTNs as inhalable sonosensitizers for bacterial pneumonia, which produces ROS under ultrasound irradiation and thus

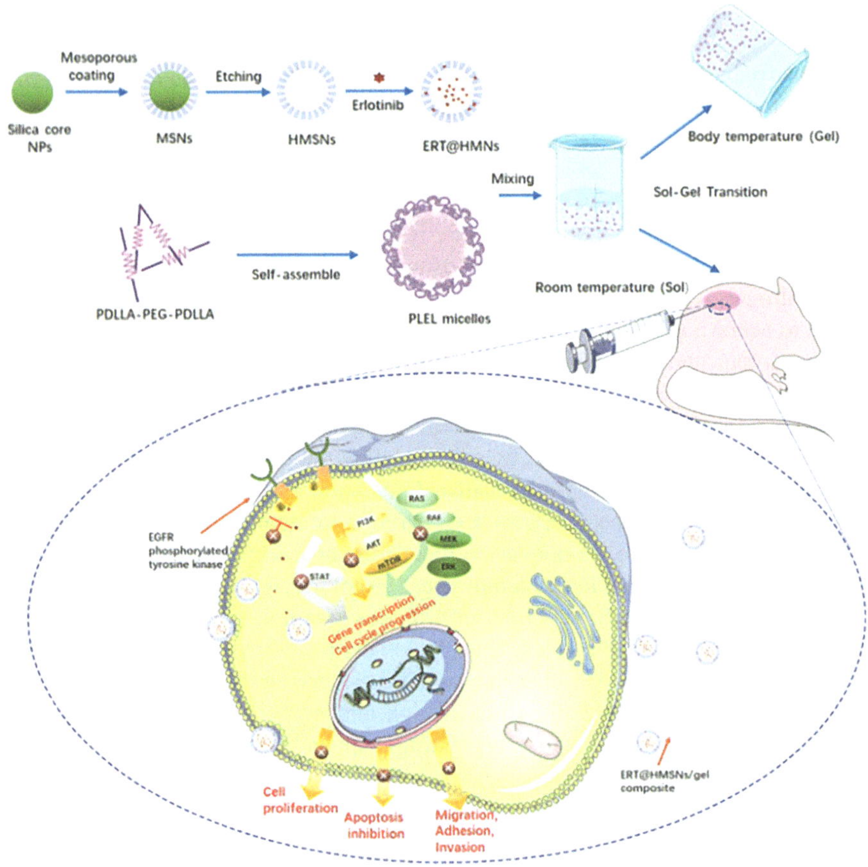

Fig. 5.13 The thermosensitive hydrogel containing erlotinib-loaded hollow mesoporous silica nanoparticles as a localized drug delivery system. (Reprinted with permission from Zhou et al. 2020)

kills Gram-negative multidrug-resistant bacteria (Fig. 5.14). Particularly, ZTNs upon US irradiation give a 100% survival rate in the severely immunodeficient mice with a lethal bacterial pneumonia. Furthermore, ultrasound-sensitive polymers change shape or volume when exposed to ultrasound, a property that could be exploited in ultrasound-driven drug delivery systems. Lin et al. (2023) synthesized an ultrasound-responsive block copolymer that can produce micelles through self-assembly behavior. These ultrasound-responsive glycopolymer micelles can undergo irreversible structural damage under high-intensity focused ultrasound to enable controlled drug release of nanoparticles.

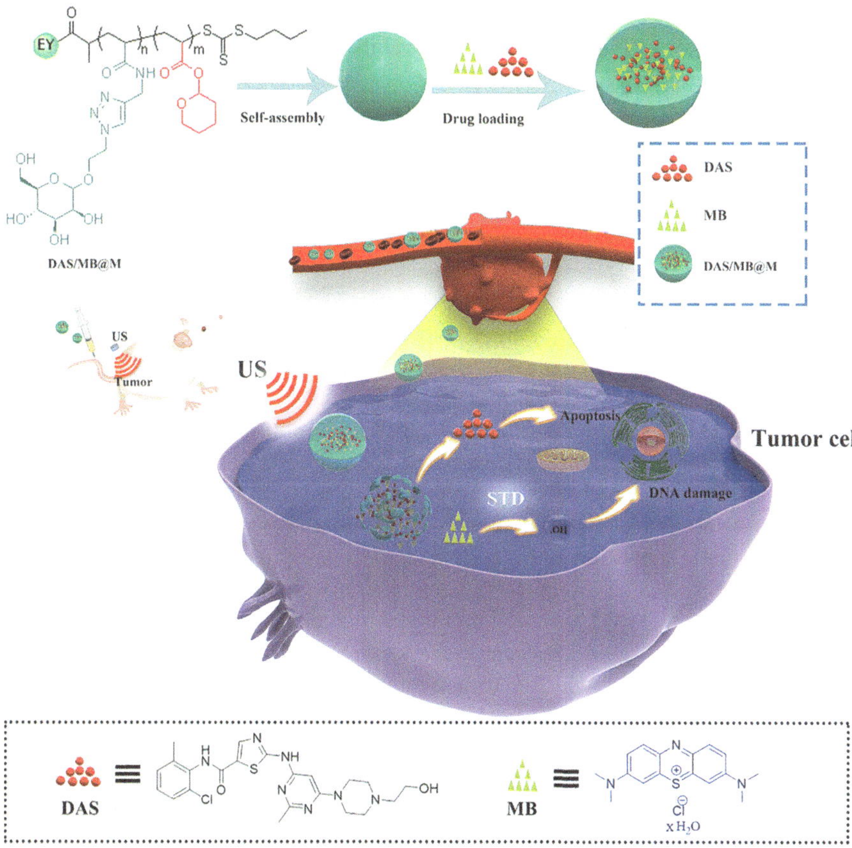

Fig. 5.14 A schematic illustration of the preparation of ultrasound-responsive glycopolymer micelles for controlled drug release. (Reused with permission from Pan et al. 2022)

5.2.2.5 Electricity

Electric fields, acting as external stimuli, can also be used to manipulate the properties of responsive biomaterials. Firstly, some biomaterials are designed to release drugs in response to electric field stimulation. Such systems could be used to enable targeted, precise drug delivery, such as the delivery of drugs via electrically-responsive implants for the treatment of localized diseases (Yasin 2016). Secondly, some electrically responsive biomaterials have electrothermal properties and can be used for localized thermal therapy, such as killing tumor cells in cancer treatment by electric-induced thermal effects (Wagstaff et al. 2016). Gu and colleagues reported that platinum nanoparticles (PtNPs) can be activated by an electric field to occur Faraday cage effect on the surface, which subsequently triggers the dissociation of water molecules to produce ·OH, thus proposing electrodynamic therapy (EDT) (Gu et al. 2019). Since then, EDT has attracted significant interest from researchers as an emerging electrotherapy method for solid tumors. In the field of neuroscience, electrically responsive biomaterials have the ability to simulate neuronal excitation, which has important applications in neuroscience research, neural repair, and brain-computer interface technologies. These materials could be used to model the growth or repair of neural tissue, as well as for the development of implantable neuroregulatory devices (Lin et al. 2020; He et al. 2020).

5.2.2.6 X-Ray

X-ray is a commonly used imaging and therapeutic tool that penetrates biological tissue and is detected to produce diagnostic images or for therapeutic purposes. At the same time, X-ray can also drive some responsive biomaterials to produce changes in physical properties, such as fluorescence emission, ionization production or energy release or chemical properties, such as color changes or charge accumulation. Scintillators are a class of materials that emit light when excited by X-rays and can be detected and used to make X-ray imaging systems (Wang et al. 2023b). More importantly, based on scintillator nanoparticles capable of down-converting X-rays to fluorescence, X-ray photodynamic therapy (X-PDT) has been derived for anti-tumor therapy. Unlike radiation therapy, which uses X-rays to kill cancer cells directly, X-PDT uses scintillator nanoparticles to convert the energy of X-rays into ultraviolet or visible light, which in turn stimulates photosensitizers to produce ROS that can damage cancer cells (Fig. 5.15) (He et al. 2022). At the same time, the X-ray dose required for X-PDT is small and it avoids damage to healthy tissue (Wang et al. 2022). In terms of imaging, some composites can be used as contrast agents for multimodal imaging, and effective T_1/T_2-weighted MRI and X-ray computed tomography multimodal imaging can be performed simultaneously (Dong et al. 2017a). In addition, X-ray-responsive biomaterials can also monitor radiation

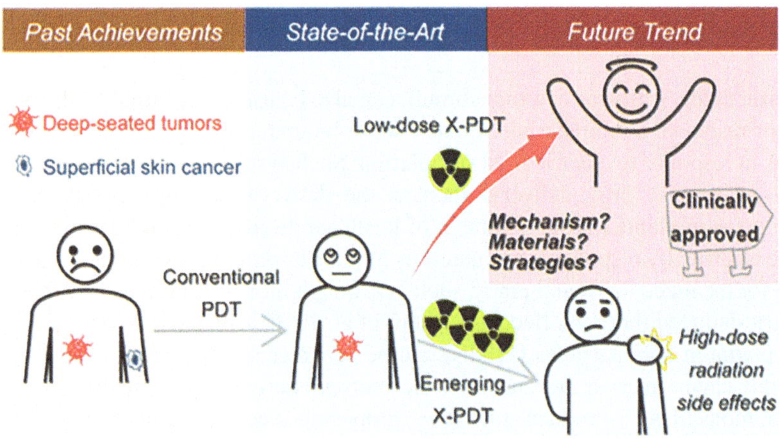

Fig. 5.15 A schematic illustration of X-PDT. (Reprinted with permission from He et al. 2022)

dose through discoloration, ensuring that the radiation dose received by the patient is within a safe range, which is essential for accurate dose control in radiation therapy (Yang et al. 2022).

5.2.2.7 Microwaves

Microwave-responsive biomaterials can interact with microwave radiation and produce specific biological effects. These biomaterials exhibit some special properties in the microwave frequency range, giving them important functions in biomedical applications. Dielectric materials have specific dielectric constant and dielectric loss, good absorption and conduction ability to microwave radiation, and can be used for microwave imaging or biosensing (Xiao et al. 2022; Shi et al. 2020). As a special kind of microwave-response biomaterials, microwave-absorbing materials can effectively absorb microwave radiation and convert it into heat energy. Such materials are useful in microwave hyperthermia and can be used to treat tumors or kill bacteria. For example, a nano-immunomodulator (BMCPH) proposed by Li et al. is able to generate heat under microwave irradiation to mediate tumor ablation while reversing immunosuppression and reactivating anti-tumor immune effects to enhance microwave hyperthermia (Fig. 5.16) (Li et al. 2023b). Another ionic liquid-based hydrogel exhibits favorable microwave thermal conversion properties and facilitates transdermal drug delivery, demonstrating effective antimicrobial properties and facilitating rapid healing of infected wounds (Gao et al. 2022). In recent years, researchers have increasingly focused on the development of versatile microwave-response biomaterials to meet the needs of different biomedical applications. Some multifunctional microwave-sensitive materials can produce both thermal effects and ROS in response to microwave stimuli, combining microwave hyperthermia and microwave dynamic therapy for maximum therapeutic benefit (Qin et al. 2023).

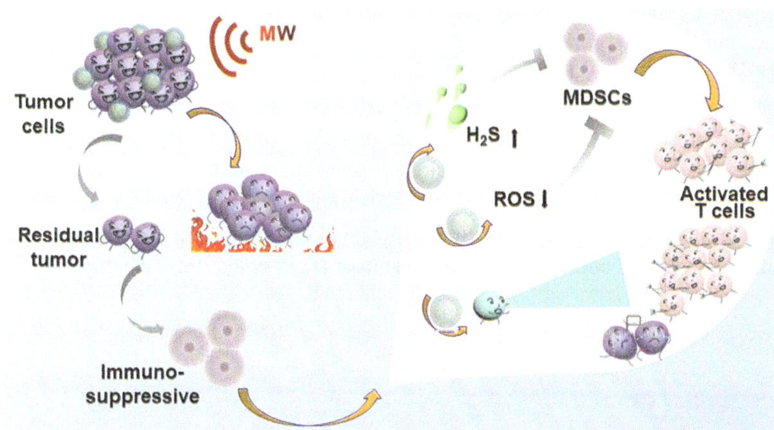

Fig. 5.16 A schematic diagram of BMCPH for mediating tumor microwave thermal therapy. (Reprinted with permission from Li et al. 2023b)

5.2.2.8 Pressure

Pressure-response biomaterials are a class of materials that sense, measure, and transmit external pressure changes, and play an important role in medical diagnosis, monitoring, and treatment. The capacitance or resistance values of pressure-responsive biomaterials can change under the influence of external pressure. Such materials tend to be able to respond to small deformations with high sensitivity. They can be integrated with sensors to monitor pressure changes in internal organs, such as pressure in the heart, arteries or bladder, or to monitor physiological parameters such as posture and breathing, providing vital data for disease diagnosis and treatment (Chen and Yan 2020). In addition, piezoelectric materials, as another important pressure-responsive biomaterial, produce changes in charge or potential when subjected to mechanical pressure or stress. Typical piezoelectric materials include barium titanate, titanium oxide (TiO_2), polyvinylidene fluoride (PVDF), etc., which can be used to manufacture pressure sensors. Recently, piezodynamic therapy (PZDT) based on piezoelectric nanoparticles has been pioneered for the treatment of diseases. In PZDT, piezoelectric materials can be wirelessly activated by ultrasound-based external pressure to promote electron and hole separation, which triggers a series of redox reactions to produce ROS (Wang et al. 2023c).

5.2.2.9 Humidity

Humidity-responsive biomaterials play an important role in the biomedical field, and their unique properties make them key components in medical devices, sensors, and implant materials. These materials exhibit measurable or observable responses to changes in ambient humidity, which can be achieved through a variety of mechanisms. A common class of humidity-responsive materials are hydrogels, such as polyethylene glycol (PEG), polyvinyl alcohol (PVA), and polyacrylamide (PAM). These materials expand after absorbing water, and this hygroscopic expansion property makes them suitable for the manufacture of humidity sensors (Jung et al. 2022). Several studies have reported the direct use of hydrogel-based humidity sensors in practical applications such as smart masks, electronic skin, breath analysis, and wireless monitoring, demonstrating their great potential for applications in healthcare and other fields (Luo et al. 2023). Another class of humidity-responsive biomaterials is based on changes in electrical properties. Conductive materials such as polypyrrole (PPy) and graphene oxide (GO) may exhibit changes in electrical conductivity under changes in humidity, which makes them ideal for the manufacture of humidity sensors and electrochemical sensors (Lei et al. 2022; Zhang et al. 2008).

5.3 Biomedical Applications

5.3.1 Drug Delivery

Drug delivery refers to the effective delivery of drugs to specific cells or tissues to improve therapeutic effectiveness and reduce adverse reactions. The functional groups and wettability of drugs are important considerations in the design of drug delivery vectors, as they affect drug-carrier interactions and, ultimately, drug delivery efficiency. In addition, the size and shape of drug molecules also play a role in determining a suitable carrier. In general, selecting the right carrier for a drug molecule requires consideration of a variety of factors, including the size, shape, functional group, and wettability of the drug molecule, as well as the nature of the carrier itself. Stimuli-responsive materials can precisely control the drug release in response to various stimuli and have attracted widespread attention in the field of drug delivery. Among them, external stimuli such as temperature, pH, and light are common regulatory factors. Temperature-responsive materials are a class of materials that adjust their structure and properties according to changes in ambient temperature. In drug delivery, this material can achieve targeted release of drugs by responding to local temperature changes in the body. For example, for tumor treatment, temperature-sensitive nanomaterials can apply heat therapy locally to increase the concentration of drugs within the tumor, thereby enhancing the therapeutic effect. Han et al. (2020) proposed a new method for lipophilic drug release based on dual pH-responsive hydrogel actuators (Fig. 5.17). The drug was encapsulated within it and could be released via a dual pH-controlled capsule switch. Photosensitive

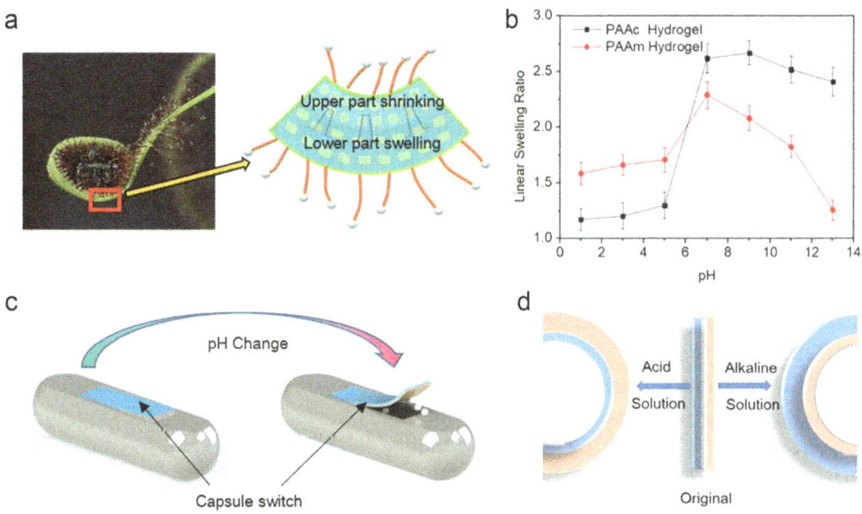

Fig. 5.17 Design of the capsule switch based on a dual pH-responsive hydrogel actuator. (Reprinted with permission from Han et al. 2020)

materials can respond to changes in light and are often used in photodynamic therapy. This material releases drugs through light activation, achieving highly targeted treatment of disease lesions while minimizing the impact on surrounding normal tissues. Boruah et al. (2023) developed a nanocomposite system synthesized from phosphatidylcholine vesicles (liposomes) and azobenzene nanoclusters as a light-responsive drug delivery vehicle. This complex exhibited sustained drug release in the presence of ultraviolet light, with a release amount of up to 77.33%. Overall, stimuli-responsive materials have shown great potential advantages in drug delivery, providing new ways to improve therapeutic effects and reduce side effects. However, its practical application still needs to face a series of challenges, including the biocompatibility, stability, and preparation complexity of the material, which require more in-depth research and optimization.

5.3.2 Tissue Engineering

Tissue engineering is a comprehensive science and engineering discipline that aims to use the principles and methods of biology, engineering, and materials science to design, construct, and maintain artificial tissues and organs with specific physiological functions. The goal of this field is to develop methods that can replace or repair damaged tissues and organs, providing new avenues for disease treatment and tissue repair. In tissue engineering, stimuli-responsive materials have important applications. These materials respond reversibly to internal or external stimuli, providing precise control and regulation methods for tissue engineering.

Stimuli-responsive materials can be used as cell carriers, designing and constructing carriers that can be used to carry, support, and guide cell growth to simulate and promote tissue regeneration. Such materials can provide supportive structures, delivery of bioactive molecules, and platforms for cell interaction. These materials can adjust their physical properties under temperature changes or other stimuli in the body to provide a suitable microenvironment for cell attachment, growth, and differentiation. Developments in this field not only provide new therapeutic avenues for treating various tissue defects and injuries, but also provide unique tools and insights for biomedical research, drug development, and basic scientific research. In bone tissue engineering, carriers made of bioceramics and biodegradable polymers can serve as bone substitutes, provide support for bone cell attachment and growth, and gradually degrade into biological tissues. Huang et al. (2022b) synthesized an injectable methylcellulose/polyvinyl alcohol/polyvinylpyrrolidone-based thermo-sensitive hydrogel, which delivered kartogenin-loaded mesoporous bioactive glass nanoparticles, revealed phase transition temperatures of 35 °C and its ability to induce chondrogenesis and osteogenesis differentiation of tendon-derived stem cells.

In soft tissue engineering, such as skin, muscle, etc., cell carriers can provide supportive structures and promote repair and regeneration of damaged areas by carrying growth factors and cells. In cardiovascular tissue engineering, vectors can be used to construct artificial blood vessels and cardiac tissue, providing an effective way to repair the cardiovascular system. Stimuli-responsive biomaterials can be combined with responses to specific heart conditions or even cardiovascular disease to enhance the delivery of targeted therapeutic components to areas of need. Stimuli-responsive biomaterials are attractive vehicles for delivering stem cells and drugs to treat heart disease. Many animal studies have demonstrated the functionality of stimuli-responsive biomaterials in improving local cell or drug retention in the injured heart to promote cardiac regeneration. Garbern et al. (2011) reported a pH- and temperature-responsive injectable hydrogel provided sustained, local delivery of bFGF, improved angiogenesis, and achieved therapeutic effects in regional blood flow and cardiac function. This polymer was a liquid at pH 7.4 and 37 °C but formed a physical gel at pH = 6.8 and 37 °C. Stimuli-responsive biomaterials also play a key role in neural tissue engineering, promoting nerve regeneration and repair by providing supportive structures and guiding the growth of nerve cells. In this regard, the combination of biodegradable polymers and bioceramics is often used to prepare carriers for neural repair that can provide appropriate mechanical properties and delivery of bioactive molecules to promote neuronal regeneration and connection. Gao et al. (2023) developed a pro-regeneration system for olfactory ensheathing cells, namely miR-26a@SPIONs-OECs, a preferred choice for promoting nerve regeneration. These entities show high responsiveness to external magnetic fields (MF), and exhibit a tendency to migrate and orientate along the direction of the MF, thereby potentially facilitating neuronal reconnection through directional neurite elongation. miR-26a-enriched EVs from miR-26a@SPIONs-OECs can interact with host astrocytes, thereby diminishing inhibitory cues for neurite growth (Fig. 5.18). By integrating gene carriers into stimuli-responsive materials, gene delivery and expression can be achieved, providing a new means of treating genetic diseases, cancer and other diseases. Sato et al. (2019) developed a pH-sensitive cationic lipid and found hydrophilic head significantly affected the apparent pKa of the final product, a key factor in both intrahepatic distribution and endosomal escape. Its role in gene silencing is more effective than Onpatto, a drug approved by the US FDA for the treatment of hereditary epithymic amyloidosis polyneuropathy in adults. This provides strong support for the development of personalized treatment and precision medicine. Bioprinting is a technology that prints cells, cell-carrying materials, and bioactive substances into three-dimensional tissue engineering structures in a predetermined structure. Stimuli-responsive materials can be used in the bioprinting process to achieve structural adjustability and in vivo biomimetic behavior through external stimulation. Dong et al. (2017b) reported the F/AT-PEI scaffolds can be considered as candidates for application in electrical stimulation of tissue regeneration such as repair of muscle and cardiac nerve tissue.

Through these applications, stimuli-responsive materials provide a higher degree of precision and controllability for tissue engineering, which is expected to improve the efficiency and success rate of engineered preparation of artificial tissues and organs.

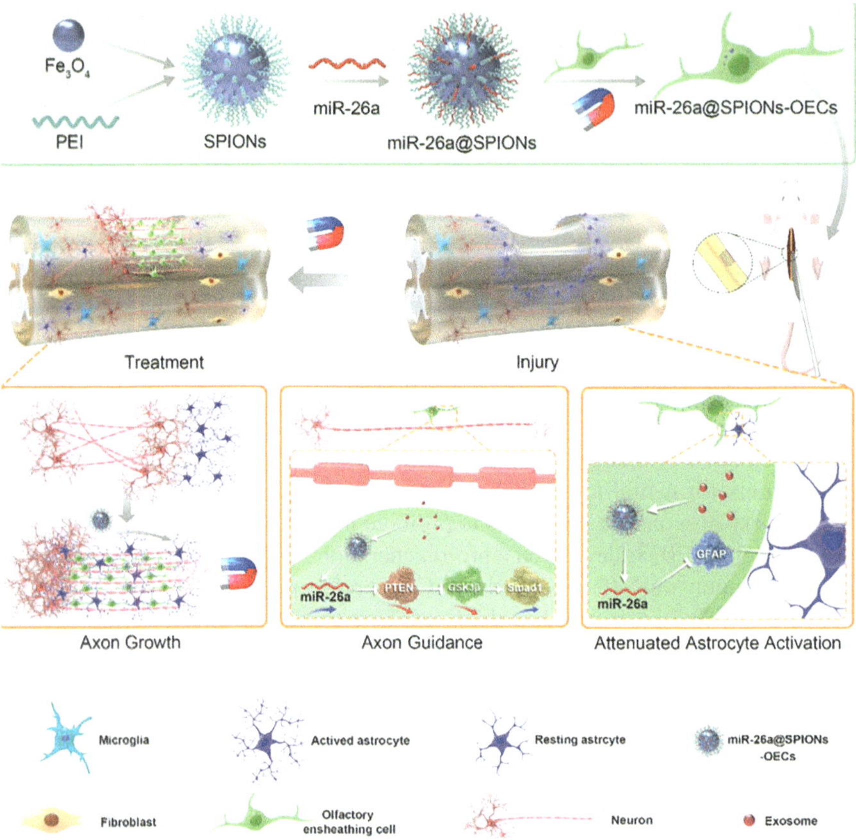

Fig. 5.18 A schematic representation of miR-26a@SPIONs-OECs for spinal cord injury repair. (Reprinted with permission from Gao et al. (2023)

5.3.3 Biosensing

Biosensors are an important class of technical tools in the biomedical field. They are designed to detect and measure biomolecules, physiological parameters, or cellular activities in organisms or the surrounding environment, and convert this information into readable signals. The core goal of these sensors is to achieve real-time and accurate monitoring of the status of living organisms, thereby providing key information for medical diagnosis, treatment, and research. The design of biosensors integrates knowledge from multiple disciplines such as physics, chemistry, biology, and engineering to meet the needs of different fields. In the structure of biosensors, biometric components are crucial components. These elements are usually highly specific biomolecules, such as enzymes, antibodies, nucleic acids, etc., which can interact with specific biomolecules or cells. Selection of appropriate biometric components is critical to ensure sensor sensitivity and specificity.

The applications of biosensors in the biomedical field are manifold. The first consideration in the design of stimuli-responsive materials is their responsiveness to stimuli from the organism's internal or external environment. These stimuli can be changes in physiological parameters, such as temperature, pH, ion concentration, etc., or they can be specific molecules or signals produced by organisms, such as biomarkers, cell secretions, etc. stimuli-responsive materials undergo measurable physical or chemical changes when exposed to these stimuli, which are captured and recorded by sensors. In the application of biosensors, the selection of stimuli-responsive materials is a critical step. These materials not only need to be highly sensitive to specific stimuli, but also need to be biocompatible to ensure that they do not cause uncomfortable reactions in organisms during implantation. Common stimuli-responsive materials include hydrogels, polymers, nanomaterials, etc. Hydrogels are often used to prepare stimuli-responsive biosensors due to their highly tunable hydration properties and biocompatibility.

In medical diagnosis, biosensors are widely used to measure biochemical indicators in blood, such as blood sugar, cholesterol, blood oxygen, etc., providing important help for the early diagnosis of diseases. For example, the continuous glucose monitoring system can monitor patients' blood sugar levels in real time by implanting subcutaneous sensors, providing diabetic patients with a more convenient and accurate blood sugar management method. Sawayama et al. (2020) described a hydrogel glucose sensor with in vivo stability based on boronic acid-based fluorescence intensity, integrating two antioxidant enzymes (Fig. 5.19). These protected the aryl boronic acid from being degraded by hydrogen peroxide in vitro, preserved the boronic acid-based fluorescence intensity of the hydrogel glucose sensors in rats for 28 days, and also allowed the hydrogel glucose sensor attached to a homemade semi-implantable CGM device to trace blood glucose concentrations in rats for 5 h with the accuracy required for clinical settings. Similarly, biosensors have also played a key role in the early diagnosis of cancer, by detecting tumor markers or extracellular vesicles in the blood, enabling early monitoring of patients' health

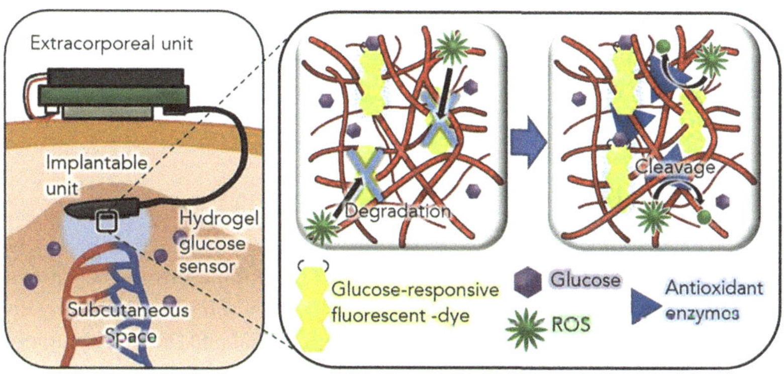

Fig. 5.19 A schematic illustration of the in vivo stable fluorescent-based hydrogel glucose sensor. (Reprinted with permission from Sawayama et al. (2020)

status, and providing strong support for timely intervention and treatment. Heydari-Bafrooei et al. reported a densely packed gold nanoparticles used as the basis for an ultrasensitive label-free electrochemical aptasensor to detect the biomarker prostate-specific antigen (PSA) in serum (Heydari-Bafrooei and Shamszadeh 2017).

On the therapeutic side, biosensors provide support for personalized treatment. Biosensors in drug delivery systems can monitor the drug concentration in the patient's body and achieve precise control of the drug treatment process. This not only improves the effectiveness of the treatment but also reduces the risk of adverse reactions and side effects that patients may face. In addition, biosensors also play a positive role in the management of chronic diseases. For example, patients with heart disease can monitor heart function through implanted heart sensors, detect abnormalities in time, and take corresponding treatment measures. Shen et al. (2023) developed piezoresistive conductive polymer composites (CPCs)-based strain sensor which can accurately monitor the pulse rate and respiration rate under different sports states and the simulated breathing modes and undoubtedly provide strong guarantee for the real-time health monitoring during the treadmill exercise and assessment of cardiac status.

Biosensors also provide powerful tools for research in fields such as biology, medicine, and pharmacology. Researchers use biosensors to monitor biological processes such as cellular activities and molecular interactions in real time and gain an in-depth understanding of the basic mechanisms of life. Optical sensors can be used to study protein folding dynamics, Sheahan et al. (2016) molecularly engineered a bioluminescence-based biosensor composed of rationally split firefly luciferase reporter fragments flanking a test protein, and used this in a protein-fragment complementation assay to quantitatively measure folding of the test protein.

Electrochemical sensors can be used to explore dynamic changes in intracellular ion concentrations. Turhan et al. (2023) developed novel fluorescent sensors based on coumarin-hydrazide-hydrazone hybrid for the detection of CN^-, Co^{2+}, and Ni^{2+} ions. The sensors showed recognition properties of CN^-, Co^{2+}, and Ni^{2+} ions. After interaction with the cyanide anion, the coumarin-hydrazide-hydrazone sensors exhibited a color change visible to the naked eye, from yellow to colorless under daylight and from blue to green under UV light. These studies not only help reveal the pathogenesis of the disease, but also provide important reference for the development of new drugs.

5.3.4 Bioimaging

Stimuli-responsive biomaterials play a key role in bioimaging in the biomedical field. In bioimaging, these materials usually produce measurable signals by changing specific signals or environmental stimuli inside or outside the organism, providing researchers with important information about the structure, function, and disease status of the organism. The design and application of stimuli-responsive materials cover a wide range of fields, including optical imaging, magnetic resonance imaging, ultrasound imaging, and other imaging technologies.

In terms of optical imaging, stimuli-responsive fluorescent probes are a common application, producing visible or near-infrared light signals through excitation by external light sources for labeling and tracking biomolecules, cells, and tissues. The luminescent properties of these probes can be regulated by environmental changes in organisms, such as changes in pH and temperature, allowing biological imaging to more accurately reflect the microenvironment in organisms. For example, some fluorescent probes are sensitive to pH changes in biological tissues and can be used to monitor the degree of acidification in the tumor microenvironment, Li et al. (2017) synthesized a rhodamine-based lysosomal pH probe, the probe exhibited favorable "on-off" reversibility between pH 3.25 and 7.20 for at least five cycles and could be employed to distinguish cancer cells from normal cells on the basis of different fluorescence response. This is important for cancer research and the assessment of treatment response. Its luminescent properties can be changed according to changes in temperature, which has potential applications in research such as hyperthermia, and can also be used to monitor temperature changes inside living organisms. Shen et al. (2021) reported on a thermosensitive rhodamine B (RhB)-derived fluorogenic probe that enables fluorescent labeling of cell mitochondria at concentrations as low as 1 μM and demonstrated that this probe exhibits a temperature-dependent response in cell mitochondria (Fig. 5.20).

In the field of magnetic resonance imaging, the application of stimuli-responsive biomaterials is mainly reflected in the design of contrast agents. Some magnetic nanoparticles respond to external magnetic fields by changing their magnetic properties to enhance or weaken the contrast of magnetic resonance images, thereby enabling high-resolution imaging of biological tissues. The stimuli-responsive nature of this contrast agent makes magnetic resonance imaging more sensitive and accurate in detecting lesions and physiological states in organisms. Some magnetic nanoparticles can be used as contrast agents in magnetic resonance imaging to enhance image contrast through the regulation of external magnetic fields, and are used to detect and locate disease foci, such as tumors or inflammation.

In terms of ultrasound imaging, the application of stimuli-responsive biomaterials is mainly reflected in the research on ultrasound contrast agents. Some microbubble-type contrast agents can change their morphology or acoustic properties when stimulated by ultrasound, thereby generating contrast signals and are widely used in imaging of cardiovascular, liver, and other organs. This stimuli-responsive characteristic enables ultrasound imaging to make significant progress in real-time monitoring and quantitative analysis in living organisms. Microbubbles respond to ultrasound waves, change their acoustic properties, and are used to enhance the contrast of ultrasound images. This has broad applications in cardiovascular and oncology imaging. Li et al. (2022) prepared folate-targeted nanodroplets and showed that the nanodroplets could enhance tumor imaging in mouse models when they were vaporized using low-intensity focused ultrasound.

Overall, stimuli-responsive biomaterials have important definitions in bioimaging in the biomedical field. Through their sensitivity to external stimuli, these materials enable imaging technology to better adapt to microenvironmental changes inside and outside the organism, provide researchers with more accurate and comprehensive information, and promote the development of biomedical imaging

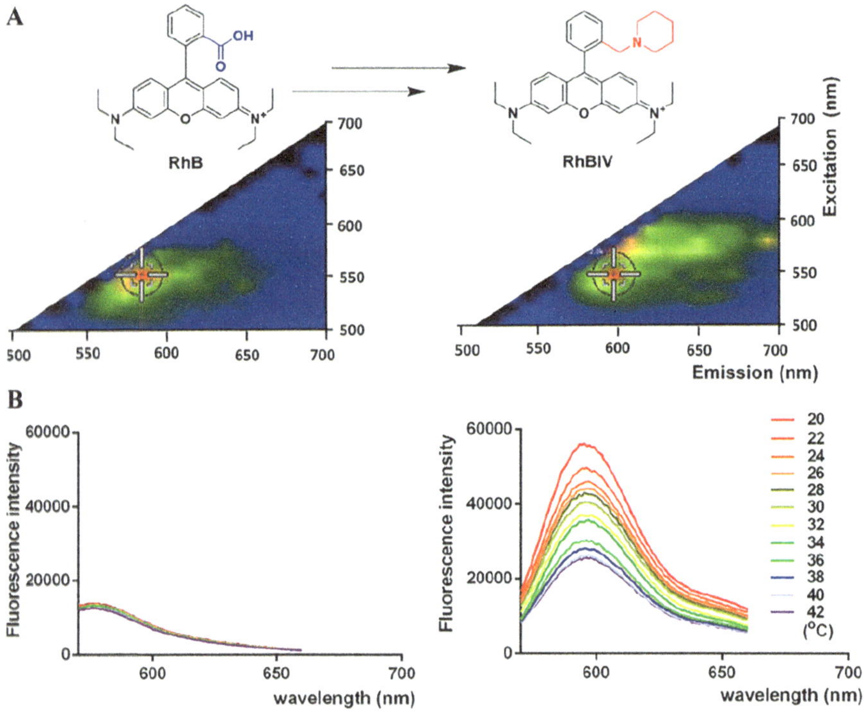

Fig. 5.20 Fluorescence spectra of RhB and RhBIV. (**a**) Reaction scheme and full-wavelength scan of RhB (10 μM) and RhBIV (10 μM). The spectra were measured with samples in phosphate-buffered saline (PBS). (**b**) Fluorescence spectra of RhB (10 μM) and RhBIV (10 μM) as a function of temperature at the optimal excitation wavelength (PBS, Ex = 520 nm). (Reprinted with permission from Shen et al. 2021)

technology. In the future, with in-depth research and innovation in stimuli-responsive biomaterials, bioimaging technology is expected to further improve its understanding of the complex structure and function of organisms, providing new possibilities for early diagnosis and treatment of diseases.

References

Akbar M, Cagli E, Erel-Göktepe I (2019) Layer-by-layer modified superparamagnetic iron oxide nanoparticles with stimuli-responsive drug release properties. Macromol Chem Phys 220(4):1800422

Bai, Y.; Liu, C.-p.; Xie, F.-y.; Ma, R.; Zhuo, L.-h.; Li, N.; Tian, W., Construction of β-cyclodextrin-based supramolecular hyperbranched polymers self-assemblies using AB2-type macromonomer and their application in the drug delivery field. Carbohydr Polym 2019**,** 213, 411

Bhole R, Gonsalves D, Murugesan G, Narasimhan MK, Srinivasan N, Dave N, Varadavenkatesan T, Vinayagam R, Govarthanan M, Selvaraj R (2023) Superparamagnetic spherical magnetite nanoparticles: synthesis, characterization and catalytic potential. Appl Nanosci 13(9):6003

Bian Q, Xue Z, Sun P, Shen K, Wang S, Jia J (2019) Visible-light-triggered supramolecular valves based on β-cyclodextrin-modified mesoporous silica nanoparticles for controlled drug release. RSC Adv 9(30):17179

Chen W, Yan X (2020) Progress in achieving high-performance piezoresistive and capacitive flexible pressure sensors: a review. J Mater Sci Technol 43:175

Chen W-H, Luo G-F, Vázquez-González M, Cazelles R, Sohn YS, Nechushtai R, Mandel Y, Willner I (2018) Glucose-responsive metal–organic-framework nanoparticles act as "smart" sense-and-treat carriers. ACS Nano 12(8):7538

Chen J, Song L, Wu Y, Zhao B, Deng J (2022) Aggregation induced emissive silicone elastomer with multiple stimuli responsiveness. ACS Appl Polymer Mater 4(6):4264

Dong L, Zhang P, Lei P, Song S, Xu X, Du K, Feng J, Zhang H (2017a) PEGylated GdF3: Fe nanoparticles as multimodal T 1/T 2-weighted MRI and X-ray CT imaging contrast agents. ACS Appl Mater Interfaces 9(24):20426

Dong SL, Han L, Du CX, Wang XY, Li LH, Wei Y (2017b) 3D printing of aniline tetramer-grafted-polyethylenimine and pluronic F127 composites for electroactive scaffolds. Macromol Rapid Commun 38(4):1600551

Du N, Ye F, Sun J, Liu K (2022) Stimuli-responsive natural proteins and their applications. Chembiochem 23(6):e202100416

Frath D, Massue J, Ulrich G, Ziessel R (2014) Luminescent materials: locking π-conjugated and heterocyclic ligands with boron (III). Angew Chem Int Ed 53(9):2290

Gao G, Jiang Y-W, Jia H-R, Wu F-G (2019) Near-infrared light-controllable on-demand antibiotics release using thermo-sensitive hydrogel-based drug reservoir for combating bacterial infection. Biomaterials 188:83

Gao Y, Hao Y, Zhang W, Wei Y, Shu Y, Wang J (2022) Microwave-triggered ionic liquid-based hydrogel dressing with excellent hyperthermia and transdermal drug delivery performance. Chem Eng J 429:131590

Gao X, Li S, Yang Y, Yang S, Yu B, Zhu Z, Ma T, Zheng Y, Wei B, Hao Y (2023) A novel magnetic responsive miR-26a@ SPIONs-OECs for spinal cord injury: triggering neural regeneration program and orienting axon guidance in inhibitory astrocytic environment. Adv Sci 10(32):2304487

Garbern JC, Minami E, Stayton PS, Murry CE (2011) Delivery of basic fibroblast growth factor with a pH-responsive, injectable hydrogel to improve angiogenesis in infarcted myocardium. Biomaterials 32(9):2407

Grønvold F, Westrum EF (1959) A-Ferric oxide: low temperature heat capacity and thermodynamic functions. J Am Chem Soc 81(8):1780

Gu T, Wang Y, Lu Y, Cheng L, Feng L, Zhang H, Li X, Han G, Liu Z (2019) Platinum nanoparticles to enable electrodynamic therapy for effective cancer treatment. Adv Mater 31(14):1806803

Han Z, Wang P, Mao G, Yin T, Zhong D, Yiming B, Hu X, Jia Z, Nian G, Qu S (2020) Dual pH-responsive hydrogel actuator for lipophilic drug delivery. ACS Appl Mater Interfaces 12(10):12010

Harada A (2001) Cyclodextrin-based molecular machines. Acc Chem Res 34(6):456

He L, Xiao Q, Zhao Y, Li J, Reddy S, Shi X, Su X, Chiu K, Ramakrishna S (2020) Engineering an injectable electroactive nanohybrid hydrogel for boosting peripheral nerve growth and myelination in combination with electrical stimulation. ACS Appl Mater Interfaces 12(47):53150

He L, Mu J, Gang O, Chen X (2021) Rationally programming nanomaterials with DNA for biomedical applications. Adv Sci 8(8):2003775

He L, Yu X, Li W (2022) Recent progress and trends in X-ray-induced photodynamic therapy with low radiation doses. ACS Nano 16(12):19691

Heydari-Bafrooei E, Shamszadeh NS (2017) Electrochemical bioassay development for ultrasensitive aptasensing of prostate specific antigen. Biosens Bioelectron 91:284

Hu X, Li F, Xia F, Guo X, Wang N, Liang L, Yang B, Fan K, Yan X, Ling D (2019a) Biodegradation-mediated enzymatic activity-tunable molybdenum oxide nanourchins for tumor-specific cascade catalytic therapy. J Am Chem Soc 142(3):1636

Hu M, Butt H-JR, Landfester K, Bannwarth MB, Wooh S, Thérien-Aubin H (2019b) Shaping the assembly of superparamagnetic nanoparticles. ACS Nano 13(3):3015
Huang Y, Morozova SM, Li T, Li S, Naguib HE, Kumacheva E (2022a) Stimulus-responsive transport properties of nanocolloidal hydrogels. Biomacromolecules 24(3):1173
Huang K, Du J, Xu J, Wu C, Chen C, Chen S, Zhu T, Jiang J, Zhao J (2022b) Tendon-bone junction healing by injectable bioactive thermo-sensitive hydrogel based on inspiration of tendon-derived stem cells. Mater Today Chem 23:100720
Jafari S, Derakhshankhah H, Alaei L, Fattahi A, Varnamkhasti BS, Saboury AA (2019) Mesoporous silica nanoparticles for therapeutic/diagnostic applications. Biomed Pharmacother 109:1100
Ji L, Griesbeck S, Marder TB (2017) Recent developments in and perspectives on three-coordinate boron materials: a bright future. Chem Sci 8(2):846
Jiang Y, Zhou H, Zhao W, Zhang S (2022) ATP-triggered drug release of self-assembled 3D DNA nanostructures for fluorescence imaging and tumor therapy. Anal Chem 94(18):6771
Jung C, Kim S-J, Jang J, Ko JH, Kim D, Ko B, Song YM, Hong S-H, Rho J (2022) Disordered-nanoparticle–based etalon for ultrafast humidity-responsive colorimetric sensors and anti-counterfeiting displays. Sci Adv 8(10):eabm8598
Kim C, Kim H, Park H, Lee KY (2019) Controlling the porous structure of alginate ferrogel for anticancer drug delivery under magnetic stimulation. Carbohydr Polym 223:115045
Kim H, Baek Y, Ha T, Choi D, Lee WJ, Cho Y, Park J, Kim S, Doh J (2023) Gold nanoparticle-carrying T cells for the combined Immuno-Photothermal therapy. Small 19(47):2301377
Lavrador P, Esteves MR, Gaspar VM, Mano JF (2021) Stimuli-responsive nanocomposite hydrogels for biomedical applications. Adv Funct Mater 31(8):2005941
Lei D, Zhang Q, Liu N, Su T, Wang L, Ren Z, Zhang Z, Su J, Gao Y (2022) Self-powered graphene oxide humidity sensor based on potentiometric humidity transduction mechanism. Adv Funct Mater 32(10):2107330
Li B, BoGe G, Wen L, Yuan Y, Zhang R, Peng X, Liu F, Sun S (2017) A lysosomal probe for monitoring of pH in living cells and ovarian tumour. Dyes Pigments 139:318
Li Z, Zhao Y, Liu H, Ren M, Wang Z, Wang X, Liu H, Feng Y, Lin Q, Wang C (2021a) pH-responsive hydrogel loaded with insulin as a bioactive dressing for enhancing diabetic wound healing. Mater Des 210:110104
Li K, Zang X, Cheng M, Chen X (2021b) Stimuli-responsive nanoparticles based on poly acrylic derivatives for tumor therapy. Int J Pharm 601:120506
Li M, Luo H, Zhang W, He K, Chen Y, Liu J, Chen J, Wang D, Hao L, Ran H (2022) Phase-shift, targeted nanoparticles for ultrasound molecular imaging by low intensity focused ultrasound irradiation [retraction]. Int J Nanomedicine 17:2751
Li Z, Wan W, Bai Z, Peng B, Wang X, Cui L, Liu Z, Lin K, Yang J, Hao J (2023a) Construction of pH-responsive nanoplatform from stable magnetic nanoparticles for targeted drug delivery and intracellular imaging. Sensors Actuators B Chem 375:132869
Li S, Xu F, Ren X, Tan L, Fu C, Wu Q, Chen Z, Ren J, Huang Z, Meng X (2023b) H2S-reactivating antitumor immune response after microwave thermal therapy for long-term tumor suppression. ACS Nano 17(19):19242
Liang Z, Pang H, Zeng G, Chen T (2022) Bioorthogonal light-up fluorescent probe enables wash-free real-time dynamic monitoring of cellular glucose uptake. Anal Chem 94(23):8293
Liang HF, Zou YP, Hu AN, Wang B, Li J, Huang L, Chen WS, Su DH, Xiao L, Xiao Y (2023) Biomimetic structural protein based magnetic responsive scaffold for enhancing bone regeneration by physical stimulation on intracellular calcium homeostasis. Adv Healthc Mater 12(32):2301724
Lin C-C, Chang J-J, Yung M-C, Huang W-C, Chen S-Y (2020) Spontaneously micropatterned silk/gelatin scaffolds with topographical, biological, and electrical stimuli for neuronal regulation. ACS Biomater Sci Eng 6(2):1144
Lin S, Zhu L, Li Z, Yue S, Wang Z, Xu Y, Zhang Y, Gao Q, Chen J, Yin T (2023) Ultrasound-responsive glycopolymer micelles for targeted dual drug delivery in cancer therapy. Biomater Sci 11(18):6149

Liu J, Ma X, Jin S, Xue X, Zhang C, Wei T, Guo W, Liang X-J (2016) Zinc oxide nanoparticles as adjuvant to facilitate doxorubicin intracellular accumulation and visualize pH-responsive release for overcoming drug resistance. Mol Pharm 13(5):1723

Liu Q, Wang H, Li G, Liu M, Ding J, Huang X, Gao W, Huayue W (2019) A photocleavable low molecular weight hydrogel for light-triggered drug delivery. Chin Chem Lett 30(2):485

Liu JY, Zhang X, Tian B (2020) Selective modifications at the different positions of cyclodextrins: a review of strategies. Turk J Chem 44(2):261

Liu J, Zhang C, Cao L, You M, Li Z, Guo H, Yao C, Lou J, Zhang P, Xu F (2022a) Ultrasensitive multiplexed detection of small molecules and enzymes using stimuli-responsive nucleic acids. Chem Eng J 440:135797

Liu Y, Zhao L, Zhao L, Xu B, Wang C, Li S, Xu B (2022b) Multi-stimuli-responsive supramolecular hydrogel based on an oxidized glutathione derivative. Dyes Pigments 205:110552

Luo Y, Yin X, Yin X, Chen A, Zhao L, Zhang G, Liao W, Huang X, Li J, Zhang CY (2019) Dual pH/redox-responsive mixed polymeric micelles for anticancer drug delivery and controlled release. Pharmaceutics 11(4):176

Luo Y, Li J, Ding Q, Wang H, Liu C, Wu J (2023) Functionalized hydrogel-based wearable gas and humidity sensors. Nano-Micro Lett 15(1):136

Mellerup SK, Wang S (2019) Boron-based stimuli responsive materials. Chem Soc Rev 48(13):3537

Mukherjee S, Thilagar P (2016) Stimuli and shape responsive 'boron-containing'luminescent organic materials. J Mater Chem C 4(14):2647

Nakal-Chidiac A, García O, García-Fernández L, Martín-Saavedra FM, Sánchez-Casanova S, Escudero-Duch C, San Román J, Vilaboa N, Aguilar MR (2020) Chitosan-stabilized silver nanoclusters with luminescent, photothermal and antibacterial properties. Carbohydr Polym 250:116973

Nie J, Zhang X, Liu Y, Schroer MA, Wang W, Ren J, Svergun DI, Zeng AP (2021) A non-structural pure enzyme protein forms a LCST type of stimuli-responsive and reversible hydrogel with novel structure and catalytic activity. bioRxiv 2021.02.07.430034

Nikravan G, Haddadi-Asl V, Salami-Kalajahi M (2019) Stimuli-responsive DOX release behavior of cross-linked poly (acrylic acid) nanoparticles. E-Polymers 19(1):203

Ortega-García A, Martínez-Bernal BG, Ceja I, Mendizábal E, Puig-Arévalo JE, Pérez-Carrillo LA (2022) Drug delivery from stimuli-responsive poly (N-isopropylacrylamide-co-N-isopropylmethacrylamide)/chitosan Core/Shell Nanohydrogels. Polymers 14(3):522

Pafiti K, Cui Z, Adlam D, Hoyland J, Freemont AJ, Saunders BR (2016) Hydrogel composites containing sacrificial collapsed hollow particles as dual action pH-responsive biomaterials. Biomacromolecules 17(7):2448

Pan X, Wu N, Tian S, Guo J, Wang C, Sun Y, Huang Z, Chen F, Wu Q, Jing Y (2022) Inhalable MOF-derived nanoparticles for sonodynamic therapy of bacterial pneumonia. Adv Funct Mater 32(25):2112145

Perigo EA, Hemery G, Sandre O, Ortega D, Garaio E, Plazaola F, Teran FJ (2015) Fundamentals and advances in magnetic hyperthermia. Appl Phys Rev 2(4):041302

Pourmadadi M, Ahmadi MJ, Dinani HS, Ajalli N, Dorkoosh F (2022) Theranostic applications of stimulus-responsive systems based on Fe2O3. Pharmaceut Nanotechnol 10(2):90

Qin Q, Yang M, Shi Y, Cui H, Pan C, Ren W, Wu A, Hu J (2023) Mn-doped Ti-based MOFs for magnetic resonance imaging-guided synergistic microwave thermal and microwave dynamic therapy of liver cancer. Bioact Mater 27:72

Qiu X, Cui Q, Guo Q, Zhou T, Zhang X, Tian M (2022) Strong, healable, stimulus-responsive fluorescent elastomers based on assembled borate dynamic nanostructures. Small 18(13):2107164

Regenold M, Kaneko K, Wang X, Peng HB, Evans JC, Bannigan P, Allen C (2023) Triggered release from thermosensitive liposomes improves tumor targeting of vinorelbine. J Control Release 354:19

Rial-Hermida MI, Rey-Rico A, Blanco-Fernandez B, Carballo-Pedrares N, Byrne EM, Mano JF (2021) Recent progress on polysaccharide-based hydrogels for controlled delivery of therapeutic biomolecules. ACS Biomater Sci Eng 7(9):4102

Roy E, Patra S, Madhuri R, Sharma PK (2016) Stimuli-responsive poly (N-isopropyl acrylamide)-co-tyrosine@ gadolinium: iron oxide nanoparticle-based nanotheranostic for cancer diagnosis and treatment. Colloids Surf B: Biointerfaces 142:248

Ruszkiewicz JA, Pinkas A, Ferrer B, Peres TV, Tsatsakis A, Aschner M (2017) Neurotoxic effect of active ingredients in sunscreen products, a contemporary review. Toxicol Rep 4:245

Ryabova A, Romanishkin I, Skobeltsin A, Moskalev A, Makarov V, Loschenov V, Nikiforov N, Orekhov A, Orekhova V, Sukhorukov V (2019) P4498 subcellular anti-atherosclerotic therapy: an approach to elimination of dysfunctional mithohondria by photodynamic therapy. Eur Heart J 40(Suppl 1):ehz745.0891

Sato Y, Hashiba K, Sasaki K, Maeki M, Tokeshi M, Harashima H (2019) Understanding structure-activity relationships of pH-sensitive cationic lipids facilitates the rational identification of promising lipid nanoparticles for delivering siRNAs in vivo. J Control Release 295:140

Sawayama J, Okitsu T, Nakamata A, Kawahara Y, Takeuchi S (2020) Hydrogel glucose sensor with in vivo stable fluorescence intensity relying on antioxidant enzymes for continuous glucose monitoring. Iscience 23(6):101243

Serpico L, Dello Iacono S, Cammarano A, De Stefano L (2023) Recent advances in stimuli-responsive hydrogel-based wound dressing. Gels 9(6):451

Shang H, Yang X, Liu H (2023) Temperature-responsive hydrogel prepared from carboxymethyl cellulose-stabilized N-vinylcaprolactam with potential for fertilizer delivery. Carbohydr Polym 313:120875

Sheahan AV, Sekar TV, Chen K, Paulmurugan R, Massoud TF (2016) A molecular imaging biosensor detects in vivo protein folding and misfolding. J Mol Med 94:799

Shen F, Yang W, Cui J, Hou Y, Bai G (2021) Small-molecule fluorogenic probe for the detection of mitochondrial temperature in vivo. Anal Chem 93(40):13417

Shen Y, Yang W, Hu F, Zheng X, Zheng Y, Liu H, Algadi H, Chen K (2023) Ultrasensitive wearable strain sensor for promising application in cardiac rehabilitation. Adv Compos Hybrid Mater 6(1):21

Shi X, Rathod VT, Mukherjee S, Udpa L, Deng Y (2020) Multi-modality strain estimation using a rapid near-field microwave imaging system for dielectric materials. Measurement 151:107243

Suenaga K, Uemura K, Tanaka K, Chujo Y (2020) Stimuli-responsive luminochromic polymers consisting of multi-state emissive fused boron ketoiminate. Polym Chem 11(6):1127

Tian B, Liu Y, Liu J (2021) Smart stimuli-responsive drug delivery systems based on cyclodextrin: a review. Carbohydr Polym 251:116871

Turhan O, Yaman M, Dikmen G, Nural Y, Sarıboyacı AE, Tasa BA, Soykan MN, Seferoğlu Z (2023) Novel fluorescent sensors based on coumarin-hydrazide-hydrazone hybrid for the detection of CN−, Co2+ and Ni2+ ions: DFT and bioimaging in living cells. J Mol Liq 392:123440

Vallet-Regi M, Rámila A, Del Real R, Pérez-Pariente J (2001) A new property of MCM-41: drug delivery system. Chem Mater 13(2):308

Vural M, Pena-Francesch A, Bars-Pomes J, Jung H, Gudapati H, Hatter CB, Allen BD, Anasori B, Ozbolat IT, Gogotsi Y (2018) Inkjet printing of self-assembled 2D titanium carbide and protein electrodes for stimuli-responsive electromagnetic shielding. Adv Funct Mater 28(32):1801972

Wagstaff PG, Buijs M, van den Bos W, de Bruin DM, Zondervan PJ, de La Rosette JJ, Laguna Pes MP (2016) Irreversible electroporation: state of the art. Onco Targets Ther 9:2437

Wang W, Wang Y, Wang Y, Gong H, Zhu H, Liu M (2019) Redox/pH dual stimuli-responsive ZnO QDs-gated mesoporous silica nanoparticles as carriers in cancer therapy. IET Nanobiotechnol 13(6):640

Wang Y, Shahi PK, Wang X, Xie R, Zhao Y, Wu M, Roge S, Pattnaik BR, Gong S (2021a) In vivo targeted delivery of nucleic acids and CRISPR genome editors enabled by GSH-responsive silica nanoparticles. J Control Release 336:296

Wang W, Hu J, Zhang R, Yan C, Cui L, Zhu J (2021b) A pH-responsive carboxymethyl cellulose/chitosan hydrogel for adsorption and desorption of anionic and cationic dyes. Cellulose 28:897

Wang X, Sun W, Shi H, Ma H, Niu G, Li Y, Zhi J, Yao X, Song Z, Chen L (2022) Organic phosphorescent nanoscintillator for low-dose X-ray-induced photodynamic therapy. Nat Commun 13(1):5091
Wang Z, Feng Z, Du F, Xiang X, Tang X, Qiu L, Qian Z (2023a) Recent progress in theranostic microbubbles. Chin Chem Lett 34(9):108137
Wang H, Wang J-X, Song X, He T, Zhou Y, Shekhah O, Gutiérrez-Arzaluz L, Bayindir M, Eddaoudi M, Bakr OM (2023b) Copper organometallic iodide arrays for efficient X-ray imaging scintillators. ACS Central Sci 9(4):668
Wang Q, Tian Y, Yao M, Fu J, Wang L, Zhu Y (2023c) Bimetallic organic frameworks of high piezovoltage for sono-piezo dynamic therapy. Adv Mater 35(41):2301784
Weng Z, Xu Y, Gao J, Wang X (2023) Research progress of stimuli-responsive ZnO-based nanomaterials in biomedical applications. Biomater Sci 11(1):76
Wu X, Liu Q, Liu F, Wu T, Shang Y, Liu J, Ding B (2021) An RNA/DNA hybrid origami-based nanoplatform for efficient gene therapy. Nanoscale 13(30):12848
Xiao H, Yan S, Chen J (2022) Microwave polarization sensing for dielectric materials based on a twisted dual-layer meta-surface. Materials 15(19):6655
Xu Q, Pan Y, Liu X, Gao Y, Luan X, Zeng F, Zhou D, Long W, Wang Y, Song Y (2022) Hypoxia-responsive platinum supernanoparticles for urinary microfluidic monitoring of tumors. Angew Chem 134(14):e202114239
Yan H, Jiang Q, Wang J, Cao S, Qiu Y, Wang H, Liao Y, Xie X (2021) A triple-stimuli responsive supramolecular hydrogel based on methoxy-azobenzene-grafted poly (acrylic acid) and β-cyclodextrin dimer. Polymer 221:123617
Yang Z, Hu J, Van der Heggen D, Feng A, Hu H, Vrielinck H, Smet PF, Poelman D (2022) Realizing simultaneous X-ray imaging and dosimetry using phosphor-based detectors with high memory stability and convenient readout process. Adv Funct Mater 32(31):2201684
Yasin MN (2016) Conducting polymer implant for on-demand ocular drug delivery. The University of Auckland
Yin J, Yao D, Yin G, Huang Z, Pu X (2019) Peptide-decorated ultrasmall superparamagnetic nanoparticles as active targeting MRI contrast agents for ovarian tumors. ACS Appl Mater Interfaces 11(44):41038
Yin B, Zhang Q, Xia X, Li C, Ho WKH, Yan J, Huang Y, Wu H, Wang P, Yi C (2022) A CRISPR-Cas12a integrated SERS nanoplatform with chimeric DNA/RNA hairpin guide for ultrasensitive nucleic acid detection. Theranostics 12(13):5914
Yougbaré S, Chou H-L, Yang C-H, Krisnawati DI, Jazidie A, Nuh M, Kuo T-R (2021) Facet-dependent gold nanocrystals for effective photothermal killing of bacteria. J Hazard Mater 407:124617
Yu J, Zhao F, Gao W, Yang X, Ju Y, Zhao L, Guo W, Xie J, Liang X-J, Tao X (2019) Magnetic reactive oxygen species nanoreactor for switchable magnetic resonance imaging guided cancer therapy based on pH-sensitive Fe5C2@ Fe3O4 nanoparticles. ACS Nano 13(9):10002
Yu X, Lin F, Li P, Yan S, Zhang K, Cui W, Yin J (2022) Porous scaffolds with enzyme-responsive Kartogenin release for recruiting stem cells and promoting cartilage regeneration. Chem Eng J 447:137454
Zhang T, He Y, Wang R, Geng W, Wang L, Niu L, Li X (2008) Analysis of dc and ac properties of humidity sensor based on polypyrrole materials. Sensors Actuators B Chem 131(2):687
Zhang K, Meng X, Cao Y, Yang Z, Dong H, Zhang Y, Lu H, Shi Z, Zhang X (2018) Metal–organic framework nanoshuttle for synergistic photodynamic and low-temperature photothermal therapy. Adv Funct Mater 28(42):1804634
Zhang S, Ye J, Liu X, Wang Y, Li C, Fang J, Chang B, Qi Y, Li Y, Ning G (2021) Titanium carbide/zeolite imidazole framework-8/polylactic acid electrospun membrane for near-infrared regulated photothermal/photodynamic therapy of drug-resistant bacterial infections. J Colloid Interface Sci 599:390
Zhang Y, Dong L, Liu L, Wu Z, Pan D, Liu L (2022) Recent advances of stimuli-responsive polysaccharide hydrogels in delivery systems: a review. J Agric Food Chem 70(21):6300

Zhang Z, Xia T, Ran P, Wei J, Meng J, Zhang G, Li X (2023) Persistent luminescence-activated Janus nanomotors with integration of photodynamic and photothermal cancer therapies. Chem Eng J 457:141226

Zhou X, He X, Shi K, Yuan L, Yang Y, Liu Q, Ming Y, Yi C, Qian Z (2020) Injectable thermosensitive hydrogel containing erlotinib-loaded hollow mesoporous silica nanoparticles as a localized drug delivery system for NSCLC therapy. Adv Sci 7(23):2001442

Zhu L, Song Q, Ma H (2023) Synthesis of hyperbranched polysiloxane/poly (N-isopropylacrylamide) microgel, its stimulus responsive behavior and study for drug release. J Macromol Sci A 60(1):18

Chapter 6
Biomimetic Materials

Yu Shi and Zhe Liu

Abstract Nature has evolved from diverse systems, objects, structures, materials, and processes with individual characteristics and multiple functions. Biomimetic and bioinspired materials were born as a bridge between nature and engineering. Natural materials and systems, deemed as exhaustive and detailed "solution manual", have performed assorted functions including self-actuation, self-cleaning, anti-fogging, anti-reflection, osmotic barrier, camouflage, aerodynamics, sensing, or combinations of two or more. The understanding of underlying mechanisms drives us to develop bioinspired nanomaterials, sophisticated nanodevices, and customized engineered products for various techniques and commercial applications. Biomimetic micro/nano-scale structures and interfaces are utilized nowadays for drug delivery, tissue engineering, regenerative medicine, etc. Furthermore, dynamical biological systems and their ability of controllable changes in structures and properties are necessary for living organisms to adapt unpredictable environments, and thus based on this principle, the reversible mechanical deformation in response to specific stimulus has become a kind of intelligent biomaterials. Bioinspiration is an inexhaustible treasure for designers, researchers, and engineers. As biomimetics matured, it has been an emerging field in manufacturing given that continuously surprising discoveries inspire infinite varieties of advanced biomaterials coupled with personalized functions and properties. Pioneering fabrication techniques combined with synthetic biology build responsive and adaptive integrated systems at all spatial scales. Moreover, interdisciplinary research, multiple tools and approaches, and revolutionary technologies might depict greater blueprint with the assistance of biomimetics. More importantly, nanotechnology incorporated with biomimetics will have significant influences on human healthcare in the future.

Y. Shi · Z. Liu (✉)
Academy of Medical Engineering and Translational Medicine, Tianjin University, Tianjin, China

Tianjin Key Laboratory of Brain Science and Neural Engineering, Tianjin University, Tianjin, China
e-mail: zheliu@tju.edu.cn

Z. Liu (ed.), *Intelligent Biomaterials*, https://doi.org/10.1007/978-981-96-8646-9_6

Keywords Biomimetics · Bioinspiration · Wettability · Self-cleaning · Micro/nanomotors

6.1 Overview of Biomimetics

Nature has transformed materials and objects with high performances, and the most important lesson from the biological world is the adaption of structure and function. Complex configuration with special surface and content renders biological objects and materials unique physicochemical and mechanical property as well as miraculous functions to adapt diversified environment, while manifold biological molecules have special interaction and communications with targeted regions or biological body (Bhushan 1893). From macroscale objects to nanoscale surface and structures, so many examples as the source of inspiration have been found of research or commercial interest. The following is the overview of common objects and materials to their selected functions, which covers the scope of bacteria, plants, and animals (Fig. 6.1).

In bacteria, their flagella motor as the stereotype of biological molecular machine possesses a rotation speed of over 10,000 rpm driven by proton flow across the membrane (Jones and Aizawa 1991). Distinct surface structures and morphology of plant leaves make nanomaterials diversity (Koch et al. 2008, 2009a). For instance, waxy leaves of Eucalyptus macrocarpa with super-hydrophobicity, and glossy

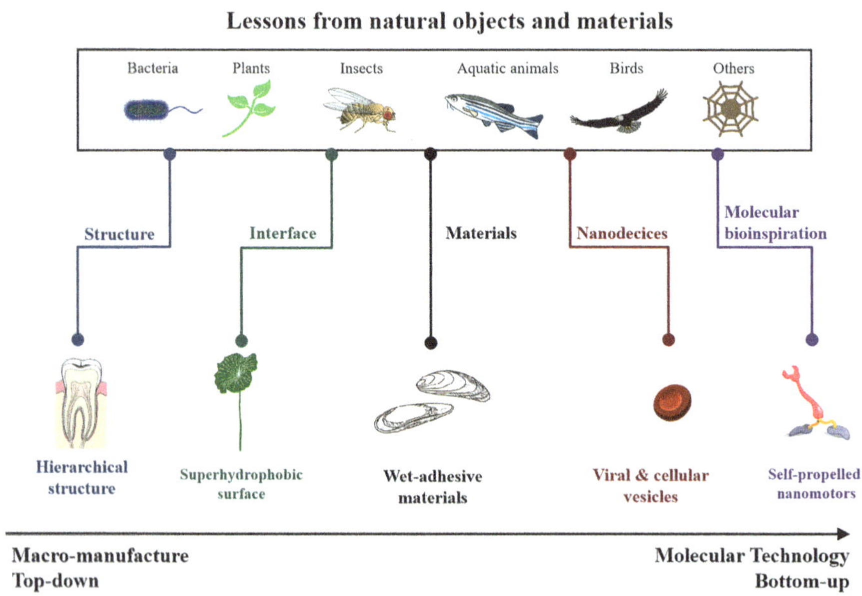

Fig. 6.1 An overview of common objects and typical bioinspired examples

leaves of Magnolia grandiflora with hydrophilic, rough but velvety surfaces of the flower leaves (petals) of Dahlia. For insects, mosquito eyes possess super-hydrophobic anti-fogging properties for outstanding vision given numerous micro-spheres "ommatidia" as individual sensory units (Gao et al. 2007). Elytra of darkling beetles collect dew with bumpy surface made by alternating hydrophobically waxy and hydrophilically non-waxy regions severed as anti-fogging coatings (Parker and Lawrence 2001). Anti reflective eyes of nocturnal moths are constituted of countless hexagonally organized nanoscopic pillars (Genzer and Efimenko 2006), which have potential for electronic displays that work in strong light with less bleakness and smaller energy consumption compared to existing displays with built-in devices of light source. For aquatic animals, mucus of their skin serves as osmotic barrier for salinity resistance, anti-infection, and drag reduction (Bechert et al. 2000). The boxfish (*Ostracion meleagris*) with a streamlined figure renders low drag in water for economical Mercedes Benz's car with a similar figure (Mueller 2008). Feathers of birds contribute waterproofer, camouflage, and insulation in addition to aerodynamic lift and inspired the creation of aircraft (Boyer 1992). Stiff and stretchable spiderweb possesses huge bearing capacity and the resistances of rain and wind in that one-dimensional fiber structures containing structure domains both of β-sheets and flexible helices to capture inspects in harsh environment (Jin and Kaplan 2003). Furthermore, sensory-aid devices have been designed for the detention of sound, frequency, heat, chemicals, and so on (Barth et al. 2003; Schütz et al. 1999).

6.1.1 Origin, Development, and Concept of Biomimetics

Objects and materials with mysterious properties and functions in nature provide substantial source of bioinspiration including raw materials, routes, designs, or ideas to guide the generation of engineering products with custom-made function, which is generally called biomimetics, mimicry of nature or biology. The term biomimetics derived from the combination of two Ancient Greek words 'bios' for life and 'mimesis' for imitation, was coined by polymath Otto Schmitt who made a physical device mimicked by nerval electrical action in 1957 (Vincent et al. 2006). Subsequently, bionics, that is the systematical sciences related to biological phenomenology for physical or composite bio-physical systems owing to natural inspiration, was proposed by Jack Steele of Wright-Patterson Air Force Base at a meeting in Dayton, Ohio in 1960 (Aziz and El Sherif 2016), and the word of biomimetics was recorded by Webster's dictionary in 1974, defined as the study of the formation, structure or function of biologically produced substances and materials (as enzymes or silk) and biological mechanisms and processes (as protein synthesis or photosynthesis) especially for the purpose of synthesizing similar products by artificial mechanisms which mimic natural ones (Bhushan 1893). Afterward, biomimicry and bioinspiration were created respectively in 1982 and 2012 (Swiegers 2012).

Every biological body is composed of a variety of biological materials that allow cells, tissues, and organs to function and has access to nature-inspired materials and devices. Biological structures serve as technique library of architecture, aerodynamics, and mechanical engineering, while natural materials are exploited to synthesize all kinds of polymers and minerals (Jeronimidis and Atkins 1995). Ancestors have developed many materials and devices long time ago, and the earliest natural inspiration dates back to man-made artificial silk more than 3000 years ago (Philip 2002). The advancement of biomimetics can be divided into model biomimetics and the combination of technology and biology phases. Initially, bioinspiration based on configuration was applied to the fabrication of transportation and engineering products in the twentieth century. There are some iconic examples of bioinspiration. Velcro of zip fastener was designed from the hooked seeds of the burdock plant by George de Mestralthe in 1955 (Velcro 1955). Clement Ader designed a steam-powered aircraft (Eole) with a flight of around 300 m inspired by wing of bats in 1987 (Coineau and Kresling 1987). The self-cleaning leaves of the lotus gave birth to the paint coating of Lotusan in 1997 (Barthlott and Neinhuis 1997). Dry adhesive tape derived from the adhesive single gecko foot hair (seta) (Autumn et al. 2000). A micro-air vehicle flying low over water with oscillating wings was designed from birds (Jones and Platzer 2002). At another phase, biology and technology were combined in both practical engineering and linguistic foundation in 1960, and then biomimetics was associated with other advanced technologies, especially computer science and molecular biology since 1980s (Bhushan 1893). The mechanism of human brain was mimicked for the construction of artificial intelligence and neural networks. Molecular biomimetics occurred for self-assembled functional materials and devices derived from DNA, protein, and cells (Grunwald et al. 1894). In the 1990s, natural materials were investigated from the angle of nanoscale with the advancement of nano-science and nanotechnology, and biomimetics has spread from macroscale, microscale to nanoscale by nanofabrication techniques, such as shark skin-inspired Fastskin swimsuits (TYR Trace Rise) with low drag, and solar bear-inspired artificial furs (Stegmaier et al. 1894; Milwich et al. 2006). Nowadays, biomimetics as an emerging field of research consists of extensive research topics and various disciplines to generate countless innovations, which has been considered as highly interdisciplinary subject involving biology, structure, physics, chemics, engineering science, electronics, and material sciences that has significant influence on science, society, and daily life with great potential in innovation of all fields (Lepora et al. 2013).

In regard to lessons from nature, the strategies or approaches of bioinspired and biomimetic system contain bioinspired configuration as well as biomimetic technologies. Several terms are coined to describe the concept including bionics, biomimicry, bioinspiration, and biomimetics, and other synonymous like biomimetics, biognosis, biologically inspired design, nature inspiration, and nature mimicry. Though these terms are presented as synonyms in the literature that mean invention or design made by taking inspiration from the knowledge of nature in terms of creative approaches, there is some discrimination among these terms in respect of the analogical specificity and fields shown in Fig. 6.2 (Shu et al. 2011; Vincent et al. 2006). Initially, the prefixes of nature involve living and nonliving natural systems, while the prefixes of Bio only refer to living organisms within the scope of nature.

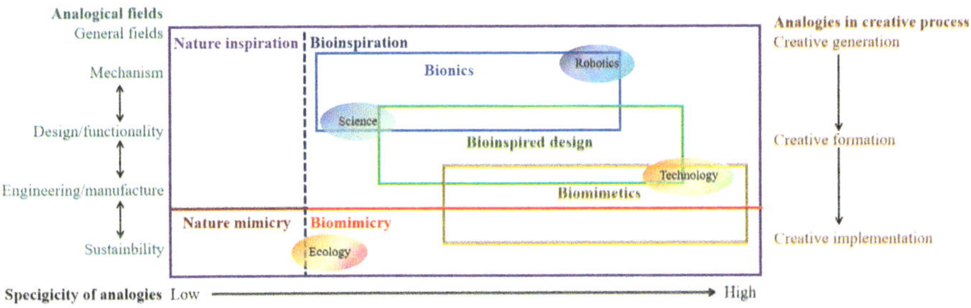

Fig. 6.2 Different conceptions of bioinspiration

Secondly, the inspiration from nature covers from divergent stage of creative generation, creative emergent stage of creative formation, to creative convergent stage of creative implementation, and the analogical fields could be mechanics (bionics), general fields (bioinspiration) and sustainability (biomimicry), based on the principles and materials extracted from biology (biologically inspired design) or specific biological system (biomimetics) (Fayemi et al. 2017).

Currently, terms such as bioinspiration, biomimicry, and biomimetics are often used synonymously in the literature according to the ISO/TC 266 Biomimetics Committee (ISO/TC266 2015).

Bioinspiration: 'Creative approach based on the observation of biological systems'. It lays stress on the primitive thinking phase during the observation of natural appearance or function resulting in the creation of idea.

Bionics: 'Technical discipline that seeks to replicate, increase, or replace biological functions by their electronic and/or mechanical equivalents'.

Biomimetics (ISO 18458): 'Interdisciplinary cooperation of biology and technology or other fields of innovation with the goal of solving practical problems through the function analysis of biological systems, their abstraction into models and the transfer into and application of these models to the solution'. It implies the integration of engineering technologies and manufacture deprived from nature, and usually described as a concept and framework for biomimetic terminology. Additionally, biomimetic system emphasizes adaptive similar/identical natural/biological systems for distinct engineering.

Biomimicry (ISO 18459): 'Philosophy and interdisciplinary design approaches taking nature as a model to extend the life and reduce the mass of components for sustainable development (social, environmental, and economic)'. It is devoted to sustainability.

Generally, biomimetics used in this chapter is regarded as the inventions and innovations of research and development approaches gained or inspired by knowledge of biological systems, and the distinctive advantage is the optimization of application by the transfer of biological principles to technology. In other words, underlying fundamental mechanisms behind desired functions in the natural model should be mined for the development of artificial system (Fratzl 2007).

6.1.2 Principle of Biomimetic Materials

Commonly, the developing approach of biomimetics is based on functional analogies, and assorted bio-systems as potential stereotype are introduced and chosen to explore adapted engineering of products with necessary interpretation or translation of mechanism and concept because of differences between living organisms and non-living engineering (Vincent et al. 2006). Biology is a descriptive subject on the ground of category analysis as living systems go through the long evolutionary process interfered with natural selection, while engineering is a prescriptive industry in line with rules and regularities since decision-making and practical demands are crucial to product production (Vincent et al. 2006). Therefore, simple replica of prototype and direct utilization of materials may be not achieved. "Theory of Inventive Problem Solving" developed by Genrich Altshuller and Rafik Shapiro could be a suitable tool for the transfer between biology and engineering (Al'tshuller 1999). Further, logical framework was established (Bogatyreva et al. 2004): things (substance, structure) do things (requiring energy and information) somewhere (space, time), which includes three objects with six dimensions. The methodology of biomimetic avails design process-related engineering. Solution-based and problem-driven approaches were introduced into biomimetics practice (Fayemi et al. 2017). Solution-based approaches start off with biological knowledge or natural inspiration-based fundamental science research for technical design, and describe the process that biological function could be emulated in technology (Goel et al. 2014). Such process is closely associated with knowledge transfer between scientific technology and industry organization involving scientific research, patent for invention as well as transfer and license to companies (Siegel et al. 2004). Conversely, the problem-driven approach starts from the biomimetic development process for problematic solution, and improvement or functions performed by biological system could be extracted to be applied technologies as the up-streaming of solution-based approaches (Goel et al. 2014). Its process includes identification, alternation by screening, and implementation with tests (Massey and Wallace 1996).

At the final stage of biomimetic design, the solution has been recommended to conform with *Life's Principle* proposed by the Biomimicry Institute for Sustainable Ecosystems (Terrier et al. 2019). Specifically, minimum of resources along with minor environment impacts and energy consumption are utilized for product or project; local resource should be preferred and form the closed circulation as nature to keep dynamic equilibrium of nature without pollution or waste; diversify cooperation and shared information with feedback loop are required for the implement of industrial ecology or circular economy approaches. In addition, there are three points in the practical development of biomimetic materials: a) selected species with objective function should be of commercial interest, or selected materials themselves derived from nature are economic and easily attained; b) modeling by nature's route are exploited to structure optimization based on the well-grounded understanding of mechanism behind the function and verification from the lab; c) high efficient smart materials could be fabricated conform to nature principle, that is a few light elements are utilized to construct adaptive biological materials with hierarchical structures by controlled self-assembly (Fratzl 2007). In the biomimetic system, the basic strategies contain surface, shape and texture, actuation, and preparation. Surface as the barrier of organisms and environment provides the communication of these two systems as the prerequisite of organic protection. Shape and texture can enhance the performance and function. Self-actuation and preparation make intelligent materials impossible. Current biomimetic study majorly focuses on the following aspects: biomimetic structure and mechanics, biomimetic interface, biomimetic materials, biomimetic functional device, molecular biomimetics, in addition to bioinspired artificial intelligent and bionic swarm intelligence optimization.

6.2 Biomimetic Structure: Hierarchical Biomineralization of Bone

6.2.1 Fundamentals

The performance and function of materials hinge on the composition and structure concerned with the processing conditions and techniques (Lei et al. 2021). Biological organism utilizes limited elements and the most common materials to construct biomaterials, and these biomaterials adhere to recipes guided by genetic code to grow, differentiate, or control self-assembly. Biological objects are created by varieties of soft and hard tissues with multi-layer hierarchical composite structures to achieve multiple function of biology and adapt changing environmental conditions (Panneerselvam et al. 2019; Fratzl and Weinkamer 2007). Precise hierarchical assembly, consisting of nanostructures in most cases, renders sophisticated and graded structure from bottom to top at all scales, which allows biological building blocks to construct tissues and organs, as well as optimization and adjustment of biomaterials at each level of hierarchy (Gupta et al. 2006). In nature, there are major five representative hierarchical structures (Wei et al. 2023): lamellar arrangement for enhanced strength and toughness of shells (Yazdani Sarvestani et al. 2022), columnar alignment for anti-vibration and support of teeth (Liu et al. 2018), coaxial layered arrangement for bending resistance and mass transport of bone and trunk (Wegst et al. 2015), Bouligand structure for toughness of carapace and fish scales, and array structure for structural color and hydrophobicity of some insects like butterfly (Fig. 6.3). Among these hierarchical structures, coaxial layered arrangement of hard tissue represented by bone has attracted great attention by researchers, which make bone substitute possible.

Hierarchical structure tends to rely on biomineralization. In order to enhance the mechanical strength of tissues, inorganic solids are generated and incorporated into tissues to form hierarchical organo-mineral nanocomposites, and such a highly controllable and dynamic process is named as biomineralization (Elsharkawy and Mata

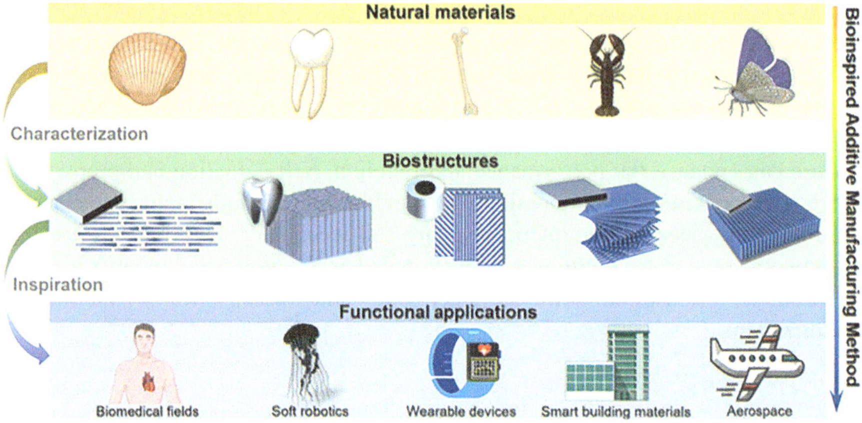

Fig. 6.3 Bioinspired-additive manufacturing techniques applied in multiple domains. (Reprinted with permission from Wei et al. 2023)

2018). Biomineralization, based on mineral templating in the preformed organic matrix or the interaction between biomacromolecules and forming mineral phases, is more often exploited to develop hard tissues, and the macromolecules hold the dominant position to control and sustain mineral phases (Weiner and Addadi 1997). Bone with mineralized collagen fibers as a coaxial layered arrangement possesses superior material transport capacity along with enough mechanical strength (Wegst et al. 2015). Its components are about 62% of inorganic mineral nanoscale-hydroxy-apatite crystals, and 38% of water and organic substances like collagen type I and glycoproteins (Salgado et al. 2004; Hamed et al. 2010). The mechanical strength of the bone ascribes to dissipate deformation energy from different layers and sacrificial bonds among individual collage fibrils in hierarchical structure (Fratzl 2008; Fantner et al. 2005). In its hierarchical structure, the former two layers are fibrous collagen and fiber bundle by helical structure, and then calcium phosphate grows from the gaps of fiber bundle along the parallel orientation of collagen fibers that generate megapascals of prestress (Ping et al. 2022). Next, mineralized collagen fiber bundles are formed by hydroxyapatite-wrapped fiber bundle, and are arranged into hollow structures, named as Havers' canals for nerve and blood transportation of bone cells. As the final sixth level of bone, porous cancellous bone accesses to bone marrow and blood vessels, while cortical bone with thick outer layer for supporting functions (Rho et al. 1998).

What's more, gluing interfaces facilitate the formation of hierarchical hybrid materials for multi-functions, for instance, integrated muscles and connective tissues for simultaneous movement and support, which provides a new avenue for active materials (Sidorenko et al. 2007).

6.2.2 *Fabrication*

Fabrication from down to up of novel biomimetic materials with large span is still a great challenge owing to large span of scales and exquisite integrity, which is far beyond present technology and process. Significant importance should be attached to the fundamental knowledge from every dimension and advanced processing routes. Biomimetic hierarchical materials with multiscale entail appropriate different processing and strategies including biomimetic mineralization, freeze casting, and additive manufacturing.

6.2.2.1 Biomimetic Mineralization

Nearly all biological hard tissues originate from mineralization of inorganic components including calcium carbonate, silica, and calcium phosphate. Concretely, committed cells could harden and stiffen minerals by depositing to soft protein matrix, the most common polymeric matrix (Wegst et al. 2015). Whereas, in practical process, only micrometer-sized specimens have been produced but with poor mechanical performance, yet applied in growing particles, capsules, and films (Ethirajan et al. 2008). For example, magnetic fields and field gradients can control the distribution of anisotropic reinforcing microparticles within the polymer matrix. These reinforcing microparticles with minimal concentrations coatings of superparamagnetic nanoparticles could be functionalized under applied magnetic fields to produce synthetic composites with complex arrangements with desired mechanical response (Erb et al. 2012). Besides, pH-responsive polymers could reversibly functionalize the particles in the assembly. Janus particles in hierarchical arrangements can be acquired to perform different chemical properties on different sides of the surface (Walther and Muller 2013).

6.2.2.2 Freeze Casting

Freeze casting, as an inexpensive technique, is utilized to fabricate bioinspired metal/ceramic, model polymer, and hybrid materials with high toughness, which guides ceramic-based suspensions to freeze in water to form lamellar ice crystals with natural structure (Deville et al. 2006). Following the sublimation of water, layered homogeneous ceramic scaffolds are formed, and then the scaffolds are filled with a second soft phase for hard-soft layered composites, or are compressed before the filling of a soft phase for brick-and-mortar structures. The structure and ductility of scaffolds are regulated by many factors, for instance, the composition and the freezing kinetics of the suspensions for micro-structure with mechanical response, the thickness of the lamellae for the control of practical sizes, the corporation of additives for solid-liquid interfacial tension, the ice crystals as bridges between lamellae (Wegst et al. 2015). Moving beyond, efficient cellular structures also could be constructed by freeze casting, for instance, platelet-based slurries were exploited to create porous honeycomb-like scaffolds with nacre-like cell walls by directional solidification during freeze casting (Hunger et al. 2013).

6.2.2.3 Additive Manufacturing

Additive manufacturing refers to a family of technologies for the construction of layer-by-layer structures with the avail of computer designs, including 3D inkjet printing, robocasting, selective laser sintering, stereolithography, and two-photon polymerization (Wegst et al. 2015). Additive manufacturing techniques involve all-scale processing according to different structural demands: nano-additive manufacturing for accurate nanoscale lattice structures but difficult-to-scale production, micro-additive manufacturing for microstructures but with limited available materials, macro-additive manufacturing for exquisite figures but limited printed materials. Such multiscale technologies have been applied for cellular functional networks, implantable scaffolds, and biomedical devices (Wei et al. 2023). For instance, Zhang et al. (2018a). designed a DNA origami with 3D rhomboid crystal structure that can be molecularly arranged and assembled in the predetermined lattice by nano-additive manufacturing; micro-additive manufacturing was exerted to fabricate abiotic enamel materials with high strength and low density based on the microstructure of natural tooth enamel (Yeom et al. 2017); resorted to macro-additive manufacturing, McAlpine's research team constructed a bionic ear based on 3D printing a cell-seeded hydrogel matrix incorporated with intertwined conducting polymer of silver nanoparticles, which enhanced auditory sensing toward radio frequency reception, and has access to stereo audio music (Mannoor et al. 2013). Therefore, additive manufacturing as a promising technology has access to the design of bioinspired structural materials of the future.

6.2.3 Biomedical Applications

Bone autografts or allografts as the gold-standard treatment are exploited for bone defects and regeneration, but are limited to potential disease transmission. Pioneering biomaterials including bioceramics, shape memory metals, and polymers have evolved in recent years though with constrained mechanics, bio-degradation, and bioactivities (Koons et al. 2020). Biomineralization-inspired materials for bone repair usually require some aspects: superior biocompatibility of biomaterials with as less chemical agents as possible, regulable structure and mechanics for motor function, and facile construction method of bone nanostructures with low cost (Tang et al. 2021). Biomimetic hierarchical bone scaffold with similar compounds of natural bone has been achieved using hydroxyapatite/collagen composite nanofibers by Bian et al (2019). They utilized electrospinning technique and rolling method to synthesize bone scaffold composed of microscopic fibers from bottom-up as shown

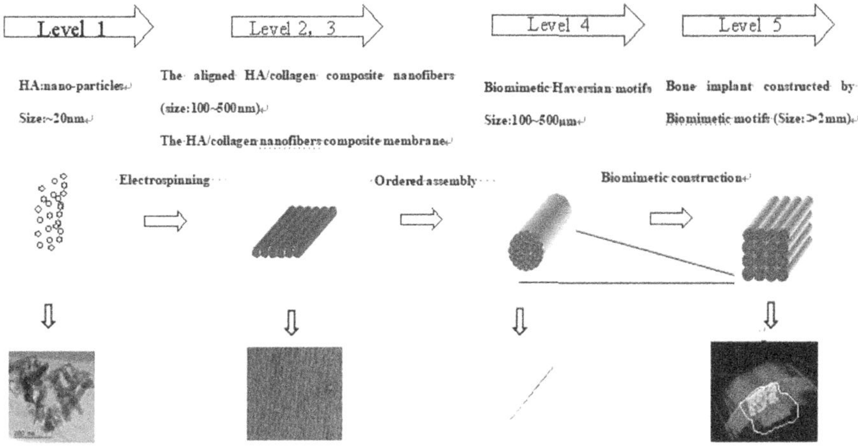

Fig. 6.4 The construction of bionic bone Haversian hierarchical hydroxyapatite/collagen composite scaffold. (Reused with permission from Bian et al 2019)

in Fig. 6.4, which was confirmed by MicroCT scanning and scanning electron microscopy for potential application in bone tissue engineering.

Based on the conception of intrafibrillar mineralization of type-I collagen with hydroxyapatite in bone tissue, intrafibrillarly mineralized scaffolds in collagen sponges were first proposed in 2003 by Gower et al., and they utilized poly (acrylic acid) to achieve mineralization of calcium carbonate and calcium phosphate by the polymer-induced liquid-precursor (PILP) process (Olszta et al. 2003). Following collagen substrates infiltrated into calcium carbonate or calcium phosphate, acid, and dilute hypochlorite removed superficial mineral and collagen matrix respectively, resulting in hierarchically mineralized collagen scaffolds for bone repair. Intrafibrillar silicification is another strategy to prepare mineralized collagen scaffolds (Sun et al. 2018). Tay et al. prepared intrafibrillar silicified collagen scaffolds (SCSs) for monocyte immunomodulation-based angiogenesis and bone regeneration via infiltration of intrafibrillar amorphous silica in collagen matrices. The formational capability of bone and related cytokines of SCSs was verified in a murine calvarial defect model (Sun et al. 2017). Hence, apart from common calcium phosphate and calcium phosphate mentioned above, silica, yttria-stabilized zirconia can be also developed scaffolds, and multiple mixed minerals may also be future trends given the diverse bioactivities of different minerals (Zhou et al. 2014). 3D printing technology provides more possibility for biomimetically bioactive scaffolds owing to simple and practical process with little impurities. Calcium phosphate (β-TCP)/calcium silicate ($CaSiO_3$) composite bio-ceramic scaffold (β-TCP/$CaSiO_3$) with hydroxyapatite surface layer was prepared by precise and accurate 3D printing technique for capillary formation, bone-augmentation, and new bone matrix formation in the model of back defection and calvarial extramembranous defection of SD rat (Fig. 6.5) (Liu et al. 2022a). In addition, the interconnected porous bio-scaffolds with hydroxyapatite coating are beneficial to up-regulation of the cellular

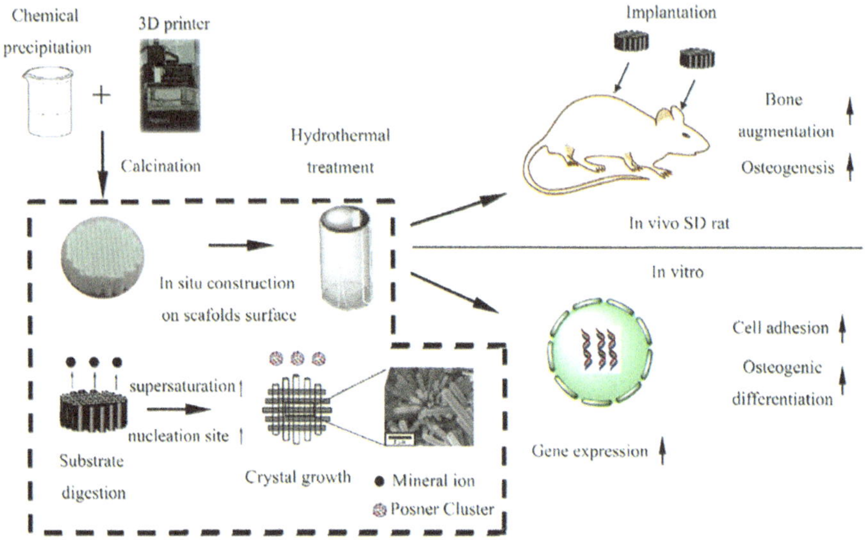

Fig. 6.5 The scheme of biomimetic ceramics scaffolds with hydroxyapatite surface as subcutaneous implants of back and extramembranous implants near the calvaria of SD rats. (Reused with permission from Liu et al. 2022a)

differentiation of mBMSCs and gene expression of HUVECs with preferable biocompatibility.

Natural extracellular matrix (ECM)-based fibrous scaffolds also possess hierarchical structures for tissue regeneration. Inspired by the fibrillar structure of ECM, nano-submicron biomimetic fibrous scaffold with integrated symmetrical nanostructure and interconnected porosity of above 90% was fabricated by hybridization of bacterial cellulose nanofibers and cellulose acetate submicrofibers using electrospinning and biosynthesis method. The fibrous scaffold has improved mechanical and biological performances that are facilitated with cell migration, proliferation, alkaline phosphatase activity, calcium depositions, and even the expression of osteogenic gene (ALP mRNA and Runx2 mRNA) and protein (ALP) (Luo et al. 2020). Considering the osteogenic effects of collagen with hierarchical architecture in ECM, collagen-biomimic anisotropically hierarchical topography was designed by silicone stretch-oxidation-release method and imprinting lithography for improvement of osteogenic differentiation of human mesenchymal stem cells (Fig. 6.6) (Yang et al. 2020). Such multiscale hierarchical structure has influence on vinculin expression, cell contractility, regulation of YAP-TAZ signaling pathway, and osteogenic differentiation contributing to the highest degree of osteogenesis, which was certified by different scales of substrates single scale (Flat, W25), double (W3||W25), and triple scale (W0.5⊥W3||W25 and W0.5||W3||W25).

On the foundation of bionic bone scaffold, the introduction of electrical stimulation will accelerate the process of bone repair. For example, a 3D biomimetic

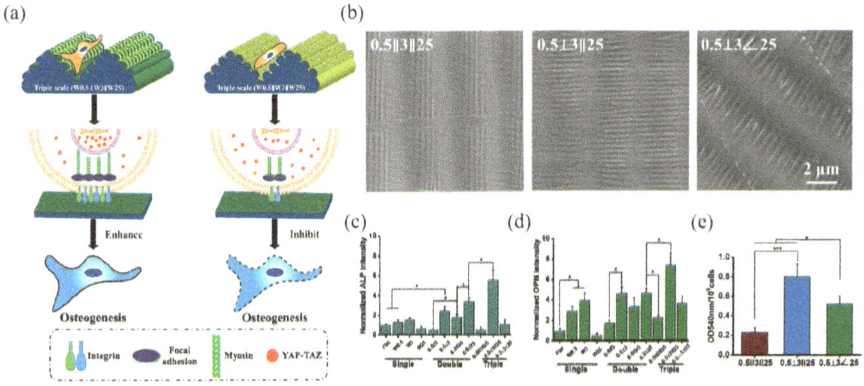

Fig. 6.6 (**a**) Scheme of hierarchical structure of collagen for stem cells adhesion and osteogenic differentiation. (**b**) Representative SEM images of the triple scale substrates in collagen with synthetic material (PDMS). (**c**, **d**) Quantification of the expression of alkaline phosphatase (**c**) and osteopontin (**d**) in cells cultured in osteogenic induction medium at day 14. (**e**) Quantitated mineralization by Alizarin Red staining. (Reprinted with permission from Yang et al. 2020)

optoelectronic scaffold integrated with thin-film silicon (Si)–based microstructures was prepared for the repair of cranial defects (Wang et al. 2023). Besides biomimetic scaffold supporting cell proliferation and differentiation, silica structures also electrically modulate membrane potentials, intracellular calcium dynamics of stem cells under the remote control of infrared illumination, which possess great prospect for biomedicine.

Bioinspired biomineralization provides a new path for the development of materials, which would render ecofriendly biocompatible biomaterials for more extensive biomedical applications, contributing to the sustainable development of natural resources. A functional protein silicatein possesses similar sequence with cathepsin-L, and can biomineralize tetraethyl orthosilicate precursor to generate biosilica. (6,7) This strategy could be exploited to fabricate mesoporous biosilica as drug delivery system. For instance, bacillus-shaped mesoporous silica nanoparticles (B-MSs) were prepared by mineralization of *Escherichia coli* (*E. coli*) biotemplates with mutant silicatein (mSα) expression, and exhibit hierarchical porous structure with original morphology of *E. coli* following biotemplates removing. Amino-modified B-MSs absorbing ConA antigens were harnessed as vaccine adjuvants for immune enhancement including the secretion levels of both IgG and CD8$^+$ T cells (Meng et al. 2023). Moreover, 3D-printed biomimetic bone structures can also boost piezoelectric performance, which may have potential in piezocatalysis biomedicine. Li et al. fabricated polyvinylidene fluoride energy harvester with bioinspired bone structures via a simple and fast in-situ chemical foaming-assisted fused deposited modeling method. The hierarchical porous structure renders superior capacity of power-to-electricity conversion for high-performance piezoelectric energy harvesting (Meng et al. 2023).

6.3 Biomimetic Surface: Biomimetic Superhydrophobic Nanostructure

6.3.1 Fundamentals

Distinctive surface nanostructure and morphology endow plant leaves with unique multi-functional properties including superhydrophobicity and low adhesion for self-cleaning, antifouling, drag reduction, and even plant motion (Bhushan 2019). As for water-repellent plants, water droplets on the surfaces of leaves lightly settle on the peak of microbumps owing to air bubbles filling into the hollows of the nanostructure under the liquid droplet, resulting in rolling of droplet. The structure of these leaves is divided into hydrophobic wax coating and roughness layer with microbump nanostructure (Bhushan et al. 1894; Bhushan and Jung 2011). On one hand, waxed cuticle as superhydrophobic coating could reduce water and ions loss, and protect organisms from the adhesion of bacterial cells and fouling. On the other hand, arranged micro-structure of surface could cause bacterial cells physically damage including lysis and death. Surface of lotus leaves could be as a model (Koch et al. 2008).

Various wetting states are accessed by the static contact angle that refers to the tangent angle of the liquid-vapor interface on the substrates described as Young Eq. (6.1) (Fig. 6.7) (Thomas 1805):

$$\cos\theta = \frac{\gamma_{SV} - \gamma_{SL}}{\gamma_{LV}} \tag{6.1}$$

where γ_{LV}, γ_{SV}, and γ_{SL} mean the interfacial tension between liquid-vapor, solid-vapor, and solid-liquid, respectively.

The value of 90°is a watershed in liquiphilic and liquiphobic, and liquiphobic surfaces require static contact angle more than 90° in comparison to that of superliquiphobic surfaces above 150° (Bhushan 2016). Superliquiphobic substrates commonly require a suitable roughness distribution, which was coined as "roughness-induced" superliquiphobicity. Besides that, low contact angle hysteresis (CAH) is another critical characteristic of a solid-liquid interface that refers to the difference of the advancing and the receding contact angle for the evaluation of

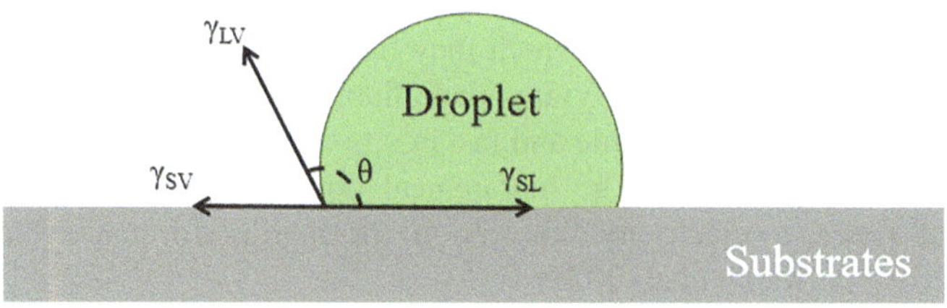

Fig. 6.7 Scheme of contact angle of a liquid droplet depositing on a substrate

self-cleaning and drag in fluid flow (Bhushan 2016). Considering the low value of CAH leading to the droplet roll and even slide, CAH less of 10° would possess self-cleaning properties to remove contaminant. The surface of lotus leaves, with a contact angle of 164° and a CAH of 3°, is made up of wax, a mixture of long-chain hydrocarbon compounds, and possesses superimposed asperities on microstructure: bundles of epicuticular wax tubules on different forms of papillary papillose epidermal cells contributing to rolled water droplets and removal of contaminant particles (Koch et al. 2009a; Bhushan et al. 1894; Koch et al. 2009b).

6.3.2 *Fabrication and Characterization*

6.3.2.1 Fabrication

The techniques of the preparation are classified into three approaches: top-down (lithography, template), bottom-up (layer-by-layer, chemical deposition), and their combination (electrospinning, sol-gel) (Celia et al. 2013).

1. Top-down.

Lithographic has access to designed structure (micro-, nano-, and hierarchical structures) and customized morphology of surface (such as circular pillars and star-shaped pots). Colloidal lithography, as a soft lithography technique, has attracted increasing attention for fabrication of ordered 2D or 3D structures (Zhang and Yang 2010). For instance, superhydrophobic poly(methyl methacrylate) surfaces with microscale pillared and nanofibrillar pattern were prepared by optical lithography and deep reactive ion etching (Accardo et al. 2010). Microfluidic channels also could be fabricated by lithographic to modify superhydrophobic surfaces for the capture and collection of water as the desert beetle, *Stenocara* sp (Kang et al. 2010).

Template can replicate pattern for surface structure by printing, growing, or pressing materials, in accompany with removed template with the inverse of pattern as repeated template for replica of the original (Victor et al. 2012). Anodic aluminum oxide (AAO), commonly applied template and membrane in the researches, could attain aligned nanofibers with controllable spacing by electrodeposition of nanostructure into the pores of the template, for example, superhydrophobic high-density polyethylene nanofiber surfaces by the regulation of AAO templates, and polystyrene nanopillars with different tip prepared on the aluminum substrate with AAO membrane (Qu et al. 2008; Cheng et al. 2010).

2. Bottom-up.

Layer-by-layer (LbL) technique is the deposition of thin films, particularly with opposite charge in each other like polyelectrolytes, by means of alternative dip in the solutions to absorb sequentially different charged species (Zhang et al. 2007; Bravo et al. 2007). LbL is also applied in nanoparticles assembly to obtain functionally hierarchically organized composite surfaces with different sizes, such as silica nanoparticles on honeycombs-patterned poly (dimethylsiloxane) substrates for alternate high/low adhesion superhydrophobic surfaces (Amigoni et al. 2009; Sun et al. 2010).

Chemical vapor deposition (CVD) is a deposition process established on decomposition or reaction of chemical precursors as vapor phase on heated substrates (Martin 2009). CVD is suitable for rough surfaces on thin film or a thin layer of hydrophobic compound deposing on rough surface, for instance, dual hierarchical structure of branched nanostructures and nanorod top on Au-coated silicon substrates, and oxidized metal (aluminum and titanium) coating on particulate films for difference of wetting properties (Yao et al. 2011; Hozumi et al. 2011).

3. The combination of top-down and bottom-up.

Electrospinning supports continuous nano-size fibers of polymers, nanoparticles, active agents, and even metals and ceramics (Greiner and Wendorff 2007). Polymers are common objects of electrospinning technique to structure superhydrophobic surfaces (Michielsen and Lee 2007). Adding additives could modulate surface hydrophobicity, such as low molecular weight additives leading to increase in hydrophobicity. Intriguingly, polymers would obtain a maximum water contact angle with all additives of 4% may be attributed to aggregates of additives molecules at high concentration (Hardman et al. 2011).

Sol-gel methods are appropriate for superhydrophobic surfaces on textiles, wood, and other substrates often with great resistance to temperature, where different precursors are transformed into glassy state under adjustable condition via hydrolysis and polycondensation with controlling surface roughness (Yang et al. 2010). For instance, glass plates of methyl trimethoxy silane with sol dip-coating were modified by trimethylchlorosilane following sintering at 150 °C for superhydrophobic surface (Mahadik et al. 2010). The functionality of the particles could enhance the adherence to the substrates owing to generated covalent bond in each other. Modified copper wafers with hydroxyl groups could react with silanol groups of coating using vinyl trimethoxy silane as precursor for superhydrophobic surfaces with water contact angle of 155°and expected corrosion resistance (Fan et al. 2012).

6.3.2.2 Characterization

The various substrates to different liquids have different wettabilities and properties, and deionized water and hexadecane are usually chosen as representatives of water and oil respectively. Superliquiphobic surfaces are usually characterized by wettability, self-cleaning, wear resistance, anti-biofouling, and anti-fogging, along with other special capability, such as resistance to hard environments (Bhushan 2019).

1. *Wettability*. Wettability with water and hexadecane can be evaluated by the measure of contact and tilt angles with a standard automated goniometer. Substrates with diversified coatings would gain the different property combination of superhydrophilic, superhydrophobic, and superoleophobic, while superoleophobic substrates include PDMS, nylon, PU, PE, and PC contributing to removal

contaminants with droplets both of water and hexadecane rolling (Bhushan and Multanen 2019).

2. *Self-cleaning*. In this test, silicon carbide is usually severed as contaminant, and images of samples were taken and quantitatively analyzed before and after water and hexadecane cleaning with particles removed by water droplets.
3. *Wear resistance*. Wear resistance of samples on substrates was examined by ball-on-flat tribometer for wear tests or scratching with tweezers, and wear tracks were observed in optical images.
4. *Anti-fogging*. Condensed droplets form into a thin and continuous water layer with transparency on superhydrophilic surfaces for rapid evaporation (Brown and Bhushan 2015). The anti-fogging properties were evaluated by the samples on substrates placed above boiling water for 5 s, and optical images were taken to appraise their transparency.
5. *Resistance to hard environments*. Some materials are required to work on hard environment including elevated temperature, humidification, soaking, or chemical ambiance, hence liquiphobicity of the coating was examined on heating substrates or under humid or soaking, and acidic/alkaline environment.

The nanoparticle/binder coating is composed of high hardness nanoparticles and durable resin to create hierarchical roughness for higher contact angles and lower tilt angles (Martin et al. 2017; Bhushan and Martin 2018). This coating with enjoyable adhesion to the substrates would render greater durability for substrates by fluorosilane deposition and one-step spray technique. Due to the highlight of facile fabrication steps and flexibility of substrate, the nanoparticle/binder-coating is appealing for superhydrophobic surfaces with more functions including resistance to special environment (Bhushan 2019).

6.3.3 Biomedical Applications

Superhydrophobic properties exhibit numerous functionalities given the rough micro/nanostructure of a low surface energy material with stable non-wetted state compared to traditional biomaterials with metastable wetted state in psychological environment. Bioinspired biomaterials with superhydrophobic surfaces have received great consideration in biomedical application, such as medical coating with antibacterial activity, drug/gene carriers, disease theranostics, and cell/biomaterials screening (Falde et al. 2016). Moreover, some superhydrophobic biomaterial could construct cellular microarrays to control superhydrophobic and superhydrophilic areas, and support droplets or fluid flow on disposable microfluidic devices.

6.3.3.1 Medical Coating

Superhydrophobic materials possess reduced wetted area and rough morphology against bacterial adsorption and cell attachment applied for medical coating. Hasan et al. designed a nanostructured superhydrophobic "super surface" with a static water contact angle of 154.0° and contact angle hysteresis of 8.3° by deep reactive ion etching technique (Fig. 6.8) (Hasan et al. 2015). The surface is constituted by randomly arranged nanopillars with height of 4 μm and diameter of 220 nm which mechanically rupture bacteria including gram-negative (*Escherichia coli, E.coli*), gram-positive (*Staphylococcus aureus, S. aureus*), and even mammalian cells (mouse osteoblasts) for potential surfaces of microfluidics, surgical instruments, and other self-cleaning and anti-bacterial devices.

The insufficient hemocompatibility of medical devices is a critical problem that may lead to life threat, especially for important organ assistants or disease treatment, and thus, inhibiting thrombosis induced by surface of blood-contacting medical devices could alleviate chronic inflammation around implanted sites avoiding device failure. Platelet adhesion to devices depends largely on the superhydrophobicity of surface. Noticeably, superhydrophobic surface, usually used fluorocarbon coatings, can also be exerted as anti-thrombotic coating to the decrease of platelet

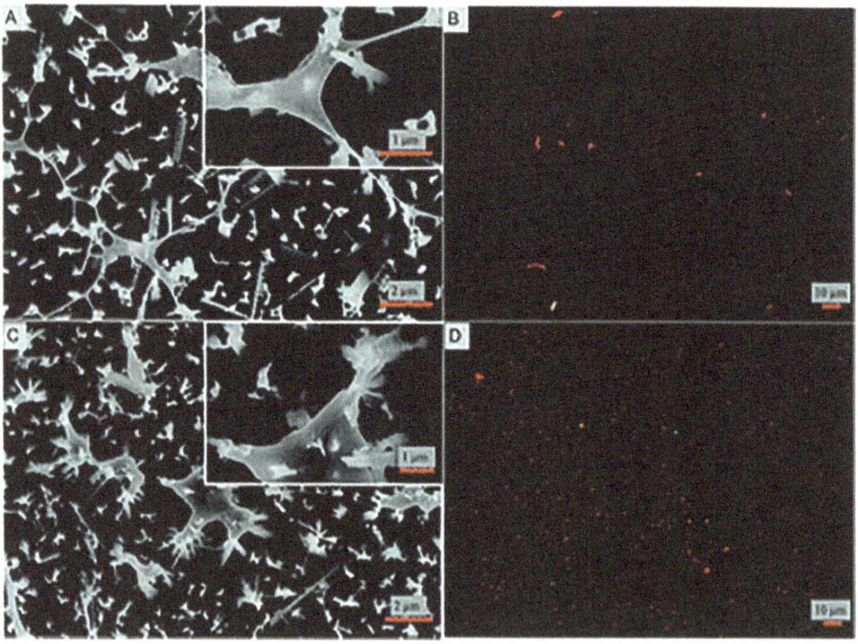

Fig. 6.8 Nanopillar-constructed surface could rupture bacteria. (**a** and **c**) Scanning electron micrographs and (**b** and **d**) fluorescent microscopic images of bacterial attachment, **a**, **b** for *E. coli* and **c**, **d** for *S. aureus*, green for viable cell and red for non-viable cells. (Reprinted with permission from Hasan et al. 2015)

attachment and coagulation in that highly hydrophobic surfaces with low surface energy can initially promote the adsorption of proteins and then easily desorption (Kumar et al. 2010). Based on the principle, Hasebe et al. fabricated a biomimetic surface for blood-contacting medical devices. The surface is coated with fluorine-incorporated diamond-like carbon (F-DLC) film, and possesses dual roughness with hydrophobicity like lotus-leaf. F-DLC has great prospect for coating clinically temporary blood-contacting medical devices due to its non-thrombogenicity that has been evaluated by the assessment of incubation with platelet-rich plasma isolated from human whole blood (Hasebe et al. 2013). In another case, superhydrophobic surface was fabricated by the graphene and titanium dioxide nanoparticles with antibacterial and antibiofilm properties (Wu et al. 2023). Poly (lactic acid) (PLA) film with the superhydrophobic coating could induce the low surface wettability against *S. aureus*, Klebsiella pneumoniae, and Pseudomonas aeruginosa. In addition to the inhibition of the formation of biofilm, the PLA film with superhydrophobic coating possesses great biocompatibility with hemolysis less than 2% and benefit to heart cells and kidney cells, which has the potential for single-use blood-contacting medical devices. Moreover, pH-responsive smart superhydrophobic surface was designed with unique controllability between dual bactericidal and releasing properties (Song et al. 2021). The smart surface was fabricated by electroless silver deposition on a copper mesh with modification of a mixed thiol solution of 1-dodecanethiol and 11-mercaptoundecanoic acid so as to form rough pine branch structures depositing on the copper mesh. The surface with exposed silver ions possesses the switchable wettability between superhydrophobicity with low adhesion to kill bacteria (pH = 7) and hydrophobicity with high adhesion (pH = 9) to release dead bacteria.

6.3.3.2 Drug Carriers

In respect of theranostics, hydrophobic drug carriers could slow dissolution both of hydrophobic and hydrophilic drugs for sustained release. For instance, 3D superhydrophobic electrospun meshes were prepared by poly (ε-caprolactone) (PCL) with poly(glycerol monostearate-co-ε-caprolactone) (PGC-C18) as a drug carrier of anticancer drugs (CPT-11 and SN-38) (Yohe et al. 2012). With the addition of PGC-C18, the hydrophobicity of electrospun mesh increased to delay the drug release for more than 90 days, and possess preferable mechanical properties for surgical buttressing of the anastomosis. Besides, several superhydrophobic surfaces (polystyrene, aluminum, and copper) were utilized to prepare temperature-responsive hydrogel beads as drug carriers (Lima et al. 2011). Hydrogel beads were fabricated by dextran-methacrylated, poly (N-isopropylacrylamide) incorporated with bioactive substance (insulin or albumin) via photo-crosslinking followed by dropping on the superhydrophobic substrates. These smart hydrogel particles as protein carriers with almost 100% encapsulate rate possess great promise for pharmaceutical or other biomedical applications.

6.3.3.3 Functional Cell Configuration and Others

Based on the inspiration of superhydrophobic surface of lotus leaf, a biomimetic fabrication process was proposed by Song et al. to construct hydrogel and polymeric spheres using superhydrophobic substrate for cell expansion, cell encapsulation, and controlled release of molecule. Potential advantages of the fabrication process include easy operation to prepare sphere, high encapsulation efficiency and loading, low cost with eco-friendly technology, and mild processing conditions (Song et al. 2010). For instance, microspheres of dextran-methacrylate (dextran-MA) and γ-cyclodextrins (γ-CD) loaded with dexamethasone were prepared by photopolymerization on biomimetic superhydrophobic substrates (Lima et al. 2014). Due to copolymerization of dextran-MA with γ-CD-NMA, microspheres possess enhanced loading capacity and sustained release character for osteogenic differentiation and bone regeneration that is verified by alkaline phosphatase staining. Thus, dextran-MA/γ-CD spherical microgels implemented by photopolymerization of droplets onto superhydrophobic substrates as cytocompatible implantable spheres could be applied in bone tissue engineering. Similarly, collagen/platelet lysate gel beads were prepared by the same method for skin ulcers treatment (Lima et al. 2015). Resorting to the biomimetic superhydrophobic surfaces, easy-to-handle formulations were obtained to encapsulate platelet lysate into collagen millimetric gel beads for sustained release of multiple growth factors supporting skin regeneration. Intriguingly, the gel particles enhance hASCs cell proliferation, migration, and connection from the particles contributing to the spontaneous formation of membrane in the target site, which provide the possibility of efficiently handleable autologous product rather than expensive recombinant growth factors. Additionally, superhydrophobically polymeric surfaces were harnessed as array-based screening flat platform for combinatorial cell-3D biomaterials screening in the fields of tissue engineering (Salgado et al. 2012). Under the UV/ozone radiation, arrays of hydrophilic regions were patterned with combinatorial hydrogel spots depositing onto the devices, and binary alginate-based polymeric arrays were also attained with different proportions of three different polymers (chitosan, collagen, and hyaluronic acid). The value of cell viability and number reflects the different cytocompatibility of different substrates to different cells.

6.4 Material Biomimetics: Mussel Foot Proteins-Inspired Biomaterials with Bioadhesion

6.4.1 Fundamentals

Marine mussels, belonging to the Mytilidae family, are mollusks that prefer to live in cold ocean. Despite being constantly washed by waves and tides, mussels can adhere to various surfaces including rocks and piles in the ocean due to their strongly adhesive secreta generated by the mussel foot (Lee et al. 2007). Recently, the mechanism of underwater adhesion of these adhesive secreta has attracted considerable interests by researchers. Their underwater adhesive property resorts to mussel foot proteins (mfps), and five proteins (mfp-2/3/4/5/6) have been found in mussel plaques, two of which mfp-3 and mfp-5 rich in 3,4-dihydroxyphenylalanine (DOPA) containing catechol functional groups play significant role in adhesive mechanism(Fig. 6.9) (Guo et al. 2020). The varieties of interactions (e.g., oxidation of catechol oxidase and complexation of ions) of DOPA render crosslinking density and adhesive strength, allowing the under-water adhesion to a wide range of surfaces (Stewart et al. 2011). Specifically, unstable DOPA is often easily oxidized to DOPA-quinone by strongly irreversible covalent bonds to enhance the adhesive strength with organic surfaces, but in the accompaniment with extremely weak interactions with inorganic surfaces (Han et al. 2017a; Lee et al. 2006). Beyond hydrophobic interactions or electrostatic shielding provided by nonpolar amino acids (e.g., glycine) of DOPA groups, acidity-alkalinity of environment could control the oxidation degree of DOPA for the balance between adhesion and cohesion (Cui and Liu 2021). DOPA is reposited in weak-acidic adhesion glands along with inactivated catechol oxidase in ventral grooves, and secreted DOPA would be oxidized after catechol oxidase activated by hydrolysis once encountering to weakly alkaline environment

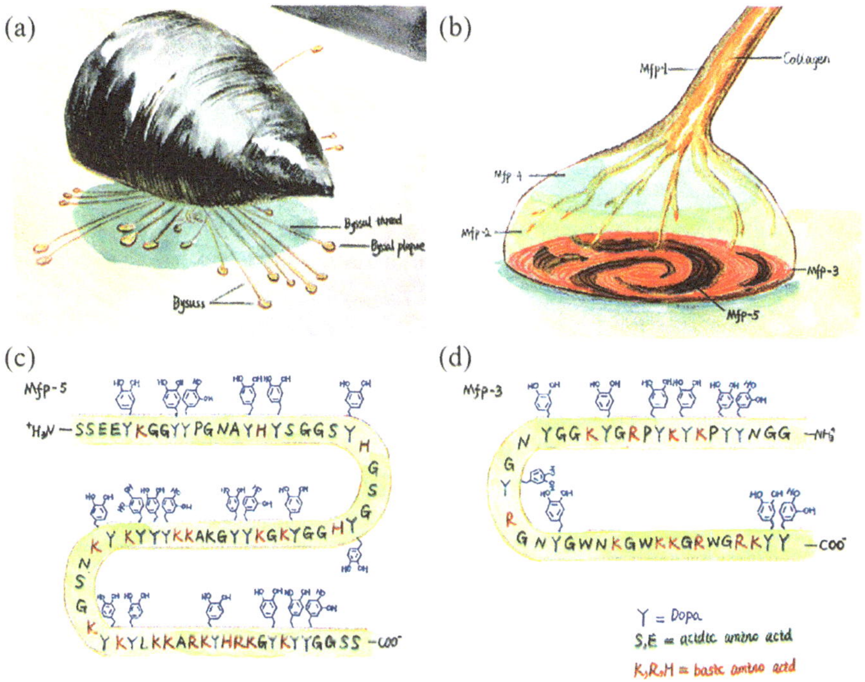

Fig. 6.9 The scheme of mussel foot and its secreted protein. (**a**) Schematic byssal threads of mussel. (**b**) Schematic byssal plaques adhering to substrate. (**c, d**) Primary sequence of mfp-5 (**c**) and mfp-3 (**d**). (Reused with permission from Zhang et al. 2020a)

of seawater and sea coasters. Under seawater conditions (pH = 8.2), the persistence of unoxidized catechols and oxidized DOPA-quinones has access to attachment to both inorganic and organic substrates given that catechols offer non-covalent interactions such as p-p electron interactions, cation-p electron interactions, and coordination with metal ions or metal oxides to adhesion to inorganic substrates, and DOPA-quinones involved with covalent cross-linking reactions of functional group of proteins and reversible catechol-Fe^{3+} ligand interactions for attachment to organic surfaces (Nicklisch and Waite 2012). Additionally, hydration film as a barrier between the liquid and the substrate surface is another critical resistance to wetted adhesive interactions, however, they can be penetrated by cationic residues adjacent to DOPA in mfps (Wang et al. 2017; Kamino et al. 2012).

Hydrogel as a kind of classically soft material applied has been extensively developed in biomedicine and soft electronics, but limited in some specific working environment, for example, harsh and humid environments (Hu et al. 2018; Yuk et al. 2019). The adhesive mechanism of marine mussels provides inspiration for wet-adhesive materials, and myriads of natural or artificial materials have been produced from the repository of polyphenolic compounds based on mussel-inspired chemistry, such as dopamine, DOPA, proanthocyanidins, pentagalloylglucose, epigallocatechin-3-gallate, tannic acid, and gallic acid (Yang et al. 2018; Faure et al. 2013). Polyphenolic compounds with similar structure of mussel proteins including quinone and catechol moieties also possess material-independent and wet adhesion given covalent bonding with some functional molecules and non-covalent self-assembly, thus multiple forms such as nanoparticles, coating, and polymers have exploited as building blocks for hydrogel (Saiz-Poseu et al. 2019). In practical design, moving beyond superior adhesion mentioned above, tough and stretchable hydrogel with self-healing ability could be easily realized due to the reversibly broken catechol-based dynamic bonds for energy dissipation (Han et al. 2017b). Additionally, diverse building blocks with multifold interactions and bio-functions render more dimensions and possibilities to hydrogel supporting further scenarios (Zhang et al. 2018b). Intriguingly, mussel-inspired hydrogels more often possess the capacity of cross-link with collagen (He et al. 2011), inhibition of protein hydrolysis (Busenlehner and Armstrong 2005), and resistance to calcification of elastin in vascular grafts (Katiyar 2006). Their biocompatible compounds have also been studied for photothermal, antioxidative, antimicrobial, anti-cancer activities, as well as their potential therapeutic effects on various diseases. Bearing cationic and zwitterion, polyphenols can be dedicated to elimination of bacteria (Zhang et al. 2020b). Proanthocyanidins derived from cocoa beans and grape seeds and epigallocatechin-3-gallate from green tea are leveraged to scavenge free radicals for anti-inflammation and anti-oxidation in the treatment of liver fibrosis, gastric ulcers and colorectal cancer (Fine 2000; Ariga 2004; Tipoe et al. 2010). Pentagalloylglucose without pro-oxidant activity in aqueous media compared to gallic acid or tea polyphenols could induce tumor apoptosis (Lee et al. 2011a; Zhang et al. 2009). Thus, these mussel-inspired materials become promising candidates for biomaterial applications, and especially come into play in soft and hard tissue repair like dental restorations (Zhang et al. 2020a).

6.4.2 Mechanism, Design, and Characterization

Bioadhesion is a rather common phenomenon in nature resulting from different superficial physical forces implemented by bioadhesives consisting of proteins, polysaccharides, and lipids (Alegre-Requena et al. 2016). Bioadhesion has been coined to the adhesion to biological tissues since 1970s (Pandey et al. 2020a). Noticeably, adhesion emphasizes the interaction of dissimilar atoms or molecules at the interface between materials and substrates rather than similar atoms or molecules within the material termed as cohesion. The mechanisms of bioadhesion could be divided into intermolecular bonding, chain entanglement induced by penetrating interactions of adhesives, electrostatic bonding and mechanical interlocking at the interface, and surface tension. Among these mechanisms, intermolecular bonding is deemed as the principal adhesive force including covalent interaction and van der Waals forces (Pandey et al. 2020b). The superficial adhesive forces contain chemical bond-based primary forces (cationic, ionic, and metallic chemical bonds) and molecular secondary forces (hydrogen bonding and van der Waals forces). In mussel biochemistry, dihydroxy groups can bind with mucosal tissues through hydrogen bonds, while the benzene ring possesses π–π interactions, cation–π interactions, robust reversible bonds and interfacial bonds with other aromatic rings, positively charged ions, metal ions, and metal oxide respectively (Lu et al. 2013). Mechanical interlocking is dictated by the microscopic roughness of tissue surface in the consequence of superficially interlocking bond of adhesive infiltrating in the defects and crevices of tissue surface, while electrostatic interaction occurs at the electrically layered interface between adhesive and tissue driven by their electronic transfer under attractive force (Modaresifar et al. 2016; Khanlari and Dubé 2013). Mfp-3 s with hydrophobic tryptophan side chains could eliminate hydration layer at the interface to promote the contact and attachment of adhesive (Maier et al. 2015). Moreover, modifications of electron withdrawing groups also can significantly increase their adhesive capacity (Cencer et al. 2015). Chain entanglement will generate semi-permanent bonds caused by the interpenetration between the adhesive and tissue in relationship with their crosslinking density (Marshall et al. 2010). Relative lower surface tension of adhesives compared to the adherent provides the foundation of adhesion. Expected adhesiveness is required for small contact angle between adhesive and substrates to contact surface area as much as possible by wetting surface. The formation of coacervate in mussel avails to improve surface wettability for underwater adhesion (Kord Forooshani and Lee 2017).

As for the general synthesis strategies of mussel-inspired tissue adhesives, there are polyphenol modification of polymer, polymerization of monomers with polyphenol group, and synthesis of peptides. Diversity of polymers has been certified that they can be modified or grafted with manifold polyphenol groups including DOPA and catechol via the linkages of amide, urethane, and ester (Mehdizadeh et al. 2012), and their adhesive property could adjust by molecular weight and structure of polymer as well as types of polyphenols (Lee et al. 2011b). Polyphenol groups also can conjugate onto the chain of block copolymers (Brodie et al. 2011), and covalently crosslink with biopolymers, for instance, chitosan, and hyaluronic acid (Ryu et al. 2015; Lee et al. 2013; Hong et al. 2013). Mentioned polymerization, dopamine methacrylamide (DMA) as a typical catechol-containing monomer has crosslinked with other monomers including common polyacrylamide (PAAm) to generate dopamine-grafted polymers (Lee et al. 2007). The last strategy is the synthesis of catechol-containing peptides by artificial technique or genetic engineering, and its unique advantage is customizablity (Jeon et al. 2015).

The adhesion measurement of adhesives can be characterized by adhesive strength for adhesive force per unit area and adhesive toughness for the energy per unit area to separation. There are several common adhesion tests of adhesive: tensile test and lap shear test for adhesive bonding strength along normal, shear direction respectively, peeling test for adhesion energy and average peel force at the defined angle and speed, and probe-on-flat indentation and flat punch indentation for loading-holding-unloading contact adhesion analysis (Zhang et al. 2020b). The tensile test and lap shear are general techniques for measurement of adhesion strength, while the peeling test as the evaluation indexes of adhesive toughness is suitable for hydrogels peeling off the substrate. Indentation tests offer more information on mechanical properties including elastic modulus and relaxation behavior of adhesives, and their probes are classified into hemispherical (or curved) probe and flat punch, among which plat punch is appropriate for hydrogel in underwater given the unobservable view of contact area induced by refractive indices (Cholewinski et al. 2017).

6.4.3 Biomedical Applications

Mussel biochemistry confers mussel-inspired biomaterials, especially hydrogel, numerous unique physical/chemical properties including superior interfacial adhesion, biocompatibility, and diverse physiological-adaptive mechanical properties. Increasing attention is drawn to mussel-inspired biomaterials in biomedical engineering, from tissue engineering, antibacterial application, to drug carriers and more biological application.

6.4.3.1 Tissue Adhesion

Medical tissue adhesives are utilized for adhesion and repair of tissues. Polyphenol-functionalized hydrogel possesses unique wet-adhesion and self-healing and hemostasis advantages, so can be prepared as underwater soft tissue adhesives. For example, Zheng et al. prepared a tissue adhesive-based poly (ethylene glycol) diacrylate (PEGDA) hydrogel interpenetrated with N-hydroxy succinimide (NHS) conjugated alginate (Alg) (Fig. 6.10) (Zheng et al. 2023). Tannic acid (TA)/Iron (III) chloride (Fe^{3+}) as the source of polyphenol potentiates the wet tissue adhesion (>400 kPa in a wound closure test) and hemostasis due to the high affinity of TA Fe^{3+}

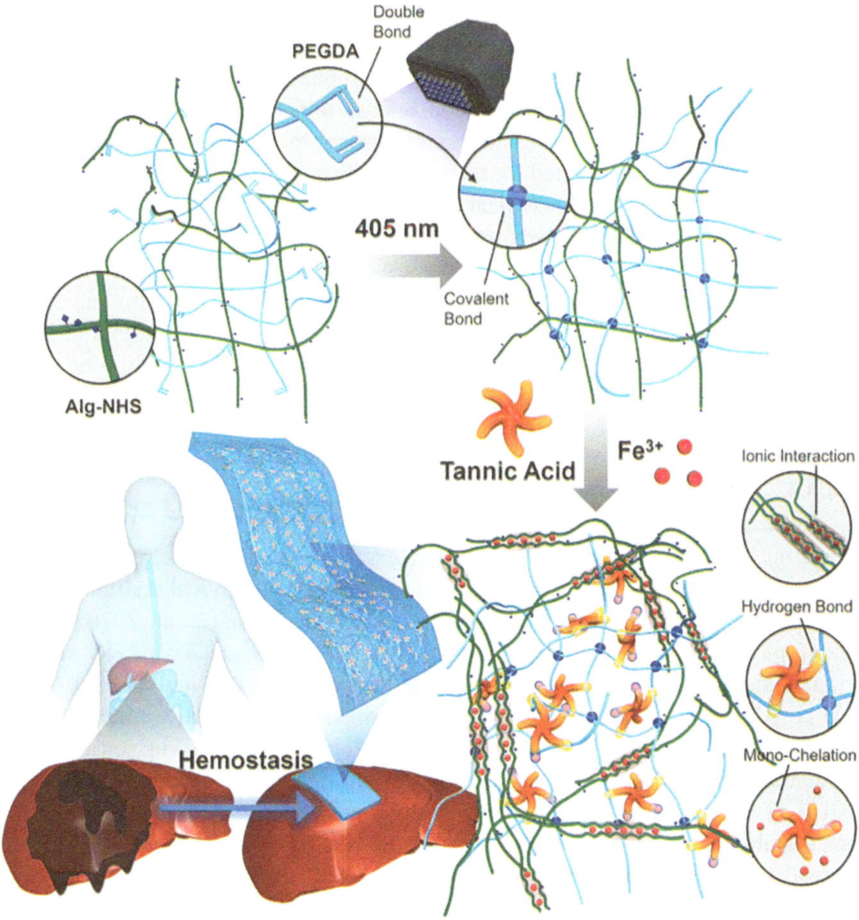

Fig. 6.10 Schematic illustration of the preparation of APTF hydrogel based on Alg-NHS, PEGDA, and TA/Fe^{3+}. (Reused with permission from Zheng et al. 2023)

with blood. The ready-to-use hemostatic tissue adhesive also possesses expected elasticity and is tough to adapt to physiological action, and thus they are suitable for both internal and external injuries. In another case, artificial-synthesis catechol derivatives with acrylic acid and other monomers formed on-demand detachable wet tissue adhesive by photo-induced copolymerization (Chen et al. 2023). The adhesion strength of the adhesive to wet porcine skin is up to 153 kPa, and the adhesive can easily peel off porcine skin in 5 wt% aqueous urea solution.

6.4.3.2 Drug Delivery

Mussel-inspired adhesive hydrogels loading with therapeutic agents are conducive to reach targeting incision for sustained release (Khanlari and Dubé 2013). For instance, catechol-functionalized chitosan hydrogel loading with genipin and sulfasalazine (SSZ) was fabricated for ulcerative colitis. The hydrogel could adhere to mucous to prompt the pharmaceutical efficiency. In the animal models of ulcerative colitis, the adhesive hydrogel performed better therapeutic efficiency in contrast by oral administration of SSZ, and also possesses the potential in colon cancer, or related diseases (Xu et al. 2017). Hence, adhesive drug delivery systems can adhere to biological tissue to improve delivery efficiencies and extend retention time without potential systematic side effect, which provides the possibility of clinic translation.

6.4.3.3 Other Applications

Some mussel-inspired hydrogels have been exerted as flexible electronics and soft actuators given their unique advantages of skin-like physiological and mechanical properties, conductivity, and stimulus-responsiveness (Zhang et al. 2020c). They are employed as skin-like sensors and implantable devices to record electromyographic signal by adhering to human skin. For instance, polyvinyl alcohol (PVA) and polydopamine (PDA) supramolecular hydrogel with conductive functionalized single-wall carbon nanotube (FSWCNT) coating severed as skin sensor could reversibly adhere to skin surface for monitoring of human activities (Liao et al. 2017). The mussel-inspired conductive hydrogel possesses extraordinary self-healing ability and robust adhesiveness, in addition to wireless remote monitoring of real-time human motion in both large and tiny scale, which has great prospect in soft electronics for personal healthcare. Furthermore, a catechol-based hydrogel was prepared by hyaluronic acids and plant-derived polyphenol (i.e., gallol) for detection of electromyographic signal of myocardium and skeletal muscles (Fig. 6.11) (Shin et al. 2019). With metal nanoparticles, the granular hydrogels obtained higher conductivity (≈ 0.05 S cm^{-1}) than other bulk hydrogels, microgels, or microgels, and could restore electrical conduction by bridging the separated muscle tissues.

On top of that, mussel-inspired hydrogels with gradient structures can be leveraged as soft actuators through diverse dynamic interactions. For example, dopamine/poly(vinyl alcohol) adhesive with metal ions (Fe^{3+}) was adapted as interface to develop a soft, ultrathin (~50 μm), self-adhesive, and skin-compliant sensor. The interface with tailorable modulus could bind two active layers (biaxially oriented polypropylene and paper) for actuation responding to heat, NIR light, and voltage along with self-locking and unlocking by removing stimuli and humidity respectively (Chen et al. 2019).

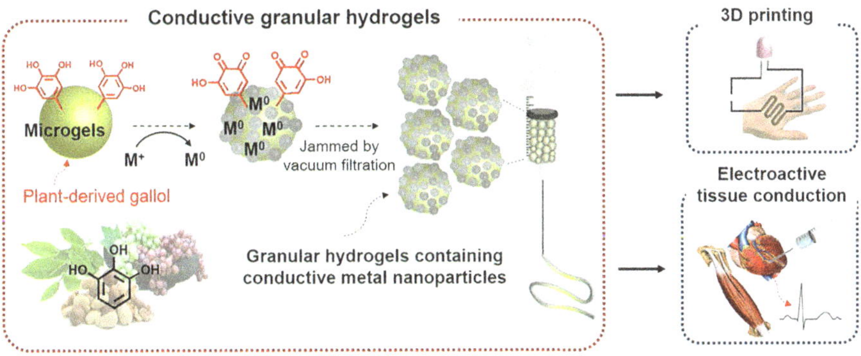

Fig. 6.11 Schematic illustration of injectable and conductive granular hydrogels for 3D printing and electroactive tissue support. (Reprinted with permission from Shin et al. 2019)

6.5 Biomimetic Micro/Nano-Devices: Viral, Cell Membrane, and their Derivatives as Biomimetic Delivery Platforms

6.5.1 Fundamentals

In biological bodies, manifold cell membranes constituted of biological macromolecules (such as protein, peptidoglycan) are responsible for wide biological functions including energy conversion, material transport, and information transition, which render more inspiration for biomimetic delivery platforms (Li et al. 2020a). Vesicles as diverse, nanoscale membrane derivative from cell, especially cancer cells and host cells of mammal, can be classified as exosomes, micro-vesicles, and apoptotic bodies. In the past, extracellular vesicles (EVs) have been thought to be "cell dust", dispose of cellular components, whereas, are esteemed stable sources of circulating biomarkers of intercellular communication and circulating for disease diagnoses and prognosis (Théry et al. 2009). In normal physiological environments, biofluids carried copious EVs with macro/micro-molecules to other sites, while in cancer settings, some host cells can remove EVs of tumor for tumor suppression although EVs are often protumorigenic (Keerthikumar et al. 2016; Pucci et al. 2016). EVs, as new targets for bioassays with the size in the range of 10–30 μm, initially are exerted as repeatedly and minimally invasive cancer monitors for time-dependent assessment of tumor molecular information such as tumor type and stage, molecular driver mutations, and treatment progression (Van der Pol et al. 2012). Exosomes as the darling of drug delivery system are a kind of small membrane vesicles (30–150 nm) secreted during cellular exocytosis process. Exosomes comprised of peripheral and transmembrane proteins also can stimulate immune responses by endocytic pathway (Rashed et al. 2017). These exosomes are usually generated from secretory endosome of cytoplasmic membrane, and undergo the formation of intraluminal vesicles within multi-vesicular body, maturation, and extracellular release (Tamura et al. 2021). They are burden with the important job of cell-cell communication and transportation with specific molecules on the surface.

6.5.2 Classification, Preparation, and Characterization

Initially, some polymersome-inspired cell membrane could self-assemble into lipid bilayer-like structure to deliver cancerous drugs with stability, responsibility, and certain bio-functionality (Wong et al. 2017). Whereas, these primarily biomimetic delivery systems lack specificity and immune properties. Nowadays, increasing advanced biomimetic delivery systems have developed owing to pioneering nanotechnology, and biomimetic delivery systems are classified into three major groups: cell membrane, vesicles, and membrane-coating nanoparticles. They fuse the advantages both of biological devices and synthetic therapeutic agents with less obstacle induced by immune system along with inherent properties and functions of the original cells (Fang et al. 2018). Bioinspired surfaces provide the possibility of the conjugation of passivating agents and targeting moieties to the nanoparticles. Importantly, lipid bilayers are fabricated by biomimicking cell membrane to transport the substances from nanoparticle core, accompanied by the incorporation of ion channel protein for controllable release of the therapeutic agents (Liu et al. 2009). Similar cell-membrane cloaking nanovesicles have been applied as therapeutic cargoes. Cell-derived vesicles including exosomes and extracellular vesicles from red blood cells (RBCs), cancer cells, immune cells, and so on (Fig. 6.12) (Xia et al. 2019).

1. **RBC and platelet membrane**. RBC membrane, namely erythrocyte membrane, as the most abundant blood cell in vertebrate blood is responsible for transportation of O_2 and CO_2. Owing to superior biological compatibility and longer lifespan than others (more than 120 days), they are usually exerted as stealth coating or cargoes to delivery vaccine/therapeutic agents, and transfuse blood with decreased risk of the rejection immune system (Xia et al. 2019).

   Platelets, anucleate blood cells originating from fragmentation of megakaryocytes for blood coagulation and even tumor metastasis can evade the immune system owing to CD47 receptors of their surfaces, and target CD44-receptors overexpressed cancer cells owing to the expression of P-selectin (Gay and Felding-Habermann 2011; Wang et al. 2019).
2. **Tumor cell membrane**. Malignant tumor more often possess immunogenic tumor-specific antigens but are eliminated by negative selection leading to inactivity of immunity, and thus tumor cell membrane carried with rich antigenic materials of tumor cells would be a potential tumor vaccine to rekindle immune system against malignant tumors (Liu et al. 2022b). Besides, surface adhesion molecules on the surface tumor cells lead to their aggregation and homologous targeting, which has access to cargoes of therapeutic agents (Li et al. 2020b).
3. **Immune cell membranes**. Immune cells with expression of immune-related receptors and immune-modulating proteins also serve as a membrane source for

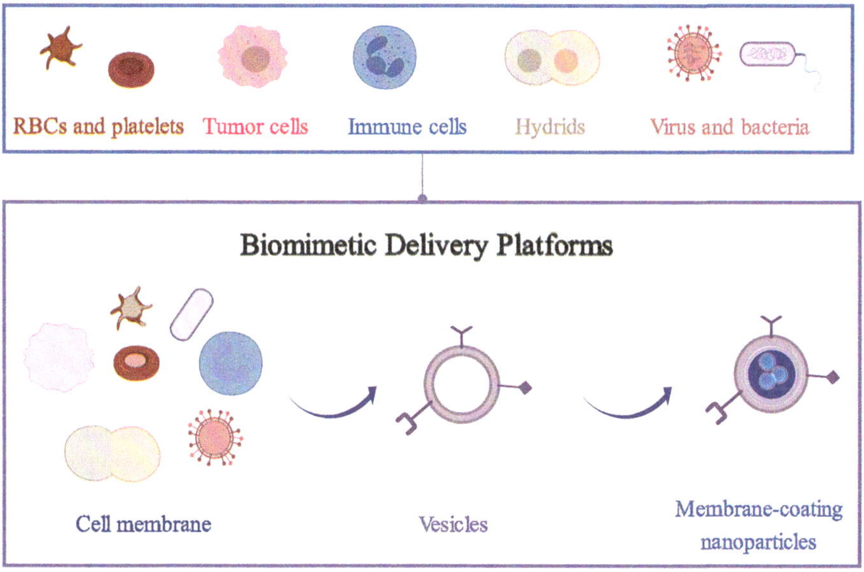

Fig. 6.12 Biomimetic delivery platform classified into cell membrane, vesicles, and membrane coating nanoparticles generated from diverse types of cells and virus

the delivery of therapeutic agents, for instance, neutrophils, T cells, macrophages, dendritic cells (DCs), and natural killer cells. However, these populations of inflammatory cells are also related to progression of the tumor, and even become tumor accomplices (Chai et al. 2019; Chen et al. 2016). Reversely, these immune cells could be as cloak of therapeutic agents to tumor targeting owing to their inflammation chemotaxis.

4. **Hybrid cell membranes**. For combined with functions of different cell membranes, different types of cell membrane fusing technology have been developed (Xu et al. 2021).
5. **Virus and bacteria-inspired systems**. Bacteria with pathogen-associated molecular pattern could activate innate immune response of host followed by improved adaptive immunity resembling immune adjuvants, which commonly selectively colonize in necrotic/hypoxic tumor areas, hence, they are utilized as targeted cancerous drug delivery systems or therapeutic agents (Zhou et al. 2018). Furthermore, engineered bacteria may be required by genetic or chemical modification modifications to further enhance anti-tumor immunity.

Analogously, on the one hand, oncolytic viruses could selectively destroy tumor cells followed by systematic antitumor immune responses; on the other hand, viruses with naturally immunogenic can also express tumor antigens by genetic engineering applied for cancer vaccines. Viral capsids also could self-assemble into virus-like particles to naturally induce immune responses as the same as virus (Noad and Roy 2003).

6. **Vesicles**. Exosomes are created by the endosomal membrane pathway accompanied by the presence of multivesicular body for the degradation of vesicular contents fused with the lysosomes or releasing vesicles considered as exosomes to extracellular matrix fused with plasma membrane (Kowal et al. 2014). The immunity of the host to EVs with hydrophilic polymers could be escaped by biomimicking to achieve long circulation in the bodies and improved treatment efficiency (Reintjes et al. 2008). Additionally, targeting ligand-receptor recognition could actively target tumor, while cell-penetrating peptide facilitates the delivery of EVs or particles across biological barrier (Chrastina et al. 2011; Said Hassane et al. 2010). Compared to inorganic nanocarriers, soft EVs could change their structure to control the release of therapeutic agents to targeted sites along with enhanced performance under the external stimuli or in the target microenvironment (Branco and Schneider 2009).
7. **Membrane-coating nanoparticles**. Membrane coatings as a dynamic and active carrier are also developed to enhance nanoparticles biocompatibility (Gulati et al. 2018). For instance, membranes of red blood cells (RBCs) coatings with lipid bilayer and membrane proteins exhibit longer blood circulation and retention times, higher uptake by the tumor cells, and deceleration in blood clearance, compared to conventional polyethylene glycol coatings (Piao et al. 2014; Ren et al. 2016).

Behind these vesicles, surfaces and structure are harnessed as biomimetic approaches of preparation and development. Preparation of cell membranes and their derivatives involves three basic types: lysis of the parental cells with a hypotonic buffer, centrifugal separation of membranes and other cellular components, collection of micro/nanovesicles with suitable size by disruption methods such as sonication, extrusion, homogenization, and nitrogen cavitation (Gao et al. 2016). Additionally, viral and cellular vesicles are commonly obtained by means of density-based ultracentrifugation, sucrose-gradient centrifugation, co-precipitation, as well as molecular weight-based chromatography, fluidic systems (Prasad and Conde 2022). Among these techniques, ultracentrifugation and sucrose-gradient centrifugation are deemed as gold standard in the preparation techniques though with time-consuming (>4 h), large size, and low recovery (Shao et al. 2018). Further, functional cell membrane and their derivatives can be pre-modified and post-modified (Le et al. 2021). Specifically, pre-modification is genetic or metabolomic manipulation on parental cells, while post-modification is introduced by functional molecules like lipids, proteins, and synthetic components with more convenience and selectivity. Not only that, two different parent cell membranes can be hybridized to prepare vesicles using a co-extrusion technique, and these vesicles with hollow core structure could be severed as coating materials of nanoparticles or cargoes of therapeutic agents. There are several approaches to coating nanoparticles: co-extrusion, the mainstreaming method to obtain controllable size of the final particles; ultrasonication, the simple method with possibility of large-scale production; electroporation with membrane integrity (Liu et al. 2019). Furthermore, genetically engineered pathogens have also been applied in cellular vesicles for drug delivery.

As a phase I clinical trial, the genetically modified Lactococcus lactis bacteria are responsible for mucosal deliver of immunomodulatory proteins with less systemic side effects to patients with Crohn's disease (Braat et al. 2006). Genetically modified Salmonella typhimurium has also been utilized as vehicles to deliver anticancer multidrug-resistant gene MDR1 siRNA or transcription factor STAT3 SiRNA, and the tumor growth was inhibited with prolonging the lifespan (Jiang et al. 2007).

As for the characterization of vesicles, electron microscope and atomic force microscope are used for the substructures, morphology and constitutive elements (e.g., lipid, protein), while scanning electron microscopy and transmission electron microscopy (TEM) for the micro/nano-analysis including 3D surface topography, elemental composition, and molecular characterization of dehydrated EVs, in addition to cryo-electron microscopy for morphology, isolated substructure and constitutive elements in the frozen form rather than dehydration (Shao et al. 2018). Moreover, dynamic light scattering is exploited to determine the hydrodynamic diameter, when nanoparticle tracking analysis and tunable resistive pulse sensing are used for concentration and size distribution of EVs (Mao et al. 2019). Last but not least, light microscopic-based single EV analysis (SEA) technique is utilized to measure protein biomarkers in distinct vesicles with higher signal-to-noise ratio than free vesicles in flow liquids (Yu et al. 2023).

6.5.3 Biomedical Applications

Owing to the advantages with natural targeting ability, preferable biocompatibility, high cargo capacity for therapeutics agents, long circulation time as well as escape from immune clearance, these biomimetic delivery systems, especially EVs, deprived from the cellular membrane and cytosol rather than intracellular organelles has been utilized for varieties drug delivery system according to their diverse protein and bio-markers. It is noteworthy that EVs have been applied for the delivery of anticancer drug in that they easily bind with tumor microenvironment (Yu et al. 2020). Moving beyond the high permeability and retention (EPR) effect for enhanced accumulation in solid tumors, these biomimetic delivery systems modified with diverse targeting proteins as "camouflage" commonly work on hindrance of cell-cell communication or bio-marker secretion of cancer cells, in addition to actively delivery of therapeutics agents due to specific targeting and selective attachment to tumor cells (Caro et al. 2021). Cellular vesicles are separated from organelles by ultracentrifugation and adsorbed on the NP surface to incorporate biological properties into the nanostructures, thereby increasing biocompatibility, decreasing allergic reaction in vivo, and enhancing their effectiveness against cancer. The resultant nanostructure carries the full array of cancer cell antigens, thereby offering a robust and innovative platform that applies to multiple modalities of anticancer therapy. These engineered biomimetic features are currently under exploration as advanced drug delivery systems in which natural components are used for theragnostic purposes, bypassing macrophage uptake and systemic clearance, improving therapeutic outcomes.

6.5.3.1 Tumor Theranostics

Although nanodrugs targeting tumor with EPR effect, intratumoral distribution and delivery still limit the drug potency still limited in solid tumor owing to abnormal extracellular matrix. The camouflage of RBC membrane facilitates nanodrug with effective accumulation, immune escape for prolonged blood circulation. A metal-organic framework (MOF) nanodrug (ZIF-8-DOX-LY-RM) with RBC membrane coating was developed for both remodel of tumor microenvironment and improvement of tumor chemotherapy (Qiao et al. 2022). As for the nanodrug, zeolitic imidazolate framework-8 (ZIF-8) as a representative MOF nanocarrier possesses the capacity of desired drug loading and sustained release, while doxorubicin (DOX) and TGFBR1 inhibitor LY364947 (LY) were encapsulated due to the strong interaction with ZIF-8 for enhanced chemotherapy, among which LY was utilized for collagen depletion and hypoxia alleviation. In another case, cancer cells-adhesive platelet exosome combined with photothermal sensitive liposomes was exploited as carriers of glucose oxidase (GOx, G) and ferric ammonium (FAC, F) to construct laser-responsive nanosystem (FG@PEL) (Ma et al. 2022). The nanosystem can realize the cascade synergetic strategy of targeting photothermal therapy (PTT) for the vascular damage and cascaded chemodynamic therapy involved with GOx oxidization and catalysis of FAC, at the same time, thermal generated by PTT also can boost the production hydroxyl radicals for CDT by producing more.

Some malignant tumors including breast cancer and melanoma could adhere to brain endothelial cells for further trans-endothelial migration may be attributed to specific surface receptors and endothelial cell adhesion molecules on the surfaces of these tumor cells (Preusser et al. 2012). Owing to invasiveness of these malignant tumors, the tumor membranes could be extracted as drug delivery carriers. Liu group designed a biomimetic nanocarrier of brain metastatic tumor cell membrane for fluorescence imaging and PTT of early brain tumors (Wang et al. 2020). The nanocarrier can across brain-blood barrier (BBB) resorting to the camouflage of B16F10 mouse melanoma cell membranes and 4 T1 mammary breast cell membranes, and its core is indocyanine green (ICG)-loaded polymeric nanoparticles for both tumor imaging and therapy. Similarly, according to homologous targeting properties, activated fibroblast cell membranes (AF) could homologously target cancer-associated fibroblasts severed as nanocarriers. With the shell of AF, semiconducting polymer nanoparticles (SPNs) were prepared by semiconducting polymer (SP) and poly(cyclopentadithiophene-alt-benzothiadiazole) (PCPDTBT) for targeting fluorescence and photoacoustic imaging as well as combined photodynamic therapy (PDT) and PTT (Li et al. 2018).

Combination of artificial receptor and synthesized receptor could achieve bioorthogonal targeting. Activated T cell membrane with specific immune recognition proteins was labeled on synthesized azide (N_3) via glycometabolism, and the

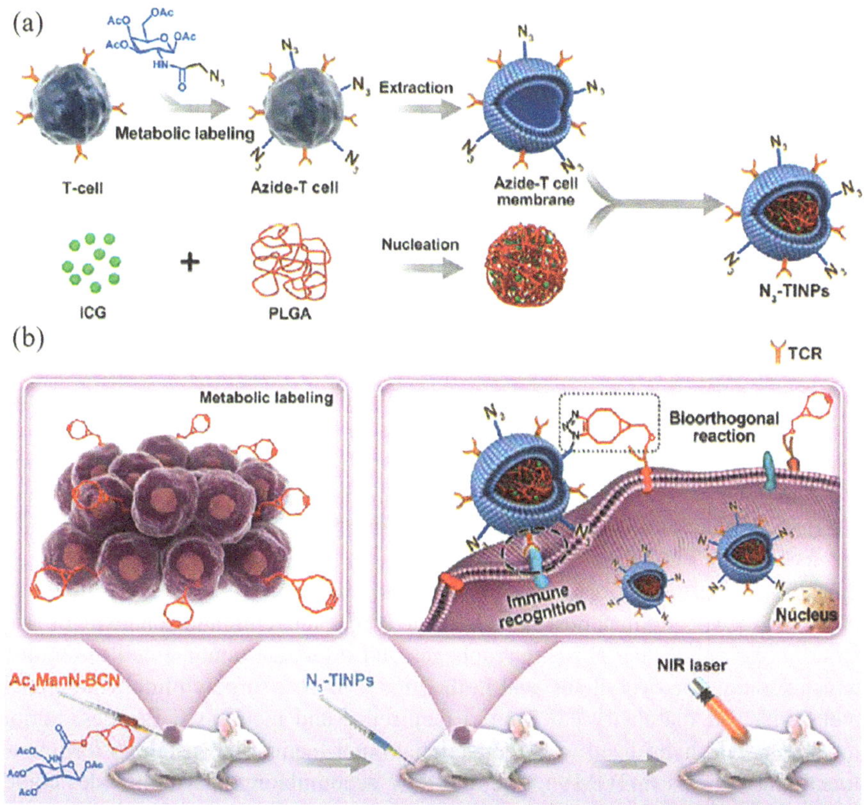

Fig. 6.13 Schematic illustration of T cell-membrane biomimicking nanoparticles (N_3-TINPs) for enhanced PTT. (**a**) The scheme of the preparation of N_3-TINPs. (**b**) The scheme of N_3-TINPs for bioorthogonal targeting and immune recognition mediated PTT. (Reprinted with permission from Han et al. 2019)

modified cell membrane (N_3-T) possess the bioorthogonally targeting capacity (Fig. 6.13) (Han et al. 2019). Taking N_3-T as shell, poly (lactic-co-glycolic acid) (PLGA) loaded ICG as ICG-PLGA polymeric core can form biomimicking nanoparticles (N_3-TINPs) for immune recognition/tumor anchor mediated-enhanced PTT, which can be severed as a novelty targeting drug delivery platform.

6.5.3.2 Tumor Vaccine and Immunotherapy

Tumor cell components like lysates or cell membranes involving tumor-associated or tumor-specific antigens could be severed as tumor antigens for personalized tumor vaccines (Ochyl et al. 2018). In terms of tumor cell membranes with tumor antigens, they can load or coat nanoparticles along with immune adjuvant, and then target antigen-presenting cells and T cells for the activation of the subsequent antitumor immunity (Zeng and Pu 2020). Taking B16 OVA melanoma cell membrane vesicles (CMVs) as example, CpG oligonucleotides, a toll-like receptor 9 (TLR9) agonist as the immune adjuvant were easily inserted onto CMVs followed by the modification of cholesterol (Liu et al. 2021). With cholesterol-modified DC-SIGN-targeted aptamer targeting dendritic cell, the nanovaccines could concentrated in the lymph nodes accompanied by strong antitumor immune responses for suppression of the tumor growth and long-term immune memory effect, which provide an approach for personalized tumor vaccines.

Moreover, genetically programmed cell vesicles combined with an immune checkpoint blockade (ICB) could further enhance the therapeutic responses with long-term immune memory avoiding tumor recurrence. As for ICB therapy, on the one hand, the interruption of programmed cell death-ligand (PD-L1) and PD-1 receptor could restore T cell activation for adaptive immune response (Tsushima et al. 2007); on the other hand, the interruption of CD47 and signal regulatory protein alpha (SIRPα) could improve macrophage phagocytosis for innate immune (Advani et al. 2018). The genetically programmable fusion cellular vesicles (Fus-CVs) were constructed by SIRPα variant-overexpressed 4 T1 murine mammary carcinoma and PD-1-overexpressed B16F10 murine melanoma cell lines though repeated sonication and extrusion techniques with a mini extruder (Meng et al. 2021). Fus-CVs simultaneously block innate and adaptive immune pathway to promote antigen presentation of macrophages and boost antitumor T-cell immunity, which prolong the overall survival in the model of post-surgery tumor (malignant melanoma and mammary carcinoma) recurrence and metastasis. Besides, tumor membrane was hybridized with the bacterial outer membrane vesicle to form new functional vesicles (mTOMV). The mTOMV accumulating in lymph nodes could evoke adaptive immune response of homologous tumor for inhibition of lung metastasis and potential personalized immunotherapy (Zou et al. 2021).

The introduction of contrast agents into vesicle-based tumor vaccine is lacing on the cake. Plasma membranes of activated dendritic cells membrane incorporated into ultrasound contrast microbubbles (MBs) as immunotheranostic microbubbles (DC-iMBs) were developed for induction anti-tumor immune response together with real-time visualization of MBs (Fig. 6.14) (Jugniot et al. 2022). DC-iMBs with T-cell-mediated antitumor effect severed as a novel immunotherapeutic cell-free vaccine realized inhibition of tumor and prolongation of animal survival.

6.6 Molecular Biomimetics: Self-Propelled Nanomotors

From the above elaboration mentioned, biomimetics and bionics have merged into materials design in every respect with the enormous advancement of science and engineering. Further, at the molecular level, bionics means the construction of micro-system with bionic or life-like characteristics, such as dynamism, homeostasis, self-healing, self-reproduction, communication, and responsiveness, which is an integrated and interdisciplinary study containing engineering, natural science

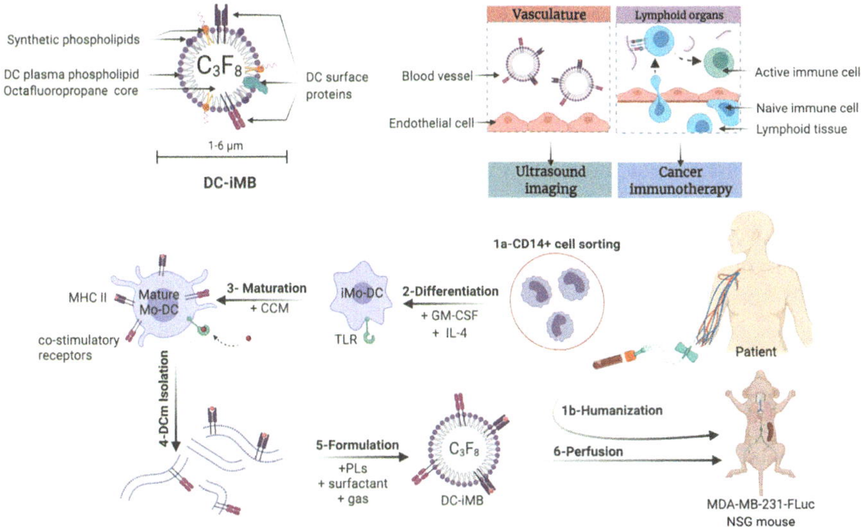

Fig. 6.14 Dendritic cell-immunotheranostic microbubbles (DC-iMB) as vaccine for ultrasound imaging and breast cancer immunotherapy as well as their preparation. *CCM* cancer cell membrane, *DCm* DC membrane, *GM-CSF* granulocyte-macrophage colony-stimulating factor, *LNs* lymph nodes, *MHC* major histocompatibility complex, *PLs* phospholipids, *TLR* Toll-like receptor. (Reprinted with permission from Jugniot et al. 2022)

and life science (Rodríguez-Arco et al. 2019). Precise supramolecular arrangements in different forms build up organisms and biological units with dedicated bio-functions.

Typically, biomolecular motors, also termed as biological nanomotors are constituted of protein units in the size of about tens of nanometers severed as energy generator, intracellular cargo, or muscle actuator (He et al. 2009). For instance, ATP synthase that can synthesis in an artificial environment is command for the catalytic synthesis of ATP molecules with rotary motion (He et al. 2008), while Kinesin with two each heavy and light chains obtained from purification or genetic engineering is responsible for intracellular transport and mitosis (Li et al. 2014). Biomolecular motors with the exceptional performance (controllable and fast motion as well as high cargo loading) and feasible acquisition and modification have been introduced into artificial system, and the self-assembly construction of micro/nanomotors involves templated-assisted layer-by-layer assembly (LbL), self-assembly of polymers with biodegradation, and cell membrane-modified micro/nanomotors with immune escape and biocompatibility (Wu et al. 2016).

To date, these assembled nanomotors have been applied for disease theranostics and healthcare, especially tumor theranostics. Yu et al. prepared microsensor-based antibody(Ab1)-modified self-propelled gold nanoparticle/polyaniline/Pt (AuNP/PANI/Pt) micromotor by templated-assisted layer-by-layer assembly for fast and convenient visualized detection of cancer biomarker (Yu et al. 2014). When testing with target protein presenting, the glycidyl methacrylate microspheres (GMA) modified with secondary antibody (Ab2-GMA) can bind with the moving microsensor resulting in slow movement of the motor, which has access to quick immunoassay method for the target protein with high sensitivity and accuracy. The self-assembly of different polymers (hyperbranched polyamide, chitosan, polylysine, and heparin/folic acid with amino-group) with L-arginine provide general avenues for molecular motor driven by nitric oxide (NO) (Fig. 6.15) (Wan et al. 2019). Meanwhile, NO also renders more biological effects including anticancer and angiogenesis, while fluorescent polymers could monitor the movement of molecular motors, which realize the development of zero-waste and self-destroyed and self-imaging molecular motors.

Surface immunosuppressive RBCs were conferred active propulsion to be severed as autonomous micromotor (Wu et al. 2014). Under the irradiation ultrasound, RBCs loading with iron oxide nanoparticles could self-propel due to generated pressure gradient along with external magnetic field as the switch of motion, which hold potential for drug carrier with long circulation and in-susceptibility to macrophages uptake. Furthermore, RBC-based multicargo-loaded micromotor system was prepared for cancer theranostics (Wu et al. 2015). In the micromotor, in

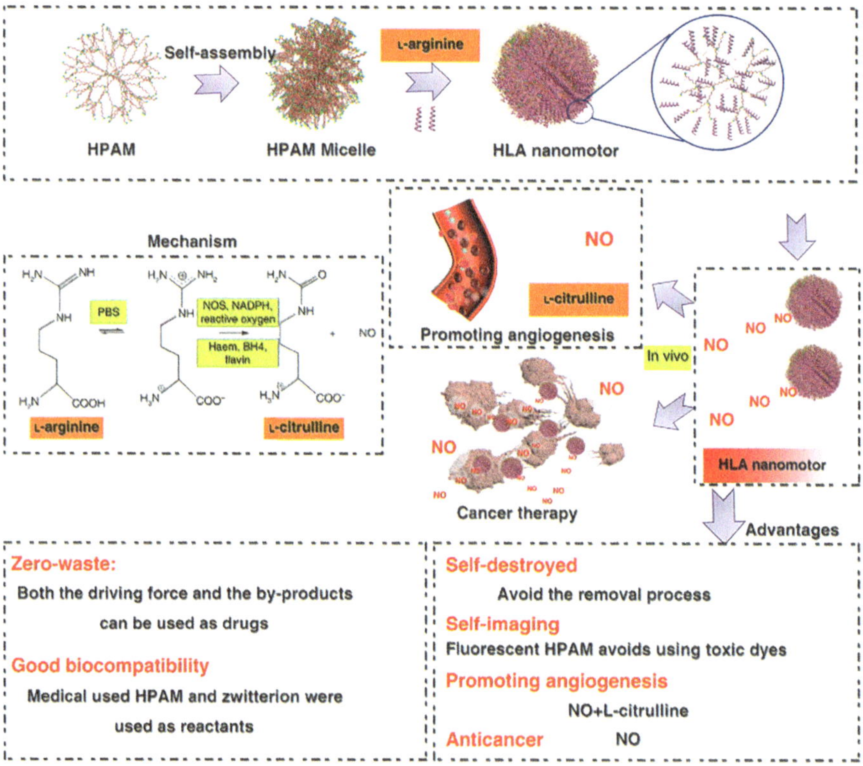

Fig. 6.15 Schematic illustration of the preparation of bio-inspired nitric-oxide-driven nanomotor for cancer imaging and therapy. (Reprinted with permission from Wan et al. 2019)

addition to iron oxide nanoparticles for propulsion and magnetic resonance imaging, quantum dots (QDs) and doxorubicin (DOX) allow the visualization of motors and cancer chemotherapy that offers a novelty and mobile all-in-one platform for imaging, delivering, and monitoring.

6.7 Conclusion and Outlook

Though, as an old topic, biomimetic materials research is still booming attributing to the synergy of biology and material sciences accompanied by the comprehension of structure-function relationships. Originating from designs of organisms, biomimetics that combine with science, medicine, and engineering have become lively impulsion of biomedical development, covering regenerative medicine and disease theranostics. According to different sources of biology, diverse biomimetic principles have been proposed and carried into practice in various forms. Numerous strategies based on simple raw materials from nature have been evolved for adaptive multi-functions, including structure construction, interface design, material simulation, micro/nano-device extraction, and even molecular biomimetics. Therefore, biomimetic biomaterials provide more prospects for devising new strategies and the creation of intelligent biomaterials via a portfolio of bioinspired elements and processes.

The occurrence of nanotechnology undoubtedly provides a new perspective for the application of biomimetics, which avail to the understanding of materials and structures as well as the development of innovative techniques. The prime example is that the discovery and application of hierarchical structural composites achieve inorganic-organic hybrids similar to brick-and-mortar structure.

Liquid-repellent and self-cleaning surfaces of plant leaves provide more inspiration for self-cleaning, antifouling, and antibacterial biomaterials. The development of diversified superhydrophobic surface area of attractive in various applications, especially biomedicine. Nanofabrication techniques provide more abundant and various superhydrophobic surfaces for different demands. Unique micro/nanostructure could either prevent bacterial attachment or inhibit platelet adhesion for coating of medical devices. Additionally, superhydrophobic surface of polymeric materials with fascinating biomolecular interaction could provide more microenvironments to support cell growth in living conditions, and also be exploited as drug carriers to control drug release.

Mussel foot proteins-inspired adhesives have access to superior material-independent and underwater adhesion by biomimicking biochemical mechanism of catechol adhesion, hence they have attracted increasing attraction in healthcare and biomedical engineering in terms of adhesives with poly-phenol motifs suitably adhering to biological tissue. With technological advancement and the deepening of research, more natural adhesive strategies have been biomimicked and combined with artificial technique to create multifunctional and high-performance biomaterials. Further, material science integrated with nature-inspired strategies would be a tendency to develop more biomaterials applied in multiple clinical setting.

The introduction of biomimetic micro/nano-device breaks a brand-new approach in drug delivery systems. These bioinspired targeting carriers with or without potential immunogenicity are commonly prepared through direct extraction or genetic engineering techniques. The assembly of synthetic system of drug and biological system of carrier possesses complementary advantages of greater efficiency and targeting delivery with defense against phagocytosis and prolonged circulation, which has become research spotlight of drug delivery systems.

Molecular bionics provide more possibilities for intelligent biomaterials. Inspired by biological motors, micro/nanomotor as a self-propelled biomedical device performs miscellaneous biomedical functions under the drive both of internal and external energy. Recently, sundry micro/nanomotor could realize navigation to targeting sites across biological barriers for disease screen, diagnosis, treatment, monitor, and so on, which facilitates nanomedicine research as well as clinically multi-tasking and challenging tasks for disease theranostics.

Given the increasing demand of biomimetics and their broadening applied fields, many countries have greatly invested in related research (Hwang et al. 2015), such as BIONIS network for the connection between businesses and universities in England, BIOKON established by twenty-eight collaborators of research centers related to biomimetics in Germany, and Biomimicry Institute, the Biomimicry Guild, and the website of Asknature.org established by Janine M Benyus in the USA. The forefront of biomimetics is being vigorously advanced with the involvement of modern science and technique engineering, from initially mere imitation of biological appearance, to the biomimicry of micro/nano-scale surface of insects and plants, and now to the bioinspiration of cell organelles, single cells, in addition to associated micro/nano-structure, material science and molecular biology, resulting in the revolution of structure, materials, and manufacture techniques for expected performance and functions. So far, we can create brand-new biomaterials inspired by natural structures, substances, and functions. In future, in the premise of compatibility with nature, biomimetics would be critical to the interdisciplinary evolution and technological upgradation, biomaterials inspired by nature organisms would be developed to crack the hard nuts.

References

Accardo A, Gentile F, Mecarini F, De Angelis F, Burghammer M, Di Fabrizio E, Riekel C (2010) In situ X-ray scattering studies of protein solution droplets drying on micro-and nanopatterned superhydrophobic PMMA surfaces. Langmuir 26(18):15057–15064

Advani R, Flinn I, Popplewell L, Forero A, Bartlett NL, Ghosh N, Kline J, Roschewski M, LaCasce A, Collins GP (2018) CD47 blockade by Hu5F9-G4 and rituximab in non-Hodgkin's lymphoma. N Engl J Med 379(18):1711–1721

Al'tshuller GS (1999) The innovation algorithm: TRIZ, systematic innovation and technical creativity. Technical Innovation Center, Inc.
Alegre-Requena JV, Häring M, Herrera RP, Díaz DD (2016) Regulatory parameters of self-healing alginate hydrogel networks prepared via mussel-inspired dynamic chemistry. New J Chem 40(10):8493–8501
Amigoni S, Taffin de Givenchy E, Dufay M, Guittard F (2009) Covalent layer-by-layer assembled superhydrophobic organic– inorganic hybrid films. Langmuir 25(18):11073–11077
Ariga T (2004) The antioxidative function, preventive action on disease and utilization of proanthocyanidins. Biofactors 21(1–4):197–201
Autumn K, Liang YA, Hsieh ST, Zesch W, Chan WP, Kenny TW, Fearing R, Full RJ (2000) Adhesive force of a single gecko foot-hair. Nature 405(6787):681–685
Aziz MS, El Sherif AY (2016) Biomimicry as an approach for bio-inspired structure with the aid of computation. Alex Eng J 55(1):707–714
Barth FG, Humphery JAC, Secomb TW (2003) Sensors and sensing in biology and engineering. Springer, New York
Barthlott W, Neinhuis C (1997) Purity of the sacred lotus, or escape from contamination in biological surfaces. Planta 202:1–8
Bechert DW, Bruse M, Hage W, Meyer R (2000) Fluid mechanics of biological surfaces and their technological application. Naturwissenschaften 87(4):157–171
Bhushan B (1893) Biomimetics: lessons from nature–an overview. Philos Trans R Soc A Math Phys Eng Sci 2009(367):1445–1486
Bhushan B (2016) Biomimetics: bioinspired hierarchical-structured surfaces for green science and technology. Springer
Bhushan B (2019) Lessons from nature for green science and technology: an overview and bioinspired superliquiphobic/philic surfaces. Phil Trans R Soc A 377(2138):20180274
Bhushan B, Jung YC (2011) Natural and biomimetic artificial surfaces for superhydrophobicity, self-cleaning, low adhesion, and drag reduction. Prog Mater Sci 56(1):1–108
Bhushan B, Martin S (2018) Substrate-independent superliquiphobic coatings for water, oil, and surfactant repellency: an overview. J Colloid Interface Sci 526:90–105
Bhushan B, Multanen V (2019) Designing liquid repellent, icephobic and self-cleaning surfaces with high mechanical and chemical durability. Phil Trans R Soc A 377(2138):20180270
Bhushan B, Jung YC, Koch K (1894) Micro-, nano-and hierarchical structures for superhydrophobicity, self-cleaning and low adhesion. Philos Trans R Soc A Math Phys Eng Sci 2009(367):1631–1672
Bian T, Zhao K, Meng Q, Tang Y, Jiao H, Luo J (2019) The construction and performance of multilevel hierarchical hydroxyapatite (HA)/collagen composite implant based on biomimetic bone Haversian motif. Mater Des 162:60–69
Bogatyreva O, Shillerov A, Bogatyrev N (2004) Patterns in TRIZ contradiction matrix: integrated and distributed systems. 4th ETRIA symposium
Boyer P (1992) Visions of a flying machine: the Wright brothers and the process of invention. Peter L Jakab Isis 83(1):165–166
Braat H, Rottiers P, Hommes DW, Huyghebaert N, Remaut E, Remon JP, Van Deventer SJ, Neirynck S, Peppelenbosch MP, Steidler L (2006) A phase I trial with transgenic bacteria expressing interleukin-10 in Crohn's disease. Clin Gastroenterol Hepatol 4(6):754–759
Branco MC, Schneider JP (2009) Self-assembling materials for therapeutic delivery. Acta Biomater 5(3):817–831
Bravo J, Zhai L, Wu Z, Cohen RE, Rubner MF (2007) Transparent superhydrophobic films based on silica nanoparticles. Langmuir 23(13):7293–7298
Brodie M, Vollenweider L, Murphy JL, Xu F, Lyman A, Lew WD, Lee BP (2011) Biomechanical properties of Achilles tendon repair augmented with a bioadhesive-coated scaffold. Biomed Mater 6(1):015014
Brown PS, Bhushan B (2015) Bioinspired, roughness-induced, water and oil super-philic and super-phobic coatings prepared by adaptable layer-by-layer technique. Sci Rep 5(1):14030

Busenlehner LS, Armstrong RN (2005) Insights into enzyme structure and dynamics elucidated by amide H/D exchange mass spectrometry. Arch Biochem Biophys 433(1):34–46

Caro C, Avasthi A, Paez-Muñoz JM, Leal MP, García-Martín ML (2021) Passive targeting of high-grade gliomas via the EPR effect: a closed path for metallic nanoparticles? Biomater Sci 9(23):7984–7995

Celia E, Darmanin T, de Givenchy ET, Amigoni S, Guittard F (2013) Recent advances in designing superhydrophobic surfaces. J Colloid Interface Sci 402:1–18

Cencer M, Murley M, Liu Y, Lee BP (2015) Effect of nitro-functionalization on the cross-linking and bioadhesion of biomimetic adhesive moiety. Biomacromolecules 16(1):404–410

Chai Z, Ran D, Lu L, Zhan C, Ruan H, Hu X, Xie C, Jiang K, Li J, Zhou J (2019) Ligand-modified cell membrane enables the targeted delivery of drug nanocrystals to glioma. ACS Nano 13(5):5591–5601

Chen Z, Zhao P, Luo Z, Zheng M, Tian H, Gong P, Gao G, Pan H, Liu L, Ma A (2016) Cancer cell membrane–biomimetic nanoparticles for homologous-targeting dual-modal imaging and photothermal therapy. ACS Nano 10(11):10049–10057

Chen S, Zeng S, Liu S, Liu H, Zheng R, White KL, Smith AT, Liu L, Sun L (2019) A biomimetic interface with high adhesion, tailorable modulus for on-skin sensors, and low-power actuators. Chem Mater 31(21):8708–8716

Chen Y, Ni P, Xu R, Wang X, Fu C, Wan K, Fang Y, Liu H, Weng Y (2023) Tough and on-demand detachable wet tissue adhesive hydrogel made from catechol derivatives with a long aliphatic side chain. Adv Healthc Mater 12:2301913

Cheng Z, Gao J, Jiang L (2010) Tip geometry controls adhesive states of superhydrophobic surfaces. Langmuir 26(11):8233–8238

Cholewinski A, Yang FK, Zhao B (2017) Underwater contact behavior of alginate and catechol-conjugated alginate hydrogel beads. Langmuir 33(34):8353–8361

Chrastina A, Massey KA, Schnitzer JE (2011) Overcoming in vivo barriers to targeted nanodelivery. Wiley Interdiscip Rev Nanomed Nanobiotechnol 3(4):421–437

Coineau Y, Kresling B (1987) Les inventions de la nature et la bionique. Hachette

Cui C, Liu W (2021) Recent advances in wet adhesives: adhesion mechanism, design principle and applications. Prog Polym Sci 116:101388

Deville S, Saiz E, Nalla RK, Tomsia AP (2006) Freezing as a path to build complex composites. Science 311(5760):515–518

Elsharkawy S, Mata A (2018) Hierarchical biomineralization: from nature's designs to synthetic materials for regenerative medicine and dentistry. Adv Healthc Mater 7(18):1800178

Erb RM, Libanori R, Rothfuchs N, Studart AR (2012) Composites reinforced in three dimensions by using low magnetic fields. Science 335(6065):199–204

Ethirajan A, Ziener U, Chuvilin A, Kaiser U, Cölfen H, Landfester K (2008) Biomimetic hydroxyapatite crystallization in gelatin nanoparticles synthesized using a miniemulsion process. Adv Funct Mater 18(15):2221–2227

Falde EJ, Yohe ST, Colson YL, Grinstaff MW (2016) Superhydrophobic materials for biomedical applications. Biomaterials 104:87–103

Fan Y, Li C, Chen Z, Chen H (2012) Study on fabrication of the superhydrophobic sol–gel films based on copper wafer and its anti-corrosive properties. Appl Surf Sci 258(17):6531–6536

Fang RH, Kroll AV, Gao W, Zhang L (2018) Cell membrane coating nanotechnology. Adv Mater 30(23):1706759

Fantner GE, Hassenkam T, Kindt JH, Weaver JC, Birkedal H, Pechenik L, Cutroni JA, Cidade GA, Stucky GD, Morse DE (2005) Sacrificial bonds and hidden length dissipate energy as mineralized fibrils separate during bone fracture. Nat Mater 4(8):612–616

Faure E, Falentin-Daudré C, Jérôme C, Lyskawa J, Fournier D, Woisel P, Detrembleur C (2013) Catechols as versatile platforms in polymer chemistry. Prog Polym Sci 38(1):236–270

Fayemi P-E, Wanieck K, Zollfrank C, Maranzana N, Aoussat A (2017) Biomimetics: process, tools and practice. Bioinspir Biomim 12(1):011002

Fine AM (2000) Oligomeric proanthocyanidin complexes: history, structure, and phytopharmaceutical applications. Alternat Med Rev 5(2):144–151
Fratzl P (2007) Biomimetic materials research: what can we really learn from nature's structural materials? J R Soc Interface 4(15):637–642
Fratzl P (2008) When the cracks begin to show. Nat Mater 7(8):610–612
Fratzl P, Weinkamer R (2007) Nature's hierarchical materials. Prog Mater Sci 52(8):1263–1334
Gao X, Yan X, Yao X, Xu L, Zhang K, Zhang J, Yang B, Jiang L (2007) The dry-style antifogging properties of mosquito compound eyes and artificial analogues prepared by soft lithography. Adv Mater 19(17):2213–2217
Gao J, Chu D, Wang Z (2016) Cell membrane-formed nanovesicles for disease-targeted delivery. J Control Release 224:208–216
Gay LJ, Felding-Habermann B (2011) Contribution of platelets to tumour metastasis. Nat Rev Cancer 11(2):123–134
Genzer J, Efimenko K (2006) Recent developments in superhydrophobic surfaces and their relevance to marine fouling: a review. Biofouling 22(5–6):339–360
Goel AK, Vattam S, Wiltgen B, Helms M (2014) Information-processing theories of biologically inspired design. In: Biologically inspired design: computational methods and tools, pp 127–152
Greiner A, Wendorff JH (2007) Electrospinning: a fascinating method for the preparation of ultrathin fibers. Angew Chem Int Ed 46(30):5670–5703
Grunwald I, Rischka K, Kast SM, Scheibel T, Bargel H (1894) Mimicking biopolymers on a molecular scale: nano (bio) technology based on engineered proteins. Philos Trans R Soc A Math Phys Eng Sci 2009(367):1727–1747
Gulati NM, Stewart PL, Steinmetz NF (2018) Bioinspired shielding strategies for nanoparticle drug delivery applications. Mol Pharm 15(8):2900–2909
Guo Q, Chen J, Wang J, Zeng H, Yu J (2020) Recent progress in synthesis and application of mussel-inspired adhesives. Nanoscale 12(3):1307–1324
Gupta HS, Seto J, Wagermaier W, Zaslansky P, Boesecke P, Fratzl P (2006) Cooperative deformation of mineral and collagen in bone at the nanoscale. Proc Natl Acad Sci 103(47):17741–17746
Hamed E, Lee Y, Jasiuk I (2010) Multiscale modeling of elastic properties of cortical bone. Acta Mech 213(1–2):131–154
Han L, Lu X, Liu K, Wang K, Fang L, Weng L-T, Zhang H, Tang Y, Ren F, Zhao C (2017a) Mussel-inspired adhesive and tough hydrogel based on nanoclay confined dopamine polymerization. ACS Nano 11(3):2561–2574
Han L, Yan L, Wang K, Fang L, Zhang H, Tang Y, Ding Y, Weng L-T, Xu J, Weng J (2017b) Tough, self-healable and tissue-adhesive hydrogel with tunable multifunctionality. NPG Asia Materials 9(4):e372–e372
Han Y, Pan H, Li W, Chen Z, Ma A, Yin T, Liang R, Chen F, Ma Y, Jin Y (2019) T cell membrane mimicking nanoparticles with bioorthogonal targeting and immune recognition for enhanced photothermal therapy. Adv Sci 6(15):1900251
Hardman SJ, Muhamad-Sarih N, Riggs HJ, Thompson RL, Rigby J, Bergius WN, Hutchings LR (2011) Electrospinning superhydrophobic fibers using surface segregating end-functionalized polymer additives. Macromolecules 44(16):6461–6470
Hasan J, Raj S, Yadav L, Chatterjee K (2015) Engineering a nanostructured "super surface" with superhydrophobic and superkilling properties. RSC Adv 5(56):44953–44959
Hasebe T, Nagashima S, Kamijo A, Moon M-W, Kashiwagi Y, Hotta A, Lee K-R, Takahashi K, Yamagami T, Suzuki T (2013) Hydrophobicity and non-thrombogenicity of nanoscale dual rough surface coated with fluorine-incorporated diamond-like carbon films: biomimetic surface for blood-contacting medical devices. Diam Relat Mater 38:14–18
He Q, Duan L, Qi W, Wang K, Cui Y, Yan X, Li J (2008) Microcapsules containing a biomolecular motor for ATP biosynthesis. Adv Mater 20(15):2933–2937
He Q, Cui Y, Li J (2009) Molecular assembly and application of biomimetic microcapsules. Chem Soc Rev 38(8):2292–2303

He L, Mu C, Shi J, Zhang Q, Shi B, Lin W (2011) Modification of collagen with a natural cross-linker, procyanidin. Int J Biol Macromol 48(2):354–359

Hong S, Yang K, Kang B, Lee C, Song IT, Byun E, Park KI, Cho SW, Lee H (2013) Hyaluronic acid catechol: a biopolymer exhibiting a pH-dependent adhesive or cohesive property for human neural stem cell engineering. Adv Funct Mater 23(14):1774–1780

Hozumi A, Cheng DF, Yagihashi M (2011) Hydrophobic/superhydrophobic oxidized metal surfaces showing negligible contact angle hysteresis. J Colloid Interface Sci 353(2):582–587

Hu B, Owh C, Chee PL, Leow WR, Liu X, Wu Y-L, Guo P, Loh XJ, Chen X (2018) Supramolecular hydrogels for antimicrobial therapy. Chem Soc Rev 47(18):6917–6929

Hunger PM, Donius AE, Wegst UG (2013) Platelets self-assemble into porous nacre during freeze casting. J Mech Behav Biomed Mater 19:87–93

Hwang J, Jeong Y, Park JM, Lee KH, Hong JW, Choi J (2015) Biomimetics: forecasting the future of science, engineering, and medicine. Int J Nanomedicine:5701–5713

Jeon EY, Hwang BH, Yang YJ, Kim BJ, Choi B-H, Jung GY, Cha HJ (2015) Rapidly light-activated surgical protein glue inspired by mussel adhesion and insect structural crosslinking. Biomaterials 67:11–19

Jeronimidis G, Atkins AG (1995) Mechanics of biological materials and structures: nature's lessons for the engineer. Proc Inst Mech Eng C J Mech Eng Sci 209(4):221–235

Jiang Z, Zhao P, Zhou Z, Liu J, Qin L, Wang H (2007) Using attenuated Salmonella typhi as tumor targeting vector for MDR1 siRNA delivery: an experimental study. Cancer Biol Ther 6(4):555–560

Jin H-J, Kaplan DL (2003) Mechanism of silk processing in insects and spiders. Nature 424(6952):1057–1061

Jones CJ, Aizawa S-I (1991) The bacterial flagellum and Flagellar motor: structure, assembly and function. In: Rose AH, Tempest DW (eds) Advances in microbial physiology, vol 32. Academic Press, pp 109–172

Jones K, Platzer M (2002) On the design of efficient micro air vehicles. WIT Trans Ecol Environ 57

Jugniot N, Dahl JJ, Paulmurugan R (2022) Immunotheranostic microbubbles (iMBs)-a modular platform for dendritic cell vaccine delivery applied to breast cancer immunotherapy. J Exp Clin Cancer Res 41(1):299

Kamino K, Nakano M, Kanai S (2012) Significance of the conformation of building blocks in curing of barnacle underwater adhesive. FEBS J 279(10):1750–1760

Kang SM, You I, Cho WK, Shon HK, Lee TG, Choi IS, Karp JM, Lee H (2010) One-step modification of superhydrophobic surfaces by a mussel-inspired polymer coating. Angew Chem Int Ed 49(49):9401–9404

Katiyar SK (2006) Matrix metalloproteinases in cancer metastasis: molecular targets for prostate cancer prevention by green tea polyphenols and grape seed proanthocyanidins. Endocr Metab Immune Disord Drug Targets 6(1):17–24

Keerthikumar S, Chisanga D, Ariyaratne D, Al Saffar H, Anand S, Zhao K, Samuel M, Pathan M, Jois M, Chilamkurti N (2016) ExoCarta: a web-based compendium of exosomal cargo. J Mol Biol 428(4):688–692

Khanlari S, Dubé MA (2013) Bioadhesives: a review. Macromol React Eng 7(11):573–587

Koch K, Bhushan B, Barthlott W (2008) Diversity of structure, morphology and wetting of plant surfaces. Soft Matter 4(10):1943–1963

Koch K, Bhushan B, Barthlott W (2009a) Multifunctional surface structures of plants: an inspiration for biomimetics. Prog Mater Sci 54(2):137–178

Koch K, Bhushan B, Jung YC, Barthlott W (2009b) Fabrication of artificial lotus leaves and significance of hierarchical structure for superhydrophobicity and low adhesion. Soft Matter 5(7):1386–1393

Koons GL, Diba M, Mikos AG (2020) Materials design for bone-tissue engineering. Nat Rev Mater 5(8):584–603

Kord Forooshani P, Lee BP (2017) Recent approaches in designing bioadhesive materials inspired by mussel adhesive protein. J Polym Sci A Polym Chem 55(1):9–33

Kowal J, Tkach M, Théry C (2014) Biogenesis and secretion of exosomes. Curr Opin Cell Biol 29:116–125

Kumar V, Pulpytel J, Rauscher H, Mannelli I, Rossi F, Arefi-Khonsari F (2010) Fluorocarbon coatings via plasma enhanced chemical vapor deposition of 1H, 1H, 2H, 2H-perfluorodecyl acrylate-2, morphology, wettability and antifouling characterization. Plasma Process Polym 7(11):926–938

Le Q-V, Lee J, Lee H, Shim G, Oh Y-K (2021) Cell membrane-derived vesicles for delivery of therapeutic agents. Acta Pharm Sin B 11(8):2096–2113

Lee H, Scherer NF, Messersmith PB (2006) Single molecule mechanics of mussel adhesion. Proc Natl Acad Sci 103(35):12999–13003

Lee H, Lee BP, Messersmith PB (2007) A reversible wet/dry adhesive inspired by mussels and geckos. Nature 448(7151):338–341

Lee H-J, Seo N-J, Jeong S-J, Park Y, Jung D-B, Koh W, Lee H-J, Lee E-O, Ahn KS, Ahn KS (2011a) Oral administration of penta-O-galloyl-β-D-glucose suppresses triple-negative breast cancer xenograft growth and metastasis in strong association with JAK1-STAT3 inhibition. Carcinogenesis 32(6):804–811

Lee BP, Messersmith PB, Israelachvili JN, Waite JH (2011b) Mussel-inspired adhesives and coatings. Annu Rev Mater Res 41:99–132

Lee C, Shin J, Lee JS, Byun E, Ryu JH, Um SH, Kim D-I, Lee H, Cho S-W (2013) Bioinspired, calcium-free alginate hydrogels with tunable physical and mechanical properties and improved biocompatibility. Biomacromolecules 14(6):2004–2013

Lei C, Xie Z, Wu K, Fu Q (2021) Controlled vertically aligned structures in polymer composites: natural inspiration, structural processing, and functional application. Adv Mater 33(49):2103495

Lepora NF, Verschure P, Prescott TJ (2013) The state of the art in biomimetics. Bioinspir Biomim 8(1):013001

Li J, Jia Y, Dong W, Feng X, Fei J, Li J (2014) Transporting a tube in a tube. Nano Lett 14(11):6160–6164

Li J, Zhen X, Lyu Y, Jiang Y, Huang J, Pu K (2018) Cell membrane coated semiconducting polymer nanoparticles for enhanced multimodal cancer phototheranostics. ACS Nano 12(8):8520–8530

Li J, Wang H, Wang Y, Gong X, Xu X, Sha X, Zhang A, Zhang Z, Li Y (2020a) Tumor-activated size-enlargeable bioinspired lipoproteins access cancer cells in tumor to elicit anti-tumor immune responses. Adv Mater 32(38):2002380

Li H, Lu J, Yan C, Xu L (2020b) Tumor cell membrane-coated biomimetic nanoplatform for homologous targeted therapy of colorectal carcinoma. Int J Polym Mater Polym Biomater 69(18):1157–1166

Liao M, Wan P, Wen J, Gong M, Wu X, Wang Y, Shi R, Zhang L (2017) Wearable, healable, and adhesive epidermal sensors assembled from mussel-inspired conductive hybrid hydrogel framework. Adv Funct Mater 27(48):1703852

Lima AC, Song W, Blanco-Fernandez B, Alvarez-Lorenzo C, Mano JF (2011) Synthesis of temperature-responsive dextran-MA/PNIPAAm particles for controlled drug delivery using superhydrophobic surfaces. Pharm Res 28:1294–1305

Lima A, Puga AM, Mano J, Concheiro A, Alvarez-Lorenzo C (2014) Free and copolymerized γ-cyclodextrins regulate the performance of dexamethasone-loaded dextran microspheres for bone regeneration. J Mater Chem B 2(30):4943–4956

Lima AC, Mano JF, Concheiro A, Alvarez-Lorenzo C (2015) Fast and mild strategy, using superhydrophobic surfaces, to produce collagen/platelet lysate gel beads for skin regeneration. Stem Cell Rev Rep 11:161–179

Liu J, Jiang X, Ashley C, Brinker CJ (2009) Electrostatically mediated liposome fusion and lipid exchange with a nanoparticle-supported bilayer for control of surface charge, drug containment, and delivery. J Am Chem Soc 131(22):7567–7569

Liu Z, Zhang Z, Ritchie RO (2018) On the materials science of nature's arms race. Adv Mater 30(32):1705220

Liu X, Sun Y, Xu S, Gao X, Kong F, Xu K, Tang B (2019) Homotypic cell membrane-cloaked biomimetic nanocarrier for the targeted chemotherapy of hepatocellular carcinoma. Theranostics 9(20):5828

Liu B, Yang Y, Chao Y, Xiao Z, Xu J, Wang C, Dong Z, Hou L, Li Q, Liu Z (2021) Equipping cancer cell membrane vesicles with functional DNA as a targeted vaccine for cancer immunotherapy. Nano Lett 21(22):9410–9418

Liu X, Miao Y, Liang H, Diao J, Hao L, Shi Z, Zhao N, Wang Y (2022a) 3D-printed bioactive ceramic scaffolds with biomimetic micro/nano-HAp surfaces mediated cell fate and promoted bone augmentation of the bone–implant interface in vivo. Bioact Mater 12:120–132

Liu J, Liew SS, Wang J, Pu K (2022b) Bioinspired and biomimetic delivery platforms for cancer vaccines. Adv Mater 34(1):2103790

Lu Q, Oh DX, Lee Y, Jho Y, Hwang DS, Zeng H (2013) Nanomechanics of cation–π interactions in aqueous solution. Angew Chem 125(14):4036–4040

Luo H, Gan D, Gama M, Tu J, Yao F, Zhang Q, Ao H, Yang Z, Li J, Wan Y (2020) Interpenetrated nano-and submicro-fibrous biomimetic scaffolds towards enhanced mechanical and biological performances. Mater Sci Eng C 108:110416

Ma Y, Zhang Y, Han R, Li Y, Zhai Y, Qian Z, Gu Y, Li S (2022) A cascade synergetic strategy induced by photothermal effect based on platelet exosome nanoparticles for tumor therapy. Biomaterials 282:121384

Mahadik SA, Kavale MS, Mukherjee S, Rao AV (2010) Transparent Superhydrophobic silica coatings on glass by sol–gel method. Appl Surf Sci 257(2):333–339

Maier GP, Rapp MV, Waite JH, Israelachvili JN, Butler A (2015) Adaptive synergy between catechol and lysine promotes wet adhesion by surface salt displacement. Science 349(6248):628–632

Mannoor MS, Jiang Z, James T, Kong YL, Malatesta KA, Soboyejo WO, Verma N, Gracias DH, McAlpine MC (2013) 3D printed bionic ears. Nano Lett 13(6):2634–2639

Mao W, Wen Y, Lei H, Lu R, Wang S, Wang Y, Chen R, Gu Y, Zhu L, Abhange KK (2019) Isolation and retrieval of extracellular vesicles for liquid biopsy of malignant ground-glass opacity. Anal Chem 91(21):13729–13736

Marshall SJ, Bayne SC, Baier R, Tomsia AP, Marshall GW (2010) A review of adhesion science. Dent Mater 26(2):e11–e16

Martin PM (2009) Handbook of deposition technologies for films and coatings: science, applications and technology. William Andrew

Martin S, Brown PS, Bhushan B (2017) Fabrication techniques for bioinspired, mechanically-durable, superliquiphobic surfaces for water, oil, and surfactant repellency. Adv Colloid Interf Sci 241:1–23

Massey AP, Wallace WA (1996) Understanding and facilitating group problem structuring and formulation: mental representations, interaction, and representation aids. Decis Support Syst 17(4):253–274

Mehdizadeh M, Weng H, Gyawali D, Tang L, Yang J (2012) Injectable citrate-based mussel-inspired tissue bioadhesives with high wet strength for sutureless wound closure. Biomaterials 33(32):7972–7983

Meng QF, Zhao Y, Dong C, Liu L, Pan Y, Lai J, Liu Z, Yu GT, Chen X, Rao L (2021) Genetically programmable fusion cellular vesicles for cancer immunotherapy. Angew Chem Int Ed 60(50):26320–26326

Meng F, Ye X, Chen W, Liu L, Hu F, Qi Y, Yu W, Lin Z, Yan J, Guo Z (2023) A mild bioinspiration route to bacillus-shaped silica with enhanced immune responses. ACS Sustain Chem Eng 11(4):1324–1332

Michielsen S, Lee HJ (2007) Design of a superhydrophobic surface using woven structures. Langmuir 23(11):6004–6010

Milwich M, Speck T, Speck O, Stegmaier T, Planck H (2006) Biomimetics and technical textiles: solving engineering problems with the help of nature's wisdom. Am J Bot 93(10):1455–1465

Modaresifar K, Azizian S, Hadjizadeh A (2016) Nano/biomimetic tissue adhesives development: from research to clinical application. Polym Rev 56(2):329–361

Mueller T (2008) Biomimetics design by natures. Natl Geogr 213:68–90
Nicklisch SC, Waite JH (2012) Mini-review: the role of redox in Dopa-mediated marine adhesion. Biofouling 28(8):865–877
Noad R, Roy P (2003) Virus-like particles as immunogens. Trends Microbiol 11(9):438–444
Ochyl LJ, Bazzill JD, Park C, Xu Y, Kuai R, Moon JJ (2018) PEGylated tumor cell membrane vesicles as a new vaccine platform for cancer immunotherapy. Biomaterials 182:157–166
Olszta MJ, Douglas EP, Gower LB (2003) Intrafibrillar mineralization of collagen using a liquid-phase mineral precursor, vol 774. MRS Online Proceedings Library (OPL)
Pandey N, Soto-Garcia LF, Liao J, Philippe Z, Nguyen KT, Hong Y (2020a) Mussel-inspired bioadhesives in healthcare: design parameters, current trends, and future perspectives. Biomater Sci 8(5):1240–1255
Pandey N, Soto-Garcia LF, Liao J, Zimmern P, Nguyen KT, Hong Y (2020b) Mussel-inspired bioadhesives in healthcare: design parameters, current trends, and future perspectives. Biomater Sci 8(5):1240–1255
Panneerselvam R, Anandhan N, Sivakumar G, Ganesan K, Marimuthu T, Sugumar V (2019) Role of annealing temperatures on mechanical, optical, electrical and magnetic properties of nanohydroxyapatite biomaterial. J Nanosci Nanotechnol 19(8):4366–4376
Parker AR, Lawrence CR (2001) Water capture by a desert beetle. Nature 414(6859):33–34
Philip B (2002) Natural strategies for the molecular engineer. Nanotechnology 13(5):R15
Piao J-G, Wang L, Gao F, You Y-Z, Xiong Y, Yang L (2014) Erythrocyte membrane is an alternative coating to polyethylene glycol for prolonging the circulation lifetime of gold nanocages for photothermal therapy. ACS Nano 8(10):10414–10425
Ping H, Wagermaier W, Horbelt N, Scoppola E, Li C, Werner P, Fu Z, Fratzl P (2022) Mineralization generates megapascal contractile stresses in collagen fibrils. Science 376(6589):188–192
Prasad R, Conde J (2022) Bioinspired soft nanovesicles for site-selective cancer imaging and targeted therapies. Wiley Interdiscip Rev Nanomed Nanobiotechnol 14(4):e1792
Preusser M, Capper D, Ilhan-Mutlu A, Berghoff AS, Birner P, Bartsch R, Marosi C, Zielinski C, Mehta MP, Winkler F (2012) Brain metastases: pathobiology and emerging targeted therapies. Acta Neuropathol 123:205–222
Pucci F, Garris C, Lai CP, Newton A, Pfirschke C, Engblom C, Alvarez D, Sprachman M, Evavold C, Magnuson A (2016) SCS macrophages suppress melanoma by restricting tumor-derived vesicle–B cell interactions. Science 352(6282):242–246
Qiao C, Wang X, Liu G, Yang Z, Jia Q, Wang L, Zhang R, Xia Y, Wang Z, Yang Y (2022) Erythrocyte membrane camouflaged metal–organic framework nanodrugs for remodeled tumor microenvironment and enhanced tumor chemotherapy. Adv Funct Mater 32(6):2107791
Qu M, Zhao G, Wang Q, Cao X, Zhang J (2008) Fabrication of superhydrophobic surfaces by a Pt nanowire array on Ti/Si substrates. Nanotechnology 19(5):055707
Rashed H, Bayraktar EK, Helal G, Abd-Ellah MF, Amero P, Chavez-Reyes A, Rodriguez-Aguayo C (2017) Exosomes: from garbage bins to promising therapeutic targets. Int J Mol Sci 18(3):538
Reintjes T, Teßmar J, Göpferich A (2008) Biomimetic polymers to control cell adhesion. J Drug Deliv Sci Technol 18(1):15–24
Ren X, Zheng R, Fang X, Wang X, Zhang X, Yang W, Sha X (2016) Red blood cell membrane camouflaged magnetic nanoclusters for imaging-guided photothermal therapy. Biomaterials 92:13–24
Rho J-Y, Kuhn-Spearing L, Zioupos P (1998) Mechanical properties and the hierarchical structure of bone. Med Eng Phys 20(2):92–102
Rodríguez-Arco L, Poma A, Ruiz-Pérez L, Scarpa E, Ngamkham K, Battaglia G (2019) Molecular bionics–engineering biomaterials at the molecular level using biological principles. Biomaterials 192:26–50
Ryu JH, Hong S, Lee H (2015) Bio-inspired adhesive catechol-conjugated chitosan for biomedical applications: a mini review. Acta Biomater 27:101–115
Said Hassane F, Saleh A, Abes R, Gait M, Lebleu B (2010) Cell penetrating peptides: overview and applications to the delivery of oligonucleotides. Cell Mol Life Sci 67:715–726

Saiz-Poseu J, Mancebo-Aracil J, Nador F, Busqué F, Ruiz-Molina D (2019) The chemistry behind catechol-based adhesion. Angew Chem Int Ed 58(3):696–714
Salgado AJ, Coutinho OP, Reis RL (2004) Bone tissue engineering: state of the art and future trends. Macromol Biosci 4(8):743–765
Salgado CL, Oliveira MB, Mano JF (2012) Combinatorial cell–3D biomaterials cytocompatibility screening for tissue engineering using bioinspired superhydrophobic substrates. Integr Biol 4(3):318–327
Schütz S, Weissbecker B, Hummel HE, Apel K-H, Schmitz H, Bleckmann H (1999) Insect antenna as a smoke detector. Nature 398(6725):298–299
Shao H, Im H, Castro CM, Breakefield X, Weissleder R, Lee H (2018) New technologies for analysis of extracellular vesicles. Chem Rev 118(4):1917–1950
Shin M, Song KH, Burrell JC, Cullen DK, Burdick JA (2019) Injectable and conductive granular hydrogels for 3D printing and electroactive tissue support. Adv Sci 6(20):1901229
Shu L, Ueda K, Chiu I, Cheong H (2011) Biologically inspired design. CIRP Ann 60(2):673–693
Sidorenko A, Krupenkin T, Taylor A, Fratzl P, Aizenberg J (2007) Reversible switching of hydrogel-actuated nanostructures into complex micropatterns. Science 315(5811):487–490
Siegel, D. S.; Waldman, D. A.; Atwater, L. E.; Link, A. N., Toward a model of the effective transfer of scientific knowledge from academicians to practitioners: qualitative evidence from the commercialization of university technologies. J Eng Technol Manag 2004, 21 (1–2), 115–142
Song W, Lima AC, Mano JF (2010) Bioinspired methodology to fabricate hydrogel spheres for multi-applications using superhydrophobic substrates. Soft Matter 6(23):5868–5871
Song YY, Zhang LH, Dong LM, Li HT, Yu ZP, Liu Y, Lv GJ, Ma HL (2021) pH-responsive smart wettability surface with dual bactericidal and releasing properties. ACS Appl Mater Interfaces 13(38):46065–46075
Stegmaier T, Linke M, Planck H (1894) Bionics in textiles: flexible and translucent thermal insulations for solar thermal applications. Philos Trans R Soc A Math Phys Eng Sci 2009(367):1749–1758
Stewart RJ, Ransom TC, Hlady V (2011) Natural underwater adhesives. J Polym Sci B Polym Phys 49(11):757–771
Sun W, Shen L, Wang J, Fu K, Ji J (2010) Netlike knitting of polyelectrolyte multilayers on honeycomb-patterned substrate. Langmuir 26(17):14236–14240
Sun J-L, Jiao K, Niu L-N, Jiao Y, Song Q, Shen L-J, Tay FR, Chen J-H (2017) Intrafibrillar silicified collagen scaffold modulates monocyte to promote cell homing, angiogenesis and bone regeneration. Biomaterials 113:203–216
Sun J-L, Jiao K, Song Q, Ma CF, Ma C, Tay FR, Niu L-N, Chen J-H (2018) Intrafibrillar silicified collagen scaffold promotes in-situ bone regeneration by activating the monocyte p38 signaling pathway. Acta Biomater 67:354–365
Swiegers GF (2012) Bioinspiration and biomimicry in chemistry: reverse-engineering nature. Wiley, Canada, pp i–xxvii
Tamura T, Yoshioka Y, Ochiya T (2021) Extracellular vesicles as the "magic bullet" for fighting threats to humanity. Extracell Vesicl Circulat Nucl Acids 2(3):224–227
Tang S, Dong Z, Ke X, Luo J, Li J (2021) Advances in biomineralization-inspired materials for hard tissue repair. Int J Oral Sci 13(1):42
Terrier P, Glaus M, Raufflet E (2019) BiomiMETRIC assistance tool: a quantitative performance tool for biomimetic design. Biomimetics 4(3):49
Théry C, Ostrowski M, Segura E (2009) Membrane vesicles as conveyors of immune responses. Nat Rev Immunol 9(8):581–593
Thomas Y (1805) An essay on the cohesion of fluids. Philos Trans R Soc London 95:65–87
Tipoe GL, Leung TM, Liong EC, Lau TYH, Fung ML, Nanji AA (2010) Epigallocatechin-3-gallate (EGCG) reduces liver inflammation, oxidative stress and fibrosis in carbon tetrachloride (CCl4)-induced liver injury in mice. Toxicology 273(1–3):45–52
Tsushima F, Yao S, Shin T, Flies A, Flies S, Xu H, Tamada K, Pardoll DM, Chen L (2007) Interaction between B7-H1 and PD-1 determines initiation and reversal of T-cell anergy. Blood J Am Soc Hematol 110(1):180–185

Van der Pol E, Böing AN, Harrison P, Sturk A, Nieuwland R (2012) Classification, functions, and clinical relevance of extracellular vesicles. Pharmacol Rev 64(3):676–705

Velcro, S. A. Improvements in or relating to a method and a device for producing velvet type fabric. 1955

Victor J, Facchini D, Erb U (2012) A low-cost method to produce superhydrophobic polymer surfaces. J Mater Sci 47:3690–3697

Vincent JFV, Bogatyreva OA, Bogatyrev NR, Bowyer A, Pahl A-K (2006) Biomimetics: its practice and theory. J R Soc Interface 3(9):471–482

Walther A, Muller AH (2013) Janus particles: synthesis, self-assembly, physical properties, and applications. Chem Rev 113(7):5194–5261

Wan M, Chen H, Wang Q, Niu Q, Xu P, Yu Y, Zhu T, Mao C, Shen J (2019) Bio-inspired nitric-oxide-driven nanomotor. Nature Commun 10(1):966

Wang R, Li J, Chen W, Xu T, Yun S, Xu Z, Xu Z, Sato T, Chi B, Xu H (2017) A biomimetic mussel-inspired ε-poly-l-lysine hydrogel with robust tissue-anchor and anti-infection capacity. Adv Funct Mater 27(8):1604894

Wang H, Wu J, Williams GR, Fan Q, Niu S, Wu J, Xie X, Zhu L-M (2019) Platelet-membrane-biomimetic nanoparticles for targeted antitumor drug delivery. J Nanobiotechnol 17:1–16

Wang C, Wu B, Wu Y, Song X, Zhang S, Liu Z (2020) Camouflaging nanoparticles with brain metastatic tumor cell membranes: a new strategy to traverse blood–brain barrier for imaging and therapy of brain tumors. Adv Funct Mater 30(14):1909369

Wang H, Tian J, Jiang Y, Liu S, Zheng J, Li N, Wang G, Dong F, Chen J, Xie Y (2023) A 3D biomimetic optoelectronic scaffold repairs cranial defects. Sci Adv 9(7):eabq7750

Wegst UG, Bai H, Saiz E, Tomsia AP, Ritchie RO (2015) Bioinspired structural materials. Nat Mater 14(1):23–36

Wei J, Pan F, Ping H, Yang K, Wang Y, Wang Q, Fu Z (2023) Bioinspired additive manufacturing of hierarchical materials: from biostructures to functions. Research 6:0164

Weiner S, Addadi L (1997) Design strategies in mineralized biological materials. J Mater Chem 7(5):689–702

Wong CK, Mason AF, Stenzel MH, Thordarson P (2017) Formation of non-spherical polymersomes driven by hydrophobic directional aromatic perylene interactions. Nat Commun 8(1):1240

Wu Z, Li T, Li J, Gao W, Xu T, Christianson C, Gao W, Galarnyk M, He Q, Zhang L (2014) Turning erythrocytes into functional micromotors. ACS Nano 8(12):12041–12048

Wu Z, De Avila BE-F, Martín A, Christianson C, Gao W, Thamphiwatana SK, Escarpa A, He Q, Zhang L, Wang J (2015) RBC micromotors carrying multiple cargos towards potential theranostic applications. Nanoscale 7(32):13680–13686

Wu Z, Lin X, Si T, He Q (2016) Recent Progress on bioinspired self-propelled micro/Nanomotors via controlled molecular self-assembly. Small 12(23):3080–3093

Wu XH, Liew YK, Lim WM, Mai CW, Then YY (2023) Blood compatible and noncytotoxic superhydrophobic graphene/titanium dioxide coating with antibacterial and antibiofilm properties. J Appl Polym Sci 140(11):e53629

Xia Q, Zhang Y, Li Z, Hou X, Feng N (2019) Red blood cell membrane-camouflaged nanoparticles: a novel drug delivery system for antitumor application. Acta Pharm Sin B 9(4):675–689

Xu J, Tam M, Samaei S, Lerouge S, Barralet J, Stevenson MM, Cerruti M (2017) Mucoadhesive chitosan hydrogels as rectal drug delivery vessels to treat ulcerative colitis. Acta Biomater 48:247–257

Xu C, Jiang Y, Han Y, Pu K, Zhang R (2021) A polymer multicellular nanoengager for synergistic NIR-II photothermal immunotherapy. Adv Mater 33(14):2008061

Yang H, Pi P, Cai Z-Q, Wen X, Wang X, Cheng J, Yang Z-R (2010) Facile preparation of super-hydrophobic and super-oleophilic silica film on stainless steel mesh via sol–gel process. Appl Surf Sci 256(13):4095–4102

Yang HC, Waldman RZ, Wu MB, Hou J, Chen L, Darling SB, Xu ZK (2018) Dopamine: just the right medicine for membranes. Adv Funct Mater 28(8):1705327

Yang L, Ge L, Zhou Q, Mokabber T, Pei Y, Bron R, van Rijn P (2020) Biomimetic multiscale hierarchical topography enhances osteogenic differentiation of human mesenchymal stem cells. Adv Mater Interfaces 7(14):2000385

Yao L, Zheng M, He S, Ma L, Li M, Shen W (2011) Preparation and properties of ZnS superhydrophobic surface with hierarchical structure. Appl Surf Sci 257(7):2955–2959

Yazdani Sarvestani H, Aranguren van Egmond D, Esmail I, Genest M, Paquet C, Ashrafi B (2022) Bioinspired stochastic design: tough and stiff ceramic systems. Adv Funct Mater 32(6):2108492

Yeom B, Sain T, Lacevic N, Bukharina D, Cha S-H, Waas AM, Arruda EM, Kotov NA (2017) Abiotic tooth enamel. Nature 543(7643):95–98

Yohe ST, Herrera VL, Colson YL, Grinstaff MW (2012) 3D superhydrophobic electrospun meshes as reinforcement materials for sustained local drug delivery against colorectal cancer cells. J Control Release 162(1):92–101

Yu X, Li Y, Wu J, Ju H (2014) Motor-based autonomous microsensor for motion and counting immunoassay of cancer biomarker. Anal Chem 86(9):4501–4507

Yu H, Yang Z, Li F, Xu L, Sun Y (2020) Cell-mediated targeting drugs delivery systems. Drug Deliv 27(1):1425–1437

Yu X, Dong M, Wang L, Yang Q, Wang L, Han W, Dong J, Liu T, Kong Y, Niu W (2023) Nanotherapy for bone repair: milk-derived small extracellular vesicles delivery of icariin. Drug Deliv 30(1):2169414

Yuk H, Lu B, Zhao X (2019) Hydrogel bioelectronics. Chem Soc Rev 48(6):1642–1667

Zeng Z, Pu K (2020) Improving cancer immunotherapy by cell membrane-camouflaged nanoparticles. Adv Funct Mater 30(43):2004397

Zhang J, Yang B (2010) Patterning colloidal crystals and nanostructure arrays by soft lithography. Adv Funct Mater 20(20):3411–3424

Zhang L, Chen H, Sun J, Shen J (2007) Layer-by-layer deposition of poly (diallyldimethylammonium chloride) and sodium silicate multilayers on silica-sphere-coated substrate—facile method to prepare a superhydrophobic surface. Chem Mater 19(4):948–953

Zhang J, Li L, Kim S-H, Hagerman AE, Lü J (2009) Anti-cancer, anti-diabetic and other pharmacologic and biological activities of penta-galloyl-glucose. Pharm Res 26:2066–2080

Zhang T, Hartl C, Frank K, Heuer-Jungemann A, Fischer S, Nickels PC, Nickel B, Liedl T (2018a) 3D DNA origami crystals. Adv Mater 30(28):1800273

Zhang C, Wu M-B, Wu B-H, Yang J, Xu Z-K (2018b) Solar-driven self-heating sponges for highly efficient crude oil spill remediation. J Mater Chem A 6(19):8880–8885

Zhang J, Zhao Y, Tian Z, Zhu J, Shi Z, Cui Z, Zhu S (2020a) Enhancement performance of application mussel-biomimetic adhesive primer for dentin adhesives. RSC Adv 10(20):12035–12046

Zhang W, Wang R, Sun Z, Zhu X, Zhao Q, Zhang T, Cholewinski A, Yang FK, Zhao B, Pinnaratip R (2020b) Catechol-functionalized hydrogels: biomimetic design, adhesion mechanism, and biomedical applications. Chem Soc Rev 49(2):433–464

Zhang C, Wu B, Zhou Y, Zhou F, Liu W, Wang Z (2020c) Mussel-inspired hydrogels: from design principles to promising applications. Chem Soc Rev 49(11):3605–3637

Zheng Y, Shariati K, Ghovvati M, Vo S, Origer N, Imahori T, Kaneko N, Annabi N (2023) Hemostatic patch with ultra-strengthened mechanical properties for efficient adhesion to wet surfaces. Biomaterials 301:122240

Zhou B, Niu LN, Shi W, Zhang W, Arola DD, Breschi L, Mao J, Chen JH, Pashley DH, Tay FR (2014) Adopting the principles of collagen biomineralization for intrafibrillar infiltration of yttria-stabilized zirconia into three-dimensional collagen scaffolds. Adv Funct Mater 24(13):1895–1903

Zhou S, Gravekamp C, Bermudes D, Liu K (2018) Tumour-targeting bacteria engineered to fight cancer. Nat Rev Cancer 18(12):727–743

Zou M-Z, Li Z-H, Bai X-F, Liu C-J, Zhang X-Z (2021) Hybrid vesicles based on autologous tumor cell membrane and bacterial outer membrane to enhance innate immune response and personalized tumor immunotherapy. Nano Lett 21(20):8609–8618

Chapter 7
Interactive Biomaterials

Zhe Liu, Xinru Kong, and Yu Shi

Abstract With the continuing progress in the biomedical field, the development of novel intelligent biomaterials has attracted more and more attention. Interactive biomaterials, a term used innovatively in this review, refers to multi-functional materials developed on the basis of biosensors and responsive biomaterials for disease surveillance, prediction, diagnosis, and treatments. By combining the benefits of biosensors and responsive components in interactive biomaterials, more intelligent, precise, and controllable disease monitoring and treatment could be achieved. So far, promising candidates for such interactive biomaterials have been developed, including digital pills, robotic capsule endoscopy, smart wound dressings, smart contact lenses, and theranostic microneedle array patches. The utilization of these interactive biomaterials simplifies procedures for disease treatment, reducing the need for multiple materials. At the same time, interactive biomaterials can seamlessly connect with patients, doctors, and medical devices to deliver more personalized and precise medical services. Starting from sensors and responsive biomaterials, this chapter will take interactive biomaterials as a starting point, explain the categories and application advantages of interactive biomaterials, discuss the challenges and prospects of interactive biomaterials, and provide new ideas for the design of personalized medical platforms.

Z. Liu (✉)
Division of Intelligent Biomaterials, Department of Medical Sciences and Engineering, Medical School, Tianjin University, Tianjin, China

Academy of Medical Engineering and Translational Medicine, Tianjin University, Tianjin, China

Tianjin Key Laboratory of Brain Science and Neural Engineering, Tianjin University, Tianjin, China
e-mail: zheliu@tju.edu.cn

X. Kong · Y. Shi
Academy of Medical Engineering and Translational Medicine, Tianjin University, Tianjin, China

Tianjin Key Laboratory of Brain Science and Neural Engineering, Tianjin University, Tianjin, China

Z. Liu (ed.), *Intelligent Biomaterials*, https://doi.org/10.1007/978-981-96-8646-9_7

Keywords Interactive biomaterials · Biosensors · Responsive biomaterials · Personalized medicine

7.1 Introduction

In the past decades, the emergence of sensors has made disease diagnosis and evaluation more convenient, while the advent of responsive biomaterials has provided opportunities for precise and efficient disease treatment. Every innovation and breakthrough in functional materials in the medical field is closely related to technological advances and society needs. Nowadays, with the advancement of medical technology and the development of material sciences, there is a growing requirement for convenient and personalized healthcare models. In this context, interactive biomaterials with dual functions of sensing and treating have emerged, combining sensors and responsive biological components. By harnessing the unique properties of interactive biomaterials, novel diagnostic and therapeutic strategies can be implemented to provide patients with a comprehensive care plan.

In the fields of biomedicine and healthcare, sensors can be designed to detect various physical or biochemical signals related to human health, such as blood pressure, body temperature, respiration, motion, as well as biochemical indicators like blood glucose, metabolism and nutrient levels, cancer biomarkers, etc. providing doctors with information about disease status and patient health conditions (Fig. 7.1a) (Zheng et al. 2021b; He et al. 2023; Khan et al. 2023). The use of these sensors not only provides accurate diagnostic results but also provides real-time and noninvasive monitoring, resulting in better medical experiences for patients. With highly sensitive and highly specific detection capabilities, sensors can recognize subtle changes in the early stages of disease, facilitating early intervention and potentially improving treatment outcomes (Tan et al. 2022; Li et al. 2023b). Leveraging these encouraging benefits, sensors are integrated with interactive biomaterials to enable real-time monitoring and detection of disease-related physiological indicators, thereby enabling disease detection or medication-adherent measurement capabilities in interactive biomaterials.

The functional properties of interactive biomaterials are further enhanced by the characteristics of the responsive biological components. Responsive biomaterials are a sort of materials with intelligent responsiveness properties, capable of specifically responding to exogenous or endogenous stimuli (Fig. 7.1b) (Badeau and Deforest 2019; Kowalski et al. 2018). Traditional approaches of disease treatments often suffer from shortcomings such as poor targeting, premature drug leakage, normal tissues damage, and suboptimal treatment outcomes, whereas responsive biomaterials can automatically adjust and respond to changes in the environment for controllable and efficient therapeutic purposes. Responsive biomaterials can not only be designed to be sensitive to physiological or pathological changes at the

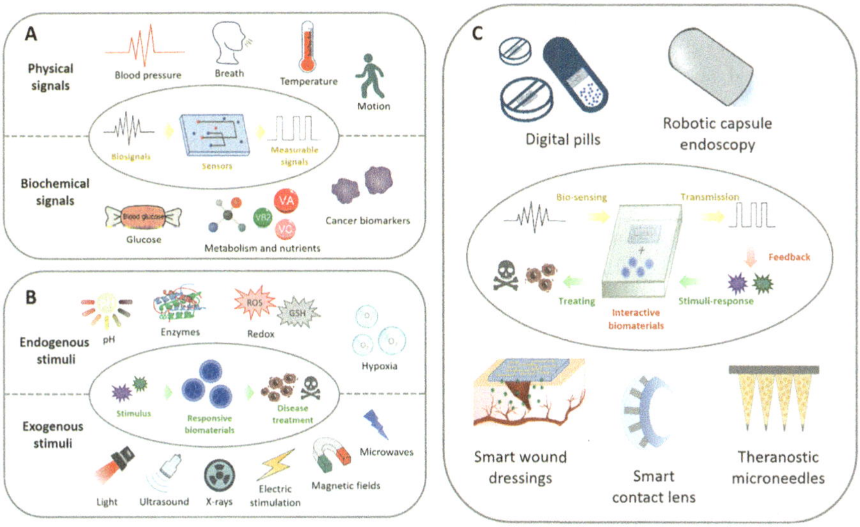

Fig. 7.1 A summary for disease monitoring or treatment with different biomaterials. (**a**) sensors, (**b**) responsive biomaterials, (**c**) interactive biomaterials

diseased site, but also can be driven to release drugs or generate biological effects under exogenous stimuli such as light, ultrasound (US), electric stimulation, microwaves, and magnetic fields (Battistella et al. 2021; Cook and Decuzzi 2021; Armenia et al. 2022). Inspired by the unique properties of responsive biomaterials, which have potential and promising applications in biomedicine, interactive biomaterials for controlled treatment based on individual patient needs can be created by integrating responsive biomaterials with diagnostic and therapeutic frameworks.

The utilization of interactive biomaterials in disease diagnosis and treatment represents a paradigm shift in healthcare. Conventional approaches often rely on separate diagnostic tests and therapeutic interventions, resulting in increased patient costs, time, and inconvenience. The integration of sensors and responsive materials facilitates the development of compact and versatile systems that can perform drug adherent measurements or trigger treatment patterns on demand based on disease detection results (Fig. 7.1c). In recent years, interactive biomaterials are being studied and increasingly applied, such as smart wound dressings that detect wounds and release drugs on demand (Xu et al. 2021), digital pills approved by the FDA for clinical treatment (Asher 2017), smart contact lenses for closed-loop diagnostic treatment of eye diseases (Kim et al. 2022c), etc. The application of these interactive biomaterials not only simplifies the medical processes, but also offers new opportunities for early detection, precise intervention, and improved patient outcomes by providing real-time monitoring, personalized treatment responses, and simplified healthcare solutions.

In this chapter, we firstly split the sensors used for disease diagnosis into two categories based on their differences in the detection signals, and then mention responsive biomaterials that can be triggered by multiple endogenous or exogenous signals. Finally, we aim to focus on interactive biomaterials that integrate the characteristics of sensors and responsive biomaterials, introduce the application and related research on this versatile material, and discuss the challenges and possible future directions. Despite its infancy in clinical applications, it is clear that interactive biomaterials play an important role and are bound to revolutionize areas such as personalized medicine, precision therapy, and early-stage diagnosis. We intend to elicit interactive biomaterials through the introduction of sensors and responsive biomaterials, and to gain an in-depth understanding of the unique advantages and recent developments of interactive biomaterials. We hope that this strategy of combining sensing and responsive components will provide more ideas for the design of interactive biomaterials, and pave the way for a more convenient and personalized medical era.

7.2 Sensors

Sensors play a key role in disease diagnosis and assessment. They are capable of detecting, measuring, and recording the physical and biochemical signals of organisms, thus providing valuable data for early diagnosis, treatment, and monitoring of diseases, enabling personalized medicine. According to the different detection signals, the sensors used for disease diagnosis can be divided into two categories: sensors that detect physical signals, and sensors that detect biochemical signals.

7.2.1 Physical Signals

Physical parameters such as blood pressure, body temperature, respiration, and movement are the main health indicators that reflect the physiological activity of the human body. The application of sensors to measure and analyze changes in these physical signals and physiological features enables healthcare professionals to obtain accurate data on physical parameters, thereby supporting decisions for early diagnosis, treatment, and monitoring of diseases.

7.2.1.1 Blood Pressure Sensors

For blood pressure monitoring, flexible and wearable blood pressure sensors that are portable compared to traditional sphygmomanometers can continuously measure an individual's blood pressure level, providing an important assessment of cardiovascular system function and helping physicians diagnose cardiovascular problems in patients (Zhao et al. 2023b; Yi et al. 2022). Min et al. developed a wearable piezoelectric blood-pressure sensor (WPBPS) with high flexibility, which can sense the pressure generated by arterial pulses and output it as electrical pulse signals for detecting tiny pulse waves in the wrist arteries. Clinically validated results in 35 subjects showed that WPBPS had a mean difference in systolic blood pressure (SBP) and diastolic blood pressure (DBP) compared with commercial sphygmomanometers of -0.89 ± 6.19 and -0.32 ± 5.28 mmHg, respectively (Min et al. 2023). In addition to piezoelectric pressure sensors, capacitive pressure sensors can be used for blood pressure measurement (Wan et al. 2023). Another supercapacitive-type pressure sensor (sCPS), using the piezoelectric effect and supercapacitive-iontronic characteristics of supercapacitor, detects slight changes in pressure in the carotid artery and displays a pulse waveform that can be used to calculate blood pressure for early diagnosis of cardiovascular diseases (Hye-Jun and Jin-Woo 2023). Besides, there are other portable sensors for continuous blood pressure measurement, such as an optical-fiber-sensor-assisted smartwatch (Fig. 7.2) (Liangye et al. 2023), and smart rings with embedded ring sensors (Sel et al. 2023). Blood pressure sensors can not only detect pressure changes caused by blood vessels in a wearable form, but also can be used in an implantable form for hemodynamic monitoring of patients with cardiovascular disease to guide post-operative management. Recently, an implantable wireless, battery-free hemodynamic monitoring device has been developed. The system, which consists of a sensor-based sensing module and a wireless electronic module for communication, supports implantation into the heart or blood vessels to monitor cardiovascular pressure, blood flow, and temperature in real-time and transmit the measurements to an external device (Kwon et al. 2023). Another biocompatible tissue-adhesive piezoelectric soft sensor (TPSS) can adhere to carotid artery of a pig to monitor blood pressure during surgery (Chan et al. 2023).

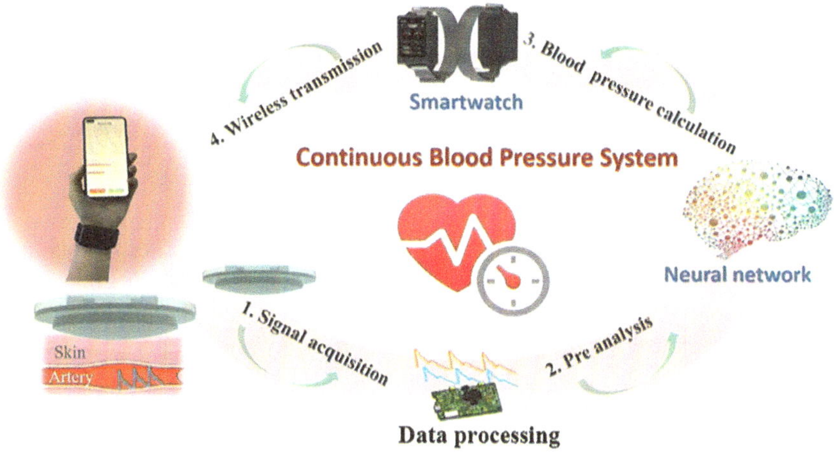

Fig. 7.2 A schematic illustration of a blood pressure monitoring system consisting of an optical-fiber-sensor-assisted smartwatch. (Reprinted with permission from Liangye et al. 2023)

7.2.1.2 Temperature Sensors

Measuring body temperature can provide clinical information relevant to the diagnosis and healing of many diseases, including the prevention of heat-related illnesses such as heat stroke from fever, and aid in the diagnosis of diseases such as infections. Currently, wearable temperature sensors are designed to simplify the process of measuring body temperature in daily life and manage personal health in a more advanced and convenient manner, as demanded by personalized medicine. For instance, flexible temperature sensors made from conductive polymer composites and high-precision temperature sensors utilizing metal-doped nanocomposite membranes present promising possibilities for monitoring body temperature and improving personal healthcare (Ryu et al. 2023; Geng et al. 2022; Sang et al. 2021; Phadkule and Sarma 2022). Furthermore, to reduce the strain cross-sensitivity of the sensor and increase its temperature sensitivity, Yu et al. developed a kirigami-capacitive sensor for continuous monitoring of skin temperature changes. In particular, the kirigami mode makes the kirigami-capacitive sensor less cross-sensitivity to reduce the interference of strain on the sensor, and the choice of capacitive sensor material allows the kirigami-capacitive sensor to have a higher temperature sensitivity and to be able to detect small temperature variations (Yu et al. 2021). Skin temperature sensors are used to detect external temperatures, which are easily affected by ambient temperature. Thus, in clinical settings, it is necessary to precisely measure the temperature of the human core. Based on this, a study conducted temperature measurements for 48 hours in 26 patients with fever using different methods, including taking armpit temperature with a thermometer, measuring forehead temperature with an infrared thermometer, monitoring skin temperature with an ETACT patch, and assessing core temperature with an ingestible capsule (Huang et al. 2020). The study analyzed the correlation and consistency among these different measurements and revealed that the ingestible capsule is a reliable method for measuring core body temperature in clinical practice, as it is less invasive for the patient and provides real-time measurement results as well.

7.2.1.3 Breath Sensors

Breathing is an important vital sign, and real-time monitoring of respiratory status is of great significance for health assessment and disease diagnosis. In clinical practice, commonly used respiratory monitoring systems, which consist of a mask and breathing tube covering the mouth and nose, are bulky and not portable, and may cause false alarms due to air leakage (Liao et al. 2017). Therefore, there is a need to develop flexible and portable respiratory sensors for dynamic monitoring of respiratory signals. Humidity can act as a link between respiratory signals and sensing signals, and inspired by this, materials that are sensitive to humidity are used for the preparation of breath sensors (Bošković et al. 2020; Yi et al. 2020; Bao et al. 2021). Silver vanadate nanowires synthesized by hydrothermal methods are highly humidity-sensitive semiconductors. By exploiting this property, a flexible breath sensor has been designed to monitor dynamic respiration by detecting humidity (Yan et al. 2023). With the integration of an apnea alarm module, the breath sensor can also monitor sleep disordered breathing and provide early warnings of sleep apnea syndrome. In addition to detecting changes in humidity, breath sensors have the capability to measure respiration rate and volume by detecting variations in the movements of the chest and abdomen (Zhang et al. 2019; Chu et al. 2019). Moreover, to facilitate the observation of the breathing processes, smart masks that measure key respiratory parameters have been designed. These smart masks can recognize various respiratory states and are critical for daily and clinical monitoring. The face masks based on a LSMO/Mica sensor are capable of identifying various breathing states by analyzing the changes in electrical resistance caused by thermal diffusion during breathing, which is critical for daily and clinical monitoring (Ye et al. 2023). Another smart face mask based on an ultrathin pressure sensor developed by Zhong et al. generates output voltage when triggered by low airflow pressure. Subsequently, the breath signal can be wirelessly transmitted via WIFI to a portable electronic device for recording and subsequent analysis (Zhong et al. 2021).

7.2.1.4 Motion Sensors

Motion sensors are another crucial type of sensors that detect physical signals. They have the ability to sense and record human body movements, distinguish movement patterns, and provide vital biological information, thus offering valuable data for medical professionals (Amit et al. 2019; Wang et al. 2022b; Zhao and Li 2020). On one hand, motion sensors have been extensively studied in everyday healthcare. An intelligent cushion based on the flexible piezoelectric sensor can detect the user's sitting position to improve the health of sedentary people (Ma et al. 2023), an all-nanofiber iontronic tactile sensor can detect the health condition of the knee joint and prevent knee joint injury (Wang et al. 2022e), and an in-shoe sensor pad (ISSP) can effectively track sports data and guide athletes' training (Fig. 7.3a) (Yang et al. 2022a). In addition, breathable and waterproof flexible sensors have been developed to enable motion detection and wireless communication in special underwater environments, taking into account factors such as comfortable wear and specific requirements for underwater use (Sun et al. 2023b; Guan et al. 2023; Wei et al. 2023).

On the other hand, motion sensors are of special interest in disease assessment and rehabilitation. Li and colleagues developed a multimodal sensor glove that captures hand movements to assess the severity of symptoms in Parkinson's patients to provide personalized rehabilitation (Fig. 7.3b) (Li et al. 2023a). In addition to gloves, an in-shoe monitoring system and a self-powered tremor sensor have been designed for gait recognition and low-frequency motion evaluation in Parkinson's patients, respectively (Wang et al. 2022a; Kim et al. 2021). For the management of oropharyngeal swallowing disorders, Kim et al. developed a flexible submental sensor patch that remotely monitors muscle activity and laryngeal movement during swallowing (Kim et al. 2019). Similarly, another piezoelectric smart patch attached to the neck converts tiny throat movements into electrical outputs that are

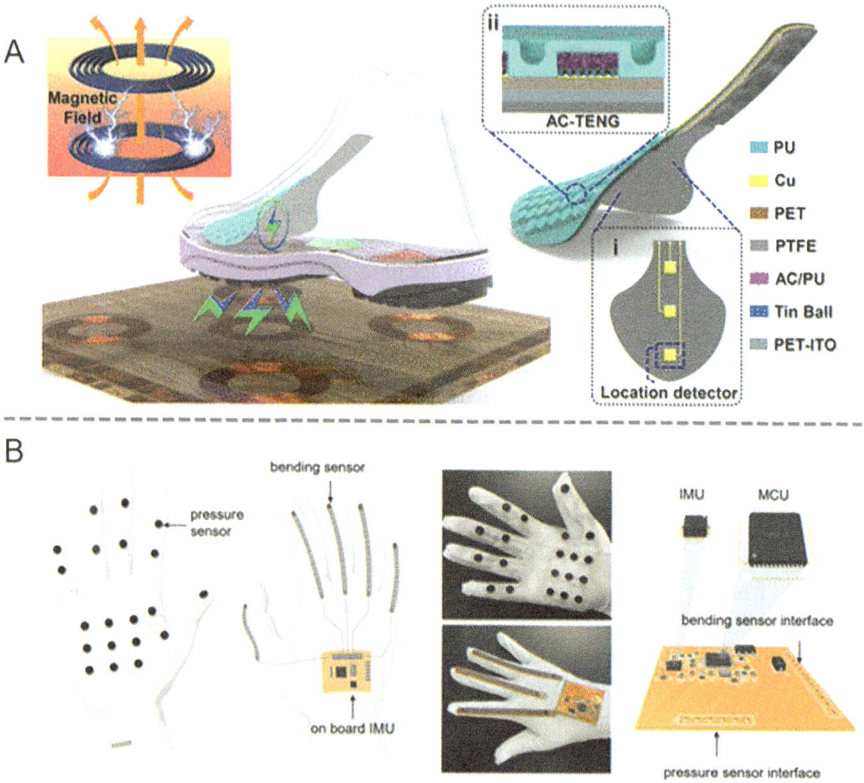

Fig. 7.3 Representative motion sensors of (**a**) construction drawing of the ISSP. (Reprinted with permission from Yang et al. 2022a) and (**b**) multimodal sensor glove structure. (Reprinted with permission from Li et al. 2023a)

transmitted via signal amplification and signal transmission modules to a smartphone for remote viewing and real-time assessment by doctors (Natta et al. 2021).

7.2.2 *Biochemical Signals*

Sensors that detect biochemical signals also have important applications in disease diagnosis. These sensors are designed to detect specific chemical molecules or biomolecules in organisms with biological response and signal transduction mechanisms, providing physicians with critical information on disease-related indicators, enabling early diagnosis and efficient treatment of diseases.

7.2.2.1 Glucose Sensors

Daily blood glucose monitoring is essential for people with diabetes. While traditional glucose monitoring requires fingertip blood sampling, glucose sensors typically employ non-invasive techniques that replace blood with non-invasive biological fluids for continuous glucose measurement, effectively avoiding pain and discomfort (Saha et al. 2023). Various glucose sensors, such as glucose sensing patches attached to the skin (Fig. 7.4a) (Dervisevic et al. 2021), pacifier biosensors and mouthguard glucose sensors (Fig. 7.4b) that monitor glucose in saliva (García-Carmona et al. 2019; Arakawa et al. 2020), and glucose biofuel cells attached to nappies detecting urine glucose, offer convenient ways to measure glucose levels in the body (Fig. 7.4c) (Shitanda et al. 2021; Zhang et al. 2021b). Besides, smart contact lens composed of porous nanohydrogels containing gold-platinum bimetallic nanocatalysts can be used for continuous glucose monitoring in tears through a REDOX reaction based on the glucose oxidase glucose oxidase (GOx) (Kim et al. 2022b). Another type of contact lens that serves as a platform for monitoring glucose in tears uses fluorescent signals generated by a glucose fluorescence probe to monitor glucose concentrations (Deng et al. 2022). Furthermore, the glucose sensor can be connected to an intelligent warning system that sends a sound or vibration when blood glucose levels reach a dangerous or preset range, alerting the patient to take timely control measures to prevent health risks caused by abnormal glucose levels (Zhao et al. 2023c).

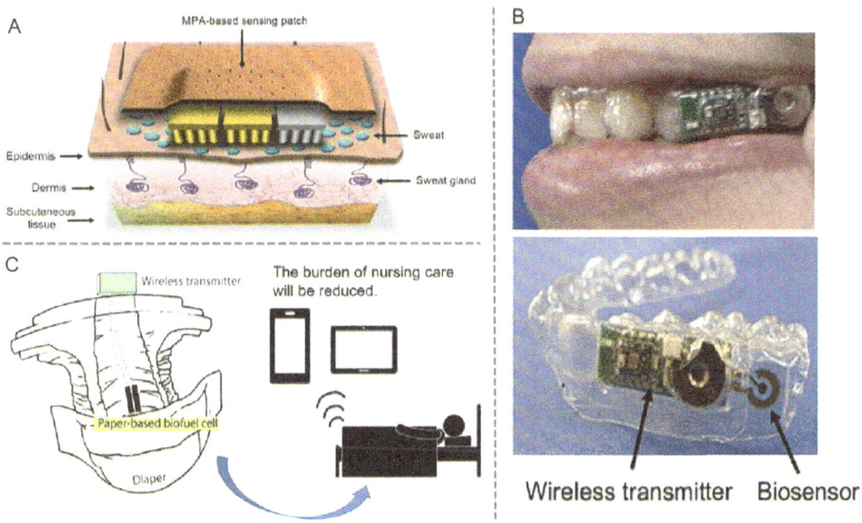

Fig. 7.4 Representative glucose sensors of (**a**) glucose sensing patch. (Reprinted with permission from Dervisevic et al. 2021), (**b**) a mouthguard glucose sensor for salivary glucose measurement. (Reprinted with permission from Arakawa et al. 2020), and (**c**) a self-powered diaper sensor. (Reprinted with permission from Shitanda et al. 2021)

7.2.2.2 Metabolite and Nutrient Monitoring Sensors

Apart from glucose, which is used for blood glucose calibration, other metabolites and nutrients in body fluids, such as amino acids, vitamins, urate, adenosine triphosphate (ATP), and nucleotides, are also closely related to human metabolism and health (Lu et al. 2022). Therefore, sensors that detect these metabolites and nutrients provide insightful access to help people understand and manage their own health. As one of the most important body fluids, sweat is often used as a detection medium for sensors (Mukasa et al. 2023; Zhu et al. 2023; Wei et al. 2021b; Cheng et al. 2019). A laser-engraved sensor constructed by Yang et al. can achieve sensitive detection of uric acid and tyrosine in sweat (Yang et al. 2019b). Similarly, Wang and colleagues designed a laser-engraved graphene-molecularly imprinted polymer (LEG MIP) sensor that performs both direct and indirect detection strategies by distinguishing the different electrochemical properties of target molecules in sweat, thus allowing for accurate and efficient monitoring of metabolite and nutrient levels (Wang et al. 2022g). Additionally, saliva and urine, as common non-invasive biological fluids, also contain a great deal of information about the body's metabolic and nutritional status. Based on this, various sensors have been designed to analyze metabolites and nutrients in these fluids (Raj et al. 2019; Liu et al. 2022c). A supramolecular cucurbit(Battistella et al. 2021)uril-based rotaxane chemosensor could be used for optical identification of tryptophan in urine (Krämer et al. 2023), while another solar cell-type photoelectrochemical (PEC) sensor could perform high throughput and multiple detections of saliva metabolites (Liu et al. 2023c). Moreover, integrated sensors for non-invasive monitoring of two or more body fluids have also been investigated (Zhang et al. 2021d). Bi et al. prepared fully integrated wearable sensor arrays (FIWSAs) composed of microfluidic, sensing, and electronic modules for the detection of metabolites in sweat, saliva, and urine. The microfluidic module collects and stores body fluids, the sensing module sensitively detects electrolytes and metabolites, and the electronic module wirelessly transmits processed signals to the user interface via Bluetooth for personalized, non-invasive metabolic and nutritional monitoring (Bi et al. 2023).

7.2.2.3 Sensors That Detect Cancer Biomarkers

Cancer biomarkers are molecules that are produced or expressed abnormally during the development of cancer, can be present in humoral tissue or cells, and are often associated with the presence of cancer progression and response to treatment (Sarhadi and Armengol 2022). Detecting biomarkers or indicators associated with cancer at an early stage is crucial for early detection and subsequent treatment of the disease. Among various monitoring methods, biosensors have the advantages of convenience, high sensitivity, high accuracy, and hold great potential in detecting and identifying cancer cells and cancer biomarkers (Ranjan et al. 2017; Deng et al. 2021b). These biosensors are equipped with biometric elements on their surfaces, which can interact with specific cancer biomarkers, resulting in color changes, readable light, or electrical signals. Xiao et al. developed a colorimetric biosensor based on Au nanoparticle-decorated Bi_2Se_3 (Au/Bi_2Se_3) nanosheets capable of determining the concentration of carcinoembryonic antigens based on color changes (Fig. 7.5a) (Xiao et al. 2017). In addition, an exosome protein microRNA one-stop (Exo-PROS) biosensor and a perception-based nano-sensor have been developed to detect multiple cancer biomarkers by exploiting changes in the fluorescence signal caused by binding of the identifying element to the target biomarker (Hsu et al. 2023; Yaari et al. 2021). These sensors have been successfully validated using clinical samples. Compared to colorimetric biosensors and fluorescence biosensors, electrochemical biosensors have been widely studied due to their higher response speed, higher sensitivity, and more reliable detection of cancer biomarkers (Khan et al. 2023; Rashid et al. 2021; Meng et al. 2022). Li et al. reported a "biphasic replacement" electrochemical aptamer-based (BRE-AB) sensing platform to achieve highly sensitive detection of biological macromolecules by reducing steric hindrance at the interface (Fig. 7.5b) (Li et al. 2022d). Similarly, to improve detection sensitivity, another electrochemical immunosensor uses Fe-Co metal-organic

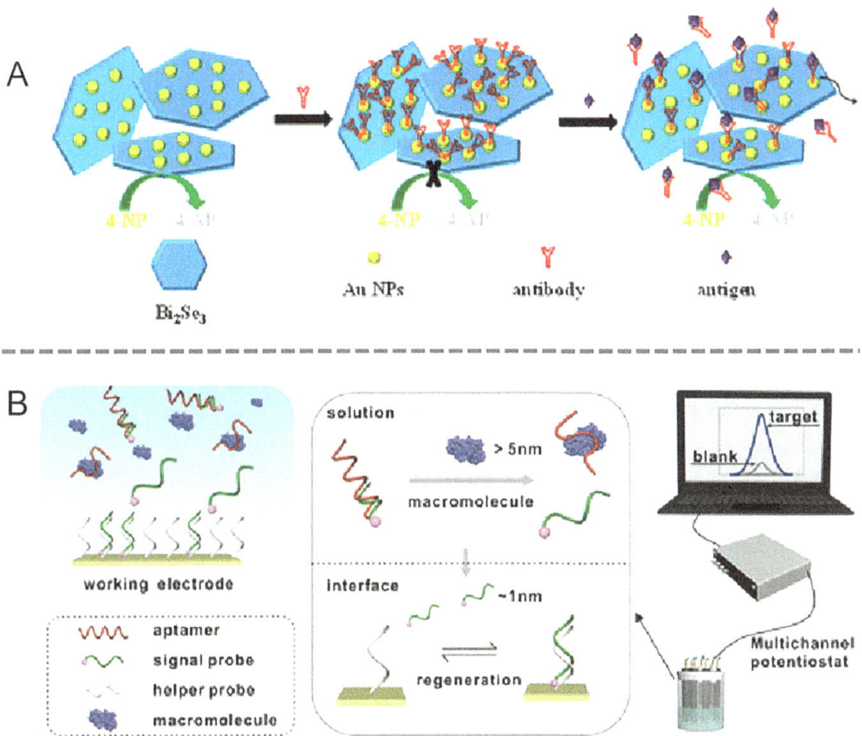

Fig. 7.5 Sensors detecting cancer biomarkers. (**a**) Sensing mechanism of colorimetric biosensors based on Au/Bi2Se3 nanosheets. (Reprinted with permission from Xiao et al. 2017). (**b**) A schematic illustration of the BRE-AB sensing platform. (Reprinted with permission from Li et al. 2022d)

framework as signal amplifier to realize ultrasensitive detection of tumor biomarkers (Shang et al. 2023).

The application of sensors in disease diagnosis and assessment encompasses the detection of both physical and biochemical signals. They provide doctors with diverse information to facilitate early diagnosis, treatment, and disease monitoring. Ongoing technological advances in sensing technology will continue to enhance the accuracy and efficiency of disease detection, leading to further innovations and breakthroughs in the medical fields (Table 7.1).

7.3 Responsive Biomaterials

Similar to biosensors, responsive biomaterials have great potential for personalized and precise medical treatments. The main difference is that sensors that detect physical or biological signals are mainly used for disease diagnosis and detection, while responsive biomaterials are mostly used for disease treatment. These materials possess the ability to sense changes in endogenous stimuli (e.g., pH, enzyme, redox, hypoxia) or exogenous stimuli (e.g., light, ultrasound, electricity, microwaves, magnetic fields), enabling them to release drugs or produce bioactive substances for precise drug delivery and treatment (Badeau and Deforest 2019; Mu et al. 2018).

7.3.1 Endogenous Stimuli

Biomaterials that respond to endogenous stimuli can mediate disease therapy by sensing physiological or pathological stimuli within the body to regulate drug release, induce tissue repair, or trigger specific physiological responses. These materials have the potential to improve the effectiveness of existing therapies while promising the development of personalized therapeutic systems.

7.3.1.1 pH

Complex environmental and pathological changes in the body result in pH variations in different organs or disease sites. In this context, the emergence of biomaterials capable of responding to pH changes provides an efficient route for drug delivery and release. On one hand, the acidic environment of the human stomach influences the effectiveness of oral drugs and even causes drug degradation. To ensure the drug reaches its target location without any problems, pH-responsive biomaterials have been designed to navigate through the stomach (Peng et al. 2023). Zhao et al. prepared a ferulic acid-derived lignin nanoparticle (FALNP) that remained stable in a strongly acidic environment due to the presence of abundant weak acidic groups (Zhao et al. 2022). As a result, FALNP is able to bypass the gastric acid

Table 7.1 A summary of various sensors for disease detection

Detection mode	Type	Example	References
Physical signals	Blood pressure sensors	1. WPBPS	Min et al. (2023)
		2. sCPS	Hye-Jun and Jin-Woo (2023)
		3. Optical-fiber-sensor-assisted smartwatch	Liangye et al. (2023)
		4. Battery-less wireless implant	Kwon et al. (2023)
		5. TPSS	Chan et al. (2023)
	Temperature sensors	1. Polymer–nanocomposite–fiber temperature sensor	Ryu et al. (2023)
		2. Au-doped SiNM temperature sensor array	Sang et al. (2021)
		3. Nano-composite film doped with silver	Phadkule and Sarma (2022)
		4. Kirigami-capacitive sensor	Yu et al. (2021)
		5. Ingestible capsule sensor	Huang et al. (2020)
	Breath sensors	1. Humidity sensor based on microknot resonator	Yi et al. (2020)
		2. All-fiber humidity sensor	Bao et al. (2021)
		3. Breath sensor based on $AgVO_3$ NWs	Yan et al. (2023)
		4. Waist-wearable respiration sensor	Zhang et al. (2019)
		5. Face mask based on LSMO/mica sensor	Ye et al. (2023)
	Motion sensors	1. ISSP	Yang et al. (2022a)
		2. Multimodal sensing glove	Li et al. (2023a)
		3. In-shoe monitoring system	Wang et al. (2022a)
		4. Flexible submental sensor patch	Kim et al. (2019)
		5. Piezoelectric smart patch	Natta et al. (2021)

(continued)

Table 7.1 (continued)

Detection mode	Type	Example	References
Biochemical signals	Glucose sensors	1. Glucose sensing patch	Dervisevic et al. (2021)
		2. Pacifier biosensor	García-Carmona et al. (2019)
		3. Mouthguard glucose sensor	Arakawa et al. (2020)
		4. Self-powered diaper sensor	Shitanda et al. (2021)
		5. Smart contact lens	Kim et al. (2022b)
	Metabolite and nutrient monitoring sensors	1. A laser-engraved sensor	Yang et al. (2019b)
		2. LEG–MIP sensors	Wang et al. (2022g)
		3. A cucurbit (Battistella et al. 2021)uril-based rotaxane chemosensor	Krämer et al. (2023)
		4. Solar cell-type PEC sensor	Liu et al. (2023c)
		5. FIWSAs	Bi et al. (2023)
	Cancer biomarker sensors	1. Colorimetric biosensors based on Au/Bi_2Se_3 nanosheets	Xiao et al. (2017)
		2. Exo-PROS biosensors	Hsu et al. (2023)
		3. Perception-based nanosensors	Yaari et al. (2021)
		4. BRE-AB sensing platform	Li et al. (2022d)
		5. PDMS/Au film-based electrochemical immunosensor	Shang et al. (2023)

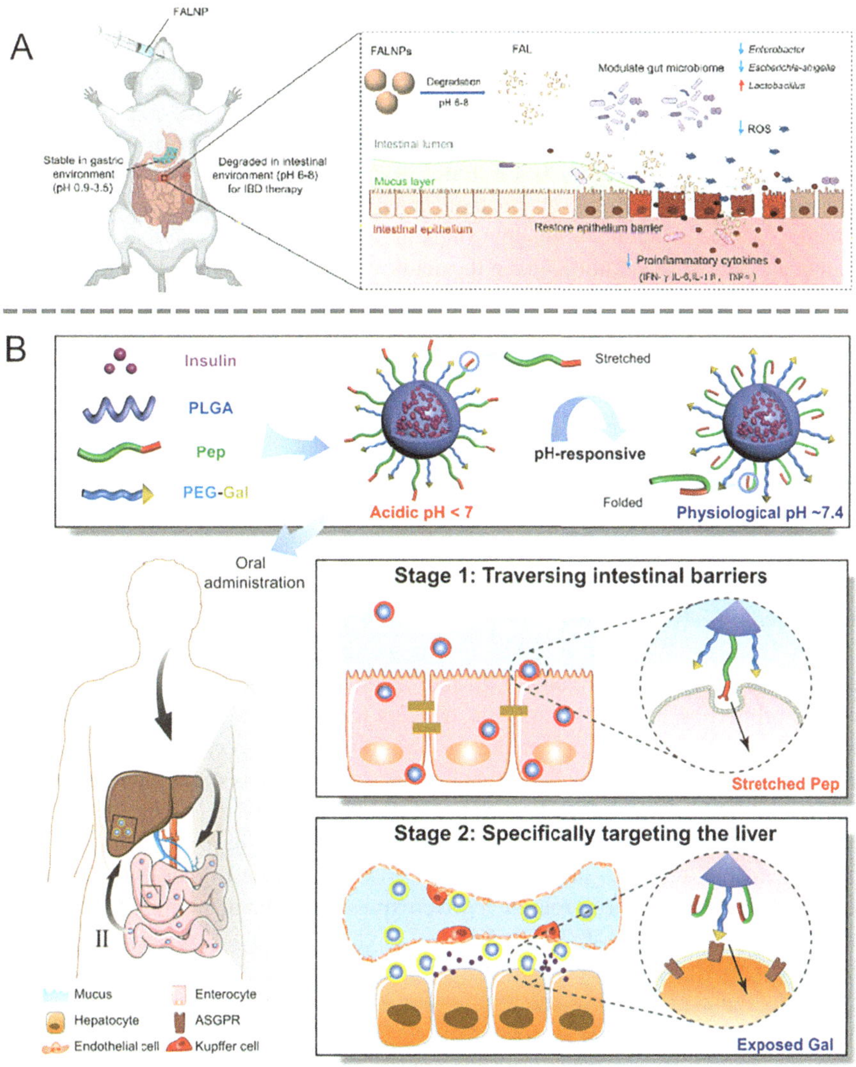

Fig. 7.6 pH-responsive biomaterials. (**a**) FALNP as an orally administrated therapeutic reagent for the therapy of IBD. (Reprinted with permission from Zhao et al. 2022). (**b**) The composition and mechanism of Pep/Gal-PNPs for oral insulin delivery. (Reprinted with permission from Yang et al. 2022b).

environment, reach the intestines, and degrade to release drugs for the treatment of inflammatory bowel disease (IBD) (Fig. 7.6a). Biomaterials that respond to pH also play a key role in the delivery of oral insulin (Hu et al. 2022; Ren et al. 2023). Nanoparticles modified with a pH-responsive cell-penetrating peptide (Pep) and a liver-targeting moiety (Gal) (Pep/Gal-PNPs) enable intestinal absorption and liver deposition of insulin (Fig. 7.6b) (Yang et al. 2022b). On the other hand, changes in pH can indicate a variety of pathological changes. For instance, the acidic tumor microenvironment, resulting from the hyperglycolysis of tumor cells, has been identified as a hallmark of solid tumors. Based on this observation, several strategies have been developed for the specific delivery of therapeutic agents to tumor sites (Liu et al. 2019; Jia et al. 2020; Li et al. 2022b). Liu and his team synthesized an isoguanosine-phenylboronic-guanosine (isoGPBG) hydrogel that degraded and released isoG in an acidic tumor environment to treat oral squamous cell carcinoma (Liu et al. 2023b). In addition, given the acidic environment caused by bacteria during infection, controlled drug release systems are used to increase efficacy and reduce resistance (Liu et al. 2022b; Song et al. 2018b). While for alkaline environments caused by certain bacterial infections, Chen et al. prepared a polydiacetylene-peptide (PDA-Pep) hydrogel that released encapsulated drugs to kill bacteria when pH was elevated (Chen et al. 2022a).

7.3.1.2 Enzymes

Enzymes, as biological macromolecules, possess a wide range of intrinsic properties, including substrate specificity, high reaction rates, and mild reaction conditions (Mu et al. 2018; Wang et al. 2022c). It is worth noting that changes in enzyme expression are highly correlated with the occurrence of diseases. Based on these special properties, enzymes can act as triggers for responsive biomaterials. Taking advantage of the fact that enzyme levels at tumor sites are very different from normal tissues, drug delivery systems that can respond to matrix metalloproteinases or hyaluronidase have been developed to target drug delivery to enhance anti-tumor efficacy (Jiang et al. 2021; Chen et al. 2018b). Bai et al. synthesized self-immolative prodrug nano-assemblies targeting tumor cells that can be triggered for intramolecular hydrolysis and drug release by overexpressed endonuclear histone deacetylases in the cells (Bai et al. 2023). Besides, in cardiovascular disease, studies have shown that increased thrombin levels are associated with the formation of blood clots (Mackman 2008). Inspired by this fact, a platform using thrombin as a stimulus signal was designed to achieve controlled drug release in the local thrombus and improve thrombolysis efficiency (Zhang et al. 2021c). Wang and colleagues used amino acid sequences that can be cut by thrombin to invent a peptide-templated manganese dioxide nanozyme (PNzyme/MnO_2) for thrombin-activated thrombolytic peptide release (Wang et al. 2023b). Enzymes also play a crucial role in the targeted treatment of bacterial infections. A hierarchical nano-platform actively targets bacteria through the dual action of mannose and galactose, which is subsequently triggered by lipase secreted by the bacteria to release antibiotics (Yu et al. 2022).

7.3.1.3 Redox

Redox homeostasis mediated primarily by reducing species such as glutathione (GSH) and oxidizing species such as reactive oxygen species (ROS) is essential for the maintenance and regulation of normal physiological functions in organisms (Zhang et al. 2023b; Wu et al. 2022). In fact, changes in the redox environment in organisms are closely related to the development of cancer, wound healing, and systemic diseases, providing a basis for the design of redox-responsive biomaterials for disease therapy that can undergo reversible structural or functional changes under redox conditions. Due to the abnormal metabolism of tumor cells, GSH and ROS at tumor sites are very different from normal tissues. In general, GSH is highly expressed at tumor sites, and based on this, disulfide bonds that can be cleaved in the presence of GSH were selected to initiate tumor therapy for redox imbalance (Lou et al. 2021; Liu et al. 2023d; Yu et al. 2018). In order to improve the therapeutic effect, Liu et al. utilized docetaxel (DTX) as a drug model to create DTX-STES-DTX NPs connected with sulfur-tellurium-sulfur (STeS) hybrid chalcogen bonds, which exhibit ultra-high redox dual responsiveness. These NPs can be activated by ROS or GSH, resulting in the release of DTX and playing an anti-tumor role (Liu et al. 2022a). In terms of wound healing, the imbalance of redox homeostasis can increase the difficulty of skin repair. To address this issue, redox-sensitive poly(N-isopropylacrylamide-acrylic acid) (PNA)-doped nanofibrous membranes have been developed to regulate the level of ROS at the wound site and thus promote wound healing (Zhang et al. 2021a). In addition, experimental and clinical data demonstrate that vascular oxidative stress is associated with cardiovascular disease and can lead to arteriosclerosis (Larocca et al. 2014). With this in mind, Gao et al. synthesized macrophage cell membrane-coated ROS-responsive NPs loaded with atorvastatin (AT) (MM-AT-NPs) that can target the lesion site and release AT in response to high levels of ROS stimulation, thereby reducing atherosclerosis (Gao et al. 2020a).

7.3.1.4 Hypoxia

Oxygen (O_2) is important for proper cell function and survival, as well as being a key component of cellular respiration and productivity. Hypoxia occurs when tissues or organs are unable to get an adequate supply of oxygen (Stoica et al. 2022). Hypoxia directly affects cell productivity and metabolism, leading to impaired cell function, and is often associated with diseases such as vascular disease, cancer, and ischemia. At the tumor site, hypoxia not only enhances the aggressiveness of tumor cells and causes tumor metastasis, but also reduces the effectiveness of tumor therapy. However, from another perspective, the anoxic tumor microenvironment also provides the conditions for a hypoxia-responsive tumor treatment platform. Azobenzene has excellent hypoxia sensitivity and can be used as a carrier to synthesize and encapsulate anti-tumor drugs, allowing the drugs to be released in oxygen-depleted cancer cells (Kulkarni et al. 2016). Yang and colleagues used an azobenzene linker to construct human serum albumin (HSA)-based nanosystem (HCHOA), which is structurally stable at normal oxygen partial pressure but separates into therapeutic particles at hypoxic tumor sites to enhance penetration into deep tumor tissue and enhance anti-tumor effects (Yang et al. 2019a). In addition, obligate anaerobes may be used in the treatment of hypoxic tumors, where they proliferate in the anoxic tumor microenvironment, competing for and depleting nutrients needed for tumor growth, leading to tumor cell death (Xia et al. 2022). For example, researchers demonstrated that *Clostridium novyi* (*C. Novyi-nt*) can perform accurate tumor localization in laboratory mice and dogs, and used *C. Novyi-NT* to treat a human patient with advanced leiomyosarcoma (Roberts et al. 2014). The results showed a reduction in tumors within and around the patient's bones after injection of *C. Novyi-NT* spore. In the treatment of ischemic cardiomyopathy, taking advantage of the characteristics that red blood cells can store oxygen in an oxygen-rich environment and release it in an anoxic environment, Shi et al. carefully designed a engineered nanosponge (Mn_3O_4@nanoerythrocyte-T7, MNET) to self-regulate the oxygen content during the treatment of ischemic stroke for better therapeutic effects (Shi et al. 2019).

7.3.2 *Exogenous Stimuli*

Compared to endogenous stimuli, exogenous physical stimuli such as light, ultrasound, electricity, microwaves, and magnetic fields have the advantage of being easy to tune and control, and have attracted much attention in disease treatment. Exogenous stimuli can act on biological materials in a non-invasive manner. In response to stimuli, biological materials can release drugs or produce thermal effects or ROS to induce necrosis at the lesion site. The current development of this treatment model is very encouraging due to its advantages of good tissue penetration depth, non-invasive, precision targeting capabilities, and negligible side effects.

7.3.2.1 Light

Light sources, as typical external physical stimuli, have become an important ingredient in therapeutic modalities mediated by light-sensitive materials, such as photodynamic therapy and photothermal therapy. Among them, PDT involves the use of light and photosensitizers to produce ROS to destroy lesion sites, and PTT uses photothermal materials to generate heat under the action of light to kill cancer cells or bacteria (Deng et al. 2021a). To date, many PDT or PTT-related clinical trials have been conducted, including employing Chlorin e6 as a photosensitizer treating patients with early cervical cancer (Ivanova et al. 2022), using aminolevulinic acid (ALA) treating patients with actinic keratosis (Berman et al. 2021), and using gold-silica nanoshells (GSNs) to perform photothermal ablation of prostate tumors (Rastinehad et al. 2019). Although some clinical studies have been conducted, PDT and PTT have not been widely used in the clinical treatment of patients. This is mainly due to the lack of ROS production caused by hypoxia at the lesion site, which limits the therapeutic effect of PDT. Additionally, PTT inevitably leads to thermal damage to the normal site, and the tissue penetration depth of light is also insufficient. To cope with these challenges, several oxygen self-supply platforms have been designed specifically to deal with the problem of hypoxia at the lesion site (Huang et al. 2021). Using endogenous hydrogen peroxide (H_2O_2) as a source of O_2, responsive biomaterials with oxygen self-supply function can enhance the killing effect of PDT on cancer cells or be used in the treatment of anaerobic infection (Wei et al. 2022; Sun et al. 2021). Similarly, the oxygen-generating capabilities of algae through photosynthesis have been harnessed for developing cyanobacteria-based PDT platforms, aiming to overcome the problem of insufficient oxygen at tumor or infected sites (Chang et al. 2022; Wang et al. 2022h). Further, to improve the therapeutic efficacy while mitigating the side effects of PTT, carefully designed responsive materials containing heat shock protein (HSP) inhibitors have been prepared to achieve mild PTT (Gao et al. 2020b; Liu et al. 2023a). In addition to their individual effects, the combination of PDT and PTT offers a novel approach to disease treatment by addressing the limitations of each approach and enhancing the safety and efficacy of the treatment. Therefore, responsive biomaterials capable of generating both ROS and heat upon light stimulation have gained momentum in recent years and are increasingly being used for the treatment of tumors, bacterial infections, atherosclerosis, and other conditions (Kim et al. 2023; Yuan et al. 2023; Dai et al. 2022). Due to limited tissue penetration of light, PDT and PTT are unsatisfactory in treating deep tumors or focal sites. To overcome this challenge, researchers have conducted studies using ultrasound or X-rays instead of light sources to drive responsive materials for mediating disease therapy, as will be described below.

7.3.2.2 Ultrasound (US)

Ultrasound-based imaging has long been popular in clinical applications due to its advantages of low cost, real-time imaging, and good safety (Yoon et al. 2017). More importantly, with the advent of US-responsive materials, ultrasound offers the opportunity for precision disease treatment. First of all, similar to light, ultrasound can trigger sonosensitizers to produce ROS to treat disease, which is known as sonodynamic therapy (SDT). Although SDT is derived from PDT, ultrasound has a higher penetration than light, as it can penetrate up to 11 cm deep into tissue, making it suitable for treating deep disease sites (Ma et al. 2018). Currently, the low quantum yield of ROS remains a major challenge hindering the applications of SDT (Wang et al. 2020). Various strategies have been developed to enhance ROS production, including optimizing the structure of sonosensitizers, consuming GSH to relieve hypoxia, and combining SDT with other therapies (Sun et al. 2023a; Song et al. 2022; Cheng et al. 2023a; Huang et al. 2022). In recent years, the combination of piezoelectric materials and US has provided a novel approach for alleviating the deficiency of ROS production due to hypoxia, and has also formed a new paradigm for disease treatment – piezoelectric dynamic therapy (PZDT) (Fig. 7.7) (Hu et al.

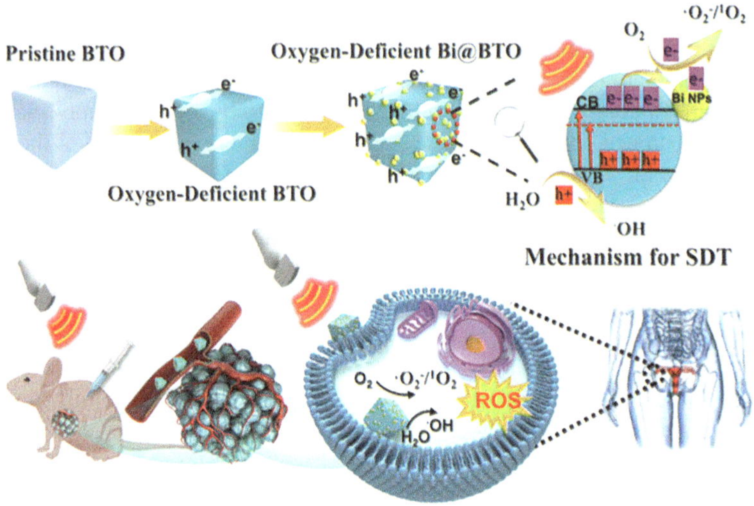

Fig. 7.7 The development and working mechanism of bismuth-doped oxygen-deficient $BaTiO_3$ for improved electron-hole pair separation and enhanced sono-piezocatalytic therapy. (Reused with permission from Cheng et al. 2022)

2021; Dong et al. 2021; Cheng et al. 2022; Truong Hoang et al. 2022; Zhu et al. 2020). Secondly, responsive biomaterials capable of generating electrical stimuli under ultrasound have been developed using piezoelectric materials. These materials, which convert mechanical deformations into direct current electrical output when subjected to physical stimuli, have been used to repair spinal cord injuries, stimulate peripheral nerves, and normalize tumor blood vessels (Chen et al. 2022c; Chen et al. 2021; Li et al. 2022c). For instance, Kim et al. report a NO-releasing and piezoelectric nanoparticle (BTNP–pDA–BNN6) for electrical stimulation of deep tissue in the brain (Kim et al. 2022a). Thirdly, ultrasound can be used in conjunction with microvesicles for thrombolysis. The principle of sonothrombolysis is to facilitate the penetration and diffusion of thrombolytic agents induced by ultrasound radiation forces and microvesicles, and to cause the mechanical destruction of the clot by inertial cavitation of the microvesicles, thus dissolving and removing the clot that has formed in the vessel (El Kadi et al. 2022; Zheng et al. 2022; Pan et al. 2022; Zhong et al. 2019). To explore the effect of different shaped particles on sonothrombolysis, Choi et al. prepared recombinant tissue plasminogen activator (rtPA)-functionalized gold nanospheres (NPt) and gold nanostars (NSt), then performed in vitro and in vivo thrombolysis experiments (Choi et al. 2022). The results revealed that asymmetric NSt enhanced the potential for thrombolysis through the ultrasound-induced cavitation effect compared to the symmetric NPt.

7.3.2.3 X-Rays

As has been mentioned earlier, the clinical application and conversion of conventional PDT, which uses ultraviolet (UV), visible (Vis), or near-infrared (NIR) light as activation sources, has been greatly hindered by limited light tissue penetration. To solve this problem, X-rays, as a source of ionizing radiation with infinite tissue penetration depth, have recently been developed for PDT. This is often referred to as X-ray Induced PDT (X-PDT) or RDT (Chong et al. 2021; He et al. 2022). Unlike conventional radiotherapy, X-PDT utilizes scintillator nanoparticles to convert the energy of X-rays into UV or Vis light, which stimulates photosensitizers and generates cytotoxic ROS. This innovative approach enhances treatment efficacy at lower X-ray doses and minimizes damage to normal tissue surrounding the tumor (Wang et al. 2016). Recently, there has been a growing interest in scintillating materials that demonstrate persistent luminescence under X-ray irradiation, making them particularly promising for low-dose X-PDT applications (Song et al. 2018a). Chen et al. loaded LGO:Cr nanoparticles and a photosensitizer, 2,3-naphthalocyanine (NC), into mesoporous silica nanoparticles to prepare a novel nanosensitizer (NC-LGO:Cr@mSiO_2) to achieve X-PDT (Chen et al. 2017). After X-ray irradiation, LGO:Cr can emit NIR light lasting for several minutes or even hours. To reduce the significant energy loss that occurs during the conversion of X-rays to Vis using inorganic scintillators, and to enhance the performance of X-PDT, researchers have created organic scintillators capable of directly emitting photons when exposed to X-ray radiation. The organic phosphorescent nanoscintillators (ITC-NPs) designed by Wang and colleagues can generate more than 75% triplet excitons to produce singlet oxygen ($^1O_2$) under a dose of 0.4 Gy X-ray irradiation (Wang et al. 2022f). In addition, a recent study prepared the first purely organic nanoscintillator (TBDCR NPs) capable of simultaneously generating $^1O_2$ and hydroxyl radicals (·OH) by direct X-ray irradiation for X-PDT in hypoxic tumors (Yu et al. 2023).

7.3.2.4 Electrical Stimulation

Electric stimulation has many applications in the treatment of diseases, including treating epilepsy, tumor treating fields (TTFields) treatment, and skin wound therapy (Fisher and Velasco 2014; Fabian et al. 2019; Luo et al. 2023a). All of these therapies work with electrical stimulation alone, while a typical use of electrical stimulation in conjunction with responsive biomaterials to treat disease is electrodynamic therapy (EDT). In 2019, Gu and colleagues reported that platinum nanoparticles (PtNPs) can be activated by an electric field to occur Faraday cage effect on the surface, which subsequently triggers the dissociation of water molecules to produce ·OH, thus proposing EDT (Gu et al. 2019). Since then, EDT has attracted significant interest from researchers as an emerging electrotherapy method for solid tumors. To achieve better tumor suppression, a growing number of studies are combining EDT with other therapies. For example, loading GOx or doxorubicin (DOX) in PtNPs enables synergy of EDT with starvation therapy or chemotherapy,

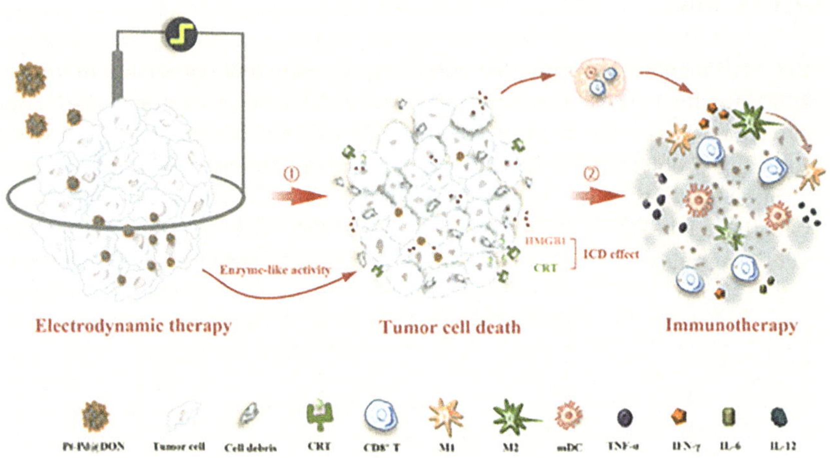

Fig. 7.8 An illustrative scheme of a Pt–Pd@DON for synergistic electrodynamic immunotherapy. (Reprinted with permission from Chen et al. 2022b)

respectively (Lu et al. 2020; Chen et al. 2020). In addition, taking into account the immunosuppressive tumor microenvironment, Chen et al. designed an advanced anti-tumor platform (Pt-Pd@DON) using Pt-Pd nanoflowers and glutamine antagonist 6-diazo-5-oxo-l-norleucine (DON) to synergize electrodynamic therapy and immunotherapy (Chen et al. 2022b). Among them, Pt-Pd nanoflowers produce ROS for EDT under square-wave current and induce immunogenic cell death (ICD), and DON enhance protective immune response by promoting dendritic cells (DCs) maturation and recruiting $CD8^{+}$ T cells (Fig. 7.8). Treatment results in mouse tumor models demonstrated that Pt-Pd@DON is effective in inducing tumor ablation under the action of a square wave current, enhancing the infiltration of $CD8^{+}$ T cells in the tumor microenvironment and the M1 phenotypic polarization of tumor-associated macrophages.

7.3.2.5 Magnetic Fields

Magnetic fields can induce heat generation through magnetothermal conversion mediated by magnetic nanoparticles (MNPs), which is known as magnetic hyperthermia. In this therapy, MNPs absorb energy from an external magnetic field and convert it into heat. Since the movement of MNPs can be manipulated by an external magnetic field, it has the ability to target the diseased site. Currently, magnetic hyperthermia has been used in clinical trials for the treatment of solid tumors, including for prostate cancer alone or as an adjunct to radiation therapy for glioblastoma (GBM) (Johannsen et al. 2007; Maier-Hauff et al. 2010). Magnetic hyperthermia mediated by MNPs also plays an important role in the treatment of bone defects.

For example, magnetic hyperthermia promotes osteogenesis and angiogenesis by upregulating HSP expression, and it can also treat infected bone defects by killing bacteria (Wang et al. 2022d; Zhao et al. 2023a). In addition, magnetic composite stents that can be induced by alternating magnetic fields (AMF) to produce thermal effects and accelerate their own degradation provide reliable guidance for controlled degradation of implants (Hao et al. 2021). Although the deep penetration of magnetic fields in the human body can achieve deep tissue therapy, the low thermal efficiency of magnetothermal materials limits the application of magnetic hyperthermia (Wang et al. 2018). To address this issue, Pan et al. attempted to enhance the magnetothermal efficiency of MNPs by doping of Zn^{2+} and increasing the size, and thus synthesize ZCMF, which was demonstrated to have excellent magnetothermal therapeutic properties by in vitro and in vivo experiments (Pan et al. 2021). Alternatively, considering the disadvantage of easy aggregation of nanoparticles, magnetotactic bacteria can be developed to replace traditional MNPs to mediate magnetic hyperthermia for killing *Staphylococcus aureus* or inhibiting the proliferation of cancer cells (Chen et al. 2016; Gandia et al. 2019).

7.3.2.6 Microwaves

Microwaves, as the electromagnetic spectrum that causes the temperature of the lesion site to rise, are used in thermal ablation therapy. However, the recurrence rate of traditional microwave ablation (MWA) is still high due to the limited ablation area (Groeschl et al. 2014). To improve the ablation effect and selectively deliver heat to the lesion site, microwave-sensitive materials capable of converting microwaves into heat have been developed for targeted enhancement of the heating effect (Chen et al. 2018a). Furthermore, recent studies have found that some microwave sensitizers can also generate ROS under the trigger of microwaves to synchronously start microwave thermal therapy (MTT) and microwave dynamic therapy (MDT) (Fu et al. 2017; Li et al. 2022a). The manganic-doped titanium metal-organic frame (Mn-Ti MOFs) nanosheets constructed by Qin and colleagues can generate both heat and ROS under low-power microwave radiation and have good biocompatibility. Treatment experiments in vivo showed that Mn-Ti MOFs were effective in inducing tumor elimination in tumor-bearing mice after 14 days of treatment (Qin et al. 2023). In the treatment of bacterial-infected osteomyelitis, responsive biomaterials that can significantly heat up and induce ROS production driven by microwaves play an important role (Zhu et al. 2022; Wei et al. 2021a). Li et al. designed a microwave-responsive system, the $Zn^{2+}/g\text{-}C_3N_4$ composite system (BZCN). Under microwave action, the electrons produced by BZCN collide with Na^+ to generate energy that is used to react with water and oxygen to produce (Zheng et al. 2021b) O_2 and $\cdot O_2^-$, while polar molecules move and generate frictional heat. Finally, the released Zn^{2+} stimulates wound tissue healing, which is conducive to osteogenic repair (Li et al. 2023c).

Biomaterials that respond to endogenous or exogenous stimuli provide valuable insights for precise disease treatment. These responsive biomaterials can recognize

Table 7.2 A summary of responsive biomaterials for disease treatment

Stimulus mode	Type	Application	References
Endogenous stimuli	pH	1. FALNP (IBD treatment)	Zhao et al. (2022)
		2. PEP/gal-PNPs (insulin delivery)	Yang et al. (2022b)
		3. isoGPBG hydrogel (treating OSCC)	Liu et al. (2023b)
		4. PDA-pep hydrogel (antimicrobial delivery)	Chen et al. (2022a)
	Enzymes	1. HA-6-PTX (targeted PTX delivery)	Chen et al. (2018b)
		2. Self-immolative prodrug nanoassemblies (cancer-selective drug delivery)	Bai et al. (2023)
		3. PNzyme/MnO_2 (thrombolytic)	Wang et al. (2023b)
		4. Hierarchical nanoplatform (therapy against bacterial infections)	Yu et al. (2022)
	Redox	1. DTX-STeS-DTX NPs (treatment of tumors)	Liu et al. (2022a)
		2. PNA-doped nanofibrous membranes (accelerated wound healing process)	Zhang et al. (2021a)
		3. MM-AT-NPs (treatment of atherosclerosis)	Gao et al. (2020a)
	Hypoxia	1. HCHOA nanosystem (deep tumor penetration and treatment)	Yang et al. (2019a)
		2. *C. novyi*-NT (antitumor therapy)	Roberts et al. (2014)
		3. MNET (salvaging ischemic stroke)	Shi et al. (2019)

Exogenous stimuli	Light	1. Ir-g-C_3N_4 (hypoxic tumor therapy)	Wei et al. (2022)
		2. F@Ce6-M (against anaerobe-induced periodontal disease)	Sun et al. (2021)
		3. Cyanobacteria-based PDT platform (PDT for tumors)	Chang et al. (2022)
		4. GOIGLs (mild-temperature PTT for tumor)	Gao et al. (2020b)
		5. Black TiO_2 nanoprobe (combination PDT and PTT for atherosclerosis)	Dai et al. (2022)
	Ultrasound	1. $MnVO_3$ NPs (SDT of cancer)	Sun et al. (2023a)
		2. PtCu NPs (sonodynamic bacterial elimination)	Cheng et al. (2023a)
		3. Bismuth-doped oxygen-deficient BaTiO3 (sono-piezocatalytic therapy)	Cheng et al. (2022)
		4. BTNP–pDA–BNN6 (treating Parkinson's disease)	Kim et al. (2022a)
		5. NSt (sonothrombolysis)	Choi et al. (2022)
	X-rays	1. NC-LGO:Cr@mSiO_2 (X-PDT for tumor)	Chen et al. (2017)
		2. ITC-NPs (X-PDT for tumor)	Wang et al. (2022f)
		3. TBDCR NPs (hypoxia-tolerant X-PDT)	Yu et al. (2023)
	Electric stimulation	1. PtNPs (EDT for tumors)	Gu et al. (2019)
		2. pPt@GOx (synergistic starvation/EDT)	Lu et al. (2020)
		3. DOX@pPt-PEG (synergistic chemo–/electrodynamic therapy)	Chen et al. (2020)
		4. Pt-Pd@DON (synergistic electrodynamic immunotherapy)	Chen et al. (2022b)
	Magnetic fields	1. Iron-oxide nanoparticles (thermotherapy for GBM)	Maier-Hauff et al. (2010)
		2. MRA hydrogels (induced bone regeneration)	Wang et al. (2022d)
		3. Magnetotactic bacteria (killing of *Staphylococcus aureus*)	Chen et al. (2016)
	Microwaves	1. MZCNs (MWA of tumor)	Chen et al. (2018a)
		2. Mn-Ti MOFs (combination of MDT and MTT for tumors)	Qin et al. (2023)
		3. BZCN (treating osteomyelitis)	Li et al. (2023c)

the difference between diseased sites and normal tissue, or be controlled by external physical fields, thereby releasing drugs or producing biological effects at specific sites to mediate precise disease treatment, improving therapeutic efficacy while avoiding damage to normal tissue (Table 7.2).

7.4 Interactive Biomaterials

Interactive biomaterials, which integrate the features of sensors that sense pathological changes and materials that respond to stimuli, can not only collect information for disease assessment, but also control the release of drugs or generate biological effects to treat diseases. The advantage of interactive biomaterials is that they simplify the diagnosis and treatment of diseases, avoiding multiple procedures and the use of multiple materials. At the same time, they can provide more precise and targeted treatment, reducing unnecessary trauma.

7.4.1 Smart Pills

Smart pills, as typical interactive biomaterials, can be defined as ingested tablets or capsules containing drug delivery systems and sensors or mechanical elements that can reach the gastrointestinal tract for the purposes of drug release and in-situ detection. The pills are easily swallowed by most people, reducing discomfort and improving patient acceptance. Depending on the purpose of detecting, smart pills can be divided into two categories, namely digital pills for drug compliance measurement and robotic capsule endoscopy for gastrointestinal disease detection.

7.4.1.1 Digital Pills

The digital pill is an innovative interactive biomaterial consisting of small sensors and drugs (Goodman et al. 2022). Once the digital pill is dissolved by gastric acid, the encapsulated drug is released and sensors inside the pill generate a signal that can be detected by an external wearable device (El Hadidi and Rosano 2020). Subsequently, the wearable device converts the collected signals into data and transmits it to a portable terminal, allowing for accurate tracking of drug intake. This technology provides objective information about patients' medication, enabling timely remote intervention by doctors (Goodman et al. 2022; El Hadidi and Rosano 2020). As of this review, there are two digital pills that have been approved by the Food and Drug Administration (FDA): Abilify MyCite, approved in 2017, and ID-Cap System, approved in 2019 (Asher 2017; Goodman et al. 2022). Abilify MyCite, which consists of aripiprazole tablets embedded with a sensor, is used to treat psychiatric disorders and provide information about objective medication patterns (Cosgrove et al. 2019). More recently, a Phase IIIb clinical trial found that

Abilify MyCite reduced hospitalization rates in schizophrenia patients (Cohen et al. 2022). Results from other clinical trials on schizophrenia patients showed that treatment with Abilify MyCite more effectively improved clinical symptoms than oral medication alone (Cochran et al. 2022). The ID-Cap System, the second digital pill approved by the FDA, consists of an ID-Capsule, an ID-Cap Reader, and associated software for measurement and reporting of medication adherence. The ID-Capsule that is made up of a standard drug capsule shell containing an ID-Tag (ingestible sensor) can be dissolved by gastric acid to release the drug and trigger the ID-Tag (Goodman et al. 2022). A study in which 17 volunteers were recruited for a validation test of the ID-Cap System's usability demonstrated that the ID-Cap System is easy to use and readily accepted by patients for accurate medication adherence measurements (Baumgartner et al. 2021).

7.4.1.2 Robotic Capsule Endoscopy

Developed from traditional wireless capsule endoscopy, robotic capsule endoscopy consists of an imaging device and a controlled drug delivery system that can be passed through the gastrointestinal tract for diagnostic, therapeutic, sampling, and other purposes (Iddan et al. 2000). Wireless capsule endoscopy was first introduced in 2000 for painless endoscopic imaging of the small intestine. Because it can minimize patient discomfort and trauma, capsule endoscopy has developed rapidly since its introduction, becoming a valuable tool for examining the gastrointestinal tract and revolutionizing the diagnostic procedure of the gastrointestinal tract (Tang et al. 2004; Melmed and Lo 2005). However, capsule endoscopy has several limitations, including lack of therapeutic capability, passive passage through the gastrointestinal tract, and lack of ability to obtain biopsy samples (Singeap et al. 2016; Son et al. 2020). To this end, robotic capsule endoscopy that can be controlled by magnetic fields have emerged, which can be used not only as imaging devices, but also for active diagnosis and treatment. Yim et al. designed and constructed a magnetically actuated soft capsule endoscope (MASCE) for medical applications in the diagnosis and treatment of diseases in the stomach (Yim and Sitti 2011). The MASCE consists of a camera module on top, an internal permanent magnet at each end, and a drug chamber formed by two internal magnets. After the patient swallows the MASCE, the doctor manipulates the MASCE through an external permanent magnet while viewing camera images of the capsule in real time. When abnormal tissue is observed, the doctor will drive the MASCE to the target location to release the drug for treatment. Furthermore, in order to improve treatment efficiency and reduce medical costs, an innovative multimodule capsule robot (MCR) composed of main module and functional module was proposed (Zheng et al. 2021a). The main module can capture intestinal images for initial disease detection, while the functional module can dock with the main module for further diagnosis and treatment. At the end of the treatment, all the robotic modules are separated from each other and then discharged through the anus. In this mode, different modules can be controlled independently, reliably connected and easily separated, making the work of diagnosis and treatment modules more specific and efficient.

7.4.2 *Noninvasive Closed-Loop Systems*

Noninvasive closed-loop systems are interactive biomaterials with great potential to integrate medicine and information technology to provide more comprehensive and intelligent medical services to patients. A major advantage of noninvasive closed-loop systems is their ability to monitor and record physiological parameters in real time in a noninvasive manner, and to administer medications on demand based on the monitoring results. This closed-loop approach to diagnosis and treatment not only provides personalized medical advice, but also simplifies the diagnosis and treatment process and improves the efficiency of medical treatment.

7.4.2.1 Smart Wound Dressings

Wound healing is a complex and dynamic process with many factors that can disrupt normal wound repair, cause wound infections, and even lead to the development of chronic wounds (Liang et al. 2021; Dong and Guo 2021). Traditional wound dressings lack the ability to monitor the wound in real time and adjust treatment modalities accordingly (Pang et al. 2020). In this context, smart wound dressings that can monitor the wound environment and administer drugs on demand in a closed-loop manner have emerged as interactive biomaterials. Smart wound dressings are composed of various components, including sensors, analog-to-digital converters (ADCs), wireless electronics, controlled drug delivery systems, or stimulators (Xu et al. 2021). Among these components, the sensor layer plays a crucial role in detecting physical and chemical indicators of wounds, such as temperature, pH, lactic acid, and uric acid. It uses an ADC to convert the detected information into a digital signal which is then sent to an external electronic device for processing. When a wound undergoes changes or becomes infected, the external device wirelessly sends a treatment command and activates a drug delivery electrode or stimulator, facilitating wound healing and enabling effective care. The battery-free and wireless wound dressing developed by Xu et al. provides electrically controlled on-demand antibiotic delivery based on the state of the wound, providing an integrated solution for monitoring and repairing damaged tissue (Xu et al. 2021). In addition, in order to realize multiple monitoring and combination treatment of chronic wounds, Sani et al. constructed a wearable patch (Shirzaei Sani et al. 2023). As a proof-of-concept, the infection-excised wound model of diabetic mice was used for in vivo multi-sensor studies and treatment evaluation. The results showed that the wearable patch was able to monitor dynamic changes in physical and chemical indicators of the wound and was able to fully close the chronic wound after 14 days.

7.4.2.2 Smart Contact Lenses

Contact lenses that come into direct contact with the eye can provide a variety of disease-related information and have the potential to deliver drugs to the eye

(Dennyson Savariraj et al. 2021). Inspired by this, researchers have incorporated biosensors or drug delivery systems into contact lenses for real-time diagnosis or treatment of diseases (Kim et al. 2017; Sun et al. 2017). It is important to note that this type of contact lens cannot be called an interactive biomaterial, as it does not integrate both detection and treatment of the disease. In recent years, novel smart contact lenses capable of closed-loop diagnosis and treatment of eye diseases have been developed with continuous research advances (Kim et al. 2022d). A wireless theranostic contact lens (WTCL) prepared by Yang and colleagues could be used for monitoring intraocular pressure (IOP) levels in glaucoma and for on-demand administration (Yang et al. 2022c). The sensing module in WTCL can super-sensitively detect fluctuations in the IOP and generate signals for wireless transmission. When intraocular pressure rises, wireless power transfer (WPT) triggered iontophoresis allows in-situ anti-glaucoma drug release to control the IOP (Fig. 7.9a). In the monitoring and treatment of diabetic retinopathy, a smart electrochemical contact lens has been proposed for non-invasive glucose monitoring and control of drug delivery (Keum et al. 2020). An evaluation of in-vivo sensing and

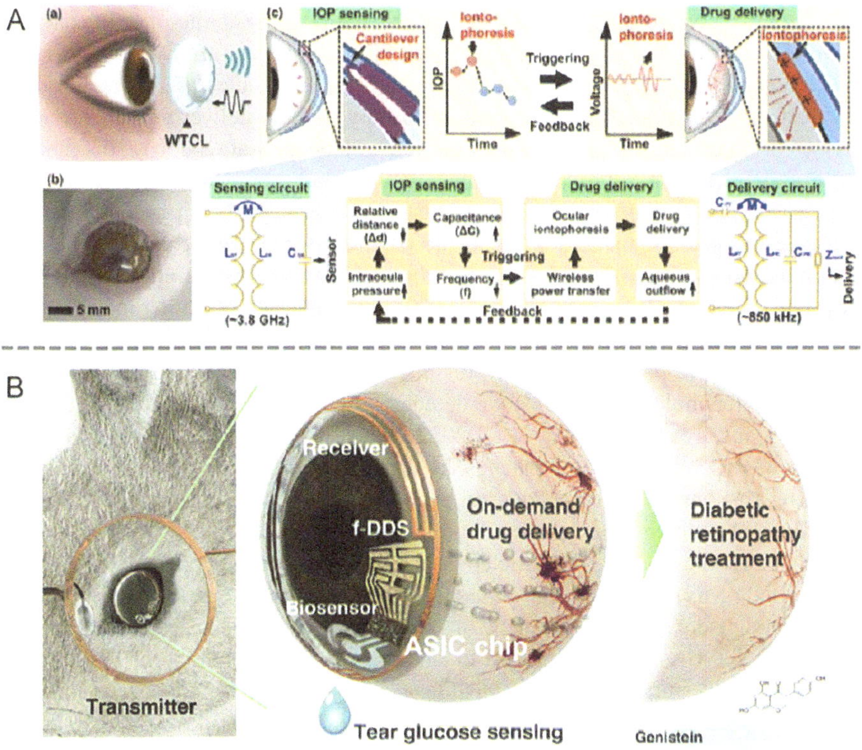

Fig. 7.9 Representative smart contact lens of (**a**) WTCL for real-time and in-situ IOP monitoring and on-demand administration. (Reprinted with permission from Yang et al. 2022c). and (**b**) A smart electrochemical contact lens for in-vivo diabetes diagnosis and treatment. (Reused with permission from Keum et al. 2020)

drug delivery applications in a diabetic rabbit model showed that glucose concentration in tears measured by contact lens strongly correlates with blood glucose concentration, and that contact lens enable electronically controlled drug release for the treatment of diabetic retinopathy (Fig. 7.9b).

7.4.3 Smart Minimally Invasive Devices

As convenient interactive biomaterials, smart minimally invasive devices can form micro-pores on the skin through the cuticle to collect samples or monitor physiological parameters, while enabling accurate drug delivery. It is gentler in nature, avoiding skin nerves and blood vessels, reducing or eliminating pain and making operations easier for medical staff and even patients. As a representative of smart minimally invasive devices, theranostic microneedle array patches have been studied and developed in recent years.

The microneedle patch consists of a series of microneedle arrays with a height range of 500–800 μm that allow penetration of the skin with minimal pain (Jin et al. 2019). By incorporating sensors onto microneedle arrays, microneedle patches enable direct and rapid measurement of specific biomarkers in vivo (Downs et al. 2023; Zeng et al. 2020). In addition, the use of microneedles as convenient drug delivery devices enables accurate drug release by directly delivering drugs to local tissues or blood circulation (Zhang et al. 2023a; Cheng et al. 2023b). This method of drug delivery can improve drug bioavailability and therapeutic effectiveness, as well as reduce side effects and drug waste. In recent years, in order to further improve the performance of microneedle patches so that they can meet the dual requirements of physiological parameter detection and treatment, efforts have been devoted to the development of a new generation of theranostic microneedle array patches. Today, theranostic microneedle array patches not only efficiently deliver drugs to multiple parts of the body for disease treatment, but also combine with biosensing technologies for sensitive detection, enabling the integration of disease detection and treatment (Yang et al. 2022d). For intelligent wound management, theranostic microneedle array patches are capable of monitoring wound status and enabling responsive subcutaneous drug delivery, which determines their unique position in wound environment regulation (Guo et al. 2021; Wang et al. 2023a; Gao et al. 2021). Shan et al. constructed a polymer nanodot-integrated microneedle patch (FND-MN) with recognition and treatment capabilities to enable intelligent synchronous sensing and treatment of wound infections (Shan et al. 2022). The color of FND-MN is closely related to changes in pH and H_2O_2 at the wound site, which makes it suitable for distinguishing normal wounds from bacterial-infected wounds and achieving on-demand therapeutic effects. Besides, the theranostic microneedle array patches can also be used for intelligent monitoring of diabetes and transdermal drug delivery. Smart microneedle array patches, which integrate sensors and insulin delivery systems, can accurately track blood glucose levels and quickly release insulin based on glucose levels. To accelerate the diffusion of insulin in the subcutaneous tissue and further improve the accuracy of the sensor and the efficiency of lowering blood glucose, Luo et al. constructed an ultrasonic pump-enabled closed-loop microneedle to enhance insulin diffusion through acoustic flow, microbubble cavitation, and induced perforation generated by an ultrasound pump (Fig. 7.10a) (Luo et al. 2023b). In addition to glucose levels, which are directly related to diabetes, ionic concentrations of Na^+ and K^+ are important for diabetes treatment (Meisinger et al. 2013; Yang et al. 2022e). With this fact in mind, the microneedle theranostic platform (MNTP) designed by Yang and colleagues can monitor glucose, K^+, and Na^+ levels in situ and enables on-demand insulin delivery based on direct feedback from the detected physiological signals (Fig. 7.10b) (Yang et al. 2022e).

The aforementioned interactive biomaterials, including digital pills, robotic capsule endoscopy, smart wound dressings, smart contact lenses, and theranostic microneedle array patches, integrate diagnosis and treatment and enrich the connotation of traditional medical drugs. The emergence of interactive biomaterials has

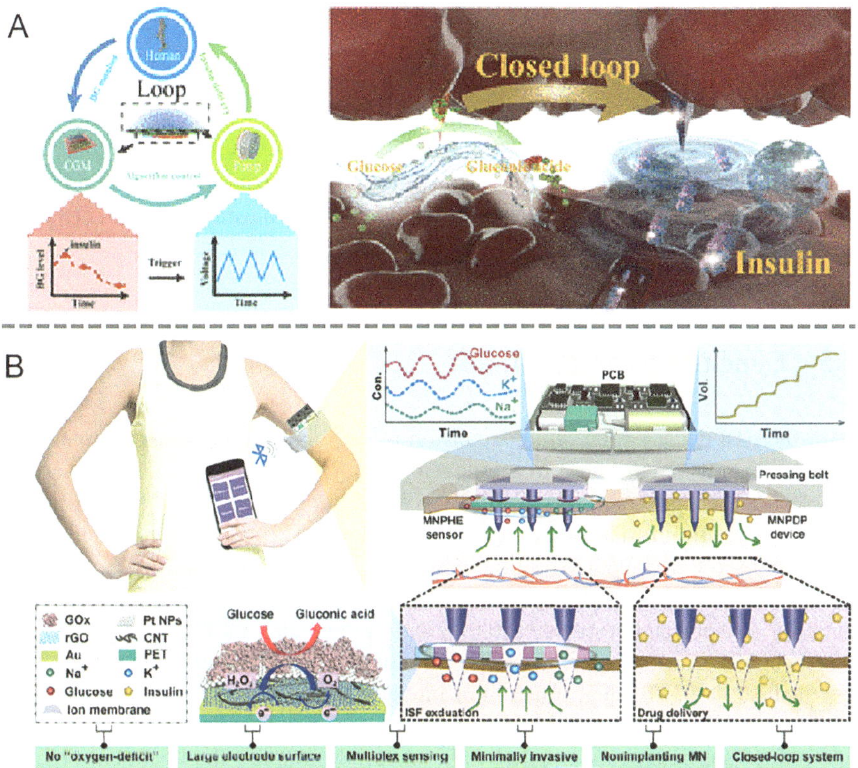

Fig. 7.10 Representative theranostic microneedle array patches. (**a**) Principle and structure of a microneedle-based closed-loop system for the treatment of diabetes. (Reprinted with permission from Luo et al. 2023b). (**b**) Illustration of the MNTP for intelligent and precise diabetic management. (Reused with permission from Yang et al. 2022e)

Table 7.3 A summary of interactive biomaterials

Model	Type	Example	References
Smart pills	Digital pills	1. Abilify MyCite	Asher (2017)
		2. D-Cap System	Goodman et al. (2022)
	Robotic capsule endoscopy	1. Soft capsule endoscope	Yim and Sitti (2011)
		2. Multimodule capsule robot	Zheng et al. (2021a)
Noninvasive closed-loop systems	Smart wound dressings	1. Battery-free and wireless wound dressing	Xu et al. (2021)
		2. Wearable patch	Shirzaei Sani et al. (2023)
	Smart contact lens	1. WTCL	Yang et al. (2022c)
		2. Smart electrochemical contact lens	Keum et al. (2020)
Smart minimally invasive devices	Theranostic microneedle array patches	1. FNDs-MNs	Shan et al. (2022)
		2. Ultrasonic pump-enabled closed-loop system	Luo et al. (2023b)
		3. MNTP	Yang et al. (2022e)

transformed the way that detect and treat diseases, breaking down the traditional barrier between detection and treatment systems. This development is crucial for continuous disease monitoring and timely treatment (Table 7.3).

7.5 Conclusions and Outlook

Over the past decades, the rapid expansion of sensors and responsive biomaterials has significantly contributed to the development of the biomedical field. Sensors enable the real-time monitoring of signals and biomarkers in the body, offering important health information. Responsive biomaterials provide more efficient and accurate disease treatment, avoiding unnecessary damage. Furthermore, interactive biomaterials for integrated diagnosis and treatment have been developed by incorporating sensors and responsive biological components into the diagnostic and therapeutic framework, showing great potential and broad application prospects in the medical field. The dual function of interactive biomaterials opens up new possibilities for disease diagnosis, treatment, and monitoring. By providing tailored and immediate treatments, interactive biomaterials simplify the treatment process, reduce the need for multiple procedures and materials, and help reduce healthcare costs and resource consumption, thereby promoting sustainable healthcare. Additionally, interactive materials can enhance patient comfort, improve treatment outcomes, and enhance quality of life.

While interactive biomaterials have shown great advantages in personalized medicine, providing more convenient and efficient solutions for disease monitoring and treatment, several challenges remain to be addressed. First, one of the main challenges is the complexity of the physiological environment. Due to individual variation, the presentation of the same disease can vary from person to person. Therefore, it is necessary to improve the sensitivity and specificity of sensor modules in interactive biomaterials. Second, interactive biomaterials may impact individual autonomy and compromise privacy. Taking Abilify MyCite as an example, when a patient takes a digital pill, their medication data may be monitored by doctors or family members, potentially exposing their privacy (De Miguel and Morla 2020). Therefore, patients' experiences should be taken into account when applying interactive biomaterials, keeping them informed and allowing them to choose who can access their medical data. Third, the prolonged use of interactive biomaterials may lead to potential side effects. For example, smart contact lenses used for disease monitoring and treatment can cause decreased oxygen permeability and increased lens protein adsorption if worn for a long time, leading to eye discomfort (Xu et al. 2018). Therefore, when designing interactive materials, it is essential to consider their long-term usability, as well as optimize their biocompatibility and comfort. Fourth, the development of interactive biomaterials is in its infancy, with limited relevant studies, most of which are in the pre-clinical research stage. There is still a long way to go before these materials can be used by patients in their daily lives. Addressing this problem requires interdisciplinary collaboration and continuous research efforts. With the joint efforts of the experts in medicine, materials science, and engineering, we can expect to further advance the development of interactive materials, leading to more significant breakthroughs and innovations in medical science and technology.

Overall, interactive biomaterials derived from sensors and responsive biomaterials bring new opportunities and challenges to the biomedical field. These materials possess characteristics of intelligence, precision, and personalization, which can improve the processes of diagnosis, treatment, and monitoring, ultimately enhancing the quality of life for patients. With the continuing breakthroughs in R&D, it can be anticipated that the development and application of more advanced and precise interactive biomaterials will lead to revolutionary advances in personalized medicine, precision medicine, and early diagnosis, contributing significantly to human health and well-being.

References

Amit M, Chukoskie L, Skalsky AJ et al (2019) Flexible pressure sensors for objective assessment of motor disorders [J]. Adv Funct Mater 30(20)

Arakawa T, Tomoto K, Nitta H et al (2020) A wearable cellulose acetate-coated Mouthguard biosensor for in vivo salivary glucose measurement [J]. Anal Chem 92(18):12201–12207

Armenia I, Cuestas Ayllón C, Torres Herrero B et al (2022) Photonic and magnetic materials for on-demand local drug delivery [J]. Adv Drug Deliv Rev 191:114584

Asher M (2017) Fda approves first digital pill [J]. Nat Rev Drug Discov

Badeau BA, Deforest CA (2019) Programming stimuli-responsive behavior into biomaterials [J]. Annu Rev Biomed Eng 21:241

Bai H, Wang H, Zhou Z et al (2023) Histone deacetylase – triggered self-immolative peptide-cytotoxins for cancer-selective drug delivery [J]. Adv Funct Mater 33(16)

Bao W, Chen F, Lai H et al (2021) Wearable breath monitoring based on a flexible fiber-optic humidity sensor [J], vol 349. Chemical, Sensors Actuators B, p 349

Battistella C, Liang Y, Gianneschi NC (2021) Innovations in disease state responsive soft materials for targeting extracellular stimuli associated with cancer, cardiovascular disease, diabetes, and beyond [J]. Adv Mater 33:2007504

Baumgartner SL, Buffkin DE Jr, Rukavina E et al (2021) A novel digital pill system for medication adherence measurement and reporting: usability validation study [J]. JMIR Hum Factors 8(4):e30786

Berman B, Bhatia N, Piacquadio D et al (2021) 25988 acceptability of 20% aminolevulinic acid photodynamic therapy for treatment of actinic keratoses on the upper extremities in a phase 3, randomized, vehicle-controlled trial [J]. J Am Acad Dermatol 85(3):AB79

Bi Y, Sun M, Wang J et al (2023) Universal fully integrated wearable sensor arrays for the multiple electrolyte and metabolite monitoring in raw sweat, saliva, or urine [J]. Anal Chem 95(16):6690–6699

Bošković MV, Sarajlić M, Frantlović M et al (2020) Aluminum-based self-powered hyper-fast miniaturized sensor for breath humidity detection [J]. Sensors Actuators B Chem 321:321

Chan W, Yiran H, Ying L et al (2023) Tissue-adhesive piezoelectric soft sensor for in vivo blood pressure monitoring during surgical operation [J]. Adv Funct Mater 2303696

Chang M, Feng W, Ding L et al (2022) Persistent luminescence phosphor as in-vivo light source for tumoral cyanobacterial photosynthetic oxygenation and photodynamic therapy [J]. Bioactive Mater 10:131–144

Chen C, Chen L, Yi Y et al (2016) Killing of Staphylococcus aureus via magnetic hyperthermia mediated by Magnetotactic bacteria [J]. Appl Environ Microbiol 82(7):2219–2226

Chen H, Sun X, Wang GD et al (2017) LiGa5O8:Cr-based theranostic nanoparticles for imaging-guided X-ray induced photodynamic therapy of deep-seated tumors [J]. Mater Horizons 4(6):1092–1101

Chen X, Fu C, Wang Y et al (2018a) Mitochondria-targeting nanoparticles for enhanced microwave ablation of cancer [J]. Nanoscale 10(33):15677–15685

Chen Y, Peng F, Song X et al (2018b) Conjugation of paclitaxel to C-6 hexanediamine-modified hyaluronic acid for targeted drug delivery to enhance antitumor efficacy [J]. Carbohydr Polym 181:150–158

Chen T, Gu T, Cheng L et al (2020) Porous Pt nanoparticles loaded with doxorubicin to enable synergistic chemo−/electrodynamic therapy [J]. Biomaterials 255:120202

Chen P, Wu P, Wan X et al (2021) Ultrasound-driven electrical stimulation of peripheral nerves based on implantable piezoelectric thin film nanogenerators [J]. Nano Energy 86:106123

Chen W, Hazoor S, Madigan R et al (2022a) Alkaline-responsive polydiacetylene-peptide hydrogel for pH-sensing and on-demand antimicrobial release [J]. Mater Today Adv 16:100288

Chen G, Xu Q, Feng Z et al (2022b) Glutamine antagonist synergizes with electrodynamic therapy to induce tumor regression and systemic antitumor immunity [J]. ACS Nano 16(1):951–962

Chen P, Xu C, Wu P et al (2022c) Wirelessly powered electrical-stimulation based on biodegradable 3D piezoelectric scaffolds promotes the spinal cord injury repair [J]. ACS Nano 16(10):16513–16528

Cheng X, Wang B, Zhao Y et al (2019) A Mediator-Free Electroenzymatic Sensing Methodology to Mitigate Ionic and Electroactive Interferents' Effects for Reliable Wearable Metabolite and Nutrient Monitoring [J]. Adv Funct Mater 30(10)

Cheng S, Luo Y, Zhang J et al (2022) The highly effective therapy of ovarian cancer by bismuth-doped oxygen-deficient BaTiO3 with enhanced sono-piezocatalytic effects [J]. Chem Eng J 442:136380

Cheng S, Chen L, Gong F et al (2023a) PtCu Nanosonosensitizers with inflammatory microenvironment regulation for enhanced Sonodynamic bacterial elimination and tissue repair [J]. Adv Funct Mater 33(22)
Cheng X, Hu S, Cheng K (2023b) Microneedle patch delivery of PROTACs for anti-cancer therapy [J]. ACS Nano 17(12):11855–11868
Choi W, Key J, Youn I et al (2022) Cavitation-assisted sonothrombolysis by asymmetrical nanostars for accelerated thrombolysis [J]. J Control Release 350:870–885
Chong LM, Tng DJH, Tan LLY et al (2021) Recent advances in radiation therapy and photodynamic therapy [J]. Appl Phys Rev 8(4)
Chu M, Nguyen T, Pandey V et al (2019) Respiration rate and volume measurements using wearable strain sensors [J]. NPJ Digital Med 2(1):8
Cochran JM, Fang H, Le Gallo C et al (2022) Participant engagement and symptom improvement: aripiprazole tablets with sensor for the treatment of schizophrenia [J]. Patient Prefer Adherence Volume 16:1805–1817
Cohen EA, Skubiak T, Boskovic DH et al (2022) Phase 3b multicenter, prospective, open-label trial to evaluate the effects of a digital medicine system on inpatient psychiatric hospitalization rates for adults with schizophrenia [J]. J Clin Psychiatry 83(3):40541
Cook AB, Decuzzi P (2021) Harnessing endogenous stimuli for responsive materials in Theranostics [J]. ACS Nano 15:2068–2098
Cosgrove L, Cristea IA, Shaughnessy AF et al (2019) Digital aripiprazole or digital evergreening? A systematic review of the evidence and its dissemination in the scientific literature and in the media [J]. BMJ Evid Based Med 24(6):231–238
Dai T, He W, Tu S et al (2022) Black TiO2 nanoprobe-mediated mild phototherapy reduces intracellular lipid levels in atherosclerotic foam cells via cholesterol regulation pathways instead of apoptosis [J]. Bioact Mater 17:18–28
De Miguel BI, Morla GM (2020) 'Digital pills' for mental diseases: an ethical and social analysis of the issues behind the concept [J]. J Law Biosci 7(1)
Deng X, Shao Z, Zhao Y (2021a) Solutions to the drawbacks of Photothermal and photodynamic cancer therapy [J]. Advanced. Science 8(3)
Deng Y, Sun Z, Wang L et al (2021b) Biosensor-based assay of exosome biomarker for early diagnosis of cancer [J]. Front Med 16(2):157–175
Deng M, Song G, Zhong K et al (2022) Wearable fluorescent contact lenses for monitoring glucose via a smartphone [J]. Chemical, Sensors Actuators B, p 352
Dennyson Savariraj A, Salih A, Alam F et al (2021) Ophthalmic sensors and drug delivery [J]. Acs Sensors 6(6):2046–2076
Dervisevic M, Alba M, Esser L et al (2021) Silicon micropillar Array-based wearable sweat glucose sensor [J]. ACS Appl Mater Interfaces 14(1):2401–2410
Dong R, Guo B (2021) Smart wound dressings for wound healing [J]. Nano Today 41:101290
Dong Y, Dong S, Liu B et al (2021) 2D piezoelectric Bi2MoO6 nanoribbons for GSH – enhanced sonodynamic therapy [J]. Adv Mater 33(51)
Downs AM, Bolotsky A, Weaver BM et al (2023) Microneedle electrochemical aptamer-based sensing: real-time small molecule measurements using sensor-embedded, commercially-available stainless steel microneedles [J]. Biosens Bioelectron 236:115408
El Hadidi S, Rosano G (2020) Evidence beyond the digital medication pill [Z], vol 6. Oxford University Press, pp 72–74
El Kadi S, Porter TR, Verouden NJW et al (2022) Contrast ultrasound, sonothrombolysis and sonoperfusion in cardiovascular disease [J]. JACC Cardiovasc Imaging 15(2):345–360
Fabian D, Guillermo Prieto Eibl M, Alnahhas I et al (2019) Treatment of glioblastoma (GBM) with the addition of tumor-treating fields (TTF) a review [J]. Cancers 11(2)
Fisher RS, Velasco AL (2014) Electrical brain stimulation for epilepsy [J]. Nat Rev Neurol 10(5):261–270

Fu C, Zhou H, Tan L et al (2017) Microwave-activated Mn-doped zirconium metal–organic framework nanocubes for highly effective combination of microwave dynamic and thermal therapies against cancer [J]. ACS Nano 12(3):2201–2210
Gandia D, Gandarias L, Rodrigo I et al (2019) Unlocking the potential of Magnetotactic bacteria as magnetic hyperthermia agents [J]. Small 15(41)
Gao C, Huang Q, Liu C et al (2020a) Treatment of atherosclerosis by macrophage-biomimetic nanoparticles via targeted pharmacotherapy and sequestration of proinflammatory cytokines [J]. Nature Commun 11(1)
Gao G, Jiang YW, Guo Y et al (2020b) Enzyme-mediated tumor starvation and phototherapy enhance mild-temperature photothermal therapy [J]. Adv Funct Mater 30(16)
Gao B, Guo M, Lyu K et al (2021) Intelligent silk fibroin based microneedle dressing (i-SMD) [J]. Adv Funct Mater 31(3):2006839
García-Carmona L, Martín A, Sempionatto JR et al (2019) Pacifier biosensor: toward noninvasive saliva biomarker monitoring [J]. Anal Chem 91(21):13883–13891
Geng Y, Cao R, Innocent MT et al (2022) A high-sensitive wearable sensor based on conductive polymer composites for body temperature monitoring [J]. Composit Part A Appl Sci Manufact 163:163
Goodman GR, Vaz C, Albrechta H et al (2022) Ingestible electronic sensors for monitoring real-time adherence to HIV pre-exposure prophylaxis and antiretroviral therapy [J]. Curr HIV/AIDS Rep 19(5):433–445
Groeschl RT, Pilgrim CHC, Hanna EM et al (2014) Microwave ablation for hepatic malignancies [J]. Ann Surg 259(6):1195–1200
Gu T, Wang Y, Lu Y et al (2019) Platinum nanoparticles to enable electrodynamic therapy for effective cancer treatment [J]. Adv Mater 31(14):e1806803
Guan M, Liu Y, Du H et al (2023) Durable, breathable, sweat-resistant, and degradable flexible sensors for human motion detection [J]. Chem Eng J 462:142151
Guo M, Wang Y, Gao B et al (2021) Shark tooth-inspired microneedle dressing for intelligent wound management [J]. ACS Nano 15(9):15316–15327
Hao L, Li J, Wang P et al (2021) Spatiotemporal Magnetocaloric microenvironment for guiding the fate of biodegradable polymer implants [J]. Adv Funct Mater 31(15)
He L, Yu X, Li W (2022) Recent Progress and trends in X-ray-induced photodynamic therapy with low radiation doses [J]. ACS Nano 16(12):19691–19721
He T, Chen J, He B-G et al (2023) Toward wearable sensors: advances, trends, and challenges [J]. ACM Comput Surv 55(14s):1–35
Hsu C-C, Yang Y, Kannisto E et al (2023) Simultaneous detection of tumor derived Exosomal protein–MICRORNA pairs with an Exo-PROS biosensor for cancer diagnosis [J]. ACS Nano 17(9):8108–8122
Hu H, Feng W, Qian X et al (2021) Emerging nanomedicine – enabled/enhanced nanodynamic therapies beyond traditional photodynamics [J]. Adv Mater 33(12)
Hu Y, Gao S, Lu H et al (2022) Acid-resistant and physiological pH-responsive DNA hydrogel composed of A-motif and i-motif toward Oral insulin delivery [J]. J Am Chem Soc 144(12):5461–5470
Huang F, Magnin C, Brouqui P (2020) Ingestible sensors correlate closely with peripheral temperature measurements in febrile patients [J]. J Infect 80(2):161–166
Huang X, Zhang W, Peng Y et al (2021) A multifunctional layered nickel silicate Nanogenerator of synchronous oxygen self-supply and superoxide radical generation for hypoxic tumor therapy [J]. ACS Nano 16(1):974–983
Huang D, Zhao C, Wen B et al (2022) Oxygen-carrying microfluidic microcapsules for enhancing chemo-sonodynamic therapy on patient-derived tumor organoid models [J]. Chem Eng J 435:134871
Hye-Jun K, Jin-Woo P (2023) Carotid artery monitoring patch using a supercapacitive pressure sensor with piezoelectricity [J]. Nano Energy 114:108636

Iddan G, Meron G, Glukhovsky A et al (2000) Wireless capsule endoscopy [J]. Nature 405(6785):417
Ivanova VA, Nikitina VP, Verenikina EV (2022) Photodynamic therapy for early cervical cancer [J]. J Clin Oncol 40(16):17528
Jia Q, Li Z, Guo C et al (2020) PEGMA-modified bimetallic NiCo Prussian blue analogue doped with Tb(III) ions: efficiently pH-responsive and controlled release system for anticancer drug [J]. Chem Eng J 389:124468
Jiang Z, Feng X, Zou H et al (2021) Poly(l-glutamic acid)-cisplatin nanoformulations with detachable PEGylation for prolonged circulation half-life and enhanced cell internalization [J]. Bioactive Mater 6(9):2688–2697
Jin Q, Chen HJ, Li X et al (2019) Reduced graphene oxide nanohybrid–assembled microneedles as mini-invasive electrodes for real-time transdermal biosensing [J]. Small 15(6):1804298
Johannsen M, Gneveckow U, Thiesen B et al (2007) Thermotherapy of prostate cancer using magnetic nanoparticles: feasibility, imaging, and three-dimensional temperature distribution [J]. Eur Urol 52(6):1653–1662
Keum DH, Kim S-K, Koo J (2020) Wireless smart contact lens for diabetic diagnosis and therapy [J]. Sci Adv 6(17):eaba3252
Khan H, Shah MR, Barek J et al (2023) Cancer biomarkers and their biosensors: a comprehensive review [J]. TrAC Trends Anal Chem 158:158
Kim J, Kim M, Lee M-S et al (2017) Wearable smart sensor systems integrated on soft contact lenses for wireless ocular diagnostics [J]. Nat Commun 8(1):14997
Kim MK, Kantarcigil C, Kim B (2019) Flexible submental sensor patch with remote monitoring controls for management of oropharyngeal swallowing disorders [J]. Sci Adv 5(12):eaay3210
Kim J-N, Lee J, Lee H et al (2021) Stretchable and self-healable catechol-chitosan-diatom hydrogel for triboelectric generator and self-powered tremor sensor targeting at Parkinson disease [J]. Nano Energy 82:105705
Kim T, Kim HJ, Choi W et al (2022a) Deep brain stimulation by blood–brain-barrier-crossing piezoelectric nanoparticles generating current and nitric oxide under focused ultrasound [J]. Nature Biomed Eng 7(2):149–163
Kim SK, Lee GH, Jeon C et al (2022b) Bimetallic nanocatalysts immobilized in nanoporous hydrogels for long-term robust continuous glucose monitoring of smart contact lens [J]. Adv Mater 34(18)
Kim TY, Mok JW, Hong SH et al (2022c) Wireless theranostic smart contact lens for monitoring and control of intraocular pressure in glaucoma [J]. Nature. Communications 13
Kim TY, Mok JW, Hong SH et al (2022d) Wireless theranostic smart contact lens for monitoring and control of intraocular pressure in glaucoma [J]. Nat Commun 13(1):6801
Kim H, Yoon J, Kim HK et al (2023) Upconverting nanoparticle-containing erythrocyte-sized hemoglobin microgels that generate heat, oxygen and reactive oxygen species for suppressing hypoxic tumors [J]. Bioact Mater 22:112–126
Kowalski PS, Bhattacharya C, Afewerki S et al (2018) Smart biomaterials: recent advances and future directions [J]. ACS Biomater Sci Eng 4:3809
Krämer J, Grimm LM, Zhong C et al (2023) A supramolecular cucurbit[8]uril-based rotaxane chemosensor for the optical tryptophan detection in human serum and urine [J]. Nature Commun 14(1)
Kulkarni P, Haldar MK, You S et al (2016) Hypoxia-responsive Polymersomes for drug delivery to hypoxic pancreatic cancer cells [J]. Biomacromolecules 17(8):2507–2513
Kwon K, Kim JU, Won SM et al (2023) A battery-less wireless implant for the continuous monitoring of vascular pressure, flow rate and temperature [J]. Nature. Biomed Eng 7:1215
Larocca TJ, Hearon CM, Henson GD et al (2014) Mitochondrial quality control and age-associated arterial stiffening [J]. Exp Gerontol 58:78–82
Li S, Chen Z, Tan L et al (2022a) MOF@COF nanocapsule for the enhanced microwave thermal-dynamic therapy and anti-angiogenesis of colorectal cancer [J]. Biomaterials 283:121472

Li J, Wang Z, Han H et al (2022b) Short and simple peptide-based pH-sensitive hydrogel for antitumor drug delivery [J]. Chin Chem Lett 33(4):1936–1940
Li C, Xiao C, Zhan L et al (2022c) Wireless electrical stimulation at the nanoscale interface induces tumor vascular normalization [J]. Bioactive Mater 18:399–408
Li F, Yang W, Zhao B et al (2022d) Ultrasensitive DNA-biomacromolecule sensor for the detection application of clinical cancer samples [J]. Adv Sci 9(6)
Li Y, Yin J, Liu S et al (2023a) Learning hand kinematics for Parkinson's disease assessment using a multimodal sensor glove [J]. Adv Sci 10(20)
Li S, Zhang H, Zhu M et al (2023b) Electrochemical biosensors for whole blood analysis: recent Progress, challenges, and future perspectives [J]. Chem Rev 123:7953–8039
Li B, Zhu H, Lv Y et al (2023c) Metal ion coordination improves graphite nitride carbon microwave therapy in antibacterial and Osteomdayelitis treatment [J]. Small 19
Liang Y, He J, Guo B (2021) Functional hydrogels as wound dressing to enhance wound healing [J]. ACS Nano 15(8):12687–12722
Liangye L, Shunfeng S, Yunfei L et al (2023) Automatic and continuous blood pressure monitoring via an optical-fiber-sensor-assisted smartwatch [J]. PhotoniX
Liao F, Zhu Z, Yan Z et al (2017) Ultrafast response flexible breath sensor based on vanadium dioxide [J]. J Breath Res 11(3)
Liu G, Zhao X, Zhang Y et al (2019) Engineering biomimetic platesomes for pH-responsive drug delivery and enhanced antitumor activity [J]. Adv Mater 31
Liu T, Li L, Wang S et al (2022a) Hybrid chalcogen bonds in prodrug nanoassemblies provides dual redox-responsivity in the tumor microenvironment [J]. Nature Commun 13(1)
Liu B, Li J, Zhang Z et al (2022b) Ph responsive antibacterial hydrogel utilizing catechol–boronate complexation chemistry [J]. Chem Eng J 441:135808
Liu M, Yang M, Wang M et al (2022c) A flexible dual-Analyte electrochemical biosensor for salivary glucose and lactate detection [J]. Biosensors 12(4)
Liu W, Di J, Ma Y et al (2023a) Mitochondria-mediated HSP inhibition strategy for enhanced low-temperature Photothermal therapy [J]. ACS Appl Mater Interfaces 15(22):26252–26262
Liu T, Du Y, Yan Y et al (2023b) pH-responsive dual-functional hydrogel integrating localized delivery and anti-cancer activities for highly effective therapy in PDX of OSCC [J]. Mater Today 62:71–97
Liu M, He D, Liu Z et al (2023c) Disposable solar microcell array-based addressable photoelectrochemical sensor for high-throughput and multiplexed analysis of salivary metabolites [J]. Biosens Bioelectron 232:115312
Liu J, Zhu H, Lin L et al (2023d) Redox imbalance triggered Intratumoral Cascade reaction for tumor "turn on" imaging and synergistic therapy [J]. Small 19(16)
Lou X, Zhang D, Ling H et al (2021) Pure redox-sensitive paclitaxel–maleimide prodrug nanoparticles: endogenous albumin-induced size switching and improved antitumor efficiency [J]. Acta Pharm Sin B 11(7):2048–2058
Lu Z, Gao J, Fang C et al (2020) Porous Pt nanospheres incorporated with GOx to enable synergistic oxygen-inductive starvation/electrodynamic tumor therapy [J]. Adv Sci 7(17)
Lu Y, Lin L, Ye J (2022) Human metabolite detection by surface-enhanced Raman spectroscopy [J]. Mater Today Bio 13:13
Luo R, Shi B, Luo D et al (2023a) Self-powered electrical stimulation assisted skin wound therapy [J]. Sci Bull 68:1740
Luo X, Yu Q, Yang L et al (2023b) Wearable, sensing-controlled, ultrasound-based microneedle smart system for diabetes management [J]. ACS Sensors 8(4):1710–1722
Ma A, Chen H, Cui Y et al (2018) Metalloporphyrin complex – based nanosonosensitizers for deep-tissue tumor theranostics by noninvasive sonodynamic therapy [J]. Small 15(5)
Ma X, Qi Y, Niu Y et al (2023) Highly sensitive, ultra-reliable flexible piezoelectret sensor for non-contact sitting motion tracking and physiological signal monitoring [J]. Nano Energy 111:108424
Mackman N (2008) Triggers, targets and treatments for thrombosis [J]. Nature 451(7181):914–918

Maier-Hauff K, Ulrich F, Nestler D et al (2010) Efficacy and safety of intratumoral thermotherapy using magnetic iron-oxide nanoparticles combined with external beam radiotherapy on patients with recurrent glioblastoma multiforme [J]. J Neuro-Oncol 103(2):317–324

Meisinger C, Stöckl D, Rückert I et al (2013) Serum potassium is associated with prediabetes and newly diagnosed diabetes in hypertensive adults from the general population: the KORA F4-study [J]. Diabetologia 56:484–491

Melmed GY, Lo SK (2005) Capsule endoscopy: practical applications [J]. Clin Gastroenterol Hepatol 3(5):411–422

Meng F, Yu W, Chen C et al (2022) A versatile electrochemical biosensor for the detection of circulating microrna toward non – small cell lung cancer diagnosis [J]. Small 18(22)

Min S, Kim DH, Joe DJ et al (2023) Clinical validation of a wearable piezoelectric blood-pressure sensor for continuous health monitoring [J]. Adv Mater 35(26):2301627

Mu J, Lin J, Huang P et al (2018) Development of endogenous enzyme-responsive nanomaterials for theranostics [J]. Chem Soc Rev 47(15):5554–5573

Mukasa D, Wang M, Min J et al (2023) A computationally assisted approach for designing wearable biosensors toward non-invasive personalized molecular analysis [J]. Adv Mater 35

Natta L, Guido F, Algieri L et al (2021) Conformable ALN piezoelectric sensors as a non-invasive approach for swallowing disorder assessment [J]. ACS Sensors 6(5):1761–1769

Pan J, Xu Y, Wu Q et al (2021) Mild magnetic hyperthermia-activated innate immunity for liver cancer therapy [J]. J Am Chem Soc 143(21):8116–8128

Pan Y, Li Y, Li Y et al (2022) Nanodroplet – coated microbubbles used in sonothrombolysis with two – step cavitation strategy [J]. Adv Healthc Mater 12(6)

Pang Q, Lou D, Li S et al (2020) Smart flexible electronics-integrated wound dressing for real-time monitoring and on-demand treatment of infected wounds [J]. Adv Sci 7(6):1902673

Peng P, Feng T, Yang X et al (2023) Gastrointestinal microenvironment responsive Nanoencapsulation of probiotics and drugs for synergistic therapy of intestinal diseases [J]. Acs Nano 17:14718

Phadkule SS, Sarma S (2022) High-performance flexible temperature sensor from hybrid nanocomposite for continuous human body temperature monitoring [J]. Polym Compos 44(2):1381–1391

Qin Q, Yang M, Shi Y et al (2023) Mn-doped Ti-based MOFs for magnetic resonance imaging-guided synergistic microwave thermal and microwave dynamic therapy of liver cancer [J]. Bioact Mater 27:72–81

Raj M, Moon J-M, Goyal RN et al (2019) Simultaneous detection of ATP metabolites in human plasma and urine based on palladium nanoparticle and poly(bromocresol green) composite sensor [J]. Biosens Bioelectron 126:758–766

Ranjan R, Esimbekova EN, Kratasyuk VA (2017) Rapid biosensing tools for cancer biomarkers [J]. Biosens Bioelectron 87:918–930

Rashid S, Nawaz MH, Rehman IU et al (2021) Dopamine/mucin-1 functionalized electro-active carbon nanotubes as a probe for direct competitive electrochemical immunosensing of breast cancer biomarker [J], vol 330. Chemical, Sensors Actuators B, p 330

Rastinehad AR, Anastos H, Wajswol E et al (2019) Gold nanoshell-localized photothermal ablation of prostate tumors in a clinical pilot device study [J]. Proc Natl Acad Sci 116(37):18590–18596

Ren C, Zhong D, Qi Y et al (2023) Bioinspired pH-responsive microalgal hydrogels for Oral insulin delivery with both hypoglycemic and insulin sensitizing effects [J]. ACS Nano 17(14):14161–14175

Roberts NJ, Zhang L, Janku F (2014) Intratumoral injection of clostridium novyi-NT spores induces antitumor responses [J]. Sci Adv 6(249):249ra111

Ryu WM, Lee Y, Son Y et al (2023) Thermally drawn multi-material fibers based on polymer nanocomposite for continuous temperature sensing [J], vol 5. Adv Fiber Mater, p 1712

Saha T, Del Caño R, Mahato K et al (2023) Wearable electrochemical glucose sensors in diabetes management: a comprehensive review [J]. Chem Rev 123(12):7854–7889

Sang M, Kang K, Zhang Y et al (2021) Ultrahigh sensitive Au-doped silicon nanomembrane based wearable sensor arrays for continuous skin temperature monitoring with high precision [J]. Adv Mater 34(4)
Sarhadi VK, Armengol G (2022) Molecular biomarkers in cancer [J]. Biomol Ther 12(8)
Sel K, Osman D, Huerta N et al (2023) Continuous cuffless blood pressure monitoring with a wearable ring bioimpedance device [J]. NPJ Digital Med 6(1):59
Shan J, Zhang X, Kong B et al (2022) Coordination polymer nanozymes-integrated colorimetric microneedle patches for intelligent wound infection management [J]. Chem Eng J 444:136640
Shang Z, Su T, Jin D et al (2023) An integrated and flexible PDMS/au film-based electrochemical immunosensor via Fe–Co MOF as a signal amplifier for alpha fetoprotein detection [J]. Biosens Bioelectron 230:115245
Shi J, Yu W, Xu L et al (2019) Bioinspired Nanosponge for salvaging ischemic stroke via free radical scavenging and self-adapted oxygen regulating [J]. Nano Lett 20(1):780–789
Shirzaei Sani E, Xu C, Wang C et al (2023) A stretchable wireless wearable bioelectronic system for multiplexed monitoring and combination treatment of infected chronic wounds [J]. Sci Adv 9(12):eadf7388
Shitanda I, Fujimura Y, Takarada T et al (2021) Self-powered diaper sensor with wireless transmitter powered by paper-based biofuel cell with urine glucose as fuel [J]. ACS Sensors 6(9):3409–3415
Singeap A-M, Stanciu C, Trifan A (2016) Capsule endoscopy: the road ahead [J]. World J Gastroenterol 22(1):369
Son D, Gilbert H, Sitti M (2020) Magnetically actuated soft capsule endoscope for fine-needle biopsy [J]. Soft Robot 7(1):10–21
Song L, Li P-P, Yang W et al (2018a) Low-dose X-ray activation of W(VI)-doped persistent luminescence nanoparticles for deep-tissue photodynamic therapy [J]. Adv Funct Mater 28(18)
Song Z, Wu Y, Cao Q et al (2018b) Light-triggered on-demand antibiotic release from functional metal-organic framework for bacterial infection combination therapy [J]. Adv Funct Mater 28(23)
Song Y, Li W, Jing H et al (2022) Spatiotemporal sonodynamic therapy for the treatment of rheumatoid arthritis based on Z-scheme heterostructure sonosensitizer of HO-1 inhibitor jointed bismuth nanotriangle [J]. Chem Eng J 438:135558
Stoica SI, Bleotu C, Ciobanu V et al (2022) Considerations about hypoxic changes in Neuraxis tissue injuries and recovery [J]. Biomedicines 10(2)
Sun J, Lei Y, Dai Z et al (2017) Sustained release of brimonidine from a new composite drug delivery system for treatment of glaucoma [J]. ACS Appl Mater Interfaces 9(9):7990–7999
Sun X, Sun J, Sun Y et al (2021) Oxygen self-sufficient nanoplatform for enhanced and selective antibacterial photodynamic therapy against anaerobe-induced periodontal disease [J]. Adv Funct Mater 31(20)
Sun L, Cao Y, Li W et al (2023a) Perovskite – type manganese vanadate sonosensitizers with biodegradability for enhanced sonodynamic therapy of cancer [J]. Small 19(27)
Sun G, Wang P, Jiang Y et al (2023b) Bioinspired flexible, breathable, waterproof and self-cleaning iontronic tactile sensors for special underwater sensing applications [J]. Nano Energy 110:108367
Tan AYS, Lo NW, Cheng F et al (2022) 2D carbon materials based photoelectrochemical biosensors for detection of cancer antigens [J]. Biosens Bioelectron 219:114811
Tang S-J, Zanati S, Dubcenco E et al (2004) Diagnosis of small-bowel varices by capsule endoscopy [J]. Gastrointest Endosc 60(1):129–135
Truong Hoang Q, Ravichandran V, Nguyen Cao TG et al (2022) Piezoelectric au-decorated ZNO nanorods: ultrasound-triggered generation of ROS for piezocatalytic cancer therapy [J]. Chem Eng J 435:135039
Wan Q, Chen Q, Freithaler MA et al (2023) Toward real-time blood pressure monitoring via high-fidelity iontronic tonometric sensors with high sensitivity and large dynamic ranges [J]. Adv Healthc Mater 12(17):2202461

Wang GD, Nguyen HT, Chen H et al (2016) X-ray induced photodynamic therapy: a combination of radiotherapy and photodynamic therapy [J]. Theranostics 6(13):2295–2305
Wang B, Chan KF, Yu J et al (2018) Reconfigurable swarms of ferromagnetic colloids for enhanced local hyperthermia [J]. Adv Funct Mater 28(25)
Wang X, Zhong X, Gong F et al (2020) Newly developed strategies for improving sonodynamic therapy [J]. Mater Horizons 7(8):2028–2046
Wang Z, Bu M, Xiu K et al (2022a) A flexible, stretchable and triboelectric smart sensor based on graphene oxide and polyacrylamide hydrogel for high precision gait recognition in parkinsonian and hemiplegic patients [J]. Nano Energy 104:107978
Wang S, Deng W, Yang T et al (2022b) Body-area sensor network featuring micropyramids for sports healthcare [J]. Nano Res 16(1):1330–1337
Wang M, Gao B, Wang X et al (2022c) Enzyme-responsive strategy as a prospective cue to construct intelligent biomaterials for disease diagnosis and therapy [J]. Biomater Sci 10(8):1883–1903
Wang L, Hu P, Jiang H et al (2022d) Mild hyperthermia-mediated osteogenesis and angiogenesis play a critical role in magnetothermal composite-induced bone regeneration [J]. Nano Today 43:101401
Wang P, Liu J, Yu W et al (2022e) Flexible, stretchable, breathable and sweatproof all-nanofiber iontronic tactile sensor for continuous and comfortable knee joint motion monitoring [J]. Nano Energy 103:107768
Wang X, Sun W, Shi H et al (2022f) Organic phosphorescent nanoscintillator for low-dose X-ray-induced photodynamic therapy [J]. Nature Commun 13(1)
Wang M, Yang Y, Min J et al (2022g) A wearable electrochemical biosensor for the monitoring of metabolites and nutrients [J]. Nature Biomed Eng 6(11):1225–1235
Wang B, Zhou L, Guo Y et al (2022h) Cyanobacteria-based self-oxygenated photodynamic therapy for anaerobic infection treatment and tissue repair [J]. Bioactive Mate 12:314–326
Wang Y, Gao B, He B (2023a) Toward efficient wound management: bioinspired microfluidic and microneedle patch [J]. Small 19(3):2206270
Wang Z, Zhao Y, Hou Y et al (2023b) A Thrombin-activated peptide-templated nanozyme for remedying ischemic stroke via thrombolytic and neuroprotective actions [J]. Adv Mater 36
Wei S, Qiao Y, Wu Z et al (2021a) Na+ inserted metal-organic framework for rapid therapy of bacteria-infected osteomyelitis through microwave strengthened Fenton reaction and thermal effects [J]. Nano Today 37:101090
Wei X, Zhu M, Li J et al (2021b) Wearable biosensor for sensitive detection of uric acid in artificial sweat enabled by a fiber structured sensing interface [J]. Nano Energy 85:106031
Wei F, Karges J, Shen J et al (2022) A mitochondria-localized oxygen self-sufficient two-photon nano-photosensitizer for ferroptosis-boosted photodynamic therapy under hypoxia [J]. Nano Today 44:101509
Wei Y, Shi X, Yao Z et al (2023) Fully paper-integrated hydrophobic and air permeable piezoresistive sensors for high-humidity and underwater wearable motion monitoring [J]. NPJ Flexib Electron 7(1)
Wu Y, Li Y, Lv G et al (2022) Redox dyshomeostasis strategy for tumor therapy based on nanomaterials chemistry [J]. Chem Sci 13(8):2202–2217
Xia Y, Duan S, Han C et al (2022) Hypoxia-responsive nanomaterials for tumor imaging and therapy [J]. Front Oncol 12:12
Xiao L, Zhu A, Xu Q et al (2017) Colorimetric biosensor for detection of cancer biomarker by au nanoparticle-decorated Bi2Se3 nanosheets [J]. ACS Appl Mater Interfaces 9(8):6931–6940
Xu J, Xue Y, Hu G et al (2018) A comprehensive review on contact lens for ophthalmic drug delivery [J]. J Control Release 281:97–118
Xu G, Lu Y, Cheng C et al (2021) Battery free and wireless smart wound dressing for wound infection monitoring and electrically controlled on-demand drug delivery [J]. Adv Funct Mater 31
Yaari Z, Yang Y, Apfelbaum E (2021) A perception-based nanosensor platform to detect cancer biomarkers [J]. Sci Adv 7(47):eabj0852

Yan Z, Zhang S, Gao M et al (2023) Ultrasensitive and wide-range-detectable flexible breath sensor based on silver vanadate nanowires [J]. ACS Appl Electron Mater 5(1):520–525

Yang G, Phua S Z F, Lim W Q, et al. A Hypoxia-Responsive Albumin-Based Nanosystem for Deep Tumor Penetration and Excellent Therapeutic Efficacy [J]. Adv Mater, 2019a, 31(25)

Yang Y, Song Y, Bo X et al (2019b) A laser-engraved wearable sensor for sensitive detection of uric acid and tyrosine in sweat [J]. Nat Biotechnol 38(2):217–224

Yang P, Shi Y, Li S et al (2022a) Monitoring the degree of comfort of shoes in-motion using triboelectric pressure sensors with an Ultrawide detection range [J]. ACS Nano 16(3):4654–4665

Yang T, Wang A, Nie D et al (2022b) Ligand-switchable nanoparticles resembling viral surface for sequential drug delivery and improved oral insulin therapy [J]. Nature Commun 13(1)

Yang C, Wu Q, Liu J et al (2022c) Intelligent wireless theranostic contact lens for electrical sensing and regulation of intraocular pressure [J]. Nature Commun 13(1)

Yang J, Yang J, Gong X et al (2022d) Recent progress in microneedles-mediated diagnosis, therapy, and theranostic systems [J]. Adv Healthc Mater 11(10):2102547

Yang J, Zheng S, Ma D et al (2022e) Masticatory system–inspired microneedle theranostic platform for intelligent and precise diabetic management [J]. Sci Adv 8(50):eabo6900

Ye L, Wu F, Xu R et al (2023) Face mask integrated with flexible and wearable manganite oxide respiration sensor [J]. Nano Energy 112:108460

Yi Y, Jiang Y, Zhao H et al (2020) High-Performance Ultrafast Humidity Sensor Based on Microknot Resonator-Assisted Mach–Zehnder for Monitoring Human Breath [J]. ACS Sensors 5(11):3404–3410

Yi Z, Liu Z, Li W et al (2022) Piezoelectric dynamics of arterial pulse for wearable continuous blood pressure monitoring [J]. Adv Mater 34(16):2110291

Yim S, Sitti M (2011) Design and rolling locomotion of a magnetically actuated soft capsule endoscope [J]. IEEE Trans Robot 28(1):183–194

Yoon YI, Tang W, Chen X (2017) Ultrasound-mediated diagnosis and therapy based on ultrasound contrast agents [J]. Small Methods 1(8)

Yu L, Chen Y, Lin H et al (2018) Ultrasmall mesoporous organosilica nanoparticles: morphology modulations and redox-responsive biodegradability for tumor-specific drug delivery [J]. Biomaterials 161:292–305

Yu Y, Peng S, Sha Z et al (2021) High-precision, stretchable kirigami-capacitive sensor with ultra-low cross-sensitivity for body temperature monitoring [J]. J Mater Chem A 9(44):24874–24886

Yu Y, Li J, Zhang Y et al (2022) A bioinspired hierarchical nanoplatform targeting and responding to intracellular pathogens to eradicate parasitic infections [J]. Biomaterials 280:121309

Yu Y, Xiang L, Zhang X et al (2023) Pure organic AIE nanoscintillator for X-ray mediated Type I and Type II photodynamic therapy [J]. Adv Sci 10

Yuan Z, Wu J, Xiao Y et al (2023) A photo – therapeutic nanocomposite with bio-responsive oxygen self-supplying combats biofilm infections and inflammation from drug-resistant bacteria [J]. Adv Funct Mater 33

Zeng Y, Wang J, Wang Z et al (2020) Colloidal crystal microneedle patch for glucose monitoring [J]. Nano Today 35:100984

Zhang H, Zhang J, Hu Z et al (2019) Waist-wearable wireless respiration sensor based on triboelectric effect [J]. Nano Energy 59:75–83

Zhang S, Li Y, Qiu X et al (2021a) Incorporating redox-sensitive nanogels into bioabsorbable nanofibrous membrane to acquire ROS-balance capacity for skin regeneration [J]. Bioactive Mater 6(10):3461–3472

Zhang J, Liu J, Su H et al (2021b) A wearable self-powered biosensor system integrated with diaper for detecting the urine glucose of diabetic patients [J], vol 341. Chemical, Sensors Actuators B, p 341

Zhang H, Qu H, He Q et al (2021c) Thrombus-targeted nanoparticles for thrombin-triggered thrombolysis and local inflammatory microenvironment regulation [J]. J Control Release 339:195–207

Zhang X, Xia Y, Liu Y et al (2021d) Integrated wearable sensors for sensing physiological pressure signals and β-Hydroxybutyrate in physiological fluids [J]. Anal Chem 94(2):993–1002
Zhang X, Gan J, Fan L et al (2023a) Bioinspired adaptable indwelling microneedles for treatment of diabetic ulcers [J]. Adv Mater 35(23):2210903
Zhang H, Mao Z, Kang Y et al (2023b) Redox regulation and its emerging roles in cancer treatment [J]. Coord Chem Rev 475:214897
Zhao J, Li G (2020) Study on real-time wearable sport health device based on body sensor networks [J]. Comput Commun 154:40–47
Zhao C, Yang J, Chen M et al (2022) Synthetic lignin-derived therapeutic Nano reagent as intestinal pH-sensitive drug carriers capable of bypassing the gastric acid environment for colitis treatment [J]. ACS Nano 17(1):811–824
Zhao Y, Hu L, Dai H et al (2023a) Smart trap-capture-kill antibacterial system for infected microenvironment improvement and vascularized bone regeneration via magnetic thermotherapy [J]. Adv Funct Mater 33(23)
Zhao L, Liang C, Huang Y et al (2023b) Emerging sensing and modeling technologies for wearable and cuffless blood pressure monitoring [J]. NPJ Digital Med 6(1):93
Zhao H, Zhang L, Deng T et al (2023c) High-performance sensing, breathable, and biodegradable integrated wearable sweat biosensors for a wireless glucose early warning system [J]. J Mater Chem A 11(23):12395–12404
Zheng L, Guo S, Wang Z et al (2021a) A multi-functional module-based capsule robot [J]. IEEE Sensors J 21(10):12057–12067
Zheng Y, Tang N, Omar R et al (2021b) Smart materials enabled with artificial intelligence for healthcare wearables [J]. Adv Funct Mater 31(51):2105482
Zheng X, Pan Y, Zhang Y et al (2022) Interventional microbubble enhanced Sonothrombolysis on left ventricular assist devices [J]. Adv Sci 9(21)
Zhong Y, Zhang Y, Xu J et al (2019) Low-intensity focused ultrasound-responsive phase-transitional nanoparticles for thrombolysis without vascular damage: a synergistic nonpharmaceutical strategy [J]. ACS Nano 13(3):3387–3403
Zhong J, Li Z, Takakuwa M et al (2021) Smart face mask based on an ultrathin pressure sensor for wireless monitoring of breath conditions [J]. Adv Mater 34(6)
Zhu P, Chen Y, Shi J (2020) Piezocatalytic tumor therapy by ultrasound –t riggered and BaTiO3-mediated piezoelectricity [J]. Adv Mater 32(29)
Zhu H, Li B, Liu X et al (2022) Interfacial Mo, W-conjugated polarization, and oxygen vacancies of MoO2/WO3 in enhanced microwave therapy for MRSA-induced osteomyelitis [J]. ACS Nano 16(12):21098–21110
Zhu C, Bing Y, Chen Q et al (2023) Nonenzymatic flexible wearable biosensors for vitamin C monitoring in sweat [J]. ACS Appl Mater Interfaces 15(15):19384–19392

Chapter 8
Bio-sensing Materials

Jie Fu, Fangying Xiong, Qiya Gao, Shouzhe Deng, Jiawang Wang, and Shuang Li

Abstract Materials used in biosensors are organized into five primary categories: physical, chemical, biological, flexible electronic, and nano-sensing materials.

Physical sensing materials convert physical stimuli into quantifiable signals. This section delves into four main types: semiconductor, ceramic, packaging, and transistor materials. It emphasizes their role in physical sensing, discussing their diverse applications and the innovative materials that have emerged from their development. For instance, contemporary ceramic-based composite materials, created by integrating ceramic reinforcements, are extensively utilized.

Chemical sensing materials interact with specific chemicals to produce measurable signals. This discussion centers on metal oxide semiconductor, electrochemical, optical, and acoustic sensing materials, categorizing them by structural characteristics. It outlines their principal uses and the advancements in material technology during their evolution.

Biological sensing materials engage with biomolecules or organisms to generate quantifiable signals. The narrative elaborates on enzymes, metabolic biochemistry, and immune binding as key modes of interaction. These materials are pivotal in human monitoring and the development of implantable devices, enabling the detection of biomarkers like proteins, genes, or metabolites, and the creation of devices such as cardiac pacemakers that monitor heart rhythms.

Flexible electronic materials, characterized by their bendability, stretchability, and adaptability, are crucial for producing flexible and wearable electronic devices. Their applications in modern medicine are vast, enabling the creation of wearable devices like smart wristbands and patches that monitor vital signs and physical activity. The discussion focuses on three material types: flexible electronics, substrates, and nanomaterials.

Nano-sensing materials, known for their nano-sized structures, are instrumental in developing nanosensors and detection devices for analyzing physical, chemical,

J. Fu · F. Xiong · Q. Gao · S. Deng · J. Wang · S. Li (✉)
Academy of Medical Engineering and Translational Medicine, Medical College, Tianjin University, Tianjin, China
e-mail: lishuangv@tju.edu.cn

Z. Liu (ed.), *Intelligent Biomaterials*, https://doi.org/10.1007/978-981-96-8646-9_8

or biological parameters on a minute scale. Since their advent in the 1980s, these materials have been recognized as among the most promising of the twenty-first century. The section categorizes nanomaterials by their properties—signal enhancement, catalytic effects, and specificity—detailing their unique attributes, sensor configurations, and applications.

Keywords Physical sensing materials · Chemical sensing materials · Bio-sensing materials · Flexible electronic materials · Nano-sensing materials

8.1 Introduction

With the continuous progress of scientific technology, the development of medical sensors has become increasingly important. Medical sensors are devices capable of perceiving and detecting internal changes in the human body, with a high degree of sensitivity and accuracy. By reliably monitoring physiological indicators and disease biomarkers, medical sensors can provide timely diagnostic and therapeutic feedback, effectively improving medical outcomes and quality of life for patients. Medical sensors are a technology used to perceive internal information in the human body, with a wide range of applications including, but not limited to, disease diagnosis, drug monitoring, and vital sign monitoring. Medical sensors can achieve monitoring and assessment of patient conditions by measuring physiological parameters such as blood pressure, body temperature, and heart rate (Meng et al. 2022a; Chen et al. 2020; Bent et al. 2020). Figure 8.1 illustrates the widespread application of physical, chemical, and biological sensing materials, as well as emerging flexible and nanomaterials in the development of sensors in the medical field.

Physical sensing materials are a crucial component in the construction of physical sensors. The selection of materials has important effects on the sensitivity, stability, and response time of the sensors, which in turn affect their performance and application range. Common physical sensor materials include semiconductor and ceramic materials (Ahmad et al. 2021; Fang et al. 2022). Semiconductors are conductive materials that have controllable conductivity, possessing conductivity between conductors and insulators at room temperature. Ceramic materials are usually prepared by sintering inorganic and non-metallic solid components, and due to the high melting points of their raw materials, high-temperature conditions are often required in their processing and application. High-temperature ceramic materials are used for manufacturing high-temperature sensors, whereas special coating materials are used for manufacturing chemical sensors. With the development of materials science and nanotechnology, new types of physical sensor materials continue to emerge, providing new possibilities for the improvement and innovation of physical sensor technology. Future research will focus on improving and innovating materials to enhance the sensitivity, stability, and response time of sensors, driving the development of physical sensor technology.

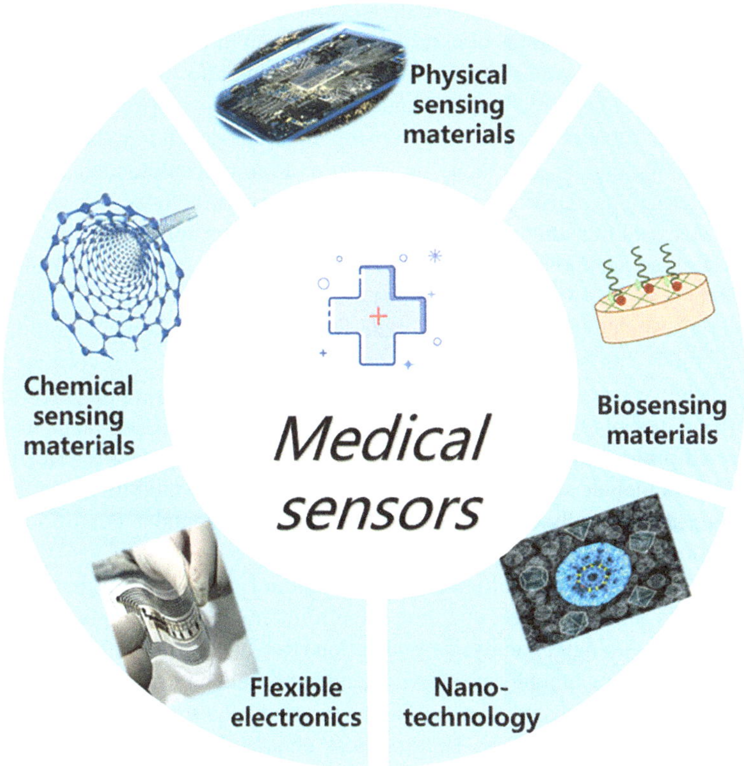

Fig. 8.1 Medical sensing materials

Chemical sensing materials are a class of materials that are capable of sensitive responses to specific chemical analyses or physical changes. They enable quantitative or qualitative detection of target substances through various mechanisms such as molecular recognition, optics, electrochemistry, and surface-enhanced effects (Berger et al. 2020; Zhang et al. 2020a; Singh et al. 2021). They find extensive applications in fields such as environmental monitoring, medical diagnosis, food safety, and bioanalysis. The design and performance of chemical sensing materials depend on their basic principles, which mainly involve molecular recognition and signal transduction. Molecular recognition is achieved through interactions with target substances, such as coordination bonds, hydrogen bonds, and electrostatic attractions. Signal transduction involves the conversion of the recognition process into signal formats suitable for measurement and analysis, typically involving optical, electrochemical, and bio-sensing technologies. The performance of chemical sensing materials depends on various factors, such as the selection of materials, preparation methods, and structure. Key performance factors include selectivity, sensitivity, response speed, stability, and reproducibility. Additionally, adaptability to environmental conditions and multifunctionality are important considerations.

Biological sensing materials are materials that achieve quantitative or qualitative detection of the internal status or external environment of living organisms by interacting with them. They can convert biological signals into measurable signals by utilizing the specificity recognition and interactions between biomolecules. The design and performance of biological sensing materials depend on their basic principles, which include the specificity recognition of biomolecules, such as antibody-antigen interactions, DNA hybridization, and enzyme activity, as well as the methods of signal transduction and amplification, such as optical, electrochemical, and bio-sensing technologies (Zhang et al. 2020b; Foroughi and Jahani 2022; Cavalcante et al. 2021). There are various preparation methods for biological sensing materials, including biosynthesis, material synthesis, and surface modification. These methods can select appropriate materials and structures according to the needs to achieve specific biological recognition and signal conversion performance. The performance of biological sensing materials depends on various factors, such as the selected biological materials, preparation methods, and structures. Key performance factors include selectivity, sensitivity, stability, reproducibility, and biocompatibility. Additionally, adaptability to environmental conditions and multifunctionality are also important considerations.

Flexible and nanoscale biological sensing materials are a class of materials with high flexibility and small size, enabling them to interact with living organisms and achieve specific biological analysis or detection (Rao et al. 2023; Aftab et al. 2023). Compared to traditional materials, flexible and nanoscale biological sensing materials have unique advantages and potentials in biomedical, bio-imaging, bio-sensing, and cellular engineering fields. They possess characteristics such as bendability, stretchability, deformability, and wearability, as well as large surface area-to-volume ratio, high sensitivity, and high selectivity. These characteristics allow them to come into contact with the biological body and achieve highly sensitive biological detection and imaging. Flexible and nanoscale biological sensing materials have broad application prospects in bio-sensing and life sciences. They can be applied in areas such as biomolecular detection, cellular imaging, microfluidic systems, biomimetic sensing, tissue engineering, and bio-delivery systems. Particularly in fields such as cancer diagnosis, drug delivery, and gene editing, flexible and nanoscale biological sensing materials demonstrate tremendous potential.

In terms of medical applications, these sensing materials play a crucial role. Physical sensing materials can be used to monitor vital signs of patients, such as body temperature, heart rate, and blood pressure, providing important diagnostic and treatment references for doctors (Kuzubasoglu and Bahadir 2020; Hinde et al. 2021; Kim et al. 2019). Chemical sensing materials can be used to detect biochemical indicators in the blood, such as blood glucose, blood oxygen saturation, and tumor markers, aiding in early diagnosis and treatment monitoring (Yao et al. 2021; Suprayitno et al. 2019; Dukle et al. 2022). Biological sensing materials have applications in medical fields including bio-imaging, cellular detection and drug delivery, which can be used for early diagnosis and treatment of cancer (Mehta et al. 2021; Chen et al. 2019; Panwar et al. 2019). Flexible sensing materials can be fabricated into flexible skin patches for monitoring patient's respiration, electrocardiogram, and drug release, while nanoscale sensing materials can be used in disease molecular labeling and drug delivery systems, offering potential and prospects for precision medicine.

This article presents the principles of physical, chemical, biological, flexible, and nanoscale sensing materials. It provides a detailed analysis of their applications in medicine. The use of sensing materials in medicine offers new means and methods for clinical diagnosis, monitoring, and treatment with the potential to drive advancements in medicine and the development of personalized healthcare in the future.

8.2 Physical Sensing Materials

8.2.1 Semiconductor Materials

Semiconductor is a material with controllable conductivity that falls between that of conductors and insulators, exhibiting conductivity at room temperature. Semiconductors are widely utilized in various fields such as integrated circuits, photovoltaic power generation, communications, and high-power power conversion. For instance, the high-density transistors used in modern integrated circuits are fabricated using semiconductor materials. The discovery of semiconductor properties traces back to 1833 when British scientist Michael Faraday observed a decrease in the electrical resistance of silver sulfide with increasing temperature. This marked the first observation of the temperature-dependent nature of semiconductor resistivity. Subsequently, various distinctive properties of semiconductors, including the photoelectric voltage effect, photoconductivity, and rectification effect, were discovered. However, it wasn't until the 1930s that the concept of semiconductors gained academic recognition. In 1947, Bell Labs summarized the four key characteristics of semiconductors. Common semiconductor materials encompass silicon, germanium, and gallium arsenide, among others. Silicon, derived from sand and refined to high purity, holds particular significance in semiconductor applications. Its capacity to form a mask during device fabrication enhances semiconductor device stability and streamlines automated industrial production. Silicon has emerged as the most impactful semiconductor material due to its extensive usage. In conclusion, semiconductors play a pivotal role in diverse technological applications, thanks to their unique properties. The translation of these properties into technological advancements has significantly shaped fields such as electronics, communications, and energy conversion.

8.2.1.1 Silicon-Based Semiconductor

Silicon wafers serve as the primary substrates for semiconductors, forming a single thin film. These wafers find extensive use in manufacturing integrated circuits and photovoltaic/solar cells. The fabrication process involves several steps, including photolithography patterning, thin film deposition, doping and ion implantation, etching, and other techniques. In the late 1960s, silicon materials began to be employed as sensing matrices for in vivo electrophysiological measurements in integrated sensors (Zias et al. 1963; Angell et al. 1970). Subsequently, Bregveld developed silicon-based devices, known as pH field-effect transistors (FETs), capable of detecting pH values in minute volumes with high precision (Bergveld 1970). The commercialization of pH sensors in the mid-1980s marked the maturation of silicon materials and device manufacturing technologies, laying the foundation for the ongoing development of devices like ion-selective and biofield-effect transistors. When silicon dioxide (SiO_2) or other metal oxides are used as chemical recognition sensing units alongside nitrides (e.g., Al_2O_3, Ta_2O_5, SnO_2, and Si_3N_4), resulting sensors exhibit high selectivity towards hydrogen ions due to active hydrogen bonds in the oxide layer. These sensors capitalize on the reversible coordination of hydrogen ions with oxide layers, influencing the surface potential of the sensing element. Changes in surface potential impact the gate potential, akin to the operation of metal gate FETs, thereby altering the signal of pH field-effect transistors. This innovative silicon technology allows the direct integration of sensing elements with readout circuits, providing independent microsensors with high sensitivity and signal-to-noise ratio. This characteristic innovation positions it prominently in the field of microsensors, particularly in biosensor chip development applications.

8.2.1.2 New Type of Semiconductor

Silicon carbide belongs to the third generation of semiconductor materials and is characterized as a wide bandgap semiconductor, distinguishing itself primarily through its bandgap width when compared to silicon. The bandgap width is a pivotal indicator for assessing the breakdown voltage of a semiconductor material, with a larger bandgap width indicating a greater ability to withstand high voltages.

Representing the third generation of semiconductor materials, silicon carbide typically possesses a wider bandgap. This feature allows devices crafted from silicon carbide to be one-tenth the size of their silicon-based counterparts while maintaining comparable performance. This reduction results in a three-quarters decrease in energy loss, establishing silicon carbide as a new substrate material for the production of high-voltage and high-frequency devices.

In addition to silicon carbide, other materials like gallium nitride (GaN), gallium arsenide (GaAs), indium nitride (InN), and diamond are gradually finding integration into the manufacturing processes of various electro-optical sensor devices (Grant et al. 2005). These materials bring superior sensing characteristics to the devices, including increased voltage tolerance, enhanced thermal conductivity, overall higher efficiency, and improved long-term operational stability, thus enhancing reliability.

Moreover, these materials exhibit excellent reverse recovery characteristics, enabling a relatively short return to baseline during operation. They are also less susceptible to electromagnetic interference, facilitating operation at higher frequencies. The unique chemical, physical, and mechanical stability of these novel semiconductor materials positions them ideally for the development of specialized sensors.

8.2.1.3 Semiconductor Classification

Intrinsic Semiconductor: An intrinsic semiconductor, completely devoid of impurities and lattice defects, relies on intrinsic excitations for its electrical conductivity. Examples include silicon (Si), germanium (Ge), and gallium arsenide (GaAs). While semiconductor materials can't be entirely pure, those with minimal impurities are often termed impurity semiconductors. Intrinsic semiconductors exist in two physical forms: polycrystalline and monocrystalline with the latter being essential in modern semiconductor device manufacturing. At a specific temperature, the generation and recombination of electron-hole pairs coexist, reaching dynamic equilibrium. At this point, the semiconductor exhibits a specific carrier density, resulting in a particular resistivity. With increasing temperature, more electron-hole pairs are generated, leading to an increased carrier density and decreased resistivity. This temperature-dependent effect in semiconductors is an intuitive explanation. Due to the relatively high resistivity of pure semiconductors without lattice defects, their practical applications are limited.

Elemental Semiconductor: An elemental semiconductor is composed of a single element, with early studies centering on silicon and selenium. It is a solid material consisting solely of one element, showcasing semiconductor characteristics. However, it is susceptible to variations influenced by trace impurities and external conditions. Presently, silicon and germanium stand out as the most extensively used elemental semiconductors owing to their favorable performance. Inorganic Synthetic Semiconductor: Inorganic synthetic semiconductors primarily encompass semiconductor materials composed of a single element though compounds with multiple elements are also included. However, not all compounds meet the criteria for semiconductor materials due to the characteristics of the elements and fabrication methods. These semiconductors are mainly employed in high-speed devices, with indium phosphide (InP) transistors demonstrating superior speeds compared to other materials. They play a vital role in optoelectronic integrated circuits and devices designed to withstand nuclear radiation.

Organic Synthesis Semiconductor: Organic compounds, characterized by carbon-carbon bonds in their molecular structure, form the basis of organic synthesis semiconductors. By arranging organic compounds perpendicularly to the carbon-carbon bonds and overlaying them, a conduction band is established. Chemical doping enables these compounds to enter energy bands, facilitating electrical conductivity and the creation of organic semiconductor materials. In contrast to traditional semiconductors, organic semiconductors offer advantages such as lower cost, good solubility, and ease of material processing. The electrical conductivity of these materials is controllable through manipulation of the molecular structure, making them versatile and applicable in various fields, including organic films and organic lighting.

Organic synthesis semiconductor: Organic compounds refer to compounds containing carbon-carbon bonds in their molecular structure. By orienting organic compounds perpendicular to the carbon-carbon bonds and overlaying them, a conduction band can be formed. Through chemical doping, these compounds can enter energy bands, allowing for electrical conductivity, thereby forming organic semiconductor materials. In comparison to traditional semiconductors, organic semiconductors are characterized by lower cost, good solubility, and ease of material processing. The electrical conductivity can be controlled by manipulating the molecular structure, making them versatile and applicable in various fields such as organic films and organic lighting.

Amorphous semiconductor; Also known as amorphous semiconductor or glass semiconductor, it belongs to a category of semiconducting materials. It is primarily achieved by altering the relative positions of atoms, disrupting the original periodic arrangement, and forming amorphous silicon.

8.2.2 *Ceramic Materials*

Ceramic materials are typically crafted through sintering and processing of inorganic, non-metallic solid components in powder form. Given their high melting points, both processing and applications often necessitate elevated temperatures. These materials comprise compounds or combinations of various metallic and non-metallic elements, primarily O, B, C, and N. They manifest as oxides, carbides, borides, or silicides, further classified based on their end-use into categories such as structural ceramics, electrical ceramics (encompassing dielectric, piezoelectric, and pyroelectric properties), optical ceramics, chemical processing ceramics, ceramic coatings, bio-ceramics, and ceramic superconductors (Ayode Otitoju et al. 2020; Lakhdar et al. 2021).

Their exceptional physical, chemical, and mechanical properties, characterized by high-temperature resistance, good hardness, high stiffness, low friction coefficient, and chemical inertness, derive from their atomic structure grounded in rigid ions and covalent bonds.

8.2.2.1 Oxide and Non-Oxide Ceramic Materials

Currently, the most extensively studied oxide ceramics include Al_2O_3 (valued for its versatility, low cost, and relatively low sintering temperature) and ZrO_2 (renowned for its high toughness, relatively low sintering temperature, and widespread industrial applications) (Drouet et al. 2017). While SiO_2 and silicates may not always be categorized as traditional ceramic materials, they serve as economically effective alternatives to Al_2O_3 and other advanced ceramics in the initial stages of biomedical applications.

Non-oxide ceramics, such as silicon carbide (SiC), tungsten carbide (WC), boron carbide (B_4C), silicon nitride (Si_3N_4), aluminum nitride (AlN), and zirconium diboride (ZrB_2), typically exhibit higher temperature resistance, strength, and fracture toughness. Oxide ceramics are relatively easy to process, with sintering temperatures rarely exceeding 1550 °C, while non-oxide ceramics, due to their high hardness, often demand sintering temperatures above 1700 °C, making processing more challenging (frequently requiring pressure-assisted ignition in a controlled inert atmosphere). Oxide and non-oxide ceramic materials are typically processed and shaped using powder processing methods, forming and sintering ceramic components from dry or wet ceramic powder mixtures, or obtained through the pyrolysis of pre-ceramic polymers (Riedel and Chen 2008).

8.2.2.2 Polymer-Derived Ceramic Materials

Polymer-derived ceramic materials are produced by pyrolyzing polymer precursors into ceramic materials under inert gas conditions at temperatures exceeding 800 °C. This method can be applied to create binary ceramic materials like SiC, Si3N4, BN, and AlN, as well as complex ternary, quaternary, or even quinary ceramic materials such as SiCN, SiCO, SiBCN, and SiAlCN (Colombo et al. 2010). The pre-mixing polymer processing route provides a novel alternative to traditional powder processing.

One significant advantage lies in the fact that pre-mix polymer precursors can be processed using commonly employed molding techniques in the plastics industry, such as injection molding and resin transfer molding. High-temperature pyrolysis results in the release of a substantial amount of volatile gases (H_2O, CO_2, H_2, CH_4), leading to significant mass loss and noticeable dimensional shrinkage ranging from 40% to 70%. This phenomenon aids in preventing the formation of pores and cracks in ceramic components larger than a few millimeters, particularly in the case of pre-amine polymers. It ensures volumetric integration at the interface between biological tissue and porous ceramic scaffolds.

8.2.2.3 Ceramic Matrix Composites

Despite possessing superior physical, chemical, and thermomechanical properties, ceramics suffer from a common drawback: brittleness, lack of toughness, and susceptibility to catastrophic failure. In many engineering applications, this flaw significantly limits advanced ceramics from replacing more ductile metal counterparts.

Ceramic matrix composites, incorporating ceramic reinforcing bodies (such as continuous fibers, whiskers, or nanoparticles) within a ceramic matrix, effectively enhance the toughness of monophasic ceramics, addressing their reliability deficiencies. Certain ceramic matrix composites exhibit substantially improved thermomechanical properties, particularly in thermal shock resistance and fracture toughness with higher fracture elongation rates. Consequently, they can function as high-performance thermal structural components under demanding operating conditions (Klocke et al. 2014).

Synthesizing composite ceramic powders can be achieved through various processing routes, including solid-state, sol-gel, co-precipitation, and hydrothermal methods. Examples include carbon fiber/carbon matrix (Cf/C), carbon fiber/silicon carbide (Cf/SiC), and silicon carbide/silicon matrix (SiCf/SiC) (Neera Singh et al. 2017). For instance, embedding piezoelectrically active ceramic particles as fillers in a non-active polymer matrix holds great potential for manufacturing low-cost, flexible, and efficient piezoelectric sensors.

Diverse compositions and advanced manufacturing techniques, such as infiltration, injection molding, sacrificial mold, electrophoresis, and 3D printing, have been employed to manufacture ceramic composite materials. These techniques aim to enhance their piezoelectric performance for applications like energy harvesting and storage, sensing and detection, and tissue engineering (Eltouby et al. 2021).

8.2.3 Encapsulation Materials

Electronic encapsulation involves packaging semiconductor chips to shield them from external environmental influences. Electronic encapsulation materials play a crucial role in supporting electronic components and their interconnections, offering mechanical support, environmental sealing, signal transmission, heat dissipation, and shielding. The key components of these materials encompass substrates, lead frames, interlayer dielectrics, and sealing materials.

8.2.3.1 Encapsulation Substrate

High-density multilayer encapsulation substrates serve as crucial electrical interfaces between semiconductor chips and conventional PCBs (printed circuit boards). These substrates play a vital role in providing electrical transition, protection, support, and efficient heat dissipation for the chips. The main materials employed in their construction include ceramics, epoxy glass, diamond, metals, and metal matrix composites.

Encapsulation substrate materials must demonstrate high thermal conductivity to prevent thermal damage to electronic components. Additionally, they should feature a coefficient of thermal expansion matching the chip to avoid failure due to thermal stress. Ensuring good high-frequency characteristics is essential to meet the demands of high-speed conduction. These substrates should possess high mechanical performance, effective electrical insulation, chemical stability (resistant to corrosion from electroplating solutions and metal materials used for wiring), and ease of processing.

Commonly used encapsulation substrate materials comprise ceramic-based, plastic-based, and metal-based materials. Ceramic-based encapsulation materials exhibit characteristics such as a low dielectric constant, high insulation, high mechanical strength, high thermal conductivity, and chemical stability. However, due to higher production costs, they are predominantly utilized in advanced micro-electronic devices within aerospace and military applications. Plastic-based encapsulation materials offer advantages such as low production costs, lightweight construction, and miniaturization. Nevertheless, they may exhibit mismatched thermal expansion coefficients and high dielectric losses. Metal-based encapsulation materials, while conducive to heat conduction and possessing high strength, are less suitable for lightweight chip encapsulation due to their higher density and cost.

8.2.3.2 Lead Frame

The lead frame is composed of two primary components: the chip pad and the pins. Its role is to act as the carrier for integrated circuit chips, facilitating electrical connections between the internal circuit terminals (bonding points) and external leads through internal and external wires. Consequently, lead frame materials must possess attributes such as good electrical and thermal conductivity, high strength, and a low coefficient of thermal expansion.

An ideal lead frame material should have a tensile strength exceeding 600 MPa, conductivity above 80%, and a softening temperature exceeding 500 °C. Copper alloys, including copper-iron-phosphorus, copper-nickel-silicon, copper-chromium-zirconium, copper-silver, and copper-tin, are widely utilized due to their excellent properties.

As modern terminal devices become more intelligent, diverse, and compact, encapsulation materials, including lead frames, undergo continuous evolution. The focus is on achieving high density, high reliability, efficient heat dissipation, low power consumption, and cost-effectiveness to meet the demands of contemporary technology.

8.2.3.3 Interlayer Dielectric

Interlayer dielectric materials play a crucial role in integrated circuit packaging by serving as insulating layers between multiple metal wiring layers. As integrated circuits continue to miniaturize and increase in density, the performance of these interlayer dielectric materials becomes a significant factor affecting the functionality of integrated circuits.

Several polymers exhibit desirable properties for interlayer dielectric materials, including high-temperature resistance, low dielectric constant, and high dimensional stability. Among these polymers, notable choices include:

1. Polyimide: Known for its high-temperature resistance and excellent dimensional stability. It is a popular choice in interlayer dielectric applications.
2. Polyphenylene Oxide: Offers low dielectric constant properties, making it suitable for minimizing signal interference in densely packed integrated circuits.
3. Poly(butylene terephthalate): This polymer also possesses high-temperature resistance and low dielectric constant characteristics, making it a favorable option for interlayer dielectric materials.

These polymers help meet the evolving demands of integrated circuit technology as they enable miniaturization, enhance performance, and maintain the required insulation between metal wiring layers.

8.2.3.4 Sealing Material

Sealing materials, also known as encapsulants or molding compounds, play a crucial role in the electronics packaging industry. A notable example is epoxy molding compounds (EMC), which fall under the category of thermosetting plastics. EMC comprises epoxy resin, phenolic resin, latent curing agents, silicon micro-powder, stress modifiers, wax, and carbon black among other components.

In response to the evolving trends in plastic packaging technology, such as thinning, high density, and low power consumption, cured encapsulation materials are required to possess specific characteristics. These include low coefficients of thermal expansion, low dielectric constants, low moduli, low water absorption, and high fluidity. EMC, being a thermosetting plastic, dominates the encapsulation materials landscape accounting for over 97% usage.

The encapsulation process involves extruding EMC into the mold cavity using a transfer molding method. This process includes embedding the semiconductor chip and simultaneously crosslinking and curing to form a semiconductor device with a defined structural shape. The functions of sealing materials encompass protecting the chip from external environmental influences, resisting external factors like humidity, solvents, and impact. They also provide electrical insulation between the chip and the external environment, ensure good installation performance, resist thermal shock and mechanical vibration during installation, and facilitate heat dissipation. Sealing materials thus play a critical role in safeguarding semiconductor devices and optimizing their performance in various environmental conditions.

8.2.4 Transistors

The term "transistor" refers to all individual components made from semiconductor materials, which includes devices such as diodes, transistors, field-effect transistors, and thyristors. As a semiconductor device, it serves as the essential building block for integrated circuits. Due to their rapid response times and high precision,

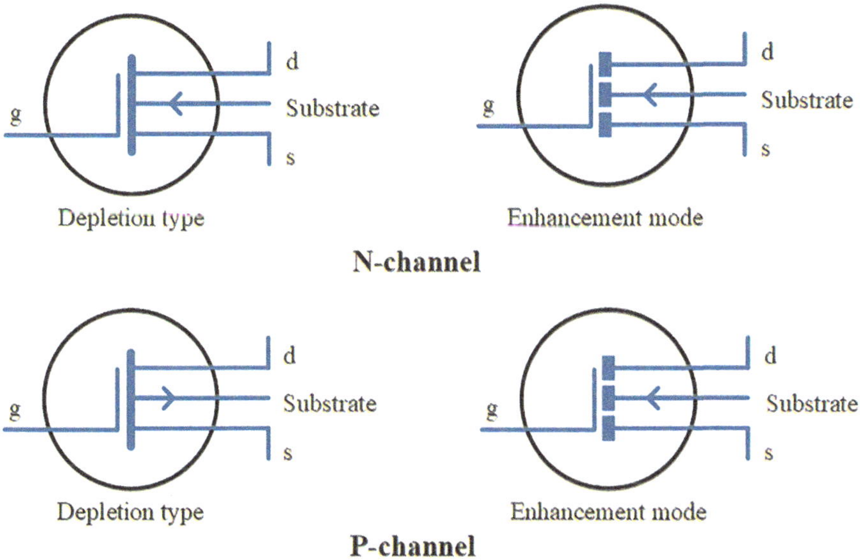

Fig. 8.2 Four different types of MOSFETs

transistors are extensively utilized in both digital and analog functions, encompassing amplification, switching, voltage regulation, signal modulation, and oscillation. Figure 8.2 illustrates four distinct MOSFET structures.

On December 16, 1947, William Shockley, John Bardeen, and Walter Brattain at Bell Labs achieved a significant milestone by creating the first germanium transistor. Fast forward to 2022, where a team led by Prof. Ren Tianling from Tsinghua University's Institute of Microelectronics accomplished a groundbreaking feat. They achieved a record gate length of 0.34 nm using a single layer of graphene as the gate, leading to the development of a "side-wall" transistor. The research, titled "Vertical MoS_2 transistors with sub-1-nm gate lengths," was published in the latest issue of Nature.

In the subsequent year of 2023, Dutch scientists made strides in material innovation by developing the first two-dimensional (2D) topological insulator called germanene composed of a single element. These transistors have the potential to replace traditional counterparts in electronic devices, offering the advantage of eliminating heat generation. This chapter delves into the exploration of bipolar junction transistors (BJTs) and field-effect transistors (FETs), highlighting their significance in the ever-evolving landscape of transistor technology.

Bipolar junction transistors, also known as bipolar transistors or semiconductor junction transistors, involve two types of charge carriers with different polarities in conduction. Through different doping methods on the same silicon wafer, three doped regions are created, forming two PN junctions that constitute the transistor. Depending on the order of these junctions, they are classified as PNP or NPN transistors. The semiconductor material used categorizes them as germanium or silicon transistors. Transistors function in cut-off, amplification, and saturation states, with cut-off and saturation commonly employed as switches in digital circuits for high-level on-off control with low-level signals. The amplification function is crucial for processing weak currents in analog circuits, making transistors essential in various detection circuits. In practical applications, weak signals from sensors often require amplification for further processing or providing sufficient energy to drive actuators. Transistors serve as the core components in amplification circuits, enabling distortion-free amplification of small input variations into significant outputs. However, due to temperature effects, the amplification factor of transistors can be unstable. Therefore, simple transistor amplification circuits may not meet the precision requirements of signal processing circuits. Integrated operational amplifier chips, incorporating transistors as units, are commonly used in current amplification circuits, offering stable gain and precision.

Field-effect transistors control output circuit current using the electric field effect in the input circuit, earning them the name unipolar transistors as they rely solely on majority carriers for conduction. In comparison to bipolar junction transistors, field-effect transistors exhibit high input circuit resistance (10^7 to 10^{12} ohms), resulting in low power consumption, excellent thermal stability, and robust radiation resistance. These advantages have led to their widespread use in electronic circuits since the 1960s, serving purposes like signal amplification, current regulation, and acting as switches.

Two main types of unipolar transistors are Metal-Oxide-Semiconductor Field-Effect Transistors (MOSFETs) and Metal-Semiconductor Field-Effect Transistors (MESFETs). While their structures differ slightly, both include three main parts: source, drain, and gate. The most common type is the MOSFET, featuring an insulating layer, an oxide layer, and a semiconductor layer. Applying voltage to the gate forms an electric field, influencing the channel's conductivity between the source and drain. Adjusting the gate voltage allows control over the current flow. MESFETs, known for high transport speed and frequency characteristics, find common use in radio frequency and microwave devices.

8.3 Chemical Sensing Materials

Chemical sensing materials comprise a category of materials uniquely capable of specifically interacting with target molecules and translating this interaction into quantifiable signals. The fundamental principle underlying their operation involves eliciting physical or chemical changes within the material due to its specific interaction with target molecules. These changes are measurable and can be employed to identify and quantify the target molecules. Chemical sensing materials are characterized by their selectivity and sensitivity. The selectivity of these materials is defined by their capacity to specifically interact with certain chemical substances while remaining inert to others, a critical attribute for the efficacy of chemical sensors. Sensitivity, on the other hand, denotes the material's ability to recognize and detect target molecules, with high sensitivity ensuring the detection of these substances even at low concentrations. Chemical sensing materials are pivotal in developing various sensor types, such as gas sensors, liquid sensors, and biosensors, for detecting a wide range of chemicals. Prominent examples of chemical sensing materials include metal oxide semiconductors, electrochemical materials, optical materials, and acoustic materials, each demonstrating superior performance in their respective application scenarios.

8.3.1 Metal Oxide Semiconductor Sensing Materials

Metal-oxide-semiconductor (MOS) sensor materials have become a staple in sensor technology, renowned for their ability to monitor changes in environmental parameters, including gases, humidity, and temperature. In recent years, MOS materials, as a category of chemiresistive sensors, have garnered significant interest in fields such as environmental monitoring, automotive emissions control, and food safety testing. This surge in popularity is attributed to their low manufacturing costs, high sensitivity, ease of use, and versatility in detecting a wide array of gases (Zhang et al. 2017). These materials possess distinctive electrical characteristics, rendering them crucial in sensor technology. Their operational principle hinges on the modulation of charge transport and resistance alterations on the material's surface. Exposure of the sensing material to target gases or environmental parameters leads to gas adsorption or surface reactions. Such interactions prompt a shift in the charge state at the material's surface, thereby modifying its resistance. This change in resistance is quantifiable and can be utilized to detect and monitor environmental alterations. Metal oxide semiconductor gas sensors are employed across various sectors and applications due to their cost-effectiveness relative to alternative sensing technologies, durability, lightweight design, longevity, and their high sensitivity coupled with rapid response times. Their extensive application in measuring and monitoring trace levels of environmentally significant gases, such as carbon monoxide and nitrogen dioxide, underscores their importance (Fine et al. 2010). Sensors utilizing metal oxide semiconductors primarily function to identify target gases via redox reactions that occur between these gases and the oxide surface. This detection process unfolds in two stages: Firstly, redox reactions take place where oxygen ions ($O-$) on the material's surface engage with the target gas molecules, leading to a change in the electronic state of the oxide surface. Secondly, this electronic alteration is converted into a variation in the sensor's electrical resistance. This resistance change can be identified through measurements of alterations in capacitance, work function, mass, optical properties, or reaction energy (Liu et al. 2012). Widely recognized metal oxide semiconductor materials for sensing include tin dioxide (SnO2), zinc oxide (ZnO), indium tin oxide (ITO), and iron oxide (Fe2O3), among others. Each of these materials exhibits unique sensitivities to particular gases or parameters, highlighting the importance of selecting the appropriate material for specific applications.

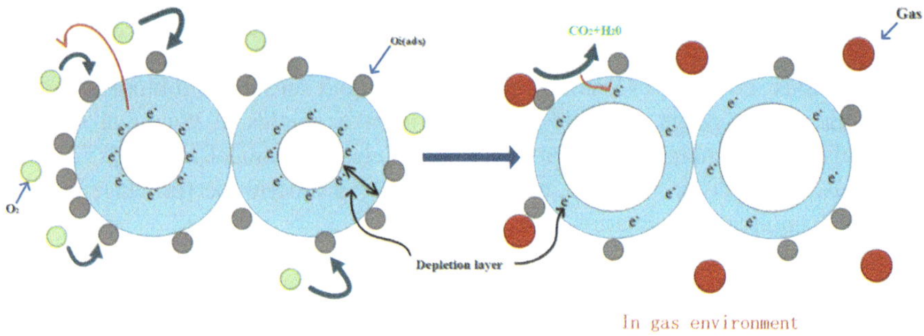

Fig. 8.3 Schematic diagram of the sensing mechanism of n-type semiconductor metal oxide nanostructures for reducing gases

8.3.1.1 Tin Dioxide

Tin dioxide (chemical formula: SnO2) stands as a pivotal metal oxide semiconductor material. As the cornerstone of solid-state gas sensors, SnO2 is among the most extensively researched metal oxides (Berger et al. 2009). Tin dioxide is frequently utilized in sensors, electronic devices, and conductive materials. This substance is characterized by its white, colorless, and transparent crystalline form, possessing a high melting point and significant thermal stability. Notably, tin dioxide exhibits both morphological and chemical stability. In n-type semiconductors like tin oxide, electrons serve as the primary charge carriers (Nikolic et al. 2020). Tin dioxide is a semiconductor material boasting a wide energy bandgap of approximately 3.6 electron volts. It functions as an insulator at room temperature, yet transitions to a conductive material at elevated temperatures. Various preparation techniques, such as chemical vapor deposition, the sol-gel method, and hydrothermal synthesis, are employed to produce tin dioxide. These methods result in diverse forms of tin dioxide, including nanoparticles, films, and fibers, among others. Figure 8.3 illustrates the schematic diagram of the sensing mechanism of tin dioxide for reducing gases..

Tin dioxide is commonly utilized in gas sensors, particularly for detecting hazardous gases like carbon monoxide (CO), sulfur dioxide (SO2), and methane (CH4). Within a gas sensor, tin dioxide's resistance fluctuates based on the adsorption or reaction with the target gas. These variations in resistance can be quantified and leveraged to ascertain the concentration of the target gas. Additionally, tin dioxide finds application as a transparent conductive film in the production of electronic devices, including solar cells, touch screens, and liquid crystal displays. It is also employed in catalytic processes, glass coatings, and materials sensitive to heat.

8.3.1.2 Zinc Oxide

Zinc oxide (ZnO) is a significant inorganic compound, appearing as a white powder or in colorless, transparent crystalline or block solid form typically white. At room temperature, it acts as an insulator, yet its electrical properties can be modified through methods like doping. Zinc oxide is synthesized through various techniques, including chemical vapor deposition, solution processes, and heat treatment. The methods for preparing nanoscale zinc oxide have been thoroughly explored, highlighting its importance in nanomaterial applications. Doping zinc oxide alters its electrical properties, transforming it into a semiconductor. Through controlled doping, zinc oxide can acquire specific conductivity and optical characteristics, making it suitable for use in photoelectric devices. Its direct band gap of 3.37 eV and high exciton binding energy, which contribute to efficient ultraviolet emission, have made zinc oxide (ZnO) a subject of extensive research for applications in areas such as gas sensing (Nikolic et al. 2020).

Zinc oxide boasts a broad spectrum of applications across various sectors. In the realm of electronic devices, it serves as a transparent conductive layer essential for the fabrication of solar panels, touch screens, and liquid crystal displays. In the cosmetics industry, zinc oxide's UV-absorbing properties make it a key ingredient in sunscreen formulations and various beauty products. Within the medical field, it is incorporated into creams and powders as a skin protectant for treating skin disorders and eczema. In the rubber industry, zinc oxide enhances the wear resistance and anti-aging characteristics of rubber goods. For ceramics and glass, it is utilized in the creation of ceramic pigments and glass materials, imparting specific colors and properties. Moreover, zinc oxide nanomaterials find significant applications in optoelectronic devices, including light-emitting diodes (LEDs), where their unique properties are highly valued.

8.3.1.3 Copper Oxide

Copper oxide (CuO) is a form of copper oxide, manifesting as a black or dark red powdery substance with notable resistance to electricity and heat. It functions as a semiconductor material, with its conductivity adjustable through doping and temperature regulation. As a p-type semiconductor, copper oxide has been extensively researched for various applications particularly in gas sensing. Intriguingly, it behaves as an insulator at ambient temperatures but transitions to a semiconductor or conductor when exposed to elevated temperatures. Copper oxide is synthesized using diverse techniques, including chemical deposition, thermal decomposition, and the hydrothermal method. The development of nanoscale copper oxide has been thoroughly investigated, underscoring its significance in the field of nanomaterials.

Copper oxide has a wide range of applications in many fields, such as electronic devices: Copper oxide is used as a semiconductor material for the manufacture of electronic devices such as diodes, solar cells, photodetectors, etc. Ceramics and glass: Copper oxide is used in the preparation of ceramic pigments and glass colorants, giving them color and special optical properties. Chemical catalysts: Copper oxide is used as a catalyst to participate in chemical reactions such as gas cleaning and organic synthesis. Drugs: Copper oxide nanoparticles are used to prepare drug delivery systems that help control the rate of drug release. Superconductor research: Copper oxide is a component of some high-temperature superconductors used to study superconducting properties. The research of copper oxide materials is still evolving, especially in the fields of semiconductor devices, catalysts, and materials science.

8.3.1.4 Nickel Oxide

Nickel oxide (NiO), a nickel derivative, presents as a solid ranging in color from green to black, typically found in powder form, though available in bulk as well. Characterized by its insulating properties, it exhibits high resistivity, inherently impeding electrical conduction. Despite its default insulating nature, nickel oxide's conductivity can be modified through doping and temperature adjustments, allowing it to transition into a semiconductor or conductor at elevated temperatures. This adaptability hints at its potential utility in batteries and sensors. As a p-type semiconductor with a substantial band gap (3.6–4.2 eV), nickel oxide has attracted considerable attention in research, particularly in the development of 0–3D nanostructures or heterostructures, for its application in gas sensing, especially targeting hazardous and toxic gases (Nikolic et al. 2020). Nickel oxide can be prepared by chemical deposition, thermal decomposition, solid phase reaction, and other methods.

The preparation method of nanometer nickel oxide has also been widely studied for the application of nanomaterials. Nickel oxide has a wide range of applications in several fields, such as electronic devices: Although it is itself an insulator, nickel oxide can be used as part of magnetic storage media, such as the magnetic layer of disk drives. Battery material: Nickel oxide can be used to prepare Ni-MH batteries and lithium-ion batteries of the positive electrode material. Catalyst: Nickel oxide is used as a catalyst in chemical reactions to facilitate processes such as gas cleaning and hydrogen production. Ceramics and glass: Nickel oxide is used in the preparation of ceramic pigments, ceramic glass, amber glass, and other materials. Thermistors: Nickel oxide is used in thermistors whose resistance values change with temperature and can be used for temperature measurement. Research on nickel oxide materials is still evolving, especially in the areas of battery technology, catalysts, and electronic devices.

8.3.1.5 Iron Oxide

Iron oxide encompasses a range of compounds including FeO, Fe2O3, and Fe3O4, each distinct in composition and application. FeO or ferrous oxide is a black powder predominantly seen at high temperatures and under reducing conditions, associated with the ferrous ion (Fe^2+). Fe2O3, known as triiron oxide, exhibits a common reddish-brown hue, representing a prevalent form of iron oxide found naturally, for instance in ochre, and is comprised of ferric (Fe^{3+}) ions. Fe3O4, or magnetite, is characterized by its black color and magnetic properties, containing both Fe^{2+} and

Fe^{3+} ions, making it pivotal in magnetic endeavors and magnetic resonance imaging (MRI) in the medical field. The synthesis of iron oxide can be achieved through various methodologies, including chemical synthesis, thermal decomposition, and the hydrothermal method, highlighting its versatility and utility across multiple domains.

Iron oxide serves a multitude of purposes across various domains. Commonly employed as a pigment, it imparts red, brown, yellow, and other hues to materials, finding extensive use in construction, coatings, ceramics, and colored concrete. Fe_3O_4, known for its magnetic qualities, plays a crucial role in magnetic storage media, magnetic separation, and medical magnetic resonance imaging (MRI). In the realm of chemistry, iron oxide acts as a catalyst, facilitating gas purification and organic synthesis. In medical applications, magnetic Fe_3O_4 is instrumental in drug delivery, magnetothermal therapy, and more. Additionally, certain iron oxide variants are being explored as potential positive electrode materials in lithium-ion batteries, noted for their high specific capacity and longevity.

8.3.1.6 Chromium Oxide

Chromium oxide, with the chemical formula Cr_2O_3, is a significant inorganic compound manifesting as a green solid, often presenting in shades from dark to vibrant green. It functions as a semiconductor material, characterized by relatively low conductivity, which can transition to conductive at elevated temperatures. Its conductivity is adjustable through doping and temperature modulation. Additionally, Cr_2O_3 is recognized as a p-type wide bandgap semiconductor, boasting a bandgap of 3.4 eV, and has been extensively researched for its gas sensing capabilities. Recent studies have particularly highlighted its effectiveness in the enhanced detection of toluene, a neurotoxic volatile organic compound (VOC), utilizing Cr_2O_3 microspheres (Nikolic et al. 2020). Chromium oxide can be prepared by chemical synthesis, thermal decomposition, REDOX, and other methods.

Chromium oxide boasts a versatile array of applications across numerous fields, highlighting its importance and multifunctionality. As a key component in magnetic materials, it is utilized in the creation of tapes, disks, and similar items, valued for its strong magnetic properties. In the realm of coloration, dark green chromium oxide serves as a pigment, widely used in paints, plastics, and ceramics, contributing vibrant hues and durability. For magnetic storage solutions, it plays a critical role in the magnetic layers of disk drives, enhancing data storage capabilities. In chemical processes, chromium oxide is employed as a catalyst, facilitating organic synthesis and gas purification, showcasing its chemical versatility. Additionally, in the steel manufacturing industry, it is instrumental in smelting processes, effectively removing impurities such as sulfur. This wide-ranging utility of chromium oxide, from magnetic materials and pigments to catalysts and beyond, underscores its significant role in various technological and industrial applications.

8.3.2 Electrochemical Sensing Materials

Electrochemical sensing materials represent a crucial category of substances extensively utilized in electrochemical sensors. These materials are distinguished by their unique electrochemical properties, enabling them to detect, measure, and monitor the concentration of chemical substances in liquids or gases as well as reaction rates and other critical parameters. Their wide-ranging applications span scientific research, industrial processes, and medical diagnostics. Electrochemical sensors offer a significant advantage over alternative methods; they allow the electrodes to sense materials present within the host system without inflicting any damage. This non-invasive sensing capability is pivotal for applications requiring integrity and preservation of the host system (Yogeswaran and Chen 2008). The operational principle of electrochemical sensing materials hinges on their electrochemical interactions with the target molecule or ion. These interactions trigger alterations in electrochemical signals, such as current, voltage, or resistance. Such changes are directly linked to the concentration or reaction extent of the target analyte, providing precise measurements. The primary categories of electrochemical sensing materials encompass ion-selective electrodes, working electrode materials, and electron conduction materials. Notably, common materials used include durable film electrodes, REDOX mediators, metals and alloys, and carbon-based substances, such as carbon nanotubes and graphene. These materials find extensive application across a broad spectrum of fields, from environmental monitoring and biomedicine to food safety and industrial process control. For instance, electrochemical sensors are adept at detecting heavy metal ions in water, measuring glucose levels, identifying biochemical molecules in blood, and more, showcasing their critical role in modern analytical and diagnostic processes.

8.3.2.1 Solid Electrolyte

Solid electrolytes play a pivotal role in the development of devices like ion-selective electrodes, solid-state batteries, and electrochemical sensors. In these applications, the solid electrolyte undergoes a reaction with a gas, leading to a change in current. The spectrum of solid electrolytes encompasses oxide electrolytes, sulfide electrolytes, oxide-sulfide mixed electrolytes, phosphate electrolytes, boron oxide electrolytes, nitride electrolytes, and composite electrolytes.

Oxide electrolytes, for example, include alumina, known for its use in high-temperature fuel cells and sensors due to its high thermal stability. Zirconia is another oxide electrolyte, favored in solid-state oxygen sensors and high-temperature battery applications for its excellent ionic conductivity. Magnesium oxide is utilized in high-temperature sensors and batteries as well.

Sulfide electrolytes feature materials like sodium sulfide, which are employed in sodium-ion batteries and can function at medium to high temperatures and calcium sulfide, known for its high ionic conductivity and use in solid-state batteries and sensors.

Oxide-sulfide hybrid electrolytes, such as magnesium selenide oxide mixtures, are used in solid-state batteries, merging the benefits of both oxides and sulfides.

In the category of phosphate electrolytes, lithium phosphate glass solid electrolytes stand out for their use in lithium-ion batteries, attributed to their high chemical stability. Boron oxide electrolytes include lithium boron oxide glass solid electrolytes, which find applications in lithium-ion batteries and sensors, offering high ionic conductivity. Lastly, nitride electrolytes, like boron nitride, are utilized in high-temperature batteries and sensors, celebrated for their high thermal stability, showcasing the diverse and crucial role of solid electrolytes in advancing electrochemical device technology.

8.3.2.2 Catalytic Electrode

In electrochemical sensing, the catalytic electrode plays a critical role in facilitating the electrochemical reaction, essential for the detection of specific molecules or ions. The platinum electrode, renowned as a highly common catalytic electrode material, is instrumental in oxygen reduction reactions (ORR), hydroxide reduction reactions (HER), and various other electrochemical processes. Gold electrodes are preferred in biosensors for their biocompatibility and stability, making them ideal for detecting biomolecules like glucose, DNA, and proteins. The glassy carbon electrode stands out as a versatile option, employed across a wide range of electrochemical reactions, praised for its excellent chemical stability.

Carbon-based nanomaterials, including graphene and carbon nanotubes, are frequently utilized in catalytic electrodes due to their high surface area and electrical conductivity, enhancing the efficiency of the electrochemical reactions. Transition metal compounds, such as tungsten oxide and cobalt oxide, are also employed in oxygen reduction reactions among other processes, owing to their effective catalytic properties. Conductive polymers, like polyaniline and polycarbazole, have found their place in constructing catalytic electrodes for use in sensors and energy storage devices, showcasing the diverse array of materials that contribute to the advancement of electrochemical sensing technology.

8.3.3 Optical Sensing Materials

Optical sensing materials are a class of materials used in optical sensors and optical devices that utilize optical properties to detect, measure, and monitor changes in various physical, chemical, and biological parameters. These materials are widely used in many fields such as scientific research, medical diagnosis, environmental monitoring, communication technology, and industrial fields. Their working principle is based on their interaction with light. The properties of these materials such as absorption, scattering, transmission, and refraction of light are affected by changes in physical or chemical parameters, resulting in changes in the properties

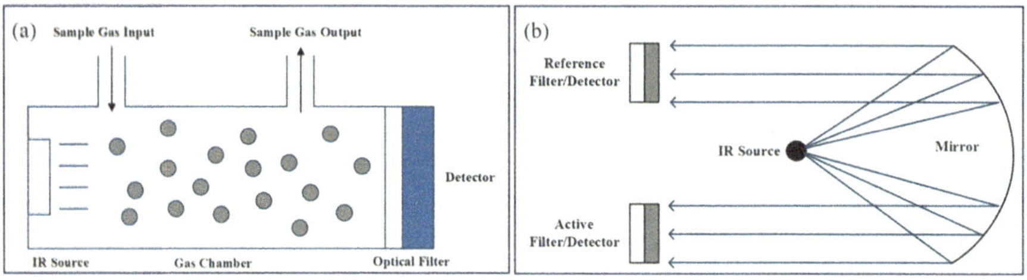

Fig. 8.4 IR-source gas sensors (**a**) based on the basic absorption spectrometry and (**b**) with reference filter/detector

of light. These changes can be measured and correlated with changes in the parameters. There are many types of optical sensing materials, including fluorescent materials, optical fibers, gratings, Surface Plasmon Resonance (SPR) materials, photonic crystals, etc. Each type of material has different properties and applications. Optical sensing materials play an important role in various applications. For example, they are used in biosensors that can detect the presence and concentration of biomolecules such as proteins, DNA, etc. In environmental monitoring, it is used to detect water quality, atmospheric composition, etc. In the field of communication, it is used for signal detection in optical fiber communication systems. Gas sensing by optical methods is usually straightforward and could achieve higher sensitivity, selectivity, and stability than non-optical methods with much longer lifetime. Their response time is relatively short, which enables online real-time detection. The performance will not be deteriorated by the changing environment or catalyst poisoning caused by specific gases, etc. Optical methods for gas sensing are mostly based on spectroscopy (Liu et al. 2012). Figure 8.4 shows a schematic diagram of the sensing mechanism of infrared light sources in gas sensors.

8.3.3.1 Absorbent Dyes

Absorbent dyes are a class of compounds that have the ability to absorb specific wavelengths of light and convert their energy into other forms, such as thermal energy or fluorescence. Light-absorbing dyes usually exhibit absorption properties over a specific wavelength range. This property allows them to selectively absorb light in specific optical applications. Light-absorbing materials are widely used in life science, chemical analysis, optoelectronics, and materials science.

Common light-absorbing dyes include Rhodamine B, a fluorescent light-absorbing dye commonly used in biomarkers, fluorescence microscopy, and cell imaging; Bromophenol Blue, a light-absorbing dye used in biological experiments, often used in protein electrophoresis gel staining; Methyl Orange, an organic dye sensitive to acid-base changes, is widely used as an acid-base indicator; Nile Red, a fluorescent light-absorbing dye commonly used for labeling and imaging lipids and oils; Photovoltaic Blue, an organic semiconductor light-absorbing dye used in organic solar cells; Lissamine Rhodamine B, a fluorescent light-absorbing dye commonly used in biomarkers and cell imaging; Malachite Green, an organic dye widely used in biological staining and tissue microsectioning; Fresh Orange G (Acridine Orange), a fluorescent light-absorbing dye, can be used to stain nucleic acids and cells; New Methylene Blue, a light-absorbing dye used for staining blood cell counts; Rhodamine 6G (Rhodamine 6G), a strong fluorescent absorbent dye, is commonly used in Raman spectroscopy and laser labeling.

8.3.3.2 Fluorescent Dyes

Fluorescence involves the absorption of ultraviolet (UV) or visible light and emission of light at higher wavelengths. Fluorescent dyes are defined as compounds which both absorb and emit strongly in the visible region, and which owe their potential for application to their intense fluorescence properties (Christie 2011). Fluorescent dyes are a class of compounds that can absorb light and emit fluorescence, and are often used in biomarkers, cell imaging, molecular probes, fluorescence microscopy, and other fields. Through the fluorescence effect, when the gas interacts with the material, the fluorescence signal changes. This material can be used for highly sensitive and selective gas sensing.

Common fluorescent dyes include FITC (fluorescein isothiocyanate), a green fluorescent dye widely used in immunofluorescence labeling and cell imaging; Rhodamine (Rhodamine series), including Rhodamine B, Rhodamine 6G, etc., with red to orange fluorescence, often used in fluorescence microscopy and cell staining; Cy series dyes (Cyanine dyes), including Cy3, Cy5, etc., have different wavelengths of fluorescence, suitable for molecular probes and biomarkers; DAPI (4′,6-diamidi no-2-phenylindole), a blue fluorescent dye used for DNA staining and nucleic acid labeling; SYBR Green, a nucleic acid fluorescent dye commonly used in gel electrophoresis and real-time fluorescence quantitative PCR; Fluorescein, a yellow-green fluorescent dye, is widely used in biomarkers, cell tracking and fluorescence microscopy; The Alexa Fluor series, including Alexa Fluor 488, Alexa Fluor 594, is an excellent series of fluorescent dyes for multi-color labeling and imaging. GFP (Green Fluorescent Protein), a fluorescent protein marker, is widely used in gene expression and cell tracking; Texas Red, a red fluorescent dye used for cell staining and fluorescence microscopy; Calcein, a green fluorescent dye, is commonly used for cell viability detection and cell labeling.

8.3.3.3 Quantum Dots

Quantum dots are nanoscale semiconductor particles with special optical and electrical properties. Quantum dots (QDs) are nanometer-scale semiconductor crystals composed of groups II–VI or III–V elements, and are defined as particles with physical dimensions smaller than the exciton Bohr radius. When a photon of visible light hits such a semiconductor, some of their electrons are excited into higher energy states. When they return to their ground state, a photon of a frequency characteristic of that material is emitted (Jamieson et al. 2007). They typically consist of a few hundred to a few thousand atoms and are between 1 and 10 nanometers in size. Quantum dots are very small in size, so they exhibit a size effect that causes their optical and electrical properties to differ greatly from those of larger materials. The electrons in a quantum dot are confined to three-dimensional space and exhibit quantum effects. This includes discretization of energy levels and quantization of electronic transitions. Quantum dots have strong absorption and emission of specific wavelengths of light and have controllable optical properties. They can emit a specific color of fluorescence, depending on their size. By adjusting the size of the quantum dot, the wavelength of its light can be adjusted. This makes them ideal materials for multi-color labeling and imaging. Quantum dots have semiconductor properties and can be used in electronic devices. They have potential applications in solar cells, light-emitting diodes (LEDs), and more.

8.3.3.4 Fluorescent Protein

Fluorescent proteins are a class of proteins that have the property of emitting light spontaneously and are commonly used for biomarkers and cell imaging. These proteins are able to fluoresce when excited by specific wavelengths of light, allowing researchers to visualize cells, molecules, and biological processes.

Common fluorescent proteins include GFP (green fluorescent protein), a protein derived from jellyfish that emits green fluorescence and is widely used for gene expression and cell tracking; RFP (Red fluorescent protein), including DsRed, mCherry, etc., emits red fluorescence. For multi-color labeling and deep tissue imaging; YFP (Yellow fluorescent protein), which emits yellow fluorescence, is often used in multi-color labeling experiments with other fluorescent proteins; BFP (Blue fluorescent protein), which emits blue fluorescence for multi-color labeling and fluorescence imaging at specific wavelengths; CFP (Cyan fluorescent protein), which emits cyan fluorescence and is often used in multi-color labeling experiments with other fluorescent proteins; mTurquoise, a fluorescent protein that emits blue to green and has a high brightness; TagRFP, a red fluorescent protein, has rapid maturation and high stability; mNeonGreen, a protein that emits green fluorescence and has high brightness and light stability; tdTomato, a red fluorescent protein widely used in cell tracking and imaging studies; Kaede, which has unique reversible fluorescent discoloration properties, can change from green to red.

8.3.3.5 Photosensitive Materials

Photosensitive materials are a class of materials that are sensitive to light, and they undergo physical or chemical changes after exposure to light. This variation can be used for a variety of applications, including in the fields of optics, optoelectronics, medicine, and printing.

Common photosensitive materials are silver salt photosensitive materials, including silver bromide film, silver halide, and so on. Commonly used in photography and printing, silver salts undergo chemical changes after light irradiation to form images; Photochromic materials, including photochromic glass, photochromic coatings, etc., used in solar-regulated glass and optical devices to change color or transparency after exposure to light; Photosensitive polymer: A photosensitive polymer that can be crosslinked or depolymerized by ultraviolet or visible light. It is widely used in lithography and 3D printing. Organic photosensitive materials, including organic photosensitizers, photosensitive dyes, etc. Used in printing, laser printing, and photosensitive paper; Indium selenide photosensitive material, used for laser printer photosensitive drum, after laser irradiation to change the charge properties, used for image formation; Photosensitive ceramics, ceramic materials with phototelegraphic properties, used in laser communication and optoelectronics devices; Photosensitive film, a photosensitive thin film material used in photoelectric sensors, solar cells. and display technology; Silicon photosensitive devices, silicon chips with photosensitivity, used in applications such as cameras, photodiodes, and photoresistors; Organic photosensitive films, including organic semiconductor materials, for use in organic optoelectronics and flexible electronic devices; Photosensitive fiber, a photosensitive optical fiber material, used in optical fiber communication and sensor technology.

8.3.4 Acoustic Sensing Materials

Acoustic sensing materials are a class of materials used in acoustic sensors and acoustic devices that are capable of sensing the propagation, transfer, and change of sound waves for detection, measurement, and monitoring of changes in sound, vibration, pressure, and acoustic parameters. These materials have a wide range of applications in many fields, including acoustics, communications, medical, and industrial fields. It works based on their sensitivity to sound waves. When the sound wave propagates to the surface of the material, the physical properties of the material, such as resistance, piezoelectric, piezoresistive, or surface wave propagation speed, will change accordingly with the change of the pressure, frequency, or amplitude of the sound wave. These changes can be measured and used to detect acoustic parameters. The main types of acoustic sensing materials include piezoelectric materials, piezoresistive materials, Surface Acoustic Wave (SAW) materials, acousto-optic materials, etc. Acoustic sensing materials are widely used in various application fields. For example, piezoelectric materials are often used in ultrasonic sensors and sound sensors for medical diagnosis, distance measurement, etc. Piezoresistive materials are used in sound sensors and vibration sensors for structural health monitoring and industrial automation; Surface acoustic wave materials are used in wireless communications and frequency controllers. Gas sensors based on acoustic transduction can be manufactured in submillimeter size and detect the mass of the analytes. Furthermore, acoustic sensors are fast, highly sensitive, cost-effective (cost per sample test), and label-free sensors. Besides, acoustic sensors can effortlessly integrate with wireless communication systems and all these inherent characteristics make them universal in use (Zhang et al. 2020b).

8.3.4.1 Piezoelectric Materials

The Curie brothers (Pierre and Jacques Curie) discovered direct piezoelectric effect in single-crystal quartz in 1880. Under pressure, quartz generated electrical charge/voltage from quartz and other materials (Uchino 2017). Piezoelectric materials are a class of materials that have the piezoelectric effect, which is capable of generating electric charge or the movement of electric charge when subjected to mechanical stress or pressure. This effect makes piezoelectric materials widely used in sensors, vibrators, ultrasonic equipment, and other fields. Piezoelectric materials also reflect changes in gas concentration by measuring changes in the speed or frequency of sound waves.

Common piezoelectric materials are crystalline piezoelectric materials, including quartz, zirconia, ammonium borate, and so on. These crystals have good piezoelectric properties and can be used in high-frequency vibration and ultrasonic equipment. Ceramic piezoelectric materials, such as lead zirconate titanate (PZT), and lithium niobate ($LiNbO_3$) are common ceramic piezoelectric materials, widely used in sensors and piezoelectric transducers; Polymer piezoelectric materials, polyvinylidene fluoride (PVDF), and its copolymers are polymers with piezoelectric properties, suitable for flexible sensors and biomedical devices; The composite material, which is composed of piezoelectric ceramics and polymer materials, has certain flexibility and high sensitivity; Organic piezoelectric materials, some organic compounds, such as piezoelectric polymers, exhibit piezoelectric properties and are suitable for flexible electronic devices and biosensors; Borate piezoelectric materials, such as ammonium borate, lithium borate, and other borate materials show good piezoelectric properties in the low frequency range, and are often used in sensors and vibration devices; Piezoelectric glass, some special composition of glass materials also show piezoelectric properties, can be used in some special applications; Single crystal piezoelectric materials, materials with a single crystal structure by growing, such as lead zirconate (PZ) single crystals, have high sensitivity and temperature stability.

8.3.4.2 Piezoresistive Materials

Piezoresistive materials are defined by their ability to change resistance when mechanical pressure or deformation is applied. This characteristic is widely leveraged in the development of sensors for detecting pressure, touch, and force. These materials are engineered for a diverse array of applications, including but not limited to touchscreen technologies, medical devices, human-computer interactions, and flexible electronics. The operational principle behind piezoresistive sensors involves the piezoresistive effect, where the conductance of the sensing material alters in response to applied strain, and the resistance of the sensor assembly changes due to geometric modifications upon deformation. The design of high-performance piezoresistive sensors takes into account a multitude of parameters such as stretchability, sensitivity, dynamic range, detection threshold, accuracy, response time, stability, durability, along with the cost and simplicity of fabrication (Irani et al. 2022).

Common piezoresistive materials are conductive rubber, rubber materials containing conductive particles or conductive polymers, which are widely used in flexible sensors, touch screens, and health monitoring; Carbon black composite material, an elastic matrix containing carbon black particles, changes in resistance due to pressure changes. Commonly used in pressure sensors and touch technology; Silicon piezoresistive materials, materials containing silicon particles or silicon micro-nano structures. Applications in flexible electronic devices and human-computer interaction; Conductive foam, a material with a conductive foam structure, which can be used to manufacture flexible sensors and pressure-sensitive materials; Polymer piezoresistive materials, including materials with conductive particles added to a polymer matrix, suitable for flexible pressure sensors and touch technologies; Conductive paper, a paper-like material consisting of conductive fibers or conductive particles for flexible sensors and pressure sensitive applications. Metal nanowire network, a network structure formed by metal nanowire, can sense the change of pressure through the change of resistance between the nanowire; Pressure-sensitive film, film material with pressure-sensitive function, used in the manufacture of flexible pressure sensors and touch screens. Aerogel piezoresistive material, piezoresistive material with aerogel structure, which can be used in flexible sensors and pressure-sensitive equipment; Flexible conductive fabrics, fabrics containing conductive fibers or conductive coatings, are used to manufacture flexible sensors and wearable devices.

8.3.4.3 Surface Wave Materials

Surface wave materials constitute a category of materials capable of supporting the propagation of surface waves. They are pivotal in the realms of acoustic and optical sensing, communications, and signal processing. Predominantly made up of quartz, lithium tantalate ($LiTaO_3$), lithium niobate ($LiNbO_3$), and other crystalline substances, surface acoustic wave (SAW) materials find extensive applications in acoustic sensors, radio frequency controllers (RF filters), and non-destructive evaluation techniques.

8.3.4.4 Acousto-Optic Crystals

Acousto-optic crystal is a kind of crystal material which can realize acousto-optic interaction. This interaction usually involves energy conversion between sound and light waves. Acousto-optic crystals have important applications in the field of optics and communication, and are often used in the manufacture of acousto-optic modulators, acousto-optic interferometers, and other devices. In acousto-optic crystals, the energy of sound wave and light wave is converted through the coupling of lattice vibration and electromagnetic wave. This coupling can be achieved through mechanisms such as the piezoelectric effect or the photoelastic effect. Common acousto-optic crystal materials include lithium niobate ($LiNbO_3$), manganese niobate (Mn: $LiNbO_3$), sodium niobate ($NaNbO_3$), and so on. These materials have specific crystal structures and acousto-optic properties. Acousto-optic crystals are widely used in communication systems to make acousto-optic modulators. They are also used to build optical interferometers, frequency-selective elements in lasers, and other optical and optoelectronics devices.

8.3.4.5 Multi-iron Materials

Multiferroic materials refer to materials with multiple ferroelectric and ferromagnetic properties at the same time. These materials exhibit simultaneous ferroelectric and ferromagnetic phases, giving them unique potential in the regulation of magnetic and electrical properties. Multiferroic materials are ferroelectric, that is, they can be electrically polarized under an external electric field. This makes their application in electronic devices potentially advantageous. Polyferric properties are usually achieved by controlling the crystal structure and chemical composition. Regulating these factors can change the ferroelectric and ferromagnetic properties of materials, providing the possibility of achieving multifunctional properties. The unique properties of multiferroic materials make them potentially useful in many fields, including electronic devices, sensor technology, magnetic storage, and energy conversion. Some common multiferroic materials include ferroelectric ferromagnetic oxides, ferroelectric ferromagnet fluoride, etc. These materials are usually composed of transition metal ions and have a complex crystal structure. A standout among the most concentrated and promising MF materials to date are the rare-earth ferrites and manganites which exhibit extraordinary properties like giant dielectric effect and colossal magnetoresistance. This makes them suitable for sensing, memory devices, and other ME applications (Kumar et al. 2020).

8.4 Bio-sensing Materials

Biological sensing materials are biological recognition elements that act on the sensor substrate. They can be enzymes, DNA, immune cells, metabolic biochemical substances, etc. With the continuous development of sensing technology, the advancement in synthetic techniques has led to increasingly diverse designs of biosensors. The incorporation of biorecombinant technology has further expanded the infinite possibilities for biosensor design in the laboratory (Fig. 8.5).

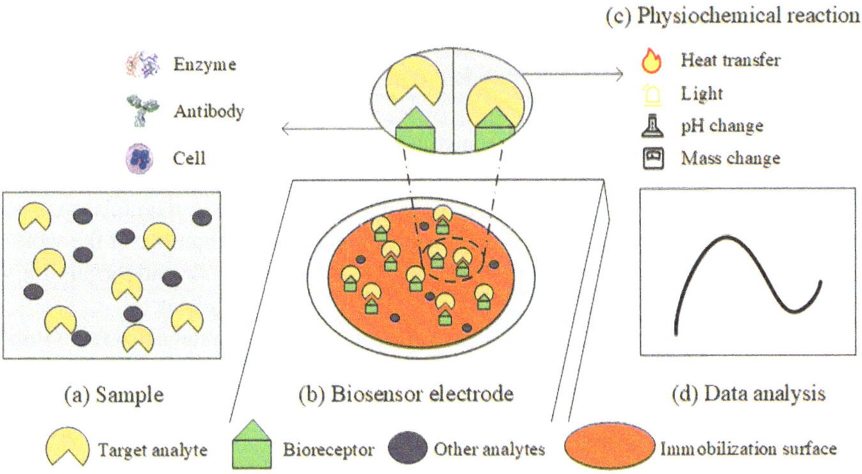

Fig. 8.5 Represents a schematic diagram of a typical biosensor system architecture. (**a**) Sample; (**b**) Biosensors electrode; (**c**) Physiochemical reaction; (**d**) Data analysis

8.4.1 Enzyme

Enzymes, serving as natural biocatalysts, exhibit remarkable selectivity and specificity, rendering them exemplary materials for bio-sensing applications. Their deployment spans numerous fields, and they are classified into three main categories according to their functional roles in sensors: oxidase, reductase, and peroxidase. This classification underscores the versatility and critical importance of enzymes in the realm of biosensor technology, highlighting their unique capabilities in facilitating accurate and sensitive detection processes.

8.4.1.1 Oxidase

Oxidase catalyzes substrate oxidation, facilitating the transfer of electrons from the substrate to the oxidant. This process often involves the concurrent reduction of oxygen molecules. Enzymes in this category include sugar oxidase, amino acid oxidase, alcohol oxidase, and amine oxidase, classified based on their fundamental types.

Sugar oxidase facilitates the oxidation reaction of the CH-OH bond in a substance, leading to the substance's oxidation. Common sugar oxidases include glucose oxidase, lactose oxidase among others. In a study by Liu et al. (2005), a glucose sensor was proposed. By coupling glucose oxidase (GOD) with a carbon nanotube/chitosan composite material, the GOD undergoes a direct electron transfer reaction with the electrode. This results in a significantly improved electron transfer rate of GOD in the system, reaching 7.73 s^{-1}, more than twice that of flavin adenine dinucleotide (3.1 s^{-1}) adsorbed on carbon nanotubes.

Moreover, the biological activity of the modified GOD within the electrode surface composite material was retained. As a result, the prepared electrode can function as a glucose biosensor with higher sensitivity (0.52 $A \cdot mM^{-1}$) and improved stability.

In the study conducted by Aila Riany de Brito et al., an electrochemical method for the quantitative detection of lactose in skim milk was introduced (de Brito et al. 2021). The approach involved immobilizing lactase in a carbon nanotube matrix-carbon paste. The resulting paste was deposited in the lower chamber of the working electrode, which, when combined with the reference electrode and the auxiliary electrode, formed an electrochemical cell operated by a potentiostat.

The system exhibited a sensitivity of 1.06 $\mu A\ cm^2\ mmol\ L^{-1}$ and a detection limit of 0.15 $mmol\ L^{-1}$ for lactose. It demonstrated ease of manufacturing, operation, low detection limit, and high reliability.

Amino acid oxidase plays a role in catalyzing the oxidation of amino acids, contributing to the decomposition of amino acids and participating in the energy production process. Common amino acid oxidases include branched-chain amino acid oxidase, cysteine dioxygenase, lysine oxidase, and phenylalanine hydroxylase.

In the study by Tian et al., a non-meso electrochemical biosensor based on DAAO (D-amino acid oxidase) was constructed for the chiral recognition of DAA (D-amino acids) without involving a complicated electron transfer process (Tian et al. 2020). Chiral recognition was achieved through the specific catalysis of DAAO to DAA. This catalytic effect led to a linear decrease in the reduction peak current with an increase in DAA concentration. Furthermore, the biosensor exhibited high resistance to LAA (L-amino acids) interference, indicating that the combination of stereospecificity and catalytic reaction significantly improved chiral recognition efficiency.

Alcohol oxidase catalyzes the oxidation reaction of alcohols, aiding in alcohol decomposition and participating in energy production. Common alcohol oxidases include alcohol dehydrogenase, glycerol dehydrogenase, and ethylene glycol dehydrogenase. In the study by Eimantas Ramonas et al., a highly sensitive amperometric glycerol biosensor was developed (Ramonas et al. 2019). Pseudomonas putida alcohol dehydrogenase was immobilized on a graphite electrode modified with carbon nanotubes and the redox mediator tetrathiafulvalene. The designed biosensor exhibited exceptional sensitivity (29.2 ± 0.9 $\mu A\ mM^{-1}\ cm^{-2}$), a low detection limit (18 μM), a linear range of 0.05 to 1.0 mM, high selectivity, and satisfactory stability for glycerol. It was successfully employed to determine glycerol concentrations in buffer solutions and human urine samples.

Amine oxidases catalyze the oxidation of amines. Common amine oxidases include phenylethanolamine oxidase, adrenaline oxidase, and phenylethylamine oxidase. In the study conducted by Ceren Kaçar et al., a biogenic amine biosensor was reported. This biosensor, based on Prussian blue (PB) and indium tin oxide nanoparticles (ITONP) modified screen-printed carbon electrode (SPCE), was further modified by diamine oxidase (DAO) or monoamine oxidase (MAO) enzymes (Kacar et al. 2020). Histamine in the concentration range of 6.0×10^{-6} to 6.9×10^{-4} M exhibited the highest response with a sensitivity of 1.84 $\mu A\ mM^{-1}$.

8.4.1.2 Reductase

Reductases constitute a class of enzymes facilitating chemical reduction reactions, playing pivotal roles in various life processes. They contribute to redox reactions, aiding in cell metabolism, energy production, and detoxification pathways to eliminate harmful oxidizing substances. Common reductases include nitrate reductase, sulfate reductase, and NADH-dehydrogenase.

Nitrate reductase is involved in reducing nitrate to nitrite or ammonia. In the human body, it primarily participates in nitrate metabolism. Nitrate is converted by nitrate reductase into nitrite or ammonia, holding significance for the body. Nitrite, when reacting with amine compounds, can form carcinogenic nitrosamines. Thus, nitrate reductase plays a detoxification role in nitrate metabolism. Additionally, nitrate reductase contributes to the synthesis of nitric oxide, which has vital physiological functions such as regulating vascular tone and preventing thrombosis. Palraj Kalimuthu et al. proposed a nitrate biosensor based on nitrate reductase operating at physiological pH (Kalimuthu et al. 2015). The catalytic current showed linearity within the nitrate concentration range of 10–400 nM with a correlation coefficient of 0.989. Through amperometry, a low detection limit of 0.76 nM ($S/N = 3$) was achieved. The electrochemical biosensor's practical application demonstrated accurate nitrate concentration determination in natural water samples, aligning well with standard spectroscopic methods.

Sulfate reductase, a vital enzyme in living organisms, plays a crucial role in the reduction reaction of sulfate, converting it to sulfide. In the human body, sulfate reductase is involved in the sulfur metabolism pathway, contributing to the reduction of sulfate to sulfide. Sulfide, with essential biological functions, participates in the formation of amino acids, coenzyme A, glutathione, and other compounds. Additionally, sulfate reductase is integral to detoxification pathways, aiding in the elimination of harmful sulfur compounds. Consequently, sulfate reductase is instrumental in maintaining the balance and metabolism of sulfur in the human body.

NADH-dehydrogenase, a pivotal enzyme within cells, participates in the oxidation reaction of NADH to NAD^+, releasing hydrogen atoms and electrons. This process is a key step in the cellular respiratory chain, where the produced NAD^+ can continue to engage in other metabolic reactions, providing the necessary energy for the cell. Therefore, NADH-dehydrogenase is indispensable for maintaining intracellular energy balance and normal metabolic processes. P.N. Bartlett et al. demonstrated the direct electrochemical oxidation of h-nicotinamide adenine dinucleotide (NADH) on a clean electrode, highlighting the formation of a radical cationic intermediate at high overpotentials and rapid fouling (Bartlett et al. 2002) .

8.4.1.3 Peroxidase

Peroxidase is a type of enzyme that catalyzes reduction reactions and plays a variety of important roles in organisms. It is mainly involved in the decomposition of hydroperoxides, reducing them to water and oxygen, thereby removing harmful peroxides. In addition, peroxidase is also involved in the regulation of intracellular redox balance and plays an important role in the stability of the intracellular environment. Common peroxidases include horseradish peroxidase and red blood cell peroxidase.

Horseradish peroxidase (HRP) is an enzyme found in plants such as horseradish. It acts as an antibacterial and antifungal agent in plants, helping them fight off pathogens. This enzyme is also used in the food industry to prevent food from spoiling. In humans, horseradish peroxidase may be helpful as an antibacterial and anti-inflammatory agent. Some research suggests horseradish peroxidase may help the body's immune system, but more research is needed to determine its exact role. In the study of Yang et al. (2006), a layer-by-layer self-assembly of horseradish peroxidase (HRP) multilayers constructed on thiol-modified gold surfaces was proposed and used to detect phenolic compounds. The response time of this biosensor to phenolic compounds is very short, reaching 95% of the maximum response in about 2 s.

Glutathione peroxidase is an enzyme present in red blood cells. It is mainly involved in regulating the concentration of intracellular hydrogen peroxide, helping to protect red blood cells from oxidative stress and maintaining intracellular redox balance. In the research of Somaye Cheraghi et al. (2022), a new enzyme biosensor based on glutathione peroxidase (GSH-Px), graphene oxide (GO), and perfluorosulfonic acid was proposed for use in body fluids Electrochemical sensing of glutathione (GSH). And successfully monitored GSH in the range of 0.003–370.0 μM, with a detection limit of 1.5 nM using the differential pulse voltammetry (DPV) method.

8.4.2 *Metabolic Biochemistry*

8.4.2.1 Carbohydrate

Carbohydrates are a class of organic compounds that include sugars, cellulose, and starch. They are generally divided into three broad categories: monosaccharides, disaccharides, and polysaccharides. Monosaccharides are carbohydrates composed of one sugar molecule, including glucose, fructose, and galactose; disaccharides are carbohydrates composed of two sugar molecules, including sucrose, lactose, maltose, etc.; polysaccharides: are composed of multiple sugar molecules Composed of carbohydrates, including starch, cellulose and glycogen. Carbohydrate sensors typically utilize specific biorecognition elements or chemical recognition elements combined with sensor technology to convert the presence of carbohydrates into a measurable signal. Taking glucose as an example, Bai et al. (2008) developed an enzyme-free glucose sensor using a three-dimensional inverse opal gold film electrode (3DGFE), which was obtained by electrochemical deposition of gold in the gaps of a polystyrene template. This sensor achieves the purpose of glucose detection by detecting the decomposition of glucose at a voltage of −0.3 V.

8.4.2.2 Lipids

Lipids play a variety of roles in organisms. For example, fatty acids are oxidized and decomposed under aerobic conditions to produce a large amount of energy, which is used as an energy storage tool; lipids are an important part of cell membranes and participate in the construction and maintenance of organelles; in the skin A lipid film is formed on the surface to protect the skin, insulate and keep warm; as an important raw material for hormone synthesis, it synthesizes hormones such as adrenocortical hormones and sex hormones; as a vitamin absorption solute, it helps fat-soluble vitamins (such as vitamins A, D, E, K) absorption and transport. These effects indicate that lipids have important physiological functions in organisms and are essential for maintaining life activities and health. In the work of Xiao WU et al. (Wu et al. 2020), an electronic tongue with a lipid membrane was proposed for taste sensing, using optical methods to analyze the structure of the membrane surface to better understand the response mechanism.

8.4.2.3 Protein

Proteins are composed of amino acids, are the building blocks of tissue, and are involved in regulating and supporting life activities. As an important component of cells and tissues, proteins give cells and tissues their shape and structure; they can transport substances in the body, such as hemoglobin transports oxygen, apolipoprotein transports lipids, etc.; they can identify and neutralize pathogens and play an immune defense role; Muscle protein is an important component of muscle contraction and participates in body movements and activities. As a sensing material, proteins have diverse structures and functions and can be used in sensors through their specific biorecognition properties. Their advantages as sensing materials lie in their natural biocompatibility, high specificity, and reliability. Regulatory, these characteristics make protein sensors have broad application prospects in medical diagnosis, food safety, environmental monitoring, and other fields. In the research of Liyun Ma et al. (2021), a flexible humidity sensor based on wool keratin and a fully biomaterial was proposed and successfully used to detect human respiration.

8.4.2.4 Nucleic Acid

Nucleic acids play various roles in organisms. Nucleic acids are mainly divided into two categories: deoxyribonucleotides (DNA) and ribonucleotides (RNA). DNA is the carrier of genetic information in cells. It stores the genetic information of organisms through encoding of base sequences and passes it to the next generation of cells when cells divide. RNA (ribonucleic acid) plays an important role in the protein synthesis process, mRNA (messenger RNA) transcribes the genetic information in DNA into RNA, and then tRNA (transfer RNA) transports the amino acids to where the protein is being synthesized. The specific pairing and molecular recognition capabilities of DNA are used to detect specific molecules or intermolecular interactions. A common DNA sensor is based on the complementary pairing principle of DNA. By designing specific DNA sequences, they can be specifically complementary to target molecules, thereby producing measurable signal changes. This principle is widely used in genetic testing, pathogen detection, drug screening, and other fields. In addition, DNA can also be used as a nanomaterial to build nanosensors. The special properties of DNA nanostructures allow them to be used to design nanosensors with specific functions for detecting small molecules, ions, or other biological molecules.

Similar to DNA, RNA can also be used for biorecognition and signal transmission in sensors. RNA sensors utilize the specific pairing and molecular recognition capabilities of RNA to detect specific molecules or intermolecular interactions. Based on the principle of complementary pairing of RNA, specific RNA sequences are designed to specifically complementary pair with target molecules, thereby producing measurable signal changes. This principle is widely used in genetic testing, pathogen detection, drug screening, and other fields. In addition, RNA can also be used as a nanomaterial to construct nanosensors. The special properties of RNA nanostructures allow them to be used to design nanosensors with specific functions for detecting small molecules, ions, or other biomolecules.

8.4.3 *Immunization*

8.4.3.1 Antibody

Antibodies are important immune materials that can bind to specific antigens to form antigen-antibody complexes for detecting target molecules. Antibodies can be divided into monoclonal antibodies and polyclonal antibodies. Monoclonal antibodies are produced by a single B cell clone and are highly specific and consistent, while polyclonal antibodies are produced by multiple B cell clones and can bind to different parts of the target molecule.

Monoclonal antibodies are antibodies produced by a single B cell clone, which can specifically bind to a specific part of the antigen. Such antibodies are produced by the same B cell clone, so they have the same antigen-binding site, that is, specific antigen binding (Ecker et al. 2015). Monoclonal antibodies can be used to treat cancer, autoimmune diseases, and infectious diseases. Additionally, they can be used to diagnose and prevent disease. The specificity of monoclonal antibodies allows them to bind precisely to specific antigens, allowing them to modulate immune responses in a targeted manner or directly intervene in the development of disease.

Polyclonal antibodies are a mixture of antibodies produced by multiple different B cell clones (Haurum 2006). These antibodies can bind to different parts of the antigen because they come from different B cell clones and have different antigen-binding sites. Polyclonal antibodies are typically produced in immune responses and may vary in specificity and affinity. Polyclonal antibodies can recognize and bind to multiple different antigens, thereby helping the body fight infections and diseases; they can be used in diagnostic kits to help detect specific antigens or pathogens, for disease diagnosis and monitoring, etc.

8.4.3.2 Antigen

Certain biosensors use specific antigens to recognize and bind corresponding antibodies to detect target molecules. Antigens can be proteins, peptides, sugars, or other biological molecules and play an important role in vaccine development, immunotherapy, and disease diagnosis.

Proteins serve as antigens in the sense that they are recognized by the immune system and trigger an immune response. Proteins often have complex structures and diverse functions, so they may be recognized by the immune system as foreign substances or abnormal cells. When a protein is recognized as an antigen, the immune system produces antibodies that bind to it and initiate an immune response against the protein. Proteins as antigens play an important role in vaccine development, immunotherapy, and disease diagnosis (Pollet et al. 2021).

Peptides as antigens are short protein fragments composed of amino acids that can be recognized by the immune system and trigger an immune response. Polypeptides are usually components of exogenous antigens (such as bacteria, viruses, etc.) or endogenous antigens (such as tumor cells, abnormal cells, etc.). Once a polypeptide is recognized as an antigen, the immune system produces antibodies against the polypeptide and initiates a corresponding immune response. As antigens, peptides play an important role in vaccine development, immunotherapy, and disease diagnosis (Zhang et al. 2019).

Carbohydrates play an important role in the immune system as antigens. Carbohydrate antigens are usually present on the surface of cells and can be recognized by the immune system and trigger an immune response. These carbohydrate antigens are of great significance in disease diagnosis, vaccine development, and immunotherapy. For example, sugar structures on the surface of some pathogens can be recognized by the immune system as foreign substances, triggering an immune response. In addition, some carbohydrate antigens on the surface of tumor cells can also become targets for tumor vaccine development and immunotherapy (Mantuano et al. 2020).

Other biomolecules that serve as antigens include polysaccharides, lipids, nucleic acids, etc. These biomolecules may be recognized by the immune system and trigger an immune response. For example, bacterial polysaccharides and lipids can be recognized by the immune system as exogenous antigens, triggering corresponding immune responses. In addition, the abnormal structure or expression of some self-molecules may also make them targets of the immune system and cause autoimmune diseases. Therefore, not only proteins and peptides, but other biological molecules also play an important role in the immune system as antigens.

8.4.3.3 Biocompatible Polymer

Biocompatible polymers are a class of polymer materials that are compatible with biological tissues and do not cause significant immune rejection. These polymers generally have good biocompatibility and biodegradability. For example, hydrogel polymers, biocompatible polymer materials, and biomimetic materials are all biocompatible polymers.

Hydrogel polymers (Choudhury et al. 2009) have good water solubility and hydrogel properties, and can form a three-dimensional network structure in water. Common hydrogel polymers include gelatin, alginate, polyethylene glycol, etc., which are often used for the biorecognition layer or immobilization of biomolecules in biosensors.

Biocompatible polymer materials include polylactic acid (PLA), polylactic acid-co-glycolic acid (PLGA), polyvinyl alcohol (PVA), etc., which have good biocompatibility and degradability is often used to prepare carriers or embedding materials for biosensors (Chen et al. 2008).

Biomimetic materials are inspired by naturally occurring substances in living organisms and have similar structures and properties to biological tissues, such as biomimetic membranes, biomimetic nanoparticles, etc., and are often used for biorecognition and signal transduction in biosensors.

8.5 Flexible Electronic Materials

8.5.1 Flexible Electronics

Flexible electronics refers to a technology that fabricates electronic devices using organic/inorganic materials on flexible/stretchable substrates. Figure 8.6 shows the stretchability of flexible electronic materials. This emerging technology aims to produce electronic devices in flexible forms to adapt to surfaces with diverse shapes and curvatures.

The realization of flexible electronics relies on innovative materials and manufacturing technologies. Key aspects of flexible electronics technology encompass flexible substrate materials, bendable electronic devices, nanoelectronic materials, and manufacturing techniques. Technical principles involve the use of substrates with high flexibility, including flexible polymers, thin films, and novel nanomaterials. This encompasses the design and manufacture of bendable electronic devices, such as flexible batteries, sensors, and circuit boards. Leveraging nanotechnology enhances the elasticity and conductivity of materials. The development of manufacturing processes for flexible electronics includes printed electronics technology, flexible film preparation, and 3D printing.

In the realm of wearable devices, flexible electronics plays a crucial role in designing and manufacturing electronic devices that seamlessly conform to the

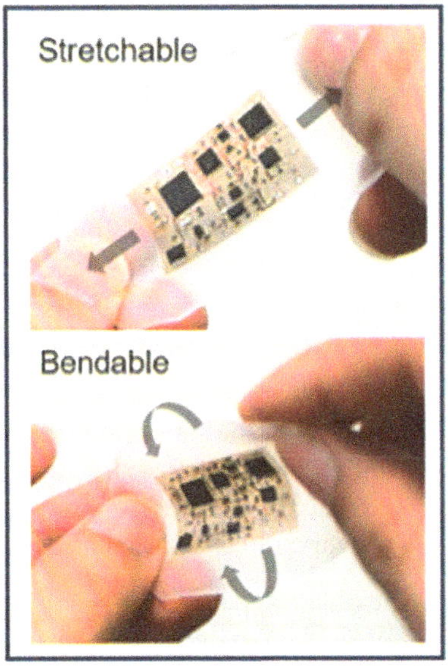

Fig. 8.6 Shows the stretchability of flexible electronic materials. (Reused with permission from Kim et al. 2021)

curves of the human body. The emergence of products like smart wristbands and smart glasses underscores the significance of flexible electronics in the wearable technology sector. In terms of flexible displays, curved, foldable, and rollable displays are pivotal developmental directions in next-generation display technologies, offering new possibilities in portability and user experience. In the field of medical health monitoring, flexible electronics find applications in medical sensors, wearable health monitoring devices, etc. The flexibility of these electronics allows them to conform to the human body surface, enabling real-time monitoring of physiological parameters and providing more accurate data for medical diagnosis and rehabilitation.

In the realm of smart skin, flexible sensors and electronic skin technology are utilized to develop smart skin with tactile sensing and adaptability. This technology is anticipated to find applications in fields like robotics and bionics. Future innovations in flexible electronics may focus on four key aspects:

1. Further development of materials with high conductivity, elasticity, and repeatability to meet the stringent requirements of flexible electronic devices.
2. Continuous optimization of manufacturing processes, including printed electronics technology and flexible film preparation technology to enhance production efficiency and reduce costs.
3. Improving the performance of flexible batteries to extend battery life for the prolonged usage needs of flexible electronic devices.
4. Developing more compact, highly integrated flexible electronic systems to reduce device size and enhance overall performance.

Flexible electronics, as a cutting-edge technology, opens avenues for the design and manufacturing of electronic products. Ongoing research and development of new materials, coupled with continuous optimization of manufacturing processes, will enable seamless integration of flexible electronics into people's daily lives. This integration promises greater flexibility and innovation in the design and manufacturing of electronic devices. From wearable technology to medical health applications and environmental monitoring, flexible electronics is poised to showcase its unique application value across various fields.

8.5.2 Flexible Electronic Substrate Materials

Flexible electronic substrate materials are integral to flexible electronic technology, serving as the supportive structure for these devices. They provide a soft, bendable, or stretchable foundation, enabling electronic devices to adapt to diverse shapes and surface curvatures. The main functions of these materials include:

1. Flexible Support: Comprising materials with high flexibility and bending properties, like flexible polymers and film materials, these substrate materials ensure the stability of electronic devices under bending or stretching conditions, preventing cracking or failure. This facilitates practical applications in wearable technology, bendable displays, and other fields.
2. Surface Adaptability: The design of flexible electronic substrate materials allows electronic devices to conform to various surfaces, including body curves, wearable device surfaces, or irregular surfaces. This flexibility enables the seamless integration of flexible electronic technology into diverse application scenarios, providing designers with increased creativity and freedom.
3. Electronic Component Integration Support: Serving as the base for electronic components, flexible electronic substrates facilitate the integration of various electronic devices such as flexible sensors, batteries, and circuit boards. This integrated capability enables the development of more compact and lightweight flexible electronic systems.
4. Bendable Electronic Devices: Electronic devices developed on flexible electronic substrates can operate normally when bent or curled. This opens up possibilities for the creation of curved electronic products, including folding mobile phones and curved displays.
5. Support for Printing and Manufacturing Processes: Flexible electronic substrate materials are generally compatible with flexible manufacturing processes such as printed electronics technology. This compatibility enhances production efficiency, reduces manufacturing costs, and supports the commercialization and large-scale application of flexible electronic technology.

In summary, flexible electronic substrate materials play a pivotal role in achieving flexibility, wearability, and bendability in flexible electronic technology. The selection of suitable substrate materials allows for better meeting the requirements of diverse application scenarios, thereby promoting the development of flexible electronics technology across various fields.

8.5.2.1 Polymer Films

Polymer films are thin film structures made from polymer materials, typically consisting of polymer compounds known for their flexibility, lightweight nature, transparency, and plasticity. These films are widely utilized in diverse fields, including flexible electronics, packaging materials, solar cells, and coating technologies. Commonly used polymer films encompass polyimide (PI), polyester (PET), and polymethyl methacrylate (PMMA). Among these, PI is frequently preferred in the fabrication of flexible electronic circuits, particularly in situations demanding high-temperature soldering of electronic components.

8.5.2.2 Flexible Glass

Flexible glass is a versatile material derived from borosilicate glass, seamlessly merging the transparency and optical properties inherent to glass with a remarkable degree of flexibility. Primarily utilized in applications demanding transparency and high-quality optical performance, flexible glass finds its niche in the realm of flexible display screens. Despite these advantages, it's worth noting that flexible glass may exhibit relative brittleness, making it susceptible to mechanical damage when compared to alternative materials.

8.5.2.3 Metal Films

Metal films, including copper and aluminum, are frequently employed as materials for flexible electronic substrates due to their excellent conductivity, making them well-suited for manufacturing flexible circuits. These films offer a cost-effective solution with a relatively straightforward manufacturing process. However, challenges may arise concerning oxidation and corrosion, necessitating measures to enhance the stability of metal films in flexible electronic applications.

8.5.2.4 Nanofiber Films

Nanofiber films, usually crafted from materials like nanocellulose or nanofiber membranes, boast excellent flexibility and transparency. This makes them well-suited for applications where both of these qualities are paramount. Despite their advantages, nanofiber films may exhibit sensitivity to environmental factors, particularly humidity and temperature, in certain conditions. Careful consideration of these factors is essential for optimizing their performance in specific applications.

8.5.2.5 Thin Film Silicon

Thin film silicon represents a form of silicon material that can be fashioned into a flexible film. Leveraging the excellent electrical properties inherent to silicon, it proves to be well-suited for the creation of high-performance flexible electronic devices. Additionally, thin film silicon exhibits commendable stability and heat resistance. However, it's important to note that the manufacturing process can be relatively intricate, contributing to potential higher costs.

Each substrate material, whether thin film silicon, polymer films, metal films, or nanofiber films, comes with its distinct set of advantages and drawbacks. The choice of material is often dictated by specific application requirements and the intricacies of manufacturing processes. In the dynamic landscape of flexible electronics development, researchers persistently explore new materials, aiming to enhance the performance and stability of flexible electronic devices. This ongoing exploration propels innovation and opens doors to novel possibilities in the field.

8.5.3 Nanomaterials

Nanomaterials showcase remarkable properties, including strong absorption bands in the visible region, high conductivity, and excellent mechanical characteristics. This diverse category encompasses various forms such as metal nanoparticles (NPs), magnetic beads, quantum dots (QD), nanotubes, nanowires, nanorods, nanofibers, nanocomposites, nanofilms, nanopolymers, and nanosheets. Their distinctive morphologies make them ideal for applications in electrochemical and optical sensors.

In the realm of biomedical applications, nanomaterials offer key advantages, including a high surface-to-volume ratio, exceptional functionalization capabilities, and numerous surface atoms providing both high surface energy and abundant anchoring sites. Diverse types of nanomaterials, including graphene, carbon nanotubes (CNT), gold nanoparticles (AuNP), and nanomotors, have been harnessed to create a myriad of nanostructured sensors across various domains.

Renowned Italian scholar Fabiana Arduini underscores the rising prominence of graphene among nanomaterials. Carbon nanotubes have experienced sustained growth, while nanogold shows a modest increase, and nanomotors remain at the conceptual verification stage according to Arduini's insights (Arduini et al. 2020). This dynamic landscape reflects the continuous exploration and application of nanomaterials in advancing sensor technologies.

8.5.3.1 Graphene

In 1962, Boehm and colleagues (Boehm et al. 1962) discovered graphene, initiating the development of nanomaterials. Subsequent decades focused on novel nanomaterials. The identification of single-layer graphene by Geim and Novoselov in 2004 (Novoselov et al. 2004) increased interest in 2D nanomaterials. Graphene, a high-quality 2D crystal with a one-atom thickness, exhibits unique electronic properties and holds significant potential in field-effect transistors, photovoltaic devices, and biosensors. Novoselov, in his Nobel Prize acceptance speech on December 8, 2010, referred to graphene not only as a "two-dimensional crystal" but also as a rare material with properties similar to massless relativistic particles due to the abundance of charge carriers. This characteristic positions graphene as an optimal candidate for designing electrochemical and optical bio-sensors. In electrochemical transduction, graphene serves as a highly functionalizable and modifiable electrode for marking/loading biomolecules and nanomaterials due to its high surface area. In optical transduction, it facilitates fluorescence quenching at any wavelength through energy transfer.

Graphene-modified screen-printed electrodes are well-suited for diverse sensing applications, spanning from medical clinical diagnostics to environmental pollution monitoring. Various forms of graphene, including graphene oxide (GO) and reduced graphene oxide (rGO), are commonly used. Additionally, composites with ZnO nanorods, carbon nanotubes (CNT), gold nanoparticles (AuNP), polyaniline, and pyrrole have been extensively explored. Graphene derivatives possess versatile characteristics beneficial for designing optical bio-sensing systems, such as the ability to undergo fluorescence quenching at any wavelength through energy transfer. This depends on factors like the number of graphene layers and the degree of oxidation. The widely utilized derivatives in optical bio-sensors are graphene oxides, which undergo electron-hole pair recombination embedded in the sp2 carbon domain, resulting in characteristic photo-induced luminescence. Adjusting oxygen content, lateral size, and oxidation degree allows tuning the maximum fluorescence emission wavelength from blue to near-infrared regions. These features facilitate in-depth studies of DNA sensing principles, focusing on the effective interaction between hexagonal cells of graphene derivatives and the cyclic structure of nucleic acids through hydrophobic and $\pi - \pi$ stacking interactions. Fluorescent probe-labeled single-stranded DNA complexed with graphene derivatives exhibits quenching in the absence of the target analyte. In the presence of the target analyte, weaker interactions occur between graphene and nucleic acid rigid structures, leading to the separation of single-stranded DNA/graphene complexes and fluorescence restoration.

8.5.3.2 Carbon Nanotubes

In 1991, Iijima reported the synthesis of carbon nanotubes (CNTs), nano-sized and needle-shaped structures of graphite carbon, in the journal "Nature" (Iijima 1991). These tubular structures, now known as carbon nanotubes, are composed of sp2 carbon units and fall into two main categories: single-walled carbon nanotubes (SWCNTs) and multi-walled carbon nanotubes (MWCNTs). They showcase diverse characteristics, including electrocatalytic performance at edge defect locations and in the presence of metal impurities. With their high surface area, carbon nanotubes are well-suited for immobilizing nanomaterials and/or biological components, making them applicable in microelectronic circuits, biological systems, and sensing devices. Carbon nanotubes are frequently combined with other nanomaterials, compounds, and/or biological components to enhance their functionality.

Jing et al. introduced an innovative approach involving the development of a glassy carbon electrode modified with a nanocomposite consisting of multi-walled carbon nanotubes (MWCNTs), cerium dioxide (CeO2), and gold nanoparticles (AuNPs) for the detection of methyl parathion. The outstanding adsorption capacity of MWCNTs facilitates solid-phase extraction of herbicides, leading to heightened sensitivity (Dong et al. 2013). Furthermore, the inclusion of CeO2 and AuNPs contributes to a high surface area and specific catalytic activity, thereby significantly improving the sensitivity of detecting methyl parathion in river and soil samples.

Carbon nanotubes possess a strong capability to immobilize biological receptors, enhancing biomolecule loading and improving electrochemical analytical performance. Additionally, their unique optical properties enable the modulation of luminescence, fluorescence, and color changes in various optical applications. The emission wavelength range, extending from 700 to 1400 nm, provides advantages such as high photostability, low autofluorescence background, and deep penetration capability for in vivo imaging.

Key optical sensing mechanisms utilizing carbon nanotubes include fluorescence spectrum changes (resulting from interactions between analytes and carbon nanotubes or encapsulated materials causing fluorescence intensity or wavelength shifts), fluorescence quenching (possible when close to other organic fluorophores), and Raman scattering, where carbon nanotubes serve as labels exhibiting extremely high Raman scattering. The diverse chiralities of carbon nanotubes enable a multiplexing platform, with each single-walled carbon nanotube functioning as a distinct color. However, combining different chiralities, lengths, and impurities may reduce sensitivity and selectivity. Addressing the lower fluorescence quantum yield compared to high quantum yield fluorophores (such as quantum dots) remains an area for improvement.

8.5.3.3 Noble Metal Nanoparticles

The utilization of nanostructures derived from noble metals is prevalent in the fabrication of electrochemical and colorimetric sensing platforms. Moreover, various other nanomaterials, including inorganic crystals, oxides, and semiconductors, contribute to the enhanced performance of sensing devices. These materials, synthesized at the nanoscale, demonstrate notable advancements in detecting trace analytes within intricate matrices such as surface water, blood, and urine.

Gold nanoparticles (AuNPs) play a crucial role in the development of screen-printed strips designed for the electrochemical detection of single-stranded and double-stranded DNA sequences associated with HIV in serum samples. The dual functionality of AuNPs involves acting as anchor points for thiol-terminated DNA probes and contributing significant enhancements in electrochemical performance compared to unmodified graphite-based printed electrodes (Cinti et al. 2018a). Furthermore, AuNPs can serve as electrochemical labels. For instance, Escosura et al. utilized magnetic beads coupled with AuNPs for the isothermal amplification of Leishmania DNA (de la Escosura-Muñiz et al. 2016). The inclusion of magnetic beads facilitates the purification/concentration of analytes, while AuNPs are employed to label oligonucleotide primers and amplify electrochemical signals, enabling rapid quantification of DNA on screen-printed carbon electrodes.

Silver demonstrates sensitivity to oxidation in the presence of halide ions when a slightly positive potential is applied, making the use of silver nanoparticles (AgNPs) particularly effective for electrochemical labeling when required. For example, AgNPs, in combination with graphene, have been employed to create a novel micro-label for clinical immunosensors designed to detect the H7 avian influenza virus. A chitosan solution was introduced to provide carboxyl groups (after NaOH treatment) for binding with H7 polyclonal antibodies, forming biohybrid conjugates (Huang et al. 2016).

AgNPs can also serve as electrochemical labels, Russo et al. utilized hollow Au-Ag nanoshells for the detection of Escherichia coli and Salmonella on screen-printed carbon-based electrodes (Russo et al. 2018). Bacterial detection is facilitated by nonspecific affinity interactions between bacterial cell walls and the surface of Au-Ag nanoshells. The electrochemical mechanism involves the corrosion of Ag + cations produced by galvanic corrosion when the Au-Ag nanoshells are exposed to high concentrations of halide ions and dissolved oxygen. This process preserves particle structural stability and enables anodic stripping analysis.

In sensor development, copper nanoparticles (CuNPs) have found application in modifying electrochemical surfaces for the detection of nitrite and nitrate ions in food and beverage samples. CuNPs have also been utilized to emulate enzyme activity, with metal-organic frameworks (MOFs) serving as protective shells for copper nanoparticles in non-enzymatic glucose sensing in an alkaline medium (Shi et al. 2016). The inclusion of MOFs enhances the stability of CuNPs, effectively preventing their dissolution and aggregation during electrocatalytic analysis. The use of transition metals like copper offers advantages over gold-based platforms, as they are easier to control during synthesis, exhibit higher activity, and are more cost-effective.

Another instance of electrocatalytic metal encapsulation involves platinum; following a similar approach, platinum nanoparticles (PtNPs) have been encapsulated in MOFs for the detection of telomerase activity (Ling et al. 2016). In an alkaline medium with the presence of $NaBH_4$ oxidation, PtNP electrocatalysis has been observed in cancer cells. This process involves up to eight electrons in the oxidation process, which is significantly higher compared to processes involving only one or two electrons.

8.5.3.4 Inorganic Nanoparticles

Not only are metal nanoparticles utilized to enhance sensor performance, but a variety of nano-modifiers, including inorganic crystals and metal oxides, have also been employed in the development of sensing platforms. Prussian Blue (ferric hexacyanoferrate), recognized as an "artificial peroxidase" for its selectivity toward hydrogen peroxide, has been extensively employed in hydrogen peroxide determination and bio-sensing structures involving oxidases (e.g., alcohol, glucose, lactate, cholesterol).

Cinti et al. demonstrated a unique approach using the structure of regular filter paper to synthesize Prussian Blue nanoparticles' precursors, without external power or reducing agents. This synthesis, termed "paper blue," resulted in nanoparticles of approximately 20 nm attributed entirely to the composition and structure of the paper (Cinti et al. 2018b). This method has been applied in glucose detection in whole blood, showcasing high selectivity, even in the presence of common blood interferents like ascorbic acid, uric acid, and acetaminophen. The electrocatalytic properties of Prussian Blue in hydrogen peroxide reduction and thiol oxidation are attributed to the redox pair: Prussian Blue and Prussian White, representing oxidized and reduced forms. Beyond electrochemical properties, Prussian Blue undergoes a color change between different oxidation states; when reduced to Prussian White, its color changes from blue to colorless.

Bagheri et al. reported a 96-well paper-based chromatography platform highlighting the fading of Prussian Blue for detecting BChE enzyme activity in human serum (Bagheri et al. 2019). The detection principle relies on the reaction between the enzyme product thiocholine and Prussian Blue; the presence of thiocholine can reduce Prussian Blue, forming the reduced form, i.e., Prussian White.

In the development of an ochratoxin A (OTA) aptamer sensor, Rivas et al. (2015) created a carbon-printed electrode by electropolymerizing a polythionine film and assembling IrO_2NPs. This label-free platform immobilized the OTA recognition aptamer through electrostatic interactions. The interaction resulted from the attraction between negatively charged citrate groups around IrO_2NPs and positively charged amino groups on the amino-modified aptamer.

Nano-cerium oxide, a rare earth oxide nanomaterial with unique physical and chemical properties, was utilized in a study by Liu et al. (2016). The Ce^{3+}/Ce^{4+} ratio in nano-cerium oxide significantly influenced its peroxidase-mimicking activity. Liu et al. prepared a nano-composite material, $H_2TCPP–CeO_2$, consisting of cerium oxide nanorods and H_2TCPP. This nano-composite exhibited peroxidase-like activity towards hydrogen peroxide and TMB, resembling horseradish peroxidase.

Many inorganic nanomaterials leverage their artificial enzyme-like properties, often combined with co-reactants like TMB, leading to color changes after specific reactions.

8.5.3.5 Nanomotors

Sensing technology has evolved from "classical" substrates to "active" substrates, incorporating autonomous micro/nanomachines capable of independent movement within a sample. These machines, ranging from nanometers to micrometers, can be constructed using diverse materials such as polymers, metals, and semiconductors (Wang and Gao 2012; Wang et al. 2013). While applications are still in the conceptual verification stage, these tiny autonomous objects demonstrate the ability to perform various tasks autonomously, including transporting cargo, cell destruction, and pollution remediation.

In a study by Cinti et al., Mg-Au-Ni-Janus microengines were employed to accelerate the degradation of neurotoxins and subsequently detect harmless remediation products (p-nitrophenol) (Cinti et al. 2015). Smart microengines generated OH^- to increase the medium's pH, promoting the oxidative degradation of organophosphorus into easily detectable p-nitrophenol. The produced H_2 facilitated rapid and reproducible degradation of neurotoxins without external stirrers or mixing devices. Rojas et al. electrochemically detected diphenyl phthalate in biological and food samples using a platform based on Au-Mg microengines, where microengines increased the pH, enabling the conversion of diphenyl phthalate into phenol (an electroactive molecule) (Rojas et al. 2016).

Maria et al. developed a colorimetric method using tubular microengines composed of a mixed SWCNT-Fe_2O_3 outer layer powered by MnO_2 catalysts to detect and differentiate isomers of phenylenediamine (Maria-Hormigos et al. 2018). The catalytic decomposition of hydrogen peroxide induced by the MnO_2 catalytic layer resulted in O_2 bubbles and the generation of hydroxyl radicals for analyte dimerization.

Employing nanomaterials from carbonaceous, metallic, or inorganic sources enhances device characteristics such as cost-effectiveness, high sensitivity and selectivity, low detection limits, and long-term stability. These nanomaterials can be mutually combined and/or combined with different biological elements, leading to the creation of novel nanocomposites with consistently enhanced performance.

8.6 Nano-sensing Materials

Nanomaterials are prepared from inorganic and organic materials and are basically defined as materials with a size range of approximately 1–100 nm. Since the discovery of nanomaterials by German physicist Gleiter in the 1980s, they have received widespread attention from the academic community. Nanomaterials are also known as "the most promising materials in 21st century. Since the theories of classical mechanics and quantum mechanics are no longer valid at the nanometer scale, its unique properties, such as transparency, hydrophobicity, photoluminescence, toughness and hardness, and good conductivity, make it useful in various fields. Excellent internal performance. It acts in coatings to give them long-term antimicrobial activity; it acts in electromagnetic interference to produce nanometers and nanotube arrays with a specific gravity of only one-sixth of steel, and in shielding and absorbing electromagnetic radiation, It performs better than traditional metals in terms of field emission and thermal conductivity (Bhattacharjee and Bose 2021); in the field of action and sensors, its high sensitivity and selectivity also provide more solutions in the field of electrochemical sensors. The traditional method of nanoparticle preparation is "top-down", which is reduction by abrasion and various crushing in the traditional sense; but in the past few years, "bottom-up" has been increasingly used Technical production method, this method can make the nanomaterials produced have smoother surfaces and more stable properties (Baer et al. 2013). Due to the unique properties exhibited by nanomaterials and their promotion of detection performance, such as speed, simplicity, specificity, sensitivity, selectivity, and robustness, in terms of signal amplification/generation and signal readout, etc. It is expected to serve as the core of nanoparticle-based sensors. The specific role of new nanomaterials in sensing will be introduced below.

8.6.1 Signal Enhancement

Nanomaterials can enhance signals in a variety of ways. For example: it has a larger specific surface area. Compared with macroscopic materials, nanomaterials can provide more active surfaces, thereby enhancing signals. The fluorescence signal can be enhanced by surface modification or doping with specific fluorescent substances; as a SERS substrate, through the surface plasmon resonance effect enhances the Raman scattering signal. This method can improve the detection sensitivity of molecules and is commonly used in chemical analysis, biomolecular detection, etc.

8.6.1.1 Electronic Transport

The electron transport performance of nanomaterials refers to the transport characteristics of electrons in materials at the nanometer scale. Due to its size effect and surface effect, nanomaterials exhibit many unique properties in electron transport. First, nanomaterials usually have a larger specific surface area, which means that electrons can be transported over a larger surface in nanomaterials, thereby increasing the electron transmission rate. Secondly, the energy band structure of nanomaterials may change, causing their electron transport properties to be different from those of macroscopic materials. For example, nanomaterials such as carbon nanotubes and graphene have excellent electron transport properties, exhibiting high carrier mobility and low resistivity. In addition, the electron transport performance of nanomaterials is also affected by factors such as lattice defects, impurities, and interfaces. These factors may affect the scattering and transmission of electrons, thereby affecting the overall electron transport performance of the nanomaterial.

8.6.1.2 Localized Surface Plasmon Resonance (LSPR)

The local field enhancement effect of nanomaterials means that when a local electromagnetic field exists in the nanostructure, the field can be enhanced to a higher degree than the surrounding environment. This effect often occurs in nanomaterials such as nanoparticles and nanostructured surfaces, and has important applications in optics, spectroscopy, sensors, and other fields. In terms of optics, the local field enhancement effect can cause the local electromagnetic field on the surface of nanomaterials to be enhanced, thereby affecting the absorption, scattering, and emission of light. This effect is widely used in technologies such as surface-enhanced Raman scattering (SERS) and localized surface plasmon resonance (LSPR) to detect trace molecules, biomolecules, etc. (Dynich et al. 2009). In the sensor field, the local field enhancement effect can be exploited to improve the sensitivity and detection limit of the sensor. By introducing nanomaterials into the sensor structure, the local field enhancement effect can be exploited to enhance the sensor's detection ability of target molecules or substances.

8.6.1.3 Surface-Enhanced Raman Scattering (SERS)

Surface-Enhanced Raman Scattering (SERS) is a spectroscopic technology based on the local field enhancement effect on the surface of nanostructures. In SERS, when molecules are adsorbed on the surface of nanostructures with appropriate surface enhancement effects, their Raman scattering signals are significantly enhanced, allowing the detection of extremely low concentrations of molecules. The principle of SERS technology is to enhance the Raman scattering signal of molecules by about 106 to 1014 times through the local field enhancement effect on the surface of nanostructures, thereby achieving highly sensitive detection of trace molecules. This technology is widely used in biomedicine, environmental monitoring, It has broad application prospects in food safety and other fields. In the field of biomedicine, SERS technology can be used to detect biomolecules, cell markers, etc., and is helpful in disease diagnosis, drug testing, etc.

8.6.2 Catalysis

Nanomaterials can act as catalysts in a variety of ways. For example: it has a large specific surface area, so under the same mass, it has a larger surface area than traditional materials and can provide more active sites; its surface effect and size effect can change its electronic structure and chemical properties, thereby enhancing catalytic activity; the reaction rate and selectivity can also be improved by regulating the adsorption and diffusion process of reactants; nanomaterials can also regulate the reaction environment, such as adjusting pH value, temperature and atmosphere, etc. In short, nano-sensing materials play a catalytic role through the above methods, improve reaction rate and selectivity, and are widely used in the field of catalytic reactions. Several nanomaterials with catalytic effects are summarized below.

8.6.2.1 Nanozymes

The term nanozyme was first proposed by Pasquato, Scrimin, and colleagues in 2004 to describe the transphosphorylation reactivity of triazacyclononane-functionalized gold nanoparticles. Then in 2007, Peixia Wang and others discovered that nanoparticles have peroxidase properties (Gao et al. 2007), and various nanoparticles with enzymatic properties were discovered and proposed. Compared with traditional biological enzymes, nanozymes, artificial enzymes, not only have the catalytic activity of enzymes, but also have the characteristics of high surface area, stability, fluorescence, and paramagnetism of nanomaterials. Due to these excellent physical and chemical properties of nanozymes, they have been used in fields such as medical imaging, immunoassays, pathological analysis, and disease treatment. From the usage scenarios of nanozymes, they are divided into: medical imaging, immunoassays, and disease treatment.

Because nanozymes can be remotely controlled through different stimuli, including magnetic fields, light, heat, etc., the purpose of imaging can be achieved. Different biological tissues can be imaged through different substances. Due to different imaging technologies, they are divided into fluorescence imaging, nuclide imaging, and magnetic resonance imaging. Some nanozymes have fluorescent properties and can be used for biofluorescence imaging. Zhao et al. reported copper ion-doped ZIF-8 loaded with gold nanozymes and doxorubicin hydrochloride (DOX) (Zhao et al. 2021). Preparation for synergistic treatment of tumors with chemotherapy and CDT assisted by activated fluorescence imaging of the tumor microenvironment (TME). Nanozymes can carry radioactive isotopes for use in positron emission tomography (PET) or single photon emission computers Tomography (SPECT) imaging helps doctors observe the biological process of disease. Wang et al. detailed the synthesis of CuS-Fn nanocages (CuS-Fn NCs), featuring ultra-small CuSNPs within the FnNC cavity. These nanocages exhibit excellent biocompatibility, robust near-infrared absorbance, and pronounced photoacoustic contrast, illustrating their potential for advanced applications in the field (Wang et al. 2016). In quantitative ratiometric photoacoustic imaging (PAI), real-time in vivo photoacoustic imaging of the entire tumor by CuS-Fn NCs Imaging features excellent photoacoustic tomography improvements. With the incorporation of the radionuclide ^{64}Cu, ^{64}CuS-Fn NCs also exhibit an excellent PET imaging agent; nanozymes can be used as MRI (Magnetic Resonance Imaging, MRI) contrast agents through their distribution in the body. Doctors observe the condition of tissues and organs to diagnose and monitor diseases. These technologies make nanozymes have broad application prospects in in vivo imaging, which can help doctors perform early diagnosis of diseases, monitor treatment effects, and implement personalized treatments. In the research of Qin et al., a nanozyme that can be used as a peroxidase to catalyze glucose decomposition and a T_2-magnetic resonance imaging contrast agent for in vivo tumor detection was produced. This nanozyme will observe negative signals in the tumor area 1 h after injection and return to pre-injection levels after 4 h (Qin et al. 2021). These technologies enable nanozymes to have a wide range of application scenarios in in vivo imaging, providing modern medical technology with more comprehensive, accurate, and safe treatment and monitoring ideas.

Immunoassay nanozyme refers to a method of immunoassay using nanozyme as a marker. Nanozymes usually refer to enzyme nanoparticles prepared through nanotechnology, which have a large specific surface area and high carrier capacity, and can improve the sensitivity and stability of immunoassays. This method has been widely used in biomedical research and clinical diagnosis, and can be used to detect the content and activity of various substances such as proteins, nucleic acids, metabolites, etc. Taking the new coronavirus as the background, many effective detection methods have been proposed, and nanozymes can be used to detect its related proteins: In the study of Liang et al. (2021), a method using nanozyme immunochromatography was proposed. The sensor is used to rapidly and quantitatively detect SARS-CoV-2 nucleocapsid protein in human blood, which is an effective means to detect early SARS-CoV-2 infection and control COVID-19, and the limit of detection (LOD) of NLICS is 0.026 ng/mL NP, with a linear detection range (LDR) of 0.05 ~ 1.6 ng/mL NP, which is more sensitive than traditional ICA. Meng et al. demonstrated the application of nanozymes for the detection of human nucleic acids by integrating recombinase polymerase amplification (RPA) with FeS2 nanozyme strips. Leveraging the high peroxidase-like activity of the FeS2 nanozymes, this approach enables straightforward nucleic acid amplification followed by enhanced colorimetric signal detection (Meng et al. 2022b). The nucleic acid detection limit of this method for SARS-CoV-2 is 200 copies/mL. In addition, nanozymes can also detect metabolites for the purpose of immunoassay: in the study of Lu et al., a method was proposed to prepare oxidation-like materials through PDA surface modification-assisted pyrolysis strategy (Hu et al. 2019). Enzymatically active CoO/N-CS-rGO hybrid nanozyme and achieves highly sensitive detection of DA and UA.

Nanozymes can play a variety of roles in the field of disease treatment: they are used for targeted therapy, and are designed as targeted delivery carriers to accurately deliver drugs to disease sites. In the study by Duo Wang et al., a novel approach utilizing superparamagnetic iron oxide (SPIO) nanozymes, modified with bevacizumab (Ava), is proposed for synergistic ferroptosis and starvation therapy in treating triple-negative breast cancer (TNBC). This method effectively inhibits tumor angiogenesis, thereby severing the tumor's energy supply and manipulating the tumor microenvironment toward a state of nutrient deprivation (Wang et al. 2022). For drug sustained release, it is designed as a nanocarrier with sustained release function, regarding the release rate and location of the drug, Hu et al. (2023) [Self-intensified synergy of a versatile] proposed a multifunctional nanozyme based on endogenous H_2O_2 and O_2 to trigger reactions in cancer cells, and verified PZIF67. The self-enhancing synergistic effect of nanozymes and DOX on PLA electrospun fibers inhibits postoperative tumor recurrence and metastasis. For integration of diagnosis and treatment, real-time monitoring during the treatment process, or simultaneous treatment during the diagnosis process, Liu et al. (2023) proposed $ZnMnFe_2O_4$-PEG-FA nanozyme, which can reach the ideal temperature to damage tumor cells and its strong catalytic ability greatly alleviates the limitations of traditional PTT and catalytic therapy. At the same time, it has excellent photoacoustic imaging and magnetic resonance imaging capabilities to enable monitoring. And guide cancer treatment; used for immune regulation, it is designed as a nanocarrier with immunomodulatory function to regulate the function of the immune system. Zeng et al. proposed a polypyrrole nanoenzyme as a tumor microenvironment modulator to reprogram macrophages and enhance immunotherapy (Zeng et al. 2022). Based on the excellent photothermal conversion ability of CuPP nanozymes, the local temperature increase at the tumor site after laser irradiation further enhances redox by enhancing trienzyme-like activity. Hemostatic destruction, in turn, initiates the reversal of the innate immune-suppressive TME, allowing macrophages to be recultured from the pro-tumor M2 phenotype to the anti-tumor M1 phenotype to enhance immunity. These effects make nanozymes a potential therapeutic tool that can help improve therapeutic efficacy, reduce side effects, and enable personalized treatment.

8.6.2.2 Metal Nanoparticle Catalysis

Metal nanoparticles were first prepared by the sol-gel method discovered by American scientist Richard Zsigmondy. Due to its high specific surface area, surface activity, controllability, and biocompatibility, it has good research and application value in biomedicine, catalysis, energy, and other fields, especially in the medical field, its photosensitivity. Its controllable properties can sense light energy in biological therapy and be used in photothermal therapy; its biocompatibility properties can be used as a carrier for drug transportation and used in drug release; its surface activity and high specific surface area can sense biomolecules and amplify signals used in the field of bio-sensing.

Photothermal therapy uses the thermal effect produced by metal nanoparticles to absorb light of specific wavelengths to destroy tumor cells or other abnormal cells. Specifically, metal nanoparticles (such as gold nanoparticles) can be injected into the affected area, and then lasers or other light sources are used to illuminate these particles, causing them to absorb light energy and convert it into heat energy, thereby locally heating up and destroying the surrounding tissue, lesion tissue. This photothermal therapy method has the advantages of local treatment, minimal invasion, and strong controllability, and has potential application prospects in tumor treatment. In addition, based on the controllability of metal nanoparticles, its photothermal conversion efficiency can be optimized and the therapeutic effect can be improved.

The application of metal nanoparticles in drug release is to achieve controlled release of drugs by combining drug carriers with metal nanoparticles. This method can improve the bioavailability of the drug, reduce the dose and toxic side effects of the drug, thereby improving the therapeutic effect. Specifically, metal nanoparticles can achieve drug release through different mechanisms, such as through pH response, temperature response, etc. For example, some metal nanoparticles such as iron oxide nanoparticles dissolve and aggregate under different pH environments, making them potentially useful in biomedical fields.

Metal nanoparticles can be designed into highly sensitive biosensors for detecting specific signals from biomolecules or cells. These nanoparticles can achieve recognition of specific biomolecules through surface modification or functionalization, thereby achieving highly sensitive detection of biomolecules. Specifically, metal nanoparticles can specifically interact with specific biomolecules, such as antibodies, DNA, or small molecule ligands, through surface modification or functionalization, thereby enabling the detection of biomolecules. When target biomolecules interact with nanoparticles, it results in changes in the properties of the nanoparticles.

8.6.2.3 Catalytic Nanocomposites

Catalytic nanocomposites refer to a new type of material formed by combining nanomaterials with catalysts. Such composite materials usually consist of nanoparticles or nanostructured catalysts and other nanomaterials (such as carbon nanotubes, nanoparticles, etc.). This combination can give full play to the advantages of nanomaterials and catalysts, improve catalytic performance and stability, and can be used to promote chemical reactions and improve reaction efficiency and selectivity. Since nanomaterials have high specific surface area and size effect, they can provide abundant active sites, and catalysts have specific catalytic activity. Therefore, the combination of the two can form a synergistic effect and improve catalytic performance.

8.6.2.4 Core/Shell-Like Nanostructure

Core-shell nanostructure refers to a nanoparticle structure composed of a core and an outer shell. In this structure, the core is usually made of one substance, while an outer shell wraps around the core and can be made of different substances. Nanoparticles with this structure usually have an interface between the core and the shell, which can give the nanoparticles special properties and functions. Such structures often have excellent optical, electronic, magnetic, and other properties, which make them widely used in biomedicine, catalysis, sensing, energy, and other fields. Since the core-shell structure can provide more active surface area and increase the contact between the catalyst and the reactants, thereby improving the catalytic activity. At the same time, by rationally designing the core-shell structure, precise control of the catalyst surface can be achieved, further improving catalytic activity and selectivity. The core-shell structure can improve the stability of the catalyst. The outer shell protects the inner core from environmental influences and extends the service life of the catalyst. In addition, the core-shell structure also enables the regeneration and recycling of catalysts. By rationally designing the core-shell structure, the catalyst can be more easily separated and recycled, thereby reducing costs and environmental impact.

8.6.3 Specificity

The main reason why nano-sensing materials play a specific role is the special properties and structure of their surfaces. Nanomaterials have large specific surface areas and highly tunable surface properties, which enable them to specifically interact with target molecules or organisms. A common nano-sensing material is gold nanoparticles, whose surfaces are rich in free electrons and can react with chemical bonds between molecules or interactions between biomolecules. By regulating the size, shape, and surface modification of gold nanoparticles, highly selective recognition and detection of target molecules can be achieved. For example, specific ligands can be used to modify the surface of gold nanoparticles to specifically coordinate with target molecules, thereby achieving detection of target molecules. In addition, nano-sensing materials can also achieve specific detection through the Surface-Enhanced Raman Scattering (SERS) effect. SERS is a spectroscopic technique based on the localized electromagnetic field enhancement effect on the surface of gold or silver nanoparticles. When a target molecule comes into contact with gold or silver nanoparticles, its Raman scattering signal is significantly enhanced, enabling highly sensitive and specific detection of the target molecule. In short, nano-sensing materials can specifically interact with target molecules or organisms by regulating their surface properties and structure, thereby achieving highly selective and sensitive detection of target molecules.

8.6.3.1 Ligand-Receptor Interaction

The receptor-ligand binding interaction of nanoparticles refers to the binding effect between receptors and ligands on the surface of nanoparticles. In this interaction, the receptor is usually a functionalized molecule on the surface of the nanoparticle, and the ligand is the molecule that interacts with the receptor. This interaction has important implications in the biomedical field. For example, nanoparticles can be designed to have specific receptors that can bind to specific ligands in organisms, enabling the recognition and detection of biomolecules. This application can be used in fields such as disease diagnosis and drug delivery. In addition, the receptor-ligand binding interactions of nanoparticles can also be used for the separation and enrichment of biomolecules. By designing nanoparticles with specific receptors, the enrichment and separation of specific molecules in biological samples can be achieved, which is helpful for the analysis and detection of biological samples.

8.6.3.2 Chemical Bonding

Chemical bonding of nanoparticles refers to the formation and breaking of chemical bonds in materials at the nanoscale. In nanomaterials, the nature of chemical bonds may change due to size effects and surface effects, so chemical bonding of nanoparticles has some special properties and applications. Chemical bonding of nanoparticles can be used to prepare functionalized nanomaterials, for example, by fixing molecules or groups on the surface of nanoparticles through chemical bonding, thereby endowing nanomaterials with specific properties and functions. This functionalized nanomaterial can be used in catalysis, biomedicine, sensors and other fields. In addition, chemical bonding of nanoparticles can also be used in the assembly and construction of nanomaterials. By controlling the way and position of chemical bonding, the self-assembly and directional assembly of nanoparticles can be achieved, thereby constructing nanomaterials with specific structures and properties.

8.6.3.3 Enzyme-Substrate Interaction

Nanomaterial enzyme-substrate interaction refers to the interaction between nanomaterials and substrate molecules. In this interaction, nanomaterials can mimic the function of enzymes and undergo specific chemical reactions or recognition with substrate molecules. This interaction has potential applications in biomedicine and biochemistry. For example, nanomaterials can be designed to have specific catalytic activity to promote specific chemical reactions of substrate molecules. This application can be used in fields such as drug synthesis and bio-sensing. In addition, nanomaterial enzyme-substrate interactions can also be used for the recognition and detection of biomolecules. By designing nanomaterials with specific structures, highly sensitive detection of specific substrate molecules can be achieved, which is helpful for biomedical diagnosis and biomolecule analysis.

8.6.3.4 Immune Reaction

The role of nanomaterials in immune responses is an area of intense research interest. Nanomaterials can influence immune responses in a variety of ways, including as vaccine vectors, immunomodulators, and tools for immune recognition and therapy. As vaccine carriers, nanomaterials can improve the stability of the vaccine, extend its existence time in the body, and enhance its stimulating effect on the immune system, thereby improving the immunogenicity of the vaccine. In terms of immunomodulation, nanomaterials can be designed to have specific immunomodulatory effects, such as by regulating antigen presentation, regulating the activation status of immune cells, etc., thus affecting the response of the immune system. In addition, nanomaterials can also be used for immune recognition and treatment. By functionalizing biomolecules on the surface of nanomaterials, identification and targeted treatment of specific immune cells or molecules can be achieved, which helps to improve the accuracy and effectiveness of treatment.

Acknowledgments This work was funded by the National Key Research and Development Program of China (Grant No. 2022YFF1202700), the National Natural Science Foundation of China (Grant No. 82001922), and the Hong Kong Scholars Program (Grant No. XJ2021034).

References

Aftab S, Hussain S, Al-Kahtani AA (2023) Latest innovations in two-dimensional flexible nanoelectronics. Adv Mater 2301280

Ahmad R, Joshi AM, Boolchandani D, Varma T (2021) Design of potentiostat and current mode read-out amplifier for glucose sensing, 2021 IEEE International Symposium on Smart Electronic Systems (iSES), IEEE2021, pp 64–69

Angell JB, Wise KD, Starr A (1970) An integrated-circuit approach to extracellular microelectrodes. IEEE Trans Biomed Eng BME-17:238

Arduini F, Cinti S, Scognamiglio V, Moscone D (2020) Nanomaterial-based sensors, handbook of nanomaterials in analytical chemistry. Elsevier, pp 329–359

Ayode Otitoju T, Ugochukwu Okoye P, Chen G, Li Y, Onyeka Okoye M, Li S (2020) Advanced ceramic components: materials, fabrication, and applications. J Ind Eng Chem 85:34–65

Baer DR, Engelhard MH, Johnson GE, Laskin J, Lai J, Mueller K et al (2013) Surface characterization of nanomaterials and nanoparticles: Important needs and challenging opportunities. J Vac Sci Technol A 31: 050820

Bagheri N, Cinti S, Caratelli V, Massoud R, Saraji M, Moscone D et al (2019) A 96-well wax printed Prussian Blue paper for the visual determination of cholinesterase activity in human serum. Biosens Bioelectron 134:97–102

Bai Y, Yang W, Sun Y, Sun C (2008) Enzyme-free glucose sensor based on a three-dimensional gold film electrode. Sens Actuators B Chem 134:471–476

Bartlett P, Simon E, Toh CS (2002) Modified electrodes for NADH oxidation and dehydrogenase-based biosensors. Bioelectrochem 56:117–122

Bent B, Goldstein BA, Kibbe WA, Dunn JP (2020) Investigating sources of inaccuracy in wearable optical heart rate sensors. NPJ Digital Med 3:18

Berger F, Sanchez J-B, Heintz O (2009) Detection of hydrogen fluoride using SnO2-based gas sensors: understanding of the reactional mechanism. Sens Actuators B Chem 143:152–157

Berger G, Frangville P, Meyer F (2020) Halogen bonding for molecular recognition: new developments in materials and biological sciences. Chem Commun 56:4970–4981

Bergveld P (1970) Development of an ion-sensitive solid-state device for neurophysiological measurements. IEEE Trans Biomed Eng

Bhattacharjee Y, Bose S (2021) Core-shell nanomaterials for microwave absorption and electromagnetic interference shielding: a review. ACS Appl Nano Mater 4:949–972

Boehm H-P, Clauss A, Fischer G, Hofmann U (1962) Das adsorptionsverhalten sehr dünner kohlenstoff-folien. Z Anorg Allg Chem 316:119–127

Cavalcante FT, de Falcão IRA, da Souza JES, Rocha TG, de Sousa IG, Cavalcante AL et al (2021) Designing of nanomaterials-based enzymatic biosensors: synthesis, properties, and applications. Electrochem 2:149–184

Chen H, Yuan L, Song W, Wu Z, Li D (2008) Biocompatible polymer materials: role of protein-surface interactions. Prog Polym Sci 33:1059–1087

Chen Y, Ji W, Yan K, Gao J, Zhang J (2019) Fuel cell-based self-powered electrochemical sensors for biochemical detection. Nano Energy 61:173–193

Chen H-Y, Chen A, Chen C (2020) Investigation of the impact of infrared sensors on core body temperature monitoring by comparing measurement sites. Sensors 20:2885

Cheraghi S, Taher MA, Karimi-Maleh H, Karimi F, Shabani-Nooshabadi M, Alizadeh M et al (2022) Novel enzymatic graphene oxide based biosensor for the detection of glutathione in biological body fluids. Chemosphere 287:132187

Choudhury N, Sampath S, Shukla A (2009) Hydrogel-polymer electrolytes for electrochemical capacitors: an overview. Energy Environ Sci 2:55–67

Christie R (2011) Fluorescent dyes, pp 562–587. Woodhead Publishing Ltd

Cinti S, Valdés-Ramírez G, Gao W, Li J, Palleschi G, Wang J (2015) Microengine-assisted electrochemical measurements at printable sensor strips. Chem Commun 51:8668–8671

Cinti S, Proietti E, Casotto F, Moscone D, Arduini F (2018a) Based strips for the electrochemical detection of single and double stranded DNA. Anal Chem 90:13680–13686

Cinti S, Cusenza R, Moscone D, Arduini F (2018b) Based synthesis of Prussian Blue nanoparticles for the development of whole blood glucose electrochemical biosensor. Talanta 187:59–64

Colombo P, Mera G, Riedel R, Soraru Gdjjotacs (2010) Polymer-derived ceramics: 40 years of research and innovation in advanced ceramics. J Am Ceram. Soc 93:1805–1837

de Brito AR, de Jesus RS, de Carvalho Tavares IM, Silva FN, Santana NB, Ferrao SPB et al (2021) Application of the electrochemical biosensor in the detection of lactose in skimmed milk. Surf Interfaces 22:100839

de la Escosura-Muñiz A, Baptista-Pires L, Serrano L, Altet L, Francino O, Sánchez A et al (2016) Magnetic bead/gold nanoparticle double-labeled primers for electrochemical detection of isothermal amplified Leishmania DNA. Small 12:205–213

Dong J, Wang X, Qiao F, Liu P, Ai S (2013) Highly sensitive electrochemical stripping analysis of methyl parathion at MWCNTs–CeO2–Au nanocomposite modified electrode. Sensors Actuators B Chem 186:774–780

C. Drouet, A. Leriche, S. Hampshire, M. Kashani, A. Stamboulis, M. Iafisco, et al., Types of ceramics, advances in ceramic biomaterials 2017, pp. 21–82

Dukle A, Nathanael AJ, Panchapakesan B, Oh T-H (2022) Role of paper-based sensors in fight against cancer for the developing world. Biosensors 12:737

Dynich R, Ponyavina A, Filippov VV (2009) Local field enhancement near spherical nanoparticles in absorbing media. J Appl Spectrosc 76:705–710

Ecker DM, Jones SD, Levine HL (2015) The therapeutic monoclonal antibody market, MAbs, vol 7. Taylor & Francis, pp 9–14

Eltouby P, Shyha I, Li C, Khaliq J (2021) Factors affecting the piezoelectric performance of ceramic-polymer composites: a comprehensive review. Ceram Int 47:17813–17825

Fang Z, Tian X, Zheng F, Jiang X, Ye W, Qin Y et al (2022) Enhanced piezoelectric properties of Sm3+-modified PMN-PT ceramics and their application in energy harvesting. Ceram Int 48:7550–7556

Fine GF, Cavanagh LM, Afonja A, Binions RJS (2010) Metal oxide semi-conductor gas sensors in environmental monitoring. SENSORS 10:5469–5502

Foroughi MM, Jahani S (2022) Investigation of a high-sensitive electrochemical DNA biosensor for determination of Idarubicin and studies of DNA-binding properties. Microchem J 179:107546

Gao L, Zhuang J, Nie L, Zhang J, Zhang Y, Gu N et al (2007) Intrinsic peroxidase-like activity of ferromagnetic nanoparticles. Nat Nanotechnol 2:577–583

Grant J, Cunningham W, Blue A, O'Shea V, Vaitkus J, Gaubas E et al (2005) Wide bandgap semiconductor detectors for harsh radiation environments. Nucl Instrum Methods Phys Res, Sect A 546:213–217

Haurum JS (2006) Recombinant polyclonal antibodies: the next generation of antibody therapeutics? Drug Discov Today 11:655–660

Hinde K, White G, Armstrong N (2021) Wearable devices suitable for monitoring twenty four hour heart rate variability in military populations. Sensors 21:1061

Hu C, Ashok D, Nisbet DR, Gautam VJB (2019) Bioinspired surface modification of orthopedic implants for bone tissue engineering. Biomater 219:119366

Hu Y, Xu Y, Mintz RL, Luo X, Fang Y, Lao Y-H et al (2023) Self-intensified synergy of a versatile biomimetic nanozyme and doxorubicin on electrospun fibers to inhibit postsurgical tumor recurrence and metastasis. Biomater 293:121942

Huang J, Xie Z, Xie Z, Luo S, Xie L, Huang L et al (2016) Silver nanoparticles coated graphene electrochemical sensor for the ultrasensitive analysis of avian influenza virus H7. Anal Chim Acta 913:121–127

Iijima S (1991) Helical microtubules of graphitic carbon, nature. 354:56–58

Irani FS, Shafaghi AH, Tasdelen MC, Delipinar T, Kaya CE, Yapici GG et al (2022) Graphene as a piezoresistive material in strain sensing applications. 13:119

Jamieson T, Bakhshi R, Petrova D, Pocock R, Imani M, Seifalian AM (2007) Biological applications of quantum dots. 28:4717–4732

Kacar C, Erden PE, Dalkiran B, Inal EK, Kilic E (2020) Amperometric biogenic amine biosensors based on Prussian blue, indium tin oxide nanoparticles and diamine oxidase- or monoamine oxidase-modified electrodes. Anal Bioanal Chem 412:1933–1946

Kalimuthu P, Fischer-Schrader K, Schwarz GN, Bernhardt PV (2015) A sensitive and stable amperometric nitrate biosensor employing Arabidopsis thaliana nitrate reductase. J Biol Inorg Chem 20:385–393

Kim J, Chou EF, Le J, Wong S, Chu M, Khine M (2019) Soft wearable pressure sensors for beat-to-beat blood pressure monitoring. Adv Healthc Mater 8:1900109

Kim H, Kim Y-S, Mahmood M, Kwon S, Epps F, Rim YS et al (2021) Wireless, continuous monitoring of daily stress and management practice via soft bioelectronics. Biosens Bioelectron 173:112764

Klocke F, Klink A, Veselovac D, Aspinwall DK, Soo SL, Schmidt M et al (2014) Turbomachinery component manufacture by application of electrochemical, electro-physical and photonic processes. CIRP Ann 63:703–726

Kumar M, Shankar S, Kumar A, Anshul A, Jayasimhadri M, Thakur OP (2020) Progress in multiferroic and magnetoelectric materials: applications, opportunities and challenges. J Mater Sci Mater Electron 31:19487–19510

Kuzubasoglu BA, Bahadir SK (2020) Flexible temperature sensors: a review. Sensors Actuators A Phys 315:112282
Lakhdar Y, Tuck C, Binner J, Terry A, Goodridge R (2021) Additive manufacturing of advanced ceramic materials. Prog Mater Sci 116:100736
Liang C, Liu B, Li J, Lu J, Zhang E, Deng Q et al (2021) A nanoenzyme linked immunochromatographic sensor for rapid and quantitative detection of SARS-CoV-2 nucleocapsid protein in human blood. Sens Actuators B Chem 349:130718
Ling P, Lei J, Jia L, Ju H (2016) Platinum nanoparticles encapsulated metal–organic frameworks for the electrochemical detection of telomerase activity. Chem Commun 52:1226–1229
Liu Y, Wang M, Zhao F, Xu Z, Dong S (2005) Bioelectronics, The direct electron transfer of glucose oxidase and glucose biosensor based on carbon nanotubes/chitosan matrix. 21:984–988
Liu X, Cheng S, Liu H, Hu S, Zhang D, Ning HJS (2012) A survey on gas sensing technology. 12:9635–9665
Liu Q, Ding Y, Yang Y, Zhang L, Sun L, Chen P et al (2016) Enhanced peroxidase-like activity of porphyrin functionalized ceria nanorods for sensitive and selective colorimetric detection of glucose. Mater Sci Eng C 59:445–453
Liu J, Shi X, Qu Y, Wang GJN (2023) Functionalized ZnMnFe2O4-PEG-FA nanoenzyme integrating diagnosis and therapy for targeting hepatic carcinoma guided by multi-modality imaging. 15:11013
Ma L, Patil A, Wu R, Zhang Y, Meng Z, Zhang W et al (2021) A capacitive humidity sensor based on all-protein embedded with gold nanoparticles@ carbon composite for human respiration detection. nanotechnol 32:19LT01
Mantuano NR, Natoli M, Zippelius A, Laubli H (2020) Tumor-associated carbohydrates and immunomodulatory lectins as targets for cancer immunotherapy. JITC 8
Maria-Hormigos R, Jurado-Sanchez B, Escarpa A (2018) Self-propelled micromotors for naked-eye detection of phenylenediamines isomers. Anal Chem 90:9830–9837
Mehta VN, Desai ML, Basu H, Singhal RK, Kailasa SK (2021) Recent developments on fluorescent hybrid nanomaterials for metal ions sensing and bioimaging applications: a review. J Mol Liq 333:115950
Meng K, Xiao X, Wei W, Chen G, Nashalian A, Shen S et al (2022a) Wearable pressure sensors for pulse wave monitoring. Adv Mater 34:2109357
Meng X, Zou S, Li D, He J, Fang L, Wang H et al (2022b) Nanozyme-strip for rapid and ultrasensitive nucleic acid detection of SARS-CoV-2. Biosens Bioelectron 217:114739
Neera Singh RM, Gupta P, Kumar D (2017) Ceramic matrix composites: processing techniques and recent advancements. Mater Environ Sci
Nikolic MV, Milovanovic V, Vasiljevic ZZ, Stamenkovic Z (2020) Semiconductor gas sensors: materials, technology, design, and application. Sens 20:6694
Novoselov KS, Geim AK, Morozov SV, Jiang D-e, Zhang Y, Dubonos SV et al (2004) Electric field effect in atomically thin carbon films, science. 306:666–669
Panwar N, Soehartono AM, Chan KK, Zeng S, Xu G, Qu J et al (2019) Nanocarbons for biology and medicine: sensing, imaging, and drug delivery. Chem Rev 119:9559–9656
Pollet J, Chen W-H, Strych U (2021) Recombinant protein vaccines, a proven approach against coronavirus pandemics. Adv Drug Deliv Rev. 170:71–82
Qin R, Feng Y, Ding D, Chen L, Li S, Deng H et al (2021) Fe-coordinated carbon nanozyme dots as peroxidase-like nanozymes and magnetic resonance imaging contrast agents. ACS Appl Bio Mater 4:5520–5528
Ramonas E, Ratautas D, Dagys M, Meskys R, Kulys J (2019) Highly sensitive amperometric biosensor based on alcohol dehydrogenase for determination of glycerol in human urine. Talanta 200:333–339
Rao Z, Wang X, Mao S, Qin J, Yang Y, Liu M et al (2023) Flexible Memristor-based Nanoelectronic devices for wearable applications: a review. ACS Appl Nano Mater 6:18645–18669
Riedel R, Chen IW (2008) Modern trends in advanced ceramics. Ceramics Science and Technology

Rivas L, Mayorga-Martinez CC, Quesada-González D, Zamora-Gálvez A, De La Escosura-Muñiz A, Merkoçi A (2015) Label-free impedimetric aptasensor for ochratoxin-a detection using iridium oxide nanoparticles. Anal Chem 87:5167–5172
Rojas D, Jurado-Sánchez B, Escarpa A (2016) "Shoot and sense" Janus micromotors-based strategy for the simultaneous degradation and detection of persistent organic pollutants in food and biological samples, Analytical chemistry. 88:4153–4160
Russo L, Leva Bueno J, Bergua JF, Costantini M, Giannetto M, Puntes V et al (2018) Low-cost strategy for the development of a rapid electrochemical assay for bacteria detection based on AuAg nanoshells. ACS Omega 3:18849–18856
Shi L, Zhu X, Liu T, Zhao H, Lan M (2016) Encapsulating cu nanoparticles into metal-organic frameworks for nonenzymatic glucose sensing. Sensors Actuators B Chem 227:583–590
Singh Y, Paswan MK, Raghuwanshi SK (2021) Sensitivity enhancement of SPR sensor with the black phosphorus and graphene with bi-layer of gold for chemical sensing. Plasmonics 16:1781–1790
Suprayitno E, Marlianto M, Mauliana M (2019) Measurement device for detecting oxygen saturation in blood, heart rate, and temperature of human body. J Phys Conf Ser 033110. IOP Publishing
Tian T, Liu M, Chen L, Zhang F, Yao X, Zhao H et al (2020) D-amino acid electrochemical biosensor based on D-amino acid oxidase: Mechanism and high performance against enantiomer interference. Biosens Bioelectron 151:111971
Uchino K (2017) The development of piezoelectric materials and the new perspective, advanced piezoelectric materials. Elsevier, pp 1–92
Wang J, Gao W (2012) Nano/microscale motors: biomedical opportunities and challenges. ACS Nano 6:5745–5751
Wang W, Duan W, Ahmed S, Mallouk TE, Sen A (2013) Small power: autonomous nano-and micromotors propelled by self-generated gradients. Nano Today 8:531–554
Wang Z, Huang P, Jacobson O, Wang Z, Liu Y, Lin L et al (2016) Biomineralization-inspired synthesis of copper sulfide-ferritin nanocages as cancer theranostics. ACS nano 10:3453–3460
Wang D, Fang W, Huang C, Chen Z, Nie T, Wang J et al (2022) MR imaging guided iron-based nanoenzyme for synergistic Ferroptosis-Starvation therapy in triple negative breast cancer. 3:159–167
Wu X, Tahara Y, Yatabe R, Toko K (2020) Taste sensor: electronic tongue with lipid membranes. Anal Sci 36:147–159
Yang S, Li Y, Jiang X, Chen Z, Lin X (2006) Horseradish peroxidase biosensor based on layer-by-layer technique for the determination of phenolic compounds. Sens Actuators B Chem 114:774–780
Yao Y, Chen J, Guo Y, Lv T, Chen Z, Li N et al (2021) Integration of interstitial fluid extraction and glucose detection in one device for wearable non-invasive blood glucose sensors. Biosens Bioelectron 179:113078
Yogeswaran U, Chen S-MJS (2008) A review on the electrochemical sensors and biosensors composed of nanowires as sensing material. sens 8:290–313
Zeng W, Yu M, Chen T, Liu Y, Yi Y, Huang C et al (2022) Polypyrrole nanoenzymes as tumor microenvironment modulators to reprogram macrophage and potentiate immunotherapy. Adv Sci 9:2201703
Zhang J, Qin Z, Zeng D, Xie CJPCCP (2017) Metal-oxide-semiconductor based gas sensors: screening, preparation, and integration. Physical Chemistry Chemical Physics 19:6313–6329
Zhang L, Huang Y, Lindstrom AR, Lin T-Y, Lam KS, Li Y (2019) Peptide-based materials for cancer immunotherapy. Theranostics 9:7807
Zhang J, Tan Y, Song W-J (2020a) Zeolitic imidazolate frameworks for use in electrochemical and optical chemical sensing and biosensing: a review. Microchim Acta 187:1–23

Zhang L, Mazouzi Y, Salmain M, Liedberg B, Boujday S (2020b) Antibody-gold nanoparticle bioconjugates for biosensors: synthesis, characterization and selected applications. Biosens Bioelectron 165:112370
Zhao D-H, Li C-Q, Hou X-L, Xie X-T, Zhang B, Wu G-Y et al (2021) Tumor microenvironment-activated theranostics nanozymes for fluorescence imaging and enhanced chemo-chemodynamic therapy of tumors. ACS Applied Materials & Interfaces 13:55780–55789
Zias CWCAR, DeMichele DJ, Emtage PR, Hartman DK, Savery JW, R. C. (1963) Thomas solid-state potentiometer. Pract Appl

Chapter 9
Intelligent Biomaterials in Diagnostic Systems

Qiya Gao, Fangying Xiong, Jie Fu, Shouzhe Deng, Jiawang Wang, and Shuang Li

Abstract This chapter introduces the practical application of diagnostic sensors within various systems. Section 9.1 delineates the taxonomy and operational principles of medical sensors, facilitating the detection and subsequent quantitative or qualitative analysis of pertinent data.

In Sect. 9.2, emphasis is placed on in vivo diagnostics, encompassing a comprehensive overview of implantable cardiac monitors such as artificial pacemakers, and auditory prostheses like cochlear implants. This section delves into their historical evolution, operational mechanisms, and pertinent application scenarios, while also elucidating ongoing developmental trajectories aimed at enhancing biocompatibility and minimizing energy consumption.

Section 9.3 pivots toward in vitro diagnostics, spotlighting cutting-edge immunodiagnostic techniques like ELISA and immunofluorescence analyses, alongside molecular diagnostic methodologies such as PCR and gene sequencing technologies. These methodologies facilitate the detection of specific biomarkers and afford precise observations via microscopy and allied instruments, thereby augmenting the accuracy of diagnostic outcomes.

Section 9.4 introduces the realm of minimally invasive diagnostics, characterized by its benefits of diminished risk, expedited recovery, and reduced trauma. This segment predominantly expounds upon the operational principles and practical applications of microneedle diagnostics, basic low-frequency electrical stimulation, and ultrasonic diagnostics. These modalities afford insights into specific anatomical functions, facilitate status assessment, and prognosticate recovery trajectories.

In Sect. 9.5, the discourse shifts toward wearable physiological index monitoring and diagnostics. Encompassing wearable technologies such as patches and wristbands, this section delineates the monitoring and diagnostic capabilities pertaining to physiological parameters like electrocardiogram (ECG), electromyogram (EMG), and blood oxygen saturation. Furthermore, it elaborates on wearable biochemical

---

Q. Gao · F. Xiong · J. Fu · S. Deng · J. Wang · S. Li (✉)
Academy of Medical Engineering and Translational Medicine, Medical College, Tianjin University, Tianjin, China
e-mail: lishuangv@tju.edu.cn

Z. Liu (ed.), *Intelligent Biomaterials*, https://doi.org/10.1007/978-981-96-8646-9_9

parameter monitoring and diagnosis, encompassing analyses of sweat, interstitial fluid, and urine. Wearable diagnostic technologies, distinguished by their swift detection capabilities and cost-effectiveness, predominantly serve in the early screening of latent health risks.

Section 9.6 elucidates the paradigm of smart integrated diagnostics, offering an exhaustive examination of multimodal diagnostics and the utilization of smart sensors within the healthcare domain. Additionally, it posits prospective developmental trajectories such as miniaturization, energy efficiency, integration with artificial intelligence, and multimodal perception.

Keywords Medical sensor · In vivo diagnosis · Minimally invasive diagnosis · Wearable biosensor

9.1 Medical Sensor Principle

Sensors, devices, or instruments, adept at discerning and identifying a variety of physical, chemical, or biological entities within their environment, possess the capability to transmute this information into measurable electrical signals or alternative forms of output signals. This conversion facilitates the subsequent transmission, processing, and analysis. Predominantly, when the output manifests as an electrical signal, a sensor is composed of three essential elements: a sensing element, a transducer, and a signal conditioning circuit. Owing to the intrinsically feeble nature of the output signal, there is a frequent necessity for auxiliary power supply and signal conditioning and conversion circuits to augment, operationalize, and modulate the signal. Sensors are stratified into physical sensors, chemical sensors, and biological sensors, contingent upon the variety of quantities they are engineered to detect, highlighting their pivotal role across a breadth of diagnostic applications.

9.1.1 Medical Physics Sensors

Physical sensors are instrumental devices employed for detecting and quantifying physical attributes within the environment. They possess the capability to perceive, gather, and transform physical phenomena—such as temperature, pressure, humidity, and light intensity—into interpretable electrical signals or alternative output formats. Temperature sensors, pressure sensors, and photodetectors are among the prevalent variants of physical sensors. These sensors are extensively utilized across diverse domains including industrial control, automation, environmental monitoring, and medical devices. Their real-time data acquisition capabilities empower individuals to comprehensively comprehend and effectively manage their surroundings.

Temperature sensors operate on the principle of correlating object temperature with a specific physical property. Common types of temperature sensors include thermistors, thermocouples, resistance temperature detectors (RTDs), infrared sensors, and semiconductor sensors. Thermistors function by responding to changes in material resistance with temperature, while thermocouples exploit the thermoelectric effect of different metals. Infrared sensors gauge temperature by detecting infrared radiation emitted by objects, and semiconductor sensors rely on the properties of semiconductor materials. These sensors can be categorized based on various factors depending on the intended application and measurement requirements. Typical classifications comprise contact and non-contact sensors, digital and analog sensors, precision sensors, and general-purpose sensors, each offering distinct advantages and suitability for specific needs. Temperature sensors have diverse applications across various fields, including medical devices, where they are employed for tasks like body temperature measurement.

Pressure sensors, on the other hand, are instruments utilized for measuring medium pressure. They convert pressure into electrical signals, which are subsequently relayed to computers, control systems, or other devices for processing, either electrically or digitally. These sensors are extensively used across industrial automation, automotive engineering, aerospace, and medical equipment. Their operational principle hinges on the impact of the medium on the internal forces of the sensor. Common pressure sensor types encompass resistive, capacitive, inductive, and piezoelectric sensors, each differing in sensitivity, measurement ranges, and anti-interference capabilities. The appropriate sensor is chosen based on specific application requirements. Pressure sensors are vital for pressure measurement and control, fluid system monitoring, liquid level measurement, safety monitoring, and more. In medical devices, they serve purposes such as blood pressure monitoring, ventilator and artificial heart control, and other equipment functions. Ensuring the accuracy, reliability, and stability of pressure sensors is paramount for precise measurement and control, often necessitating calibration and regular maintenance. Additionally, attention must be paid to applicable ranges, working environments, and safety requirements during installation and use to prevent operational errors and equipment damage.

Photodetectors constitute a pivotal type of sensing device in modern technology, capable of converting light signals into electrical signals via the photovoltaic effect. They facilitate the detection and measurement of light, objects, or other optical parameters, finding widespread applications across industrial automation, electronic devices, robotics, and medical equipment. Photodetectors can be classified into various types, including photoresistors, photodiodes, phototransistors, optocouplers, and fiber optic sensors, each with distinct characteristics and applications. For example, photoresistors regulate light by monitoring material resistance changes under illumination, while optocouplers discern the presence or absence of objects using light thresholds, and fiber optic sensors detect optical signals based on light transmission properties. Photodetectors offer numerous advantages such as high accuracy, rapid response, non-contact measurement, and high reliability, suitable for a range of applications including object detection, position measurement, flow measurement, liquid level detection, light intensity control, remote controls, and optical communications. In electronic devices, photodetectors are employed in touch screens and optoelectronic encoders, while in medical equipment, they are utilized for tasks like blood oxygen measurement and heart rate monitoring. With ongoing technological advancements, photodetector technology continues to progress, with improvements in sensitivity, stability, and reliability, alongside miniaturization and enhanced integration, offering expanded possibilities for innovative applications across industries.

Physics sensors are widely utilized in the medical sector, serving a crucial function in diagnosing, treating, and overseeing illnesses. Electrophysiology sensors, including electrocardiogram and electroencephalogram sensors, are instrumental in capturing and assessing cardiac and cerebral activities. They assist healthcare professionals in identifying conditions such as heart diseases, neurological disorders, and sleep disorders. Additionally, blood pressure sensors, blood oxygen sensors, and various other physiological sensors are employed to monitor essential signs like blood pressure and blood oxygen saturation, aiding patients in managing chronic conditions like hypertension.

Jeong et al. proposed an implantable device called the mechanical electrocardiogram (MECG) sensor that can simultaneously measure both mechanical electrocardiography and electrocardiography (Fig. 9.1a) (Oh et al. 2023). The key to success was the creation of a permeable electrode on a single low-modulus nanoporous fiber pad, which allowed the sensor to adhere to tissue immediately. The strain-insensitive electrode was used for the ECG electrode, while the strain-sensitive electrode was used for the MCG. A wearable continuous monitoring system was developed to ensure remote monitoring of human vital signs in daily life, enabling assessment of the critical condition. The use of dry woven electrodes not only improved the comfort and flexibility of the wearable continuous monitoring system but also reduced the overall manufacturing cost due to the ultra-low power technology of the microsensor. Prakash et al. evaluated the performance of ultra-low-power flexible single-lead silver-coated nylon braided dry electrodes for electrocardiography (Fig. 9.1b) (Sriraam et al. 2023). The system demonstrated excellent performance in terms of reusability, wetness, washing, treadmill, and activity tests. Liu et al. designed a multifunctional wearable sensor for the simultaneous detection of electroencephalography and sweat rate (Fig. 9.1c) (Gao et al. 2020). Conductive fabric electrodes were combined on the multifunctional sensor for measuring brain electrical signals. The conductive material on the fabric was metal Ag, which has low contact impedance and low noise.

Continuous monitoring of blood pressure in daily life is important and necessary for early detection and control of hypertension. Blood pressure monitoring is a significant concern in the healthcare field, and the utilization of flexible encapsulated sensors offers a promising solution for non-invasive and comfortable blood pressure monitoring. Chang et al. conducted research on flexible blood pressure sensors and developed a blood pressure monitoring system based on a flexible encapsulated sensor (Fig. 9.1d) (Sun and Chang 2023). To enhance wearer comfort and obtain reliable pulse signals, a flexible encapsulated sensor combining poly(para-xylylene)

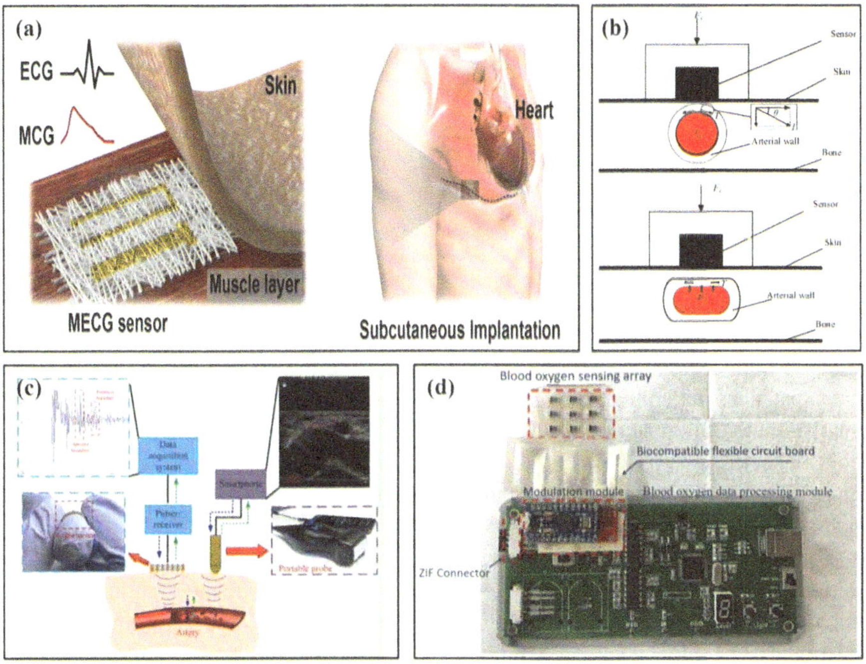

Fig. 9.1 (**a**) Schematic diagram of a fully implantable mechanical ECG sensor. Reprinted with permission from Oh et al. 2023; (**b**) Schematic diagram of flexible packaged sensor blood pressure monitoring system. Reprinted with permission from Sun and Chang 2023; (**c**) Schematic diagram of flexible piezoelectric composite ultrasonic blood pressure sensor. Reused with permission from Peng et al. 2021; (**d**) Schematic diagram of flexible blood oxygen sensing system. Reprinted with permission from Kuo et al. 2022

and PDMS materials was fabricated. Furthermore, to address the issue of low accuracy in blood pressure measurement, various machine learning algorithms were compared and analyzed, ultimately identifying the random forest model as the optimal regression model. Similarly, Jiang et al. reported, for the first time, a flexible piezoelectric composite ultrasound sensor for continuous blood pressure measurement by tracking the ultrasonic motion of the vessel wall (Fig. 9.1e) (Peng et al. 2021). The performance of portable or wearable pulse oximeters can be affected by improper movement or wearing, leading to erroneous blood oxygen concentration calculations. Such errors stem from external stray light or light leakage caused by improper coupling of the sensor with the skin. Wang's study aimed to develop a flexible blood oxygen sensing system that utilizes a reflective-type blood oxygen sensing chip and sequentially measures the oxygen levels at nine positions through time-division pulse modulation (Fig. 9.1f) (Kuo et al. 2022). Each sensing chip consists of an optical transmission and reception part. Inverse chip packaging was employed to integrate the sensing chips, and a flexible poly(para-xylylene) substrate was used, accommodating the curvature of the wrist and positioning the photodiode array around the radial artery. By scanning the sensor array during dynamic behavior, accurate light intensities can be extracted to obtain blood oxygen concentrations and prevent errors caused by improper fitting or sensor movement during motion.

9.1.2 Medical Chemical Sensors

Chemical sensors are devices that can detect, perceive, and convert chemical information into usable signals. They are often used to monitor gases, ions, molecules, and other chemical substances, converting them into measurable forms such as electrical or optical signals, enabling quantitative or qualitative analysis of these chemical substances. Chemical sensors have wide applications in various fields, including environmental monitoring, food safety, medical diagnostics, and industrial process control. They can provide rapid, sensitive, cost-effective, and portable chemical analysis solutions, which are of great significance in protecting the environment and human health.

Chemical sensors typically consist of two main components: the sensing element and the transducer. The sensing element is the core part of the chemical sensor, capable of undergoing specific chemical reactions or interactions with the target chemical substance. These reactions or interactions often lead to changes in the properties of the sensing element, such as conductivity, optical properties, electrochemical properties, etc. The design and selection of the sensing element are closely related to the properties of the target chemical substance, ensuring high selectivity and sensitivity. The transducer is the part that converts the changes in the sensing element into measurable signals. Commonly used transducers include electrochemical transducers, optical transducers, electronic transducers, etc. For example, in electrochemical sensors, changes in the sensing element can result in variations in current, voltage, or resistance, and the transducer converts these changes into measurable electrical signals. In the case of optical sensors, changes in the fluorescent, absorption, or scattering properties of the sensing element are utilized to obtain measurement signals. In practical applications, chemical sensors usually need to be combined with data acquisition systems or signal processors to obtain and analyze measurement signals, and to derive the final chemical analysis results. Additionally, the selection of suitable detection techniques, sensing elements, and transducers is crucial for the performance and application range of chemical sensors.

Chemical sensors offer a fast, accurate, sensitive, economical, and portable means of obtaining real-time analytical results. However, chemical sensors also face challenges in terms of selectivity, sensitivity, stability, and lifespan, which require continual technical improvement and optimization. With the continuous development of nanotechnology, flexible electronics, and smart materials, chemical sensors are becoming more advanced and multifunctional. For example, the application of nanomaterials can improve the sensitivity and selectivity of sensors; the development of flexible electronics allows electrochemical sensors to be wearable and capable of continuous monitoring and analysis of chemical components in the human body and surrounding environment; and intelligent chemical sensors can be connected to data processing systems or the internet, enabling remote monitoring and data sharing.

Nanomaterials provide numerous advantages for electrochemical biosensors used in environmental, food safety, and clinical diagnostic applications, including achieving lower detection limits. Due to their large surface area and high loading capacity, a variety of signal enhancement strategies have been reported utilizing nanomaterials as carriers of multiple signal molecules and biomolecular recognition elements. As many nanomaterials, such as metal nanoparticles, exhibit electrochemical activity, they can also be utilized as electroactive tracers to generate nanostructured electrochemical sensors with improved performance and as catalysts possessing readily accessible active sites on their surfaces. Gold nanoparticles may be the most widely used nanomaterial in biosensors due to their rapid and simple chemical synthesis, shape control, narrow size distribution, biocompatibility, and easy biomolecule binding. In the field of electrochemical biosensors, gold nanoparticles have garnered particular interest because of their unique electrochemical properties, which make the development of various biosensors possible. Therefore, among the numerous nanomaterials, gold nanoparticles are the most intensively studied and hold great promise for applications in bio-sensing, disease diagnosis, and medical treatment.

In recent years, there has been an increasing demand for personalized healthcare, leading to exploration of wearable chemical sensors. Due to their unique ability to detect target biomarkers, wearable chemical sensors enable non-invasive on-site measurement for obtaining more comprehensive information about the wearer's health. With the advent of cloud computing and the Internet of Things (IoT), the obtained biochemical information can be transmitted to processing centers for remote diagnosis and medical research. Human biological fluids, such as blood and bodily fluids, contain rich physiological information that reflects health. Therefore, they are considered satisfactory research subjects for the development of wearable chemical sensors. Compared to traditional blood testing, bodily fluids, including sweat, tears, and saliva, have the advantages of non-invasiveness, ease of access, and convenience. The sampling points make sweat a particularly attractive medium for non-invasive biomarker monitoring. Traditional surface plasmon resonance (SPR), fluorescence spectroscopy, mass spectrometry, and gas chromatography require bulky or expensive instruments that are not suitable for wearable applications (Singh et al. 2022). Therefore, several common electrochemical sensing strategies are used for wearable chemical sensors, such as current and potential methods, due to their fast response, high sensitivity, and portability (Sempionatto et al. 2020).

Among them, the current method is generally used to detect metabolites (such as glucose and lactate) by measuring current changes. Typically, a constant voltage is applied between the working electrode and the reference electrode, and the oxidation-reduction reaction of the metabolic product is induced using the corresponding oxidase to induce current changes (Tuteja et al. 2018). Neethirajan et al. designed a sensitive and specific immunosensor using label-free electrochemical timed-ampere technology, which was used for detecting cortisol and lactate hormones in human or animal fluids, such as sweat and saliva (Fig. 9.2a) (Tuteja et al. 2018). Electrochemically reduced graphene oxide was used as a synergistic platform for signal amplification, and a biological coupling template for silk screen printing electrodes was used. Cortisol and lactate salt antibodies were coupled with e-RGO using covalent carbon diimide chemistry. Unlabeled electrochemical timed-current detection was used to analyze the response of the desired biomolecules over a wide detection range. This immunosensor showed significant specificity, sensitivity, and non-invasiveness, and had caregiving capabilities. Wang et al. reported a cost-effective silk screen-printed tattoo platform for measuring alcohol in sweat and glucose in ISF (Fig. 9.2b) (Kim et al. 2018a). Ag/AgCl ink was used to create ion electroporative, WE, RE, and contact pads, while Prussian blue (PB) was used to create CE and patterned WE. Enzyme bio-receptor was immobilized in chitosan. After meals and drinking alcohol, blood sugar and alcohol levels were measured, providing future potential for monitoring blood sugar and related alcohol effects in large-scale studies on diabetes and pre-diabetes subjects.

The potential method detects chemicals by measuring changes in potential (voltage), and its working principle is based on the potential changes caused by chemical reactions. The sensor usually consists of two electrodes: a working electrode and a reference electrode. The working electrode is in contact with the chemical being measured, and when the chemical reaction occurs, potential changes occur on the

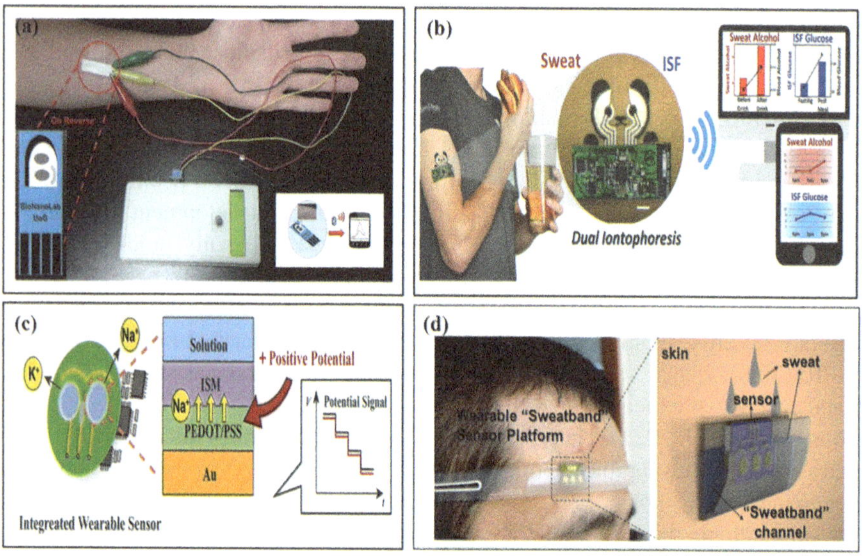

Fig. 9.2 (**a**) Immunosensors for detecting cortisol and lactate hormones in human or animal body fluids. Reprinted with permission from Tuteja et al. 2018; (**b**) Screen-printed tattoo platform for measuring alcohol in sweat and glucose in ISF. Reprinted with permission from Kim et al. 2018a; (**c**) Wearable solid-contact sodium ion ISE. Reused with permission from Liu et al. 2022; (**d**) New all-solid-state sodium ion ISE wearable sweat sensor platform. Reprinted with permission from Wang et al. 2017

working electrode. The reference electrode is used to provide a stable reference potential to make the measurement more accurate. Common potential-based chemical sensors include ion-selective electrodes (ISE) for measuring the concentration of specific ions, such as sodium and chloride ions, based on the selective reaction between specific ions and electrodes. Xiu et al. proposed a wearable solid-contact ISE that is easy to prepare and highly integrated, allowing multiple repeated measurements using an open circuit potential by applying a positive potential to the working electrode (Fig. 9.2c) (Liu et al. 2022). The integrated sensor was manufactured on a printed circuit board, eliminating external hardware. This wearable platform has been tested on humans by monitoring ion levels in real sweat. Zhang et al. proposed a new all-solid-state sodium ion ISE wearable sensor platform based on a gold nanobranch array electrode as a solid contact and a polyvinyl acetate/inorganic salt film-coated all-solid-state RE (Fig. 9.2d) (Wang et al. 2017). A sweat strap sensor platform was developed for effective collection of sweat and real-time analysis of sweat sodium during indoor exercise. This all-solid-state ISE and RE-integrated sensing platform provides a very simple and reliable method for constructing various portable and wearable devices.

In summary, chemical sensors, through the coordinated interaction between sensing elements and transducers, can convert information about the target chemical substances into measurable signals. They hold promising applications in real-time monitoring and detection of chemical substances and play a crucial role in various fields. With continuous innovation and progress, chemical sensors will provide us with more convenient, reliable, and efficient solutions for chemical analysis.

9.1.3 Medical Biosensors

Biosensors are devices capable of identifying specific biological elements that interact with substances in a sample, resulting in physicochemical changes that can be measured as signals. Since the proposal of the glucose enzyme amplifying electrode by Leland Clark Jr. in 1962, biosensors have found widespread applications in clinical, industrial, environmental, and agricultural analysis. According to the definition provided by the International Union of Pure and Applied Chemistry (IUPAC), biosensors are independent, integrated analytical devices in which the biorecognition element (biochemical receptor, including enzymes, antibodies, antigens, peptides, DNA, aptamers, or living cells) is in direct spatial contact with the transducer element(s) (such as electrochemical, optical, or bioluminescent signals). Depending on the type of biorecognition element and mode of application, biosensors can be categorized into various types, such as enzyme sensors, immunosensors, DNA sensors, and cell sensors, among others. Biosensors can also be classified into electrochemical biosensors, optical biosensors, and other types based on the signal transduction methods employed. Each type of sensor has its specific recognition mechanism and application range. Biosensors were initially developed for point-of-care testing (POCT) of biomolecular targets, with the aim of extending clinical analysis from specialized laboratories to public settings, including hospitals, non-hospital healthcare settings, or even home environments. Although various biosensors have been developed for sensitive and selective detection of a range of disease-related molecules, the clinical translation of biosensors remains limited due to challenges in integrating and miniaturizing biosensors into portable devices.

Among different bio-sensing platforms, electrochemical biosensors that integrate the biorecognition elements within the electrochemical transducer(s) (e.g., electrodes or field-effect transistors) are particularly suitable for device integration, as they can be easily miniaturized, mass-manufactured, and integrated with electronic readout modules on a single chip. Furthermore, electrochemical signals, such as current and potential, can be collected with simple, portable, and low-cost peripheral instruments and consume low power. Additionally, the signal generated by the affinity recognition of the biorecognition element to the target analyte can be amplified through physical, chemical, or biological strategies, greatly enhancing the detection sensitivity. Therefore, electrochemical biosensors hold great potential for the development of POC diagnostic devices. According to the World Health Organization (WHO), POC biosensors should be affordable, sensitive, specific, user-friendly, rapid, robust, equipment-free, and deliverable to end users for on-site testing and diagnosis in the routine lives of individual patients and consumers. As such, POC diagnostics are expected to play a critical role in the diagnosis and treatment of major diseases worldwide. For example, the electrochemical glucose meter, the most successful commercial POC bio-sensing device, has been widely used worldwide to assist diabetes patients.

Yaacob et al. reported the development of a DNA biosensor for the detection of hooked-end Holliday junction DNA using carbon quantum dots (CQDs) (Fig. 9.3a) (Zainuddin et al. 2023). The biosensor system employed a tapered optical fiber as the sensing transducer. Prior to incubating the DNA probe with 16S ribosomal RNA gene probe DNA, commercial chemicals were used to functionalize the CQDs via in situ deposition (drop-casting) onto the tapered region. The detection was achieved through the hybridization between the probe DNA and complementary DNA at different concentrations. Vakilian et al. developed an electrochemical biosensor for the determination of arsenite concentrations using arsenite oxidase (Fig. 9.3b) (Tabibi et al. 2022). The study utilized gold nanoparticles (AuNPs) to improve the analytical characteristics of the biosensor by providing a suitable nanoenvironment for enzymatic activity. Cadavid et al. proposed a novel, compact, and highly sensitive bovine serum albumin biosensor based on etched single-mode-multimode-single-mode optical fiber structure (Fig. 9.3c) (Patiño-Jurado et al. 2022). This sensitive label-free biosensor has the potential to be widely applied in various fields such as disease detection, medical diagnosis, and food safety testing. Zhao et al. presented a microfluidic biosensor for the immunodetection of Toxoplasma gondii using a thin fiber-core microfiber coated with molybdenum disulfide (Fig. 9.3d) (Chen et al. 2023). The thin fiber-core microfiber was obtained by fusion splicing a single-mode fiber and a thin-core fiber using arc discharge and flame heating. To avoid interference and protect the sensing structure, the thin-core microfiber was encapsulated in a microfluidic chip. Molybdenum disulfide and Toxoplasma gondii antigens were

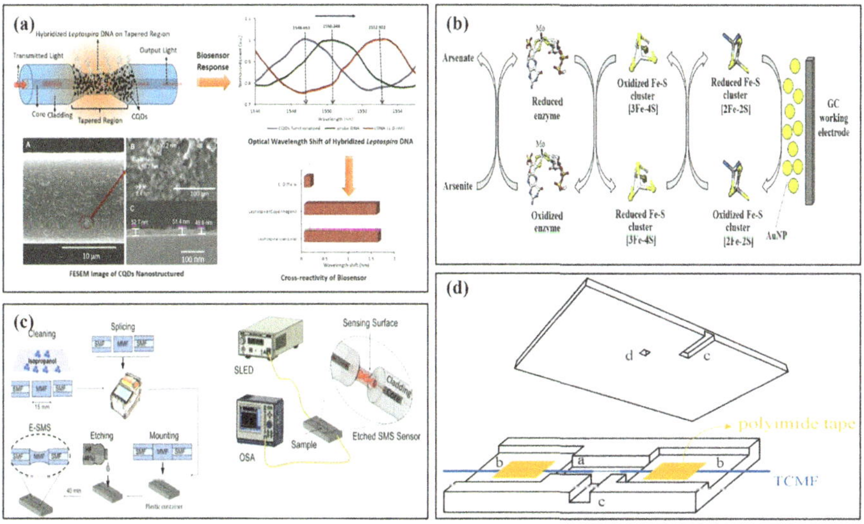

Fig. 9.3 (**a**) Carbon quantum dots (CQDs)-enhanced DNA biosensor for Leptospira DNA detection. Reused with permission from Zainuddin et al. 2023; (**b**) The proposed biosensor for arsenite concentration determination. Reused with permission from Tabibi et al. 2022; (**c**) Fabrication process of the biosensor structures and the experimental setup used to measure transmission spectra of the biosensor. Reused with permission from Patiño-Jurado et al. 2022; (**d**) Structure design diagram of microfluidic chip. Reprinted with permission from Chen et al. 2023

modified on the surface of the thin-core microfiber for the immunodetection of Toxoplasma gondii.

9.2 In Vivo Diagnosis

The advancement of in vivo diagnostic technology is primarily showcased in the rapid evolution of implantable diagnostic devices. These devices rely on electronic components with monitoring capabilities implanted within the body to conduct ongoing detection of physiological and biochemical parameters. They transmit biological information wirelessly to external devices for processing, enabling timely responses to emergencies, identification of abnormalities, and appropriate actions. Fundamental active implantable medical devices encompass implantable electronic materials, energy supply systems, integrated circuit technology, signal transmission, and reception systems, among other technologies. They offer several advantages: (1) By utilizing implantable diagnostic device systems, there is no need for surface measurements, thus reducing various sources of interference and enhancing data precision. (2) They enable the continuous measurement and analysis of various physiological and biochemical parameters under diverse physiological conditions.

(3) They facilitate direct modulation of organs and tissues involved in the healing process of injured areas. (4) They serve as a basis for assessing recovery status.

Consequently, implantable diagnostic device systems exhibit significant developmental potential and promising market prospects.

Implantable diagnostic devices constitute high-risk medical devices implanted into the human body to support bodily functions, albeit with inherent potential risks. Hence, stringent regulations are imperative to ensure their safety and efficacy. The origins of early implantable diagnostic devices trace back to the inception of pacemakers in the early 1950s. Initially designed to deliver electrical impulses for heart stimulation in patients with complete or partial heart block, these devices were bulky and powered directly by 110 V alternating current, resulting in skin burns due to technical constraints.

Through collaborative efforts among engineering, academic, and medical communities, significant progress has been made in the development of implantable diagnostic devices. Innovations include stable voltage battery power systems, myocardial wires to mitigate skin burns from external electrodes, circuit integration to reduce device size, and the use of biocompatible materials to minimize rejection.

From the introduction of implantable ECG recorders in the mid-twentieth century to implantable EEG recorders and hearing devices in the late twentieth century, and now encompassing a broader range of applications in the twenty-first century, these devices can continuously monitor indicators such as blood sugar, blood pressure, and nerve electrical stimulation. Moreover, advancements in materials have resulted in enhanced biocompatibility, signal transmission rates, and accuracy.

9.2.1 *Insertable Cardiac Monitor*

ECG monitoring is a widely used technique for screening and diagnosing heart conditions, notably arrhythmia. The effectiveness of ECG signals in identifying abrupt conditions such as arrhythmia is well-established. Traditional ECG monitoring devices encompass external loop recorders, Holter monitors, and multi-lead electrocardiograms. Nevertheless, the limited occurrence and brevity of arrhythmias pose challenges for conventional clinical monitoring approaches, often leading to overlooked diagnoses and incorrect assessments. Consequently, there's a delay in detecting and addressing emergencies such as syncope and stroke (Members et al. 2015).

Hence, the proposal is to employ an insertable Cardiac Monitor (ICM) for continuous real-time ECG monitoring, accomplished by implanting subcutaneous electrodes. This strategy mitigates the deficiencies associated with conventional ECG monitoring techniques, such as restricted monitoring periods, intermittent data collection, and vulnerability to external disturbances. ICMs offer comprehensive data support and a conceptual foundation for diagnosing conditions like unexplained syncope, monitoring and identifying atrial fibrillation, sudden strokes, unexplained palpitations, and other related ailments.

9.2.1.1 The Development History of ICMs

ICMs made their debut in the 1990s, evolving from the technology of cardiac pacemakers, which were pioneered in 1952. Dr. Senning and engineer Elmqvist in Sweden engineered the initial pacemaker suitable for human implantation, marking the advent of fully implantable pacemakers. Initially deployed for emergency treatment of patients with complete or partial heart block, these devices function by emitting electrical pulses of specific frequencies, directly stimulating the myocardium through electrode leads. This stimulation prompts corresponding excitation of myocardial cells, ultimately restoring contraction within the atrium or ventricle. However, early external pacemakers were bulky, weighing 7.2 kg, with wires extending externally to the heart's surface or internally through veins into the right ventricle, causing inconvenience for users. Furthermore, their pacing thresholds were unstable, susceptible to external influences and power fluctuations, necessitating frequent hospital visits for adjustments. Initially, nickel-chromium or zinc-mercury batteries powered these devices, typically lasting around 2 years, but often leading to lead fractures and subsequent power irregularities (Zeppenfeld et al. 2022).

The evolution of cardiac pacemakers can be categorized into four distinct phases: the first-generation fixed-rate type, the second-generation on-demand type, the third-generation physiological type, and the fourth-generation adaptive type. Following 1958, the development of implantable pacemakers experienced rapid strides. That year, Dr. Seymour Furman's extensive endocardial pacing experiments revealed a lower pacing threshold compared to epicardial pacing, leading to the advocacy for endocardial electrodes. Collaborating with Schaldach in 1959, Furman facilitated the production of endocardial pacing electrode leads, streamlining pacemaker implantation and reducing energy consumption.

In 1964, Castellanos et al. achieved a significant milestone by successfully crafting the ventricular on-demand pacemaker, ushering in the second generation of pacing technology. In 1976, American scientist M. Stanley Whittingham devised the inaugural rechargeable metal lithium battery, utilizing titanium sulfide as the positive electrode material and metallic lithium as the negative electrode material. This breakthrough marked the transition from zinc-mercury batteries to lithium batteries for pacemaker energy supply, boasting improved stability and a lifespan of up to 13 years, thus diminishing the need for surgeries and minimizing infection risks.

Further innovation occurred in 1978 when Furman introduced the pioneering dual-chamber pacemaker, implanting two electrode leads separately in the right atrium and right ventricle for atrioventricular sequential pacing to address bradycardia. Concurrently, the inaugural implantation of a DDD pacemaker underscored the entry of pacing technology into its third generation.

In 1986, pacemakers with frequency response capabilities emerged, epitomizing the automation trend in pacemaker development. The year 1995 witnessed the debut of the first pacemaker with automatic capture of pacing threshold, further enhancing the automation aspect of pacemakers. Finally, in 1997, Auricchio and Blane achieved a significant breakthrough by completing the first implantation of cardiac resynchronization therapy (CRT), marking a pivotal advancement in the field.

In recent years, implantable cardiac pacemakers have achieved notable progress in terms of extended lifespan, miniaturization, passivity, and biocompatibility. Presently, implantable cardiac pacemakers primarily consist of three components: pacemakers, implantable cardioverter defibrillators, and implantable loop recorders. These three device types are elaborated upon below:

9.2.1.2 Pacemaker

An artificial pacemaker comprises two essential components: the pulse generator and the electrode leads. Typically, the pulse generator is positioned in the subclavian region, situated anteriorly on either the left or right side of the chest. It functions by emitting a pulsed current of specific frequency through the electrode lead, which is threaded via the subclavian vein or cephalic vein (Crozier and McAlister 2011). The device stimulates the myocardium directly through an electrode, triggering excitement once it reaches the motor potential threshold of myocardial cells. This excitement propagates to adjacent myocardium via gap junctions or intercalated disc junctions between cells, resulting in sequential atrial or ventricular excitement and subsequent heart contraction. This mechanism effectively addresses cardiac dysfunction arising from arrhythmias. Additionally, pacemakers are increasingly utilized in treating tachyarrhythmias and non-cardiac electrical disorders such as preventing paroxysmal atrial tachyarrhythmias, carotid sinus syncope, and achieving biventricular synchronization in drug-refractory congestive heart failure cases. Pacemakers are categorized into single-chamber, dual-chamber, and multi-chamber configurations based on the targeted pacing chamber. For instance, a single-chamber pacemaker implants a wire into a specific heart chamber (e.g., right atrium or right ventricle) for pacing. Dual-chamber pacemakers utilize pacing electrode leads implanted into both the atrium and ventricle to address cardiac function pauses caused by arrhythmias. Single-chamber pacemakers encompass ventricular demand inhibition (VVI) and atrial demand inhibition (AAI) pacemakers. In a VVI pacemaker, the upper electrode lead is inserted into the apex of the right ventricle, specifically within the muscle trabeculae. In an AAI pacemaker, the upper electrode lead is positioned within the pectinate muscle of the right atrial appendage. Dual-chamber pacemakers, such as the DDD model, typically involve the placement of two electrode leads—one in the right atrial appendage (atrium) and the other in the right ventricular apex (ventricle)—to facilitate atrioventricular sequential pacing. Multi-chamber pacemakers encompass three-chamber and four-chamber configurations, primarily utilized for managing refractory heart failure and paroxysmal heart failure associated with interatrial conduction block, as well as addressing symptoms like atrial fibrillation and congestive heart failure. However, direct implantation of leads into cardiac chambers carries the risk of complications such as infection or hematoma, which may vary among individuals. The early complication rate for the dual-chamber DDD pacemaker (8.7%) is notably higher compared to the single-chamber VVI pacemaker group (2.9%) (Kuang et al. 2021).

The pacemaker, as an implantable device, can automatically and continuously monitor the patient's heart rhythm and the functioning of the pacemaker post-implantation. The collected data can be analyzed to assess the patient's ECG status under normal conditions. Presently, pacemakers offer features such as automatic pacing mode adjustment, pacing frequency adjustment, voltage output adjustment, and atrioventricular delay adjustment. By continuously recording and storing single-lead ECG signals, pacemakers can automatically detect arrhythmias and record ECG signals before and after abnormalities occur, aiding in diagnosis.

It's crucial to recognize that pacemakers can occasionally malfunction or incorrectly detect tachycardia, which might resemble the signal of an acute myocardial infarction. Therefore, physicians must possess a comprehensive understanding of electrocardiogram abnormalities related to pacemakers, be capable of distinguishing various types of pacemakers and their lead placements, and be ready to manage any emergencies that may arise during pacemaker assessment (Chauhan et al. 1994).

9.2.1.3 Implantable Cardioverter Defibrillator

The implantable cardioverter defibrillator (ICD) resembles a pacemaker in appearance. It's an electronic device designed to promptly identify and terminate malignant ventricular arrhythmias detected by the pacemaker's ECG. With a larger motor, the ICD offers multiple functions, including supportive pacing, anti-tachycardia pacing, low-energy cardioversion, and high-energy defibrillation. It automatically identifies malignant arrhythmias like ventricular fibrillation within 10 seconds and delivers electric shock defibrillation to address ventricular tachycardia or ventricular fibrillation. The ICD serves as a highly effective treatment method for preventing sudden death, significantly reducing its incidence and mortality rates.

Like pacemakers, ICDs consist of two components: a pulse generator and electrode leads. However, the ICD system requires a higher energy source, typically utilizing two series-connected lithium-silver barium pentoxide batteries in the pulse generator. ICD leads are distinguishable from pacemaker leads by the presence of one or two shock coils, identifiable by dense bands along specific sections.

Traditionally, ICDs are implanted below the clavicle, with wires passed through veins to stimulate myocardial cell excitement. In 2023, Medtronic introduced the Aurora EV-ICD system, implanted below the patient's left axilla (left mid-axillary area). This innovative technique, unlike traditional methods, involves minimally invasive placement of the leads under the sternum instead of within the heart and veins. This external positioning significantly reduces the risk of long-term complications.

In numerous clinical trials, implantable cardioverter-defibrillators (ICDs) have proven effective in preventing sustained ventricular tachycardia (VT), ventricular fibrillation (VF), heart failure, and severe left ventricular dysfunction in high-risk patients. The life-saving role of ICDs in preventing sudden death is firmly established, with their ability to deliver high-energy therapy to terminate fatal ventricular arrhythmias and administer shocks being pivotal in saving lives (Kuang et al. 2021).

Furthermore, ICDs are appropriate for acute conditions such as non-ischemic dilated cardiomyopathy, organic heart disease, and syncope accompanied by significant left ventricular dysfunction.

Given its implantable nature, ICDs may encounter operational malfunctions that result in deactivation and associated risks. Hence, comprehending the common causes of ICD deactivation and their respective remedies is crucial. Additionally, it's imperative for the user's family members to be educated on assessing the device for anomalies based on remote data (Williams and Poole 2013; McEvedy et al. 2018).

9.2.1.4 Implantable Loop Recorder

The implantable loop recorder (ILR) is a single-lead electrocardiogram detection device implanted beneath the skin. It's designed to identify and record abnormal electrocardiographic events like atrial fibrillation (AF), ventricular tachycardia, bradycardia, and asystole. Positioned beneath the chest skin on the left side of the sternum, the ILR functions as a continuous electrocardiogram processor, capturing irregular heart rhythms. It's commonly employed in diagnosing patients experiencing unexplained or recurring symptoms such as syncope and palpitations. For instance, it can help identify generalized tonic-clonic seizures stemming from intractable epilepsy (MacIver et al. 2016). For stratified management of arrhythmia risks, the implantable loop recorder (ILR) is especially well-suited for conditions that cannot be diagnosed using short-term external ECG recording technology or conditions requiring long-term ECG data, such as atrial fibrillation. Due to its automatic ECG data storage function, the ILR can accurately diagnose potential risk factors and eliminate redundant examinations.

To facilitate layered management of arrhythmia risks, the implantable loop recorder (ILR) incorporates an automatic function for storing ECG data, rendering it exceptionally apt for conditions that elude diagnosis with brief external ECG recordings or necessitate enduring ECG data collection, such as atrial fibrillation. This feature enables the ILR to pinpoint potential risk factors precisely, thereby obviating the need for repetitive tests.

The ILR concept originated in 1992 when Canadian physician George Klein adapted a pacemaker into a basic ECG recorder. He utilized the ILR's ECG monitoring capability to document prolonged sinus arrest in patients. Through extended electrocardiogram analysis, the underlying cause of a patient's syncope was ascertained, leading to the implantation of a pacemaker. The inaugural ILR version had a battery life of merely around 14 months and solely permitted manual ECG recordings; the second-generation ILR, introduced in 2002, incorporated an automatic trigger feature that stores the ECG automatically when detected heart rhythms align with predefined arrhythmia criteria. Subsequent ILR generations have furthered this technology, offering diagnostics for ventricular and atrial fibrillation in addition to tachycardia or bradycardia, with a battery lifespan extending to approximately 36 months, suitable for prolonged applications.

In recent years, advancements in technology have ushered in high-performance ILRs. For instance, Medtronic's LINQ II TM device is downsized to one-third of a 3A battery and weighs a mere 3.4 g, boasting the enhanced TruRhythmTM algorithm which delivers 100% sensitivity in diagnosing sinus-atrial block and maintains a low false positive rate of 4.7% in detecting atrial fibrillation. It supports continuous monitoring for up to 5 years. Furthermore, Abbott's Confirm Rx™ represents the first ICM that pairs with a smartphone via Bluetooth, utilizing Sharp Sense™ technology to refine the ECG's P wave and R wave sensitivity while reducing false positives. Additionally, BIOTRONIK's German-engineered BIOMONITOR product offers flexibility, optimizes signal detection, provides superior ECG data, and supports up to 5.5 years of continuous monitoring.

In recent years, the guidelines governing the use of implantable cardiac monitors (ICMs) have seen significant updates. For instance, in 2012, a collaborative effort by the American College of Cardiology Foundation (ACCF), the American Heart Association (AHA), and the American Heart Rhythm Society (HRS) resulted in the release of the "2012 Updated Guidelines for the Management of Arrhythmia Devices." This document introduced revisions to the criteria for cardiac resynchronization therapy (CRT) and laid out detailed implantation, follow-up, and management strategies for cases with a left ventricular ejection fraction (LVEF) under 30% for Class I indications. Then, in August 2021, the European Society of Cardiology (ESC) published its most recent guidelines on cardiac pacing and CRT in the European Heart Journal. These guidelines enhanced the protocols for implanting pacemakers under special conditions. For example, the use of leadless pacemakers is advised for patients undergoing hemodialysis or those lacking upper limb venous access, which significantly raises the risk of infection at the device pocket site. For patients undergoing surgery that might cause or has caused an atrioventricular block or those having tricuspid valve surgery, the placement of epicardial pacing leads is recommended. Furthermore, it's suggested that patients with implantable pacemakers should only be considered for MRI scans when there's a definite need, provided that specific leads are in place, and certain precautions are observed.

9.2.2 Implantable Hearing Devices

Over the last 30 years, the fields of audiology, otology, and electronic device technology have advanced significantly, leading to the clinical adoption of various implantable auditory devices (IADs), with cochlear implants being a notable example. These implantable hearing devices, surgically inserted into the human body, aim to restore hearing for individuals suffering from sensorineural, conductive, or mixed deafness. Unlike passive implants, such as those used for ossicular chain reconstruction, IADs are active devices. They work by converting sound into specifically coded electrical signals via a converter. These signals are then transmitted as energy through either skull vibration, the ossicular chain, or by directly stimulating the lymphatic vessels near the cochlea, thereby directly exciting the cochlear nerve or the brainstem nerve (Ho et al. 2006). Patients with different forms of deafness are fitted with specific types of hearing implants tailored to their condition. For instance, cochlear implants (CI) are ideal for individuals with severe to profound sensorineural hearing loss, while vibrating sound bridges (VSB) and bone-anchored hearing aids (BAHA) are recommended for those experiencing moderate to severe sensorineural, conductive, or mixed deafness. For patients who have lost bilateral auditory nerve function, resulting in total deafness, auditory brainstem implants are considered appropriate (Svrakic and Vambutas 2019). Traditional acoustic hearing aids are effective for mild hearing impairment, whereas implantable hearing devices are primarily utilized for more severe forms of hearing loss, including severe sensorineural and conductive deafness.

9.2.2.1 The History of Implantable Hearing Devices

The inception of implantable hearing devices dates back to 1800 when the Italian physicist Alessandro Volta discovered that electrical stimulation could induce hearing in normal ears. Volta experimented by inserting wires connected to a 50 V battery into his ears, creating a current path through his auditory system. However, due to the nascent stage of electrical technology and safety concerns, no further experiments were pursued at the time.

In 1957, the first implantable hearing device, capable of electrically stimulating the auditory nerve, was invented by French physiologist Andrè Djourno and otolaryngologist Charles Eyriès. They developed this technology through their work on implantable induction coils, which converted electromagnetic signals into electric currents to stimulate the skin or tissues, thus creating a conductive pathway.

For patients suffering from bilateral cholesteatoma, who had undergone extensive removal of cochlear nerves, a tiny portion of the nerve near its base was preserved. The device's wires were implanted at the base of the remaining cochlear nerve, with a ground electrode placed in the temporal muscle. Electrical stimulation from the implanted device enabled these patients to perceive intensity variations in the stimulation, though they struggled to localize sounds accurately in their environment. The placement of the stimulation wire outside the cochlea and the extensive removal of cochlear nerves hindered the effective transmission of electrical stimulation. Consequently, the implanted device had a limited lifespan and was susceptible to malfunctions.

Building on the original concept, subsequent generations have made significant advancements. American doctors William F. House and John Doyle, in 1961, expanded upon Djourno et al.'s research by implanting devices into the round windows of three subjects. This allowed patients to better perceive variations in sound intensity and pitch. In 1967, Professor Graeme Clark from the University of Melbourne, Australia, suggested the insertion of electrodes into the cochlea. This idea laid an important theoretical foundation for future wearable devices. The rapid advancement in the sealing and safety of these devices also played a crucial role. Utilizing sealing technology developed by NASA to protect electronic components from bodily fluids, electronic engineer Jack Urban designed and implemented the first wearable sound processor in 1972. Following these developments, the House/3 M single-channel implant was approved by the FDA for clinical trials in 1982 and was available for commercial distribution by 1984, marking it as the first cochlear implant device to achieve this milestone (Bond et al. 2009). Within the realm of single-channel detection, a user's ability to perceive and recognize sound frequencies is constrained, leading to a markedly superior efficacy of multi-channel treatments over their single-channel counterparts. Multi-channel configurations are designed to be gentler, mitigating the risk of cochlear damage. They also feature a pre-curved shape that optimizes the placement of the electrode array closer to the nerve. This design enhances the compatibility of surgical procedures and stimulation techniques with the cochlea's anatomical and physiological characteristics (Wolfe 2018).

In 1972, French physician Chouard spearheaded a research and development team for multi-channel cochlear implants. By 1977, he emphasized the importance of tailoring the stimulation level to the electrophysiological properties unique to each electrode's contact point. This insight established a theoretical bedrock for future cochlear implant rehabilitation techniques and underscored the significance of early implantation. Following this, in 1976, the French firm Nuerelec introduced the inaugural cochlear implant device to the market. A significant milestone was achieved a year later with the successful development and implantation of the world's first eight-channel microelectronic cochlear implant, heralding cochlear implants' advent into clinical application.

Post-1980, the quest for enhanced precision in cochlear implants gravitated toward refining coding strategies. A pivotal 1984 clinical trial by Symbion Company demonstrated that participants could recognize open-set speech sounds, with CNC monosyllabic word recognition attaining an accuracy of 73% a few months post-implantation. The CIS coding strategy, introduced in 1989, significantly improved user ability to recognize open-set speech. This was accomplished through the introduction of the current steering concept and the deployment of the Advanced Combination Encoder (ACE) coding strategy (Wolfe 2018). Designed for long-term implantable devices, nanoparticles of barium titanate (BaTiO3) and lithium niobate (LiNbO3) exhibit characteristics that preserve the viability, morphological features, distribution of cytochrome C, and levels of reactive oxygen species (ROS) in the neuronal cell line PC12 cells. These properties not only ensure the maintenance of cell health but also promote the growth and stimulation of neurite branches, highlighting their excellent biocompatibility (Lei et al. 2016). The following sections will provide a detailed overview of implantable auditory devices, which include the cochlear implant, vibrating sound bridge, and neurostimulator.

9.2.2.2 Cochlear Implant

The Cochlear Implant (CI) stands as the pioneer and most advanced technology in artificial hearing implants. It functions by converting sounds into electrical signals, which are then transmitted to the patient's cochlea, generating stimuli that enable sound perception. This technology is now capable of restoring or even reconstructing auditory functions for individuals with hearing loss. The implantation of a cochlear device is a surgical process that places a device within the inner ear. The success of this surgery in enhancing hearing capabilities hinges on various factors including the root cause and duration of the hearing loss, and the condition of the auditory nerve.

Before undergoing surgery, a thorough assessment of the patient's hearing abilities is conducted using techniques like electrical impedance evaluation and hearing aid trials, tailored to their unique physiological state. The surgical procedure entails making an incision behind the ear to reach the cochlea and the hearing organ. A non-invasive electrode is inserted into the inner ear to conserve any remaining hearing, while the other electronic components are positioned in a specially prepared bone bed on the skull's surface, subsequently concealed by skin. The surgery involves stimulating the auditory nerve with an electrical current to observe nerve reflexes, which helps in confirming the implant's functionality.

The key to successful cochlear implant surgery lies in the precise and safe insertion of electrodes into the scala tympani via the round window membrane, which effectively stimulates the nerve fibers. The implantation of the electrode is a crucial phase, demanding meticulous control over the force and timing to preserve the delicate structure of the inner ear and any residual hearing as much as possible. External forces beyond 100 mN could lead to damage, such as the rupture of the basilar membrane, which necessitates technical assistance since manual control cannot achieve the required precision. The depth, angle, and effectiveness of electrode stimulation might also vary based on individual physical conditions, underscoring the importance of preserving the inner ear's fine structure. This depends greatly on the surgeon's experience and the equipment's precision. However, the complexity of this surgery limits its widespread application, and it may lead to various complications. Mild issues like inflammation at the incision site, tympanic membrane perforation, and temporal hematoma can be resolved with conservative treatments. In contrast, more severe complications might necessitate surgical remedies, such as addressing cerebrospinal fluid leaks, implant migration, and electrode malformation.

Due to the specialized methods and high precision required for ear surgeries, traditional surgical robots cannot be directly applied to these procedures. For instance, the mechanical arm of the well-known da Vinci robot lacks the adaptability and finesse needed to navigate the intricate structure of the inner ear and does not possess the delicate touch required for such operations. Consequently, there's a pressing need for the development of otological surgery robots specifically tailored for these complex procedures. Currently, otological surgery robots are divided into two main types based on the imaging technologies utilized during surgery: those guided by endoscopes and those guided by medical imaging systems. Furthermore, these robots are categorized into active, semi-active, or passive types, depending on the level of involvement required from surgeons during the procedure (Candito et al. 2022). The endoscope-guided robotic system operates by inserting an endoscope into the inner ear to directly monitor the spatial relationship between surgical tools and tissues. This approach offers the advantage of high-resolution imaging, enhanced brightness in the field of view, and increased procedural efficiency. Furthermore, it provides visibility to areas that are typically challenging to view with a surgical microscope, such as the upper and posterior sections of the femoral compartments. On the other hand, the medical image-guided robotic system begins its process by placing calibration marks on the patient's head to gather detailed medical images. Utilizing this data, it constructs a three-dimensional representation to accurately map out the relationship between internal tissues, organs, nerves, and reference markers. This creates a precise positional coordinate system that enables surgeons to pinpoint the surgical site with accuracy by navigating the surgical path within this coordinated framework (Riojas and Labadie 2020). Several cochlear implant robots designed for clinical application are already available in the market. For example, the HEARO robot, a collaborative effort between MED-EL, an Austrian Cochlear Company, and CAScination AG, a medical technology firm, creates a tunnel directly from the skull's surface with a diameter of 1.8 mm. This allows for the insertion of cochlear implant electrodes with minimized trauma and achieves a precision of 0.14 mm ± 0.07 mm. Additionally, the iotaSOFT cochlear implant robot, developed by iotaMotion, provides assistance to surgeons by controlling the insertion speed of the electrode array during cochlear implant surgeries.

9.2.2.3 Vibrating Sound Bridge

The inaugural vibrating sound bridge implantation surgery was conducted by Professor Ugo Fisch at Zurich Hospital in Switzerland in 1996. Today, this technique is regarded as a relatively established method for artificial middle ear implantation. Differing from traditional hearing aids, which gather, amplify, and channel sound directly to the eardrum, the vibrating sound bridge is composed of an external and an implanted component. The external segment includes a microphone, battery, and an electronic device, discreetly attached beneath the patient's hair via magnetic means. This device converts external sounds into electronic signals, which are then relayed to the implanted segment. Meanwhile, the implanted section encompasses coils, magnets, conductor connections, and the Floating Magnetic Tunneling Sensor (FMT). The FMT, acting as the transducer within the vibrating sound bridge system, converts these signals into mechanical vibrations. Uniquely, it achieves this through the transmission of inertial energy, eliminating the need for an additional supporting structure typically required in inertial drive systems (Bell et al. 2014). The implanted part is directly attached to the incus in the middle ear, inducing vibrations in the ossicles to offset sensory deficits in the inner ear. Within the FMT, the coil and magnet are positioned closely together, facilitating a robust electromagnetic connection. Consequently, a displacement as minimal as 1 μm in the middle ear can generate a sound pressure level of 120 dB, leading to significant sound amplification (Ball 2010).

The Vibrating Sound Bridge (VSB) was initially recommended for adults experiencing moderate to severe sensorineural hearing loss. With the advancement of the technology, it has become applicable for cases of conductive and mixed hearing loss as well. Moreover, VSB implantation preserves the openness of the patient's external auditory canal, thus avoiding the blockage and discomfort that might result from the closure associated with cochlear implants. It is deemed suitable for individuals suffering from chronic suppurative otitis media, external auditory canal atresia, and similar conditions. In evaluating the safety of auditory surgery, preserving residual hearing is considered the primary measure for surgical safety, while the incidence of adverse events is a secondary consideration. A follow-up study involving 100 patients who received VSB implants revealed that most experienced a change in their pure tone average (PTA) of less than 10 dB over 3 months post-activation, with stability observed after 1 year, thereby affirming the safety of the VSB procedure (Goode et al. 1996).

9.2.2.4 Bone-Anchored Hearing Aid (BAHA)

Bone-anchored hearing aids are implantable bone conduction hearing systems in the 1980s. BAHA consists of four parts (Arthur 2002): sound processor, vibration driver, bone anchor, and control unit. The sound processor receives the audio input and performs digital signal processing on it; the vibration driver converts the processed sound signal into mechanical vibration, and conducts the vibration directly to the part of the skull through the bone anchor; the bone anchor is an implanted hearing aid auxiliary equipment, usually made of titanium alloy or other biocompatible materials with good durability and biocompatibility, implants bone anchors into the temporal bone so that external mechanical vibrations can be directly transmitted to the temporal bone through the bone anchors. The inner ear is used to realize auditory perception; the control unit is often located outside the human body and can be used to adjust the volume, select frequency, and so on, so that different individuals can adjust according to their own conditions.

Unlike cochlear implants, which transmit sound to the inner ear through an implanted electrode array, bone-anchored hearing aids (BAHA) receive sound waves through a sound processor outside the body, convert them into digital signals, and amplify them. The vibration driver converts the sound into vibrations and transmits them to the bone anchored in the temporal bone. By bypassing the outer and middle ears, the anchor directly transmits vibrations through the skull to the inner ear, which then reaches the auditory nerve, enabling the hearing-impaired to perceive sound. Bone-anchored hearing aids restore auditory perception to patients through bone conduction, making them suitable for certain types of hearing impairments such as conductive hearing loss or certain inner ear malformations. They can avoid ear canal obstruction and feedback problems, while providing clearer and more precise hearing assistance that offers a natural sound experience (Dun et al. 2011). In addition, since the outer and middle ears are bypassed, bone-anchored hearing aids can enhance hearing and help localize sounds in noisy situations with less distortion of sound signals compared to traditional hearing aids. For patients with unilateral sensorineural deafness, using BAHA can transmit sound through the skull from the deaf side to the hearing side, providing 360° sound perception.

9.2.2.5 Auditory Brainstem Implantation (ABI)

Auditory brainstem implantation (ABI) represents a form of central auditory prosthetics. Unlike traditional cochlear implants that necessitate a functional auditory nerve for hearing restoration, ABI enables hearing in individuals lacking auditory nerves.

The ABI system comprises sound processors, sound generators, implanted devices, and control units. The sound processor, an external apparatus, captures, processes, and encodes auditory signals into digital formats, which are relayed to the sound generator. This generator then transforms the digital inputs into electrical impulses, directed toward the implanted device. Positioned within the auditory nerve or brainstem, the implanted device receives these impulses, directly activating nerve fibers or brainstem nuclei to evoke auditory perceptions. The control unit fine-tunes settings like current strength and frequency to cater to the varied requirements of patients.

ABI is particularly beneficial for individuals with significant auditory nerve loss or profound inner ear anomalies. It circumvents the conventional auditory pathway by directly stimulating alternative neural receptors in the brainstem's cochlear nucleus. These receptors initiate auditory nerve impulses that ascend to the cerebral cortex, culminating in the perception of sound (Mylanus et al. 1998).

Initially, auditory brainstem implantation (ABI) was primarily indicated for individuals with neurofibromatosis type 2 (NF2). Over time, as the technology advanced, its application has expanded to include other auditory conditions such as cochlear ossification, stenosis of the internal auditory canal, and auditory nerve aplasia. ABI surgery, noted for its longer duration compared to cochlear implant procedures, is conducted by a neurosurgeon. The surgical process entails cranial opening, implant placement, intraoperative evaluation, securing the implant, and wound closure. The procedure demands rigorous neuromonitoring standards throughout (Sennaroglu and Ziyal 2012). The cochlear nucleus is located in the brainstem, which is the center that regulates vital life activities such as breathing and heartbeat. Therefore, the installation and fixation of the electrode device are critical focal points of the surgery (Fig. 9.4).

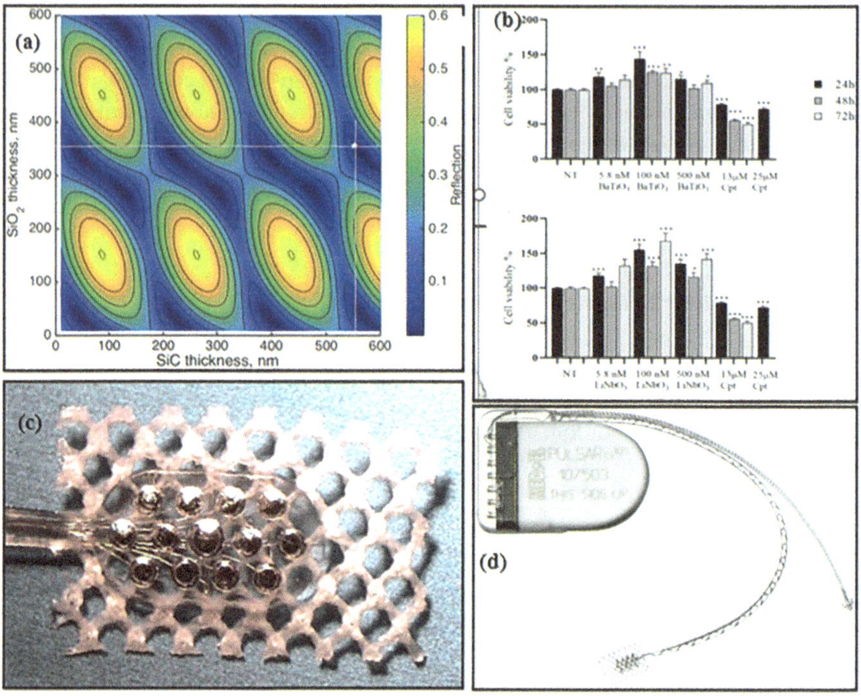

Fig. 9.4 (**a**) Investigating the optical reflectivity of a SiC and thermal SiO2 stack, it has been found that a configuration with 560 nm of SiC on top of 310–380 nm of SiO2 can reduce the reflection of 880 nm light below 10% at vertical incidence. This finding is significant for optimizing the material composition of microneedles. (Reused with permission from Lei et al. 2016. (**b**) Results of MTS assay of differentiated PC12 cells treated with three different concentrations of barium titanate ($BaTiO_3$). Reprinted with permission from Candito et al. 2022. (**c**) The receiver/stimulator has an array with 12 platinum electrodes. Reused with permission from Sennaroglu and Ziyal 2012. (**d**) ABI electrode of the Med-El Company. Reused with permission from Sennaroglu and Ziyal 2012

9.2.3 Neurostimulator

Implantable neurostimulators function by delivering specific levels of electrical pulses to targeted muscles or nerves, aiming to modify or restore functions within the brain, nerves, or muscles. This approach relies on neurofunctional modulation to stimulate different bodily regions, addressing a range of symptoms. The inception of neurostimulation traces back to 1874, when French neurologists Gustav Fritsch and Eduard Hitzig first applied electrical stimulation to the cerebral cortex of dogs, eliciting motor responses that helped delineate brain functional areas. This pioneering work demonstrated the impact of electrical stimulation on neural tissues.

Following the advent of pacemaker technology, the development of nerve stimulators for various body areas accelerated. Notably, in the late 1970s, American physician Norman Shealy introduced the first spinal cord neurostimulator, employing it to alleviate chronic pain. The 1980s saw the introduction of deep brain stimulation into neuroscience, while the 1990s marked a surge in the use of peripheral nerve stimulators to manage chronic pain, urinary incontinence, and neuromuscular disorders.

The U.S. FDA has authorized several implantable neurostimulators for clinical applications, including spinal cord stimulators for pain management, deep brain stimulators for treating Parkinson's disease, tremors, dystonia, and obsessive-compulsive disorder, vagus nerve stimulators for epilepsy and depression, and sacral nerve stimulators for urinary incontinence (Brackmann et al. 1993).

An implantable neurostimulator, classified as an Active Implantable Medical Device (AIMD), undergoes continuous enhancement in performance. This progress aligns with the development of novel medical therapies that leverage electrical stimulation of the central or peripheral nervous systems, alongside a deepening understanding of their underlying mechanism (Pelter et al. 2013). Implantable nerve stimulators, as opposed to traditional external nerve stimulators, offer prolonged operational duration, capable of functioning within a patient's body for several years due to their surgical implantation. These devices provide a significant advantage over their external counterparts by minimizing risks such as accidental damage or displacement, thus enhancing safety and reliability. Specifically for patients in need of long-term therapy, implantable nerve stimulators deliver precise localization of the nerve stimulation area, enabling stimulation at higher frequencies and intensities to notably improve treatment outcomes. Consequently, this discussion will cover three key types of implantable stimulators: Spinal Cord Stimulators, Deep Brain Stimulators, and Vagus Nerve Stimulation.

9.2.3.1 Spinal Cord Stimulator (SCS)

In 1965, American neuroscientists Melzack and Wall introduced the "gate control theory" of pain, suggesting that pain results from a complex interplay of various factors. This theory laid the foundational basis for the development of Spinal Cord Stimulation (SCS). The 1970s saw the advent of the fully implantable SCS system. By delivering electrical stimulation directly to the spinal cord area of the patient, an implantable spinal cord stimulator (SCS) effectively mitigates pain symptoms. SCS is predominantly utilized in the treatment of chronic back pain, limb pain, and neuropathic pain, and it typically comprises two main components: an electrode and a pulse generator. The electrode, connected by a slender wire, is surgically implanted near the spinal cord. Its tip features several small metal contacts that engage with the nerve fibers in the spinal cord to transmit electrical signals. Precise placement of the electrode is vital to ensure accurate delivery of stimulation signals to the designated area. The pulse generator, implanted in the patient's abdomen or buttocks, facilitates the provision of spatially selective electrical stimulations through real-time activation, directing them to the intended location via the electrodes (Hrdlicka 2002). Parameters such as frequency and intensity are adjusted through an external controller to meet the needs of different individuals.

Early SCS was mainly used to relieve chronic back pain, including neuralgia caused by spinal cord injury and complex regional pain syndrome (CRPS), among others. There are two main neurophysiological mechanisms of spinal cord stimulation: analgesia in neuropathic pain states and the release of neurotransmitters such as dopamine and GABA, which play a role in analgesia in ischemic pain. With a better understanding of the connection mechanism between the spinal cord and other nerves, SCS can now be used to treat a wider range of diseases. For example, implantable spinal cord stimulators can improve movement disorders after stroke, limb paralysis, spasticity, and intervertebral disc herniation, as well as alleviate nerve root pain symptoms. In cases where spinal cord injury disrupts communication between the brain and the areas of the spinal cord involved in walking, the Brain-Spine Interface (BSI) developed by Lorach et al. can help patients regain independent upright walking. BSI consists of a fully implantable recording and stimulation system that establishes a direct link between cortical signals and simulated modulation of epidural electrical stimulation targeting the spinal cord areas involved in walking. The FDA has approved SCS for the treatment of failed lumbar spine surgery syndrome, chronic pain, complex regional pain syndrome, refractory angina, visceral abdominal pain, perineal pain, and limb pain caused by nerve damage.

9.2.3.2 Deep Brain Stimulator (DBS)

The concept of deep brain stimulation (DBS) was first introduced in 1950. Surgical implantation of electrodes and stimulators into the deep brain has proven that stimulating the deep brain can improve the symptoms of certain neurological diseases. However, due to technical limitations, surgical positioning and intensity cannot be achieved. In 1987, microelectrode recording technology was used for the first time, resulting in improved experimental accuracy and applications for additional neurological diseases such as Parkinson's disease and obsessive-compulsive disorder. The US FDA approved DBS for the treatment of Parkinson's disease, torsades, and chronic pain in 2002, 2003, and 2005 respectively. DBS involves the implantation of electrodes into the patient's brain, and uses electrical signals generated by a pulse generator to stimulate specific nerve nuclei deep in the brain, correcting abnormal electrical circuits and alleviating neurological symptoms. During the implantation process, doctors use electroencephalogram scanning to locate targeted nerve nuclei by recording the characteristics of different depths and trajectories of the brain, observing the patient's response to stimulation, and determining the best electrode placement location while minimizing side effects.

DBS consists of three parts: implanted electrodes, positioning system, and pulse generator. Implanted electrodes are wires and electrodes surgically implanted deep into the brain. Abnormal neural activity is regulated by giving electrical stimulation to the electrodes. The diameter of the intracerebral stimulation electrodes is typically 1.27 mm, and they have four stimulation contacts spaced at either 0.5 mm or 1.5 mm intervals for stimulation selection. Metal electrodes with good biocompatibility, such as titanium alloy, are commonly used. The surgical positioning system typically includes head fixators, stereotaxic instruments, and electroencephalogram (EEG) examinations. These tools help locate the target nerve nucleus during surgery and observe the patient's reaction, allowing for optimal electrode placement. The pulse generator is a device that controls the stimulation delivered to the electrodes in the DBS system. It is usually implanted in the chest or neck area and provides regular electrical stimulation to achieve therapeutic effects. To ensure stable output current, the pulse generator is powered by either a rechargeable battery or an MRI-safe battery.

DBS enables direct measurement of pathological brain activity, providing a theoretical basis for exploring the connection between brain circuit abnormalities and dysfunction (Lorach et al. 2023). This technology has been beneficial for studying and treating various brain diseases. Initially, DBS was primarily used to treat Parkinson's disease. However, as our understanding of the relationship between brain neural structures and related diseases has improved, the causal relationship between brain stimulation and brain lesions or neuropsychiatric disorders has become clearer. Lozano et al. (2019) conducted experiments using DBS, transcranial magnetic stimulation, and other methods to identify a single depression circuit. Through their research, they found that different methods converged on this specific circuit, and the experimental group showed improved antidepressant effects. This demonstrates the potential of DBS and similar neurostimulation techniques to optimize the treatment of depression and other related conditions. DBS is a promising technology that allows for the direct measurement and regulation of pathological brain activity. It has shown potential in the treatment of various brain disorders and has contributed to our understanding of the brain's neural circuits and their relation to dysfunction.

9.2.3.3 Vagus Nerve Stimulation (VNS)

The vagus nerve is the tenth pair of the 12 pairs of cranial nerves in the human body. It is issued by the vagus nucleus located on the back and inside of the brain. It reaches the neck through the jugular foramen, then enters the chest and abdomen, and is distributed throughout the throat, heart, gastrointestinal tract, and so on. Its main function is to regulate the movement and secretion functions of the heart, lungs, gastrointestinal, and other internal organs. The implantable vagus nerve stimulator (VNS) was first used in humans in 1988. It was first used to explore the relationship between vagus nerve stimulation and epileptic seizures. The frequency of epileptic seizures in patients after vagus nerve stimulation treatment through implanted electrodes and stimulators and the severity was significantly reduced without obvious complications, laying a theoretical foundation for the application of VNS. In 1997, the FDA approved implantable vagus nerve stimulation as a selective therapy for treatment-resistant epilepsy, and in 2005, the FDA approved implantable vagus nerve stimulation as a selective therapy for treatment-resistant depression.

The vagus nerve stimulator consists of three parts: a pulse generator, a spiral electrode, and a flexible lead. The pulse generator is placed in the subcutaneous tissue below the left midclavicular line. The spiral electrode is wrapped around the vagus nerve. The flexible lead is implanted in a subcutaneous tunnel and connects the spiral electrode to the pulse generator. Stimulation parameters can be adjusted using external control equipment, and the pulse generator emits intermittent current pulses to stimulate the vagus nerve, thereby impacting physiological functions like heart rate, breathing, and blood pressure. The potential relationship between VNS and the treatment of common diseases like anxiety and Alzheimer's disease is still being explored. For instance, VNS may be used to investigate the connection between the vagus nerve and long-term loss of arm function after ischemic stroke (Siddiqi et al. 2021). Furthermore, as the vagus nerve is interconnected with numerous body systems, research into VNS for the diagnosis and treatment of other diseases is ongoing, such as exploring the link between VNS and cytokines that inhibit rheumatoid arthritis, and investigating the use of VNS as an adjuvant treatment for heart disease (Koopman et al. 2016).

9.3 In Vitro Diagnosis

9.3.1 Immunodiagnosis

Immunodiagnosis is a pivotal diagnostic approach utilized for disease identification, primarily by detecting specific proteins, antibodies, or other biomarkers present in bodily fluids. These biomarkers encompass various entities including pathogens like viruses, bacteria, fungi, tumor markers, and immunoglobulins. Among the array of in vitro diagnostic immunodiagnostic techniques are ELISA, immunofluorescence analysis, and immunochromatography. These technologies hold a wide application spectrum, spanning from diagnosing infectious diseases, detecting tumor markers, diagnosing autoimmune conditions, monitoring drug concentrations, and beyond. In vitro diagnostic immunodiagnostic technology boasts traits of rapidity, precision, and high throughput, thereby serving as a cornerstone in early disease identification and subsequent treatment. As technological progress marches on, the evolution of in vitro diagnostic immunodiagnostic technology persists, promising clinicians more precise diagnostic instruments. This advancement fosters early disease detection and treatment, consequently enhancing patient survival rates and overall quality of life.

9.3.1.1 Enzyme-Linked Immunosorbent Assay(ELISA)

The Enzyme-Linked Immunosorbent Assay (ELISA), initially outlined by Eva Engvall and Peter Perlmann in 1971, is a pivotal diagnostic technique commonly used in laboratories worldwide (Engvall and Perlmann 1972). It is a crucial analytical biochemical assay utilized for detecting specific proteins or ligands in liquid samples. This method employs antibodies that target the protein of interest, making it a valuable tool in medical diagnosis, biological research, and quality control.

In tumor marker testing, ELISA plays a vital role in detecting specific tumor markers in a patient's serum or plasma. The process involves adding the sample to a well plate containing antibodies that bind to the tumor markers if they are present. Subsequently, an enzyme-labeled antibody and a staining substrate are added, leading to a measurable color change that indicates the quantity of the tumor marker present (Molina et al. 2005). This capability aids in early tumor diagnosis, disease monitoring, and treatment evaluation.

For the diagnosis of autoimmune diseases, ELISA is instrumental in identifying specific autoantibodies in a patient's serum. These autoantibodies emerge when the immune system erroneously targets the body's own tissues. The detection of these autoantibodies through ELISA enables doctors to diagnose autoimmune conditions such as rheumatoid arthritis, systemic lupus erythematosus, and autoimmune thyroid diseases. Owing to its high sensitivity and specificity, ELISA is a staple in diagnosing autoimmune diseases and is extensively employed in clinical settings.

In allergen testing, ELISA is key to identifying specific IgE antibodies to allergens in a patient's serum. These antibodies are generated by the immune system as a reaction to certain allergens, including pollen, food, or pet dander. The ability of ELISA to detect these specific IgE antibodies allows doctors to pinpoint the allergens causing allergic reactions in patients, aiding in the diagnosis of allergy causes.

ELISA stands as an invaluable asset in medical diagnostics, offering precise detection of proteins, antibodies, and antigens across a variety of diseases and conditions. Its adaptability and dependability have cemented its role as a frequently utilized method in both research and clinical environments.

9.3.1.2 Immunofluorescence Analysis

Immunofluorescence analysis is a biochemical analysis technique that uses fluorescently labeled antibodies to detect specific molecules in samples (Hemmila 1985). In this technology, molecules in the sample such as proteins, organelles, or cell surface receptors are combined with specific fluorescently labeled antibodies to form complexes. These complexes are then viewed with a fluorescence microscope or other fluorescence detection device to determine the presence and location of specific molecules in the sample. Immunofluorescence analysis is widely used in biomedical research and clinical diagnosis, such as for the detection of viruses, bacteria, fungi, tumor markers, etc. Immunofluorescence can be classified according to its application and technical details into: direct immunofluorescence analysis, indirect immunofluorescence analysis, immunocytochemical fluorescence analysis, and flow cytometry.

Direct immunofluorescence analysis directly detects target molecules by directly observing the fluorescence signal after combining antigens in samples with fluorescently labeled specific antibodies (Aoki et al. 2010). This method is commonly employed for the detection of cell surface antigens and the study of cell morphology. In the laboratory, direct immunofluorescence analysis is commonly utilized to identify and quantify the presence and localization of cell surface proteins, organelles, or intracellular molecules.

Indirect immunofluorescence analysis, on the other hand, indirectly detects the presence of antigens by using an unlabeled primary antibody to bind the antigen, followed by a fluorescently labeled secondary antibody to bind the primary antibody (Aoki et al. 2010). This technique is frequently used for antibody titer determination, diagnosis of autoimmune diseases, and quantitative analysis of antigens in research.

Immunocytochemistry is a technique employed to identify specific proteins within cells (Burry 2011). This method involves the fixation and permeabilization of cell samples, followed by their conjugation with fluorescently labeled antibodies specific to the proteins of interest. The presence and localization of these proteins are then determined by visualizing the fluorescent signals under a fluorescence microscope.

Flow cytometry is a technique that analyzes individual cells through the use of fluorescently labeled antibodies, which identify and examine proteins present on or inside the cell surface. This method facilitates the high-throughput analysis of cells by injecting a suspension of cells into a flow cytometer and then exciting the fluorescent markers within these cells using a laser.

9.3.1.3 Immunochromatography

Immunochromatography is a commonly used biochemical separation and detection technique used to detect the presence and concentration of specific molecules in samples (Pyo and Yoo 2012). This technology combines samples with specific antibodies and then uses solid or liquid phases to separate and detect target molecules.

In solid-phase immunoassays, specific antibodies are anchored onto a solid surface, such as a microplate or a nitrocellulose membrane. The process begins with applying a sample that contains the antigen of interest to the solid surface, allowing the antigen to adhere to the fixed antibodies. The formation of a visible mark, either a line or spot, signifies the antigen's presence. The signal detection can be amplified by introducing labeled secondary antibodies or substrates, making this assay particularly useful in clinical settings for identifying specific proteins or biomolecules in serum samples.

Conversely, liquid-phase immunoassays feature both antibodies and antigens in a soluble state. Here, the antibody binds to the desired antigen in a liquid medium. The mixture is then passed through a solid support, like a column or flow cell, for separation and detection. This approach is widely adopted in clinical diagnostics to identify acute viral infections due to its efficiency.

Competitive immunochromatography represents a distinct approach within immunochromatography techniques, primarily aimed at quantifying specific molecules in a sample. In this setup, the sample's antigen and a labeled antigen vie for binding sites on antibodies fixed to a solid phase. The binding reduces the amount of labeled antigen attached to the solid phase, with the antigen concentration in the sample inferred from the decrease in labeled antigen. This method is routinely applied in clinical settings to measure specific proteins or biomolecules in serum.

Sandwich immunochromatography, or sandwich immunoassay, constitutes another variant of immunochromatography technology. This technique operates on the principle of capturing the target antigen between two antibodies – one immobilized on the solid phase and the other, a labeled detection antibody, binding to a different epitope on the antigen. The "sandwich" formation allows for the determination of the antigen's presence and quantity in the sample by assessing the label's signal intensity.

9.3.2 Molecular Diagnostics

Molecular diagnostics within the realm of in vitro diagnostics utilizes molecular biology techniques for disease diagnosis. This approach detects molecular alterations, including DNA, RNA, or protein modifications, in bodily fluids or tissue samples to identify or monitor disease progression. Its primary benefits are the high sensitivity and specificity it offers, leading to more precise diagnostic outcomes. Additionally, molecular diagnostics supports personalized medicine, enabling physicians to tailor treatment plans and devise more effective strategies based on a patient's genetic profile and molecular markers. Key technologies employed in molecular diagnostics encompass PCR (Polymerase Chain Reaction), gene sequencing, and nucleic acid hybridization, among others.

9.3.2.1 Polymerase Chain Reaction (PCR)

PCR, developed by Nobel Laureate Kary Mullis, is a technique employed to amplify specific DNA sequences (Mullis et al. 1992). By using PCR, a significant quantity of DNA can be produced from a minimal initial sample, facilitating subsequent analysis and diagnostic procedures. Applications of PCR span numerous fields, including:

1. Infectious pathogen detection: PCR can detect the DNA of viruses, bacteria, and other pathogens, aiding in the diagnosis of infectious diseases like influenza, HIV, hepatitis B virus, etc.

2. Genetic disease diagnosis: PCR can be used, It is used to detect the causative gene mutations of genetic diseases and help doctors diagnose genetic diseases, such as cystic fibrosis, thalassemia, etc.
3. Tumor diagnosis: PCR facilitates the detection of causative gene mutations associated with genetic diseases, assisting doctors in diagnosing conditions such as cystic fibrosis, thalassemia, etc.
4. Drug sensitivity testing: PCR can evaluate a patient's metabolic ability or drug sensitivity to specific medications, guiding personalized drug treatment plans.
5. Disease monitoring: PCR enables the monitoring of disease progression, treatment effectiveness, and disease recurrence, including HIV load monitoring.

9.3.2.2 Gene Sequencing

Gene sequencing is a technology used to determine the sequence of DNA. DNA is the molecule that makes up an organism's genetic information, and genetic sequencing determines the exact sequence of specific genes in DNA or the entire genome. The development of this technology is of great significance for understanding genetics, studying diseases, and developing personalized medical solutions. Gene sequencing can be divided into various types, including Sanger sequencing, next-generation sequencing (NGS), and third-generation sequencing technology. The development of these technologies has made gene sequencing faster, more accurate, and more economical, thus promoting the development of genomics and biomedicine.

Through genetic sequencing, scientists can identify mutations in the genome, search for genetic variations associated with disease, study the evolutionary history of species, and develop personalized medical treatments. In addition, gene sequencing is also widely used in agriculture, biotechnology, environmental science, and other fields. The clinical applications of gene sequencing generally include:

1. Genetic disease diagnosis: Gene sequencing can identify gene mutations related to genetic diseases, helping doctors diagnose and assess the risk of genetic diseases.
2. Tumor diagnosis: By sequencing the genome of tumor cells, tumor-related genetic variations can be discovered, helping doctors determine tumor type, prognosis, and treatment options;
3. Drug sensitivity testing: Gene sequencing helps doctors determine a patient's ability to metabolize specific drugs or drug sensitivity, guiding individualized drug treatment plans.
4. Disease risk assessment: Through gene sequencing, genetic variations related to specific disease risks can be discovered, helping individuals assess their disease risk and formulate preventive measures.
5. Disease monitoring: Gene sequencing can monitor disease progression, treatment effects, and disease recurrence, such as HIV load monitoring.

9.3.2.3 Nucleic Acid Hybridization

Nucleic acid hybridization is a technique used to detect and analyze nucleic acid sequences. In nucleic acid hybridization, a labeled nucleic acid probe binds to a complementary sequence in the nucleic acid sample to be detected. This technology is often used to detect the presence, expression levels, or mutations of specific genes. It can also be used to determine the homology between two nucleic acid sequences. It is often used in clinical testing for tumor markers. These molecular diagnostic technologies play an important role in clinical diagnosis and personalized treatment, helping doctors more accurately diagnose diseases, predict disease progression, and formulate personalized treatment plans.

9.3.3 *Biochemical Diagnosis*

Biochemical diagnosis of in vitro diagnostics refers to the method of disease diagnosis using biochemical principles and techniques. This diagnostic method involves detecting biochemical components such as enzymes, proteins, metabolites, etc. in body fluid or tissue samples to help determine the presence or progression of disease. Commonly used techniques for biochemical diagnosis include: enzyme activity determination, protein analysis, and metabolite determination.

9.3.3.1 Enzyme Activity Assay

Enzyme activity assays are used to evaluate the activity levels of specific enzymes in body fluids or tissues. This technology assists doctors in diagnosing and monitoring several diseases that may cause changes in enzyme activity within the body. Typically, enzyme activity assays involve combining the sample to be tested with a specific substrate, followed by measuring the amount or concentration of the reaction product produced. By measuring the rate or concentration of reaction products, the activity level of the enzyme being tested can be determined. The use of this technology is widespread in clinical practice, including evaluating liver function, myocardial damage, osteoporosis, and other diseases. Specific enzyme activity levels detected in body fluids can aid in determining the presence of certain diseases such as abnormal liver function and myocardial infarction.

9.3.3.2 Protein Analysis

Protein analysis refers to the detection and analysis of protein levels and properties in body fluids or tissues through experimental techniques. This analysis can help doctors diagnose disease, monitor treatment effects, and understand physiological and pathological processes. Protein analysis generally includes mass spectrometry and protein electrophoresis. There are several methods commonly employed in protein analysis, including mass spectrometry and protein electrophoresis.

Mass spectrometry is a technique used to identify compounds by measuring their mass and relative abundance to determine their structure and composition. In the field of biochemistry, mass spectrometry is frequently used to identify biomolecules such as proteins, nucleic acids, and metabolites. The basic principle of mass spectrometry is to ionize the molecules in the sample to be measured and measure the mass-to-charge ratio of these ions with a mass spectrometer. Based on the mass and ionization properties of the molecule, the structure and composition of the compound can be inferred. Proteins are identified and quantified through mass spectrometry techniques, such as MALDI-TOF mass spectrometry, liquid chromatography-mass spectrometry (LC-MS), etc.

Protein electrophoresis: Protein electrophoresis is a commonly used protein separation and analysis technology. It involves the separation of proteins into different bands or ribbons based on their size and charge under the influence of electric fields. Protein electrophoresis usually includes two main types: gel electrophoresis and two-dimensional gel electrophoresis. In gel electrophoresis, proteins are separated into different bands based on their size and charge, which can be used to analyze the relative molecular mass and concentration of the protein. Two-dimensional gel electrophoresis combines two different separation dimensions, usually based on the isoelectric point and molecular weight of proteins, and can more precisely analyze complex protein mixtures for separation and quantification of proteins.

9.3.3.3 Metabolite Determination

Metabolite determination involves the quantitative analysis of metabolites in organisms. Metabolites are the products of metabolic activities in organisms, including glucose, fatty acids, amino acids, nucleotides, hormones, enzymes, and more. Measuring these metabolites provides insights into the metabolic state of an organism, aiding in disease diagnosis, health assessment, and treatment monitoring. Metabolite measurement is utilized in various areas such as diabetes diagnosis, liver disease monitoring, and tumor metabolism research.

The general process of metabolite determination includes the following steps:

1. Sample collection: Biological tissues, blood, urine, or other relevant samples are collected for analysis. Proper sample collection procedures are followed to ensure sample purity and integrity.
2. Sample preparation: The collected samples undergo appropriate preparation techniques to extract the metabolites and remove interfering substances. Sample preparation methods may include centrifugation, extraction, washing, and other necessary steps.
3. Selection of analysis technology: Based on the properties of the metabolites to be measured and the analysis requirements, suitable analysis technologies are

selected. Examples include mass spectrometry analysis, chromatography-mass spectrometry coupling, nuclear magnetic resonance, and others.
4. Sample analysis: The prepared samples are analyzed using the chosen analytical instrument to obtain quantitative data on metabolites.
5. Data analysis: The obtained data is processed and analyzed, which involves peak identification, quantitative calculation, result interpretation, and other relevant analyses.
6. Result reporting: A report is prepared based on the analysis results and provided to clinicians or researchers for further diagnosis or research purposes.

9.3.4 Microbiological Diagnosis

Microbiological diagnosis is a type of in vitro diagnosis that utilizes microbiological principles and techniques to identify the presence of microorganisms such as bacteria, viruses, and fungi in body fluids or tissue samples. This diagnostic method is crucial in determining the causative agents and characteristics of infectious diseases. Common techniques used in microbiological diagnosis include bacterial culture, viral isolation, and fungal culture.

Bacterial culture involves inoculating patient samples such as blood, urine, cerebrospinal fluid, and sputum onto appropriate culture media and cultivating them under optimal conditions to promote the growth and reproduction of potentially pathogenic bacteria. The results of bacterial culture are vital for clinical diagnosis and treatment. They help doctors determine the type of infection, establish an effective antibiotic treatment regimen, and monitor treatment effectiveness. Moreover, drug susceptibility testing can assist doctors in selecting antibiotics that are effective against the patient's specific infection.

Virus isolation is by inoculating patient samples such as blood, saliva, urine, tissue samples, etc. into cell culture, and using cell culture to allow potential pathogenic viruses to proliferate and reproduce under appropriate conditions. Doctors determine the type and quantity of pathogenic viruses by observing and identifying the morphology, growth characteristics, and biological characteristics of the virus.

Fungal culture is often used to diagnose fungal infections by inoculating samples onto a medium containing appropriate nutrients to grow fungi, and then observing and identifying the type and quantity of fungi.

9.4 Minimally Invasive Diagnosis

Minimally invasive surgery represents a technological advance over traditional surgical techniques, inflicting less trauma or leaving smaller incisions in patients. This approach offers benefits such as reduced surgical risks, quicker recovery times, and diminished pain. It is recognized as a key surgical evolution for the twenty-first century. The ongoing advancements in electronic imaging, precision medical instruments, and medical sensors, combined with their integration into clinical practices, have sharpened the visualization of internal human anatomy. This progress has led to a decrease in nerve tissue damage and postoperative infections, facilitating successful surgeries via video-assisted laparoscopy and remotely operated surgical robots, among other methods. Beyond clinical procedures, minimally invasive technologies are increasingly applied in various medical processes, including diagnostics, which has seen a rapid growth in interest and development. Particularly in the areas of spinal, gynecological, and urinary system diseases, minimally invasive diagnostics are gaining prominence. Primary diagnostic techniques involve microneedle systems, electrical stimulation, and composite methods. Research is increasingly focused on developing and analyzing microneedle and advanced sensing technologies for better biocompatibility, sensitivity to target detection, and rapid material analysis, paving the way for further advancements in this field.

9.4.1 Microneedle Diagnosis

Microneedle is a new physical penetration-enhancing technology that has emerged as an alternative to traditional hypodermic injection needles. The tip of the microneedle penetrates the skin to deliver drugs through rapid dissolution or slow degradation. The needle body is typically 10–2000 μm high and 10–50 μm wide. The appropriate material, length, size, and shape are selected based on the depth of implantation required and the patient's physical condition.

Microneedles can directly and accurately penetrate the skin, creating nanoscale mechanical channels in the epidermis or dermis. This allows for the collection or injection of drugs without passing through the stratum corneum, minimizing contact with nerve endings and blood vessels during use. This technology has advantages in efficiently and accurately absorbing skin interstitial fluid for subsequent analysis. Many studies focus on developing minimally invasive transdermal sensing diagnostic devices based on microneedles to achieve easy-to-use real-time biomonitoring for long-term health monitoring (Tasca et al. 2019). Microneedle-related diagnostic platforms primarily achieve diagnostic functions by extracting interstitial fluid and transdermal sensing for in-situ diagnosis. They improve detection efficiency and accuracy by optimizing the material and shape of microneedles. Leveraging the advantages of microneedle non-invasiveness, ease of collection, and dosage control, it can be combined with other detection methods such as polymerase chain reaction (PCR), high-performance liquid chromatography (HPLC), enzyme-linked immunosorbent assay (ELISA), proteomic analysis, etc. These combinations can achieve efficient and fast detection with higher capture efficiency and monitoring accuracy than individual detection methods. Bin Yanget al. proposed combining polymerase chain reaction (PCR) and genotyping technology with microneedle patches to directly capture tumor metastasis-related interstitial fluid (ISF) containing Epstein-Barr virus cell-free DNA (EBV Cf DNA) and quantify it (Yang et al.

2019). Zhan et al. proposed using microneedles to deliver drugs through the skin in combination with high-performance liquid chromatography (HPLC) and dissolvable microneedles (DMNs) (Zhan et al. 2018). These DMNs carry lidocaine hydrochloride and rapidly dissolve within the skin, achieving high drug loading for local anesthesia. The required anesthesia time is significantly reduced compared to market-available anesthetic creams. Keng et al. proposed applying microneedles in the field of immune diagnosis (Ng et al. 2015). They combined microneedles with a detection mechanism for multiple antigens, enabling transdermal diagnosis and signal monitoring at the microneedle tip. They constructed a depth profile of skin markers and used enzyme-linked immunosorbent assay (ELISA) for quantitative detection of signals, verifying the test results. The specific research and development process mainly includes the following three aspects:

9.4.1.1 Characteristics and Functions

So far, microneedle materials can be divided into four generations. The first-generation microneedles are made of silicon or silicon oxide. The second-generation microneedles are made of metal. The third-generation microneedles are soluble microneedles. And the fourth-generation microneedles are hydrogel microneedles. Additionally, based on different drug release mechanisms, microneedles can also be classified into five types: solid microneedles, hollow microneedles, coated microneedles, soluble microneedles, and hydrogel microneedles.

The most basic hollow microneedle is made of metal or silicon. Similar to the principle of a conventional syringe, it penetrates the skin directly to form tiny holes and then extracts tissue fluid through the holes or applies patches containing drugs to the surface. Although it does not require a driving device, it cannot control the extraction speed and dosage, etc. Moreover, because the diameter of the microneedles is in the micron range, processing is difficult and prone to deformation or breakage, which is not conducive to mass production and transportation. For this reason, some people have proposed relevant improvement directions. For example, making hollow microneedles and ceramic microneedles (CM) combined with the developed porous ceramic microneedles, using sacrificial microfabrication and improved LIGA improves the microneedles, improves the storage of thermally unstable compounds at room temperature, and reduces the difficulty of transportation (Dixon et al. 2021). Ceramic microneedles can change the porosity and strength of ceramics by adjusting the parameters of the slurry, which is used to prepare materials such as ceramic manganese (Bhatnagar et al. 2019). In addition, with the development of fine processing technology, equipment, and polymer materials, it has become possible to develop microneedle manufacturing materials into polymers and biomaterials. More new biomaterials such as glass fiber, hyaluronic acid, silk fibroin, and new nanocomposite materials have also been used to manufacture microneedle devices; at the same time, new technologies such as 3D Printing, photolithography technology, and micro-molding technology are also used in the manufacturing process of microneedles. Li et al. proposed using different mask images and X-rays to photolithography of the hollow holes at different positions to obtain hollow microneedle arrays of different specifications (Yigui et al. 2018). Breakthrough progress has been made in terms of stability, stretchability, and biocompatibility.

In addition, new types of microneedles such as hydrogel microneedles, patch microneedles, coated microneedles, etc. can also be used for the extraction of biological tissue fluids, blood, and other substances. Hydrogel microneedles are different from conventional microneedles. They are prepared from cross-linked hydrogel or super-swollen polymer and are considered to be a subtype of soluble microneedles. Soluble microneedles can be divided apical layer, middle layer, and base layer. In terms of transdermal drug delivery and tissue fluid extraction, it has the advantages of excellent delivery rate, tightness, production efficiency, and no problem of needle residue. For example, UV cross-linking and molding methods are used to prepare soluble microneedles, and the good water absorption of the soluble microneedles is used to achieve swelling and degradation of the microneedles, minoxidil sustained release of Minoxidil-methacrylated gelatin soluble microneedles (Lu et al. 2022). In addition, the construction of nanoassemblies of Chlorin e6, which are loaded into soluble microneedles, significantly improves the water solubility and tumor cell uptake of Ce6 Small molecule prodrug compared to free Ce6 molecules, thereby enabling efficient and precise delivery to tumor sites (Huang et al. 2021) (Fig. 9.5).

9.4.1.2 Detect the Fluctuation of Physiological Ions

In situ diagnosis refers to using certain substances (such as antibodies, enzymes, specific probes, lectins, etc.) to screen out target molecules from various biomolecules obtained, and using methods such as fluorescence color development or enzymatic reaction color development. Indicates the distribution range and degree of aggregation of molecules to determine the distribution and physiological effects of target molecules. Microneedles are currently used as microelectrodes in clinical diagnosis. In addition to collecting tissue fluid through the skin, they can also open minimally invasive continuous diagnostic sensors or detect corresponding biomarkers according to different detection targets. Realize POC diagnosis, enabling more convenient, fast, and low-cost diagnosis. The detection of physiological signals is to collect pure physiological signal molecules and convert them into digital signals in real time for analysis. The traditional method mainly collects blood or other secretions such as tears, sweat, saliva, urine, etc. to analyze physiological ions, which requires invasive procedures, professional equipment, and trained personnel, thus limiting its applicability for real-time and continuous monitoring. The needle punctures a vein or capillary to extract blood and moves the blood to a test paper or test tube for in vitro analysis. The entire process is invasive and prone to errors due to human operation, which is not conducive to the construction of portable home health testing equipment. The interstitial fluid of skin tissue (ISF) conducts frequent molecular exchanges with capillaries and has similar small molecule composition

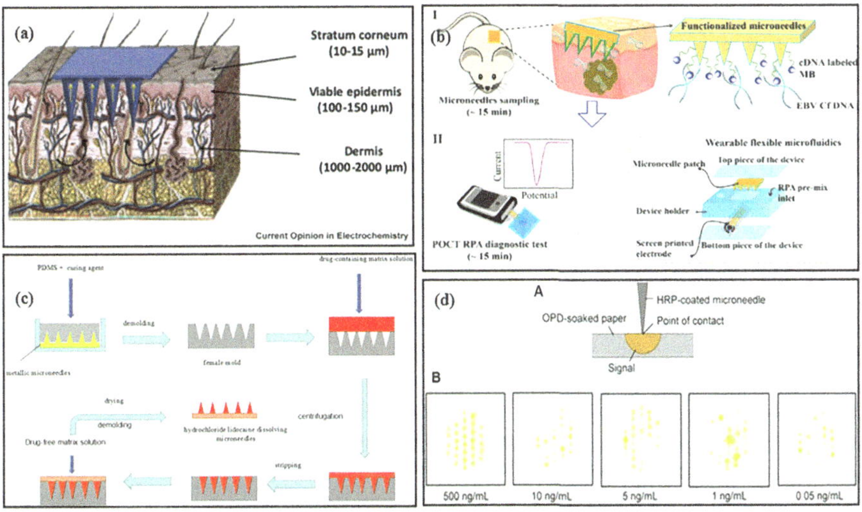

Fig. 9.5 (**a**) Schematic representation of the microneedles penetrating the skin. Reused with permission from Tasca et al. 2019; (**b**) Schematic workflow of the programmable assay, with microneedle patch sampling for 15 min (I) and diagnostic testing based on a POCT flexible wearable electrochemical microfluidic platform (II). Reprinted with permission from Yang et al. 2019; (**c**) 56-Preparation of female mold and the DMNs with lidocaine hydrochloride using two-step casting method and reverse molding method. Reused with permission from Zhan et al. 2018; (**d**) Paper-based detection of human TNF-alpha using analyte-capturing microneedles. Reprinted with permission from Dixon et al. 2021

and metabolites to blood (Tran et al. 2018), while microneedle serves as a method to extract ISFA commonly used tool that uses microneedle arrays to extract I from the user's subcutaneous tissue in vivo. ISF is efficient, accurate, and minimally invasive. In the early stage of development, some people proposed to use expandable hydrogel microneedles. The microneedles in a dry state are inserted into the skin. The microneedles begin to expand and use the pressure difference to insert the I into the skin. After capturing the hydrogel matrix, it is analyzed through colorimetry, fluorescence, and other methods. However, the analysis process requires the extracted ISF manual export and then processing is not conducive to real-time monitoring, and ISF long-term storage in the hydrogel matrix will lead to the loss of some biological macromolecules and other effects. Huang proposed an integrated system of microneedle sensor array, which uses a three-dimensional microneedle array to replace the traditional planar microneedle sheet (Huang et al. 2023). The three working electrodes of the sensor share a reference electrode and are insulated from each other, overcoming the problem of manufacturing multiple electrode channels. The limitations of the three-dimensional microneedle structure reduce the mutual interference between electrodes during measurement. Modify a specific ion-selective membrane on the surface of the microneedle working electrode so that target ions can pass through and reach the electrode, but non-target ions cannot come into contact with the electrode (Issa et al. 1994). Integrating a microneedle array and potential sensing, a dual-electrode system is used to measure the potential change signal related to ion concentration. This system enables real-time detection of changes in Ca^{2+}, K^{+}, and Na^{+} concentrations in the body. By connecting the microneedle array with the integrated potential sensor to a circuit module for recording potential, processing signals, and controlling feedback, and wirelessly transmitting the data to a customized device via Bluetooth, a wearable home monitoring system, can be constructed to analyze real-time changes in Ca^{2+}, K^{+}, and Na^{+} concentrations in the body. This information can be combined with relevant disease backgrounds such as hypercalcemia to provide reminders for emergencies like hypokalemia.

9.4.1.3 Blood Glucose Monitoring

Currently, home blood glucose tests typically involve pricking the fingertip or another body part with a needle to collect a small drop of blood for testing with test strips or instruments. This process can be painful, uncomfortable, and carries a risk of infection. Additionally, the collected blood may be affected by substances like tissue fluid, and frequent blood collection can impact hand health.

To achieve more convenient, faster, and non-invasive self-blood glucose monitoring (SMBG), the interstitial fluid (ISF) in the skin tissue can be utilized. ISF has a molecular composition and metabolites similar to that of blood. A microneedle system is used to collect ISF for biomarker analysis. Traditional hydrogel microneedles are inserted into the skin in a dry state and use volume expansion to create pressure, pushing ISF into a hydrogel matrix for subsequent analysis. However, this process may result in the loss of certain biomarkers and can be inconvenient to operate.

To address these challenges, electrochemical sensors can be incorporated into expandable microneedle tips to minimize biomarker loss. A patch system with a double-layer structure is developed, with the inner surface modified with a biomarker-specific oxidase biosensor. The outer layer consists of a polymer that can undergo volume expansion, actively extracting ISF from the skin through this expansion. The ISF is then transported directly to the inner device, which integrates the signal processing and transmission module. By combining glucose with the specific oxidase in the blood, the concentration of blood sugar can be reflected in real time and transmitted to a mobile device.

This innovative system can quickly extract skin interstitial fluid and analyze it within 30 seconds. It exhibits a rapid linear response to glucose concentrations ranging from 0 to 21 mM (Zhang et al. 2023). In addition, a micro-molar sensitive glucose biosensor can be integrated into a bioresorbable microneedle by using polyhistidine-labeled GBP biosensors, which is combined with microneedling to increase skin permeability and overcome the stratum corneum restrictions on transport. The high sensitivity of the GBP biosensor allows for an effective reduction in the required volume of extracted ISF, minimizes microneedle processing time, and reduces the errors introduced during the ISF extraction process. The collected data are processed using a normalized fluorescence response associated with a standard saturation curve. The glucose concentration is reflected by measuring the fluorescence intensity using a spectrophotometer, making it suitable for nursing points and portable sensing devices.

9.4.1.4 Detection of Infectious Diseases

During the detection of infectious diseases, it is essential to ensure the health and safety of the testing personnel to prevent infection during the testing process. Therefore, implementing personal protective measures for infectious disease testing is crucial. For those who need to collect blood or ISF samples, the non-invasive, convenient, and efficient advantages of microneedle patches have significant application value.

The use of bio-sensing devices to detect infectious diseases is an emerging method for monitoring health and the spread of infectious diseases. Although microneedle technology is still in the research and development stage in the field of infectious disease detection, there are similar devices that provide a theoretical basis. For example, solid microneedles integrated with biosensors use covalent cross-linking chemicals to fix specific identifiers such as enzymes and antibodies to the microneedle surface, preventing them from falling off during use (Moniz et al. 2012). The sensing device utilizes enzymes or antibodies with specific recognition to modify the electrode, enabling the detection of electrical signals that are proportional to the concentration of the target. These quantifiable electrical signals provide information about the relative abundance of the target (Bhalla et al. 2016). Common techniques for signal quantization include the differential pulse voltammetry method (DPV), cyclic voltammetry (CV), and square wave voltammetry (SWV), among others.

The sensors can detect infectious agents by directly detecting the entire pathogen or by measuring the metabolites or nucleic acids secreted by the pathogen. The obtained concentrations are then converted into quantitative signals to facilitate the detection process. Moreover, the selection of microneedle materials and types varies based on different applications. In transdermal in-situ sensing, microneedles are used as electrodes, employing two design strategies: the development of electroactive solid microneedles or the incorporation of traditional electrochemical sensing elements within hollow microneedles.

The manufacturing process of electroactive solid microneedles involves metalizing polymer or metal microneedles using sputtering or electron beam evaporation to create an electroactive surface. Subsequently, a coating method is used to attach the corresponding biomarkers to the electrode surface, enabling specific detection of target substances (Falup-Pecurariu 2013). Metals such as gold and platinum, which possess large electroactive surface areas, are commonly employed. However, considering the cost and processing complexity of microneedles, nanocomposite solid microneedles based on polymers and nanomaterials like carbon nanotubes are utilized for mass production. Carbon nanocomposites exhibit excellent mechanical strength, temperature adaptability, and a high aspect ratio, which increases the electroactive surface area of the microneedles. Additionally, carbon nanotubes and carbon paste electrodes have lower background currents, minimizing their impact on subsequent detection. The design of in-situ biosensors is also less challenging compared to individual microfluidic measurements, enabling easier customization by selecting different microneedles and biomarkers to achieve the desired combination.

In addition, some researchers have proposed the use of ion-selective sensors to measure interstitial fluid (ISF) pH and electrolyte concentration, among other factors, to determine an individual's health status and whether an infection has occurred. For example, Miller et al. used modified carbon paste materials to make hollow microneedle electrodes to achieve multiple detection of pH fluctuations, lactic acid, and glucose (Miller et al. 2012). Some reports suggest that skin pH and lactic acid content are related to bacterial infections, but the specificity of sensor detection cannot be guaranteed due to the potential influence of daily exercise, diet, environment, and other factors on these indicators.

9.4.2 Basic Low Frequency Electrical Stimulation

Basic electrical stimulation involves the use of low-frequency pulse currents to stimulate the nerves and muscle tissue of the user. By detecting changes in the electrophysiological activity of nerves and muscles, it is possible to determine the location, functional status, extent of damage, and recovery of specific nerves and muscles. This information can also be used to predict the recovery time and effectiveness throughout the course of the disease. In medical terminology, pulse currents with frequencies lower than 1000 Hz are referred to as low-frequency currents or low-frequency pulse currents.

Low-frequency pulse currents can produce various effects on the body depending on the frequency. For example, human motor nerves can achieve complete forced contraction of muscles under electrical pulse stimulation at a frequency of 50 Hz. Sensory nerves exhibit noticeable tremors at frequencies around 100 Hz. Electrical stimulation at a frequency of about 100 Hz can also have analgesic and sedative effects on the central nervous system. Stimulation ranging from 1 to 10 Hz activates the sympathetic nervous system, while stimulation ranging from 1 to 50 Hz excites the vagus nerve.

Implanted devices utilizing low-frequency, low-intensity currents can effectively stimulate nerves and muscles. The voltage and current intensity can be adjusted in real time to meet the specific needs of different individuals, minimizing potential damage and fatigue. This approach achieves minimally invasive or even non-invasive characteristics. For example, myoelectricity can be detected minimally invasively by placing electrode patches on the skin surface or using acupuncture electromyography. This information can serve as a basis for determining peripheral fatigue and mental fatigue in individuals (Falup-Pecurariu 2013). For example, muscle fatigue symptoms often manifest before the onset of muscle diseases, leading to changes in signal characteristics, conduction effects, and signal noise. By collecting electromyographic signals, the status of the corresponding muscle groups can be more accurately assessed. This characteristic, where fatigue appears before the onset of the disease, can be utilized for early detection and other important applications.

In the mid-twentieth century, implantable ECG recorders for detecting ECG, implantable EMG recorders for detecting EMG, and implantable EEG recorders for detecting EEG were introduced. Over time, the accuracy of detection algorithms has improved, materials have advanced, biocompatibility has been enhanced, and signal transmission rates and accuracy have increased. As a result, the damage caused during the detection process has significantly reduced. Different detection methods have been developed for various targets.

Next, the corresponding technologies will be introduced from the three detection aspects of electromyography, electrocardiography, and electroencephalography.

9.4.2.1 Electromyography

Using nerve conduction studies (NCS) and electromyography (EMG) detection, it is possible to detect indicators of muscle activation status and enable functional monitoring and diagnosis of specific body parts. For example, NCS/EMG can be used to assess the status of neck-related muscles during swallowing or to detect facial paralysis by analyzing EMG changes in facial muscles. Additionally, it can also be utilized to observe changes in biceps fatigue before and after exercise. These techniques provide valuable insights into muscle function and aid in diagnosis and monitoring of various conditions.

Surface electromyography (sEMG) analysis is indeed widely used for characterizing and assessing muscle fatigue (Venugopal et al. 2014). Surface electromyography signal is an electrical signal generated during skeletal muscle contraction, including the action potential sequence of different motor units in the contracting muscle. The corresponding signal can be obtained through electrodes attached to the skin surface (Merletti and Parker 2004). Under fatigue, factors such as the rapid twitching of muscle fibers and the synchronization of motor units within the muscle cause the spectrum of the signal to shift to the low-frequency area, and the signal amplitude increases (Dimitrova and Dimitrov 2003; Rogers and MacIsaac 2011), early diagnosis of various types of diseases can be achieved by analyzing myoelectric signals, such as detecting myoelectricity to determine whether patients have Parkinson's syndrome such as tremor, bradykinesia, stiffness, loss of postural reflexes, and freezing phenomena (PD) symptoms to determine whether the patient has symptoms of early Parkinson's syndrome (Eftaxias et al. 2015). Real-time detection of user's occurrence of epileptic symptoms can be achieved by quantitatively analyzing the characteristic surface electromyography (EMG) of bilateral (generalized) tonic-clonic seizures (TCS). It can also be used to differentiate between TCS and non-epileptic convulsions, thereby improving diagnostic accuracy (Beniczky et al. 2018). By using surface electromyography (sEMG) to detect superimposed motor unit action potentials of muscle fibers, adsorbing surface electrodes on the surface of facial skin, can enable diagnosis and treatment of temporomandibular joint disorders (TMD) in patients, and it can evaluate the degree of damage and main damage locations of patients with occlusal dysfunction in the oral and maxillofacial system through the signal of sEMG, and it is used as an important basis for monitoring subsequent orthodontic treatment (Szyszka-Sommerfeld et al. 2020). Considering the non-stationarity of surface EMG signals, Du et al. (2003) proposed to use multiple time windows (MTW) to extract electromyographic features, which can use various multi-window functions to calculate changes in signal energy relative to time, meet the needs of signal parameter changes due to signal non-stationarity, and extract multiple features to obtain a more complete information collection (MacIsaac et al. 2006).

In addition, transcutaneous electrical nerve stimulation (TENS) can also be used as a method of diagnosing diseases by connecting electrodes to the skin surface of the detection site. Intermittent direct current and induction electrical stimulation of nerves and muscles causes forced contraction of muscles. Intermittent direct current will cause the muscles to cause a flash-like contraction at the moment of power on and off. The recovery situation is judged by detecting the threshold difference between the cathode and the anode. If there is an obvious difference between the two, it can think something abnormal has occurred. This feature can be used to diagnose myasthenia gravis (MG), Lambert-Eaton myasthenic syndrome and other neuromuscular transmission diseases as the diagnostic basis for the effect of subsequent rehabilitation treatment. Martinet achieved significant improvements in pain intensity, knee joint function, and lower limb function in patients with myasthenia gravis during exercise by utilizing percutaneous electrical nerve stimulation (PENS) for diagnosis (Martin Perez et al. 2023). The diagnosis and treatment of acute and chronic myofascial pain syndrome in patients using a combination of PENS and deep dry needling can effectively determine the patient's pain intensity, pressure pain threshold, and other indicators. The use of a visual analog scale can also assist in evaluating the rehabilitation progress (Hernandez et al. 2021).

9.4.2.2 Clinical Electrophysiological Examination

Clinical electrophysiological examination is an invasive technique that utilizes the patient's electrocardiogram (ECG) and intracardiac stimulation technology to record the electrical signal sequence that triggers the heartbeat. This technique is mainly used to diagnose cardiac arrhythmias such as atrioventricular block and ectopic tachycardia, and to select antiarrhythmic drugs and cardiac pacing instruments. It is also useful for judging the degree of recovery after surgery.

Clinical electrophysiological examinations are particularly suitable for patients with chronic coronary heart disease, myocardial infarction, and ventricular aneurysms who have recurring symptoms. By stimulating specific areas of the heart and recording the resulting electrical signals, this technique can accurately diagnose the origin of ventricular tachycardia.

In recent years, wearable electrocardiograms have become increasingly popular as a non-invasive method for monitoring heart conditions in real time. These devices can detect arrhythmias by analyzing the frequency, regularity, origin, or conduction site of the heart's electrical impulses. By continuously monitoring the user's ECG and comparing it to a pre-constructed database of arrhythmia patterns, these devices can provide early warning of potential heart problems and help prevent serious complications such as stroke or sudden death.

In summary, clinical electrophysiological examination and wearable electrocardiograms are both valuable techniques for diagnosing and monitoring cardiac arrhythmias, helping to prevent serious complications and improve patient outcomes (Shen et al. 2020).

Acute coronary syndrome (ACS): Early ECG recording is a key step to ensure timely screening of patients. Short-distance transportation often leads to arrhythmia complications. Therefore, trained medical staff can diagnose whether it is st-segment elevation myocardial infarction (STEMI) by detecting patients' ECG. Detecting important transient abnormalities in the ECG allows for prompt measures to be taken, reducing the time and distance of transfer within the hospital and improving treatment effectiveness and patient survival rates (Rosiek and Leksowski 2016);

Considering the timeliness of ECG signals and the fact that cardiovascular diseases often accompany asymptomatic myocardial ischemia without obvious pain, 24-hour Holter monitoring is gaining increasing attention. Chen et al. selected a total of 206 elderly patients with suspected coronary heart disease and concurrent asymptomatic myocardial ischemia and arrhythmias as the study subjects; traditional electrocardiogram and 24 h dynamic electrocardiogram were used to measure the subjects (Chen et al. 2022).

Both the experimental group and the control group were in the same supine position, with electrode patches placed on the chest to record ST-T segment changes, duration, start and stop characteristics, heart rate variability, and other features. The accuracy of coronary angiography, which is considered the gold standard for diagnosing heart disease, was compared with the findings from the dynamic electrocardiogram. It was found that 24-hour dynamic electrocardiogram can observe the nature of coronary artery lesions through angiography. By utilizing multi-lead synchronous, long-term, and continuous monitoring of electrocardiogram changes, the scope of diagnosis can be expanded, allowing for the detection of resting and active symptoms and achieving more accurate and efficient detection. For example, simultaneous measurement of time-series data from multiple multi-channel induced electrocardiograms enables the conversion of this data into frequency-domain electrocardiogram data for each channel. This allows for the construction of three-dimensional electrocardiogram information, which can promptly detect pathological Q waves, ST-segment elevation, and occasionally irregular functions (Kwon 2022). Therefore, 24-hour Holter monitoring holds great significance in guiding treatment decisions and evaluating prognosis in clinical practice.

To further enhance detection accuracy and efficiency in 24-hour Holter monitoring, various technologies can be combined. For instance, combining ambulatory electrocardiography with a new-onset atrial fibrillation prediction model based on electrocardiographic characteristics can improve the accuracy of atrial fibrillation prediction (Nishimura et al. 2023). By using Convolutional Neural Networks (CNN) to process electrocardiograms measured from 24-hour Holter monitoring, a one-dimensional CNN trained on segmented ECG waveforms and their spectral entropy is used to extract abnormal waveforms in the electrocardiogram. A two-dimensional CNN trained on segmented ECGs is then used to recognize atrial fibrillation and determine whether it exceeds a set threshold, achieving an accuracy rate of approximately 90% (Kamozawa et al. 2023). By integrating the dynamic electrocardiogram system with a 12-lead format, continuous potential measurement using time-domain methods, and spectral analysis-based measurement of T-wave alternans in the frequency domain, it is possible to achieve long-term and efficient detection over a 24-hour period (Ikeda 2021). To ensure user comfort and usability over an extended period, a sticky dynamic electrode patch that is waterproof, wireless, and biocompatible can be utilized. This minimizes damage to the body while maintaining detection accuracy (Lang et al. 2022). Ultimately, these advancements contribute to achieving the goal of minimally invasive surgery.

9.4.2.3 EEG Signal

EEG signals provide valuable insights into the overall activity and functioning of brain nerve cells in different regions of the brain. They are closely related to neuromuscular pathways, nerve functions, sensory sensations, and more. However, EEG signals have certain limitations, such as low amplitude, high randomness, non-stationarity, and non-linearity. Additionally, during signal collection, the blood-brain barrier can influence the EEG signal. Early EEG collection techniques were invasive, involving the removal of a part of the skull and the implantation of electrodes, which carried risks and difficulties. To address these challenges and obtain EEG signals safely and efficiently, non-invasive techniques have been developed.

Common non-invasive techniques for collecting brain signals include electroencephalography (EEG), magnetoencephalography (MEG), functional magnetic resonance imaging (fMRI), functional near-infrared imaging (fNIR), and positron emission tomography (PET). Among these, EEG and MEG offer high temporal resolution, require simple experimental setups and equipment, and pose minimal harm to the human body.

Brain-Computer Interface (BCI) is a hardware-software complex used to connect traditional control systems, computing systems, and biological objects. It records the activity of various brain areas and converts biological signals into digital signals for processing. BCI holds great potential in enabling direct communication between the brain and external devices, offering promising applications in fields such as medical diagnosis, rehabilitation, and assistive technology (Hramov et al. 2021).

The brainwave recording system consists of four components: electrodes, amplifiers, Analog to Digital Converter (ADC), and a recording device. The electrodes are used to capture the user's brainwave signals, the amplifiers process the analog signals and amplify the amplitude of the brainwave signals, the ADC accurately converts the brainwave signals into digital format, and the recording device stores the data (Nicolas-Alonso and Gomez-Gil 2012). The measurement electrodes for EEG include active electrodes, reference electrodes, and ground electrodes. For good conductivity, these electrodes are mainly made of silver chloride (Sinclair et al. 2007). The EEG signal is indeed the potential difference between active electrodes and reference electrodes. Users typically wear an EEG cap with built-in electrodes that fit snugly on the scalp. Conductive gel is used to establish a conductive path and reduce impedance, ensuring the strength of the collected signals.

EEG signals reflect the superposition of postsynaptic potentials generated by a large number of neurons, which can provide insights into the functional status of specific brain regions. By analyzing the collected EEG signals from various parts of the brain, it is possible to diagnose neurological conditions such as Alzheimer's disease and depression. EEG is particularly useful in situations where patients cannot provide subjective judgments, such as in cases of hemiplegia or confusion.

To efficiently diagnose abnormal EEG conditions, advanced neural computing and machine learning techniques are combined with EEG analysis. Specific features of EEG signals from patients with certain diseases are extracted and compared to those of healthy individuals. This process involves constructing a database and undergoing a learning process that ultimately enables efficient diagnosis.

For example, deep convolutional neural networks (CNN) can be used for automatic EEG-based depression screening. Higher-order spectra (HOS) analysis can be employed to analyze normal, pre-epileptic, and epileptic EEG signals. These techniques aim to enhance the accuracy and efficiency of interpreting EEG data for diagnostic purposes (Kumar and Ajitha 2014), using convolutional neural networks to realize Parkinson's disease (PD) early diagnosis (Oh et al. 2020), diagnosing schizophrenia based on abnormal brain electrical signals automatically (Oh et al. 2019), etc.

Magnetoencephalography (MEG) is indeed a widely used neuroimaging technology that measures neuronal activity in the living brain by detecting the weak magnetic fields generated by ion currents in the neural cortex. MEG theoretically offers high temporal resolution (1 ms) and spatial resolution (1–2 mm). However, due to the limited magnetic field strength of brain tissue and the inherent limitations of equipment accuracy, the actual spatial resolution is relatively low (approximately 2 cm).

For example, the Vectorview MEG scanner is a widely used system that includes a 306A sensor array covering the entire scalp. This sensor array consists of magnetometers and gradiometers that directly measure the magnetic field. By measuring the magnetic field difference between these two types of sensors, the influence of environmental magnetic fields on brain magnetic field measurements can be minimized. Compared to electroencephalography (EEG), MEG does not require the long-term installation of electrodes or the use of conductive gel. It offers higher spatiotemporal resolution and faster data acquisition speed. MEG enables the non-invasive analysis of brain activity, making it suitable for diagnosing neurological disorders in children and infants, such as motor impairment following brain injury, cognitive dysfunction, and epileptic-like activity.

However, MEG has certain limitations that currently restrict its more widespread use in brain-computer interface (BCI) applications. These limitations include its bulkiness, high cost, requirement for a special shielded room to isolate environmental electrical signals, and inability to be used for daily data collection. Consequently, MEG cannot be widely used at present, and the field is still in its early stages of development.

9.4.3 Ultrasound Diagnosis

Ultrasound medicine is a discipline that utilizes the physical properties of ultrasound and the interaction between ultrasound and human tissues and organs to achieve diagnosis and treatment. By emitting ultrasound to the target parts that need to be monitored, the ultrasound is affected by the reflection, transmission, absorption, attenuation, etc. of sound during the propagation process in the body. The collected signals are then received, amplified, and processed to generate curves, spectra, or other representations that form the basis for disease diagnosis. This process is known as auscultation. Ultrasound imaging allows real-time monitoring of procedures such as needle punctures by placing an ultrasound probe on the skin's surface in the area of interest. By adjusting the position of the probe, different angles and views of the tissue structures can be observed. Currently, ultrasound diagnostic equipment is predominantly used to examine lesions and diseases of the digestive and circulatory systems, as well as in obstetrics and gynecology, urology, sports medicine, respiratory organs, and cranial nerves. The specific ultrasound testing methods employed vary depending on the area of the body being examined (Akiyama 2020).

9.4.3.1 Features and Applications

In 1881, French physicists Pierre Curie and Jacques Paul Curie proposed the forward and inverse piezoelectric effects, which led to the development of a method for using electronic technology to generate ultrasonic waves based on the principle of the piezoelectric effect. Early applications of ultrasound included locating objects underwater and inspecting the internal structures of opaque objects.

Although medical ultrasound diagnosis can be traced back to 1922, in the early days, its application scope was limited due to the limited power and detection sensitivity of artificial ultrasound. However, in the 1940s, American scientist Firestone proposed a pulse emission method for ultrasonic testing, which laid the theoretical foundation for subsequent development. This led to the advancement of ultrasound technology, which today plays a crucial role in non-invasive imaging and is widely used in various medical specialties. In 1942, German scientist Dussik utilized ultrasound for the diagnosis of brain tumors.

The rapid development of electronic technology has significantly improved many indicators that restrict instrument performance, such as dynamic range, sensitivity, and linear amplification range. In particular, ultrasound has benefited from these advancements and has become a widely used tool in medical diagnosis due to its strong penetration, minimal invasiveness, and fast speed.

Throughout its history, ultrasound technology has continued to evolve and improve. In the 1950s, A-mode ultrasound equipment emerged and was put into practical use. Later, in the 1970s, B-mode ultrasound technology matured, and real-time grayscale ultrasound was introduced. This breakthrough allowed for the construction of two-dimensional ultrasound images, known as sonograms, which provide cross-sectional views of human tissue and organs. Real-time grayscale ultrasound also enabled the dynamic imaging of organs, such as the heart and lungs, opening up new possibilities for medical diagnosis and treatment.

In the 1990s, the maturity of Doppler detection technology enabled the detection of hemodynamic parameters of tissues and organs, and combined with color Doppler blood flow imaging (CDFI) and two-dimensional ultrasound images can visually display the direction and speed of blood flow in tissues and organs, improving detection efficiency, etc. With the introduction of contrast-enhanced ultrasound, three-dimensional ultrasound, and other technologies as well as the performance improvement of electronic devices, the detection range and detection accuracy of ultrasonic diagnosis have been significantly improved. The descriptions of A-scan ultrasonography, B-scan ultrasonography, M-scan ultrasonography, and D-scan ultrasonography are as follows.

9.4.3.2 A-scan Ultrasonography

A-scan ultrasonography is a technique that utilizes amplitude modulation to detect the strength of echo signals by measuring the variations in amplitude. The probe receives the amplitude of the reflected or transmitted ultrasonic signal that changes with time and outputs a one-dimensional curve of the amplitude changing with time. Initially proposed during the early development of ultrasound diagnosis, A-scan ultrasonography was used to measure the size and borders of organs, the distance between tissue interfaces, and differentiate between normal and diseased tissues based on their acoustic properties.

With the maturity of two-dimensional image construction technology and the improvement of instrument performance, A-scan ultrasonography is now primarily used in ophthalmology for biometric measurements. For example, it can be used to detect retinoblastoma, a tumor in the retina that causes retinal protrusion near the optic disc and increases reflectivity. By comparing A-scan ultrasound measurements between normal and affected eyes, efficient and non-invasive detection of retinoblastoma can be achieved (Ikeda et al. 1995); A-scan ultrasonography can be used as a supportive diagnostic tool for acute optic neuron (De Bernardo et al. 2019); The combined use of A-scan ultrasonography and the Orbscan(R) device allows for measurement of ocular biometric features, such as anterior chamber depth, axial length, and corneal diameter (Allouch et al. 2001).

9.4.3.3 B-Scan Ultrasonography

B-scan ultrasonography utilizes brightness-modulated signals to produce imaging. By using multiple ultrasonic beams to scan the target area and receiving the echo signals, a two-dimensional grayscale image is generated based on depth and intensity. In this image, areas with high-intensity echoes are displayed as bright spots, while areas with low-intensity echoes appear darker. B-scan ultrasonography offers several advantages, including ease of operation, strong repeatability, non-invasiveness, and no ionizing radiation. It provides fast imaging speed and authentic results, allowing for real-time imaging and convenient diagnosis when the imaging speed exceeds 24 frames per second.

B-scan ultrasonography is currently the main ultrasound detection method used clinically. It is mainly used for ultrasound examination of superficial organs, abdominal organs, and obstetrics and gynecology. Superficial organs include the thyroid gland, mammary gland, salivary gland, lymph node, superficial mass, muscle, nerve, and joint. The detection of internal tumors and inflammatory lesions using B-mode ultrasonic beams is the preferred method. Abdominal organ ultrasound examination can be used to examine solid abdominal organs such as the liver, pancreas, spleen, bilateral kidneys, prostate, and hollow organs such as the gastrointestinal tract. For solid organs, it can assess their shape, internal structure, and the condition of surrounding tissues. It has a significant effect on the early diagnosis of inflammation or tumors. For example, morphological analysis of ovarian masses and evaluation of the possibility of transformation into tumors can be conducted (Zimmer et al. 2003). It can be used alone or in combination with computed tomography (CT) and magnetic resonance imaging (MRI) to analyze the volume, morphology, secretory function of the thyroid, as well as the volume of thyroid tumors. This can be used to evaluate the treatment process and the patient's recovery status in the treatment of thyroid diseases and parathyroid adenomas (Khati et al. 2003). For hollow organs such as the intestines, corresponding visualization aids need to be taken during the detection process to eliminate the interference of air and mucus in the gastrointestinal tract, which can effectively diagnose the location and range of lesions of tumors or gastric ulcers. Besides, B-scan ultrasonography has high accuracy and clarity in detecting fluid-containing organs such as the gallbladder and bladder in the abdominal cavity. It can effectively detect internal swelling conditions such as gallbladder tumors, bladder stones, and gallstones when there is fluid filling inside these organs. In obstetrics and gynecology, pregnant women are a special group for examination. With its advantages of non-invasiveness, repeatability, and dynamic imaging, B-scan ultrasonography has become the preferred method for screening fetal malformations. Prenatal ultrasound can show the growth and development of the fetus at different stages as well as dynamic images, accurately detecting whether the fetus has any abnormalities.

9.4.3.4 M-Scan Ultrasonography

M-scan ultrasonography, also known as ultrasound light point scanning, involves adding a slow-scanning sawtooth wave to the B-mode image. This causes the echo signals to move from left to right on the display, showing the curve of the distance between the interfaces of the tissues along a specific beam and the probe over time. The vertical axis represents the depth of the detected tissue, while the horizontal axis represents the motion curve of the tissue at different depths over a certain period of time. Similar to A-scan ultrasonography, M-scan ultrasonography is often synchronized with B-scan ultrasonography and is mainly used for cardiovascular system examination. It can observe the course of large blood vessels connected to the heart, the origin, thickness, hardening, and blood flow conditions of the blood vessels, and determine the degree of stenosis. It can also observe the internal structure, lumen thickness, blood flow direction, blood flow speed, and other conditions of arteries and veins in the head, neck, trunk, and limbs. Observing the function of venous valves can help identify abnormal conditions such as inflammatory lesions and blood clots. For instance, M-scan ultrasonography can be used to measure the thickness of the intima-media layers of arteries separately. By comparing the ratio of these measurements, the condition of the cardiovascular system can be assessed, and cardiovascular diseases can be diagnosed (Naessen 2009). The use of ultrasonography to measure carotid intima-media thickness (CIMT) and perform quantitative measurements of carotid plaques (such as plaque number, plaque thickness, plaque area, and three-dimensional evaluation of plaque volume) is employed for predicting the risk of cardiovascular disease onset (Naqvi and Lee 2014); By utilizing ultrasound to detect carotid artery plaques, early diagnosis of atherosclerosis can be achieved, and the acute rupture of plaques can be prevented etc. (Arbab-Zadeh and Fuster 2015).

9.4.3.5 Color Doppler Flow Imaging (CDFI)

By utilizing the Doppler effect of ultrasound, and further analyzing the resulting frequency shift on the basis of Doppler two-dimensional imaging, color Doppler ultrasound blood flow images can be formed, which allows simultaneous analysis of ultrasound structural images and hemodynamics. CDFI has multiple probes and has tissue harmonic imaging function to improve the clarity of the image. It is mainly used to detect abdominal and cardiovascular systems, such as digestive system, urinary system, obstetrics and gynecology, etc.

Color Doppler ultrasound blood flow mapping (CDF) or Color Doppler ultrasound imaging (CDI) utilizes the Doppler effect of ultrasound, combined with B-ultrasound, to detect moving tissues or organs (such as blood vessels and the heart) in the human body. It superimposes color-coded representations of blood flow onto the B-ultrasound image. By using Doppler two-dimensional imaging, coding is used to display the direction and flow rate of blood flow. For example, blood flow close to the ultrasound probe is displayed in red, while blood flow far away from the probe is displayed in blue. In addition to displaying the direction of blood flow, CDFI is useful for distinguishing between arteries and veins, differentiating vascular diseases from non-vascular diseases in a timely manner, detecting shunts and reflux, and performing quantitative analysis of the length, width, and area of blood flow bundles. Compared to spectral Doppler, CDFI can more accurately combine with any specific blood vessel. CDFI is primarily used in organs such as the kidneys, liver, and heart. For instance, it can be used to diagnose benign and malignant lesions in the liver, renal vascular diseases, and obtain accurate three-dimensional images of mitral valve structures and blood flow in the heart.

With the introduction of 3D color Doppler ultrasound in color Doppler flow imaging, dynamic stereoscopic displays can be achieved. This technique is particularly sensitive to uterine and ovarian blood flow, unaffected by obesity, and does not require bladder filling, making it suitable for obstetric examinations. For example, performing a 3D ultrasound examination on a pregnant woman can display the overall shape and position relationship of the umbilical cord, umbilical cord-placenta circulation, umbilical cord-fetal circulation, and fetal circulation (Xu et al. 2002). It can provide more information for obstetric examination; determine whether there is a nuchal cord in the fetus by comparing two-dimensional, color Doppler, and three-dimensional ultrasound (Hanaoka et al. 2002).

9.5 Wearable Diagnosis

The growing use of wearable diagnostics in medicine provides a convenient, real-time way to better understand a patient's physiological state. These wearable diagnostic technologies will not only provide doctors with more patient data, but also enable patients to better participate in the management of their own health. As technology continues to evolve, the use of wearable medical devices in clinical practice will continue to expand.

9.5.1 Wearable Physiological Index Monitoring and Diagnosis

9.5.1.1 Electrocardiogram Monitoring

The electrocardiogram (ECG) records the electrical activity in the heart in real time, providing an important opportunity to detecting various cardiac pathologies. The 12-lead ECG currently serves as the "standard" ECG acquisition technique for diagnostic purposes for many cardiac pathologies other than arrhythmias. However, these devices are not portable enough, expensive, not suitable for home use scenarios, so a portable detection device is needed. Thanks to the development of mobile Internet and wireless sensor networks (WSNs), wearable ECG monitoring systems have emerged which are able to detect ECG signals using a non-intrusive sensor and transmit the signal to the smartphone through wireless transmission techniques, such as Bluetooth or Zigbee (Yang et al. 2016). The core of wearable ECG diagnostic system is advanced ECG sensing technology. These sensors are typically designed as soft, comfortable, patch electrodes that are attached to the patient's chest or wrist. Some systems even use non-invasive photoelectric sensors to monitor heart activity through light reflection from the skin. The equipment itself is generally made of lightweight and soft materials to ensure comfort and convenience. These systems can monitor ECG signals in real time and capture changes in the electrical activity of the heart. Through the built-in data recording function, the system can continuously record the ECG data and form the ECG file of the patient. This real-time monitoring and recording feature provides physicians with continuous, comprehensive heart health information that can help in the early detection of potential heart problems. Wearable ECG diagnostic systems are typically equipped with a mobile app that enables patients to view ECG data anytime, anywhere via a smartphone or tablet device. At the same time, these systems often support data upload to the cloud platform, to achieve remote monitoring. Doctors can remotely access patients' electrocardiogram data to provide timely medical advice to patients, especially for those patients who cannot travel to medical institutions frequently, this remote monitoring method is particularly important. Due to the comfortable wearability of the wearable system, patients can have prolonged ECG monitoring, providing more comprehensive and continuous data. This provides doctors with a more accurate basis to support the practice of personalized medicine. By analyzing ECG data over time, doctors can better understand a patient's heart health and develop treatment plans that are more tailored to individual needs. Although wearable ECG diagnostic systems have made significant progress in heart health monitoring, there are still some challenges, such as data standardization, battery life, comfort, and more. In the future, with further innovation in technology, these systems are expected to become more intelligent and provide more reliable support for early diagnosis and personalized treatment of heart disease.

9.5.1.2 Electromyography Monitoring

With the advances in wearable sensors, wireless communication, and embedded computing technologies, we can indeed now obtain EMG data unintrusively using wearable EMG armbands. These EMG armbands typically include multiple EMG sensors positioned radially around the circumference of a flexible band, allowing ease of donning and wear in daily life (Phinyomark et al. 2018). The core of wearable EMG diagnostic system lies in its advanced EMG sensing technology. These sensors are usually attached directly to the surface of the skin and are able to pick up and record subtle changes in muscle electrical activity. Sensing technology options include patch electrodes, flexible sensors, or photoelectric sensors to ensure accurate and reliable acquisition of EMG signals. The device design of these systems focuses on lightness, flexibility, and comfort to ensure that patients can easily wear them during different activities. Possible forms of devices include patches, wristbands, or wearable devices that allow patients to perform daily activities without being bothered. Battery life, comfort, and stability are key factors to consider during the design process. Wearable EMG diagnostic systems enable real-time monitoring of muscle electrical activity, providing critical biofeedback. This not only helps to analyze movement patterns and muscle coordination, but also has important significance for muscle function assessment during rehabilitation. By monitoring EMG signals during exercise, the system can provide real-time movement analysis to help rehabilitation professionals optimize rehabilitation programs. Wearable EMG diagnostic systems are often equipped with a mobile app that allows patients to easily view the EMG data. This not only provides real-time insight into the individual's rehabilitation process, but also provides rehabilitation guidance and advice through the mobile app. This personalized rehabilitation method helps patients to participate in the rehabilitation process more actively and improve the treatment effect. Wearable EMG diagnostic systems show great potential in rehabilitation, motion analysis, and work environment improvement. However, the system still faces some challenges, including the improvement of data accuracy, the durability of the device, and integration with other medical devices. In the future, technological innovations are expected to further improve the intelligence of these systems, providing more comprehensive support for rehabilitation and improvement of quality of life.

9.5.1.3 Dermatoelectric Monitoring

The wearable electrodermal diagnostic system is based on advanced electrodermal sensing technology. These sensors are typically designed as electrodes that come in direct contact with the skin, and they're used to measure changes in the skin surface conductivity. These changes can reflect the patient's mood, stress level, and autonomic nervous system activity. The systems are designed for comfort and portability, with soft, breathable materials often being used to ensure patients can wear them comfortably for long periods of time. The device can take the form of a patch, wristband, or watch-like form, making it easy to wear daily with a long battery life.

The wearable electrodermal diagnostic system is primarily used to monitor patients' mood and stress levels. By collecting skin conductivity data in real time, the system is able to identify physiological changes in patients in different situations. This provides doctors with an intuitive way to understand the emotional state of patients and also provides timely stress feedback, which is conducive to the improvement of emotional management and mental health. Wearable electrodermal diagnostic systems also have potential applications in sleep monitoring. By monitoring skin electrical activity, the system can provide a more comprehensive analysis of sleep, including time to fall asleep, deep sleep stages, and dream stages. This helps doctors more fully assess their patients' sleep quality and can also help patients adjust their biorhythms and improve their sleep habits.

The wearable electrodermal diagnosis system has a wide application prospect in the diagnosis and treatment of mental diseases. For example, in the diagnosis of anxiety disorders, depression, and other diseases, by monitoring the patient's skin electrical activity, the system can provide objective physiological indicators to assist doctors with diagnosis. In addition, combined with psychotherapy, the system can also provide real-time biofeedback to help patients learn to better cope with emotions and stress. The wearable electrodermal diagnostic system shows great application potential in mood monitoring, mental health management, and sleep analysis. By integrating advanced sensing technology and intelligent data analysis, these systems are expected to become key tools for personalized medicine and mental health management in the future.

9.5.1.4 Blood Pressure Monitoring

Blood pressure (BP) is a vital sign that can be used to indicate the risk for atherosclerotic cardiovascular disease when elevated1,2. High blood pressure or hypertension is one of the most prevalent chronic diseases that threaten human health worldwide, regardless of age, gender, body mass index, ethnicity, or economic condition (Moon et al. 2020). The wearable blood pressure diagnostic system is mainly based on advanced blood pressure measurement technology. Sensors are often embedded in the device and are able to monitor changes in a patient's blood pressure in real time. These sensors can use photoelectric measurement, piezoelectric sensing, or other innovative technologies to ensure accurate, non-invasive blood pressure monitoring. The design of the system focuses on the lightness, comfort, and fashion of the device to promote the daily wear of the user. The wearable form can be a bracelet, watch, or patch design, so that users can easily integrate into their daily life. Wearing comfort is one of the key factors to encourage users to wear for a long time and achieve continuous, long-term blood pressure monitoring. A wearable blood pressure diagnostic system can provide real-time blood pressure monitoring and record the data on a device or cloud platform. This ability to continuously monitor helps doctors gain a more comprehensive understanding of patients' blood pressure fluctuations and provide more targeted recommendations for the management of chronic conditions such as hypertension. A wearable blood pressure diagnostic system sets a blood pressure threshold and alerts when blood pressure exceeds a safe range. This alarm and early warning function helps patients, doctors, or caregivers to take the necessary measures in a timely manner to prevent the progression of chronic diseases and the occurrence of complications. Based on continuous blood pressure monitoring data, the system is expected to enable more personalized treatment and prevention management. Through the application of deep learning algorithms, the system can identify individual health patterns, so as to provide more personalized and accurate health recommendations to help patients improve their lifestyle and control the risk of chronic diseases. Wearable blood pressure diagnosis systems have great potential to improve chronic disease management and individual health monitoring. However, the system still needs to overcome some challenges, including data accuracy, long-term wearability of the device, and user willingness to adopt.

9.5.1.5 Blood Oxygen Saturation Test

The wearable blood oxygen saturation diagnostic system is mainly based on advanced optical sensing technology. Sensors in the system typically use a combination of infrared and visible light to calculate oxygen saturation in the blood by measuring how well the light is absorbed. This non-invasive measurement technology allows users to wear it for long periods of time, enabling continuous monitoring. It is usually in the form of a bracelet, watch, or patch to facilitate users to wear it in their daily lives. Wearable oximetry diagnostic systems show great potential in respiratory, cardiovascular health, and chronic disease management. At present, a reflective blood oxygen sensor is implemented in a wearable device, which can be conveniently worn by a user and can free the user's hands, so that the user can have a better user experience (Chen and Tang 2020).

9.5.1.6 Temperature Monitoring

The core of the wearable temperature diagnosis system is the advanced temperature sensing technology. The sensor usually uses thermal elements or infrared technology to monitor the user's temperature changes in real time and accurately. This non-invasive measurement technology allows users to easily perform continuous temperature monitoring. The wearable form can be a bracelet, patch, or smart clothing, etc., to ensure that the user's body temperature is monitored in daily life without feeling. It is usually combined with the monitoring of other physiological and environmental parameters, such as heart rate, ambient temperature, etc. Wearable temperature diagnosis system has broad application prospects in epidemic surveillance, chronic disease management, and individual health monitoring.

9.5.2 Wearable Biochemical Index Monitoring and Diagnosis

The wearable biochemical index monitoring system uses advanced bio-sensing technology to monitor human biochemical indicators in real time, such as blood sugar, blood lipids, electrolytes, and so on. The selection and design of sensors is critical and often includes technologies such as optical sensing, electrochemical sensing, and biosensing to ensure accurate, sensitive measurements. This section will discuss the biochemical monitoring of sweat monitoring, interstitial fluid monitoring, and urine monitoring.

9.5.2.1 Sweat Monitoring

Sweat is an epidermally available biofluid containing metabolites, electrolytes, trace elements, and low levels of macromolecules (Kim et al. 2018b). Sweat is a prime candidate for interrogation due to its compatibility with non-invasive sampling. Several sweat sensing devices have been demonstrated in patch, watch and even temporary tattoo form factors, with the ability to measure analytes such as lactate, glucose, ammonia, ethanol, trace metals, hydration biomarkers, pH, and other metal ions present in sweat (Lu et al. 2019). The wearable sweat diagnostic system is based on advanced bio-sensing technology that enables real-time monitoring of biochemical indicators in sweat, such as glucose, electrolytes, lactic acid, and more. The selection and design of sensors are at the heart of the system, often using miniaturized sensors and nanotechnology to ensure highly sensitive detection of components in sweat. The system has real-time monitoring function, which can analyze the sweat composition in real time when the user sweats. Through Bluetooth or other wireless technology, the system can transmit the analyzed data to a smartphone or cloud platform, providing users and medical professionals with immediate feedback on their health status. Wearable sweat diagnostic systems are particularly important for the health management of patients with chronic diseases. By monitoring the biochemical indicators in sweat, the system can provide a more comprehensive health status monitoring, especially in the daily management of patients with chronic diseases such as diabetes and hypertension. Wearable sweat diagnostic systems have great potential in sports health, chronic disease management, and personalized medicine. Future developments will face challenges, including improving the stability of sensors, optimizing data algorithms, and standardizing the interpretation of sweat components. As technology continues to innovate, these systems are expected to become an important tool for health management, providing users with more comprehensive and intelligent physiological data monitoring services.

9.5.2.2 Interstitial Fluid Monitoring

The wearable interstitial fluid diagnostic system is based on advanced bio-sensing technology and is designed to monitor interstitial fluid composition in the body in real time. This may cover a variety of biochemical indicators such as proteins, metabolites, hormones, etc. The system uses miniaturized sensors and advanced analytical techniques to ensure highly sensitive detection of interstitial fluids. Wearable interstitial fluid diagnostic systems have the potential for early diagnosis, especially for certain potentially serious diseases. By monitoring biochemical changes in the interstitial fluid, the system can provide an alert for early signs of disease, providing doctors and patients with earlier, more effective treatment options. Wearable interstitial fluid diagnostic systems are expected to play a key role in personalized medicine, chronic disease management, and early disease diagnosis.

9.5.2.3 Urine Monitoring

Urine analysis is a routine clinical test technology for the diagnosis of diseases, which usually requires professional operators and relies on expensive and bulky instruments. It is difficult to collect urine of people who are suffering from urinary incontinence. Therefore, a low-cost and convenient device to detect urine in-situ is of great significance for these patients (Su et al. 2022). The wearable urine diagnostic system uses advanced bio-sensing technology to monitor biochemical indicators in urine in real time, including but not limited to proteins, sugars, metabolites, etc. The application of micro-sensors makes the system highly sensitive to detect changes in trace components in urine, providing accurate data support for disease diagnosis and health monitoring. The design of the system pays attention to the convenience and comfort of the device. Wearable forms could be wristbands, smart patches, or even sensors integrated into clothing. By monitoring multiple biochemical indicators in urine, the system enables individualized health management. This includes the management of chronic diseases such as diabetes and kidney disease. For pregnant women, the system can monitor relevant indicators in urine, providing real-time monitoring of pregnancy health and early detection of potential health risks. The system not only focuses on disease diagnosis, but also combines the user's diet and lifestyle habits, analyzes the relevant components in the urine, and provides users with personalized recommendations on diet and lifestyle, promoting long-term health maintenance.

9.6 Smart Integrated Diagnosis

9.6.1 Multimodal Diagnosis

Multimodal diagnosis is a comprehensive medical diagnostic approach that utilizes various modalities or methods to gain a holistic understanding of a patient's condition. These modalities typically include different medical imaging techniques, laboratory tests, clinical symptoms, and medical history, among other aspects of information. By combining information from multiple modalities, physicians can assess the patient's health status more comprehensively, leading to improved accuracy in disease diagnosis and treatment planning.

In contrast, single-modal diagnosis refers to the use of a specific sensor, imaging technique, or diagnostic method to obtain information about a patient's health. This diagnostic approach focuses on applying a particular technology to gather information about diseases, abnormalities, or physiological states. While single-modal diagnostic methods are effective in specific contexts, they often come with limitations such as constraints on effective information, insufficient specificity, low sensitivity, low cost-effectiveness, lack of functional information, impact of individual patient variations, and technological limitations. These limitations underscore the importance of adopting multimodal diagnosis to fully leverage the advantages of different sensors and imaging technologies, thereby enhancing the comprehensiveness and accuracy of medical diagnosis.

For instance, combining various imaging techniques such as magnetic resonance imaging (MRI), computed tomography (CT), ultrasound, and X-rays allows for a clearer observation of the internal structures and lesions within the human body. Simultaneously, integrating imaging results with laboratory tests measuring biochemical indicators in blood and urine provides a more comprehensive understanding of the patient's physiological status.

Multimodal diagnosis has extensive applications in various fields, including tumor diagnosis, assessment of neurological disorders, cardiovascular disease diagnosis, and more. By integrating information from diverse modalities, physicians can gain a more comprehensive and systematic insight into a patient's condition, thereby enhancing the accuracy and comprehensiveness of medical diagnosis. This approach allows for a more thorough evaluation of different aspects of the patient's health, leading to improved treatment planning and patient outcomes.

9.6.2 Smart Sensors

With the rapid advancement of technology, smart sensors have become a key component in areas such as the Internet of Things (IoT) and artificial intelligence (AI), leading a technological revolution. These compact yet powerful devices are now an integral part of daily life and play a crucial role in various fields, including industry, healthcare, and environmental protection.

The concept of sensors emerged as early as the early twentieth century, initially used for measuring and detecting physical quantities such as temperature, humidity, and pressure. However, with the rapid development of microelectronics and information technology, sensors have evolved into more intelligent entities. Smart sensors not only sense environmental parameters but also possess data processing and communication capabilities. This enables them to transmit information to other devices or systems for real-time monitoring and remote control.

The evolution of sensors into smart sensors has transformed them from passive data collectors into active participants in intelligent systems. One of the core features of smart sensors is their embedded processing capability, allowing them to analyze and process data locally. This reduces dependence on external systems and enhances their autonomy and intelligence. These advancements have significantly broadened the range of applications for sensors. They can now contribute to the development of intelligent systems, enabling real-time monitoring, predictive analysis, and efficient control in various domains. From industrial automation to personalized healthcare, smart sensors have revolutionized the way we interact with the environment and improve our quality of life.

The working principles of smart sensors encompass multiple key areas, including sensing technology, data processing, and communication technology. In terms of sensing technology, various types of sensors are designed to perceive different physical quantities such as light, sound, temperature, and pressure. These sensors convert physical quantities into electrical signals, transforming environmental information into readable digital or analog signals. Data processing is another critical aspect of smart sensors. The internal microprocessor of the sensor can execute various algorithms, enabling real-time processing and analysis of collected data. This autonomy allows sensors to autonomously identify specific patterns, trends, or anomalies, providing more in-depth information. Communication technology is crucial for smart sensors to achieve remote monitoring and networking. The progress in wireless communication technology allows sensors to seamlessly connect with other devices or cloud platforms, facilitating data sharing and remote control. This networking capability lays the foundation for building IoT systems.

The application of smart sensors is diverse and impacts various industries, providing innovative solutions. In the industrial sector, smart sensors are widely used for monitoring and controlling production processes. For instance, temperature sensors deployed on the production line enable real-time monitoring of equipment status, predicting potential failures and facilitating timely maintenance. This helps improve production efficiency and reduce equipment maintenance costs.

In the healthcare sector, the application of smart sensors is becoming increasingly important. Wearable health monitoring devices, such as smartwatches and health trackers, integrate various sensors to monitor users' physiological indicators such as heart rate, motion, and sleep in real-time. The collection and analysis of this data help doctors gain a more comprehensive understanding of patients' health conditions, enabling personalized medical advice.

Smart sensors also contribute to the development of smart cities in urban management. Deploying sensor networks in cities enables real-time monitoring and management of aspects such as the environment, traffic, and waste disposal. This helps improve city sustainability and quality of life.

Environmental monitoring is another crucial area of smart sensor application. By placing sensors in the atmosphere, water bodies, and soil, real-time monitoring of pollutants can be achieved, providing scientific evidence for environmental protection.

Looking to the future, the ongoing advancements in smart sensor technology are set to include miniaturization, low power consumption, integration with artificial intelligence, multi-modal sensing, and increased utilization of edge computing. These developments will enable sensors to become even more compact, lightweight, energy-efficient, and capable of executing more complex tasks locally, thus reducing dependence on cloud computing. Ultimately, smart sensors will continue to act as a catalyst for technological innovation, playing a crucial role in shaping the future. Their wide-ranging applications across an array of fields not only increase work efficiency, but also form the foundation for a more intelligent and convenient way of life.

Smart sensors have diverse applications across various industries, bringing innovative solutions to many fields. In the healthcare sector, the use of smart sensors is becoming increasingly important. Wearable health monitoring devices, such as smartwatches and fitness trackers, integrate various sensors to monitor users' real-time physiological indicators such as heart rate, activity, and sleep. The collection and analysis of this data assist doctors in gaining a more comprehensive understanding of patients' health conditions, enabling the provision of personalized medical advice. As technology continues to innovate, smart sensors remain in a continuous process of evolution. Future development trends may include the following aspects:

Miniaturization and Low Power Consumption: For scenarios such as wearable devices and drones, smart sensors will trend toward being more compact, lightweight, and require lower power consumption to extend battery life.

Integration with Artificial Intelligence: Smart sensors will become more closely integrated with artificial intelligence technologies, providing them with advanced data processing and decision-making capabilities, enabling them to execute more complex tasks locally.

Multi-Modal Sensing: Future smart sensors may integrate multiple sensors to achieve multi-modal sensing, acquiring more comprehensive and accurate environmental information.

Edge Computing: To reduce reliance on cloud computing, smart sensors will increasingly utilize edge computing to achieve faster, more real-time data processing and response.

In summary, smart sensors, as a key driver of technological innovation, will continue to play a crucial role in the future. Their widespread applications in various fields not only enhance work efficiency but also lay the foundation for creating a more intelligent and convenient way of life.

Multimodal detection in wearable sensors refers to the utilization of various built-in sensors within wearable devices, combined with multiple detection technologies, to achieve comprehensive and accurate monitoring of physiological parameters. This approach typically involves various sensors, such as motion sensors, bio-sensors, environmental sensors, etc., to gather multifaceted information. In terms of monitoring physiological parameters, wearable devices commonly use sensors such as heart rate monitors, pulse oximeters, and skin conductivity sensors to track users' cardiovascular health, respiratory health, and other physiological conditions. When it comes to motion detection, accelerometers and gyroscopes are employed to monitor users' movement and activities, including step counting, assessing activity intensity, analyzing sleep quality, and more. By combining data from physiological sensors with motion monitoring, wearable devices can provide a better understanding of users' overall health status.

In addition, some wearable devices integrate environmental sensors, such as temperature and humidity sensors, to detect the impact of the environment on health and offer personalized health recommendations. Wearable devices also utilize a combination of heart rate sensors and motion sensors for sleep monitoring, which is crucial for improving sleep habits and diagnosing sleep disorders. Moreover, some advanced wearable devices are starting to incorporate emotion recognition technology, analyzing physiological parameters, voice, and motion data to identify users' emotional states. This has potential value in psychological health management and emotional support.

By integrating these multimodal data, wearable devices can offer users more comprehensive, personalized health information, contributing to the early detection of potential health issues. This holistic monitoring approach contributes to achieving more comprehensive health management and prevention.

9.6.3 Applications

In the context of multi-modal detection using wearable sensors, the combination of physiological and biochemical indicators is common. Biochemical indicators are often derived from bodily fluids such as saliva, sweat, urine, blood, interstitial fluid, and tears. Physiological indicators include temperature, blood oxygen levels, electrocardiography (ECG), electromyography (EMG), skin conductance, and respiratory impedance, among others. A team led by Dr. Shuang Li from Tianjin University has reported a method for monitoring psychological stress based on the joint measurement of salivary cortisol and skin conductance (Bai et al. 2023). The team first developed a cortisol-sensing electrode using screen printing technology. The results demonstrated excellent detection performance with a detection range of 1 to 10,000 ng/mL and a goodness of fit for the standard curve of 0.977. Specificity tests conducted on the electrode showed strong resistance to interference. Subsequently, an integrated detection system for cortisol and skin conductance was designed. This system integrated the cortisol detection method and, simultaneously, employed alternating current impedance measurement for real-time skin conductance monitoring. Finally, an experimental paradigm was designed to validate the effective psychological stress monitoring capability of the system. This innovative joint detection method, along with the integrated system featuring wearability, electrochemical and liquid detection, and real-time analysis, holds significant potential for applications in future scenarios involving psychological stress monitoring. Tianyu Li and colleagues have designed a wearable device equipped with a multifunctional and flexible hydrogel paper patch capable of simultaneously monitoring electrocardiogram (ECG) signals and biochemical signals (glucose levels) in sweat during physical activity (Li et al. 2022). The ADS1293 chip was utilized for bioelectric potential measurement to record ECG signals from the hydrogel paper electrodes. The LMP91000 chip was employed to determine glucose concentrations. A Bluetooth chip, CC2541, was integrated to collect data from each sensor and transmit it to a receiver. A lithium battery (3.7 V) served as the power source, regulated by the TPS78233 boost converter. Lastly, a graphical interface was developed using Java, demonstrating measurement results on a mobile application. The hydrogel paper patches, integrated with a flexible printed circuit board (FPCB), can be positioned on the chest, achieving real-time monitoring of both electrophysiological and biochemical signals during physical activity. Wan-Young Chung et al. proposed a novel configuration that measures photo-plethysmography (PPG) and impedance plethysmography (IPG) signals simultaneously from the subject wrist (Rachim and Chung 2019). The proposed biosensor system enables a fully non-intrusive system that is cuff-less, and also utilizes a single measurement site for maximum wearability and convenience of the patients. The efficacy of the proposed technique was evaluated on 10 young healthy subjects. Experimental results indicate that the pulse transit time (PTT)-based features calculated from an IPG peak and PPG maximum second derivative (f14) had a relatively high correlation coefficient (r) to the reference BP, with -0.81 ± 0.08 and -0.78 ± 0.09 for systolic BP (SBP) and diastolic BP (DBP), respectively. Moreover, here they proposed two BP estimation models that utilize six- and one-point calibration models. The six-point model is based on the PTT, whereas the one-point model is based on the combined PTT and radial impedance (Z). Thus, in both models, they observed an adequate root-mean-square-error estimation performance, with 4.20 ± 1.66 and 2.90 ± 0.90 for SBP and DBP, respectively, with the PTT BP model; and 6.86 ± 1.65 and 6.67 ± 1.75 for SBP and DBP, respectively, with the PTT-Z BP model. This study suggests the possibility of estimating a subject's BP from only wrist bio-signals. Thus, the six- and one-point PTT-Z calibration models offer adequate performance for practical applications. The smart bandage based on a multi-modal wearable device enables real-time physiological monitoring and active intervention to promote the healing of chronic wounds. The team led by Zhenan Bao has developed a flexible bioelectronic system comprising wireless power supply, closed-loop sensing, and stimulation circuits (Jiang et al. 2023). The system features skin-interface hydrogel electrodes capable of adhering and detaching as needed. In mouse experiments, they demonstrated that the wound care system could continuously monitor skin impedance and temperature, providing electrical stimulation based on the wound environment. In preclinical wound models, the treatment group showed an accelerated healing rate of approximately 25% and enhanced dermal remodeling by about 50% compared to the control group. Additionally, they observed activation of pro-regenerative genes in mononuclear cells and macrophage clusters, which may enhance tissue regeneration, neovascularization, and dermal recovery. Ali Javey's team has proposed a wearable flexible integrated sensing array (FISA) designed to selectively screen a set of biomarkers in sweat simultaneously (Gao et al. 2016). By merging commercial integrated circuit technology, the team bridges the existing technological gap in wearable biosensors between signal transduction (sensor-generated electrical signals), conditioning (amplification and filtering in this case), processing (calibration and compensation here), and wireless transmission, all integrated on a flexible printed circuit board (FPCB). FISA can selectively measure a set of metabolites and electrolytes in human sweat, along with skin temperature during prolonged indoor and outdoor physical activities. The selection of target analytes and skin temperature is aimed at aiding the understanding of an individual's physiological state. For instance, excessive loss of sodium and potassium in sweat may lead to hyponatremia, hypokalemia, muscle spasms, or dehydration. Skin temperature can provide clinical information for various diseases and skin injuries, such as pressure ulcers.

9.6.4 Challenges and Prospects

Multi-modal diagnosis has extensive applications in the medical field but faces several technological challenges. Firstly, there is the complexity of algorithms and models. The analysis of multi-modal data requires sophisticated algorithms and models, including data fusion, feature extraction, and pattern recognition. Solutions for processing and analyzing large-scale data are necessary. Secondly, there are challenges related to data integration and standardization. The diverse sources of data from different modalities make it challenging to integrate these heterogeneous data and establish uniform standards, which may affect diagnostic accuracy. Lastly, there is the issue of hardware and device compatibility. Diagnostic devices from different modalities may come from various manufacturers, and ensuring compatibility between their hardware and devices may involve standardization issues.

However, continuous technological advancements are propelling progress in the field of multi-modal diagnosis. The development of deep learning techniques is increasingly prevalent and shows promise in handling multi-modal information more effectively, thus enhancing diagnostic accuracy. Additionally, the rise of personalized medicine positions multi-modal diagnosis as a key component, enabling the formulation of customized diagnostic and treatment plans based on each patient's multi-modal data. Finally, the promotion of remote healthcare and advancements in real-time monitoring technology allow patients to receive multi-modal remote diagnostic services over the internet. This enables healthcare professionals to gain a more comprehensive understanding of disease progression and the physiological status of patients.

Overall, while there are challenges, the continuous advancement of technology and the integration of multi-modal data hold great potential for improving diagnostic accuracy and personalized healthcare.

Acknowledgments This work was funded by the National Key Research and Development Program of China (Grant No. 2022YFF1202700), the National Natural Science Foundation of China (Grant No. 82001922), and the Hong Kong Scholars Program (Grant No. XJ2021034).

References

Akiyama I (2020) Basic and recent applied technologies of ultrasound in the field of clinical diagnosis. Acoustical Science and Technology 41:845–850

Allouch C, Touzeau O, Borderie V, Puech M, Scheer S, Laroche L (2001) Ocular biometric measurements with a slit-lamp method (Orbscan®). Journal Français d'Ophtalmologie 24:710–715

Aoki VR, Sousa JX, Fukumori LGM, Perigo AM, Freitas EL, Oliveira ZN (2010) Direct and indirect immunofluorescence. Anais Brasileiros de Dermatologia 85:490–500

Arbab-Zadeh A, Fuster V (2015) The myth of the "vulnerable plaque" transitioning from a focus on individual lesions to atherosclerotic disease burden for coronary artery disease risk assessment. J Am Coll Cardiol 65:846–855

Arthur D (2002) The vibrant® soundbridge™. Journal of Hearing Science 6:67–72

Bai Y, Fu J, Qin Z, Gao Q, Li S (2023) Integrated biosensing system of electrochemistry and electrophysiology for cortisol and skin conductance analysis on smartphone. Sensors Actuators B Chem 394:134368

Ball GR (2010) The vibrant soundbridge: design and development, active middle ear implants. Karger Publishers, pp 1–13

Bell B, Williamson T, Gerber N, Gavaghan K, Wimmer W, Kompis M et al (2014) An image-guided robot system for direct cochlear access. Cochlear Implants Int 15(Suppl 1):S11–S13

Beniczky S, Conradsen I, Wolf P (2018) Detection of convulsive seizures using surface electromyography. Epilepsia 59:23–29

Bhalla N, Jolly P, Formisano N, Estrela P (2016) Introduction to biosensors. Essays Biochem 60:1–8

Bhatnagar S, Gadeela PR, Thathireddy P, Venuganti VVK (2019) Microneedle-based drug delivery: materials of construction.Journal of Chemical Sciences 131:1–28

Bond M, Mealing S, Anderson R, Elston J, Weiner G, Taylor RS et al (2009) The effectiveness and cost-effectiveness of cochlear implants for severe to profound deafness in children and adults: a systematic review and economic model. Health Technology Assessment 13

Brackmann DE, Hitselberger WE, Nelson RA, Moore J, Waring MD, Portillo F et al (1993) Auditory brainstem implant: I. Issues in surgical implantation. Otolaryngology–Head and Neck Surgery 108:624–633

Burry RW (2011) Cytochemistry, controls for immunocytochemistry: an update. Journal of Histochemistry and Cytochemistry 59:6–12

Candito M, Simoni E, Gentilin E, Martini A, Marioni G, Danti S et al (2022) Neuron compatibility and antioxidant activity of barium titanate and lithium niobate nanoparticles. International Journal of Molecular Sciences 23:1761

Chauhan A, Grace A, Newell S, Stone D, Shapiro L, Schofield P et al (1994) Early complications after dual chamber versus single chamber pacemaker implantation. Pacing and Clinical Electrophysiology 17:2012–2015

Chen Q, Tang L (2020) A wearable blood oxygen saturation monitoring system based on bluetooth low energy technology. Computer Communications 160:101–110

Chen Z, Tan H, Liu X, Tang M (2022) Application of 24 h dynamic electrocardiography in the diagnosis of asymptomatic myocardial ischemia with arrhythmia in elderly patients with coronary. Heart Dis 2022:1

Chen H, Luo B, Wu S, Shi S, Dai Q, Peng Z et al (2023) Microfluidic biosensor based on molybdenum disulfide (MoS2) modified thin-Core microfiber for immune detection of toxoplasma gondii. Sensors 23

Crozier I, McAlister H (2011) History of cardiac pacing in New Zealand: the early years. Internal Medicine Journal 41:836–841

De Bernardo M, Vitiello L, Rosa N (2019) A-scan ultrasonography as a supportive diagnostic tool in case of acute optic neuritis. Mult Scler Relat Disord 36:101434

Dimitrova NA, Dimitrov GV (2003) Interpretation of EMG changes with fatigue: facts, pitfalls, and fallacies. J Electromyogr Kinesiol 13:13–36

Dixon RV, Skaria E, Lau WM, Manning P, Birch-Machin MA, Moghimi SM et al (2021) Microneedle-based devices for point-of-care infectious disease diagnostics. Acta Pharmaceutica Sinica B 11:2344–2361

Du S (2003) Feature extraction for classification of prehensile electromyography patterns: San Diego State University San Diego, CA, USA

Dun CA, Faber HT, de Wolf MJ, Cremers CW, Hol MK (2011) An overview of different systems: the bone-anchored hearing aid. Advances in Otorhinolaryngology 71:22–31

K. Eftaxias, S. Enshaeifar, O. Geman, S. Kouchaki, S. Sanei, IEEE, Detection of Parkinson's tremor from EMG signals, a singular Spectrum analysis approach, IEEE international conference on digital signal processing (DSP), Singapore, Singapore, 2015, pp. 398–402

Engvall E, Perlmann P (1972) Enzyme-linked immunosorbent assay, ELISA: III. Quantitation of specific antibodies by enzyme-labeled anti-immunoglobulin in antigen-coated tubes. The Journal of Immunology 109:129–135

Falup-Pecurariu C (2013) Fatigue assessment of Parkinson's disease patient in clinic: specific versus holistic. Journal of Neural Transmission (Vienna) 120:577–581

Gao W, Emaminejad S, Nyein HYY, Challa S, Chen K, Peck A et al (2016) Fully integrated wearable sensor arrays for multiplexed in situ perspiration analysis. Nature 529:509–514

Gao K-P, Shen G-C, Zhao N, Jiang C-P, Yang B, Liu J-Q (2020) Wearable multifunction sensor for the detection of forehead EEG signal and sweat rate on skin simultaneously. IEEE Sensors Journal 20:10393

Goode RL, Ball G, Nishihara S, Nakamura KJO (1996) Neurotology, Laser doppler vibrometer (ldv). Neurotology LWW:813–822

Hanaoka U, Yanagihara T, Tanaka H, Hata T (2002) Comparison of three-dimensional, two-dimensional and color Doppler ultrasound in predicting the presence of a nuchal cord at birth. Ultrasound Obstet Gynecol 19:471–474

Hemmila I (1985) Fluoroimmunoassays and immunofluorometric assays. Clinical Chemistry 31:359–370

Hernandez JVL, Calvo-Lobo C, Zugasti AMP, Fernandez-Carnero J, Alacreu HB (2021) Effectiveness of dry needling with percutaneous electrical nerve stimulation of high frequency versus Low frequency in patients with myofascial neck pain. Pain Physician 24:135–143

Ho RT, Wicks T, Wyeth D, Nei M (2006) Generalized tonic-clonic seizures detected by implantable loop recorder devices: diagnosing more than cardiac arrhythmias. Heart Rhythm 3:857–861

Hramov AE, Maksimenko VA, Pisarchik AN (2021) Physical principles of brain–computer interfaces and their applications for rehabilitation, robotics and control of human brain states. Physics Reports 918:1–133

G. Hrdlicka, Ieee, Ieee, implantable neurostimulators and the EMC environment, IEEE International Symposium on Electromagnetic Compatibility, Minneapolis, MN, 2002, pp. 916–919

Huang L, Wan J, Wu H, Chen X, Bian Q, Shi L et al (2021) Quantitative self-assembly of photoactivatable small molecular prodrug cocktails for safe and potent cancer chemo-photodynamic therapy. Biomaterials Research 36:101030

Huang X, Zheng S, Liang B, He M, Wu F, Yang J et al (2023) 3D-assembled microneedle ion sensor-based wearable system for the transdermal monitoring of physiological ion fluctuations. Microsystems & Nanoengineering 9:25

Ikeda T (2021) Current use and future needs of noninvasive ambulatory electrocardiogram monitoring. Intern Med 60:9–14

Ikeda T, Ogawa K, Kitanishi K (1995) A case of uncomplicated retinal astrocytoma. Nippon Ganka Gakkai Zasshi 99:1052–1055

Issa Y, Rizk M, Shoukpy A, Abdel-Aziz R, Abdel-Fattah H, Atia E (1994) Construction and performance characteristics of new tetramisole selective plastic membrane electrodes. Analytical Letters 41:135–141

Jiang Y, Trotsyuk AA, Niu S, Henn D, Chen K, Shih C-C et al (2023) Wireless, closed-loop, smart bandage with integrated sensors and stimulators for advanced wound care and accelerated healing. Nat Biotechnol 41:652–662

Kamozawa H, Muroga S, Tanaka M (2023) A detection method of atrial fibrillation from 24-hour Holter-ECG using CNN. IEEJ Trans Electr Electron Eng 18:577–582

Khati N, Adamson T, Johnson KS, Hill MC (2003) Ultrasound of the thyroid and parathyroid glands. Ultrasound Q 19:162–176

Kim J, Sempionatto JR, Imani S, Hartel MC, Barfidokht A, Tang G et al (2018a) Simultaneous monitoring of sweat and interstitial fluid using a single wearable biosensor platform. Adv Sci (Weinh) 5:1800880

Kim J, Sempionatto JR, Imani S, Hartel MC, Barfidokht A, Tang G et al (2018b) Simultaneous monitoring of sweat and interstitial fluid using a single wearable biosensor platform. Advanced Science 5:1800880

Koopman FA, Chavan SS, Miljko S, Grazio S, Sokolovic S, Schuurman PR et al (2016) Vagus nerve stimulation inhibits cytokine production and attenuates disease severity in rheumatoid arthritis. Proc Natl Acad Sci USA 113:8284–8289

Kuang RJD, Pirakalathanan J, Lau T, Koh D, Kotschet E, Ko B et al (2021) An up-to-date review of cardiac pacemakers and implantable cardioverter defibrillators. J Med Imaging Radiat Oncol 65:896–903

Kumar SSP, Ajitha L (2014) Ieee, Early detection of epilepsy using EEG signals, International conference on control, instrumentation, communication and computational technologies (ICCICCT), Kanyakumari, India, pp 1509–1514

Kuo WC, Wu TC, Wang JS (2022) Design and application of a flexible blood oxygen sensing array for wearable devices. Micromachines (Basel) 13

Kwon JM (2022) System for diagnosing and predicting heart disease, has electrocardiogram measuring unit for measuring time series data of multi-channel induced ECG, and preprocessing and heart disease prediction system for converting two-dimensional ECG, BODYFRIEND CO LTD (BODY-Non-standard) MEDICAL AI (MEDI-Non-standard)

Lang A, Basyal C, Benger M, Bhalla A, Edwards F, Farag M et al (2022) Improving stroke pathways using an adhesive ambulatory ECG patch: reducing time for patients to ECGs and subsequent results. Future Healthc J 9:64–66

Lei X, Kane S, Cogan S, Lorach H, Galambos L, Huie P et al (2016) SiC protective coating for photovoltaic retinal prosthesis. Journal of Neural Engineering 13:046016

Li T, Liang B, Ye Z, Zhang L, Xu S, Tu T et al (2022) An integrated and conductive hydrogel-paper patch for simultaneous sensing of Chemical–Electrophysiological signals. Biosens Bioelectron 198:113855

Liu HX, Gu Z, Zhao Q, Li S, Ding X, Xiao XX et al (2022) Printed circuit board integrated wearable ion-selective electrode with potential treatment for highly repeatable sweat monitoring. Sensors Actuators B-Chem 355:131102

Lorach H, Galvez A, Spagnolo V, Martel F, Karakas S, Intering N et al (2023) Walking naturally after spinal cord injury using a brain-spine interface. Nature 618:126

Lozano AM, Lipsman N, Bergman H, Brown P, Chabardes S, Chang JW et al (2019) Deep brain stimulation: current challenges and future directions, nature reviews. Neurology 15:148–160

Lu Y, Jiang K, Chen D, Shen G (2019) Wearable sweat monitoring system with integrated micro-supercapacitors. Nano Energy 58:624–632

Lu C, Gong Q, Liu X, Liu X, Liu P, Luo Z (2022) Preparation and evaluation of Minoxidil-GelMA bio-soluble microneedles. Chin Pharm J 57:823–828

MacIsaac DT, Parker PA, Englehart KB, Rogers DR (2006) Fatigue estimation with a multivariable myoelectric mapping function. IEEE Trans Biomed Eng 53:694–700

MacIver J, Tibbles A, Billia F, Ross H (2016) Patient perceptions of implantable cardioverter-defibrillator deactivation discussions: a qualitative study. SAGE Open Med 4:2050312116642693

Martin Perez SE, Martin Perez IM, Sanchez-Romero EA, Sosa Reina MD, Munoz Fernandez AC, Alonso Perez JL et al (2023) Percutaneous Electrical Nerve Stimulation (PENS) for Infrapatellar Saphenous Neuralgia Management in a Patient with Myasthenia gravis (MG). Intl J Environ Res Pub Health 20

McEvedy SM, Cameron J, Lugg E, Miller J, Haedtke C, Hammash M et al (2018) Implantable cardioverter defibrillator knowledge and end-of-life device deactivation: a cross-sectional survey. Palliat Med 32:156–163

Members ATF, Priori SG, Blomström-Lundqvist C, Mazzanti A, Blom N, Borggrefe M et al (2015) ESC Guidelines for the management of patients with ventricular arrhythmias and the prevention of sudden cardiac death: the task force for the management of patients with ventricular arrhythmias and the prevention of sudden cardiac death of the European Society of Cardiology (ESC) endorsed by: Association for European Paediatric and Congenital Cardiology (AEPC). 17(2015):1601–1687

Merletti R, Parker PJ (2004) Electromyography: physiology, engineering, and non-invasive applications. Wiley

Miller PR, Skoog SA, Edwards TL, Lopez DM, Wheeler DR, Arango DC et al (2012) Multiplexed microneedle-based biosensor array for characterization of metabolic acidosis. Talanta 88:739–742

Molina R, Barak V, van Dalen A, Duffy MJ, Einarsson R, Gion M et al (2005) Tumor markers in breast cancer-European Group on Tumor Markers recommendations. Tumor Biology 26:281–293

Moniz ARB, Michelakis K, Trzebinski J, Sharma S, Johnston DG, Oliver N et al (2012) Minimally invasive enzyme microprobes: an alternative approach for continuous glucose monitoring. Biosensors & Bioelectronics 6:479–480

Moon JH, Kang M-K, Choi C-E, Min J, Lee H-Y, Lim S (2020) Validation of a wearable cuff-less wristwatch-type blood pressure monitoring device. Scientific Reports 10:19015

K. Mullis, F. Faloona, S. Scharf, R. Saiki, G. Horn, H. Erlich (1992), Specific enzymatic amplification of DNA in vitro: the polymerase chain reaction. Cold Spring Harbor Protocols 17

Mylanus EA, van der Pouw KC, Snik AF, Cremers CW, Surgery N (1998) Intraindividual comparison of the bone-anchored hearing aid and air-conduction hearing aids. Otology & Neurotology 124:271–276

Naessen T (2009) Non-invasive method for determining cardiovascular status of individual, involves separately measuring thickness of artery intima layer and/or artery media layer. US Patent No. US20090287105A1

Naqvi TZ, Lee MS (2014) Carotid intima-media thickness and plaque in cardiovascular risk assessment, Jacc-cardiovascular imaging. 7:1025–1038

Ng KW, Lau WM, Williams AC (2015) Towards pain-free diagnosis of skin diseases through multiplexed microneedles: biomarker extraction and detection using a highly sensitive blotting method. Drug Delivery and Translational Research 5:387–396

Nicolas-Alonso LF, Gomez-Gil J (2012) Brain computer interfaces, a review. Sensors (Basel) 12:1211–1279

Nishimura T, Senoo K, Makino M, Munakata J, Tomura N, Shimoo S et al (2023) Prediction model for the new onset of atrial fibrillation combining features of 24-hour Holter electrocardiogram with 12-lead electrocardiogram. IJC Heart & Vasculature 47:101245

Oh SL, Vicnesh J, Ciaccio EJ, Yuvaraj R, Acharya UR (2019) Deep convolutional neural network model for automated diagnosis of schizophrenia using EEG signals. Appl Sci Basel 9

Oh SL, Hagiwara Y, Raghavendra U, Yuvaraj R, Arunkumar N, Murugappan M et al (2020) A deep learning approach for Parkinson's disease diagnosis from EEG signals. Neural Comput Applic 32:10927–10933

Oh J, Kim J, Moon S, Lee Y, Park D, Joo J et al (2023) Subcutaneous mechano-electrocardiogram (MECG) sensor for complementary cardiac diagnosis. Biosens Bioelectron 236:115443

Patiño-Jurado B, Cardona-Maya Y, Jaramillo-Grajales M, Montagut-Ferizzola YJ, Botero-Cadavid JF (2022) A label-free biosensor based on E-SMS optical fiber structure for anti BSA detection. Opt Fiber Technol 74:103116

Pelter MM, Kozik TM, Al-Zaiti SS, Carey MG (2013) Implantable electrical devices. Am J Critic Care 22:163–164

Peng C, Chen M, Sim HK, Zhu Y, Jiang X (2021) Noninvasive and nonocclusive blood pressure monitoring via a flexible piezo-composite ultrasonic sensor. IEEE Sensors J 21:2642–2650

Phinyomark A, Khushaba RN, Scheme E (2018) Feature extraction and selection for myoelectric control based on wearable EMG sensors. Sensors (Basel) 18:1615

Pyo D, Yoo J (2012) Immunochemistry, new trends in fluorescence immunochromatography. Journal of Immunoassay and Immunochemistry 33:203–222

Rachim VP, Chung W-Y (2019) Multimodal wrist biosensor for wearable cuff-less blood pressure monitoring system. Sci Rep 9:7947

Riojas KE, Labadie RF (2020) Robotic ear surgery, otolaryngologic clinics of North America. 53:1065

Rogers DR, MacIsaac DT (2011) EMG-based muscle fatigue assessment during dynamic contractions using principal component analysis. J Electromyogr Kinesiol 21:811–818

Rosiek A, Leksowski K (2016) The risk factors and prevention of cardiovascular disease: the importance of electrocardiogram in the diagnosis and treatment of acute coronary syndrome. Ther Clin Risk Manag 12:1223–1229
Sempionatto JR, Jeerapan I, Krishnan S, Wang J (2020) Wearable chemical sensors: emerging Systems for on-Body Analytical Chemistry. Anal Chem 92:378–396
Sennaroglu L, Ziyal I (2012) Auditory brainstem implantation. Auris Nasus Larynx 39:439–450
Shen Q, Gao HX, Li YW, Sun Q, Chen ML, Li JQ et al (2020) An open-access arrhythmia database of wearable electrocardiogram. J Med Biol Eng 40:564–574
Siddiqi SH, Schaper F, Horn A, Hsu J, Padmanabhan JL, Brodtmann A et al (2021) Brain stimulation and brain lesions converge on common causal circuits in neuropsychiatric disease. Nat Human Behav 5:1707
C.M. Sinclair, M.C. Gasper, A.S. Blum, Basic electronics in clinical neurophysiology, (2007) 3–18
Singh SU, Chatterjee S, Lone SA, Ho H-H, Kaswan K, Peringeth K et al (2022) Advanced wearable biosensors for the detection of body fluids and exhaled breath by graphene. Microchim Acta 189:236
Sriraam N, Srinivasulu A, Prakash VS, Low-Cost A (2023) Low-power flexible single-Lead ECG textile sensor for continuous monitoring of cardiac signals. IEEE Sensors J 23:20189–20198
Su H, Sun F, Lu Z, Zhang J, Zhang W, Liu J et al (2022) A wearable sensing system based on smartphone and diaper to detect urine in-situ for patients with urinary incontinence. Sensors and Actuators B: Chemical 357:131459
Sun W, Chang W (2023) Blood pressure monitoring based on flexible encapsulated sensors. Appl Sci 13
Svrakic M, Vambutas A (2019) Medical and audiological indications for implantable auditory devices. Otolaryngologic Clinics of North America 52:195–210
Szyszka-Sommerfeld L, Lipski M, Wozniak K (2020) Surface electromyography as a method for diagnosing muscle function in patients with congenital maxillofacial abnormalities. J Healthc Eng 2020:1
Tabibi Z, Massah J, Asefpour Vakilian K (2022) A biosensor for the sensitive and specific measurement of arsenite using gold nanoparticles, Measurement. 187
Tasca F, Tortolini C, Bollella P, Antiochia R (2019) Microneedle-based electrochemical devices for transdermal biosensing: a review. Science of Sensing and Biosystems 16:42–49
Tran BQ, Miller PR, Taylor RM, Boyd G, Mach PM, Rosenzweig CN et al (2018) Proteomic characterization of dermal interstitial fluid extracted using a novel microneedle-assisted technique. Journal of Proteome Research 17:479–485
Tuteja SK, Ormsby C, Neethirajan S (2018) Noninvasive label-free detection of cortisol and lactate using graphene embedded screen-printed electrode. Nano-Micro Letters 10:41
Venugopal G, Navaneethakrishna M, Ramakrishnan S (2014) Extraction and analysis of multiple time window features associated with muscle fatigue conditions using sEMG signals. Expert Systems with Applications 41:2652–2659
Wagner FB, Mignardot JB, Le Goff-Mignardot CG, Demesmaeker R, Komi S, Capogrosso M et al (2018) Targeted neurotechnology restores walking in humans with spinal cord injury. Nature 563:65
Wang S, Wu Y, Gu Y, Li T, Luo H, Li L-H et al (2017) Wearable sweatband sensor platform based on gold nanodendrite array as efficient solid contact of ion-selective electrode. Anal Chem 89:10224–10231
Williams ES, Poole JE (2013) Understanding implantable cardioverter-defibrillator shocks and mortality. J Am Coll Cardiol 62:1680–1682
Wolfe J (2018) Cochlear implants: audiologic management and considerations for implantable hearing devices. Plural Publishing
Xu H, Zhang Q, Lu M (2002) Three-dimensional color doppler ultrasound in obstetrics. Chin J Ultrason Med 18:299–302
Yang Z, Zhou Q, Lei L, Zheng K, Xiang W (2016) An IoT-cloud based wearable ECG monitoring system for smart healthcare. Journal of Medical Systems 40:1–11

Yang B, Fang XE, Kong JL (2019) In situ sampling and monitoring cell-free DNA of the Epstein-Barr virus from dermal interstitial fluid using wearable microneedle patches. ACS Appl Mater Interfaces 11:38448–38458

Yigui L, Wen-Yuan W, Huan W, Jin-Dong C, Tong L (2018) Fabrication, testing and simulation of microneedle array based on X-ray lithography. Optics and Precision Engineering 26:1156–1164

Zainuddin NH, Chee HY, Rashid SA, Ahmad MZ, Bakar MHA, Mahdi MA et al (2023) Carbon quantum dots functionalized tapered optical fiber for highly sensitive and specific detection of Leptospira DNA. Opt Laser Technol 157:108696

Zeppenfeld K, Tfelt-Hansen J, De Riva M, Winkel BG, Behr ER, Blom NA et al (2022) ESC guidelines for the management of patients with ventricular arrhythmias and the prevention of sudden cardiac death: developed by the task force for the management of patients with ventricular arrhythmias and the prevention of sudden cardiac death of the European Society of Cardiology (ESC), Endorsed by the Association for European Paediatric and Congenital Cardiology (AEPC). 43(2022):3997–4126

Zhan H, Ma F, Huang Y, Zhang J, Jiang X, Qian Y (2018) Application of composite dissolving microneedles with high drug loading ratio for rapid local anesthesia. European Journal of Pharmaceutical Sciences 121:330–337

Zhang Y, Zhao G, Zheng M, Hu T, Yang C, Xu C (2023) A nanometallic conductive composite-hydrogel core–shell microneedle skin patch for real-time monitoring of interstitial glucose levels. Nanoscale (Royal Society of Chemistry) 15:16493–16500

Zimmer Y, Tepper R, Akselrod S (2003) An automatic approach for morphological analysis and malignancy evaluation of ovarian masses using B-scans. Ultrasound Med Biol 29:1561–1570

Chapter 10
Intelligent Biomaterials and Microfluidic Technology

Cherie S. Tan, Maokun He, Yongde Guo, and Jingyu Fu

Abstract Over the past two decades, microfluidic technology has advanced significantly and is now widely used in materials science, chemistry, and biology. This chapter discusses various materials for creating microfluidic chips, their pros and cons, and highlights their applications in multiple fields, particularly organ chips. Additionally, it delves into digital microfluidics technology, a cutting-edge concept in the field of microfluidics, and provides an overview of its essential components and commonly used materials. The chapter concludes by presenting some examples of digital microfluidics applications in biology.

Keywords Microfluidics · Organ-on-a-chip · Drug screening

10.1 Microfluidic Materials

Over the past two decades, microfluidic technology has advanced significantly, offering a convenient platform for research and application in materials science, chemistry, life sciences, and biomedicine (McWilliams 1986). Microfluidic chips are widely used in these areas, and their selection is closely related to the choice of microfluidic materials. The materials used for making microfluidic chips exhibit good biochemical compatibility, which helps prevent reactions between the chips, materials, and reaction solvents. Additionally, chip materials usually have excellent modification properties, allowing for different modifications through chemical

---

C. S. Tan (✉) · M. He · Y. Guo · J. Fu
Department of Medical Sciences and Engineering, Medical School, Tianjin University, Tianjin, China

Academy of Medical Engineering and Translational Medicine, Tianjin University, Tianjin, China

Tianjin Key Laboratory of Brain Science and Neural Engineering, Tianjin University, Tianjin, China
e-mail: Cherie.tan@tju.edu.cn

Z. Liu (ed.), *Intelligent Biomaterials*, https://doi.org/10.1007/978-981-96-8646-9_10

reactions. Furthermore, most chips possess good transparency and support multiple detection methods. With the discovery of new materials and improvements in production processes, the range of microfluidic materials continues to expand. Microfluidic chips made from different materials have distinct advantages and applications. This section will discuss various types of chip materials, their pros and cons, and their applications.

10.1.1 Inorganic Materials

The concept of microfluidic chips was proposed after using technologies such as gas chromatography and capillary electrophoresis, which employed quartz or glass as the primary material for separating substances in microchannels. With the ongoing advancements in inorganic material processing technology and microfabrication technology, silicon materials have become increasingly refined in their processing techniques, making silicon the primary material for the first generation of microfluidic devices (Whitesides 2006). Silicon material, as the first material used in producing microfluidic chips, has the following advantages: (1) The thermal conductivity of silicon material is good, with a thermal conductivity of up to 157 W (m · K), which means that the temperature distribution of chips made of silicon material will be uniform and have good heat dissipation, which can adapt to various temperature environments; (2) The surface of silicon materials is prone to chemical modification, and the silicon hydroxyl groups on the surface can react with different silanization reagents to achieve varying degrees of hydrophilic and hydrophobic modification; 3. Silicon materials have good chemical inertness and organic solvent resistance, are prone to metal deposition, have low surface roughness, high purity, and stable electroosmotic mobility. Therefore, silicon materials have unique advantages in achieving fine structure chips that other materials cannot. For example, making silicon-based digital PCR chips containing millions of reaction wells (Li et al. 2019), producing pyramid microcavity arrays to achieve high-throughput capture and analysis of circulating tumor cells (Yin et al. 2019). However, silicon materials also have their drawbacks: they do not have good elasticity, are easy to break, are inconvenient to carry, have poor electrical insulation and transparency, and have complex surface modifications, which significantly limit the development and application of this material.

Due to its reliable electroosmotic flow, excellent optical properties, chemical resistance, and biocompatibility, glass has become another widely used inorganic chip material. It has unique advantages: (1) Glass has good light transmittance and can pass through ultraviolet and visible light above 250 nm, thus enabling effective ultraviolet detection; And due to the weak fluorescence background of the glass itself, it is compatible with other types of imaging observation methods in addition to ultraviolet detection; (2) Using a stable electroosmotic flow to construct an electroosmotic pump, driving the liquid flow to achieve valveless injection and complete the separation of analytes in a short time (Woolley and Mathies 1994). (3)

Similar to silicon materials, glass materials have less non-specific adsorption on the surface. They can also be subjected to subsequent biological modifications. Therefore, it has excellent application prospects in the field of biomolecular detection (Lin et al. 2015). However, due to its high production cost, high environmental design requirements, and low reuse rate, it is not easy to integrate fluid control units such as valves, poor breathability, hardness, and other reasons, so there are still many limitations in its application field.

Quartz, another commonly used inorganic material for microfluidic chips, has the following advantages: (1) Its excellent electroosmotic and optical properties can support the electroosmotic characteristics (Ou et al. 2009). (2) It is conducive to surface modification by chemical methods. (3) Photoetching and etching techniques can be used for processing, and the process is mature. (4) The raw material cost is low, easy to process and shape, and suitable for mass production. However, quartz material chips have some drawbacks: (1) Obtaining channels with a large aspect ratio is inaccessible. (2) The processing cost is relatively high. (3) Sealing is challenging. (4) It is not resistant to high temperatures, which, to some extent, limits its application in specific high-temperature environments.

However, inorganic materials are currently not mainstream, mainly because producing each inorganic material chip requires a complete process, which is relatively complex. Most manufacturing processes involve hazardous chemicals with high environmental requirements and low chip repetition rate. Therefore, it is necessary to find more suitable materials.

10.1.2 Silicone Elastomer Materials

The invention of polymer materials and the ongoing advancements in synthesis processes have led to their application in producing microfluidic chips using inorganic materials. Polymer materials offer a wider variety than inorganic materials, allowing for selecting different structures based on requirements. Additionally, due to their ability to be modified, they possess high flexibility in selection. Furthermore, polymer materials exhibit excellent processability and plasticity, with lower synthetic processing costs than inorganic materials, making them more suitable for large-scale mass production (Sollier et al. 2011). Polymer chip materials are mainly divided into four categories: silicon elastomer materials, thermosetting plastics, thermoplastics, and gel materials.

The material most commonly used in microfluidics is polydimethylsiloxane (PDMS), a liquid at room temperature. By adding polymerization initiators and heating them to 40–70 °C, PDMS can be cured into solid materials with low surface tension, creating microchannel silicon elastomer chips with nanometer-level accuracy through the pouring method. PDMS chips can bind reversibly to other PDMS, glass, silicon, and other matrix materials through simple physical contact. Additionally, PDMS chips can be surface-ionized by oxygen plasma treatment and irreversibly bonded to other substrate materials through chemical reactions. Due to

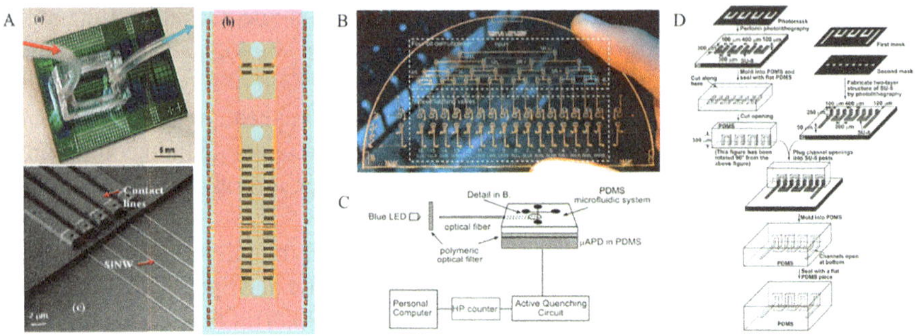

Fig. 10.1 Inorganic materials and silicon elastomer material chips (**a**) Glass chip for detecting cardiac biomarkers. Reprinted with permission from Chua et al. 2009. (**b**) Microfluidic glass chips with efficient valves. Reprinted with permission from 14. (**c**) Schematic diagrams of the encapsulation of an optical fiber and a μAPD in PDMS for detection of fluorescence in CE. Reprinted with permission from McDonald and Whitesides 2002. (**d**) Manufacturing of 3D microchannels in PDMS. Reprinted with permission from Wu et al. 2003

its affordability, fast preparation, minimal equipment requirements, and ease of use, PDMS has become a popular choice for chip materials in scientific research and industrial applications (Luo et al. 2021).

Compared to other chip materials, silicon elastic materials have better breathability, which is crucial for long-term cell culture in sealed microchannels. The PDMS surface is suitable for cell culture, allowing for gas exchange with gas environments, as well as easy gradient and oscillation as well complex medium replacement 3D microchannels can be constructed in PDMS (Wu et al. 2003), and lung function can be simulated through the deformation of PDMS structure (Huh et al. 2010). Therefore, PDMS equipment has been widely used in biological-related fields such as cell culture, cell screening, and biochemical analysis (Nge et al. 2013) (Fig. 10.1).

While PDMS possesses the benefits above and is prevalent in research, it also has notable drawbacks. Silicon elastomer materials exhibit poor resistance to organic solvents. For instance, PDMS is a porous matrix with alkyl Si-O backbones incompatible with organic solvents. Hydrophobic small molecules can be absorbed into the channel walls, resulting in swelling. As a result, its application is restricted to aqueous solution systems. Nonetheless, even within aqueous solution systems, some aqueous solutions may evaporate on the channel walls due to the material's permeability, leading to concentration-induced errors and rendering it unsuitable for specific quantitative experiments.

10.1.3 Thermosetting and Thermoplastic Materials

When silicone elastomer polymer materials are unsuitable, we often use thermosetting and thermoplastic materials as polymer chip substitutes for silicone elastomer.

Before the introduction of microfluidics, thermosetting materials were already used in negative photoresists, represented by substances such as SU-8 photoresist and polyimide (Jackman et al. 2001). Under heating or radiation conditions, thermosetting materials undergo cross-linking and solidification of thermosetting molecules, forming a rigid network. Once cured, the cured material cannot be reshaped. This material has good transparency, remains stable at high temperatures, and is resistant to most solvents. It can be used for true 3D microfabrication using photopolymerization technology, producing high aspect ratios and independent structures while maintaining high strength (Sato et al. 2006). By using appropriate bonding methods, thermosetting materials can be used to manufacture microfluidic chips (Becker and Gartner 2008). However, due to the significant rigidity characteristics of thermosetting materials after solidification, it is challenging to integrate functional units such as micropumps and valves, similar to inorganic chip materials. Moreover, the high cost of this material limits its application in the field of microfluidics.

Unlike thermosetting plastics, thermoplastic materials can be reshaped after curing because they often have a linear or branched molecular chain structure. After high-temperature heating, the polymer segments can move freely, causing the thermoplastic material to soften and undergo reshaping. Therefore, we can process near the temperature at which the thermoplastic softens, design the required microchannel pattern model according to the needs, reheat it, and then cool it to shape and solidify, forming the desired chip material with microchannels. Typical thermoplastic materials that can be used to make microchips include polymethyl methacrylate (PMMA), polycarbonate (PC), polystyrene (PS), polyethylene terephthalate (PET), and polyvinyl chloride (PVC) (Becker and Gärtner 2008). Compared to silicon elastomer materials such as PDMS, thermoplastic materials have certain advantages in the field of chip applications: 1. Thermoplastic materials, after covalent modification through reshaping, are usually more stable on the surface than PDMS materials. 2. Thermoplastic materials have slightly better solvent compatibility than PDMS and moderate resistance to alcohols, but are still incompatible with most other organic solvent substances such as ketones and hydrocarbons. After heating and softening, thermoplastic materials can be produced in large quantities through templates, making them widely used in commercial products at relatively low prices. Similarly, thermoplastic materials also have certain drawbacks that limit their application: 1. Thermoplastic materials cannot bond with other materials through simple physical contact like PDMS materials and often require additional conditions such as adding adhesives, requiring a specific temperature and pressure to bond with other substrate surfaces to form chips. 2. After forming, thermoplastic materials have high rigidity and almost no gas permeability, so their sealed microchannels and microcavities are unsuitable for long-term cell research. For example, they use a Paper/PMMA hybrid microfluidic platform to study cell crosstalk (Lei et al. 2017).

In recent years, heat-treatable perfluorinated polymers (Teflon) have been used for microfluidic structures, such as perfluoroalkoxy (Teflon PFA) and fluorinated ethylene propylene (Teflon FEP), breaking through the conventional drawbacks of thermoplastic materials (Ren et al. 2011). All Teflon is exceptionally inert to chemicals and solvents, demonstrating excellent compatibility with other organic solvents, with non-adhesion, excellent anti-fouling performance, and moderate gas permeability, making it suitable for cell culture. More importantly, the material is optically transparent and has a soft texture. It can generate fine microstructures with nano resolution through high-temperature thermoforming technology and utilize the functions of integrated membranes, microvalves, and other microfluidic devices through hotkeys and methods.

10.1.4 Gel Material

As a new type of microfluidic material, gel material, with its unique structure, characteristics, and advantages, has promoted the further development of microfluidic chips in the field of biomedicine in the application of biomedicine, such as the replication of internal microenvironment, cell culture, and so on (Ghaemmaghami et al. 2012).

The main components of gel materials are mainly composed of three-dimensional water-insoluble hydrophilic polymers, whose network structure contains the unique physical and chemical properties of gelled water-soluble compounds. This material can express characteristics similar to those of an extracellular matrix in an aqueous environment, so it has natural advantages in simulating cell culture microenvironment.

The microfluidic chips of gel materials have the following advantages: 1. Gel materials are widely available and can be extracted from natural materials or artificially synthesized, including gel extracted from animals such as matrix gel and collagen, gel extracted from plants such as sodium alginate and agarose, and artificially synthesized gel materials such as polyethylene glycol and polyacrylamide. The advantages of comprehensive sources and easy access make gel materials cheap and suitable for commercial promotion. 2. Gel materials have good biocompatibility, can interact with different kinds of cells, and show good affinity. This characteristic makes gel materials have significant advantages in biological experiments such as cell culture, which can promote the growth and reproduction of cells (Tibbitt and Anseth 2009). 3. Gel materials have good material transport and diffusion capabilities and can establish stable and adjustable gradient distribution through diffusion to reduce reagent waste caused by perfusion methods (Choi et al. 2007). 4. By combining with 3D printing technology, the rheological properties of gel materials enable them to form chips with three-dimensional structures, which avoids some limitations of other materials in template casting. Therefore, gel materials are widely used in in vitro biological simulation analysis, including constructing unstructured and personalized medical simulations (Ghaemmaghami et al. 2012).

Although gel materials have their advantages in the field of biomedicine, they still have some limitations and shortcomings: 1. Gel materials have soft characteristics, which makes it challenging to process them directly like other polymer materials. 2. It is challenging to form high-precision microchannels on the surface of gel

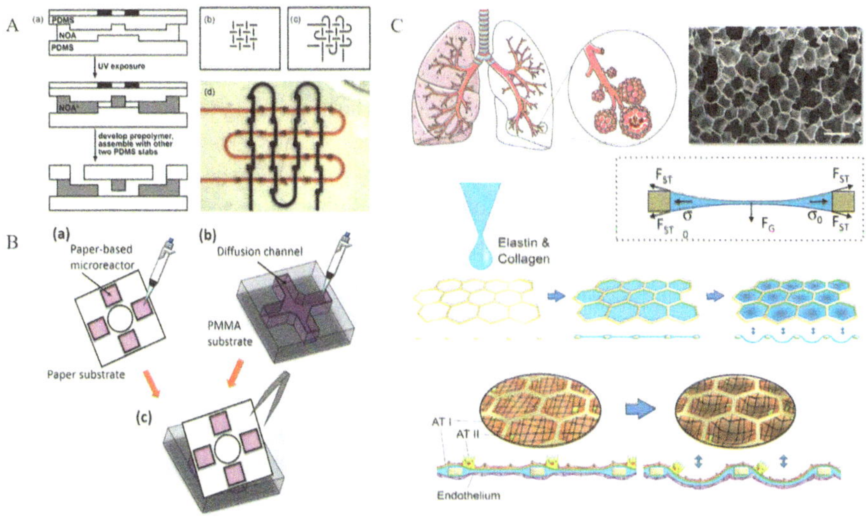

Fig. 10.2 Thermosetting, thermoplastic, and gel microfluidic chips. (**a**) 3D microfluidic channel fabricated in thermoset NOA. Reprinted with permission from Zheng et al. 2010. (**b**) Using a Paper/PMMA hybrid microfluidic platform to study cell crosstalk. Reprinted with permission from Lei et al. 2017. (**c**) Microfluidic device made of collagen I hydrogel. Reprinted with permission from Zamprogno et al. 2021

materials directly, so it is necessary to use machines to fix, specially modify or select specific photosensitive gel materials. 3. Due to the softness and water properties of gel materials, it is challenging to realize channel secretion and connection with external equipment by mechanical or chemical methods. 4. Although the gel material is similar to the extracellular matrix to a certain extent, the growth of cells on its surface is often unable to show normal morphology or function, which also has specific restrictions on its biological behavior. 5. The storage performance of gel materials is limited. Its high water property and expansibility limit its long-term storage ability. Low-temperature storage can ensure the integrity of gel materials, which increases the storage difficulty (Fig. 10.2).

10.1.5 Paper-Based Materials

With the continuous progress of technology, paper-based microfluidic chips have shown enormous development potential and become a hot research direction in microfluidic chips. This chip has unique advantages in the biomedical environment and food quality monitoring, and it has been widely used (Han et al. 2018; Jokerst et al. 2012; Yu et al. 2011).

Paper is a highly porous matrix mainly composed of cellulose, and paper-based microfluidic chips have become a popular research direction mainly due to their unique advantages: paper-based microfluidic chips exhibit high research value and application potential due to their unique material characteristics (Lin et al. 2020). Its main advantages include 1. Low cost: The raw materials for paper-based microfluidic chips are widespread and can be obtained from filter paper, cotton cloth, and wood. They come from various sources and are cheap and easy to obtain. Compared with traditional microfluidic materials such as silicon and glass, they have significant cost advantages. 2. Portable and easy to operate: The analysis system is easy to miniaturize and portable. The paper-based microfluidic chip microchannel does not require an external drive pump, and the raw material paper can be folded, making it easier to store and transport. 3. Good biocompatibility: The main component of paper, cellulose, can immobilize biological macromolecules such as enzymes, proteins, and DNA, while the porous structure and large surface area to volume ratio can effectively filter non-target particles, which is conducive to the treatment, modification, and detection of biological samples.4. Rich detection methods and precise results: It can be applied to various methods such as color reaction, electrochemical, and chemiluminescence, and the color of the paper is usually white. Colorimetric analysis on the paper chip can see the difference with the naked eye. 5. The process is relatively simple: In the microchannel processing of paper-based microfluidic chips, there are mainly two ways to achieve hydrophilic and hydrophobic treatment in different regions. Firstly, hydrophobic materials can be sprayed on the surface of hydrophilic cellulose to achieve a hydrophobic effect in some areas (Suresh et al. 2018). Secondly, the entire paper chip can be hydrophobically modified and then partially removed through photolithography and plasma oxidation to achieve hydrophilic and hydrophobic treatment of different regions. Due to the advantages of paper-based microfluidic devices have been widely used in various portable and low-cost analyses, such as clinical diagnosis and bioassay (Qin et al. 2021).

However, paper-based microfluidic chips still have the following drawbacks: 1. The hydrophilicity of paper-based chips can affect the detection of low surface tension samples, as low surface tension samples cannot be well limited in the required channels. 2. The hydrophilic and hydrophobic modified microchannels are not compatible with functional units such as microvalves. 3. Paper-based chips face the problem of liquid evaporation, which affects the accuracy of quantitative experiments. 4. The detection sensitivity is low, and fiber-like substances in the channel will block the transmission of the signal to be tested.

10.1.6 Hybrid Materials

As mentioned above, different chip materials have unique advantages and characteristics that can be utilized in different application fields. However, there may also be shortcomings and limitations that prevent them from being applied in a broader range of fields. Table 10.1 shows the advantages, disadvantages, and applications of the most current chip materials. When the existing chip materials cannot meet our social production needs, a new concept of hybrid materials emerged. Hybrid materials combine two chip materials that do not interfere with each other's performance.

Table 10.1 Advantages, disadvantages, and applications of different materials

Chip material	Representative substance	Advantages	Disadvantages	Application fields	References
Inorganic materials	Silicon	Good thermal conductivity; chemical modifiability; solvent resistance	Poor elasticity; not transparent to light; high processing costs	Processing integrated circuits; making other material chip molds	Whitesides (2006) and Li et al. (2019)
	Glass	Good transparency; good electro permeability; chemical modifiability	High processing costs; the operation process is complex	Chip electrophoresis technology; single molecule detection; UV detection;	Woolley and Mathies (1994) and Lin et al. (2015)
	Quartz	Excellent electroosmotic and optical properties, mature processing technology.	High processing costs. Sealing is challenging. Impatience to heights	Chemical analysis	Ou et al. (2009) and Cheng et al. (2012)
Polymer materials	Silicone elastomer materials	Good elasticity performance; good breathability and light transmittance	Non-specific adsorption; liquid evaporation; high specific surface area channels are prone to collapse	Cell culture; integrated micro valve functional unit; optical testing; 1.1.3 thermosetting and thermoplastic materials	Luo et al. (2021 and Nge et al. (2013)
	Thermosetting materials	High-temperature resistance; A certain degree of solvent tolerance; good transparency	Poor elasticity; processed into high	3D microchannel production	Sato et al. (2006) and Zheng et al. (2010)
	Thermoplastic materials	Stable covalent modification; easy mass production	Strict closure conditions; poor breathability; strong rigidity (except for some perfluorinated polymers)	Integrated flexible circuit electrodes; digital microfluidic liquid technology	Lei et al. (2017) and Shamsi et al. (2015)
	Gel material	Comprehensive source and low cost; good biocompatibility; porous structure allows for free diffusion of molecules	Strict storage conditions; difficulty connecting to other external devices; limited machining accuracy;	3D cell culture; personalized medical simulation; 3D chip printing and production	Ghaemmaghami et al. (2012), Tibbitt and Anseth (2009), and Choi et al. (2007)
Paper-based materials	Paper	Comprehensive source, small size, easy storage and transportation; simple processing technology; diverse detection methods, easy to commercialize production	Low surface tension samples are prone to leakage; liquids are prone to evaporation; difficulty in integrating multiple functional units	Portable analysis equipment, applied to biomedical, environmental, and food quality monitoring and detection	Lin et al. (2020) and Qin et al. (2021)

By leveraging the advantages of the two materials, we can meet our needs and apply them in a broader range of fields. Common hybrid materials include the following:

1. The glass PDMS hybrid material sandwiches the soft film between the PDMS material and the glass material, forming a diaphragm valve structure (Mu et al. 2009). This structure makes it easier for glass to form microchannels, resulting in an ideal structure for diaphragm valves, and can withstand pressure of tens of kilopascals.

2. Glass metal electrode hybrid materials: process glass into a substrate with metal electrode patterns, which can be combined with microfluidic chips of other polymer materials for electrochemical detection. Transparent electrodes can be formed through photolithography composition, especially for processing ITO-coated glass. For example, a gradient can be formed by applying the concentration difference of solution on both sides of the hydrogel. The gradient can be transferred to a free solution in the microchannel attached to the hydrogel (Qiu et al. 2003).

3. Gel hybrid material: inject gel material into other chip materials to give play to the advantages of gel material in simulating the microenvironment in vivo, which is conducive to cell culture on other chip materials (Huh et al. 2010).

4. Other additives can be added to different chips according to demand to achieve the expected functions (Siegel et al. 2010). For example, incorporating photocurable materials such as photoinitiator LAP into chip materials can achieve in situ microchannel fabrication and form complex structures (Beebe et al. 2000). Alternatively, specific nanomaterials can be doped to leverage their unique advantages and form different types of sensors for clinical biological and other applications (Medina-Sanchez et al. 2012) (Fig. 10.3).

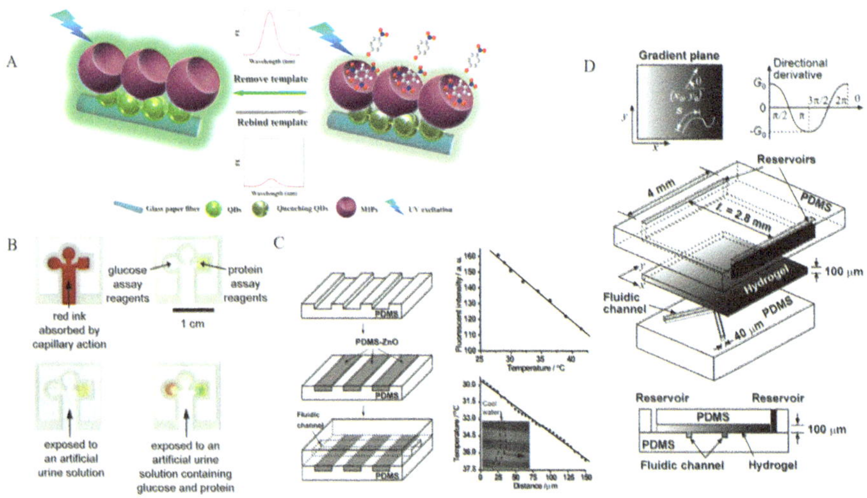

Fig. 10.3 Paper-based materials and hybrid materials. (**a**) Rotating paper-based microfluidic chips. Reprinted with permission from Qi et al. 2018. (**b**) Bioassay on paper-based microfluidic chips. Reprinted with permission from Martinez et al. 2007. (**c**) PDMS-ZnO composite as in-situ temperature sensor on a PDMS chip. Reprinted with permission from Zhou et al. 2009. (**d**) PDMS hydrogel hybrid chip for gradient generation. Reprinted with permission from Ren et al. 2013

In summary, two main development trends in chip materials are currently: powerful micro research platforms and low-cost portable analysis and applications. The powerful chip functionality is in line with laboratory research, and low cost, portable analysis, and application are the main factors that must be considered for commercial use. With the continuous development of microfluidic chip material science and technology, more microfluidic materials will be introduced, suitable for large-scale production at lower prices and have better flexibility and adaptability.

10.2 Microfluidic Applications

10.2.1 Biochemical Reaction Chip

Biochemistry studies chemical reactions and substances within living organisms, primarily focusing on the structure, function, and interactions of biological macromolecules such as proteins, nucleic acids, carbohydrates, and lipids. By examining various biochemical reactions in organisms, biochemists uncover the fundamental principles governing life processes, providing a theoretical foundation for drug development, disease diagnosis, and treatment. In recent years, integrating biochemical reactions and microfluidic chips has emerged as a significant innovation in bioengineering. By leveraging microfluidic chip technology, researchers can achieve cost-effectiveness, reduced reagent volume requirements, and increased analysis speed, enhancing experimental efficiency and opening up new possibilities for high-throughput screening, drug discovery, and disease diagnosis.

10.2.1.1 Polymerase Chain Reaction

Polymerase chain reaction (PCR) is a revolutionary molecular biology technique that can efficiently amplify specific DNA fragments (Cassinari et al. 2021; Lee et al. 2022). This technology has comprehensive applications in biomedicine, genetics, and pathogen detection. However, traditional PCR technology has limitations, such as long reaction time, large reaction volume, high energy consumption, easy generation of by-products, and difficulty in integration and portability. Scientists have shifted their innovative focus toward microfluidic technology to address these issues.

Microfluidic PCR chip is a new detection method that utilizes microfluidic technology to improve traditional PCR. Compared with traditional PCR, microfluidic PCR chips have the following advantages: 1. Low sample and reagent usage: The amount of samples and reagents required for microfluidic PCR chips is much lower than that of traditional PCR, which helps to reduce experimental costs and also reduces the demand for precious samples. 2. Fast heat transfer rate and uniform internal temperature: The microchannel structure inside the microfluidic chip can achieve rapid heat transfer, enabling the reaction system to reach the required temperature quickly. In addition, due to the smaller size of the microchannel, the heat distribution is more uniform, which is beneficial for improving the uniformity of the reaction. 3. The reaction process is easy to control: The microfluidic PCR chip can precisely control the reaction process, including parameters such as temperature, pressure, flow rate, etc. This makes the experimental results more stable and reliable while facilitating high-throughput and automated operations. 4. Short detection time, high accuracy, and intense sensitivity: Due to the above advantages, microfluidic PCR chips can significantly shorten detection time and improve detection accuracy and sensitivity. This is significant for areas such as virus testing, infectious diseases, food and water quality testing, pathogen identification (Cremonesi et al. 2022; Moron-Lopez et al. 2020), and important microbial gene testing (Jeannot et al. 2020). 5. Green analysis technology: Due to the low consumption, low pollution, and high efficiency of microfluidic PCR chip samples, it is expected to provide a green and sustainable analysis technology for environmental monitoring (Chirizzi et al. 2021) and food safety (Noh et al. 2019; Capobianco et al. 2020). At present, microfluidic PCR chips have been widely used in virus detection, infectious diseases, food and water quality detection, pathogenic bacteria identification, important microbial gene detection, detection of crop pathogens, livestock and poultry viruses, and crop gene research (Cremonesi et al. 2022; Moron-Lopez et al. 2020).

10.2.1.2 Immunological Testing

Microfluidic chip immunodetection technology has broad application prospects in clinical and biological diagnosis fields (Fitzgerald and Leonard 2017; Khalid et al. 2019). With the continuous development of monoclonal antibody research, qualitative and quantitative immune detection techniques based on antigen-antibody binding have significantly been promoted. Enzyme-linked immunosorbent assay (ELISA) has become the gold standard for detecting and quantifying protein biomarkers in clinical diagnosis. However, traditional ELISA methods are performed on a 96-well plate, requiring a larger reagent volume and longer culture time. In order to overcome these shortcomings, researchers have combined immune analysis with microfluidic chip technology in recent years and achieved significant results (Lin et al. 2020). Using microfluidic chips to improve ELISA reduces sample usage and improves reaction speed and experimental design flexibility (Sackmann et al. 2014). With the development of chip miniaturization technology, microdots' scale on a single microarray immune analysis chip can be further reduced, thereby improving detection flux and sensitivity. In order to achieve the miniaturization of microfluidic chip immune detection technology, we need to have a deep understanding of antibody immobilization, fluid control, biological matrix interference, and signal detection modes. Fluorescence technology has become the most common signal transduction platform for detecting the formation of immune complexes due to its high signal sensitivity. However, other signal transduction techniques, such as electrochemical methods, make it easier to achieve multi-target array detection while utilizing microfluidics for liquid transport. In addition to detecting proteins,

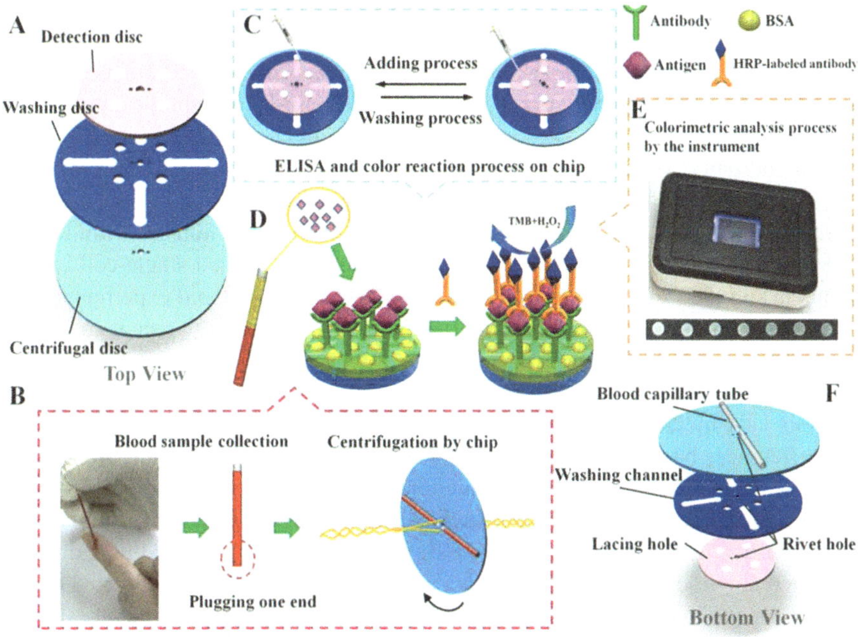

Fig. 10.4 Fully integrated manual centrifuge and analytical paper microfluidic equipment (FICA) for ELISA µ The structure and operation of PAD. Reused with permission from Lin et al. 2020. (**a**) Fully integrated manual centrifuge and analytical device (FICA) µ Exploded view of PAD. (**b**) Load and precipitate the finger-prick blood sample into a capillary tube, then centrifuge. (**c** & **d**) Then ELISA assay was performed using HRP labeled secondary antibody to catalyze 3,3′, 5,5v-tetramethylbenzidine (TMB)-H_2O_2 solution and generate a blue product (**e**). The generated signal was obtained using a portable reader. (**f**) FICA µ Bottom view of PAD, highlighting immune response areas, washing channels, and blood collection vessels

non-invasive detection of important disease markers based on metabolites in body fluids can also be considered using microfluidic chip immunoassay technology (Duan et al. 2019). This will help improve the disease's early diagnosis rate and treatment effectiveness and bring patients better quality of life (Fig. 10.4).

10.2.2 Droplet Microfluidic Chips

As the technology for generating and manipulating droplets continues to advance, microfluidic structures are being increasingly used in applications such as merging, splitting, or injecting droplets. Furthermore, an increasing number of detection technologies are being incorporated into droplet microfluidic systems, resulting in greater diversity, integration, and automation of this technology. For instance, droplet platforms can be utilized to separate single-cell droplets for secretion analysis

(Sun et al. 2020), which has widespread use across various fields. Droplet microfluidic chips present a challenge in achieving high-throughput parallel analysis of thousands of independent single cells. Nevertheless, using microdroplet chips can effectively address this challenge. The technology behind microfluidic chips allows for processing liquids the same size as cells, enabling large-scale, direct, and rapid placement, detection, manipulation, and sorting of cells. Initial research primarily focused on cell culture, freezing, thawing, dissolution, electroporation, and other operations performed on the chip, as well as culture and gene transfer within liquid and gel droplets. However, the true advantage of droplet-based single-cell experiments lies in their ability to conduct extremely detailed and rapid experiments in a highly parallel manner (Thorsen et al. 2001; Anna et al. 2003). In recent years, significant progress has been made in droplet-based single-cell experiments. For example, single-cell RNA sequencing, single-cell genome and protein expression analysis, and metabolite analysis are performed using droplet technology. Single-cell differences have attracted widespread attention in the study of molecular differences in differentiation, signal transduction, and diseases. Significant results have also been achieved in sequencing and constructing cell maps for multicellular organisms using droplet barcode technology (Guo et al. 2012). In addition, droplet-mediated single-cell experiments have also achieved success in the directed evolution of enzymes (Chiu and Stavrakis 2019). For example, based on the droplet method, it is helpful to screen horseradish peroxidase on a large scale, and the results show that new mutants with a catalytic efficiency increase of more than 10 times can be screened.

10.2.3 Drug Screening

In drug screening and new drug development, it is necessary to conduct a detailed analysis of the active ingredients, metabolism, and toxicity of many candidate drugs (Pushpakom et al. 2019). Microfluidic chips have become an ideal tool for drug screening due to their ability to simulate the internal environment and high throughput, automation, and speed (Shang et al. 2019; Uhl and Liu 2019). In recent years, drug screening platforms based on microfluidics have developed rapidly, including concentration gradient-based microfluidic devices, droplet-based microfluidic devices, and 3D cell culture-based drug screening (Wu et al. 2017). Cellular-level drug testing is one of the effective means to evaluate the sensitivity of anti-tumor drugs. There are reports of combining microfluidic chips with mass spectrometry to integrate cell culture, organ or tissue chips, metabolite collection, sample pretreatment, and detection units for drug metabolism research. The integrated microfluidic chip can also perform high-throughput cytotoxicity screening and online monitoring of metabolites simultaneously.

Concentration gradients play an essential role in life activities, and microfluidic chips can quickly form and adjust precise concentration gradients with high stability and reproducibility (Liu et al. 2013). Meanwhile, concentration gradient microfluidic chips can simulate microvessels in the body, closer to the complex microenvironment, and can perform cell growth and differentiation, cell chemotaxis, and intracellular substance analysis. Therefore, concentration gradient chips provide an effective platform for in vitro drug screening (Hu et al. 2018).

Droplet formation is an essential field in microfluidics, providing a pathway for drug screening (Zhai et al. 2019; Mastiani et al. 2018). Depending on the geometric structure of the chip, droplets of varying volume can be generated from nano-to-fly upgrades. These droplets are easy to mix, combine, capture, and transfer to the outside of the chip for incubation or sorting, which is conducive to reaction and analysis based on droplet drug screening methods for reading detection. Compared to the screening method based on a concentration gradient generator, the droplet screening method only needs to measure fluorescence intensity to quantify the results. Droplet-based drug screening analysis has advantages such as high-throughput screening ability, low reagent consumption, good reproducibility, fast response, fast mixing, precise control of cell concentration, and quantitative display of activity indicators (Trinh et al. 2023). These advantages help to measure the sensitivity of the target to drugs. Therefore, droplet-based microfluidics has been partially applied in drug screening (Fang et al. 2021). However, droplets can only encapsulate one or several cells, lacking intercellular interactions. Implementing multi-step liquid treatment in droplets makes long-term cell culture challenging and detection limited. In addition, this process requires an external professional pump, and the formation of stable droplets also requires additional treatment of oil and surfactants. Continuous flow cell culture data can be scanned, and multiple methods are available for detection, but the flux is relatively low, requiring a channel structure that can generate concentration gradients.

10.2.4 Biological and Chemical Synthesis

The application of microreactors in biology and pharmacy has achieved significant results (Stanley et al. 2012). One example of the application of microreactors is in the field of biosynthesis and biochemical processes, where they can precisely control key steps such as enzyme catalysis, biological screening analysis, and protein folding, leading to enhanced yields and purity of biological products. Furthermore, microreactors can also be employed for the synthesis of nanoparticles, enabling the preparation of complex nanomaterials like metal particles, quantum dots, and metal-organic framework materials through liquid phase synthesis methods (Fair et al. 2007).. Compared with traditional liquid phase synthesis methods, microfluidic technology has advantages such as higher yield, product size uniformity, and monodispersity.

In the synthesis of organic compounds, the high specific surface area of microreactors makes heat transfer efficiency extremely high, which can avoid the danger caused by local overheating and reduce the generation of by-products (Zhang et al. 2004; Jahnisch et al. 2004). Microreaction technology adopts continuous flow through microchannels, which can accurately control the residence time of materials and effectively reduce the generation of by-products. At the same time, the large contact-specific surface area of the microreactor significantly enhances the catalytic effect of surface modification or loading of specific catalysts within the reactor.

Microreactors have been widely used in biology and pharmacy due to their ability to control various processes precisely. In polymer synthesis, microreactors can achieve controllable multiphase microscale flow, which enhances mixing, mass transfer, and heat transfer during polymerization reactions (Munirathinam et al. 2015). By strictly controlling reaction time and combining reaction units modularly, microreactors offer significant advantages in controlling polymer molecular weight distribution, simplifying reaction environments, improving reaction selectivity, and adjusting polymer molecular structures and macroscopic morphologies. Overall, the application of microreactors has significantly supported research and technological development in these fields.

10.2.5 Medical Application POCT

Clinical biochemical testing is a commonly used method for in vitro clinical diagnosis, primarily focusing on the detection of more than ten biochemical indicators in human body fluids. To accomplish this, automated clinical biochemical instruments are required to perform a series of precise operations, including sample pretreatment, measurement, mixing reaction, detection, and cleaning. Nevertheless, as clinical demands continue to evolve, there is an urgent need for clinical, biochemical analysis to achieve multi-step serial fluid operation and multi-indicator parallel analysis.

POCT (Point of Care Testing) is a testing technique that can be performed outside the laboratory, near the test object, and obtain results in a short period of time (Gubala et al. 2012). The POCT chip based on microfluidic technology has the advantages of high integration, good accuracy, low reagent consumption, fast reaction speed, and potent portability, and has become the core chip for constructing POCT analysis instruments. The early POCT chips were mainly used to detect and analyze pathogens such as bacteria, viruses, and parasites. Later, they developed to support molecular diagnosis of substances such as blood gas, hemagglutination, nucleic acid analysis, and myocardial biomarkers, as well as biochemical diagnosis represented by metabolite analysis (Mahato et al. 2017). For example, the POCT chip integrated with nucleic acid extraction, amplification, and detection units has been applied in clinical diagnosis, and the entire operation process is carried out in a closed environment, which can reduce the burden on operators and achieve the goal of "sample in answer out". The nucleic acid extraction methods include magnetic bead separation, solid-phase extraction column based on silica gel, separation of silica gel microspheres, paper-based separation, and chemical separation of materials using imidacloprid dimethyl ester. Advanced methods such as faster and more sensitive digital PCR and isothermal amplification have become mainstream technologies and have been applied in POCT based on nucleic acid detection (Ji et al. 2020).

In recent years, applying POCT microfluidic chips has broadened from the analysis of molecules to cells and biomimetic human organs, such as those replicating smell, vision, and taste. As intelligent technology advances, future POCT analysis platforms will become more portable and user-friendly. Despite their application in immune and nucleic acid analysis, microfluidic chips are not widely used in commercial instruments and have limited adoption at the grassroots level due to the complexity of biological reaction steps and chip structure, resulting in high costs.

10.2.6 Cell and Organ Chips

Microfluidic chips have extensive applications in cell culture, separation, component analysis, and cell manipulation (Kim and Takayama 2015). Due to their unique advantages, microfluidic chips have found widespread applications in various fields. Single-cell analysis is one of the most significant applications, and it plays a crucial role in early diagnosis, treatment, drug screening, and research on cellular physiology and pathological processes of significant diseases. By integrating multiple operational units such as optical tweezers or ultrasound capture, optical perforation, cell lysis, electrophoresis separation, and cell loss counting onto a microfluidic chip and analyzing the obtained information, precise manipulation of individual cells can be achieved.

Organ chips are another promising application of microfluidic technology. They utilize the flexible manipulation capabilities of microfluidic chip systems to create microenvironments that mimic complex internal environments, including fluid shear forces, dynamic mechanical stresses, and concentration gradients. In addition, organ chips can also simulate multicellular structures, cell-cell interactions, tissue-tissue interactions, vascular perfusion, and the interaction between cells/tissues and blood (Huh et al. 2010; Huh et al. 2012).. These applications help evaluate the effectiveness and biosafety of drugs and vaccines and provide low-cost screening and research models closer to the human body's natural physiological and pathological conditions for organ physiology, tissue development, and disease etiology research.

In recent years, the field of organ chip research has achieved rapid development. Currently, most organ chips can achieve co-culture of multiple cells, simulate dynamic culture of blood flow and material transport, and provide chip organ models that can be used for drug screening experiments. These models include muscle chips, bone chips (Sun et al. 2019), lung chips (Huh et al. 2010), intestinal chips, heart chips, liver chips, intestinal kidney chips (Jang and Suh 2010), liver intestinal chips, intestinal liver cancer chips, and integrated multi-organ human chips.

Although it is currently possible to simulate specific organ tissue structures and functions on organ chips, cultivating these tissues for an extended period in vitro and preserving their physiological functions remains a challenge. For instance,

PDMS material chips may absorb the drugs being studied, while other polyester or natural hydrogel materials (such as matrix gel or collagen) can be used to construct equipment, albeit at a higher cost. These limitations make it challenging to fully replicate all the features of organs on a chip. Therefore, using microfluidic chips to create a controlled environment must be combined with ongoing advancements in other technologies to enhance the performance of organ chips and bring them closer to replicating real organs.

10.2.7 Food and Agricultural Applications

Microfluidic chip technology has broad application prospects in food and agriculture (Whitesides 2006; Hu et al. 2015). Firstly, using microfluidic chip technology for high-throughput genotyping analysis of plants can provide genotype data of plant species in a short period and at a lower cost, thereby helping to design genetically modified crops. This method can improve crop yield and disease resistance and reduce resource consumption and environmental pollution in agricultural production.

In addition to genotyping analysis, microfluidic chip technology can also be used for phenotype analysis. Using multiple chips to study the dynamic characteristics of plants under different nutrient stress conditions (such as concentration, pH, temperature, and other chemicals) over time, more precise regulatory strategies can be provided for agricultural production. For example, a three-channel paper-based microfluidic chip based on smartphones can detect Escherichia coli in field water (Park and Yoon 2015), while a smartphone-based colorimetric method can simultaneously detect pH and nitrite in soil (Lopez-Ruiz et al. 2014). Based on this, fertilization and irrigation strategies can be adjusted to improve crop yield and quality.

In addition, microfluidic chip technology provides a new way to quickly and effectively detect food pollutants. By applying microfluidic chips to detect chemical and physical properties such as pathogens, allergens, toxins, pesticide residues, heavy metals, additives, etc., a large number of samples can be analyzed quickly, improving the efficiency and accuracy of food safety monitoring. This is significant for safeguarding consumer health and promoting sustainable development of the food industry (Figs. 10.5 and 10.6).

10.3 Organ-on-a-Chip

Cells are the basic units of structure and function in the human body. Cells with similar functions combine in a particular arrangement to form tissues, and tissues combine to form organs with specific functions. Since traditional two-dimensional cell culture grows cells on a flat and hard surface, it cannot simulate the three-dimensional environment of cells in the human body, which causes differences in

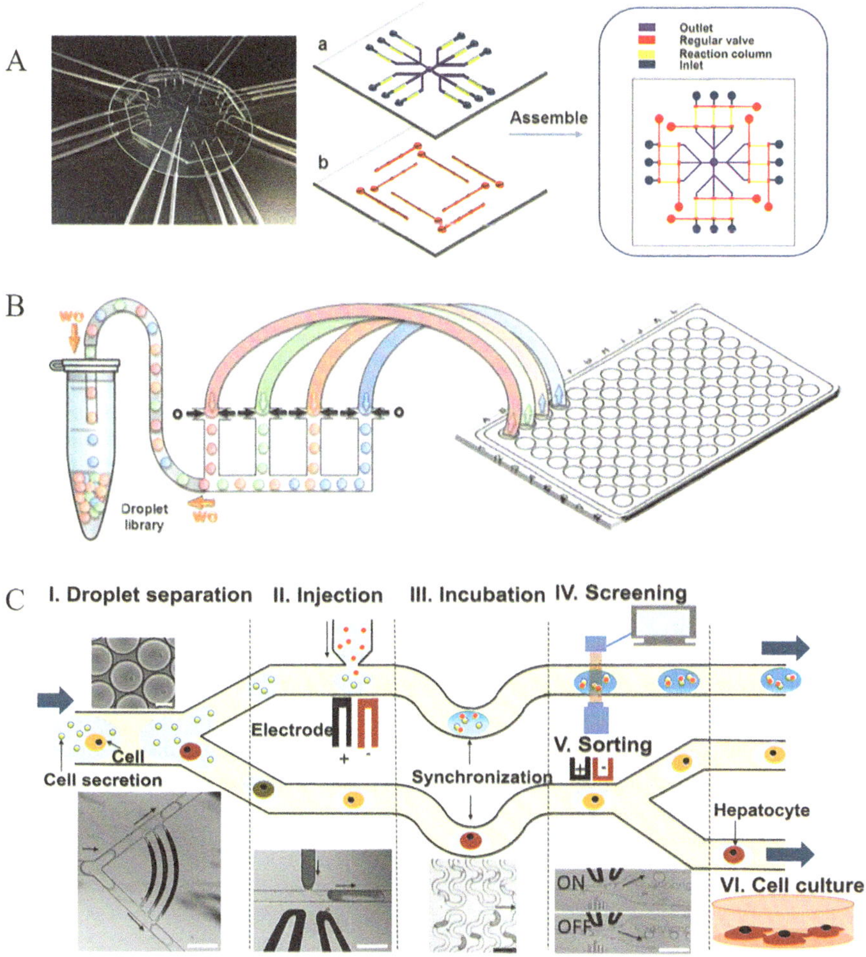

Fig. 10.5 Application of microfluidic chips in different field. (**a**) Microfluidic immunoassay system (MIAS) with integrated immunoassay and microfluidic components. Reprinted with permission from Duan et al. 2019. MIAS images: (*a*) fluid layer and (*b*) aerodynamic control layer. (**b**) A droplet sorting platform with multiple operating units. Reprinted with permission from Sun et al. 2020. Synchronize the workflow of supernatant analysis and cell sorting. After splitting into two sub-droplets, a cytotoxic chemical sensor was injected and analyzed by fluorescence spectroscopy. Then, another synchronous droplet containing cells can be sorted based on the fluorescence signal, and then cultured. (**c**) Using droplet barcode technology to achieve sequencing. Reprinted with permission from Guo et al. 2012. The holes of the microtitration plate are connected to the water sample inlet of the multiple droplet generation device (colored arrow) to simultaneously emulsify the contents of each hole. Droplets gather together to form a droplet reservoir. Each hole contains a unique barcode, represented in different colors in the figure, which is carried by the droplets in the hole and used to decode the content of each droplet

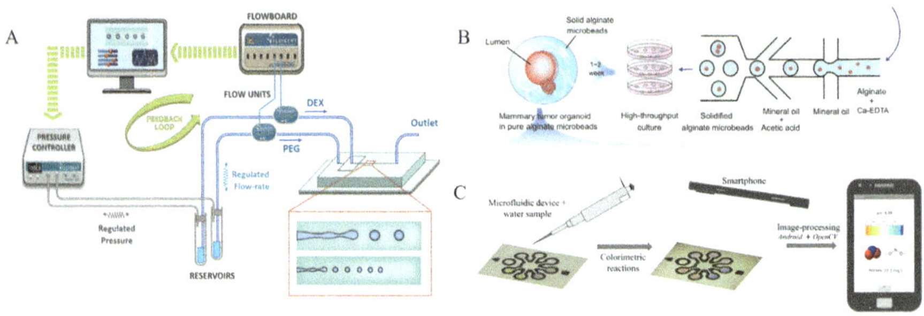

Fig. 10.6 Application of microfluidic chips in different field. (**a**) Experimental setup and schematic diagram of microfluidic chips for droplet generation. Reprinted with permission from Mastiani et al. 2018. (**b**) Application example of droplet-based microfluidics in drug screening, schematic diagram of high-throughput generation of alginate microspheres by encapsulating tumor blocks using microfluidic droplet technology. Reprinted with permission from Fang et al. 2021. (**c**) Simultaneous detection of pH and nitrite using smartphone-based colorimetric method. Reprinted with permission from Lopez-Ruiz et al. 2014

cell behavior between in vitro and in vivo cells, such as cells are flattened instead of normal spherical shape, cells lose normal function due to lack of cell-cell or cell-matrix interactions, which leads to inaccurate results of cell research.

Organ-on-a-chip (OOC) is the cultivation of cells on microfluidic chips to simulate human organs' physiological and biological characteristics. The primary objective of OOCs remains the emulation of the smallest functional unit of organ function in vitro. Consequently, in this context, certain single-layer cell-cultured microfluidic chips capable of simulating the function of the corresponding organ are still classified as OOCs.

10.3.1 Advantage and Application of OOCs

Culturing organs in microfluidic chips has several advantages: first, high fidelity. The materials and design of microfluidic chips are highly controllable and can create a more realistic microenvironment for cell culture by using biomimetics. Huh et al. cultured lung cells on a bio-membrane wrapped with extracellular matrix and covered the other side of the membrane with pulmonary capillary cells, thus simulating the structure of human alveoli (Huh et al. 2010). Second, high sensitivity. The Reynolds number of the fluid in the microfluidic chip is low, and the fluid is laminar (the adjacent layers of the laminar flow do not mix), which can achieve precise control of the fluid. Third, low pollution. The cell sources for organ culture are primarily human induced pluripotent stem cells (hiPSCs), and a critical step in inducing hiPSCs to differentiate into organoids is to form embryoid bodies (EBs), and then transfer the EBs to the matrix gel. In this process, if the cells are contaminated, it will cause differences between the batches of organoids formed, which is not conducive to subsequent research. Because PMDS has light transmittance and air permeability, the mainstream microfluidic chips are closed, and oxygen and nutrients are perfused through pipes, which can achieve a closed culture environment and avoid external pollution causing cell mutation. Zhu et al. cultured hiPSCs on PDMS with micropillar arrays (Zhu et al. 2017). The optimized micropillar arrays generated abundant EBs, which were then induced to differentiate into brain organoids in situ on the chip, thus avoiding possible contamination.

At present, the application prospect of OOCs is mainly drug screening. Preclinical experiments are a key step in drug research. However, because of the large differences between animal cells and human cells, for example, there is already research that proves that colon stimulating factor has the opposite effect on the content of class II major histocompatibility complex proteins (MHC II) in microglia cells in mice and humans (Lambert et al. 2008). Therefore, animal models have poor predictive ability, increasing the drug development cost. On the other hand, animal experiments also often encounter various ethical issues. In a word, OOCs, as a kind of artificial organ, are an ideal alternative to animal experiments because of their low cost and no ethical problems (Fig. 10.7).

10.3.2 Brain-on-a-Chip(BOC)

The human brain stands as the most intricate organ within the human body. Structurally, it comprises the telencephalon and the diencephalon, with its cellular makeup primarily consisting of neurons and neuroglia. Neurons serve as the conduits for various brain functions, whereas neuroglia predominantly offer support and protection to neurons, ensuring their metabolic needs are met.

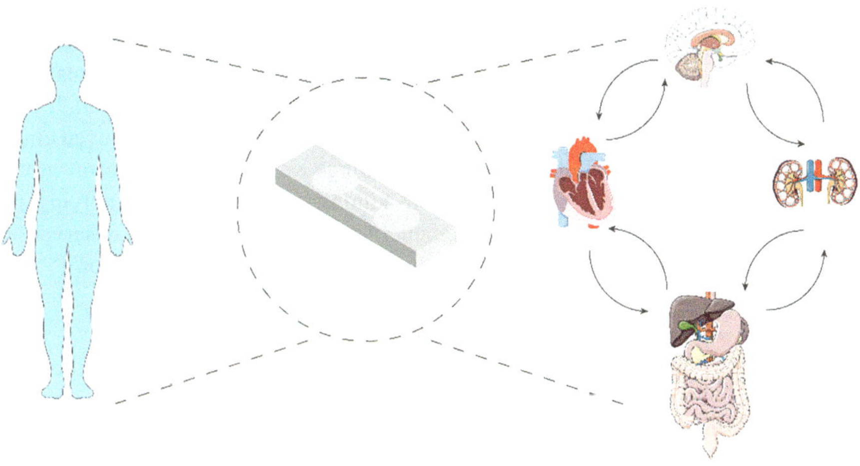

Fig. 10.7 Schematic diagram of organ-on-a-chip. As depicted in the figure, the physiological functions of distinct anatomical components are simulated by cultivating diverse tissue cells on a microfabricated chip platform

10.3.2.1 BOC for Blood-Brain Barrier

The neuroglia in the brain mainly includes astrocytes, oligodendrocytes, and microglia. Brain has three main barrier sites between blood and the brain: the blood-brain barrier (BBB), blood-cerebrospinal fluid barrier, and arachnoid barrier. Their main function is to regulate the diffusion of ions, macromolecules, neurotoxins, and other substances between the brain and the outside and to provide nutrition to the brain. Among them, the BBB is produced by tight junctions of capillary endothelial cells closest to brain cells. Once drugs pass through the BBB, the diffusion distance of drugs to nerve cells is very short, so the BBB is the preferred route for delivering drugs to brain cells (Abbott et al. 2010). It has been demonstrated that many diseases are associated with BBB dysfunction, including Parkinson's (Desai et al. 2007), Alzheimer's disease (Desai et al. 2007), and epilepsy (Remy and Beck 2006).

Due to the advantages of microfluidic chips in high throughput and fabrication efficiency, several microfluidic chip-based blood-brain barrier modeling studies have been conducted. Zakharova et al. prepared a multichannel multiplexed PDMS chip consisting of a common inlet branch to form eight parallel channels, each with a separate outlet and used soft lithography to fabricate a 2-μm-thick PDMS membrane inside the chip using soft lithography (Zakharova et al. 2020). Cell culture results showed that human cerebral microvascular endothelial cells (hCMEC) and astrocytes formed a monolayer, expressing the tight junction protein ZO-1 and the adhesion link protein VE-calmodulin in endothelial cells. Fibrillar acidic proteins in astrocytes, and molecular permeability tests showed that macromolecules crossed the barrier at a lower rate. In addition, Campisi et al. co-cultured hiPSCs-derived endothelial cells (ECs), pericytes (PCs), and astrocytes (ACs) in a fibrin gel microenvironment to form a self-assembled microvascular network (Campisi et al. 2018). Confocal microscopy results showed contact interactions between PCs and ACs, consistent with the structure of the BBB in humans.

Recently, there have been several studies applying BBB microarrays to drug screening. Zeng et al. achieved triple co-culture of endothelial cells, astrocytes, and microglia on the hydrogel surface and reproduced the BBB in both normal and abnormal states, which creates a possibility for drug screening (Zeng et al. 2023). Moreover, researchers incorporated dexmedetomidine, a frequently employed clinical anesthetic, into the cellular milieu. The findings revealed a suppression in the motility of astrocytes and microglia upon administering the drug, implying its potential efficacy in impeding the movement of endothelial cells. This observation suggests a plausible role for the drug in mitigating glial cell hyperactivation.

10.3.2.2 Microelectrode Arrays and BOC

Brain cells' functional physiological and pathological roles are widely recognized to be intricately tied to the electrical signals of neuronal cells. Numerous diseases are closely linked to disruptions in neuronal electrical activity, such as stroke (Wann et al. 2020) and epilepsy (Acharya et al. 2013). The gold standard for characterizing the electrophysiological activity of individual neurons is the membrane-clamp technique, but it is unsuitable for combining with microfluidics due to its complexity and damage to cells. With the development of various engineering techniques, a microelectrode array (MEA) technology has emerged that allows for high-throughput and non-destructive stimulation of cells.

Pelkonen et al. designed an in vitro epilepsy model combining MEA with a microfluidic chip with three independent partitions connected only by microtunnels, with 24 disk microelectrodes at the base of each partition and eight rectangular microelectrodes at the base of each microtunnel (Pelkonen et al. 2020). Neuronal cells were cultured discretely within individual partitions. Upon prolonged cultivation, the micro tunnels exhibited the presence of axons and discernible astrocytes, signifying the model's capacity to develop into a mature neuronal network. Subsequent drug stimulation in a designated partition elicits spike potential alterations within the targeted sector and across the remaining two partitions. This observation implies that the interconnected microelectrode network facilitated the synchronization of potential activities, establishing a groundwork for subsequent investigations into epileptic seizures.

Electroencephalogram (EEG) cap represents a prevalent electrode array employed for the acquisition of EEG measurements in the human brain. Huang et al. mimicked the EEG cap, developed a foldable 3D metal electrode that wraps an organoid sphere, and used it to potentially measure brain cell spheres (Huang et al. 2022). The experimental findings revealed that the electrode effectively captured and recorded the field potential emanating from the spheres upon establishing contact between the electrode and cellular spheres. Furthermore, a discernible alteration in signal amplitude was observed in response to the introduction of the neurotransmitter glutamate, indicating a pronounced impact on the recorded electrical activity. However, akin to the distinction between scalp EEG and intracranial EEG (iEEG) in human brain research, the EEG cap can only capture electrical signals from the external surface of the human brain. At the same time, needle electrodes can penetrate internally, obtaining more precise electrical signals. Zips et al. prepared an array of micrometer needle electrodes (MnEAs) (Zips et al. 2023). Measurements of electrical signals from neurospheres using MnEAs showed heightened accuracy.

Another direction of integrating BOC with MEA involves leveraging the real-time monitoring capability of MEA to acquire live data from cells on the chip, thereby alleviating the operational burden on experimenters. Ao et al. developed a one-stop brain organoid culture device to avoid heterogeneity between different brain organoids by reducing the need to change the culture medium during organoid generation (Ao et al. 2020), and measured the effect of the neuroactive cannabinoid Δ-9-tetrahydrocannabinol (THC) on the electrical activity of the organoids was measured using MEA, and the electrical activity of the organoids was significantly reduced after THC treatment (Fig. 10.8).

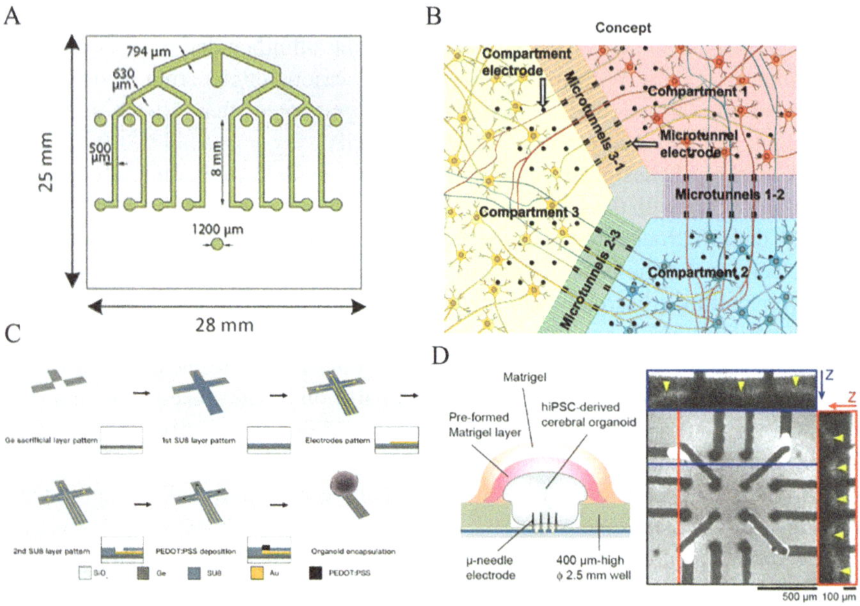

Fig. 10.8 Schematic of BOCs. (**a**) Multiplexed chip for multiplexed drug testing. Reprinted with permission from Zakharova et al. 2020. (**b**) BOC allowed axonal connections through microchannels for modelling epilepsies. Reused with permission from Pelkonen et al. 2020. (**c**) Foldable microelectrode arrays for recording electrical signals from three-dimensional cell spheroids. Reprinted with permission from Huang et al. 2022. (**d**) Microelectrode arrays equipped with high-aspect-ratio µ-needle electrodes for recording electrical signals of neurosteroids. Reprinted with permission from Zips et al. 2023

10.3.3 Heart-on-a-Chip(HOC)

The cardiac organ, situated anteriorly within the thoracic cavity, holds paramount significance in the human physiological framework by orchestrating the systemic circulation of blood. Anatomically, a typical human heart approximates the dimensions of a clenched right-hand fist. Comprising predominantly of cardiac muscle cells, the heart exhibits a structured composition characterized by four chambers: the left atrium, right atrium, left ventricle, and right ventricle. The customary hemodynamic sequence involves pumping oxygenated blood from the left atrium to the left ventricle, followed by systemic circulation throughout the body. Subsequently, deoxygenated blood returns to the right atrium through systemic circulation, enters the right ventricle under negative intraventricular pressure, and completes its pulmonary circulation before re-entering the left atrium.

10.3.3.1 Fabrication of HOC

Conventional microfluidic chips consist of PDMS, which, although they are biocompatible, may lead to various problems in cell culture due to their adsorption of proteins and other biomolecules. Hydrogels have been widely used in cell culture due to their low cost, biocompatibility, low cytotoxicity, and ease of preparation (Cao et al. 2021). There have been many studies on using hydrogels for HOC preparation. Zhang et al. prepared strip-shaped cardiomyocyte tissues by wrapping cardiomyocytes in collagen-based biomaterials (Zhang et al. 2017). After adding 3% of human neonatal dermal fibroblasts, the contractile properties of the strip-shaped cardiomyocyte tissues were improved, suggesting that the addition of auxiliary cells may contribute to the improvement of the function of cardiomyocytes, possibly due to the existence of a specific interaction between auxiliary cells and cardiomyocytes. To further investigate the factors affecting the developmental state of cardiomyocytes, Kolanowski et al. explored the effects of hemodynamics on the maturation of cardiomyocytes by culturing the cells in fluid and by varying the flow rate of the fluid (Kolanowski et al. 2020), the degree of oxygen supply. Moreover, they found that cardiomyocytes cultured under cyclic pulsatile flow conditions exhibited heightened contractile force and synchronicity. This observation implies that alterations in perfusion conditions may confer greater advantages for the maturation of cardiomyocytes.

The continuous advancement of various innovative technologies has propelled the ongoing progress in preparing HOC materials. For instance, bioprinting technology facilitates the construction of three-dimensional structures with specific shapes and functionalities by layer-by-layer deposition of biomaterials. This significantly streamlines the fabrication cost of microfluidic chips. Zhang et al. used bioprinting to prepare a multilayered hydrogel fiber scaffold and cultured endothelial cells inside the microfibers, which converged into a single layer as the fibers migrated outward, and the authors concluded that this model they prepared could be used for cardiovascular toxicity drug screening (Zhang et al. 2016). Another example of applying bioprinting technology to HOC preparation is illustrated by Liu et al. (2023). The researchers integrated bioprinting and electrospinning techniques by initially utilizing a 3D-printed polycaprolactone (PCL) scaffold to mimic the anatomical structure of human myocardial cells. Subsequently, electrospinning was employed to fabricate a directional nanofiber network, guiding cellular differentiation in a specific orientation. The study also investigated the cardiotoxicity of doxorubicin (DOX) and the cardioprotective effects of dexrazoxane (DEX).

Similar to the development of BOC, real-time inspection of the original space is also a direction of progress in HOC. Aptamers, serving as highly specific single-stranded oligonucleotides, are currently widely employed for the ex vivo monitoring of biomarkers (Li et al. 2023). Shin et al. combined a microfluidic chip with an electrochemical sensing platform to detect whether cardiomyocytes are damaged using the specific binding of aptamers to creatine kinase (CK), a biomarker of cardiomyocyte injury (Shin et al. 2016). In this study, cardiac organoid spheroids were cultured in gelatin methacryloyl (GelMA) hydrogel, and a cell damage model was constructed using the anticancer drug DOX. Subsequently, aptamer-modified electrodes were combined with the hydrogel chip, and the experimental results showed that as the concentration of DOX increased, so did the concentration of CK, which suggests that the model can be used to monitor the damage of cardiac organoids at the original space. Another type of real-time inspection at the original space utilizes the fast response properties of electrodes. For example, Ren et al. engineered a microfluidic chip featuring microgrooved structures (MGs) created through laser etching, and gold electrodes were deposited on the substrate for cell stimulation (Ren et al. 2020). The investigation revealed that the combined influence of MGs and electrical stimulation resulted in the development of aligned and synchronized cardiac organoids with beating characteristics. The introduction of anti-cancer drugs with cardiotoxic effects to the chip demonstrated that these drugs induced the demise of cardiac cells, indicating the potential applicability of the model in drug screening (Fig. 10.9).

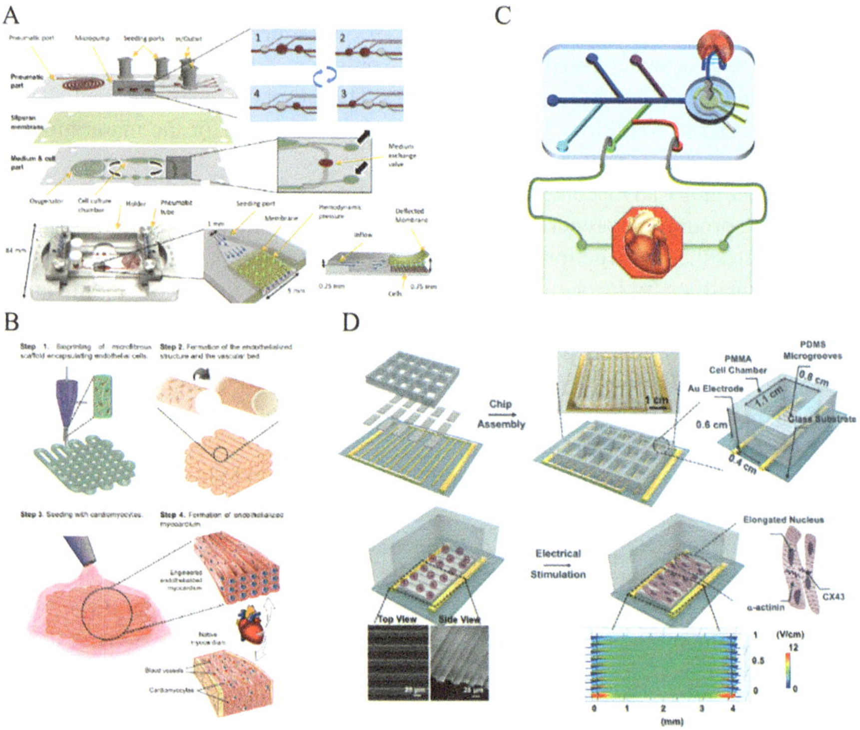

Fig. 10.9 Fabrication of Heart-on-a-chip. (**a**) HOC enhances maturation of human iPSC-derived cardiomyocytes through cyclic pulsatile hemodynamic forces. Reused with permission from Kolanowski et al. 2020. (**b**) Endothelialized myocardium through 3D bioprinting. Reused with permission from Zhang et al. 2016. (**c**) Aptamer-based compatible electrochemical biosensor for monitoring cardiac organoids. Reprinted with permission from Shin et al. 2016. (**d**) HOC incorporates microgrooves and electrical pulse stimulations for drug screening. Reprinted with permission from Ren et al. 2020

10.3.4 Gastrointestinal Tract

Gastrointestinal tract refers to the pipeline from the oral cavity to the anus, whose primary physiological function is to digest food and absorb nutrients. Structurally, it can be divided into the upper and lower digestive tracts, with the upper digestive tract including the oral cavity, pharynx, esophagus, and stomach, and the lower digestive tract including the small intestine, large intestine, and anus. Organ chips mainly study the stomach and intestines, which structurally consist of four parts: mucosa, submucosa, muscular layer, and adventitia.

10.3.4.1 Stomach-on-a-Chip

The primary physiological role of the stomach is to mechanically break down substantial food particles into smaller fragments, thereby enhancing subsequent absorption processes within the small intestine. Therefore, the key to stomach-on-a-chip is to mimic the mechanical contraction of the stomach (Carvalho et al. 2023). Lee et al. cultured human gastric organoids (hGOs) directed to differentiate from hiPSCs in Matrigel and injected nutrients into hGOs via micropipettes attached to a peristaltic pump (Lee et al. 2018). Micropipettes are used on peristaltic pumps to inject nutrients like hGO. In this way, the authors observed hGO peristalsis. Cherne and his co-workers integrated dendritic cells and gastric epithelial cells onto a microfluidic chip, enabling real-time monitoring of immune-gastric cell interactions (Cherne et al. 2021). However, both models use Matrigel as a substrate for cell culture and do not reproduce human gastric extracellular matrix (ECM) composition and architecture well. Ferreira et al. modeled the epithelial barrier, basement membrane, and supporting lamina propria of the gastric wall by a nine-layer PDMS, and the ECM of gastric cells in vivo by embedding a gastric fibroblast cell line (NST-20) in a typeIgel (Ferreira et al. 2023).

A model of 2D monolayer gastric organoids (2DMGOs) has recently been reported and claimed that there is spatial heterogeneity in the growth and differentiation of 3D organoids (Jantaree et al. 2021), and that 2D organoids can rapidly form homogeneously fused monolayers of cells, which would create model for drug development that more closely resembles the human physiological environment.

10.3.4.2 Gut-on-a-Chip

The gut is the primary digestive organ and the main route for drug absorption. The human gut contains about 10 (Chua et al. 2009)–10 (Grover et al. 2006) microbial communities, and the roles of these microorganisms in the human body include immune system regulation, absorption of nutrients, and drug metabolism (Wardman et al. 2022). In addition, the gut has been linked to many diseases, such as Alzheimer's disease (Liu et al. 2020), inflammation (Martel et al. 2022), and more. The inner surface of the small intestine is lined with a folded structure called the villous structure. The villous structure of the small intestine is able to substantially increase the surface area of the inner wall of the small intestine, thus improving the efficiency of nutrient absorption in the small intestine. Hence, the pivotal aspect of the Gut-on-a-chip (GOC) lies in the emulation of the villous structure inherent in the small intestine. Maurer et al. constructed a human gut model consisting of human umbilical cord vein endothelial cells and human epithelial colorectal cells on a microfluidic chip (Maurer et al. 2019). In this model, continuous fluid perfusion during cell culture could spontaneously form three-dimensional structures with columnar epithelium at the cell tips. The study also simulated the intestinal flora inside the human body by adding live lactobacilli, and the experimental results showed that the intestinal barrier could organize the passage of the pathogen albicans after the addition of lactobacilli. Similarly, Kim and his colleagues deformed the cells by controlling the vacuum chambers on either side of the microchannels, causing the cells to undergo rhythmic mechanical deformation (Kim et al. 2012). The result suggests that small intestinal cells can spontaneously form villus-like structures in response to mechanical forces. In addition, it has been found that culturing epithelial cells on scaffolds also produces villous structures (Costello et al. 2014). Nikoolaev et al. cultured mice, small intestinal cells in microchannels that mimicked mouse native crypts, and the cell culture results showed that the cells could recapitulate the structure of native crypts (Nikolaev et al. 2020).

As mentioned above, gut flora plays a significant role in the human body, and gut flora is a risk factor for many diseases. For example, Helicobacter pylori causes the greatest risk factor for stomach cancer (Amieva and Peek Jr. 2016). Therefore, it is crucial to understand the relationship between intestinal flora and the gastrointestinal tract. Grassart et al. used PDMS to construct a microarray with retractable vacuum channels on both sides of the central channel and introduced a wild-type strain expressing GFP (Shigella-WT-GFP) in the microarray (Grassart et al. 2019). The experimental outcomes revealed that the porous structural arrangement was more conducive to bacterial adhesion.

Additionally, peristaltic activity was observed to facilitate bacterial invasion from the apical side of the epithelium. Boquet-Pujadas et al. developed a microchip capable of real-time assessment of mechanical stresses in tissues and studied the infection dynamics associated with E. histolytica and S. flexneri (Boquet-Pujadas et al. 2022) to investigate host-pathogen interactions further. To enable real-time imaging of infection process, the study combined a programmed microscope with a microfluidic chip and algorithmically controlled the microscope's focus. The experimental results showed that cyclic peristalsis promoted S. flexneri invasion and amplified virulence. However, each of these models can only study the action of one microorganism on the small intestine at a time, and many microbial communities exist in the human body simultaneously. Hence, investigating the simultaneous presence of diverse microbial communities in the gastrointestinal tract is of paramount importance, Jalili-Firoozinezhad developed a GOC that can support the symbiosis of multiple microbial communities (Jalili-Firoozinezhad et al. 2019). This investigation incorporated fluorescent particles sensitive to oxygen quenching into the microfluidic chip to monitor real-time oxygen partial pressure continuously. Additionally, a diverse human fecal-derived microbiome was introduced into the chip. The experimental outcomes demonstrated the device's proficiency in facilitating co-cultivation of viable microbiota with small intestinal cells. Significantly, the chip demonstrated increased microbial diversity in a low-oxygen environment.

In summary, the GOCs mentioned above have simulated specific physiological characteristics of the small intestine. However, nutrient absorption is the most crucial function of the human body's small intestine. Numerous diseases are closely associated with impediments in the absorption of nutritional substances by the small intestine; for instance, compromised calcium absorption may result in osteoporosis. Huang et al. developed a microfluidic chip that can detect the small intestinal barrier for calcium ion absorption (Huang et al. 2014). The authors cultured Caco-2 cells inside the chip and utilized the calcium-sensitive dye fura-2 to determine the transported free Ca^{2+} from the top microfluidic chamber to the basolateral side of the cell monolayer (Fig. 10.10).

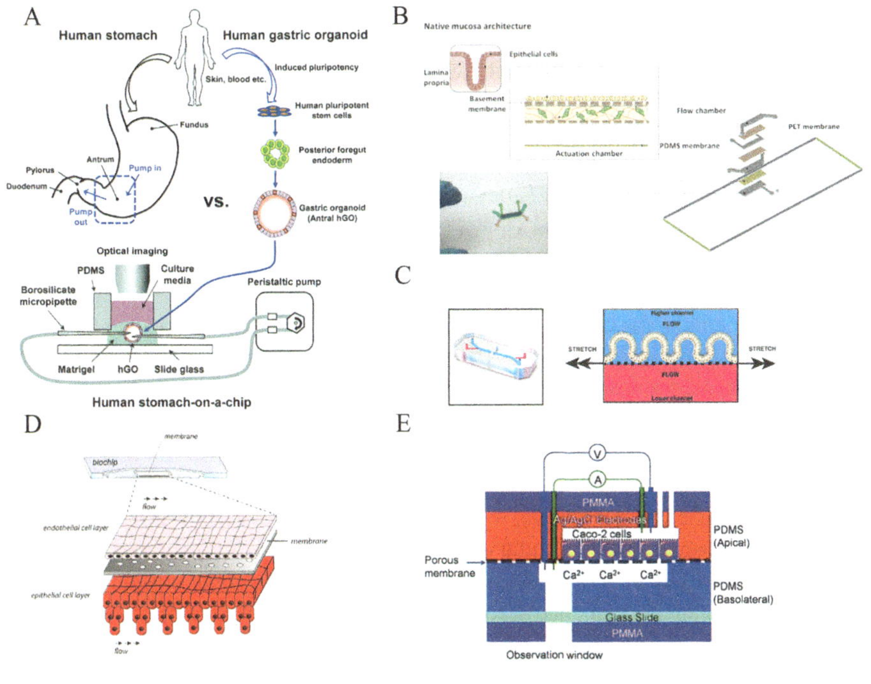

Fig. 10.10 Gastrointestinal tract-on-a-chip. (**a**) Stomach-on-a-chip with peristaltic luminal flow in vitro. Reprinted with permission from Lee et al. 2018. (**b**) Supporting peristalsis-like motion Stomach-on-a-chip. Reused with permission from Ferreira et al. 2023. (**c**) Chips capable of regulating mechanical forces in microenvironments. Reused with permission from Grassart et al. 2019. (**d**) A three-dimensional microphysiological model of the human intestine living microorganisms. Reused with permission from Maurer et al. 2019. (**e**) A microfluidic model allowing free calcium (Ca^{2+}) transport through a confluent monolayer of Caco-2 cells. Reprinted with permission from Huang et al. 2014

10.3.5 *Liver-on-a-Chip*

In the human body, the liver is an organ whose main function is metabolism. The liver's metabolic functions include glucose, lipid, protein, and drug metabolism. The human liver includes five main types of cells: hepatocytes, cholangiocytes, hepatic stellate cells, and hepatic sinusoidal endothelial cells (Banales et al. 2019). The most basic functional unit of the liver is the hepatic lobule, which is centered on the central vein, and the hepatic plates and hepatic sinusoids are arranged in a radial pattern centered on the central vein.

Microfluidic-based liver-on-a-chip (LOC) can mimic the physiological activities of the liver in the human body through suitable substrate design, thus enabling hepatocyte function studies, and drug screening.

10.3.5.1 Fabrication of LOCs

Typically, the preparation process of LOC involves selecting a suitable substrate material, using a flexible material such as PDMS to form a three-dimensional channel for cell growth, and then the implantation of different hepatocytes (Liu et al. 2022). PDMS is currently the most common material used for LOCs. However, the non-permeable property of PDMS hinders material transfer, and in order to better mimic the structure of the liver lobules, some studies have utilized permeable hydrogels to prepare LOCs with endothelial cells attached to the inner walls of the channels to mimic the capillaries in the liver (Meng et al. 2021). Meanwhile, as cell printing technology has matured, it has also been applied to preparing LOCs. Lee et al. used cell printing to prepare a 3D-LOC with two fluidic channels, where the upper channel provides nutrients to the cells, and the lower channel removes waste products generated by cell metabolism (Lee et al. 2019). Cyclic olefin copolymer (COC) is a common material used to prepare microfluidic chips with low adsorption and better biocompatibility (van Midwoud et al. 2012). However, its permeability is not as good as PDMS, which may lead to cellular hypoxia (Jellali et al. 2016). In order to maintain good permeability and low adsorption, a hybrid COC-PDMS device was prepared to culture hepatocellular carcinoma cells by Kulsharova et al. (2021).

Real-time observation of cellular activity on the LOC has been another main goal of this field. Bavli et al. achieved real-time monitoring of mitochondrial respiration by embedding phosphorescent microprobes inside tissues (Bavli et al. 2016).

10.3.5.2 LOCs for Drug Discovery

Drug-induced liver injury (DILI) is a major cause of drug withdrawal from the market (Kulsharova and Kurmangaliyeva 2021), and using in vitro liver modeling to assess the toxicity of a drug prior to its introduction to the market is one of the main application scenarios for LOCs.

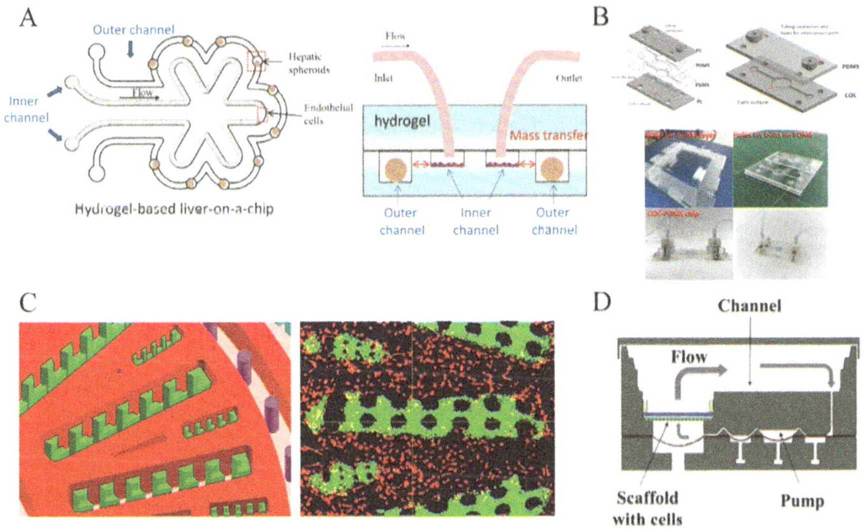

Fig. 10.11 Liver-on-a-chip. (**a**) Hydrogel-based LOC for research on nutrient perfusion. Reprinted with permission from Meng et al. 2021. (**b**) LOC in dynamic conditions. Reprinted with permission from Kulsharova et al. 2021. (**c**) a microfluidics-based biomimetic approach for fabricating an in vitro 3D liver lobule-like microtissue. Reprinted with permission from Ma et al. 2016. (**d**) A three-dimensional human microphysiological hepatocyte–Kupffer cell coculture system. Reprinted with permission from Sarkar et al. 2015

To study the metabolism of drugs in the liver, Ma et al. constructed a microfluidic chip that mimics the structure of human liver lobules, and experimental results showed that cells cultured in this device could exhibit liver function (Ma et al. 2016). Upon addition of an appropriate amount of acetaminophen (APAP) alone in this model, about 90% of the drug was metabolized to N-acetyl-p-benzoquinone imine (NAPQI), where the addition of an excessive amount of APAP resulted in hepatocellular injury. In order to study the interaction between different drugs, 50 μM omeprazole (OME), 25 μM ciprofloxacin (CPFX), respectively, followed by APAP were added, and the results showed that the death of hepatocytes was significantly increased by OME treatment. In contrast, the density of live hepatocytes was increased by CPFX treatment. Cortisol, also known as hydrocortisone, has anti-inflammatory and inhibitory effects on immune responses. Sarkar et al. co-cultured primary hepatocytes and Kupffer cells on a microfluidic chip and stimulated the cells to produce inflammatory cytokines using lipopolysaccharide (Sarkar et al. 2015). The findings demonstrated concentration-dependent anti-inflammatory effects of hydrocortisone. The authors postulated that this experimental system holds promise for investigating the efficacy of anti-inflammatory pharmaceuticals. To improve drug screening efficiency, Bircsak et al. cultured iPSC-derived hepatocytes in a microfluidic multiwell plate and tested the hepatotoxicity of 159 drugs using this platform (Bircsak et al. 2021) (Fig. 10.11).

10.3.6 Multi-Organ-on-a-Chip

The human body operates as an integrated unit, wherein organs function interdependently rather than in isolation. Hence, the imperative development of multi-organ-on-a-chip systems is required to accurately replicate the intricate interactions among organs within the human body.

10.3.6.1 MOC for Pharmacology

Since drug absorption, distribution, metabolism, and elimination occur in separate organs, they are also known as ADME. ADME is one of the most common physiological processes of multi-organ interactions in the human body. More importantly, using microchannels on the microfluidic chip makes it possible to link metabolites between different organs, thus simulating organ interactions in the human body.

The first-pass effect refers to the phenomenon that after absorption of a drug from the digestive tract, part of the drug may be metabolized and inactivated as it passes through the intestinal mucosa and the liver, resulting in a decrease in the amount of drug entering the circulation and a decrease in the efficacy of the drug. It has been shown that most of the first-pass effects occur in the gut and liver (Nakanishi and Tamai 2015). Thus, there have been studies utilizing multi-organ microarrays to reproduce the first-pass effect in the human body. Imura et al. co-cultured hepatocytes, intestinal cells, and cancer cells on a microfluidic chip, where intestinal cells Caco-2 were at the front of the chip, hepatocytes HegP2 were in the middle of the chip, and cancer cells were at the very end of the chip (Imura et al. 2010). When fluid flows inside the chip, the drug first passes through the intestinal cells, then through the hepatocytes, and finally reaches the cancer cells, by which the authors simulated the first-pass effect of the drug in the body. Transdermal drug delivery allows drugs or macromolecules to pass through the skin into the circulation at a fixed rate, avoiding first-pass effects, improving drug utilization, and maintaining plasma concentrations of the drug (Marwah et al. 2016). A team of researchers developed a heart-liver on-a-chip with a skin substitute (Strat-M membrane) and used the model to test the effects of various common drugs on heart and liver function (Pires de Mello et al. 2020). The outcomes revealed that 3% diclofenac substantially reduced cardiomyocytes' conduction rate, contractility, and spontaneous frequency. Moreover, exposure of the heart-liver system to a high concentration of hydrocortisone resulted in a significant decrease in the heart's spontaneous conduction frequency and a reduction in the liver's albumin production. These findings suggest that elevated levels of hydrocortisone adversely impact both cardiac and hepatic functions.

On the other hand, the liver is the main organ for drug metabolism in the human body too; in order to make the organ chip more suitable for drug screening, Yin et al. prepared a liver-cardiac organ chip for evaluating the safety of antidepressant drugs on the heart after metabolizing them in the liver (Yin et al. 2021). In this study, a microfluidic chip with a four-layer structure was designed, in which two layers of cavities were used to culture cells, a porous chamber in the upper layer was used to culture hepatocytes, and a micropillar array in the lower layer was used to differentiate and culture cardiac organoids. After the addition of the common antidepressant drug clomipramine to this model, significant cell death occurred in cardiac organoids, and cytoplasmic calcium ion levels decreased in cardiac cells, suggesting that the drug is cardiotoxic.

Furthermore, Skardal et al. prepared an organ-on-a-chip that can accommodate the liver, heart, lungs, blood vessels, testes, brain, and colon and investigated the differences in cytotoxicity of the FDA-recalled drug in 2D cell cultures and in 3D cell cultures (Skardal et al. 2020). The results showed that 20 μM capecitabin did not exhibit cytotoxicity when liver tissue was absent, while 20 μM capecitabin produced cardiotoxicity in the presence of liver cells, demonstrating the complexity of organ interactions. Additionally, the authors stated that when six organs were integrated onto the same chip, all six tissues maintained a high survival rate. This suggests that when multiple organs are present simultaneously, it may not be necessary to culture the cells on a specific substrate, only on serum.

10.3.6.2 MOC for Metabolism

Metabolism is a general term for the chemical reactions that sustain life activities in the body. Metabolism is usually categorized into two types: anabolic and catabolic. Meanwhile, metabolism is also a typical physiological activity that requires the joint action of multiple organs; therefore, many research teams have utilized MOC to study the interactions between different organs. For example, Slaughter et al. constructed an MOC model by culturing white adipose tissue and hepatocytes on a chip (Slaughter et al. 2021). The study added glucose and insulin to the serum to simulate diabetic patients, added fatty acids to simulate obese patients, and added the common hypoglycemic drug metformin to the microarrays; the hepatocyte steatosis was not reduced, which seems to indicate that metformin is not an effective treatment for steatosis. Similarly, Essaouiba et al. utilized rat islets of Langerhans and hepatocytes to construct a pancreas-liver organ-on-chip coculture model for probing the metabolic process of crosstalk between liver and pancreas (Essaouiba et al. 2020). The study results demonstrated a diminished expression of several genes, including *Pdx1*, *Glut2*, *App*, etc., in the pancreas-liver organ-on-chip compared to pancreatic islets cultured in isolation. The authors concluded that this model reproduces the in vivo crosstalk between organs.

However, all of these MOCs need to be individually optimized and designed for specific functions and are limited to interactions between two organs, which not only raises the cost of using MOCs, but also does not fully reproduce the entire process of substance metabolism in vivo. Vernetti et al. coupled gut-on-a-chip, kidney-on-a-chip, and the BBB-on-a-chip, and investigated vitamin D3 and its metabolic products were absorbed, transformed, and transported across the BBB in vivo (Vernetti et al. 2017) (Fig. 10.12).

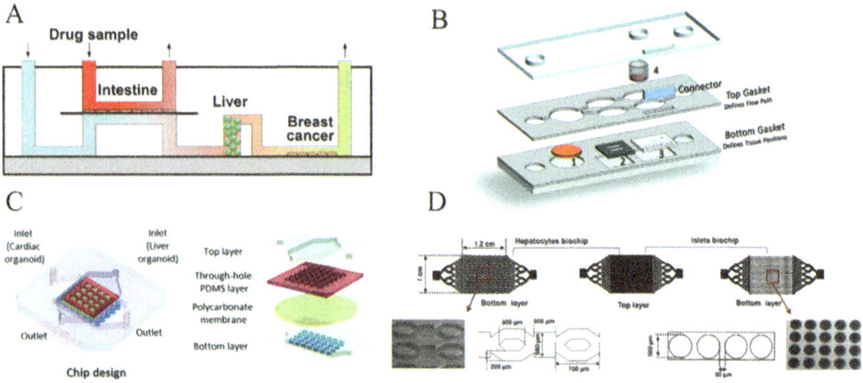

Fig. 10.12 Multi-organ-on-a-chip. (**a**) Microfluidic system containing intestinal tissue and liver tissue. Reprinted with permission from Imura et al. 2010. (**b**) Heart–liver body-on-a-chip system with a skin surrogate for the assessment of potential drug toxicity from dermal absorption. Reprinted with permission from Pires de Mello et al. 2020. (**c**) Liver-heart organoid-chip device for reflecting human organ-specific functions. Reprinted with permission from Yin et al. 2021. (**d**) Pancreas-liver coculture model in a microfluidic biochip for reproducing crosstalk between liver and pancreas. Reused with permission from Essaouiba et al. 2020

10.3.7 Summary of OOCs

The organ-on-a-chip mentioned above (OOC) has proven its significant role in the fields of drug screening and drug discovery. Unlike traditional static cell cultures, microfluidic chips enable precise control of small-scale fluids and allow simultaneous processing of multiple cell samples, providing a high-throughput experimental platform. More importantly, the design flexibility of microfluidic chips creates an ideal environment for cells to simulate the in vivo conditions. In summary, these advantages make microfluidic chips a powerful tool in biomedical research.

However, due to the complexity of human organs, current organ chips only partially mimic certain functions of human organs from specific perspectives. No organ chip has yet replicated all functions of a particular organ, so OOCs cannot wholly replace animal models. Attaining a comprehensive simulation of organs continues to stand as a primary objective within this field, involving various approaches such as changing chip materials and optimizing fluid dynamics performance. On the other hand, many cells in OOCs are derived from iPSCs, which may pose issues such as unstable cell differentiation results and cell maturity. However, directly using cells isolated from tissues may raise ethical and cost-related concerns. In conclusion, the absence of a standardized design pattern for OOCs may impede the industrial-scale production of such devices.

Overall, the future development direction of OOCs may involve the systematic integration of multiple organ systems to simulate the human body more comprehensively. This would aid in more accurate studies of cell interactions, organ interactions, and inter-organ communication. Another potential direction is high throughput and automation. With the advancement of electronic technology, integrating various electronic components into OOC could enable real-time high-throughput monitoring. OOCs may also play a crucial role in personalized medicine by using patients' own iPSCs as a cell source to produce customized organ models, thus assisting in precision medicine. These directions will further propel Organ-on-a-chip technology to become an important tool in biomedical research, advancing disease research, drug screening, and personalized treatment.

10.4 Digital Microfluidic

10.4.1 Introduction to Digital Microfluidics Technology

10.4.1.1 Definition of Digital Microfluidic Technology

Digital microfluidic technology is an emerging field within microfluidics, characterized by manipulating discrete droplets (Choi et al. 2012). It encompasses several methods for manipulating these droplets, including dielectric electrophoresis (Abdelgawad and Wheeler 2007), dielectric wetting (Zou et al. 2019), magnetic techniques (García et al. 2007), and surface acoustic waves. Electrowetting alters the surface tension between a liquid and a solid through the adjustment of electric potential, changing the contact angle between them. The advent of electrowetting on dielectric (EWOD) technology in the early twenty-first century transformed electrowetting from a relatively obscure phenomenon into a practical application, leading to significant commercial developments over the past two decades. Known as digital microfluidics (DMF), EWOD serves as a crucial foundation for various commercial applications. Digital microfluidic chips manage discontinuous droplets by using electronic circuits to control liquid surface tension, facilitating the creation, movement, splitting, and merging of droplets. This technology has extensive applications in chemical synthesis, biological analysis, and disease diagnosis.

10.4.1.2 Advantages of Digital Microfluidic Technology

Compared to traditional experimental methods, digital microfluidic technology offers unique advantages: 1. It enables precise control and manipulation of small volumes of liquids, reducing reagent consumption and improving utilization of reagents and samples. This simplifies and enhances the efficiency of complex chemical reactions, biological analyses, and diagnostic processes in laboratories; 2. The manufacturing costs of digital microfluidic chips are relatively low, with the added benefit of being easily mass-produced, thereby reducing experimental costs; 3. This technology features high throughput and can process multiple samples simultaneously, boosting experimental efficiency; 4. Digital microfluidic chips are compact and portable, allowing experiments to be conducted in various settings, offering researchers greater flexibility and convenience; 5. The closed system of digital microfluidics, particularly with its dual flat structure chip, minimizes contact between the experimenter and samples due to its excellent sealing, significantly reducing external contamination and ensuring experiments' accuracy and smooth operation; 6. Digital microfluidic platforms can achieve integrated detection and analysis, utilizing electrical signals for precise fluid control and mixing within microscale channels. This flexibility enables the integration of various analytical methods and technologies, such as spectral analysis and mass spectrometry, making it highly compatible and easy to incorporate with other devices. In summary, the advantages of digital microfluidic technology render it an essential tool in contemporary scientific research and medical diagnosis (Wang et al. 2018; Jebrail et al. 2010).

10.4.1.3 Development of Digital Microfluidic Technology

Since the introduction of microfluidic systems, this technology has experienced significant advancements. These developments can be primarily attributed to the progress in digital microfluidic structures and the enhancement of digital microfluidic control. Traditional digital microfluidic chips are designed with a dual flat plate structure, which typically consists of parallel upper and lower plates. The lower plate is equipped with a microelectrode array connected to the driving voltage, while the upper plate is connected to the ground electrode. Droplets are positioned between these plates, forming a sandwich-like structure. This design features a simple electrode and circuitry configuration with stable droplet actuation and enables various complex operations such as droplet distribution, movement, merging, and splitting.

With the increasing complexity of electrode structures, another type of open single-plate structure has emerged. This design eliminates the need for an upper flat plate by using only one flat plate to carry the driving electrode, simplifying manufacturing and enhancing droplet actuation speed. In addition to these basic single and dual flat plate structures, a digital microfluidic chip combines single and dual frequency boards. This type of chip integrates elements from single- and dual-frequency board designs, allowing droplets to switch freely between the plates (Berthier et al. 2006a).

Digital microfluidic chips are known for their simple operation and diverse methods, contributing to their rapid development and widespread use. These chips can be customized to meet the specific requirements of different functional systems by adjusting the droplet manipulation depth and width accordingly. As a result, the accuracy and volume range of liquid manipulation have always been central aspects of research in this field. Regarding operational depth, digital microfluidic chips involve a series of microdroplet manipulations, necessitating a certain level of precision in droplet handling. At present, the accuracy of liquid generation has been improved through methods such as pressure pumps (Ren et al. 2004) and liquid storage tanks (Gong and Kim 2008). Regarding manipulation breadth, digital microfluidic chips can control the driving droplet volume range between micro and nano upgrades by designing chip structure and electrode size (Song et al. 2008). In recent years, researchers have proposed the concept of multiple electrodes controlling a single droplet, which can simultaneously manipulate multiple droplets of different sizes, further promoting the development of microfluidic technology (Wang et al. 2011).

10.4.2 Processing Technology of Digital Microfluidic Chips

Digital microfluidic chips are mainly divided into two types of structures: open structure (Berthuy et al. 2016) and closed structure (Pollack et al. 2002), namely single plate structure and double plate structure, and double plate structure. Its typical structure includes parallel upper and lower plates. The lower board is a microelectrode array connected to the driving voltage, while the upper board is connected to the ground electrode. The droplets are located between the upper and lower boards, forming a sandwich structure that requires low driving voltage and no cross-infection. Due to its single flat mode, open structures are more accessible to connect to external detection equipment and have more robust control over droplets than closed structures. However, the risk of droplet evaporation and environmental pollution is also higher due to contact with the outside world. There will also be hybrid structures that combine single and double plates (Fig. 10.13).

10.4.2.1 Basic Composition of Digital Microfluidic Chips

Digital microfluidic chips typically consist of four fundamental components: a substrate, an electrode layer, a dielectric layer, and a hydrophobic layer. The following discussion will provide an overview of each of these components.

Base Material

The choice of substrate material significantly influences the fabrication and utilization of the chips. Various substrate materials offer distinct advantages and disadvantages, necessitating selection based on the application's specific requirements. Table 10.2 summarizes commonly used substrate materials for chips and their respective pros and cons. As a substrate material, glass boasts key benefits such as smooth droplet movement, high chip precision, and reusability (Moon et al. 2006; Lin et al. 2010). These attributes afford glass a competitive edge in chip manufacturing processes that demand high accuracy and stability. However, the elevated glass cost may challenge its adoption in cost-sensitive applications. On the other hand, printed circuit boards (PCBs) are characterized by their low cost and ease of

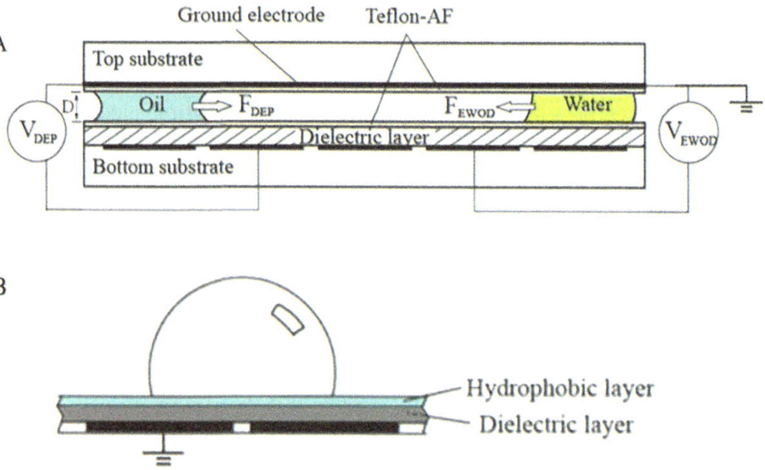

Fig. 10.13 Two structures of digital microfluidic chips. (**a**) Closed structure schematic diagram. Reprinted with permission from Wang and Jones 2015. (**b**) Open structure schematic diagram. Reprinted with permission from Wang and Jones 2015

Table 10.2 Advantages and disadvantages of commonly used substrates

Substrate material	Advantages	disadvantages	References
Glass	Smooth droplet movement, high chip accuracy, reusable	Cost	Moon et al. (2006) and Lin et al. (2010)
PCB	Low cost, easy to process	High driving voltage, poor chip accuracy	Gong and Kim (2008) and Abdelgawad et al. (2008)
ITO glass	Good transparency, high chip accuracy	Complex processing, high cost	Zhao et al. (2011)
Paper	Low cost, easy to process	Poor chip accuracy	Abbott et al. (2010)
Polyester film	Low cost, good transparency	Poor chip accuracy	Abdelgawad et al. (2008)

fabrication, making them advantageous for mass production (Gong and Kim 2008; Abdelgawad et al. 2008). However, PCBs' high driving voltage and poor chip accuracy may hinder their application in scenarios demanding high precision and low power consumption. As a substrate material, ITO glass is distinguished by its excellent transparency and high chip accuracy, making it particularly suited for applications requiring both high transparency and precision. Nonetheless, the processing of ITO glass is both complex and expensive, potentially restricting its use in applications prioritizing low cost and simplicity (Zhao et al. 2011). As a substrate material, paper offers significant advantages in terms of low cost and ease of processing (Fobel et al. 2014). These attributes make paper suitable for cost-sensitive and straightforward applications. However, paper's relatively poor chip accuracy may limit its suitability for high-precision requirements. Polyester film, noted for its low cost and good transparency, is advantageous in applications that value both affordability and high transparency (Abdelgawad et al. 2008).. Yet, like paper, the chip accuracy of polyester film is limited, posing challenges for its use in high-precision applications. The selection of substrate material should be tailored to the application's specific requirements. Glass or ITO glass may be the preferred choice for high-precision and high-stability chips. If the priority is low cost and mass production, PCBs or polyester film could be more suitable. Paper or polyester film might be considered for a balance of low cost and high precision (Fig. 10.14).

Electrode Layer Material

The electrode layer material requires good conductivity, adhesion to the substrate, and compatibility with microfabrication techniques. Among the various electrode layer materials, heavily doped polycrystalline silicon, gold (Au), copper (Cu), aluminum (Al), platinum (Pt), and indium tin oxide (ITO) are common choices (Müller et al. 2014; Fan et al. 2011; Yu et al. 2014). In the following, we will discuss the advantages and disadvantages of these materials. First, heavily doped polycrystalline silicon is compatible with microfabrication technology, allowing for high precision during its preparation. However, its manufacturing process is intricate and time-consuming, which may lead to higher production costs. Additionally, the conductivity of heavily doped polycrystalline silicon is relatively low, potentially impacting the chip's performance. Secondly, Au exhibits outstanding chemical stability and can resist corrosion in various environments. Nevertheless, its processing procedure is complex and costly, increasing overall expenses. Cu is well-established in production, and its price is relatively low, giving it certain advantages in practical applications. Nevertheless, Cu is susceptible to oxidation, which might shorten the chip's lifespan. The preparation of Al is mature and cost-effective, making it highly efficient in numerous fields. Al is prone to breakdown, which could reduce the chip's stability. Pt possesses exceptional chemical stability and can maintain excellent performance in harsh environments. However, the processing of Pt is challenging and expensive, potentially limiting its use in some applications. Finally, ITO has high conductivity and a low price tag, making it highly competitive in many areas. However, the processing of ITO is somewhat complex and may affect production efficiency (Shih-Kang et al. 2003).

Dielectric Layer Materials

The dielectric layer primarily accumulates charges, ensures droplet operation, and prevents electrode breakdown. The required operating voltage is closely related to the dielectric layer material's dielectric constant. A higher dielectric constant typically results in a lower required driving voltage. While selecting a dielectric layer material with a high dielectric constant is essential to prevent breakdown, other factors must also be considered to choose the most suitable material comprehensively. Commonly used dielectric layer materials include SiO_2 (Vallet et al. 1996), Si_3N_4 (Berthier et al. 2006b), Al_2O_3 (Chang et al. 2009), polydimethylsiloxane (PDMS) (Kuo

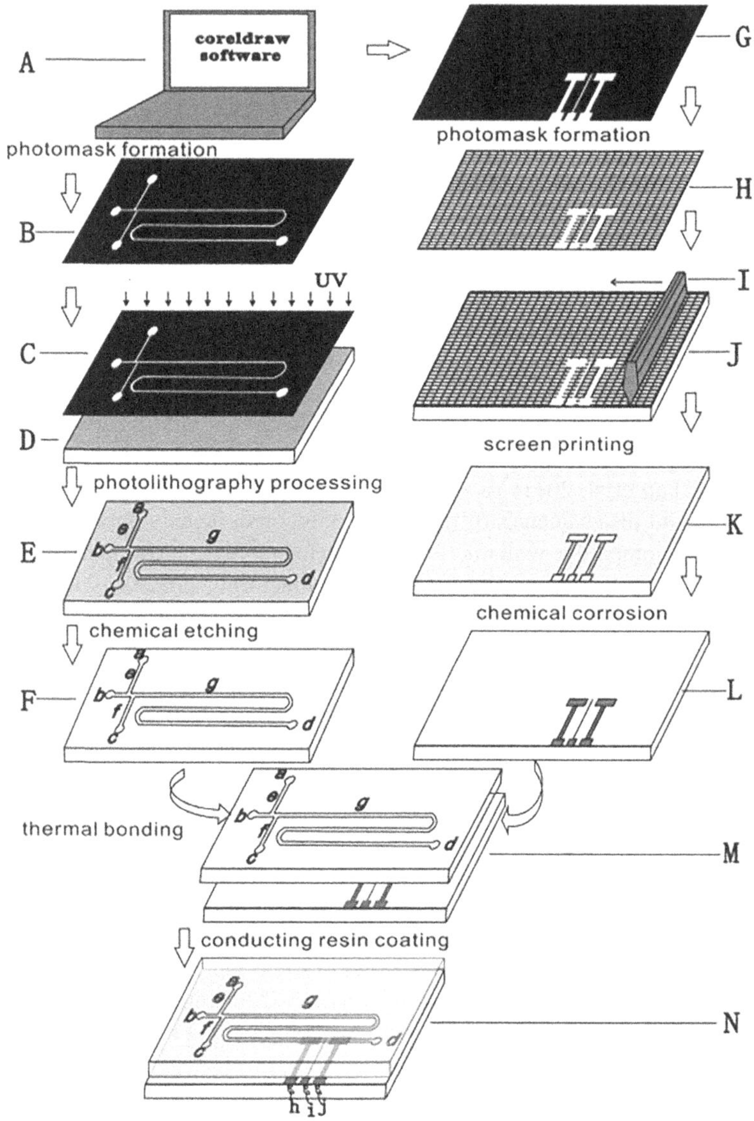

Fig. 10.14 A novel microchip based on indium tin oxide coated glass for non-contact conductivity detection. Reused with permission from Zhao et al. 2011. (**a**) computer; (**b** and **c**) photomask of chip; (**d**) chrome blank; (**e**) glass substrate with channels design; (**f**) substrate plate; (**g**) the mask of electrodes; (**h**) silk screen with planar electrodes pattern; (**i**) brush; (**j**) ITO coated glass; (**k**) ITO coated glass with shaped electrodes; (**l**) coverslip with conductivity electrodes; (**m**) the fully assembled microchip and (**n**) the microchip with conducting resin; (a) sample reservoir; (b) buffer reservoir; (c) sample waste reservoir; (d) buffer waste reservoir; (e and f) double-T injection channel; (g) separation channel; (h) input electrode terminal; (i) earth electrode terminal, and (j) output electrode terminal.

et al. 2003), SU-8 photoresist, and polyxylene (Peykov et al. 2000). SiO_2 offers excellent electrical insulation and properties, with a mature electrical technology that facilitates the formation of uniform thin films. However, its relatively low dielectric constant of 2.7 necessitates a high driving voltage, which poses significant limitations.Si_3N_4, known for its excellent insulation, mechanical wear resistance, and high-temperature resistance, has a dielectric constant of 7.8 and high dielectric strength. However, due to an incomplete preparation process, the resulting silicon nitride film may contain many particles and have an uneven distribution, leading to leakage and breakdown risks, making it less than ideal as a dielectric layer material. Al_2O_3 possesses good adhesion, excellent electrical insulation, and high-temperature resistance, meeting the requirements of a dielectric layer material. However, its complex processing process hinders its suitability for mass production and widespread use. PDMS is a non-toxic, non-flammable, simple, and fast organic silicon polymer compound with good adhesion to glass, excellent chemical inertness, and insulation properties. It has the potential to become a preferred dielectric layer material, with many scholars conducting research on PDMS-related dielectric electrowetting (Shabani and Cho 2011; Virgilio and Castaner 2009). SU-8 photoresist, a newer dielectric layer material, has gained wide usage in recent years due to its excellent mechanical, insulation, optical, and chemical properties (Chang et al. 2007).. Parylene materials, known for their excellent electrical properties, heat resistance, and chemical stability, are polymer films with a polydimethylphenylene structure, typically prepared by chemical vapor deposition. The thin film prepared by the vacuum pyrolysis vapor deposition process is very thin and is generally used as a coating.

Hydrophobic Layer Material

The hydrophobic layer is primarily used to reduce the driving resistance of droplets and increase their contact angle. Commonly chosen materials for this layer include Teflon AF (Papageorgiou et al. 2012; Pollack et al. 2002) and CYTOP (Lin et al. 2010). Teflon AF, known for its excellent chemical stability, transparency, and electrical properties, is widely used in dielectric wetting studies (Moon and Kim 2006).. Its main components are 4,5-difluoro-bis(trifluoromethyl)-1,3-dioxole (PDD) and tetrafluoroethylene (TFE), typically mixed with FC-40 to prepare Teflon AF 1600 solutions with varying mass fractions. The film formed in dielectric wetting exhibits high hydrophobicity, with the contact angle of microdroplets reaching 110° to 120°. CYTOP materials, characterized by high transmittance, good electrical properties, and high transparency, have been shown to achieve contact angles of 106° to 112° when used as hydrophobic layers (Fig. 10.15).

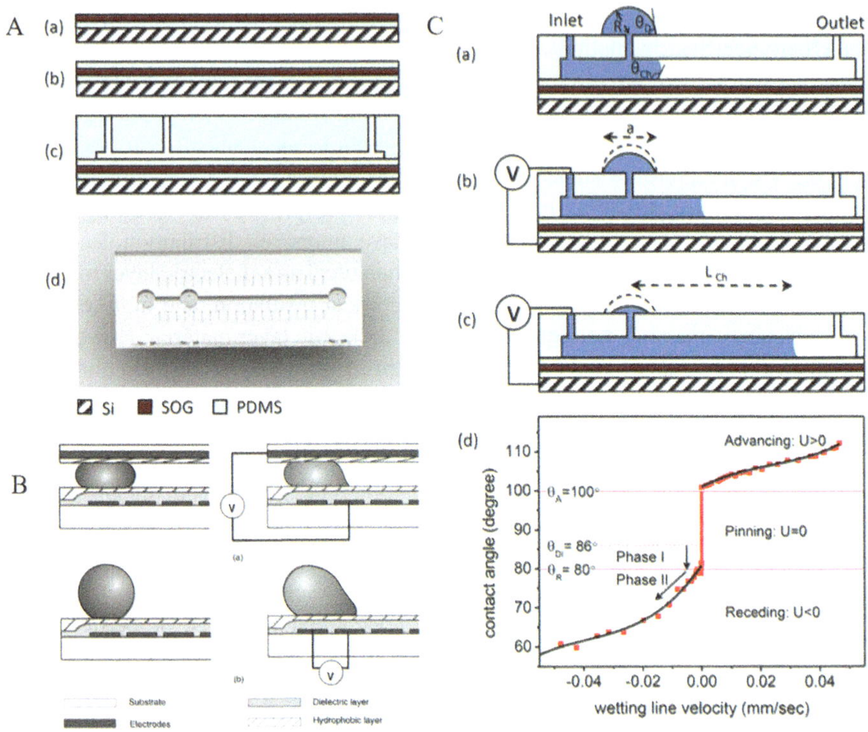

Fig. 10.15 Research on electrowetting and micropump operation. (**a**) Research on PDMS-related dielectric electrowetting, controlled by EWOD micro pump. Reprinted with permission from Shabani and Cho 2011. Micropump manufacturing steps (a–c). (a) Form an insulation layer. (b) Spin coat hydrophobic layer. (c) Use atmospheric plasma to bond PDMS layers with microchannels onto PDMS thin films. (d) Top view of a micro pump with a closed channel. (**b**) Application in the study of Teflon AF materials dielectric wetting. EWOD (electro-wetting) drive is used in the micro conveying system. The EWOD system configuration and equipment schematic. Reused with permission from Moon and Kim 2006. (a) Include an isolated additional grounding electrode, and (b) do not include an additional electrode. (**c**) Micropump operation: (a) the droplet is stable on top of the channel with the initial contact angle, θDI, of 86°. (b and c) Applying a constant voltage of 100 V drives the droplet into the channel in phases I and II. (d) Droplet contact angle, θD, vs. its wetting line velocity, *U*. Reprinted with permission from Shabani and Cho 2011

10.4.2.2 Production of Digital Microfluidic Chips

The typical fabrication process of digital microfluidic chips mainly involves the production of electrode layers, dielectric layers, and hydrophobic layers (Gao et al. 2015).. This process requires a clean room with stringent air quality, humidity, temperature, and particle density requirements. The following is a brief overview of the fabrication process for digital microfluidic chips.

Mask Making: The initial step is to create a mask for the lithography process. A mask is a layer of photosensitive material with the desired pattern, which is transferred to the substrate using a lithography machine. Precise control of the exposure time and dosage of the lithography machine is essential to ensure the accuracy and quality of the pattern.

Substrate Preparation: Select a suitable substrate as the base material for the microfluidic chips. The substrate undergoes pre-treatment processes such as cleaning and polishing to remove surface impurities and damaged layers. Additionally, the substrate is cut, ground, and processed to meet the dimensional requirements of the chip design.

Lithography process: Align the prepared mask with the substrate and place it in the lithography machine for exposure (Waldbaur et al. 2011).. During exposure, light passes through the pattern on the mask, forming the desired pattern on the substrate surface. Post-exposure processes such as developing and fixing are required to remove the unexposed photosensitive material.

Etching Process: After forming the desired pattern on the substrate, excess silicon material is removed through the etching process. Various etching methods are used, such as wet etching and dry etching. Accurate control of parameters such as the etchant's type, concentration, and temperature is necessary to ensure the accuracy and quality of the pattern.

Deposition Process: Following etching, required materials such as metals and oxides are deposited on the substrate. Various deposition methods, such as chemical vapor deposition (CVD) and physical vapor deposition (PVD), are employed. Accurate control of deposition conditions, such as temperature, pressure, and gas flow rate, is crucial to ensure the quality and performance of the deposited layer.

Packaging Process: After chip fabrication is complete, packaging protection is required. Various packaging methods are used, such as epoxy resin and glass packaging. During packaging, it is essential to ensure the chip is isolated from the external environment to prevent contamination and damage.

Finally, after packaging is complete, the microfluidic chip undergoes testing and debugging. Testing methods include electrical testing and optical testing to verify the functionality and performance of the chip. If any issues are identified, it is necessary to return to the corresponding production steps for modification and optimization (Fig. 10.16).

10.4.3 Application of Digital Microfluidic Technology

Compared to traditional biological analysis methods, digital microfluidic technology exhibits advantages such as reduced sample consumption, shorter analysis times, and enhanced containment. These features have led to its widespread adoption in the field of biology (Yu et al. 2014). This section provides a concise overview of the application of digital microfluidics in biological analysis.

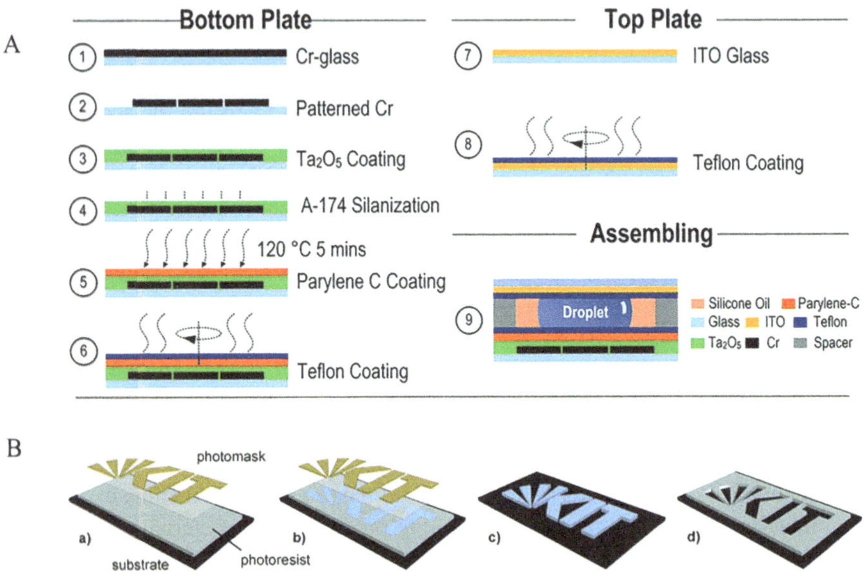

Fig. 10.16 Manufacturing of digital microfluidic chips. (**a**) The manufacturing process of digital microfluidic chips. Reprinted with permission from Gao et al. 2015. (**b**)Classical photolithography process. Reprinted with permission from Waldbaur et al. 2011. (a) A substrate is coated with a thin layer of photoresist, usually applied via spin coating. The coated substrate is placed under a photomask consisting of a radiation transparent material with a radiation absorbing material locally confined to it. (b) By exposition to radiation (usually UV light) the pattern is translated to the resist layer. Depending on the type of resist the exposed area will either be more soluble in a designated solvent (positive resist) or less soluble (negative resist). After exposure, (c) these areas are removed in the case of a positive resist or (d) these areas will remain on the substrate in the case of a negative resist

A significant challenge in employing digital microfluidics for biological analysis is the issue of biological adhesion. This phenomenon occurs when biological macromolecules, such as nucleic acids and proteins, adhere to the chip's surface, leading to sample loss or cross-contamination and impairing chip performance. This, in turn, can adversely affect droplet movement. Several strategies have been developed to mitigate biological adhesion: 1. Filling with an oil phase can enhance the droplet's driving effect, diminishing the interaction between the biomolecules within the droplets and the chip surface, thereby reducing biological adhesion to a certain extent (Srinivasan et al. 2004). 2.The addition of surfactants to the droplets can decrease biological adhesion (Luk et al. 2008; Au et al. 2011). 3. Establishing a replaceable hydrophobic isolation layer, which is renewed for each experiment, addresses the issue of biological adhesion at its root (Yang et al. 2009) (Fig. 10.17).

Immunoassay is a technique for detecting proteins, drugs, hormones, and other substances by leveraging the specific interaction between antigens and antibodies. It is known for its excellent specificity and sensitivity, making it a popular choice in various applications (Wadkins et al. 1998; McBride et al. 2003). However, the method faces significant challenges, including its complex procedures, lengthy processing times, and high costs, which limit its practicality. In contrast, digital

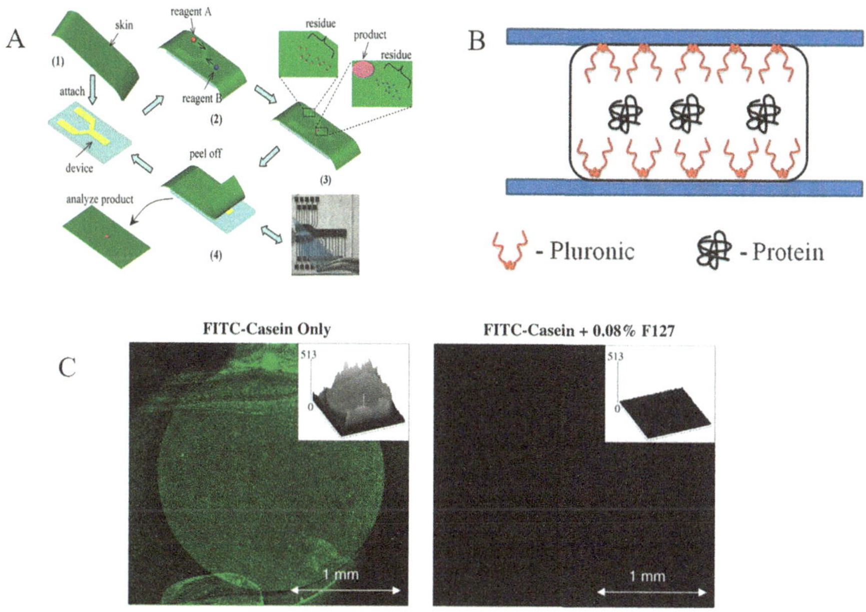

Fig. 10.17 Methods to prevent biological adhesion. (**a**) Alternative hydrophobic isolation layer to solve biological adhesion. Reprinted with permission from Yang et al. 2009. (**b**) Pluronic additive prevents bio adhesion schematic diagram. Reprinted with permission from Au et al. 2011. (**c**) Characterization of pluronic additives for preventing biological adhesion by confocal fluorescence analysis. Reprinted with permission from Luk et al. 2008. Fluorescent images of test substrates exposed to FITC-casein without (left) and with (right) 0.08% w/v F127

microfluidics emerges as a novel approach for manipulating discrete droplets, offering a solution to the complexities and inefficiencies of traditional immunoassays. This technology enables the automation of assay processes and the manipulation of droplets via electrical signals, streamlining operations and significantly cutting down on time and reagent consumption.

Consequently, it reduces the overall cost of analysis (Jebrail et al. 2010). The immunoreaction in digital microfluidics typically involves heterogeneous immunoassays, which utilize two primary methods for antibody immobilization: attachment to a solid surface, such as that of a chip (Miller et al. 2011), or to the surface of magnetic beads (Sista et al. 2008a). This advancement facilitates the detection and quantification of specific antigens or antibodies, aiding in the early diagnosis and treatment of diseases. For instance, by immobilizing antibodies on a chip's surface, rapid and precise antibodies can be detected to ascertain a patient's exposure to specific viruses or bacteria (Fig. 10.18).

Nucleic acid analysis is a widely employed molecular diagnostic technique celebrated for its accuracy and high sensitivity. It is crucial in gene expression analysis, pathogen detection, and disease diagnosis. However, traditional laboratory methods for nucleic acid analysis can be cumbersome, time-consuming, and labor-intensive. These methods often depend on precise, complex instruments and equipment,

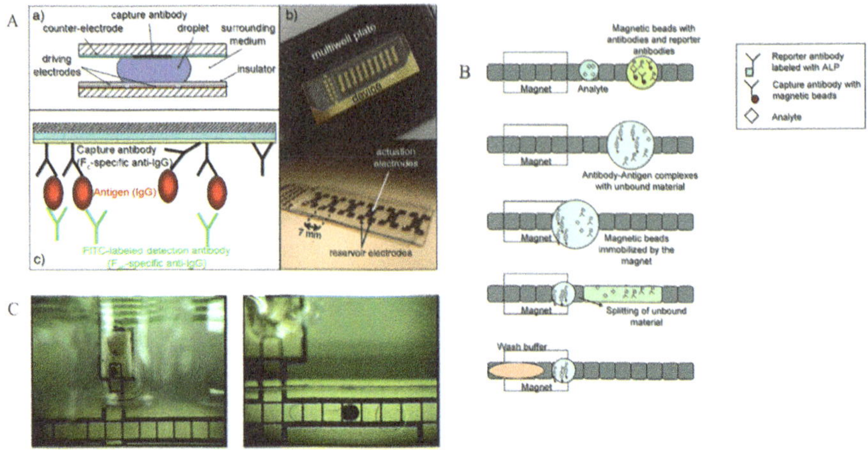

Fig. 10.18 Two types of immune detection and magnetic bead-based nucleic acid manipulation. (**a**) Immunodetection of antibodies immobilized on solid surfaces. Reprinted with permission from Miller et al. 2011. (**b**) Immune detection of antibodies immobilized on the surface of magnetic beads. Reprinted with permission from Sista et al. 2008a. (**c**) Sequence of droplet operations for concentration, extraction, and processing of DNA from a large sample volume containing infectious disease pathogens. Reprinted with permission from (Sista et al. 2008b). Concentration of DNA-carrying magnetic beads in the sample well, washing and eluting DNA from the beads over a magnet.

hindering their further development and widespread adoption. Digital microfluidics has revolutionized this field, offering programmability, automation, rapid testing, and integrated analysis capabilities. This technology holds unique advantages and applications in nucleic acid extraction and purification (Sista et al. 2008b), nucleic acid amplification (Sista et al. 2008b), and DNA sequencing (Welch et al. 2011).

Protein analysis, enzyme analysis, protein purification, and analysis are standard methodologies within digital microfluidics. This technology facilitates the study of protein structure and function and the identification of disease-related protein markers. For instance, analyzing enzyme activity of protein samples on a chip allows for the exploration of protein functional characteristics (Miller and Wheeler 2008).

Similarly, cell culture, cell sorting, and cell purification are pivotal cell research methods enabled by digital microfluidics. This technology is instrumental in examining cell growth, differentiation, and function and identifying cell markers associated with diseases. For example, cell culture on a chip provides insights into cell growth characteristics (Eydelnant et al. 2012); cell sorting on a chip aids in the study of cell differentiation (Valley et al. 2011); and cell purification from mixtures allows for the investigation of specific cell functional traits (Kumar et al. 2014) (Fig. 10.19).

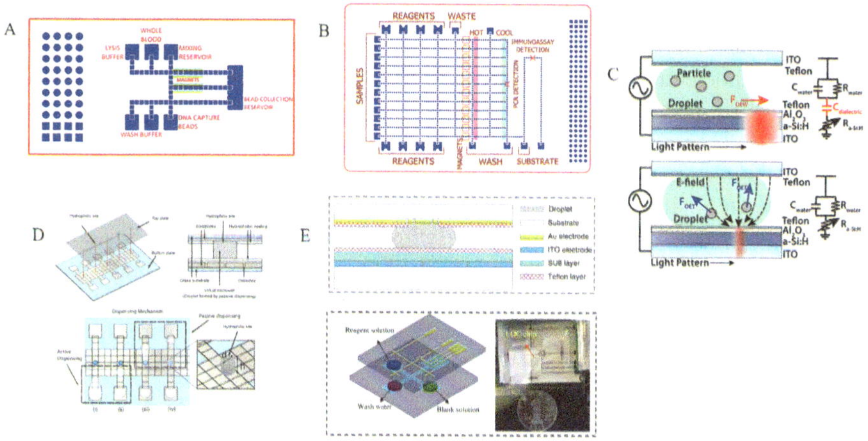

Fig. 10.19 Application of digital microfluidic chips in nucleic acid and cells. (**a**) Schematic of the chip used for extraction of human genomic DNA from whole blood. Reprinted with permission from Sista et al. 2008b. (**b**) Schematic diagram of a digital microfluidic chip with multi-region temperature control and PCR amplification. Reprinted with permission from Sista et al. 2008b. (**c**) Controlling cells on digital microfluidic chips based on optical method sorting. Reprinted with permission from Valley et al. 2011. (**d**) Cell culture on digital microfluidic chips. Reprinted with permission from Eydelnant et al. 2012. (**e**) Configuration of biochips based on digital microfluidics. It consists of a top plate with a patterned ground wire and an electrochemical cell and a bottom plate with an actuating electrode. And the assembly and pictures of the completed LOC. Reprinted with permission from Yu et al. 2014

10.5 Summary

The advent of new materials and manufacturing advancements has significantly broadened the spectrum of materials used in microfluidic chips. Initially, inorganic materials such as silicon, glass, quartz, and silicon elastomer were developed alongside thermosetting and thermoplastic materials, synthetic polymers, natural or synthetic gel polymers, paper-based materials, and functional hybrid materials. Each material type offers unique physical and chemical properties, showcasing distinct advantages and applications in various fields.

Over time, the requirements for microfluidic chips have evolved, necessitating a shift towards multifunctional materials to meet the diverse needs of different sectors. For example, the biomedical field demands materials that support various detection capabilities, including PCR, protein detection, and cell culture. Consequently, researchers are developing materials with multiple functionalities to enable a single device to serve multiple purposes. Future trends in microfluidic chip materials are expected to lean towards environmental sustainability and cost-effectiveness, driven by an increasing demand for eco friendly materials and the need for materials that support high-throughput screening and large-scale production. Innovations in materials, including two-dimensional materials and nanomaterials, are set to enhance the capabilities of microfluidics. The integration of micro/nanoelectronic technology and artificial intelligence algorithms for automation control and data analysis represents a promising direction for the future of microfluidic chip materials.

Microfluidic chip technology has found widespread applications in biomedicine, chemistry, and materials science, demonstrating significant value and potential for growth. In biomedicine, it enables cell culture, protein detection, drug screening, gene sequencing, and immune detection, facilitating a deeper understanding of cellular biology and disease mechanisms and improving the efficiency of new drug development. In chemistry, it aids in monitoring and controlling chemical reactions and constructing microreactors. The technology also holds potential in food and agriculture, environmental testing, and medical applications. As technology advances, microfluidic chips are expected to see broader application scenarios and increased market demand, with potential uses in personalized medicine, environmental monitoring, and energy applications.

Organ chips, a key application of microfluidic technology, have shown considerable promise in drug screening and discovery. They offer precise control over fluidics and the ability to process multiple cell samples simultaneously, providing a high-throughput experimental platform. Despite their advantages, organ chips can only partially simulate human organ functions and cannot wholly replace animal models. The full simulation of organ functions remains a significant goal, requiring improvements in chip materials and fluid dynamics performance. Using induced pluripotent stem cells (iPSCs) in organ chips presents challenges, such as unstable cell differentiation and ethical concerns. Overcoming the lack of standardized organ chip design patterns is crucial for industrial production.

Future developments in organ chips may include integrating multiple organ systems for a comprehensive simulation of the human body, aiding in the study of cell and organ interactions. Advancements in electronic technology could enable real-time high-throughput monitoring. Customized organ models using patients' iPSCs could play a significant role in personalized medicine, advancing precision medicine.

Digital microfluidic technology, recognized for its automation, miniaturization, integration, and high parallelism, has rapidly evolved since its inception nearly two decades ago. The development of digital microfluidic chip processing technology, driven by advances in materials science, micro/nano processing technology, and microelectronics, aims to enhance throughput and portability. Deepening the understanding of its mechanisms could lead to more stable droplet manipulation, making digital microfluidics a powerful tool for everyday life and a potent platform for personalized medical treatment, real-time detection, food safety, and more.

References

Abbott NJ, Patabendige AA, Dolman DE, Yusof SR, Begley DJ (2010) Structure and function of the blood-brain barrier. Neurobiol Dis 37:13–25. https://doi.org/10.1016/j.nbd.2009.07.030

Abdelgawad M, Wheeler AR (2007) Rapid prototyping in copper substrates for digital microfluidics. Adv Mater 19:133. https://doi.org/10.1002/adma.200601818

Abdelgawad M, Freire SL, Yang H, Wheeler AR (2008) All-terrain droplet actuation. Lab Chip 8:672–677. https://doi.org/10.1039/b801516c

Acharya UR, Sree SV, Swapna G, Martis RJ, Suri JS (2013) Automated EEG analysis of epilepsy: a review. Knowl-Based Syst 45:147–165. https://doi.org/10.1016/j.knosys.2013.02.014

Amieva M, Peek RM Jr (2016) Pathobiology of helicobacter pylori-induced gastric cancer. Gastroenterology 150:64–78. https://doi.org/10.1053/j.gastro.2015.09.004

Anna SL, Bontoux N, Stone HA (2003) Formation of dispersions using "flow focusing" in microchannels. Appl Phys Lett 82:364–366. https://doi.org/10.1063/1.1537519

Ao Z et al (2020) One-stop microfluidic assembly of human brain organoids to model prenatal cannabis exposure. Anal Chem 92:4630–4638. https://doi.org/10.1021/acs.analchem.0c00205

Au SH, Kumar P, Wheeler AR (2011) A new angle on pluronic additives: advancing droplets and understanding in digital microfluidics. Langmuir 27:8586–8594. https://doi.org/10.1021/la201185c

Banales JM et al (2019) Cholangiocyte pathobiology. Nat Rev Gastroenterol Hepatol 16:269–281. https://doi.org/10.1038/s41575-019-0125-y

Bavli D et al (2016) Real-time monitoring of metabolic function in liver-on-chip microdevices tracks the dynamics of mitochondrial dysfunction. Proc Natl Acad Sci USA 113:E2231–E2240. https://doi.org/10.1073/pnas.1522556113

Becker H, Gartner C (2008) Polymer microfabrication technologies for microfluidic systems. Anal Bioanal Chem 390:89–111. https://doi.org/10.1007/s00216-007-1692-2

Beebe DJ et al (2000) Functional hydrogel structures for autonomous flow control inside microfluidic channels. Nature 404:588–590. https://doi.org/10.1038/35007047

Berthier J, Fouillet Y, Peponnet C, Roux J-M (2006a) Modeling microdrop motion between covered and open regions of EWOD microsystems. NSTI-Nanotech

Berthier J et al (2006b) Computer aided design of an EWOD microdevice. Sensors Actuators A Phys 127:283–294. https://doi.org/10.1016/j.sna.2005.09.026

Berthuy OI, Blum LJ, Marquette CA (2016) Cancer-cells on chip for label-free detection of secreted molecules. Biosensors 6. https://doi.org/10.3390/bios6010002

Bircsak KM et al (2021) A 3D microfluidic liver model for high throughput compound toxicity screening in the OrganoPlate(R). Toxicology 450:152667. https://doi.org/10.1016/j.tox.2020.152667

Boquet-Pujadas A et al (2022) 4D live imaging and computational modeling of a functional gut-on-a-chip evaluate how peristalsis facilitates enteric pathogen invasion. Sci Adv 8:eabo5767. https://doi.org/10.1126/sciadv.abo5767

Campisi M et al (2018) 3D self-organized microvascular model of the human blood-brain barrier with endothelial cells, pericytes and astrocytes. Biomaterials 180:117–129. https://doi.org/10.1016/j.biomaterials.2018.07.014

Cao H, Duan L, Zhang Y, Cao J, Zhang K (2021) Current hydrogel advances in physicochemical and biological response-driven biomedical application diversity. Signal Transduct Target Ther 6:426. https://doi.org/10.1038/s41392-021-00830-x

Capobianco JA et al (2020) Detection of Shiga toxin-producing Escherichia coli (STEC) in beef products using droplet digital PCR. Int J Food Microbiol 319:108499. https://doi.org/10.1016/j.ijfoodmicro.2019.108499

Carvalho MR et al (2023) Gastrointestinal organs and organoids-on-a-chip: advances and translation into the clinics. Biofabrication 15:042004. https://doi.org/10.1088/1758-5090/acf8fb

Cassinari K et al (2021) Assessment of multiplex digital droplet RT-PCR as a diagnostic tool for SARS-CoV-2 detection in nasopharyngeal swabs and saliva samples. Clin Chem 67:736–741. https://doi.org/10.1093/clinchem/hvaa323

Chang Y-J, Mohseni K, Bright VM (2007) Fabrication of tapered SU-8 structure and effect of sidewall angle for a variable focus microlens using EWOD. Sensors Actuators A Phys 136:546–553. https://doi.org/10.1016/j.sna.2007.01.009

Chang J-H, Choi DY, Han S, Pak JJ (2009) Driving characteristics of the electrowetting-on-dielectric device using atomic-layer-deposited aluminum oxide as the dielectric. Microfluid Nanofluid 8:269–273. https://doi.org/10.1007/s10404-009-0511-9

Cheng YQ, Yang L, Guo CL, Zhou Y, Yang Y (2012) 2nd international conference on Advanced Engineering Materials and Technology (AEMT), pp 891–894

Cherne MD et al (2021) A synthetic hydrogel, VitroGel((R)) ORGANOID-3, improves immune cell-epithelial interactions in a tissue Chip co-culture model of human gastric organoids and dendritic cells. Front Pharmacol 12:707891. https://doi.org/10.3389/fphar.2021.707891

Chirizzi D et al (2021) SARS-CoV-2 concentrations and virus-laden aerosol size distributions in outdoor air in north and south of Italy. Environ Int 146:106255. https://doi.org/10.1016/j.envint.2020.106255

Chiu FWY, Stavrakis S (2019) High-throughput droplet-based microfluidics for directed evolution of enzymes. Electrophoresis 40:2860–2872. https://doi.org/10.1002/elps.201900222

Choi NW et al (2007) Microfluidic scaffolds for tissue engineering. Nat Mater 6:908–915. https://doi.org/10.1038/nmat2022

Choi K, Ng AH, Fobel R, Wheeler AR (2012) Digital microfluidics. Annu Rev Anal Chem (Palo Alto Calif) 5:413–440. https://doi.org/10.1146/annurev-anchem-062011-143028

Chua JH, Chee R-E, Agarwal A, Wong SM, Zhang G-J (2009) Label-free electrical detection of cardiac biomarker with complementary metal-oxide semiconductor-compatible silicon nanowire sensor arrays. Anal Chem 81:6266–6271. https://doi.org/10.1021/ac901157x

Costello CM et al (2014) Synthetic small intestinal scaffolds for improved studies of intestinal differentiation. Biotechnol Bioeng 111:1222–1232. https://doi.org/10.1002/bit.25180

Cremonesi P et al (2022) Development of quantitative real-time PCR and digital droplet-PCR assays for rapid and early detection of the spoilage yeasts Saccharomycopsis fibuligera and Wickerhamomyces anomalus in bread. Food Microbiol 101:103894. https://doi.org/10.1016/j.fm.2021.103894

Desai BS, Monahan AJ, Carvey PM, Hendey B (2007) Blood-brain barrier pathology in Alzheimer's and Parkinson's disease: implications for drug therapy. Cell Transplant 16:285–299. https://doi.org/10.3727/000000007783464731

Duan X et al (2019) Microfluidic immunoassay system for rapid detection and semi-quantitative determination of a potential serum biomarker Mesothelin. ACS Sensors 4:2952–2957. https://doi.org/10.1021/acssensors.9b01430

Essaouiba A et al (2020) Development of a pancreas-liver organ-on-chip coculture model for organ-to-organ interaction studies. Biochem Eng J 164:10778. https://doi.org/10.1016/j.bej.2020.107783

Eydelnant IA, Uddayasankar U, Li B, Liao MW, Wheeler AR (2012) Virtual microwells for digital microfluidic reagent dispensing and cell culture. Lab Chip 12:750–757. https://doi.org/10.1039/c2lc21004e

Fair RB et al (2007) Chemical and biological applications of digital-microfluidic devices. IEEE Design Test Comput 24:10–24. https://doi.org/10.1109/Mdt.2007.8

Fan SK, Hsu YW, Chen CH (2011) Encapsulated droplets with metered and removable oil shells by electrowetting and dielectrophoresis. Lab Chip 11:2500–2508. https://doi.org/10.1039/c1lc20142e

Fang G et al (2021) Mammary tumor organoid culture in non-adhesive alginate for luminal mechanics and high-throughput drug screening. Adv Sci (Weinh) **8**, e2102418. https://doi.org/10.1002/advs.202102418

Ferreira DA et al (2023) Bioinspired human stomach-on-a-chip with in vivo like function and architecture. Lab Chip 23:495–510

Fitzgerald V, Leonard P (2017) Single cell screening approaches for antibody discovery. Methods 116:34–42. https://doi.org/10.1016/j.ymeth.2016.11.006

Fobel R, Kirby AE, Ng AH, Farnood RR, Wheeler AR (2014) Paper microfluidics goes digital. Adv Mater (Deerfield Beach, Fla) 26:2838–2843. https://doi.org/10.1002/adma.201305168

Gao J et al (2015) Adhesion promoter for a multi-dielectric-layer on a digital microfluidic chip. RSC Adv 5:48626–48630. https://doi.org/10.1039/c5ra08202a

García AA et al (2007) Magnetic movement of biological fluid droplets. J Magn Magn Mater 311:238–243. https://doi.org/10.1016/j.jmmm.2006.10.1149

Ghaemmaghami AM, Hancock MJ, Harrington H, Kaji H, Khademhosseini A (2012) Biomimetic tissues on a chip for drug discovery. Drug Discov Today 17:173–181. https://doi.org/10.1016/j.drudis.2011.10.029

Gong J, Kim CJ (2008) All-electronic droplet generation on-chip with real-time feedback control for EWOD digital microfluidics. Lab Chip 8:898–906. https://doi.org/10.1039/b717417a

Grassart A et al (2019) Bioengineered human organ-on-Chip reveals intestinal microenvironment and mechanical forces impacting Shigella infection. Cell Host Microbe 26:435–444

Grover WH, Ivester RHC, Jensen EC, Mathies RA (2006) Development and multiplexed control of latching pneumatic valves using microfluidic logical structures. Lab Chip 6:623–631. https://doi.org/10.1039/B518362F

Gubala V, Harris LF, Ricco AJ, Tan MX, Williams DE (2012) Point of care diagnostics: status and future. Anal Chem 84:487–515. https://doi.org/10.1021/ac2030199

Guo MT, Rotem A, Heyman JA, Weitz DA (2012) Droplet microfluidics for high-throughput biological assays. Lab Chip 12:2146–2155. https://doi.org/10.1039/c2lc21147e

Han J et al (2018) Simple way to fabricate novel paper-based valves using plastic comb binding spines. ACS Sensors 3:1789–1794. https://doi.org/10.1021/acssensors.8b00518

Hu X, Lu L, Fang C, Duan B, Zhu Z (2015) Determination of apparent amylose content in Rice by using paper-based microfluidic chips. J Agric Food Chem 63:9863–9868. https://doi.org/10.1021/acs.jafc.5b04530

Hu ZL, Chen XY, Wang L (2018) Design and fabrication of concentration-gradient generators with two and three inlets in microfluidic chips. Chem Eng Technol 41:489–495. https://doi.org/10.1002/ceat.201700287

Huang C et al (2014) Microfluidic chip for monitoring Ca^{2+} transport through a confluent layer of intestinal cells. RSC Adv 4:52887–55289

Huang Q et al (2022) Shell microelectrode arrays (MEAs) for brain organoids. Sci Adv 8:eabq5031. https://doi.org/10.1126/sciadv.abq5031

Huh D et al (2010) Reconstituting organ-level lung functions on a Chip. Science 328:1662–1668. https://doi.org/10.1126/science.1188302

Huh D, Torisawa YS, Hamilton GA, Kim HJ, Ingber DE (2012) Microengineered physiological biomimicry: organs-on-chips. Lab Chip 12:2156–2164. https://doi.org/10.1039/c2lc40089h

Imura Y, Sato K, Yoshimura E (2010) Micro total bioassay system for ingested substances: assessment of intestinal absorption, hepatic metabolism, and bioactivity. Anal Chem 82:9983–9988. https://doi.org/10.1021/ac100806x

Jackman RJ, Floyd TM, Ghodssi R, Schmidt MA, Jensen KF (2001) Microfluidic systems with on-line UV detection fabricated in photodefinable epoxy. J Micromech Microeng 11:263–269. https://doi.org/10.1088/0960-1317/11/3/316

Jahnisch K, Hessel V, Lowe H, Baerns M (2004) Chemistry in microstructured reactors. Angew Chem Int Ed Engl 43:406–446. https://doi.org/10.1002/anie.200300577

Jalili-Firoozinezhad S et al (2019) A complex human gut microbiome cultured in an anaerobic intestine-on-a-chip. Nature Biomed Eng 3:520–553

Jang KJ, Suh KY (2010) A multi-layer microfluidic device for efficient culture and analysis of renal tubular cells. Lab Chip 10:36–42. https://doi.org/10.1039/b907515a

Jantaree P, Bakhchova L, Steinmann U, Naumann M (2021) From 3D Back to 2D monolayer stomach organoids-on-a-Chip. Trends Biotechnol 39:745–748. https://doi.org/10.1016/j.tibtech.2020.11.013

Jeannot E et al (2020) A single droplet digital PCR for ESR1 activating mutations detection in plasma. Oncogene 39:2987–2995. https://doi.org/10.1038/s41388-020-1174-y

Jebrail MJ et al (2010) Synchronized synthesis of peptide-based macrocycles by digital microfluidics. Angew Chem Int Ed Engl 49:8625–8629. https://doi.org/10.1002/anie.201001604

Jellali R, Paullier P, Fleury MJ, Leclerc E (2016) Liver and kidney cells cultures in a new perfluoropolyether biochip. Sensors and Actuators B-Chemical 229:396–407. https://doi.org/10.1016/j.snb.2016.01.141

Ji M et al (2020) Automated multiplex nucleic acid tests for rapid detection of SARS-CoV-2, influenza a and B infection with direct reverse-transcription quantitative PCR (dirRT-qPCR) assay in a centrifugal microfluidic platform. RSC Adv 10:34088–34098. https://doi.org/10.1039/d0ra04507a

Jokerst JC et al (2012) Development of a paper-based analytical device for colorimetric detection of select foodborne pathogens. Anal Chem 84:2900–2907. https://doi.org/10.1021/ac203466y

Khalid N, Arif S, Kobayashi I, Nakajima M (2019) In: Santos HA, Liu D, Zhang H (eds) Microfluidics for pharmaceutical applications. William Andrew Publishing, pp 371–422

Kim S, Takayama S (2015) Organ-on-a-chip and the kidney. Kidney Res Clin Pract 34:165–169. https://doi.org/10.1016/j.krcp.2015.08.001

Kim HJ, Huh D, Hamilton G, Ingber DE (2012) Human gut-on-a-chip inhabited by microbial flora that experiences intestinal peristalsis-like motions and flow. Lab Chip 12:2165–2174. https://doi.org/10.1039/c2lc40074j

Kolanowski TJ et al (2020) Enhanced structural maturation of human induced pluripotent stem cell-derived cardiomyocytes under a controlled microenvironment in a microfluidic system. Acta Biomater 102:273–286. https://doi.org/10.1016/j.actbio.2019.11.044

Kulsharova G, Kurmangaliyeva A (2021) Liver microphysiological platforms for drug metabolism applications. Cell Prolif 54:e13099. https://doi.org/10.1111/cpr.13099

Kulsharova G, Kurmangaliyeva A, Darbayeva E, Rojas-Solorzano L, Toxeitova G (2021) Development of a hybrid polymer-based microfluidic platform for culturing hepatocytes towards liver-on-a-Chip applications. Polymers 13:3215. https://doi.org/10.3390/polym13193215

Kumar PT et al (2014) Digital microfluidic chip technology for water permeability measurements on single isolated plant protoplasts. Sensors Actuators B-Chem 199:479–487. https://doi.org/10.1016/j.snb.2014.04.018

Kuo JS, Spicar-Mihalic P, Rodriguez I, Chiu DT (2003) Electrowetting-induced droplet movement in an immiscible medium. Langmuir 19:250–255. https://doi.org/10.1021/la020698p

Lambert C et al (2008) Dendritic cell differentiation signals induce anti-inflammatory properties in human adult microglia. J Immunol 181:8288–8297. https://doi.org/10.4049/jimmunol.181.12.8288

Lee KK et al (2018) Human stomach-on-a-chip with luminal flow and peristaltic-like motility. Lab Chip 18:3079–3085. https://doi.org/10.1039/c8lc00910d

Lee H et al (2019) Cell-printed 3D liver-on-a-chip possessing a liver microenvironment and biliary system. Biofabrication 11:025001. https://doi.org/10.1088/1758-5090/aaf9fa

Lee SS, Kim S, Yoo HM, Lee DH, Bae YK (2022) Development of SARS-CoV-2 packaged RNA reference material for nucleic acid testing. Anal Bioanal Chem 414:1773–1785. https://doi.org/10.1007/s00216-021-03846-y

Lei KF, Chang CH, Chen MJ (2017) Paper/PMMA hybrid 3D cell culture microfluidic platform for the study of cellular crosstalk. ACS Appl Mater Interfaces 9:13092–13101. https://doi.org/10.1021/acsami.7b03021

Li HA et al (2019) Versatile digital polymerase chain reaction chip design, fabrication, and image processing. Sensors Actuat B-Chem 283:677–684. https://doi.org/10.1016/j.snb.2018.12.072

Li Y et al (2023) The application of Aptamer in biomarker discovery. Biomark Res 11:70. https://doi.org/10.1186/s40364-023-00510-8

Lin YY et al (2010) Low voltage Electrowetting-on-dielectric platform using multi-layer insulators. Sens Actuators B Chem 150:465–470. https://doi.org/10.1016/j.snb.2010.06.059

Lin X et al (2015) Assay of multiplex proteins from cell metabolism based on tunable aptamer and microchip electrophoresis. Biosens Bioelectron 63:105–111. https://doi.org/10.1016/j.bios.2014.07.013

Lin D et al (2020) Low cost fabrication of microfluidic paper-based analytical devices with water-based polyurethane acrylate and their application for bacterial detection. Sensors Actuators B-Chem 303:127213. https://doi.org/10.1016/j.snb.2019.127213

Liu WW, Zheng WF, Yuan B, Jiang XY (2013) A micropatterned coculture system for axon guidance reveals that slit promotes axon fasciculation and regulates the expression of L1CAM. Integr Biol 5:617–623. https://doi.org/10.1039/c3ib20198h

Liu S, Gao J, Zhu M, Liu K, Zhang HL (2020) Gut microbiota and Dysbiosis in Alzheimer's disease: implications for pathogenesis and treatment. Mol Neurobiol 57:5026–5043. https://doi.org/10.1007/s12035-020-02073-3

Liu M et al (2022) State-of-the-art advancements in liver-on-a-chip (LOC): integrated biosensors for LOC. Biosens Bioelectron 218:114758. https://doi.org/10.1016/j.bios.2022.114758

Liu S et al (2023) Multiscale anisotropic scaffold integrating 3D printing and electrospinning techniques as a heart-on-a-Chip platform for evaluating drug-induced cardiotoxicity. Adv Healthc Mater 12:2300719

Lopez-Ruiz N et al (2014) Smartphone-based simultaneous pH and nitrite colorimetric determination for paper microfluidic devices. Anal Chem 86:9554–9562. https://doi.org/10.1021/ac5019205

Luk VN, Mo G, Wheeler AR (2008) Pluronic additives: a solution to sticky problems in digital microfluidics. Langmuir 24:6382–6389. https://doi.org/10.1021/la7039509

Luo JJ et al (2021) Durable and flexible ag-nanowire-embedded PDMS films for the recyclable swabbing detection of malachite green residue in fruits and fingerprints. Sensors Actuators B-Chem 347:130602. https://doi.org/10.1016/j.snb.2021.130602

Ma C et al (2016) On-Chip construction of liver lobule-like microtissue and its application for adverse drug reaction assay. Anal Chem 88:1719–1727. https://doi.org/10.1021/acs.analchem.5b03869

Mahato K, Srivastava A, Chandra P (2017) Paper based diagnostics for personalized health care: emerging technologies and commercial aspects. Biosens Bioelectron 96:246–259. https://doi.org/10.1016/j.bios.2017.05.001

Martel J et al (2022) Gut barrier disruption and chronic disease. Trends Endocrinol Metab 33:247–265. https://doi.org/10.1016/j.tem.2022.01.002

Martinez AW, Phillips ST, Butte MJ, Whitesides GM (2007) Patterned paper as a platform for inexpensive, low-volume, portable bioassays. Angew Chem Int Ed 46:1318–1320. https://doi.org/10.1002/anie.200603817

Marwah H, Garg T, Goyal AK, Rath G (2016) Permeation enhancer strategies in transdermal drug delivery. Drug Deliv 23:564–578. https://doi.org/10.3109/10717544.2014.935532

Mastiani M, Seo S, Mosavati B, Kim M (2018) High-throughput aqueous two-phase system droplet generation by oil-free passive microfluidics. ACS Omega 3:9296–9302. https://doi.org/10.1021/acsomega.8b01768

Maurer M et al (2019) A three-dimensional immunocompetent intestine-on-chip model as in vitro platform for functional and microbial interaction studies. Biomaterials 220:119396. https://doi.org/10.1016/j.biomaterials.2019.119396

McBride MT et al (2003) Multiplexed liquid arrays for simultaneous detection of simulants of biological warfare agents. Anal Chem 75:1924–1930. https://doi.org/10.1021/ac026379k

McDonald JC, Whitesides GM (2002) Poly(dimethylsiloxane) as a material for fabricating microfluidic devices. Acc Chem Res 35:491–499. https://doi.org/10.1021/ar010110q

McWilliams NB (1986) IV-S neuroblastoma: treatment controversy revisited. Med Pediatr Oncol 14:41–44. https://doi.org/10.1002/mpo.2950140110

Medina-Sanchez M, Miserere S, Merkoci A (2012) Nanomaterials and lab-on-a-chip technologies. Lab Chip 12:1932–1943. https://doi.org/10.1039/c2lc40063d

Meng Q, Wang Y, Li Y, Shen C (2021) Hydrogel microfluidic-based liver-on-a-chip: mimicking the mass transfer and structural features of liver. Biotechnol Bioeng 118:612–621. https://doi.org/10.1002/bit.27589

Miller EM, Wheeler AR (2008) A digital microfluidic approach to homogeneous enzyme assays. Anal Chem 80:1614–1619. https://doi.org/10.1021/ac702269d

Miller EM, Ng AH, Uddayasankar U, Wheeler AR (2011) A digital microfluidic approach to heterogeneous immunoassays. Anal Bioanal Chem 399:337–345. https://doi.org/10.1007/s00216-010-4368-2

Moon I, Kim J (2006) Using EWOD (electrowetting-on-dielectric) actuation in a micro conveyor system. Sensors Actuators A-Phys 130:537–544. https://doi.org/10.1016/j.sna.2005.12.016

Moon H, Wheeler AR, Garrell RL, Loo JA, Kim CJ (2006) An integrated digital microfluidic chip for multiplexed proteomic sample preparation and analysis by MALDI-MS. Lab Chip 6:1213–1219. https://doi.org/10.1039/b601954d

Moron-Lopez S et al (2020) Human splice factors contribute to latent HIV infection in primary cell models and blood CD4+ T cells from ART-treated individuals. PLoS Pathog 16:e1009060. https://doi.org/10.1371/journal.ppat.1009060

Mu X et al (2009) Laminar flow used as "liquid etch mask" in wet chemical etching to generate glass microstructures with an improved aspect ratio. Lab Chip 9:1994–1996. https://doi.org/10.1039/b904769g

Müller P, Kopp D, Llobera A, Zappe H (2014) Optofluidic router based on tunable liquid-liquid mirrors. Lab Chip 14:737–743. https://doi.org/10.1039/c3lc51148k

Munirathinam R, Huskens J, Verboom W (2015) Supported catalysis in continuous-flow microreactors. Adv Synthesis Catal 357:1093–1123. https://doi.org/10.1002/adsc.201401081

Nakanishi T, Tamai I (2015) Interaction of drug or food with drug transporters in intestine and liver. Curr Drug Metab 16:753–764. https://doi.org/10.2174/138920021609151201113537

Nge PN, Rogers CI, Woolley AT (2013) Advances in microfluidic materials, functions, integration, and applications. Chem Rev 113:2550–2583. https://doi.org/10.1021/cr300337x

Nikolaev M et al (2020) Homeostatic mini-intestines through scaffold-guided organoid morphogenesis. Nature 585:574. https://doi.org/10.1038/s41586-020-2724-8

Noh ES et al (2019) Quantitative analysis of Alaska Pollock in seafood products by droplet digital PCR. Food Chem 275:638–643. https://doi.org/10.1016/j.foodchem.2018.09.093

Ou J, Glawdel T, Ren CL, Pawliszyn J (2009) Fabrication of a hybrid PDMS/SU-8/quartz microfluidic chip for enhancing UV absorption whole-channel imaging detection sensitivity and application for isoelectric focusing of proteins. Lab Chip 9:1926–1932. https://doi.org/10.1039/b821438g

Papageorgiou DP, Tserepi A, Boudouvis AG, Papathanasiou AG (2012) Superior performance of multilayered fluoropolymer films in low voltage electrowetting. J Colloid Interface Sci 368:592–598. https://doi.org/10.1016/j.jcis.2011.10.035

Park TS, Yoon JY (2015) Smartphone detection of from field water samples on paper microfluidics. IEEE Sensors J 15:1902–1907. https://doi.org/10.1109/Jsen.2014.2367039

Pelkonen A et al (2020) A modular brain-on-a-chip for modelling epileptic seizures with functionally connected human neuronal networks. Biosens Bioelectron 168:112553. https://doi.org/10.1016/j.bios.2020.112553

Peykov V, Quinn A, Ralston J (2000) Electrowetting: a model for contact-angle saturation. Colloid Polym Sci 278:789–793. https://doi.org/10.1007/s003960000333

Pires de Mello CP et al (2020) Microphysiological heart-liver body-on-a-chip system with a skin mimic for evaluating topical drug delivery. Lab Chip 20:749–759. https://doi.org/10.1039/c9lc00861f

Pollack MG, Shenderov AD, Fair RB (2002) Electrowetting-based actuation of droplets for integrated microfluidics. Lab Chip 2:96–101. https://doi.org/10.1039/b110474h

Pushpakom S et al (2019) Drug repurposing: progress, challenges and recommendations. Nat Rev Drug Discov 18:41–58. https://doi.org/10.1038/nrd.2018.168

Qi J et al (2018) Rotational paper-based microfluidic-Chip device for multiplexed and simultaneous fluorescence detection of phenolic pollutants based on a molecular-imprinting technique. Anal Chem 90:11827–11834. https://doi.org/10.1021/acs.analchem.8b01291

Qin XX et al (2021) Microfluidic paper-based chips in rapid detection: Current status, challenges, and perspectives. Trac-Trends Analyt Chem 143:116371. https://doi.org/10.1016/j.trac.2021.116371

Qiu H et al (2003) Microchip capillary electrophoresis with an integrated indium tin oxide electrode-based electrochemiluminescence detector. Anal Chem 75:5435–5440. https://doi.org/10.1021/ac034500x

Remy S, Beck H (2006) Molecular and cellular mechanisms of pharmacoresistance in epilepsy. Brain 129:18–35. https://doi.org/10.1093/brain/awh682

Ren H, Fair RB, Pollack MG (2004) Automated on-chip droplet dispensing with volume control by electro-wetting actuation and capacitance metering. Sensors Actuat B-Chem 98:319–327. https://doi.org/10.1016/j.snb.2003.09.030

Ren K, Dai W, Zhou J, Su J, Wu H (2011) Whole-Teflon microfluidic chips. Proc Natl Acad Sci USA 108:8162–8166. https://doi.org/10.1073/pnas.1100356108

Ren K, Zhou J, Wu H (2013) Materials for microfluidic chip fabrication. Acc Chem Res 46:2396–2406. https://doi.org/10.1021/ar300314s

Ren L et al (2020) Combined effects of electric stimulation and microgrooves in cardiac tissue-on-a-chip for drug screening. Small Methods 4:2000438. https://doi.org/10.1002/smtd.202000438

Sackmann EK, Fulton AL, Beebe DJ (2014) The present and future role of microfluidics in biomedical research. Nature 507:181–189. https://doi.org/10.1038/nature13118

Sarkar U et al (2015) Metabolite profiling and pharmacokinetic evaluation of hydrocortisone in a perfused three-dimensional human liver bioreactor. Drug Metab Dispos 43:1091–1099. https://doi.org/10.1124/dmd.115.063495

Sato H, Matsumura H, Keino S, Shoji S (2006) An all SU-8 microfluidic chip with built-in 3D fine microstructures. J Micromech Microeng 16:2318–2322. https://doi.org/10.1088/0960-1317/16/11/010

Shabani R, Cho HJ (2011) A micropump controlled by EWOD: wetting line energy and velocity effects. Lab Chip 11:3401–3403. https://doi.org/10.1039/c1lc20440h

Shamsi A et al (2015) High throughput blood plasma separation using a passive PMMA microfluidic device. Microsyst Technol 22:2447–2454. https://doi.org/10.1007/s00542-015-2664-7

Shang M, Soon RH, Lim CT, Khoo BL, Han J (2019) Microfluidic modelling of the tumor microenvironment for anti-cancer drug development. Lab Chip 19:369–386. https://doi.org/10.1039/c8lc00970h

Shih-Kang F, Hashi C, Chang-Jin K (2003) The sixteenth annual international conference on micro electro mechanical systems. MEMS-03 Kyoto. IEEE, pp 694–697

Shin SR et al (2016) Aptamer-based microfluidic electrochemical biosensor for monitoring cell-secreted trace cardiac biomarkers. Anal Chem 88:10019–10027. https://doi.org/10.1021/acs.analchem.6b02028

Siegel AC et al (2010) Cofabrication: a strategy for building multicomponent microsystems. Acc Chem Res 43:518–528. https://doi.org/10.1021/ar900178k

Sista RS et al (2008a) Heterogeneous immunoassays using magnetic beads on a digital microfluidic platform. Lab Chip 8:2188–2196. https://doi.org/10.1039/b807855f

Sista R et al (2008b) Development of a digital microfluidic platform for point of care testing. Lab Chip 8:2091–2104. https://doi.org/10.1039/b814922d

Skardal A et al (2020) Drug compound screening in single and integrated multi-organoid body-on-a-chip systems. Biofabrication 12:025017. https://doi.org/10.1088/1758-5090/ab6d36

Slaughter VL et al (2021) Validation of an adipose-liver human-on-a-chip model of NAFLD for preclinical therapeutic efficacy evaluation. Sci Rep 11:13159. https://doi.org/10.1038/s41598-021-92264-2

Sollier E, Murray C, Maoddi P, Di Carlo D (2011) Rapid prototyping polymers for microfluidic devices and high pressure injections. Lab Chip 11:3752–3765. https://doi.org/10.1039/c1lc20514e

Song JH, Evans R, Lin YY, Hsu BN, Fair RB (2008) A scaling model for electrowetting-on-dielectric microfluidic actuators. Microfluid Nanofluid 7:75–89. https://doi.org/10.1007/s10404-008-0360-y

Srinivasan V, Pamula VK, Fair RB (2004) Droplet-based microfluidic lab-on-a-chip for glucose detection. Anal Chim Acta 507:145–150. https://doi.org/10.1016/j.aca.2003.12.030

Stanley CE, Wootton RC, de Mello AJ (2012) Continuous and segmented flow microfluidics: applications in high-throughput chemistry and biology. Chimia (Aarau) 66:88–98. https://doi.org/10.2533/chimia.2012.88

Sun W et al (2019) Organ-on-a-Chip for cancer and immune organs modeling. Adv Healthc Mater 8:e1801363. https://doi.org/10.1002/adhm.201801363

Sun G et al (2020) Functional stem cell sorting via integrative droplet synchronization. Anal Chem 92:7915–7923. https://doi.org/10.1021/acs.analchem.0c01312

Suresh V, Qunya O, Kanta BL, Yuh LY, Chong KSL (2018) Non-invasive paper-based microfluidic device for ultra-low detection of urea through enzyme catalysis. R Soc Open Sci 5:171980. https://doi.org/10.1098/rsos.171980

Thorsen T, Roberts RW, Arnold FH, Quake SR (2001) Dynamic pattern formation in a vesicle-generating microfluidic device. Phys Rev Lett 86:4163–4166. https://doi.org/10.1103/PhysRevLett.86.4163

Tibbitt MW, Anseth KS (2009) Hydrogels as extracellular matrix mimics for 3D cell culture. Biotechnol Bioeng 103:655–663. https://doi.org/10.1002/bit.22361

Trinh TN et al (2023) Droplet-based microfluidics: applications in pharmaceuticals. Pharmaceuticals 16

Uhl CG, Liu Y (2019) Microfluidic device for expedited tumor growth towards drug evaluation. Lab Chip 19:1458–1470. https://doi.org/10.1039/c8lc01250d

Vallet M, Berge B, Vovelle L (1996) Electrowetting of water and aqueous solutions on poly(ethylene terephthalate) insulating films. Polymer 37:2465–2470. https://doi.org/10.1016/0032-3861(96)85360-2

Valley JK, Pei SN, Jamshidi A, Hsu HY, Wu MC (2011) A unified platform for optoelectrowetting and optoelectronic tweezers. Lab Chip 11:1292–1297. https://doi.org/10.1039/c0lc00568a

van Midwoud PM, Janse A, Merema MT, Groothuis GM, Verpoorte E (2012) Comparison of biocompatibility and adsorption properties of different plastics for advanced microfluidic cell and tissue culture models. Anal Chem 84:3938–3944. https://doi.org/10.1021/ac300771z

Vernetti L et al (2017) Functional coupling of human microphysiology systems: intestine, liver, kidney proximal tubule, blood-brain barrier and skeletal muscle. Sci Rep 7:42296. https://doi.org/10.1038/srep42296

Virgilio VD, Castaner L (2009) Comparison of static contact angle change and relaxation in EWOD devices. Spanish Conf Electron Dev:46–49

Wadkins RM, Golden JP, Pritsiolas LM, Ligler FS (1998) Detection of multiple toxic agents using a planar array immunosensor. Biosens Bioelectron 13:407–415. https://doi.org/10.1016/s0956-5663(97)00113-9

Waldbaur A, Rapp H, Länge K, Rapp BE (2011) Let there be chip-towards rapid prototyping of microfluidic devices: one-step manufacturing processes. Anal Methods 3:2681–2716. https://doi.org/10.1039/c1ay05253e

Wang W, Jones TB (2015) Moving droplets between closed and open microfluidic systems. Lab Chip 15:2201–2212. https://doi.org/10.1039/c5lc00014a

Wang G, Teng D, Fan SK (2011) Digital microfluidic operations on micro-electrode dot array architecture. IET Nanobiotechnol 5:152–160. https://doi.org/10.1049/iet-nbt.2011.0018

Wang Y et al (2018) Highly sensitive and automated surface enhanced Raman scattering-based immunoassay for H5N1 detection with digital microfluidics. Anal Chem 90:5224–5231. https://doi.org/10.1021/acs.analchem.8b00002

Wann EG, Wodeyar A, Srinivasan R, Frostig RD (2020) Rapid development of strong, persistent, spatiotemporally extensive cortical synchrony and underlying oscillations following acute MCA focal ischemia. Sci Rep 10:21441. https://doi.org/10.1038/s41598-020-78179-4

Wardman JF, Bains RK, Rahfeld P, Withers SG (2022) Carbohydrate-active enzymes (CAZymes) in the gut microbiome. Nat Rev Microbiol 20:542–556. https://doi.org/10.1038/s41579-022-00712-1

Welch ER, Lin YY, Madison A, Fair RB (2011) Picoliter DNA sequencing chemistry on an electrowetting-based digital microfluidic platform. Biotechnol J 6:165–176. https://doi.org/10.1002/biot.201000324

Whitesides GM (2006) The origins and the future of microfluidics. Nature 442:368–373. https://doi.org/10.1038/nature05058

Woolley AT, Mathies RA (1994) Ultra-high-speed DNA fragment separations using microfabricated capillary array electrophoresis chips. Proc Natl Acad Sci USA 91:11348–11352. https://doi.org/10.1073/pnas.91.24.11348

Wu H, Odom TW, Chiu DT, Whitesides GM (2003) Fabrication of complex three-dimensional microchannel systems in PDMS. J Am Chem Soc 125:554–559. https://doi.org/10.1021/ja021045y

Wu J, Chen QS, Liu W, He ZY, Lin JM (2017) Recent advances in microfluidic 3D cellular scaffolds for drug assays. Trac-Trends in Analyt Chem 87:19–31. https://doi.org/10.1016/j.trac.2016.11.009

Yang H, Luk VN, Abelgawad M, Barbulovic-Nad I, Wheeler AR (2009) A world-to-chip interface for digital microfluidics. Anal Chem 81:1061–1067. https://doi.org/10.1021/ac802154h

Yin J et al (2019) Highly efficient capture of circulating tumor cells with low background signals by using pyramidal microcavity array. Anal Chim Acta 1060:133–141. https://doi.org/10.1016/j.aca.2019.01.054

Yin F et al (2021) HiPSC-derived multi-organoids-on-chip system for safety assessment of antidepressant drugs. Lab Chip 21:571–581. https://doi.org/10.1039/d0lc00921k

Yu J, Wang S, Ge L, Ge S (2011) A novel chemiluminescence paper microfluidic biosensor based on enzymatic reaction for uric acid determination. Biosens Bioelectron 26:3284–3289. https://doi.org/10.1016/j.bios.2010.12.044

Yu YH, Chen JF, Zhou J (2014) Parallel-plate lab-on-a-chip based on digital microfluidics for on-chip electrochemical analysis. J Micromech Microeng 24:Artn 015020. https://doi.org/10.1088/0960-1317/24/1/015020

Zakharova M et al (2020) Multiplexed blood-brain barrier organ-on-chip. Lab Chip 20:3132–3143. https://doi.org/10.1039/d0lc00399a

Zamprogno P et al (2021) Second-generation lung-on-a-chip with an array of stretchable alveoli made with a biological membrane. Commun Biol 4:168. https://doi.org/10.1038/s42003-021-01695-0

Zeng YL et al (2023) On-chip modeling of physiological and pathological blood-brain barrier microenvironment for studying glial responses to neuroinflammation. Nano Today 52:101947. https://doi.org/10.1016/j.nantod.2023.101947

Zhai J, Yi SH, Jia YW, Mak PI, Martins RP (2019) Cell-based drug screening on microfluidics. Trac-Trends Analyt Chem 117:231–241. https://doi.org/10.1016/j.trac.2019.05.018

Zhang XN, Stefanick S, Villani FJ (2004) Application of microreactor technology in process development. Org Process Res Dev 8:455–460. https://doi.org/10.1021/op034193x

Zhang YS et al (2016) Bioprinting 3D microfibrous scaffolds for engineering endothelialized myocardium and heart-on-a-chip. Biomaterials 110:45–59. https://doi.org/10.1016/j.biomaterials.2016.09.003

Zhang W et al (2017) Maturation of human embryonic stem cell-derived cardiomyocytes (hESC-CMs) in 3D collagen matrix: effects of niche cell supplementation and mechanical stimulation. Acta Biomater 49:204–217. https://doi.org/10.1016/j.actbio.2016.11.058

Zhao J, Chen Z, Li X, Pan J (2011) A novel microchip based on indium tin oxide coated glass for contactless conductivity detection. Talanta 85:2614–2619. https://doi.org/10.1016/j.talanta.2011.08.029

Zheng Y, Dai W, Ryan D, Wu H (2010) Fabrication of freestanding, microperforated membranes and their applications in microfluidics. Biomicrofluidics 4:036504. https://doi.org/10.1063/1.3491474

Zhou J, Yan H, Zheng Y, Wu H (2009) Highly fluorescent poly(dimethylsiloxane) for on-Chip temperature measurements. Adv Funct Mater 19:324–329. https://doi.org/10.1002/adfm.200801064

Zhu Y et al (2017) In situ generation of human brain organoids on a micropillar array. Lab Chip 17:2941–2950. https://doi.org/10.1039/c7lc00682a

Zips S et al (2023) Aerosol jet-printed high-aspect ratio micro-needle electrode arrays applied for human cerebral organoids and 3D Neurospheroid networks. ACS Appl Mater Interfaces 15:35950–35961. https://doi.org/10.1021/acsami.3c06210

Zou F et al (2019) Rapid, real-time chemiluminescent detection of DNA mutation based on digital microfluidics and pyrosequencing. Biosens Bioelectron 126:551–557. https://doi.org/10.1016/j.bios.2018.09.092

Chapter 11
Micro/Nanorobots

Runtan Li, Yong Kang, and Xiaoyuan Ji

Abstract Intelligent micro/nanorobots represent a breakthrough at the intersection of nanotechnology, robotics, and artificial intelligence. These minuscule robots, which can operate at micrometer and nanometer scales, are designed for precision tasks in constrained environments, including targeted drug delivery, precise surgery, environmental monitoring, and pollution control. They integrate micro-electromechanical systems (MEMS), nanomaterials, and computational intelligence, showing immense potential in both medical and industrial applications. Key components include an actuation system, a control system, and an executive mechanism, driven by various methods such as chemical propulsion, physical drive, or bio-actuation. The evolution of this technology spans from theoretical concepts to practical applications, facilitated by advancements in material science, precision manufacturing, and computational technologies. Intelligent micro/nanorobots promise significant advancements in targeted therapies, environmental remediation, and complex industrial tasks, although challenges remain in areas like precise control, biosafety, and large-scale production. Ongoing research aims to enhance their autonomy, intelligence, and multifunctionality, paving the way for revolutionary applications across diverse fields.

Keywords Micro/nanorobots · Nanotechnology · Artificial intelligence · Biomedical applications

R. Li · Y. Kang · X. Ji (✉)
Academy of Medical Engineering and Translational Medicine, Medical College, Tianjin University, Tianjin, China
e-mail: jixiaoyuan@tju.edu.cn

Z. Liu (ed.), *Intelligent Biomaterials*, https://doi.org/10.1007/978-981-96-8646-9_11

11.1 Overview of Intelligent Micro/Nanorobots

Intelligent micro/nanorobots are the product of the intersection and integration of nanotechnology, robotics technology, and artificial intelligence fields, representing a new type of robot. These robots are tiny in size and precise in movement, capable of performing tasks in narrow or complex environments at the micrometer or nanometer scale, such as targeted drug delivery, precise surgery, environmental monitoring, and pollution control. They integrate micro-electromechanical systems (MEMS), nanomaterial science, and computational intelligence, showing great potential in both medical and industrial fields, and challenging traditional robotics technology and operational concepts.

11.1.1 Definition of Intelligent Micro/Nanorobots

The concept of micro/nanorobots encompasses the technology of robots at two scales: the micrometer scale and the nanometer scale. Although there is a clear difference in size between these two scales of robots, they share similar design philosophies and technological pursuits, that is, to achieve precise operation and control at an extremely small scale.

Microrobots, as the name suggests, are usually at the micrometer level in size, that is, ranging from 1 micrometer to 1 millimeter. These robots can perform tasks at the microscopic level, such as drug delivery within the body and the assembly of microstructures in material science. The design and manufacturing of microrobots usually involve micro-electromechanical systems (MEMS) technology, which makes it possible to manufacture complex mechanical equipment at the micrometer scale.

Nanorobots further reduce the size range, and their size is usually at the nanometer level, that is, ranging from 1 nanometer to 1 micrometer. The design prototype of nanorobots is based on the biological principles at the molecular level, aiming to design and manufacture "functional molecular devices" that can operate in nanospace. These nanoscale devices simulate the function of biomolecules at the molecular level, showing extremely high operational precision and biocompatibility.

Technologically, nanorobots can be divided into two categories: one category is the nanorobots whose volume is nano-level, which can perform autonomous movement and operation at the nanoscale; the other category is the devices for nano-level operation, whose terminal actuators have a very small and precise size, capable of fine operation, while the volume of the device itself does not have to be nano-level. For example, Scanning Tunneling Microscope (STM) and Atomic Force Microscope (AFM) are representatives of such devices, which, although their overall volume is similar to that of traditional microscopes, can perform precise manipulation at the nanoscale with their terminal actuators.

Although the application prospects of micro/nano-technology in the medical field are broad, the truly nano-level volume and controllable nanorobots have not yet been realized. Current research mainly focuses on devices for micro/nano-level operation, such as micro/nano-carriers for precise drug delivery.

Broadly speaking, systems capable of operating at the micro- and nanoscales are termed micro/nanorobots. They are also known as Micro/Nanomachine or Micro/Nanomotor, reflecting the varying stages and depths of research. At present, "micro/nano machines" may be a more precise term, as these entities have yet to achieve full autonomy and intelligence in the traditional sense. However, the term 'micro/nanorobots' not only reflects the customary description people use for emerging technologies but also carries the anticipation and longing for the future development of intelligence.

The essence of micro/nanorobots lies in their intelligent potential. They can make autonomous decisions based on the internal environment and executing corresponding actions, thus being termed intelligent micro/nanorobots. Research from 2015 has already demonstrated this potential, with the development of micro/nanorobots that could detect mutations in DNA or proteins and release medication when disease criteria were met. This breakthrough signified not only the convergence of bio-computing technology with micro/nanorobotics but also heralded a new era of precision medicine.

Intelligent micro/nanorobots represent an emerging field of research that merges micro/nano-technology with robotics. Characterized by their small size, lightweight, high thrust-to-weight ratio, flexibility, and sensitivity, they are typically defined as autonomous machines capable of operating at the micro- and nanoscales and responding to environmental changes to perform complex tasks. These robots generally consist of three key components: the actuation system, the control system, and the executive mechanism:

Actuation System: The propulsion systems of intelligent micro/nanorobots often rely on chemical propulsion, electromagnetic drive, or bio-actuation mechanisms. For instance, the generation of bubbles through chemical reactions can propel the robots, or external magnetic fields can control robots equipped with magnetic materials.

Control System: The control system is central to the functionality of intelligent micro/nanorobots. It may include integrated circuits, sensors, and microprocessors that process information from internal or external environments and make autonomous decisions based on preset programs.

Executive Mechanism: The executive mechanism is the part of the intelligent micro/nanorobot that interacts directly with the environment, encompassing various micro-tools, drug delivery systems, or other functional materials. These mechanisms perform precise biochemical operations or physical interventions according to the instructions from the control system.

The evolution of micro/nanorobots dates back to the science fiction of 1966, but true scientific research began in 1982 with the invention of the scanning tunneling microscope at IBM's Zurich Laboratory, which made observation and manipulation at the micro- and nanoscales possible. Subsequently, the rise of nanotechnology in the 1990s injected new impetus into the development of micro/nanorobots. Over the past four decades, micro/nanorobots have become a cutting-edge hotspot in scientific research, attracting strategic attention and substantial investment from countries around the world.

The advancement of micro/nanorobotic technology is not just about miniaturization but also about expanding functionality and enhancing intelligence. Their potential applications in the biomedical field are especially broad, including but not limited to targeted drug delivery, disease diagnosis and treatment, and medical imaging. For example, researchers at Harbin Institute of Technology have developed micro/nanorobots capable of actively swimming and treating brain diseases, capable of crossing the blood-brain barrier, and delivering drugs precisely to the lesion area.

Moreover, the materials and actuation technologies for micro/nanorobots are also continuously advancing. From the initial metal and semiconductor materials to the current biodegradable materials, the biocompatibility and safety of micro/nanorobots have been significantly improved. In terms of actuation technology, in addition to traditional magnetic and electric actuation, light, acoustic, chemical, and bio-hybrid actuation methods are also being continuously explored and developed.

With the ongoing progress in material science, nanotechnology, artificial intelligence, and machine learning, the capabilities of intelligent micro/nanorobots will become more powerful, and their application fields will continue to expand. For instance, by integrating artificial intelligence, intelligent micro/nanorobots will be able to adjust their behavior automatically in response to environmental changes without external instructions, performing more complex decision-making and operations. Furthermore, the collaborative work of multi-robot systems is expected to play a role in complex environments, such as precise multi-point treatments within the human body or efficient search and rescue operations at disaster sites.

The application prospects for micro/nanorobots are vast, spanning healthcare, environmental science, biotechnology, and military defense. In healthcare, they can be utilized to implement targeted drug delivery systems, maximizing therapeutic effects by delivering drugs precisely to the site of disease while minimizing side effects on surrounding healthy cells. In environmental monitoring and management, intelligent micro/nanorobots can detect and break down harmful substances in the environment or play a key role in disaster response.

However, the research and application of micro/nanorobots still face many challenges. Achieving precise control while ensuring biosafety, enhancing the robots' autonomy and intelligence, and realizing large-scale production and clinical application are all focal points of current research. Additionally, as technology continues to evolve, ethical, legal, and social issues surrounding micro/nanorobots are increasingly gaining attention.

11.1.2 The Development History of Intelligent Micro/Nanorobots

The concept of intelligent micro/nanorobots was inspired by the 1966 science fiction film *Fantastic Voyage* which transported viewers to a microscopic world where brave explorers shrunk to the micrometer level and embarked on a thrilling journey inside the human body. A notable scene depicted a miniaturized submarine navigating through blood vessels to clear blood clots. Although purely fictional, this film sparked boundless imagination about micro/nanorobot technology. In reality, micro/nanorobotic technology is a genuine and rapidly advancing branch of modern science, integrating cutting-edge technologies from nanotechnology, biomedicine, mechanical engineering, and artificial intelligence (Fig. 11.1).

11.1.2.1 Initial Stage: Laying the Foundations

The idea of intelligent micro/nanorobots dates to the mid-twentieth century when scientists began contemplating the manipulation of matter at the microscopic level.

In 1959, renowned physicist Richard Feynman delivered a speech titled "There's Plenty of Room at the Bottom" at the California Institute of Technology. Feynman envisioned the potential for manipulating at the atomic and molecular levels and proposed the direct manipulation of individual atoms and molecules to create minuscule yet powerful devices. This bold concept, though considered radical at the time, is regarded as the genesis of nanotechnology, and inspired subsequent research at the nanoscale.

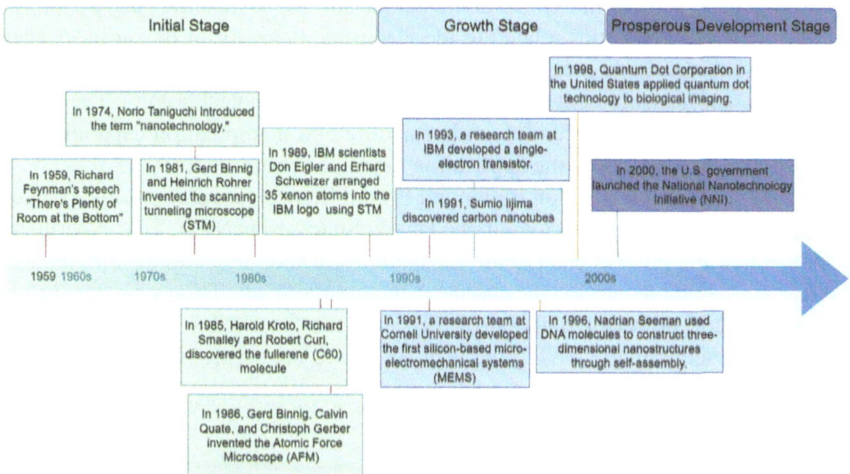

Fig. 11.1 Development timeline of micro/nanorobots

In 1974, Japanese scientist Norio Taniguchi introduced the term "nanotechnology" in his paper "On the Basic Concept of 'Nano-Technology'." Taniguchi described the processing and manufacturing techniques at the nanoscale, defining nanotechnology as "precision processing at the molecular level, based on an atom, a molecule, or larger." This definition established the terminological foundation for the field of nanotechnology, promoting clarity and systematic research directions.

In 1981, Gerd Binnig and Heinrich Rohrer invented the scanning tunneling microscope (STM) at the IBM Zurich Research Laboratory in Switzerland. The STM enabled the observation and manipulation of individual atoms at atomic-level resolution through the tunneling effect, which measures the electron density on a sample's surface. This groundbreaking invention allowed scientists to "see" and manipulate individual atoms for the first time, significantly advancing nanotechnology. For their substantial scientific contributions, Binnig and Rohrer were awarded the Nobel Prize in Physics in 1986.

In 1985, British scientist Harold Kroto, along with American scientists Richard Smalley and Robert Curl, discovered the fullerene (C60) molecule at Rice University in the United States through laser evaporation of graphite. In the experiment, they discovered the spherical molecular structure of carbon atoms, later named fullerene, also known as Buckminsterfullerene. This new form of carbon not only broadened people's understanding of the element carbon but also provided new research directions for nanomaterial science. The discovery of fullerene earned the three scientists the Nobel Prize in Chemistry in 1996.

In 1986, Gerd Binnig, Calvin Quate, and Christoph Gerber invented the Atomic Force Microscope (AFM) at Stanford University in the United States, building on the STM. The AFM achieves three-dimensional imaging at the nanoscale through the interaction between the probe and the sample surface. This type of microscope can not only observe the surface morphology of the sample but also measure the mechanical properties of the surface. Since then, humans have been able to observe the arrangement state of individual atoms on the surface of materials and the physical and chemical properties related to the electronic behavior of the surface. They can also use the probe to precisely manipulate atoms at low temperatures. They are both important measurement and processing tools for nanotechnology. The invention of AFM further enhanced the understanding and manipulation capabilities of the micro world, expanding the application range of nanotechnology.

In 1989, IBM scientists Don Eigler and Erhard Schweizer successfully arranged 35 xenon atoms into the IBM logo in the United States using STM. This experiment demonstrated the precise manipulation capability of individual atoms and marked the first-time humans arranged patterns at the atomic level. This achievement not only proved the powerful function of STM but also marked a new era in nanomanipulation technology, paving the way for future nanoscale manufacturing and processing technology.

11.1.2.2 Growth Stage: Technological Validation and Application Exploration

Entering the 1990s, with advancements in material science, precision manufacturing, and computational technologies, research on micro/nanorobots began transitioning from theoretical concepts to tangible applications.

In 1991, a research team at Cornell University developed the first silicon-based micro-electromechanical systems (MEMS), marking a significant step toward the actual manufacturing of micro/nanorobots from conceptual stages. MEMS technology enabled the fabrication of complex mechanical and electronic devices at the microscale, while nano-electromechanical systems (NEMS) further reduced this scale to the nanoscale.

Concurrently, significant progress was made in the study of nanomaterials. In 1991, Japanese scientist Sumio Iijima discovered a tubular nanostructure formed by the rolling of graphite sheets while researching carbon materials at the NEC Corporation laboratory using a high-resolution transmission electron microscope. This discovery, known as carbon nanotubes, displayed their excellent mechanical, electrical, and thermal properties, making them an ideal nanomaterial. This finding propelled the development of nanomaterial science, opening numerous new research directions and providing new possibilities for the design and manufacturing of micro/nanorobot systems.

In 1993, a research team at IBM in the United States developed a single-electron transistor. This transistor leveraged the quantum tunneling effect, controlling the flow of individual electrons to perform computations. Operating at extremely low power consumption, the single-electron transistor demonstrated the immense potential of nanoelectronics, offering new ideas for future nanoscale computing and storage devices.

In 1996, Nadrian Seeman from New York University utilized DNA molecules to construct three-dimensional nanostructures. Seeman employed self-assembly techniques, using DNA molecules as building blocks to create three-dimensional nanostructures with specific shapes and functions. This breakthrough illustrated the potential for constructing nanoscale structures using biological molecules, providing an essential foundation for the development of bio-nanotechnology.

In 1998, Quantum Dot Corporation in the United States applied quantum dot technology to biological imaging. Quantum dots, with their unique optical properties such as high brightness and narrow bandwidth emission spectra, have been widely used in biological imaging and optical detection. The application of quantum dots provided high-resolution and highly sensitive tools for biological imaging, offering new methodologies for biomedical research.

In 2000, the U.S. government launched the National Nanotechnology Initiative (NNI). This initiative aimed to promote research and development in nanotechnology, providing substantial funding and resources. The establishment of NNI marked a period of rapid development for nanotechnology, greatly facilitating its application and commercialization across various fields.

11.1.2.3 Prosperous Development Stage: Interdisciplinary Integration and Intelligent Development

Entering the twenty-first century, with breakthroughs in nanotechnology, research on micro/nanorobots has entered a new stage. Nanotechnology enables scientists to design and manufacture structures at the atomic and molecular levels, providing new possibilities for precise control and functional integration of micro/nanorobots. In addition, the progress in biomedical engineering also provides a theoretical basis and technical support for the application of micro/nanorobots in the human body.

Currently, micro/nanorobot technology is developing toward intelligence, autonomy, and multifunctionality. By integrating advanced sensors, actuators, and control systems, micro/nanorobots can achieve more complex operations and more precise control. At the same time, the integration of artificial intelligence technology also provides the possibility for micro/nanorobots to make autonomous decisions and learn, enabling them to better adapt to different environments and tasks.

11.1.3 Common Construction Strategies of Micro/Nanorobots

In the development process of intelligent micro/nanorobots, choosing the right construction strategy is crucial as it directly affects the functionality, application scope, and ultimate practicality of the robots. This section will discuss two main construction strategies—Top-Down and Bottom-Up—detailing their specific manifestations and technical details in practical applications, and showcasing these techniques' operation modes, potential advantages, and limitations through examples.

11.1.3.1 Top-Down Construction Strategy

The top-down construction strategy for micro/nanorobots is a method that gradually narrows down to the micro/nanoscale through macro-scale manufacturing methods, thereby constructing precise and versatile micro/nanorobots. The core of this strategy lies in utilizing existing macro-processing technologies and processes to gradually achieve fine processing and control of micro/nanoscale structures. This method is of great significance in the manufacturing of micro/nanorobots as it can leverage mature Micro-Electromechanical Systems (MEMS) and Nano-Electromechanical Systems (NEMS) technologies to achieve efficient and large-scale production of micro/nanorobots.

The top-down construction strategy mainly relies on several key processing technologies, including photolithography, thin-film deposition, etching, and mechanical processing technologies. Photolithography is a precision manufacturing method widely used in the microelectronics industry, which uses photosensitive materials and ultraviolet light to define patterns at the micrometer or nanometer scale. First, a layer of photosensitive material is coated on the substrate, then ultraviolet light is projected onto the photosensitive material through a mask, causing chemical changes in the exposed parts while the unexposed parts remain unchanged. Subsequently, the exposed or unexposed parts are removed through chemical solutions, forming the desired micro/nano-structures. This method is characterized by high resolution and precision, suitable for large-scale manufacturing of micro/nanorobots.

Thin-film deposition technology is another key processing method that constructs micro/nanoscale functional layers by depositing a thin layer of material onto a substrate. Common thin-film deposition methods include Physical Vapor Deposition (PVD), Chemical Vapor Deposition (CVD), and Atomic Layer Deposition (ALD). PVD involves physical processes such as evaporation or sputtering to deposit material onto the substrate, forming a uniform thin film; CVD involves chemical reactions that decompose gaseous precursors and deposit the desired material on the substrate surface; ALD achieves ultra-thin and uniform thin-film construction through a molecular layer-by-layer deposition process. These thin-film technologies provide high-quality material layers when constructing micro/nanorobots, ensuring excellent mechanical, electrical, and chemical properties.

Etching technology is an essential means of achieving fine processing of micro/nano-structures, including wet etching and dry etching. Wet etching uses chemical solutions to remove materials, achieving the construction of specific patterns through selective etching. Dry etching, on the other hand, uses plasma or reactive ions to etch materials, offering higher anisotropy and precision. Through etching technology, complex three-dimensional micro/nano-structures can be further realized based on the patterns defined by photolithography. Etching technology enables precise control of material removal during the manufacturing process of micro/nanorobots, constructing the required micro/nanoscale functional structures.

Mechanical processing techniques such as micro-milling, micro-drilling, and micro-abrasive grinding are essential methods for constructing micro/nanorobots. These techniques allow for precise machining at the micro/nanoscale through high-precision mechanical operations. Micro-milling uses miniature cutting tools to machine materials, forming precise micrometer or nanometer structures; micro-drilling creates small channels by using miniature drills to bore into materials; micro-abrasive grinding uses abrasive tools to finely grind material surfaces, achieving high precision surface smoothness. These mechanical processing techniques enable high-precision, highly flexible structural machining in the manufacturing of micro/nanorobots, especially suitable for complex shapes and high-demand functional components.

The top-down construction strategy in micro/nanorobot manufacturing relies not only on the key processing technologies but also on the integration and synergy of various technologies. By combining photolithography, thin-film deposition, etching, and mechanical processing technologies organically, complex micro/nano-structures and multifunctional integration can be achieved. For example, when constructing sensors and actuators for micro/nanorobots, electrode patterns can be defined through photolithography, followed by forming conductive layers through thin-film deposition technology, then refining the electrodes through etching technology, and finally constructing the three-dimensional structure of sensors and actuators using mechanical processing techniques. This integrated approach using multiple technologies not only improves manufacturing efficiency and precision but also enables multifunctional integration of micro/nanorobots. However, this method is relatively costly, especially when manufacturing large areas or complex three-dimensional structures. In addition, this physical etching process generates a large amount of waste, and its environmental impact should not be overlooked.

11.1.3.2 Bottom-Up Construction Strategy

In contrast to the top-down approach is the bottom-up strategy, which starts from atoms, molecules, or nanoscale basic units and gradually assembles and constructs complex micro/nano-structures. This strategy complements the top-down construction strategy, enabling the realization of high-precision and highly integrated micro/nanorobots. The bottom-up approach often utilizes chemical synthesis, self-assembly, and bio-templating technologies to gradually assemble molecular-level building blocks into micro/nanorobots with specific functions and structures.

Chemical synthesis is the foundation of the bottom-up construction strategy, synthesizing nanomaterials, and nanostructures with specific functions at the nanoscale through chemical reactions. For example, quantum dots and nanowires with excellent optoelectronic properties can be synthesized through solution chemistry methods, and high-quality carbon nanotubes and graphene can be prepared through vapor deposition. These nanomaterials have unique physical and chemical properties and can serve as basic building blocks for micro/nanorobots. The advantage of chemical synthesis methods is the precise control of material composition and structure, achieving high-purity and high-uniformity nanomaterial preparation.

Self-assembly technology is a key component of the bottom-up construction strategy, where molecular interactions such as electrostatic forces, hydrogen bonds, van der Waals forces, and hydrophobic effects are utilized to spontaneously assemble molecules or nanoparticles into nanosystems with regular structures under specific conditions. For instance, DNA self-assembly technology uses the base pairing rules of DNA molecules to precisely design and assemble various complex nanostructures. This method not only has a high degree of structural controllability but also enables multifunctional integration. Self-assembly technology in the manufacturing of micro/nanorobots can be used to construct key components such as sensors, actuators, and energy conversion devices.

Bio-templating technology is another bottom-up construction strategy that uses biological molecules or structures as templates to guide the assembly and growth of inorganic materials at the nanoscale. For example, using biological templates such as viruses, bacteria, and proteins, the directed growth and assembly of metal nanoparticles, nanowires, and nanotubes can be achieved. The advantage of bio-templating technology lies in the high specificity and precision of biological molecules, which can precisely control the morphology and size of nanomaterials. By combining bio-templating technology with chemical synthesis methods, complex-structured and multifunctional micro/nanorobots can be prepared.

An important application of the bottom-up construction strategy is the preparation of nanocomposite materials. By combining nanomaterials with different properties, multifunctional nanocomposite materials can be constructed. For example, combining conductive carbon nanotubes with high-strength ceramic nanoparticles can produce nanocomposite materials with both conductivity and mechanical strength. These nanocomposite materials can serve as structural materials for micro/nanorobots, providing excellent mechanical, electrical, and chemical properties. In addition, by regulating the composition and structure of nanomaterials, the functionality and responsiveness of nanocomposite materials can be adjusted, providing possibilities for multifunctional integration of micro/nanorobots.

In terms of energy and driving, the bottom-up construction strategy can be used to construct energy conversion and storage devices for micro/nanorobots. For example, quantum dots and nanowires prepared by self-assembly technology can achieve efficient photoelectric conversion and electrochemical energy storage. By integrating these energy conversion devices into micro/nanorobots, self-powered and wireless driving can be realized. The bottom-up construction strategy can also be used to design and prepare nanocatalysts with specific functions, using these catalysts to achieve chemical driving and reaction control of micro/nanorobots.

In summary, the bottom-up construction strategy, which starts from molecular and nano-level basic units and gradually assembles and constructs complex micro/nano-structures, has high precision and tunable functionality. This strategy plays an important role in the manufacturing of micro/nanorobots, achieving high functional integration and a variety of micro/nanorobots through chemical synthesis, self-assembly, and bio-templating technologies. In the future, with the further development and innovation of the bottom-up construction strategy, micro/nanorobots will show broader application prospects in the fields of biomedicine, environmental protection, energy, and electronics. By continuously optimizing and innovating construction strategies, micro/nanorobots will achieve higher performance and more complex functions, bringing more possibilities to scientific research and practical applications.

The core of this strategy lies in using intermolecular interactions such as hydrogen bonds, electrostatic forces, or hydrophobic interactions to guide the self-assembly of structures. It does not require expensive equipment support and is possible to operate at room temperature and atmospheric pressure, reducing production costs. However, the bottom-up construction strategy also faces its inherent challenges, such as the unpredictability and complexity of the self-assembly process. Since it relies on the subtle balance of intermolecular forces, these structures may exhibit different stabilities under different environmental conditions, which needs special attention when designing micro/nanorobots for biomedical or environmental monitoring applications.

In practical applications, researchers often combine these two strategies to make up for their respective shortcomings. For example, metal nanoparticles can be guided to self-assemble on a mini template manufactured by the top-down method.

This method not only maintains structural accuracy but also endows the material with new functions, such as changing the electromagnetic properties or optical response of the material by adjusting the composition and arrangement of nanoparticles.

11.1.4 Control Strategies for Micro/Nanorobots

Effective control of micro/nanorobots is at the core of realizing their functions. To perform precise tasks in complex environments, researchers have developed various control strategies. These strategies require not only precise manipulation of the robot's movements but also real-time response to environmental changes and task requirements. This section will delve into several main control strategies, including optical control, chemical drive, magnetic control, and electrical control strategies, and showcase the application and challenges of these technologies through specific examples.

11.1.4.1 Optical Control Strategy

The optical control strategy uses light signals to manipulate micro/nanorobots. The advantage of this method lies in its non-contact nature and high spatial resolution. Light can be used to activate specific photo-sensitive materials or directly manipulate robots through precise laser scanning.

A typical application involves using photo-sensitive polymers as the main material for robots. When these polymers are exposed to specific wavelengths of light, they undergo morphological changes that drive the robot to move or alter its structure. For instance, researchers have developed a type of micro/nanorobot that can move forward by rapidly curling and unrolling under ultraviolet light exposure. This optical control mechanism is particularly suitable for applications requiring operation in remote or inaccessible areas, such as precise drug delivery within the body.

11.1.4.2 Chemical Drive Strategy

The chemical drive strategy controls micro/nanorobots through chemical reactions, relying on the robot's materials to react with the surrounding environment's chemicals. A typical example of this strategy is the self-propelled motion of metal nanoparticles in a hydrogen peroxide solution, which occurs due to the catalytic decomposition reaction happening on the nanoparticle surface, releasing bubbles that produce thrust.

The chemical drive control method is simple and effective, suitable for environments without external energy input. However, this strategy is often limited by the types of chemical reactions and reaction rates and may face challenges regarding the toxicity of reaction byproducts or environmental compatibility.

11.1.4.3 Magnetic Control Strategy

Magnetic control is one of the most widely used methods for micro/nanorobot control. This strategy guides robots containing magnetic materials through an external magnetic field. The advantages of magnetic control include its deep penetration and strong controllability, especially the precise manipulation of robots in three-dimensional space in terms of position and attitude.

For example, micro-spiral robots made of ferromagnetic materials can perform precise rotation and linear motion under the influence of a magnetic field, enabling them to navigate complex internal environments and perform tasks such as precise drug delivery. A challenge of magnetic control strategy is designing robot structures and magnetic circuits that can stably operate in complex magnetic fields.

11.1.4.4 Electrical Control Strategy

The electrical control strategy uses electric fields or currents to control micro/nanorobots, including using electromagnetic forces to drive robot movement or change its structure. Electrical control strategies can provide extremely fast response speeds and precise control, suitable for applications requiring rapid response.

An application example is the electrophoretic driving of micro/nanorobots; these robots move by adjusting their surface charge properties under the influence of an electric field. This method is particularly suitable for operations in liquid environments, such as movement and manipulation within blood vessels or cellular spaces. Electrically controlled robots can achieve highly integrated control systems but also require power sources and circuit designs that meet strict safety and compatibility standards.

In practical applications, a single control strategy often falls short of meeting all requirements, so researchers often adopt integrated control strategies, combining the above methods to adapt to different environmental and task demands. For instance, magnetic control can be combined with optical control, using a magnetic field for rough positioning of the robot and optical control for fine manipulation.

The development of control strategies for intelligent micro/nanorobots has not only driven innovation in robot technology but also expanded their application prospects in various fields such as healthcare, environmental protection, and industry. Future research will focus more on improving the precision and efficiency of control and better integrating these robots into practical application environments.

11.1.5 Driving Strategies for Micro/Nanorobots

When intelligent micro/nanorobots operate in body fluids with low Reynolds numbers, they need to overcome their own viscous resistance and Brownian motion to achieve effective displacement and reach the target area.

In the research and application of micro/nanorobots, the design of driving strategies is key to ensuring the robots' effective operation in complex environments. A well-designed driving system can not only improve the operational accuracy of the robots but also expand their applications in clinical treatment, environmental monitoring, and other fields. This section will explore several main driving strategies currently used for micro/nanorobots, including physical driving, chemical driving, and biological driving.

The physical driving strategy involves using externally applied physical fields to control the movement and behavior of micro/nanorobots, including light-driven, electric-driven, magnetic-driven, and ultrasound-driven.

The light-driven strategy uses the radiation pressure or photothermal effect of light to drive micro/nanorobots. Micro/nanorobots driven by light fields are usually made of light-responsive materials that can generate motion under illumination. The advantage of this driving method is its non-contact nature, high resolution, and rapid response. In the biomedical field, light field-driven micro/nanorobots can be used for optical tweezer technology, capturing, and manipulating cells or particles with a focused laser beam. Additionally, light field driving can be used for photodynamic therapy, generating local photothermal effects through photosensitizers to treat tumors.

The electric-driven strategy manipulates micro/nanorobots by applying an electric field. These micro/nanorobots are typically made of dielectric or polar materials that can generate a dipole moment under the influence of an electric field, thus moving in an electric field gradient. This driving method relies on the polarization effect of dielectric materials, and the speed and direction of movement can be controlled by adjusting the frequency and intensity of the electric field. Electric field-driven micro/nanorobots have a wide range of applications in microfluidic chips and biosensors. For example, electric fields can control the opening and closing of micropumps or micro-valves, thereby regulating fluid flow, and achieving precise fluid manipulation. Additionally, electric field-driven micro/nanorobots can be used in the biomedical field for cell manipulation and tissue engineering.

The magnetic-driven strategy uses an external magnetic field to control magnetic micro/nanorobots. These robots contain magnetic materials, such as ferromagnetic or superparamagnetic nanoparticles, which can be magnetized under the influence of an external magnetic field, generating a torque for directional movement (Chung et al. 2020; Ebrahimi et al. 2020).By changing the intensity and direction of the external magnetic field, the movement state of the robots can be precisely controlled. Magnetic field-driven micro/nanorobots have a wide range of applications in the biomedical field, especially in drug delivery and minimally invasive surgery. By controlling the intensity and direction of the external magnetic field, the direction and speed of the robot's movement can be precisely adjusted, thereby delivering drugs directly to the affected area.

Ultrasonic-driven strategies utilize the mechanical vibrations of ultrasonic waves to propel micro/nanorobots. These waves can propagate within biological bodies, offering high penetration and resolution. Micro/nanorobots driven by the ultrasonic field are applicable for ultrasonic imaging and therapy. For instance, guided by the ultrasonic field, they can transport microbubbles or drugs to the affected areas, achieving synchronized ultrasonic imaging and drug delivery. Additionally, ultrasonic-driven micro/nanorobots can be employed for tissue cutting and coagulation, enhancing the precision and safety of surgical procedures.

The chemical fuel-driven strategy utilizes the biochemical fuel carried by intelligent micro/nanorobots to react with the surrounding environment, generating uneven chemical gradients (such as uneven distribution of ions, electric fields, or concentration) or unstable bubbles, achieving energy conversion, and thereby propelling the robot's own movement (Simmchen et al. 2016; Cai et al. 2023b; Esteban-Fernández et al. 2017; Cai et al. 2023a; Zhan et al. 2019). This strategy is highly flexible and diverse, and appropriate chemical reactions can be selected according to different application requirements. Chemical gradient driving is one way to implement the chemical driving strategy, which specifically refers to using the chemical concentration gradient in the environment (such as oxygen, pH value, ion concentration, etc.) for path guidance to drive micro/nanorobots.

Micro/nanorobots are equipped with sensors on their surface that can respond to specific chemical substances. When the robot is exposed to a chemical gradient field, the sensor converts the concentration difference into mechanical power, driving the robot to move toward or away from the direction of the chemical source. This mechanism is particularly important inside biological bodies, for example, in targeting tumor cells during cancer treatment. Some micro/nanorobots can use the gas pressure generated by chemical reactions to propel the robot forward, and this strategy has potential applications in micropumps or drug release systems. Chemical gradient-driven micro/nanorobots can respond to the chemical concentration gradient in the environment, achieving autonomous navigation and target detection, which is especially useful in environmental monitoring and biological detection.

Biohybrid propulsion leverages biological molecules or processes present within organisms as a source of energy. This strategy offers significant advantages in terms of biocompatibility and the utilization of energy within biological systems. For instance, certain micro/nanorobots can harness ATP, produced by cellular metabolism, as an energy source, achieving motion through the catalytic action of ATPases. Micro/nanorobots driven by bioelectric potential differences can utilize the potential differences within biological bodies for navigation, such as in neural tissues, where they can respond to local potential changes to achieve precise positioning.

The selection and optimization of these driving strategies are crucial for the efficient and safe operation of micro/nanorobots. As technology continues to advance, more innovative driving methods may emerge in the future, further expanding the application fields of micro/nanorobots. Through ongoing research and development, these driving technologies will provide dedicated support for the application of micro/nanorobots in fields such as healthcare, environmental protection, and industry (Table 11.1).

Table 11.1 Advantages and limitations of various micro-nanorobot driving methods

Driving method	Advantages	Limitations
Light-driven	High spatial resolution Non-contact driving Precise control – potential for use in optically accessible environments	Requires transparent media Limited by light source availability May have photothermal effects that could be harmful to biological tissues
Electrically driven	Deep penetration Compatible with MEMS systems Rapid response and precise control	Potential bio-rejection of electrodes Safety concerns regarding the use of electric fields in biological environments
Magnetic-driven	Deep penetration – fast driving speed – non-invasive control	Safety and compatibility of magnetic materials Potential interference with other medical devices or implants
Ultrasonic-driven	Good biocompatibility Efficient energy transfer Capable of real-time imaging and treatment	Inevitable thermal effects and cavitation
Chemical fuel-driven	Autonomous driving without external energy Flexibility in chemical reactions for propulsion	Risk of pollution from reaction byproducts – rapid fuel consumption Difficulty in controlling direction and speed of movement
Biohybrid-driven	Efficient use of biological energy Good incompatibility Adaptability to complex biological environments	Poor operability Potential tissue interference Immunological rejection

11.2 Categories of Intelligent Micro/Nanorobots

11.2.1 Light-Driven Micro/Nanorobots

Light-driven micro/nanorobots represent a cutting-edge technology in nanotechnology and micro-electromechanical systems (MEMS), showcasing immense potential in biomedical applications and environmental remediation through self-actuation and manipulation driven by light energy. These micro/nanorobots convert light energy into mechanical energy, enabling efficient movement and precise control in microenvironments (Fig. 11.2).

11.2.1.1 Principle of Light-Driven

The fundamental principle of light-driven micro/nanorobots is the absorption and conversion characteristics of photo-sensitive materials to light. In these systems, light energy is transformed into thermal, electrical, or chemical energy, generating propulsion. The most common implementation involves semiconductor materials, such as titanium dioxide (TiO_2) or perovskite, which drive robot movement through

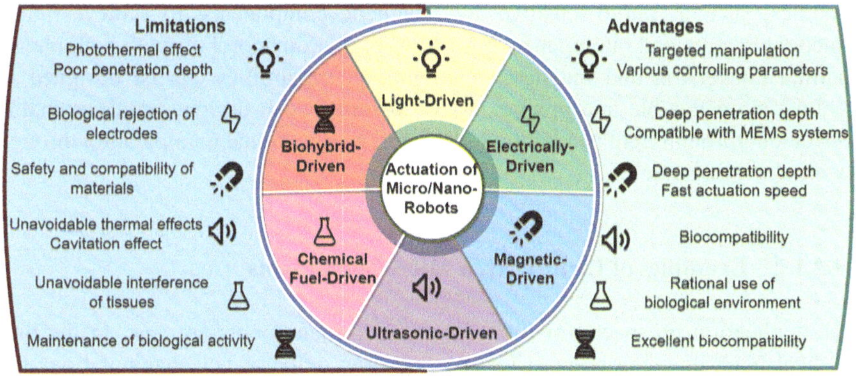

Fig. 11.2 Categories of intelligent micro/nanorobots

the generation of photogenerated electron-hole pairs. When light irradiates these materials, photon energy is absorbed, exciting electrons to the conduction band, leaving holes in the valence band. The migration and separation of these photogenerated charge carriers, under the influence of electric fields or chemical gradients, can produce propulsion, enabling micro/nanorobots to move in liquids or gas media. For instance, in titanium dioxide micro/nanorobots, light irradiation triggers photocatalytic reactions, generating electrons and holes that react with water and oxygen, creating bubbles or chemical gradients that propel the robot's movement.

In addition to titanium dioxide, metal-organic frameworks (MOFs) and quantum dots are also widely used as photosensitive materials. The choice of material not only determines the efficiency of light absorption but also affects the mode and efficiency of energy conversion. For instance, perovskite materials possess a high light absorption coefficient and excellent charge separation capabilities, making them ideal for high-efficiency light-actuated micro/nanorobots. Moreover, surface modification and functionalization of materials can further enhance their photoresponsive performance and stability. Introducing catalysts or photosensitizers on material surfaces can improve the separation efficiency of photogenerated electrons and holes, thereby increasing driving efficiency. Certain nanoparticles, such as gold nanoparticles, can increase the movement speed of micro/nanorobots by enhancing light energy conversion efficiency through localized surface plasmon resonance effects.

Current design strategies primarily include asymmetric design and structural functionalization. Asymmetric design introduces asymmetric elements, such as shape asymmetry or uneven material distribution, in the structure of micro/nanorobots. This design can produce uneven temperature fields or chemical gradients under light irradiation, achieving directional movement. For example, Janus particles, a common type of asymmetric structure, have one-half of the surface covered with photosensitive materials and the other half with inert materials. This design can form an asymmetric reaction field under light irradiation, driving particle movement. Structural functionalization involves introducing different functional modules, such as photo-sensitive and magnetic components, to achieve multi-functionalization and intelligence. For instance, micro/nanorobots with both photo-thermal conversion and magnetic manipulation capabilities can be designed to achieve more flexible and precise motion control. Such designs not only enable movement through light irradiation but also allow directional manipulation through external magnetic fields.

11.2.1.2 Examples of Light-Driven Micro/Nanorobots

The application prospects for light-driven micro/nanorobots are vast. In the biomedical field, they can be used for targeted drug delivery, cell manipulation, and minimally invasive surgery. For example, light-driven micro/nanorobots carrying drugs can achieve directed movement through light irradiation, delivering drugs accurately to the lesion site, thereby improving treatment effects and reducing side effects. VRA Ferreira (2024) explored the application of semiconductor-based light-driven micro/nanorobots in drug delivery. By optimizing photosensitive semiconductor materials such as titanium dioxide and perovskite, these micro/nanorobots can efficiently convert light energy into motion energy, achieving precise drug delivery. The potential of light-driven micro/nanorobots in treating diseases, especially targeted cancer therapy, has been demonstrated. Thanks to the strong penetration characteristics of near-infrared light, near-infrared light-driven micro/nanorobots can be used in the biomedical field. Wan et al. (2020) developed a type of platelet membrane-modified silica/platinum nanorobot that, under near-infrared light irradiation, can not only move autonomously but also control drug release according to a predetermined program. Specifically, they can rapidly release thrombolytic urokinase for the thrombolysis phase of thrombus treatment within a few hours and slowly release the anticoagulant heparin for subsequent anticoagulation therapy. This precise drug release control provides a new method for the treatment of vascular diseases such as thrombosis.

In the field of environmental remediation, light-driven micro/nanorobots can be used for water purification and pollutant degradation. For example, through photocatalytic reactions, micro/nanorobots can decompose organic pollutants in water, achieving self-cleaning and environmental remediation. Zhao et al. (2023) in 2024 developed a light-driven micro-motor based on metal-organic frameworks (MOF) for water disinfection and pollutant removal. The study showed that these micromotors have efficient driving capabilities and multi-responsive characteristics under multi-wavelength light irradiation, capable of working efficiently in complex water environments, bringing new breakthroughs in water treatment technology. Li et al. (2024) in 2024 developed an asymmetric yolk-shell structure near-infrared light-driven nano-motor for selective pollutant adsorption. The study showed that these micro/nano-motors exhibited excellent selectivity and efficiency in complex environments, providing new tools for environmental pollution control.

Optoelectronic tweezers technology offers a method distinct from other non-contact micro/nano-manipulation techniques such as magnetic field manipulation.

It can generate non-uniform electric fields and produce sufficient driving force even under low light intensity conditions, achieving independent and parallel control of micro/nano-devices. Feng Lin's team conducted in-depth research on optoelectronic tweezers technology, particularly in the parallel manipulation of large-scale micro-spiral algae. By using specific projection light patterns, they were able to independently control micro-spiral algae with multiple light points, constructing diverse micro-structure combinations such as T-shapes and micro-coils (Liang et al. 2022).

Despite significant research progress in light-actuated micro/nanorobots, they still face some challenges in practical applications. The movement speed and direction of light-actuated micro/nanorobots can be easily controlled by adjusting the intensity and direction of the light source. However, light penetration in human tissues is limited, which means that light-actuated micro/nanorobots may not function effectively in deeper environments within the body. Increasing light energy to maintain motion effects in deeper environments may cause damage to human tissues. Therefore, light-actuated micro/nanorobots have certain limitations when applied to deeper tissues in the human body, potentially restricting their application range in the biomedical field.

11.2.2 Electrically-Driven Micro/Nanorobots

Electrostatic micro-nanorobots represent a significant research direction in the field of micro-nanotechnology, achieving self-actuation and manipulation through the conversion of electrical energy. Compared to traditional driving techniques, electrostatic driving offers a more direct method of energy conversion, enabling rapid response and precise control. The advantage of electrostatic driving technology lies in its direct-action capability on biological tissues and its efficiency in specific applications.

11.2.2.1 Principle of Electrically Driven

The fundamental and core principles of electrostatically driven micro-nanorobots involve multiple aspects, including the interaction between electric fields and charges, electrophoretic and electroosmotic effects, materials science, and design strategies. These elements are key to understanding the operation of electrostatically driven micro-nanorobots.

Firstly, the core working mechanism of these robots primarily relies on the interaction between electric fields and charges. Electric fields can be directly applied through external power sources or indirectly generated by electrochemical reactions. Under the influence of an electric field, the charges on the surface of the micro-nanorobots interact with the field to produce electrophoretic forces, propelling the robots into motion. Electrophoresis, the movement of charged particles within an electric field, is one of the common driving methods for micro-nanorobots in liquid media. By adjusting the intensity and direction of the electric field, the motion trajectory and speed of the micro-nanorobots can be precisely controlled. Electroosmosis is another common driving method that utilizes the electric field to generate fluid flow on or within the surface of the micro-nanorobots, thereby achieving propulsion. The electroosmotic effect is often used to drive fluids in microfluidic systems, and the combination of electrophoresis and electroosmosis enables more complex and precise manipulation.

Commonly used materials include metals, semiconductors, and conductive polymers. Metallic materials such as gold and platinum offer excellent electrical conductivity and chemical stability but may present biocompatibility issues in certain biological environments. Semiconductor materials like titanium dioxide and silicon possess unique optoelectronic properties that can enhance driving efficiency through the photoelectric effect. Conductive polymers, such as polypyrrole and polyaniline, are flexible and highly designable, making them suitable for micro-nanorobots with complex shapes and multifunctional integration. The choice of material affects not only the electrical conductivity but also the stability and lifespan of the micro-nanorobots. To improve driving efficiency and stability, researchers often modify materials with nanostructures, such as introducing nanoparticle, nanowire, or nanotube structures, to increase surface area and active sites, thereby enhancing the electric field effect and chemical reactivity.

Asymmetric design is a common strategy that introduces asymmetric elements, such as shape asymmetry or uneven material distribution, into the structure of micro-nanorobots, creating an uneven electric field distribution and thus enabling directional movement. For instance, Janus particles are a typical example of asymmetric design, with one-half of the surface covered in conductive material and the other half in insulating material. Under the influence of an electric field, the conductive side generates electrophoretic forces while the insulating side remains force-free; this uneven force distribution drives the particle's movement. Multifunctional design involves integrating various functional modules onto the micro-nanorobots, such as sensors, drug carriers, and photo-sensitive elements, enabling collaborative work for multi-tasking. For example, micro-nanorobots with both electrophoretic and photo-responsive dual functions can be designed to achieve more flexible manipulation and applications.

The performance of micro-nanorobots depends not only on the materials and design but is also influenced by the external environment. In liquid media, micro-nanorobots are affected by fluid resistance, and fluid dynamics principles play a significant role in their movement. Under low Reynolds number conditions ($Re < 1$), the viscous forces dominate the motion of micro-nanorobots, while inertial forces are negligible. This means that they need to overcome the viscous resistance of the fluid to continue moving. Researchers optimize the shape and size of micro-nanorobots through fluid dynamics simulations and experimental validation to maximize propulsion efficiency and stability. For example, fish-shaped or spiral-shaped micro-nanorobots, due to their special geometric shapes, can generate more effective propulsion forces and higher motion stability in fluids.

11.2.3 Examples of Electrically Driven Micro/Nanorobots

Kim et al. (2014) reported on the development of an ultra-high-speed rotating nanoelectromechanical system (NEMS) device assembled from nanoscale building blocks. These nanomotors were constructed using a bottom-up assembly approach, with nanowires serving as rotors, nano-magnets as bearings, and four-pole microelectrodes as stators. The rotation angle, speed (exceeding 18,000 RPM), and chirality of these nanomotors can be controlled by an electric field. Researchers utilized analytical models to reveal the fundamental electrical, magnetic, and mechanical interactions within the nanomotor systems, which were highly consistent with experimental results, providing key insights for the design of metallic nanoelectromechanical systems. These nanomotors are capable of continuous rotation for 15 hours, over 240,000 cycles, and have been applied to control the release of biochemical substances, demonstrating the ability to precisely regulate the release rate of biochemical substances on nanoparticles through mechanical rotation. The innovation of this study, from concept, design, and actuation to application, is relevant to nanoelectromechanical systems, nanomedicine, microfluidics, and lab-on-a-chip architectures.

Yamamoto et al. (2019) reported on a novel type of micro-spiral motor that operates in a stationary direct current (DC) electrostatic field. Researchers discovered that in an oil phase containing ionic surfactants, non-spherical particles exhibit various types of regular motion, such as spinning and circular orbital movement. In this study, they observed a new type of periodic motion in micro-spirals in a specific direction, that is, helical rotation, which is dependent on the material of the spiral, whether metallic or non-metallic organic material. Experimental results showed that left-handed nickel spirals rotate clockwise when observed at the positive electrode, while organic spirals rotate in the opposite direction (counterclockwise) under the same electrode arrangement. Moreover, by changing the helicity (chirality) of the spiral, the direction of rotation can be altered. Notably, these micro-motors can maintain stable movement without any mechanical support, such as a rotating shaft or electronic switching devices. This new type of DC micro-motor may be applicable to mechanical and fluid devices as intelligent devices in micro-robots and microfluidics.

11.2.4 Magnetic-Driven Micro/Nanorobots

In the field of micro/nanorobot driving technology, magnetic driving is valued by the scientific community for its unique advantages. Compared with chemical-driven and light-driven, magnetic-driven technology provides a non-invasive and remotely controllable manipulation method. Under the action of magnetic force, micro/nanorobots can achieve precise navigation within biological bodies while avoiding thermal damage or chemical interference to surrounding tissues. The biocompatibility and safety of this driving method give it significant advantages in biomedical applications such as drug delivery, cell manipulation, and minimally invasive surgery.

11.2.4.1 Principle of Magnetic-Driven

The principles of magnetically driven micro/nanorobots include the selection and preparation of magnetic materials, the manipulation mechanism of the magnetic field, the realization of motion modes, and the design and optimization of the magnetic driving system.

Magnetic materials are a key component of magnetically driven micro/nanorobots. Commonly used magnetic materials include ferromagnetic materials (such as iron, cobalt, nickel, and their alloys) and superparamagnetic materials (such as iron oxide nanoparticles). These materials have high magnetization rates and good magnetic responsiveness, enabling micro/nanorobots to achieve efficient magnetization and movement under the action of an external magnetic field. Ferromagnetic materials can maintain high residual magnetism under an external magnetic field, suitable for applications requiring strong magnetic field driving; while superparamagnetic materials do not retain magnetism after the magnetic field is removed, suitable for safe applications in the biomedical field. By selecting the appropriate magnetic materials, the magnetic response performance and application effects of micro/nanorobots can be optimized.

Magnetic fields can be manipulated through various means, including static magnetic fields, rotating magnetic fields, and gradient magnetic fields, to control micro/nanorobots. Static magnetic fields can cause magnetic micro/nanorobots to move along the direction of the magnetic field; rotating magnetic fields can drive micro/nanorobots with specific geometric shapes to achieve rotation or spiral motion; gradient magnetic fields create spatially non-uniform magnetic field strengths, enabling micro/nanorobots to move along the direction of the magnetic field gradient. By precisely controlling the type, intensity, and direction of the magnetic field, a variety of motion modes of micro/nanorobots can be manipulated, such as linear motion, rotational motion, and three-dimensional trajectory motion.

The realization of motion modes is an important feature of magnetically driven micro/nanorobots. Under the influence of a magnetic field, micro/nanorobots can achieve various motion modes through different geometric designs and structural characteristics, such as helical structures (Ghosh et al. 2018; Patil et al. 2021), Janus microspheres (Alsaadawi et al. 2021). For example, micro/nanorobots with a rod-like structure can achieve helical propulsion under the action of a rotating magnetic field, similar to the movement of bacterial flagella; to break the symmetry in motion, the design of micro/nanorobots usually has an asymmetric shape, such as helical, fish-shaped, etc., or contains flexible components, such as a flexible tail. Micro/nanorobots with an asymmetric structure can achieve eccentric rotation under the action of a static magnetic field, generating a linear propulsive force; these designs help to achieve non-reciprocal motion, thereby improving motion efficiency. Moreover, by modifying the surface of micro/nanorobots with magnetic nanoparticles, their magnetic response performance can be enhanced, achieving complex three-dimensional motion trajectories. By optimizing the structural design and motion modes of micro/nanorobots, their motion efficiency and control accuracy can be improved.

The design and optimization of the magnetic driving system are key to achieving high-performance magnetically driven micro/nanorobots. A magnetic driving system typically includes magnetic micro/nanorobots, a magnetic field generation device, and a motion control system. The magnetic field generation device can use electromagnetic coils, permanent magnets, or rare earth magnets, etc., to generate a stable, uniform, or adjustable magnetic field to control micro/nanorobots. The motion control system can precisely adjust the magnetic field parameters, such as magnetic field strength, frequency, and direction, to achieve real-time control of the micro/nanorobot's movement. Optimizing the design of the magnetic driving system can improve the motion stability and control flexibility of micro/nanorobots, adapting to the needs of different application scenarios.

In addition, micro/nanorobots can achieve movement in the vicinity of surfaces or boundaries by interacting with the surface. This surface-assisted movement utilizes surface friction and fluid dynamics interactions to help the robot roll or slide on the surface, producing a net displacement. This mechanism is particularly effective in complex environments, such as tissue surfaces within biological bodies.

11.2.4.2 Examples of Magnetic-Driven Micro/Nanorobots

One of the most promising application areas for magnetically driven micro/nanorobots is the biomedical field. Mushtaq et al. (2019) developed a new type of nanorobot—motile piezoelectric nanoeels, which mimic the shape and function of electric eels, possessing flexible piezoelectric tails and magnetic heads. These nanoeels are composed of a piezoelectric tail made of polyvinylidene fluoride (PVDF)-based copolymer and polypyrrole (Ppy) nanowires, driven magnetically through a nickel ring. Under the application of a rotating magnetic field, the piezoelectric tail deforms, changing its polarization state, and utilizing the magnetoelectric coupling effect, it achieves the electrostatic enhancement of drug release on demand. Studies have shown that by adjusting the parameters of the magnetic field, not only can the translational movement of the nanoeels be controlled, but also the drug release kinetics can be controlled, demonstrating the ability to release drugs in a pulsed manner. By precisely adjusting the magnetic field parameters, an effective selection can be made between propulsion and drug release, providing new possibilities for the design of bio-inspired nanodevices for future biomedical applications.

Inspired by the spiral structure of bacterial flagella, researchers have developed various helical magnetic nanomaterials (Yan et al. 2015). These helical micro/nanomotors can obtain power through a continuously rotating magnetic field and achieve active movement in specific liquid environments. Zhang et al. (2009) were the first to use self-rolling technology in 2009, using Cr/Ni/Au metal films as magnetic heads and InGaAs/GaAs/Cr as spiral tails, to manufacture artificial bacterial flagella (ABFs) that can move and orient precisely in a rotating magnetic field. Wang et al. (2018) also designed a magnetic polymer helical micro-swimmer, which was coated with nickel to achieve magnetic driving, then coated with gold to prevent oxidation of the nickel and to provide a suitable surface for functionalization. The research team explored the motion characteristics of ABFs with different surface wettabilities in water and biological aqueous media under magnetic field modulation. In particular, ABFs were manipulated in a Y-shaped microchannel, demonstrating the ability to selectively control groups of ABFs under different magnetic field frequencies. Notably, ABF micro-motors can respond to the external magnetic field frequency, achieving active stirring by turning on and off multiple micro-motors.

To further improve the motion efficiency and controllability of micro/nanorobots, magnetic driving is often combined with other driving mechanisms. For example, magnetic fields can be combined with chemical driving, photo driving, or ultrasonic driving to form a hybrid power system. This combination can provide greater propulsion and more precise motion control to meet the needs of different application scenarios.

11.2.5 Ultrasonic-Driven Micro/Nanorobots

Ultrasonic waves are high-frequency sound waves with frequencies above the upper limit of human hearing (above 20 kHz), characterized by high penetration and non-invasive properties. Particularly, even at high frequencies, the harm of ultrasonic waves to biological systems is minimal, and their excellent compatibility and controllability with biological systems (Gao et al. 2018; Juanfeng et al. 2020; Peng et al. 2017) make them effectively utilized in various biomedical applications, such as targeted drug delivery, detoxification, and nano-surgery (Wang et al. 2012; Francis et al. 2020; Li et al. 2022).

11.2.5.1 Principle of Ultrasonic-Driven

The working principle of ultrasonic-driven micro/nanorobots mainly relies on the interaction between acoustic waves and micro/nanostructures. Ultrasonic waves are mechanical waves with frequencies above the audible range of human ears, characterized by high frequency and high energy density. When ultrasonic waves propagate through a liquid medium, they create alternating high-pressure and low-pressure areas, forming an acoustic pressure field. Micro/nanorobots, through the design of their geometric shape and material properties, can effectively interact with the acoustic pressure field, thereby generating propulsive force or torque to drive their movement. Specifically, ultrasonic waves can produce acoustic radiation force and acoustic streaming on the surface of micro/nanorobots through the acoustic pressure field. These forces, in conjunction with fluid dynamic effects, propel the micro/nanorobots along a predetermined trajectory (Li et al. 2021).

When ultrasonic waves propagate through a liquid, they create alternating high-pressure and low-pressure areas, forming an acoustic field. Micro/nanorobots in the acoustic field are subject to acoustic radiation force, causing compression in high-pressure areas and stretching in low-pressure areas, thereby generating a net propulsive force that moves the robot along or perpendicular to the direction of ultrasonic wave propagation. The magnitude of the acoustic radiation force depends on the frequency and intensity of the ultrasonic waves, as well as the material and shape of the micro/nanorobots.

As ultrasonic waves propagate through a liquid, they generate small fluid movements around the surface of the micro/nanorobots, known as acoustic streaming. The formation of acoustic streaming is mainly due to the viscous and thermal effects produced by the propagation of ultrasonic waves in the liquid medium. Micro/nanorobots, through interaction with acoustic streaming, gain propulsive force and torque, enabling complex motion and manipulation. The velocity and direction of the acoustic streaming depend on the frequency and intensity of the ultrasonic waves, the viscosity of the liquid, and the surface characteristics of the micro/nanorobots.

Commonly used materials include metals, polymers, and composite materials. Metallic materials such as gold and platinum have good acoustic transmission properties and mechanical strength but may have biocompatibility issues in some biological applications. Polymeric materials such as polydimethylsiloxane (PDMS) have good flexibility and designability, suitable for complex shapes and multifunctional integration of micro/nanorobots. Composite materials combine the advantages of metals and polymers, with excellent acoustic transmission properties and mechanical performance. By selecting appropriate materials and nano-structure modification, the acoustic response efficiency and motion performance of micro/nanorobots can be improved.

Microfluidic technology also plays an important role in ultrasonic-driven micro/nanorobots. Microfluidic technology manipulates fluids at a micro-scale to achieve precise control of the motion of micro/nanorobots. Under the action of ultrasonic waves, microfluidic systems can generate complex acoustic streaming fields, propelling micro/nanorobots along predetermined trajectories. Microfluidic technology can also be combined with optical and electrical technologies to achieve multimodal driving and detection, enhancing the functionality and application range of micro/nanorobots. For example, combined with optical imaging technology, the position and motion trajectory of micro/nanorobots can be monitored in real-time, achieving precise control; combined with electrical detection technology, the interaction between micro/nanorobots and the surrounding environment can be measured in real-time, improving the flexibility and reliability of operation.

11.2.5.2 Examples of Ultrasonic-Driven Micro/Nanorobots

Esteban-Fernández et al. (2016) developed an acoustically propelled nanorobot for efficient delivery of small interfering RNA (siRNA) into cells. The researchers utilized gold nanowires (AuNW) coated with rolling circle amplification (RCA) DNA chains, which anchor the siRNA therapeutic molecules. By driving the AuNW with ultrasonic (US) propulsion, rapid cellular uptake, and intracellular movement were achieved, accelerating the delivery and gene-silencing reaction of siRNA. By optimizing the effects of motion, time, and siRNA dosage, the study showed that treatment with US-propelled siRNA-AuNWs can achieve up to 94% gene silencing efficiency within minutes, and the silencing reaction is about 13 times higher compared to statically modified nanowires.

You et al. (2023) developed an ultrasonic-driven bionic micro/nanorobot for the treatment of cardiovascular and cerebrovascular diseases. These robots mimic biological structures found in nature, using ultrasonic waves to move quickly in blood vessels and achieve precise positioning.

11.2.6 Chemical Fuel-Driven Micro/Nanorobots

Chemically fueled micro/nanorobots represent an important research direction in the field of nanotechnology and micro-electromechanical systems. These types of micro/nanorobots can convert chemical energy into mechanical energy to perform specific tasks or movements based on their chemical environment. These self-driven micro/nanorobots primarily achieve propulsion through catalytic reactions, such as enzymatic reactions, hydrogen peroxide catalytic reactions, decomposing chemicals in liquids (such as hydrogen peroxide, glucose, etc.), and the chemical energy released is converted into kinetic energy to drive the robot's movement.

11.2.6.1 Principle of Chemical Fuel-Driven

The working mechanisms of these micro/nanorobots involve multiple disciplines, including chemical reaction dynamics, material science, and fluid mechanics. A deeper understanding of their fundamental principles requires examining the following aspects:

Chemical fuels that serve as the driving force include hydrogen peroxide (H_2O_2), glucose, urea, and acidic microenvironments. Among the various fuels reported for the autonomous movement of micro/nanorobots, H_2O_2 is the most commonly used (Zhang et al. 2021). The catalyst coated on the surface of the micro/nanorobots (typically metals such as platinum or silver) can accelerate the decomposition of hydrogen peroxide into water and oxygen (Liangxing et al. 2017). The generation of oxygen forms bubbles on the catalyst surface, and as these bubbles detach, the reaction force propels the micro/nanorobot forward, also producing thermal energy that can be harnessed for specific therapeutic or operational applications. Additionally, the kinetic movement of the oxygen and water molecules generated can disturb the surrounding liquid, further driving the robot's movement. This self-propulsion mechanism, achieved through chemical reactions, allows micro/nanorobots to move efficiently in liquid media.

Platinum is one of the most commonly used catalyst materials, known for its high catalytic efficiency and stability. Other metals such as silver, gold, and nickel are also widely studied as alternative catalysts. The choice of material impacts not only the catalytic efficiency but also the robot's lifespan and stability. To enhance catalytic efficiency and stability, researchers often modify the catalyst surface by introducing nano-particle, nanotube, or nanowire structures, increasing the specific surface area of the catalyst and thus improving the activity of the chemical reaction. Moreover, the surface properties of the material, such as hydrophilicity and hydrophobicity, significantly affect the progression of the chemical reaction and the generation of propulsive force. For instance, modifying the catalyst surface with hydrophobic molecules can enhance bubble formation and detachment, improving propulsion efficiency.

The structural design of chemically fueled micro/nanorobots is crucial for their efficient movement. Asymmetric design is a common strategy that introduces asymmetric elements, such as shape asymmetry or uneven material distribution, creating an asymmetric chemical reaction field. For example, Janus particles, a typical asymmetric design, have one-half of their surface covered with a catalyst and the other half with inert material. In a hydrogen peroxide solution, the catalyst side generates oxygen bubbles, while the inert side does not, causing the particle to move in the opposite direction of the catalyst. Furthermore, the design of micro/nanorobots can also include multifunctional modules, such as magnetic elements and photosensitive elements, enabling external manipulation and propulsion. For example, magnetic elements can be directed using an external magnetic field, while photosensitive elements can provide auxiliary propulsion and control through exposure to light.

Fluid dynamics principles are also essential in the movement of chemically fueled micro/nanorobots. The movement of micro/nanorobots in liquid media involves complex fluid dynamics phenomena, such as Stokes flow and boundary layer effects. Under low Reynolds number conditions ($Re < 1$), where viscous forces dominate the movement of micro/nanorobots and inertial forces are negligible, the propulsive force generated by the chemical reaction must overcome the viscous resistance of the fluid to enable sustained movement. Researchers optimize the shape and size of micro/nanorobots through fluid dynamics simulations and experimental validation to maximize propulsion efficiency and stability. For example, fish-shaped or spiral-shaped micro/nanorobots, due to their unique geometric shapes, can generate more effective propulsion forces and higher movement stability in fluids.

11.2.6.2 Examples of Chemical Fuel-Driven Micro/Nanorobots

Utilizing the tiny airflows and separation forces generated by bubble bursts, micro/nanorobots can achieve self-actuation. For instance, using hydrogen peroxide as a fuel, metal catalysts such as platinum (Pt) can promote its decomposition into oxygen and water. The oxygen bubbles produced provide propulsion for the micro/nanorobots. However, hydrogen peroxide, a common fuel, may have potential toxicity to biological organisms and is not easily adaptable to diverse tissue environments. Esteban-Fernández et al. (2017) designed a Janus-type magnesium-based micro/nanorobot that uses gastric acid as fuel to produce hydrogen gas bubbles, achieving self-actuation in the stomach for the treatment of gastric infections. Cai et al. (2023a) developed a rocket-style micro/nanorobot equipped with effervescent tablets, which can generate a large amount of carbon dioxide bubbles in water, propelling the robot to move autonomously and facilitating the penetration of the micro-needle tip through the intestinal mucosa to achieve in vivo drug release. Mathesh et al. (2021) designed a two-dimensional graphene oxide nanorobot that carries catalase enzyme, enabling self-actuation even at very low concentrations of hydrogen peroxide, providing a strategy to reduce the environmental medium dependence of intelligent micro/nanorobots.

In the study conducted by Wang et al. (2024), polysaccharide-based micro/nanorobots were developed, which utilize chemical fuel for propulsion, achieving the delivery of active food ingredients. The study showed that these micro/nanorobots can stably exist in the gastrointestinal tract environment and release the loaded nutrients through chemical reactions, significantly improving the bioavailability of dietary supplements. This research demonstrates the potential application of micro/nanorobots in food science, achieving efficient nutrient delivery by optimizing polysaccharide materials and reaction conditions.

11.2.7 Biohybrid Micro/Nanorobots

In addition to external physical field sources and chemical fuels as additional power sources, naturally occurring self-propelling organisms, such as bacteria (Li et al. 2019), sperm (Schwarz et al. 2020), and microalgae (Birgul Akolpoglu et al. 2020), are also used to develop intelligent micro/nanorobots. The natural mobility of these organisms endows intelligent micro/nanorobots with the ability to move autonomously, and by leveraging the taxis characteristics of bacteria, such as preferences for magnetism, chemical substances, light, or temperature, these micro/nanorobots can be guided to move directionally (Mestre et al. 2021), while also possessing the stress response to react to environmental changes. Moreover, the incorporation of synthetic components can endow biological hybrid micro-robots with unique characteristics, such as magnetic control, drug release, and targeted drug delivery.

11.2.7.1 Principle of Biohybrid Micro/Nanorobots

The driving mechanism of biological hybrid micro/nanorobots mainly relies on biomolecular motors. Biomolecular motors are a class of proteins that exist in nature, capable of converting chemical energy into mechanical energy for movement. Typical biomolecular motors include myosin, kinesin, and flagellar motors. Myosin generates energy through ATP hydrolysis, causing muscle fibers to contract and facilitating movement; kinesin moves along microtubules, providing energy through ATP hydrolysis for intracellular transport of materials. Flagellar motors propel bacteria forward through rotational movement. Biological hybrid micro/nanorobots utilize these biomolecular motors, fixed onto microstructures and supplied with the appropriate chemical fuel such as ATP, enabling self-propelled motion in liquid media. This drive mechanism based on biomolecular motors is efficient, precise, and controllable, capable of stable operation in complex biological environments.

Biomimicry is another common strategy, harnessing biological structures and motion mechanisms found in nature to replicate similar functions. For instance, micro/nanorobots designed with a structure mimicking bacterial flagella can achieve propulsion through rotational movement. Those designed to imitate muscle fiber structures can move by contracting and relaxing. Self-assembly technology is also an essential design strategy, where molecules spontaneously assemble into micro/nanorobots with specific structures and functions under certain conditions due to intermolecular forces such as electrostatic forces, hydrogen bonds, and hydrophobic interactions. By precisely controlling the self-assembly process, complex structures, and multifunctional integrated micro/nanorobots can be achieved.

11.2.7.2 Examples of Biohybrid Micro/Nanorobots

Bacteria, as the most diverse group in the biological world with numbers exceeding 10 to the 30th power, have been applied in the construction of biological hybrid-driven micro/nanorobots. For instance, Alapan et al. (2018) developed a novel type of bioengineered, mobile E. coli for active and controllable drug delivery. By using biotin-avidin-biotin binding technology, red blood cells loaded with superparamagnetic iron oxide nanoparticles and the anticancer drug doxorubicin were combined with bacteria to construct multifunctional micro-swimmers. The propulsion trajectories of these micro-swimmers demonstrated their active motion performance and drug payload capacity.

Algal species, which proliferate in aquatic environments and rely on photosynthesis for energy conversion, differ from traditional plants as they lack structures such as leaves, stems, and roots, as well as complex reproductive systems. Among the many types of algae, spirulina, with its unique three-dimensional helical structure, has garnered attention. Due to its ability to be cultivated on a large scale and its widespread use as a nutritional supplement in the biomedical field, Yan et al. (2017) designed a novel type of biohybrid magnetic robot by coating spirulina with magnetic Fe_3O_4 nanoparticles through a dip-coating process. These biohybrid magnetic robots, driven by a uniformly rotating magnetic field and observed through microscopy for their active motion trajectories, have shown strong navigation capabilities in various biological fluids. They also enable non-invasive tracking in superficial tissues or deep organs using spontaneous fluorescence/magnetic resonance imaging techniques. Overall, these biohybrid magnetic micro-robots exhibit ideal biodegradability, remote sensing and diagnostic capabilities, and excellent anticancer potential, presenting a new type of biohybrid micro-robot carrier with high potential for imaging-guided therapy.

Sperm, as a biological power source for hybrid-driven motors, has attracted significant attention for its exceptional motility (Magdanz et al. 2013). Schwarz et al. (2020) developed a cutting-edge sperm-hybrid micro-motor that not only possesses excellent biocompatibility but also a high drug payload capacity, making it highly suitable for targeted drug delivery. To gain a deeper understanding of the drug release mechanism of the sperm-hybrid micro-motor, the research team conducted detailed observation experiments using a microfluidic channel model. During the 8-hour experiment, the release pattern of sperm cells loaded with DOX-HCl was recorded. As the fluorescence signal gradually spread within the cellular spheroids, the fluorescence intensity of the drug-loaded sperm decreased, confirming the release of the sperm-hybrid micro-motor into the HeLa cell spheroids. The target HeLa cells exhibited significant cell body contraction after 8 h, further demonstrating the initial stage of apoptosis.

Biological hybrid-driven micro/nanorobots can also integrate various propulsion mechanisms such as chemical, ultrasonic, and magnetic forces, giving them potential advantages in complex physiological environments and making them suitable for various medical applications including drug delivery and tissue repair. However, the adaptability of biological hybrid methods in in vivo applications, apart from those involving the gastrointestinal tract, still needs to be improved. If using sperm, bacteria, or other microorganisms to construct nanomotors, their immunogenicity and immune reactivity must be considered, being cautious of the potential immune responses they may trigger in the host. Unless preventive measures are taken to eliminate antigenic determinants or pathogen-associated molecular patterns, these reactions could be quite intense. In terms of immune response and biocompatibility, using cell membranes from the patient's own red blood cells or platelets to manufacture biohybrid micro/nanorobots may be a more feasible option (Ye et al. 2024).

11.3 Applications of Intelligent Micro/Nanorobots in the Medical Field

With the relentless march of scientific progress, the integration of micro/nanorobots in the medical sector has moved from theoretical concept to tangible application. These robots, though tiny in stature, demonstrate significant potential in performing specific tasks within biological systems, heralding a new era of innovation in medical technology.

Their capabilities span the spectrum of healthcare, from targeted drug delivery to tumor treatment, wound healing, disease diagnostics, and the precision of minimally invasive surgeries. The development of micro-nanorobots has opened new avenues for research, gradually overcoming the limitations of traditional methods and offering enhanced precision and efficacy in treatment.

The contributions of these studies are dual-fold: they not only accelerate the evolution of micro/nanorobotic technology but also pave the way for a broadened

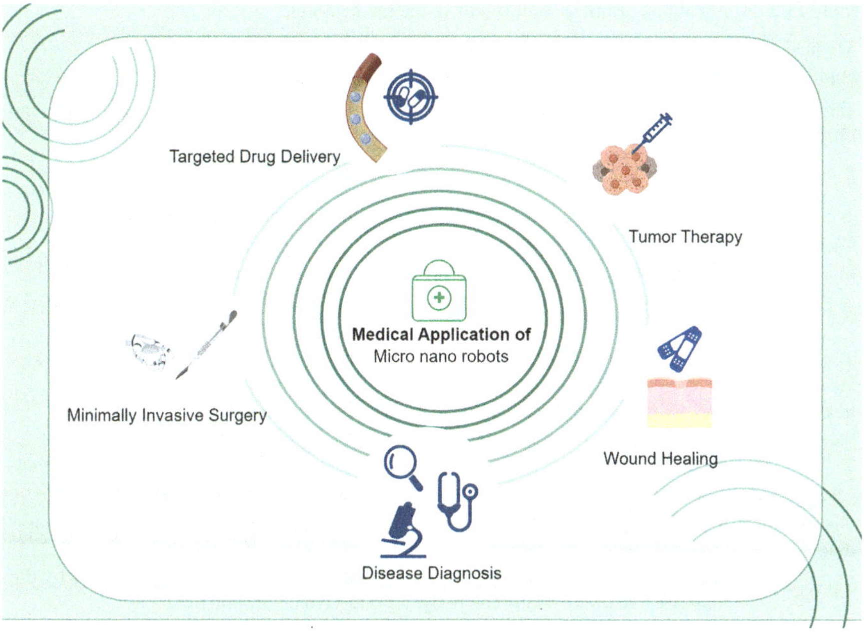

Fig. 11.3 Applications of micro/nanorobots in the medical field

spectrum of applications in biomedicine. As the field matures, the practicality and reliability of micro/nanorobots are expected to grow, bringing forth a future where their role in medical applications is both substantial and widespread.

Nanotechnology can be integrated with medicine in three main aspects:

1. The nano-ization of therapeutic drugs and the development of novel pharmaceutical formulations.
2. The detection technology of highly sensitive and precise biological nanostructures and properties, such as nanosensor systems for early disease diagnosis.
3. The combination with minimally invasive medical procedures for delicate therapeutic surgeries, such as intravascular nanorobotic surgeries (Fig. 11.3 and Table 11.2).

11.3.1 Targeted Drug Delivery

Targeted drug delivery systems are vital means of enhancing the efficacy and safety of drug treatments using micro/nanotechnology. Over the past few decades, the application of micro/nanorobots in targeted drug delivery has shown tremendous potential, especially in treating stubborn diseases such as cancer and cardiovascular diseases.

Table 11.2 Applications of micro/nanorobots in the medical field

Medical application	Advantages	Limitations
Targeted drug delivery	Micro/nanorobots can achieve precise in vivo targeting, delivering drugs directly to the lesion site, reducing side effects on healthy tissues, and improving therapeutic outcomes	The preparation process is complex, and the cost is high; Further research and verification are needed regarding biocompatibility and immune reactions within the body.
Tumor therapy	Micro/nanorobots capable of carrying and releasing anticancer drugs or thermotherapy materials can target tumor cells for treatment, reducing the side effects of traditional therapies	The complexity and diversity of the tumor microenvironment need to be overcome to ensure that micro robots can accurately reach and effectively target tumor sites
Wound healing therapy	Micro/nanorobots carrying growth factors or antimicrobial agents can promote wound healing, prevent infection, and increase the speed of healing	The potential issues of biological degradation and local inflammatory responses that may arise from the prolonged presence of micro/nanorobots in the body require further research
Disease diagnosis	By carrying diagnostic probes or contrast agents, micro/nanorobots are capable of reaching otherwise inaccessible areas within the body, providing high-resolution diagnostic information	The movement control and navigation precision within the body require optimization and may be affected by the in vivo environment
Minimally invasive surgery	Micro/nanorobots can perform delicate operations such as vascular recanalization and stone removal, reducing surgical trauma and improving surgical accuracy	Operations are complex, requiring high-precision control systems and imaging technology support, with high costs involved

These micro/nanorobots are typically made from biocompatible materials, with drug molecules either physically adsorbed, chemically bonded, or encapsulated within degradable nanocarriers. They can specifically target the affected area through the bloodstream or actively release the drug upon sensing the biochemical environmental characteristics, thereby improving the therapeutic efficiency and reducing toxic side effects on normal tissues. These robots, ranging from micrometers to nanometers in size, can penetrate biological barriers to deliver drugs precisely, enhancing efficacy and minimizing side effects. The main mechanisms include recognition and binding to specific biomarkers and responding to changes in the in vivo environment (such as alterations in pH, temperature, or enzyme activity) to release drugs.

A common strategy for targeted drug delivery is the use of surface modification techniques. By modifying antibodies, peptides, or other molecular recognition elements on the surface, micro/nanorobots can specifically recognize and bind to diseased tissues or tumor cells, achieving precise drug release. For instance, using antibody-drug conjugate (ADC) technology, antibodies can be linked to drugs, with the antibody portion responsible for recognizing tumor cells. Once bound, the drug is released at the target site, maximizing the killing of cancer cells while minimizing toxic side effects on normal cells. In this process, it is essential to ensure the stability and biocompatibility of micro/nanorobots, which involves not only the selection of materials but also the assessment of their biodegradability, immunogenicity, and long-term in vivo behavior. Additionally, optimizing their surface properties to reduce non-specific adsorption and enhance targeting capabilities is a current research focus.

Cai et al. (2021) developed a suction cup-inspired bubble-driven micro/nanorobot with a hemispherical bottom loaded with magnesium powder. This robot leverages the high volumetric ratio of porous hydrogel to efficiently carry drugs. In the gastric fluid environment, the robot achieves self-propulsion by generating hydrogen gas. With its unique suction cup shape, it can effectively adhere to the surface of gastric ulcers and release therapeutic drugs, showing significant efficacy in treating gastric ulcers.

Some naturally tropic bacteria can naturally locate tumor tissues, using the hypoxic conditions of solid tumors and specific molecular signals released by the tumor microenvironment, these bacteria can be guided to the tumor site. To enhance the targeting and lethal effect of these bacteria on tumors, Nie et al. (2023) used genetically engineered E. coli and modified them with ferromagnetic nanoparticles, developing a magnetic bio-micro/nanorobot with a signal processing function module. This robot, after colon-specific administration, can actively seek out and locate tumors, using alternating magnetic fields to cause local temperature increases for fluorescence imaging, and release immunotherapeutic drugs after bacterial lysis, thus achieving precise immunotherapy for tumors while avoiding the potential blood toxicity issues associated with traditional immunotherapeutic drugs.

The various biological barriers within living organisms, while playing an important role in preventing the invasion of foreign harmful substances, also pose obstacles to effective drug delivery. Therefore, intelligent micro/nanorobots have also been designed to penetrate these biological barriers to achieve efficient drug delivery (Wang et al. 2022a). For instance, the blood-brain barrier (BBB) prevents conventional drug molecules from entering the brain, which is a significant barrier to treating brain tumors and neurodegenerative diseases. Kim et al. (2022) constructed an ultrasound-controlled piezoelectric nanorobot that releases nitric oxide under the action of high-intensity focused ultrasound, capable of breaking through the blood-brain barrier to aggregate in the brain parenchyma, and outputting electricity under the action of ultrasound to achieve deep brain stimulation, stimulating dopaminergic neuronal cells to release dopamine, thereby alleviating the symptoms of neurodegenerative diseases.

Despite the high potential demonstrated by targeted drug delivery technology in theoretical and laboratory research, there are still many challenges in clinical application. These include but are not limited to:

1. The uncertainty and variability of biodistribution, how to ensure that drugs effectively reach and accumulate in target tissues.
2. Issues of biocompatibility and safety, with long-term in vivo studies still ongoing.
3. The technical difficulties of production costs and large-scale production, how to achieve large-scale production while ensuring quality.

11.3.2 Tumor Treatment

Tumor therapy is one of the most extensively applied and promising medical fields for micro/nanorobot technology. These miniature robots, harnessing their unique mechanical, chemical, and biological properties, can achieve targeted therapy, maximizing the destruction of cancer cells while minimizing damage to normal cells.

The role of micro/nanorobots in tumor therapy is primarily based on their ability to deliver therapeutic drugs precisely to cancer cells. This process typically involves three fundamental steps: identification, targeting, and drug release. Initially, micro/nanorobots are designed to recognize specific biological markers unique to tumors, such as specific proteins or receptors. These robots enhance their binding affinity to cancer cells by altering the chemical structure of their surface molecules.

Once positioned within tumor tissue, to achieve precise drug release, micro/nanorobots often integrate stimulus-responsive mechanisms into their design. These mechanisms can respond to external stimuli (such as pH, temperature, light, magnetic fields) or changes in internal biomarkers, enabling on-demand drug release. For example, some robots are designed to release drugs under specific pH levels or temperature conditions, or under the action of specific enzymes, conditions commonly found in the tumor microenvironment.

The design and synthesis of nanomaterials are the cornerstone of micro/nanorobot development. The selected nanomaterials must possess excellent biocompatibility and controllable degradation within biological bodies. Nanomaterials commonly used in research include metallic nanoparticles (such as gold, iron), polymeric nanoparticles (such as poly(lactic-co-glycolic acid) PLGA), and liposomes. These materials must ensure the safe transport and effective release of drugs and exhibit the expected responsive behavior under specific in vivo conditions.

Next is the development of payload encapsulation and release mechanisms. The encapsulation technology for micro/nanorobots requires the efficient loading of drugs, genes, or other therapeutic molecules into nanostructures while ensuring that the drug does not leak before reaching the lesion site. This involves the close integration of nanotechnology and pharmaceutical chemistry, addressing the balance between the physicochemical stability of the payload and its release kinetics. Common encapsulation techniques include chemical bonding, physical encapsulation, and covalent bonding.

The surface chemistry of nanocarriers significantly impacts their distribution within the body, cellular uptake, and ultimate therapeutic efficiency. Through surface modification, the targeting of nanocarriers can be enhanced, improving their specific binding ability to cancer cells while reducing toxicity to normal cells.

Zhong et al. (2020) have developed a type of micro/nanorobot that uses superparamagnetic iron oxide nanoparticles coated on microalgae as a living scaffold through a dipping process, "dressing" them in a magnetic coating. These magnetically engineered microalgae can be controlled by an external magnetic field to target delivery to tumor tissues. Through the photosynthesis of microalgae, oxygen is produced, improving the hypoxic microenvironment of tumors, and effectively achieving tumor diagnosis and treatment guided by magnetic resonance/fluorescence/photoacoustic multimodal medical imaging.

Cong et al. (2022) designed a magnetic propulsion motor that uses oncolytic adenovirus (OA)-infected 293 T cells, modified with a cyclic arginine-glycine-aspartic acid peptide (cRGD), and then asymmetrically fixed Fe_3O_4 nanoparticles on the cell surface to form 293 T-R-Fe@OA. This novel medical robot platform has demonstrated precise steering capabilities, capable of actively delivering OAs to bladder cancer cells through biorecognition of cRGD and strong propulsion by an external rotating magnetic field, thereby enhancing the penetration of viral therapy. This study combines rotating magnetic fields with oncolytic virus therapy to develop a medical robot with good biocompatibility, providing new possibilities for efficient viral treatment of bladder cancer and paving a new way for the application of magnetically propelled nano microspheres in the field of biomedicine.

Despite the great potential of micro/nanorobots in the field of tumor therapy, they still face many challenges in clinical applications. Among them, the uncertainty of biodistribution and biocompatibility are two major challenges. The movement and positioning of micro/nanorobots in the body need to be very precise to avoid damage to healthy normal tissues. In addition, potential immune responses and long-term biosafety issues of nanomaterials also need to be carefully evaluated in preclinical and clinical trials.

Furthermore, due to the heterogeneity of tumors and differences between patients in terms of biomarker expression levels, tumor microenvironment, genetic background, etc., the same micro/nano-treatment plan may show different therapeutic effects in different patients, that is, individual differences in treatment effects. This requires future research to focus on the development of personalized treatment strategies, by in-depth analysis of patients' pathological characteristics and molecular markers, to design personalized treatment plans. The regulation of the treatment process is also a key issue. How to precisely control the timing, dosage, and duration of drug release to adapt to the pathological state and treatment response of different patients is also an important part of achieving personalized treatment.

11.3.3 Wound Healing Treatment

Wound healing has long been a significant subject in the medical field. For instance, severe trauma or surgery may lead to excessive bleeding, posing a threat to patients' lives. Open wounds are susceptible to infection, which can delay the healing process and potentially lead to serious complications. Certain diseases, such as diabetes, can make chronic wounds difficult to heal. In the evolution of medical technology, the application of micro/nanorobots in wound healing has demonstrated innovation and potential, especially in dealing with complex wounds like chronic wounds and burns. These miniature robots can provide precise drug delivery, infection control, and promoting tissue regeneration, thereby accelerating the wound healing process.

The core mechanism relies on the robots' ability to precisely release drugs locally at the wound site. These sophisticated robots can carry a variety of therapeutic agents, such as growth factors that promote cell proliferation and differentiation, antimicrobial drugs, and other therapeutic compounds, supplying the necessary bioactive substances for wound healing.

In design, the release system of micro/nanorobots can be set for timed release or respond to specific biochemical signals unique to the wound area. For example, changes in pH levels or the activity of specific enzymes at the wound site are often closely related to infection, inflammation, or other pathological processes. Micro/nanorobots can sense these changes and release drugs at the appropriate time to effectively inhibit infection and promote tissue repair and regeneration at the wound site. Moreover, some innovative micro/nanorobot designs incorporate photosensitive materials, allowing the drug-release process to be controlled by external light sources. This light-triggered release mechanism offers greater flexibility and precision in treatment, allowing physicians to adjust treatment strategies in real time according to the specific conditions of wound healing, optimizing therapeutic outcomes.

Drug loading and release strategies are central to the realization of micro/nanorobot technology in wound healing therapy. Biomaterials such as biodegradable polymers, hydrogels, and nanofibers have attracted attention for their potential advantages in drug carrier functions and biocompatibility. Researchers need to precisely control the loading amount and release kinetics of drugs to achieve continuous and uniform drug coverage at the wound site. This often involves fine-tuning the chemical and physical properties of the drug carrier and optimizing the release mechanism, such as controlling drug release in response to biochemical signals like pH, temperature, or enzyme activity at the wound site.

He et al. (2016) developed a novel type of Janus micromotor that can be guided by thermophoresis, containing gold nano-painters for infrared laser-assisted tissue welding. These micromotors, composed of multilayered polyelectrolytes and superparamagnetic iron oxide nanoparticles, are coated with a gold layer to form a Janus structure, enabling them to generate heat and achieve autonomous movement under laser irradiation. Studies have shown that these micromotors are efficient in biological tissue welding, capable of successfully closing wounds in mice and confirming the mechanical recovery of the welded biological samples through mechanical property testing. Additionally, these micromotors exhibit healing characteristics comparable to medical suturing or new nanoparticle-based tissue adhesives in tissue sealing and could guide dyes to the wound site, suitable for the treatment of actively bleeding wounds. Histological analysis has demonstrated the biocompatibility of these micromotors, and the recovery of thermal damage caused by tissue welding within 9 days.

Bacteria, especially drug-resistant bacteria, and infections caused by wound contamination, pose a significant threat to human health. These infections can hinder the normal wound healing process, leading to prolonged healing times. The presence of bacteria can trigger an inflammatory response, thereby delaying wound healing and increasing the risk of infection spread. To address this issue, Liu et al.

(2023) developed a drug-free antimicrobial nanomotor for precise treatment of multidrug-resistant bacterial infections. This nanomotor is composed of heterogeneous nanostructures with specific physicochemical properties and controllable design, notably by growing copper-rich Cu7S4 nanocrystals at the high-curvature ends of gold nanorods (AuNRs), forming a Janus-like dual plasmonic (AuNR-SiO2-Cu7S4) antimicrobial nanomotor. These nanomotors are driven by near-infrared second-window light (NIR-II) and are guided by photoacoustic imaging (PAI) for synergistic photothermal and photocatalytic treatment of bacterial infections. Due to the strong plasmonic coupling and energy transfer between the Au NRs and Cu7S4 components, the photothermal performance and photocatalytic activity of the nanomotors are significantly enhanced. The motion behavior of the nanomotors facilitates penetration through the skin and enhances interaction with bacteria. More importantly, the directed navigation and synergistic antimicrobial activity of the nanomotors can be driven synchronously by NIR-II light. In a mouse model of abscess infection, these nanomotors with active movement and enhanced antimicrobial activity have shown good antibacterial effects.

Despite the significant potential demonstrated by micro/nanorobots in wound healing treatment, there are still multiple challenges in practical application. First, the complexity of the wound environment requires the robots to possess a high degree of adaptability and stability. The biochemical state at the wound site may undergo rapid changes, necessitating that drug-carrying robots work stably in these dynamically changing environments. Additionally, the biocompatibility and long-term safety of the robots require focused consideration, especially in the treatment of chronic wounds and burns.

11.3.4 Disease Diagnosis

With the rapid advancement of medical technology, the application of micro/nanorobots in the field of disease diagnosis is gaining increasing attention. These miniature robots, capable of highly precise operations, can identify specific pathologies at the molecular level, enhancing the accuracy and efficiency of early diagnosis.

The working principle of micro/nanorobots in disease diagnosis is typically based on biosensing technology. Designed to recognize and bind to specific biological markers such as proteins, DNA segments, or particular cell types, diagnostic micro/nanorobots can integrate a variety of sensors. These sensors react chemically or physically to the captured biological markers, generating quantifiable signals such as electrical, optical, or thermal signals. These signals are then analyzed to determine the pathological state.

For instance, micro/nanorobots using magnetic nanoparticles as carriers can enhance image contrast through Magnetic Resonance Imaging (MRI) technology, achieving precise localization and diagnosis of pathological tissues like tumors. Moreover, nanoparticles marked with fluorescence can be observed under a fluorescence microscope, making the visualization of pathological tissues more distinct.

Wang et al. (2022b) have developed magnetic nanorobots coated with multiple fluorescent dyes. By leveraging the fine control of an external magnetic field, these robots can form clusters of adjustable sizes, increasing the concentration of fluorescent dyes in specific areas and thereby improving the signal-to-noise ratio. Compared to traditional cell staining techniques, this method uses less dye, reduces toxicity to cells, and achieves high-precision controllable cell imaging.

Although micro/nanorobots show great potential in disease diagnosis, they also face many challenges. The complexity and diversity of biological markers pose the first challenge, as markers from different diseases may cross-react, necessitating that micro/nanorobots exhibit selectivity and specificity. Secondly, the complexity of the in vivo environment also presents challenges to the stability and functional realization of micro/nanorobots. Factors such as enzymatic degradation and immune rejection reactions can affect the accuracy of diagnostic results.

11.3.5 Minimally Invasive Surgical Procedures

Minimally invasive surgery, as a significant branch of modern surgery, represents a major leap in medical technology. Its core objective is to effectively treat diseases through the smallest possible incisions and trauma, thereby reducing patients' pain and recovery time, and enhancing the safety of surgery as well as the quality of life for patients. The introduction of micro-nanorobot technology has infused new vitality into the development of minimally invasive surgery. With their superior micromanipulation capabilities and precision, these robots are designed to perform complex surgical operations while minimizing damage to surrounding tissues, thereby optimizing surgical outcomes and shortening the recovery period for patients.

The implementation of micro-nanorobots in minimally invasive surgery involves multiple technical key points. First is the design and manufacturing of the robots, which require highly precise techniques to ensure their functionality and reliability. These robots must not only be miniaturized but also possess sufficient operational flexibility and stability to adapt to complex in vivo environments.

Secondly, the design of the control system is also crucial. This includes highly accurate navigation and real-time feedback systems to ensure that every movement of the robot is under strict control, reducing surgical risks. Moreover, the robot system must be compatible with existing medical equipment to facilitate integration into traditional surgical procedures. The compatibility of micro-nanorobot systems is also a key factor in the widespread application of these technologies. These systems need to be compatible with existing medical equipment to integrate into traditional surgical workflows, enabling effective technological integration and optimization of workflow processes. This may involve standardizing interfaces, unifying communication protocols, and coordinating operations with other devices in the operating room.

Particularly, magnetically controlled micro/nanorobots have demonstrated their unique guidance capabilities; they can be precisely positioned by adjusting external magnetic fields to reach specific surgical areas for diagnosis or treatment.

Zhiguang et al. (2018) developed a propeller-shaped magnetic micro/nanorobot, whose sub-micron dimensions perfectly match the pore structure of the vitreous body, allowing it to move autonomously under the guidance of an external magnetic field on smooth surfaces. This feature enables it to overcome the adhesive forces of biopolymers and reach the retinal area smoothly, providing an effective tool for ophthalmic minimally invasive surgery.

Lin et al. (2021) developed a flexible magnetic continuous robot (MCR) aimed at enhancing the flexibility and precision of retrograde intrarenal surgery (RIRS) by expanding the effective working space. By quantifying the impact of magnetic field non-uniformity on robot movement and combining it with heterogeneous field and Cosserat rod models, they established a kinematic model that successfully predicts the precise movement of the robot under complex magnetic fields and external constraints. Experimental results show that the effective working space of the robot can be significantly expanded, demonstrating good manipulability in simulated surgical environments, offering new possibilities for improving surgical efficiency and reducing patient pain.

Vyskočil et al. (2020) developed a magnetic micro/nanorobot capable of microsurgical operations on individual cells by sequentially depositing gold, silver, and nickel in polycarbonate membrane pores. This robot not only has high propulsion efficiency but also responds quickly to changes in the magnetic field. Under the precise control of a lateral rotating magnetic field, it can penetrate the cytoplasm of tumor cells and remove part of the cytoplasm while keeping the cell membrane intact, offering a new avenue for in-depth biological research.

Although the application of micro/nanorobots in minimally invasive surgery has great potential, it does face a series of challenges. First and foremost, the precision and reliability of robot operation are key technical issues. In minimally invasive surgery, extreme accuracy is required, as even minor operational errors can lead to unpredictable and severe consequences. Therefore, ensuring the precise control mechanism and robustness of the micro/nanorobots is crucial.

11.4 Future Prospects of Intelligent Micro/Nanorobots

11.4.1 Future Development Trends of Micro/Nanorobot Technology

The future development trends of intelligent micro/nanorobot technology can be explored from multiple aspects, including advancements in materials science, optimization of driving mechanisms, evolution of control technology, and the trend toward integration and multifunctionality. As scientific and technological progress continues, intelligent micro/nanorobots will evolve to be more efficient, intelligent, and multifunctional, providing advanced solutions for a variety of applications.

Materials science plays an essential role in the development of intelligent micro/nanorobots. Future research will focus more on the development and application of new materials, such as composite materials with high magnetic responsiveness and biocompatibility, and nanomaterials with photothermal effects and photoelectric conversion capabilities. These new materials will not only improve the driving efficiency of micro/nanorobots but also endow them with additional functionalities. For instance, the introduction of self-healing materials will allow micro/nanorobots to operate stably in complex environments over the long term, unaffected by minor damages. The multifunctionality of nanomaterials will also enable micro/nanorobots to perform multiple tasks in a single operation, such as drug delivery and lesion detection simultaneously.

Optimization of driving mechanisms is another significant trend in the development of intelligent micro/nanorobot technology. Current micro/nanorobots mainly rely on external driving mechanisms such as magnetic fields, light fields, electric fields, and chemical reactions. Future research will strive to enhance the efficiency and controllability of these driving mechanisms. For example, developing more efficient magnetic and electric driving systems will enable micro/nanorobots to achieve more efficient movement at lower power consumption. At the same time, combining various driving mechanisms, such as magneto-optical driving and magneto-chemical driving, will allow flexible switching of driving modes in different application scenarios, enhancing operational flexibility and adaptability. In addition, researchers will explore biological driving mechanisms, utilizing the motion of biomolecular motors or living cells to achieve autonomous driving of micro/nanorobots.

The development of control technology is key to achieving precise operation of intelligent micro/nanorobots. With the continuous advancement of artificial intelligence and machine learning technologies, the control systems of micro/nanorobots will become more intelligent and autonomous. By incorporating machine learning algorithms, micro/nanorobots can autonomously adjust their movement trajectories and operation modes according to environmental changes and task requirements, achieving more efficient task execution. For example, using deep learning algorithms for image recognition and path planning allows micro/nanorobots to navigate autonomously in complex biological environments and accurately reach target locations. Furthermore, developing real-time control systems based on sensor networks can achieve coordinated control of multiple micro/nanorobot groups, improving the efficiency and stability of swarm operations.

Integration and multifunctionality are important directions for the development of intelligent micro/nanorobot technology. Future micro/nanorobots will not only have a single function but will be able to integrate multiple functions, such as drug delivery, lesion detection, cell manipulation, and environmental monitoring. By integrating various sensors, actuators, and functional materials into the tiny structure of the robot, multiple tasks can be accomplished in one operation. For instance, in cancer treatment, micro/nanorobots can be designed to carry chemotherapeutic drugs, gene therapy vectors, immunostimulants, or photothermal therapeutic agents, achieving multifaceted and synergistic anti-tumor strategies through simultaneous or sequential release. They can also perform targeted drug delivery to tumors and monitor the temperature of the lesion area simultaneously, enhancing therapeutic effectiveness and safety. With modular design, different functional modules can be flexibly combined according to various application requirements, enabling the rapid realization of customized intelligent micro/nanorobots.

11.4.2 Future Application Potential of Micro/Nanorobots

Micro/nanotechnology is emerging as an innovative science of the twenty-first century, gradually transitioning from theoretical concepts to practical applications. Its interdisciplinary nature foretells a profound impact on a multitude of fields. Micro/nanorobots, as the epitome of micro/nanotechnology, are currently in a phase of rapid development.

Although micro/nanorobots are still in the research and development stage, their potential applications have garnered widespread attention. The multifunctionality of nanorobots, coupled with their high surface area-to-volume ratio, endows them with significant potential in areas such as chemical and biological sensors, medical devices, catalysts, optoelectronic materials, and nano components. Advances in material science have enabled the synthesis of various nanomaterials, including nanofilms, nanowires, nanotubes, nanobelts, nanoparticles, and nanoporous structures. These materials, with their unique physical and chemical properties, are poised to play crucial roles across diverse domains.

The prospects for the industrialization and marketization of nanorobot technology are equally promising. As technology matures and costs decrease, nanorobots are expected to find broad applications in various fields, including healthcare, environmental monitoring.

In the biomedical field, they can be utilized for targeted drug delivery, minimally invasive surgery, disease diagnosis, and biosensing. Micro/nanorobots, capable of carrying drugs, can navigate precisely to the site of disease, releasing their payload under external control to enhance therapeutic efficacy and minimize side effects. In minimally invasive surgeries, these robots can access tiny spaces within the human body to perform complex surgical operations, such as clearing blood clots, repairing tissues, and transplanting cells, thereby reducing surgical trauma and recovery time. Integrated with highly sensitive biosensors, micro/nanorobots can also monitor physiological parameters and pathological changes in real time, offering the possibility of early diagnosis and personalized treatment. They may even provide new solutions for currently incurable diseases, thereby significantly improving the quality of life and life expectancy for humans.

In the realm of environmental protection, micro/nanorobots can be employed for water purification, pollutant detection, and environmental monitoring. By designing micro/nanorobots with efficient catalytic capabilities, the degradation and removal of harmful pollutants in water can be achieved, leading to the purification of water resources. These robots can also be used for the detection of heavy metal ions and organic pollutants in the environment, enabling real-time monitoring of environmental quality through sensitive sensors. Deployed across various settings, micro/nanorobots can continuously monitor environmental changes, promptly detecting and warning of pollution issues, thus offering effective technical means for environmental conservation.

The future development trends of intelligent micro/nanorobot technology are primarily focused on the advancement of materials science, optimization of driving mechanisms, evolution of control technologies, and the trends toward integration and multifunctionality. These trends are propelling the technology of micro/nanorobots toward a direction of greater efficiency, intelligence, and versatility. In terms of application potential, micro/nanorobots are poised to play significant roles in fields such as biomedicine, environmental protection, and precision manufacturing, bringing revolutionary changes to these areas.

However, the research and application of nanorobots are not without challenges. The complexity of the technology, safety issues, and ethical considerations are all obstacles that need to be addressed. In the short term, it may be difficult for nanorobots to achieve large-scale applications. But in the long run, through sustained investment in scientific research and technological innovation, the potential of nanorobots will gradually be unearthed and realized.

To ensure the healthy and sustainable development of nanorobot technology, interdisciplinary collaboration should be emphasized during the research and development process, along with an enhanced assessment and management of potential risks. By continuous experimentation and exploration, technical challenges can be gradually overcome, advancing the technology of nanorobots and ultimately achieving its widespread application in various fields, leading to revolutionary changes for human society.

Acknowledgments This study was financially supported by grants from the National Natural Science Foundation of China (Grant No. 32071322), the National Natural Science Funds for Excellent Young Scholars (Grant No. 32122044), and the Technology & Innovation Commission of Shenzhen Municipality (Grant No. JCYJ20210324113004010).

References

Alapan Y, Alapan Y, Yaşa Ö, Yasa O, Schauer O, Schauer O, Giltinan J, Giltinan J, Tabak AF, Tabak AF, Sourjik V, Sourjik V, Sitti M, Sitti M (2018) Soft erythrocyte-based bacterial microswimmers for cargo delivery. Sci Robotics 3(17). https://doi.org/10.1126/scirobotics.aar4423

Alsaadawi Y, Alsaadawi Y, Eichler-Volf A, Eichler-Volf A, Heigl M, Heigl M, Zahn P, Zahn P, Zahn P, Albrecht M, Albrecht M, Erbe A, Erbe A (2021) Control over self-assembled janus clusters by the strength of magnetic field in H_2O_2. Eu Phys J E 44(2):23. https://doi.org/10.1140/epje/s10189-021-00010-3

Birgul Akolpoglu M, Akolpoglu MB, Dogan NO, Dogan NO, Bozuyuk U, Bozuyuk U, Ceylan H, Ceylan H, Kızılel S, Kizilel S, Sitti M, Sitti M (2020) High-yield production of biohybrid microalgae for on-demand cargo delivery. Adv Sci 7(16):2001256. https://doi.org/10.1002/advs.202001256

Cai L, Cai L, Zhao C, Zhao C, Chen H, Chen H, Liu F, Fan L, Zhao Y, Zhao Y, Qian X, Qian X, Chai R, Chai R (2021) Suction-cup-inspired adhesive micromotors for drug delivery. Adv Sci 9:2103384. https://doi.org/10.1002/advs.202103384

Cai L, Chen G, Sun L, Miao S, Shang L, Sun L, Zhao Y (2023a) Rocket-inspired effervescent motors for oral macromolecule delivery. Adv Mater 35(33). https://doi.org/10.1002/adma.202210679

Cai L, Xu D, Zhang Z, Li N, Zhao Y (2023b) Tailoring functional micromotors for sensing. Research 6:6. https://doi.org/10.34133/research.0044

Chung H-J, Chung H-J, Parsons AM, Parsons AM, Zheng L, Zheng L (2020) Magnetically controlled soft robotics utilizing elastomers and gels in actuation: a review. Adv Intell Syst 3(3):2000186. https://doi.org/10.1002/aisy.202000186

Cong Z, Cong Z, Cong Z, Tang S, Tang S, Tang S, Xie L, Xie L, Xie L, Yang M, Yang M, Yang M, Li Y, Li Y, Li Y, Lu D, Lu D, Lu D, Li J, Li J, Li J, Yang Q, Yang Q, Yang Q, Chen Q, Chen Q, Chen Q, Zhang Z, Zhang Z, Zhang Z, Zhang X, Zhang X, Zhang X, Wu S, Wu S (2022) Magnetic-powered janus cell robots loaded with oncolytic adenovirus for active and targeted virotherapy of bladder cancer. Adv Mater:2201042. https://doi.org/10.1002/adma.202201042

Ebrahimi N, Ebrahimi N, Bi C, Bi C, Cappelleri DJ, Cappelleri DJ, Ciuti G, Ciuti G, Conn AT, Conn AT, Faivre D, Faivre D, Faivre D, Habibi N, Habibi N, Hošovský A, Hošovský A, Iacovacci V, Iacovacci V, Khalil ISM, Khalil ISM, Magdanz V, Magdanz V, Misra S, Misra S, Pawashe C, Pawashe C, Rashidifar R, Rashidifar R, Soto-Rodriguez PED, Soto-Rodriguez PED, Fekete Z, Fekete Z, Jafari A, Jafari A (2020) Magnetic actuation methods in bio/soft robotics. Adv Funct Mater 31(11):2005137. https://doi.org/10.1002/adfm.202005137

Esteban-Fernández B, de Ávila BE-F, Angell C, Angell C, Soto F, Soto F, Lopez-Ramirez MA, Lopez-Ramirez MA, Báez DF, Báez DF, Xie S, Xie S, Wang J, Wang J, Chen Y, Chen Y, Chen Y (2016) Acoustically propelled nanomotors for intracellular siRNA delivery. ACS Nano 10(5):4997–5005. https://doi.org/10.1021/acsnano.6b01415

Esteban-Fernández B, de Ávila BE-F, Angsantikul P, Angsantikul P, Li J, Li J, Lopez-Ramirez MA, Lopez-Ramirez MA, Ramírez-Herrera DE, Ramírez-Herrera DE, Thamphiwatana S, Thamphiwatana S, Chen C, Chen C, de Moura Delezuk JA, Delezuk J, de Moura Delezuk JA, Samakapiruk R, Samakapiruk R, Ramez V, Ramez V, Obonyo M, Zhang L, Obonyo M, Zhang L, Wang J, Wang J (2017) Micromotor-enabled active drug delivery for in vivo treatment of stomach infection. Nat Commun 8(1):272–272. https://doi.org/10.1038/s41467-017-00309-w

Ferreira VRA, Azenha MA (2024) Recent advances in light-driven semiconductor-based micro/nanomotors: optimization strategies and emerging applications. Molecules 29. https://doi.org/10.3390/molecules29051154

Francis J, Francis J, Michael PA, Michael PA, Michael PA (2020) Investigation of micro/nano motors based on renewable energy sources. Mater Today Proceedings 27:150–157. https://doi.org/10.1016/j.matpr.2019.09.138

Gao C, Gao C, Lin Z, Lin Z, Lin X, Lin X, He Q, He Q (2018) Cell membrane–camouflaged colloid motors for biomedical applications. Adv Ther 1(5):1800056. https://doi.org/10.1002/adtp.201800056

Ghosh S, Ghosh S, Ghosh S, Ghosh S, Ghosh A, Ghosh A (2018) Mobile nanotweezers for active colloidal manipulation. Sci Robotics 3(14). https://doi.org/10.1126/scirobotics.aaq0076

He W, He W, Frueh J, Frueh J, Hu N, Hu N, Liu L, Liu L, Gai M, Gai M, Gai M, He Q, He Q (2016) Guidable Thermophoretic Janus micromotors containing gold nanocolorifiers for infrared laser assisted tissue welding. Adv Sci 3(12):1600206–1600206. https://doi.org/10.1002/advs.201600206

Juanfeng O, Ou J, Juanfeng O, Liu K, Liu K, Liu K, Liu K, Jiang J, Jiang J, Wilson DA, Wilson DA, Liu L, Liu L, Wang F, Wang F, Wang S, Wang S, Wang S, Tu Y, Tu Y, Peng F, Peng F (2020) Micro-/nanomotors toward biomedical applications: the recent progress in biocompatibility. Small 16(27):1906184. https://doi.org/10.1002/smll.201906184

Kim K, Kim K, Xu X, Xu X, Guo J, Guo J, Fan D, Fan D (2014) Ultrahigh-speed rotating nanoelectromechanical system devices assembled from nanoscale building blocks. Nat Commun 5(1):3632–3632. https://doi.org/10.1038/ncomms4632

Kim T, Kim T, Kim HJ, Kim HJ, Choi W, Choi W, Lee YM, Lee YM, Pyo JH, Pyo JH, Lee J, Lee J, Kim J, Kim C, Kim J, Kim J, Kim J-H, Kim J-H, San Kim L, Kim WJ, Kim WJ (2022) Deep brain stimulation by blood-brain-barrier-crossing piezoelectric nanoparticles generating current and nitric oxide under focused ultrasound. Nature Biomed Eng 7:149. https://doi.org/10.1038/s41551-022-00965-4

Li Q, Li Q, Chen H, Chen H, Feng X-Q, Feng X, Yu C, Yu C, Feng F, Feng F, Chai Y, Chai Y, Pan L, Lu P, Song T, Song T, Wang X, Wang X, Yao L, Yao L (2019) Nanoparticle-regulated semi-artificial magnetotactic bacteria with tunable magnetic moment and magnetic sensitivity. Small 15(15):1900427. https://doi.org/10.1002/smll.201900427

Li J, Li J, Mayorga-Martínez CC, Mayorga-Martinez CC, Ohl C-D, Ohl C-D, Pumera M, Pumera M (2021) Ultrasonically propelled micro- and nanorobots. Adv Funct Mater 32:2102265. https://doi.org/10.1002/adfm.202102265

Li H, Li H, Peng F, Peng F, Yan X, Yan X, Mao C, Mao C, Ma X, Ma X, Wilson DA, Wilson DA, He Q, He Q, Tu Y, Tu Y (2022) Medical micro- and nanomotors in the body. Acta Pharm Sin B 13:517. https://doi.org/10.1016/j.apsb.2022.10.010

Li X, Hao Q, Luan Y, Wang D, Du X (2024) The controllable fabrication of asymmetric yolk-Shell ZIF-67@PDA as NIR light-driven Nanomotor for enhanced selective adsorption. Appl Mater Today 38:102220. https://doi.org/10.1016/j.apmt.2024.102220

Liang S, Liang S, Sun J, Sun J, Zhang C, Zhang C, Zhu Z, Zhu Z, Dai Y, Dai Y, Gan C, Gan C, Cai J, Cai J, Chen H, Chen H, Feng L, Feng L (2022) Parallel manipulation and flexible assembly of micro-spiral via optoelectronic tweezers. Front Bioeng Biotechnol 10:868821–868821. https://doi.org/10.3389/fbioe.2022.868821

Liangxing H, Hu L, Miao J, Miao J, Grüber G, Grüber G (2017) Temperature effects on disk-like gold-nickel-platinum nanoswimmer's propulsion fuelled by hydrogen peroxide. Sensors Actuators B-Chem 239:586–596. https://doi.org/10.1016/j.snb.2016.08.028

Lin D, Lin D, Wang X, Wang J, Wang J, Jiao N, Jiao N, Wang Z, Wang Z, Liu L, Liu L (2021) A flexible magnetically controlled continuum robot steering in the enlarged effective workspace with constraints for retrograde intrarenal surgery. Adv Intell Syst 3:2000211. https://doi.org/10.1002/aisy.202000211

Liu L, Li S, Yang K, Chen Z, Li Q, Zheng L, Wu Z, Zhang X, Su L, Wu Y, Song J (2023) Drug-free antimicrobial nanomotor for precise treatment of multidrug-resistant bacterial infections. Nano Letters (Print) 23(9):3929–3938. https://doi.org/10.1021/acs.nanolett.3c00632

Ma X, Liang X, Li Y, Feng Q, Cheng K, Ma N, Zhu F, Guo X, Yue Y, Liu G, Zhang T, Liang J, Liu R, Zhao X, Nie G (2023) Modular-designed engineered bacteria for precision tumor immunotherapy via spatiotemporal manipulation by magnetic field. Nature Commun 14(1). https://doi.org/10.1038/s41467-023-37225-1

Magdanz V, Magdanz V, Sánchez S, Sanchez S, Schmidt OG, Schmidt OG (2013) Development of a sperm-flagella driven micro-bio-robot. Adv Mater 25(45):6581–6588. https://doi.org/10.1002/adma.201302544

Mathesh M, Mathesh M, Bhattarai E, Bhattarai E, Yang W, Yang W (2021) 2D active Nanobots based on soft Nanoarchitectonics powered by an ultralow fuel concentration. Angew Chem 61. https://doi.org/10.1002/anie.202113801

Mestre R, Mestre R, Patiño T, Patiño T, Patiño T, Patiño T, Patiño T, Sánchez S, Sanchez S, Sánchez S (2021) Biohybrid robotics: from the nanoscale to the macroscale. Wiley Interdiscip Rev Nanomed Nanobiotechnol 13(5). https://doi.org/10.1002/wnan.1703

Mushtaq F, Mushtaq F, Torlakcik H, Torlakcik H, Hoop M, Hoop M, Jang B, Jang B, Carlson F, Carlson F, Grunow T, Grunow T, Läubli NF, Läubli NF, Ferreira A, Ferreira A, Ferreira A, Chen X, Chen X-Z, Chen X-Z, Nelson BJ, Nelson BJ, Pané S, Pané S (2019) Motile piezoelectric nanoeels for targeted drug delivery. Adv Funct Mater 29(12):1808135. https://doi.org/10.1002/adfm.201808135

Patil G, Patil G, Ghosh A, Ghosh A (2021) Anomalous behavior of highly active helical swimmers. Front Phys 8:656. https://doi.org/10.3389/fphy.2020.628276

Peng F, Peng F, Peng F, Tu Y, Tu Y, Wilson DA, Wilson DA (2017) Micro/nanomotors towards in vivo application: cell, tissue and biofluid. Chem Soc Rev 46(17):5289–5310. https://doi.org/10.1039/c6cs00885b

Schwarz L, Schwarz L, Medina-Sánchez M, Medina-Sánchez M, Schmidt OG, Schmidt OG (2020) Sperm-hybrid micromotors: on-board assistance for nature's bustling swimmers. Reproduction 1(2). https://doi.org/10.1530/rep-19-0096

Simmchen J, Simmchen J, Katuri J, Katuri J, Uspal WE, Uspal WE, Popescu MN, Popescu MN, Tasinkevych M, Tasinkevych M, Sánchez S, Sanchez S (2016) Topographical pathways guide chemical microswimmers. Nature Commun 7(1):10598–10598. https://doi.org/10.1038/ncomms10598

Vyskočil J, Vyskočil J, Mayorga-Martínez CC, Mayorga-Martinez CC, Jablonská E, Jablonská E, Novotný F, Novotný F, Ruml T, Ruml T, Pumera M, Pumera M (2020) Cancer cells microsurgery via asymmetric bent surface au/ag/Ni microrobotic scalpels through a transversal rotating magnetic field. ACS Nano 14(7):8247–8256. https://doi.org/10.1021/acsnano.0c01705

Wan M, Wan M, Wang Q, Wang Q, Wang R, Wang R, Wang R, Wu R, Wu R, Li T, Li T, Fang D, Fang D, Huang Y, Huang Y, Yu Y, Yu Y, Fang L, Fang L, Wang X, Wang X, Zhang Y, Zhang Y, Miao Z, Miao Z, Zhao B, Zhao B, Zhao B, Wang F, Wang F, Mao C, Mao C, Jiang Q, Jiang Q, Xu X, Xu X, Shi D, Shi D (2020) Platelet-derived porous nanomotor for thrombus therapy. Sci Adv 6(22). https://doi.org/10.1126/sciadv.aaz9014

Wang W-J, Wang W, Wang W, Castro LA, Castro LA, Hoyos M, Hoyos M, Hoyos M, Hoyos M, Mallouk TE, Mallouk TE (2012) Autonomous motion of metallic microrods propelled by ultrasound. ACS Nano 6(7):6122–6132. https://doi.org/10.1021/nn301312z

Wang X, Wang X, Hu C, Hu C, Schurz L, Schurz L, De Marco C, De Marco C, Chen X, Chen X-Z, Chen X-Z, Pané S, Pané S, Nelson BJ, Nelson BJ (2018) Surface-chemistry-mediated control of individual magnetic helical microswimmers in a swarm. ACS Nano 12(6):6210–6217. https://doi.org/10.1021/acsnano.8b02907

Wang L, Wang L, Shi Y, Shi Y, Jiang J, Jiang J, Li C, Li C, Zhang H, Zhang H, Zhang X, Zhang X, Jiang T, Jiang T, Wang L, Wang Y, Wang Y, Feng L, Feng L (2022a) Micro-nanocarriers based drug delivery technology for blood-brain barrier crossing and brain tumor targeting therapy. Small 18:2203678–2203678. https://doi.org/10.1002/smll.202203678

Wang X, Wang X, Wang T, Wang T, Chen X, Chen X, Law JT, Law J, Shan G, Shan G, Tang W, Tang W, Gong Z, Gong Z, Pan P, Pan P, Liu X, Liu X, Yu J, Yu J, Ru C, Ru C, Huang X, Huang X, Sun Y, Sun Y (2022b) Microrobotic swarms for intracellular measurement with enhanced signal-to-noise ratio. ACS Nano. https://doi.org/10.1021/acsnano.2c02938

Wang X, Lin S, Zhang S, Yan Z, Liu W, Li F, Zhang S (2024) Polysaccharide-based micro/nanomotors for active ingredient delivery in food. ACS Appl Mater Interfaces 16:27668. https://doi.org/10.1021/acsami.4c04522

Yamamoto D, Yamamoto D, Kosugi K, Kosugi K, Hiramatsu K, Hiramatsu K, Zhang W, Zhang W, Shioi A, Shioi A, Kamata K, Kamata K, Iyoda T, Iyoda T, Yoshikawa K, Yoshikawa K (2019) Helical micromotor operating under stationary DC electrostatic field. J Chem Phys 150(1):014901. https://doi.org/10.1063/1.5055830

Yan X, Yan X, Zhou Q, Zhou Q, Yu J, Yu J, Xu T, Xu T, Deng Y, Deng Y, Tang T, Tang T, Qian F, Feng Q, Bian L, Bian L, Zhang Z, Zhang Y, Zhang Y, Ferreira A, Ferreira A, Zhang L, Zhang L (2015) Magnetite nanostructured porous hollow helical microswimmers for targeted delivery. Adv Funct Mater 25(33):5333–5342. https://doi.org/10.1002/adfm.201502248

Yan X, Yan X, Zhou Q, Zhou Q, Vincent M, Vincent M, Deng Y, Deng Y, Yu J, Yu J, Xu J, Xu J, Xu T, Xu T, Tang T, Tang T, Bian L, Bian L, Wáng YXJ, Wang Y-XJ, Kostarelos K, Kostarelos K, Zhang L, Zhang L (2017) Multifunctional biohybrid magnetite microrobots for imaging-guided therapy. Sci Robotics 2(12). https://doi.org/10.1126/scirobotics.aaq1155

Ye J, Fan Y, Niu G, Zhou B, Kang Y, Ji X (2024) Intelligent micro/nanomotors: fabrication, propulsion, and biomedical applications. Nano Today 55:102212. https://doi.org/10.1016/j.nantod.2024.102212

You Q, Shao X, Wang J, Chen X (2023) Progress on physical field-regulated micro/nanomotors for cardiovascular and cerebrovascular disease treatment. Small Methods 7(10). https://doi.org/10.1002/smtd.202300426

Zhan X, Zhan X, Wang J, Wang J, Xiong Z, Xiong Z, Zhang X, Zhang X, Zhang X, Zhang Y, Zhou Y, Zhou Y, Zhou Y, Zhou Y, Zheng J, Zheng J, Chen J, Chen J, Chen J, Chen J, Feng S-P, Feng S-P, Tang J, Tang J, Tang J (2019) Enhanced ion tolerance of electrokinetic locomotion in polyelectrolyte-coated microswimmer. Nat Commun 10(1):1–9. https://doi.org/10.1038/s41467-019-11907-1

Zhang L, Zhang L, Abbott JJ, Abbott JJ, Dong L, Dong L, Peyer KE, Peyer KE, Kratochvil BE, Kratochvil BE, Kratochvil BE, Zhang H, Zhang H, Bergeles C, Bergeles C, Nelson BJ, Nelson BJ (2009) Characterizing the swimming properties of artificial bacterial flagella. Nano Lett 9(10):3663–3667. https://doi.org/10.1021/nl901869j

Zhang Y, Zhang Y, Zhang Y, Hess H, Hess H (2021) Chemically-powered swimming and diffusion in the microscopic world. Nature Rev Chem 5(7):500–510. https://doi.org/10.1038/s41570-021-00281-6

Zhao Y, Yuan M, Yang H, Li J, Ying Y, Li J, Wang W, Wang S (2023) Versatile multi-wavelength light-responsive metal-organic frameworks micromotor through porphyrin metalation for water sterilization. Small 20:e2305189. https://doi.org/10.1002/smll.202305189

Zhiguang W, Wu Z, Troll J, Jeong H-H, Jeong H-H, Wei Q, Wei Q, Wei Q, Stang M, Stang M, Ziemssen F, Ziemssen F, Wang Z, Wang Z, Dong M, Dong M, Schnichels S, Schnichels S, Qiu T, Qiu T, Fischer P, Fischer P (2018) A swarm of slippery micropropellers penetrates the vitreous body of the eye. Sci Adv 4(11):1–10. https://doi.org/10.1126/sciadv.aat4388

Zhong D, Zhong D, Li W, Li W, Qi Y, Qi Y, He J, He J, Zhou M, Zhou M, Zhou M, Zhou M (2020) Photosynthetic biohybrid Nanoswimmers system to alleviate tumor hypoxia for FL/PA/MR imaging-guided enhanced radio-photodynamic synergetic therapy. Adv Funct Mater 30(17):1910395. https://doi.org/10.1002/adfm.201910395

Chapter 12
Future of Intelligent Biomaterials

Yandai Lin and Zhe Liu

Abstract Intelligent biomaterials including shape memory materials, self-powered biomaterials, stimuli-responsive biomaterials, interactive biomaterials, and micro/nanorobots are expected to solve key scientific and clinical problems in disease diagnosis and treatment, tissue engineering, bioimaging, wearable biosensors, and so forth. In this chapter, the situation and potentials of intelligent biomaterials for clinical translations are discussed, and the main bottleneck problems are also highlighted. The trends of future intelligent biomaterials are outlooked, including the emerging 3D/4D-printing intelligent polymers, organoid biomaterials, and flexible devices. It is noteworthy that researchers have designed next-generation intelligent biomaterials to meet the huge and increasing need for clinical theranostic solutions. Among them, the integration of intelligent biomaterials with artificial intelligence will be a very important and promising part to provide more efficient and convenient means for discovery, development, and ultimate medical translation of innovative intelligent biomaterials.

Keywords Intelligent biomaterials · Clinical translations · Artificial intelligence · Flexible biosensors

---

Y. Lin
Academy of Medical Engineering and Translational Medicine, Tianjin University, Tianjin, China

Tianjin Key Laboratory of Brain Science and Neural Engineering, Tianjin University, Tianjin, China

Z. Liu (✉)
Academy of Medical Engineering and Translational Medicine, Tianjin University, Tianjin, China

Tianjin Key Laboratory of Brain Science and Neural Engineering, Tianjin University, Tianjin, China

Division of Intelligent Biomaterials, Department of Medical Sciences and Engineering, Medical School, Tianjin University, Tianjin, China
e-mail: zheliu@tju.edu.cn

Z. Liu (ed.), *Intelligent Biomaterials*, https://doi.org/10.1007/978-981-96-8646-9_12

12.1 Current Intelligent Biomaterials and Clinical Translations

12.1.1 Current Intelligent Biomaterials

Intelligent biomaterials include shape memory materials, self-powered biomaterials, stimuli-responsive biomaterials, interactive biomaterials, biomimetic materials, micro/nanorobots, etc. which have been widely used in the diagnosis and treatment of diseases, tissue engineering, bioimaging, biosensing, and nerve engineering (Fig. 12.1) (Wu et al. 2024; Wang and Zhang 2024; Zhang et al. 2018). Up to date,

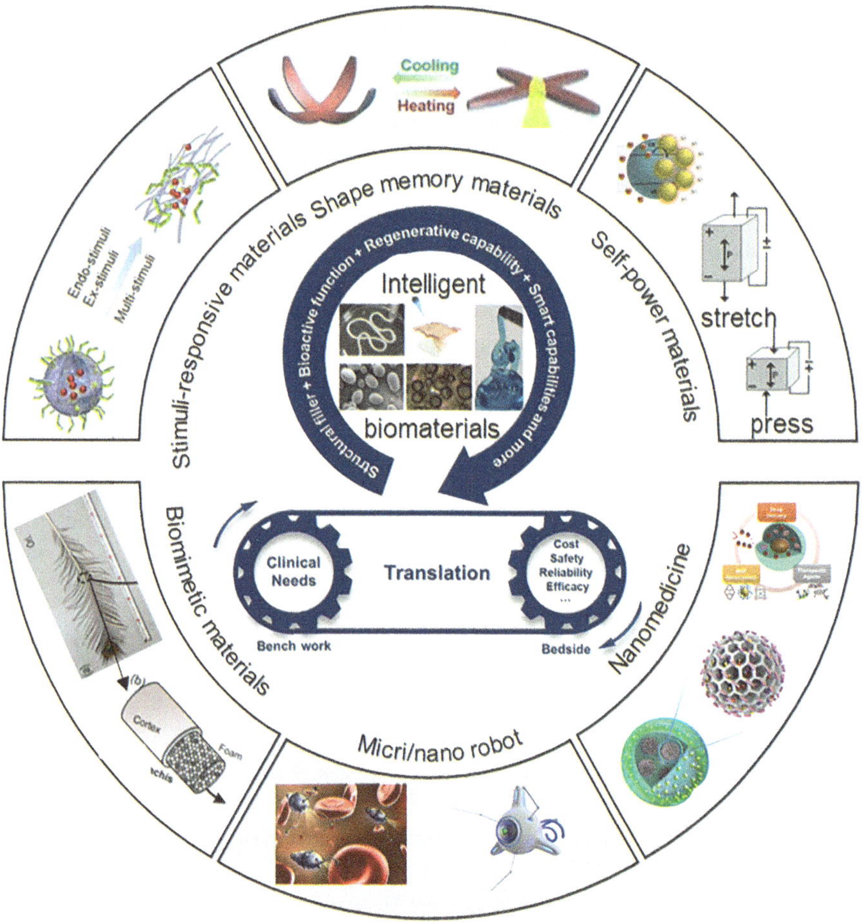

Fig. 12.1 A summary of intelligent biomaterials including shape memory biomaterials, self-powered biomaterials, stimuli-responsive biomaterials, interactive biomaterials, biomimetic materials, micro/nanorobots, and their promising potentials to meet increasing clinical needs

intelligent biological materials have been comprehensively studied, and it is expected to solve key scientific and clinical problems such as innovative degradable cardiovascular and cerebrovascular stents, valves, heart occludes, and cardiovascular thrombosis (Shodeinde et al. 2020; Dormont et al. 2018; Wang et al. 2022a).

Some highly sensitive shape memory materials can be used as flexible robotic arms to grasp and transport bioactive drugs (Xia et al. 2021). In addition, the shape memory materials can recover to original shape after transplanted into the human body through the change of shape from small to large, so as to achieve minimally invasive treatment of tiny wounds for patients and reduce the suffering of patients (Wang et al. 2022b). Because of the advantages of their nano-dimension, intelligent nanoparticles are widely used in the treatment of tumors, thrombus ablation and other diseases (Xu et al. 2021). On one hand, intelligent nanoparticles as drug carriers can avoid direct contact between drugs and normal tissues in order to reduce the toxic side effects of drugs. Secondly, nanomaterials rely on active and passive targeting to accumulate to the tumor location for improving the availability of drugs and enhancing the therapeutic effect (Sharma et al. 2022). Thirdly, the functionalized intelligent nanoparticles are able to recognize the tumor microenvironment and achieve precise in-situ release of drugs (Kong et al. 2021). It is most noteworthy that intelligent nanoparticles can integrate different treatment paradigms in one single nanoplatform, and this combination can be selected and targeted according to the characteristics of different diseases. Overall, the multifunction of intelligent nanoparticles with biocompatibility can enhance their circulation ability in the body, actively target the tumor sites for specific drug accumulation, and realize the integration of diagnosis and treatment which has attracted wide attention in the scientific community.

Intelligent hydrogels are 3D cross-linked networks with high water contents, which have excellent biocompatibility and are widely used in biological applications. In terms of tumor treatment, drug-loaded hydrogels can achieve continuous release of drugs, which can not only inhibit the growth of tumors in situ, but also fill the holes after surgery to inhibit tumor recurrences (Pertici et al. 2019). In tissue engineering, smart hydrogels can promote tissue repairing, inhibit bacterial growth, and reduce inflammation (Waters et al. 2018). In bone therapy, the bone cement prepared by smart hydrogel can fill irregular bone defects and promote osteogenic differentiation. In terms of hemostasis, the smart hydrogel has the effect of wet adhesion and can achieve rapid gumming and hemostasis in a liquid environment (Weng et al. 2022). For wearable devices, intelligent hydrogels can sensitively sense changes in human sweat indicators, and even feel changes in human blood sugar to notify patients in time (Li et al. 2023). Therefore, intelligent hydrogels have found wide applications in biological and medical areas and have made great achievements. Micro-nanorobots have become a hot research topic in recent years. They are small in size and can perform complicated tasks in small spaces through biological barriers, and thus have great potentials in disease diagnosis, sensing, microsurgery, targeted drug/cell delivery, thrombus ablation, and wound healing (Soto et al. 2020).

With advances in tissue engineering and bioelectronics, flexible electronic hydrogels that allow for tissue integration, online accurate diagnosis, and simultaneous tissue regeneration are expected to become the next-generation platform for the treatment of myocardial infarction (Chen et al. 2024; Yu et al. 2023). In the physiological environment, boric acid and carboxyl side-chains in the bonded hydrogel patch spontaneously form dynamic covalent borate bond and non-covalent hydrogen bond with polyvinyl alcohol, which induced the formation of the bonded hydrogel patch in situ. The adhesive hydrogel patch has high adhesion toughness, rapid self-healing ability, and enhanced electrochemical properties, contributing to the sensitive perception of cardiac mechanophysiology-mediated micro-deformations, while improving the electrophysiology induced by myocardial fibrosis. The adhesive hydrogel patch platform can monitor the diastolic amplitude and rhythm of infarcted myocardium online, and effectively inhibit ventricular remodeling, promote angiogenesis, and improve electrophysiological function through electrocoupling therapy.

In addition, in order to construct tissue-like prosthesis materials, soft electroactive hydrogels with physiological mechanical modulus, low resistance, and bidirectional stimulation and recording of biological tissue electrophysiological signals are the best candidate biomaterials. Researchers have proposed a neuromuscular system with instantaneous two-way electric conduction injectable tissue implants (Jin et al. 2023). This soft injectable prosthesis consists of biocompatible hydrogels with unique phenyl borate-mediated multiple crosslinks, such as irreversible but freely rearranged biphenyl bonds and reversible coordination bonds of conductive gold nanoparticles formed in situ by cross-coupling. The closed-loop robot-assisted rehabilitation achieved by injecting the prosthesis biomaterial was successfully verified in the early stage of severe muscle injury in rats, and accelerated tissue repair was achieved in the later stage. In summary, intelligent biomaterials are affecting and changing the modality of human medical treatment, and promote the prognosis and recovery quality of patients.

12.1.2 Intelligent Biomaterials in Clinical Translations

New materials for intelligent biomedical applications are revolutionizing the people's healthcare. For example, neurolink is developing a brain-computer interface system that uses a network of biocompatible electrodes implanted directly into the human brain. The system has been approved by the U.S. Food and Drug Administration (FDA) and human trials began in 2023. In addition, multi-functional biomaterials and bio-inks with biocompatibility and natural origins became cutting-edge technologies in 2024. 3D-printed surgical guide plates, bone implants, maxillofacial implants, and non-biological medical devices are common in research, but the application of 3D-printed living cell scaffolds to the clinics is still facing great challenges (Fig. 12.2). Atala et al. systematically reviewed and analyzed the opportunities and challenges of translational 3D-bioprinting (Murphy et al. 2020). In the literatures, key translational considerations of 3D-bioprinting and challenges of developing biocompatible bioinks were discussed.

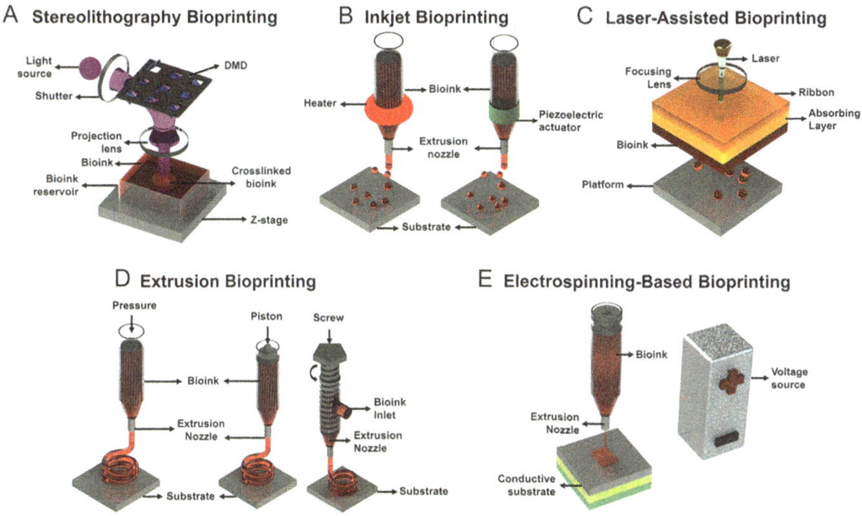

Fig. 12.2 Schematic representations of the different strategies currently used in 3D-bioprinting. (**a**) Stereolithography-based bioprinting based on DMD. (**b**) Inkjet bioprinting using (left) thermal or (right) piezoelectric actuations. (**c**) Laser-assisted bioprinting. (**d**) Extrusion-based bioprinting strategies. From left to right: pressure-based, piston-based, and screw-based cases. (**e**) Electrospinning-based bioprinting. (Reprinted with permission from (Heinrich et al. 2019))

Bioprinted tissues or organs need to reach thresholds of functional and supporting cell types (including stem cells) with biological properties similar to those of natural organs at micro/macro-scales and a complete network of blood vessels including arteries, veins, and capillaries. While some bioprinted cartilage, bone, and skin already satisfied clinical standards, there are still major challenges in printing and applying more complexed tissues. Currently, the printed cartilage tissue has histological and mechanical properties after implantation in the body. In order to better adapt the physiological functions, it is necessary to focus on the gradient printing of growth factors. As for bone healing, bioprinting technology is mainly used to large segmental defects in combination with non-printing traditional products. In addition, bioprinting is also difficult to produce bone tissues that combine formation and function. Bioprinting is only used to precisely control cells and materials to achieve in-situ skin repairing presently, but the existing techniques are still unable to fully simulate the morphology, physicochemical and physiological properties of the skin, including promoting and regulating the normal development of hair follicles, pigmentation, epidermal formation, and maturation. To achieve a breakthrough in realizing multi-scale, multi-component tissue/organ bioprinting, further research needs to be done in isolating and expanding primary and stem cell populations, developing intelligent biomaterials, and integrating complementary bioprinting technologies.

Liposomes, closed bilayer phospholipid vesicles, are ideal delivery drug systems that have been firstly applied clinically (Bulbake et al. 2017). In terms of clinical transformation of intelligent nanoparticles, Doxil liposomes were designed with doxorubicin and approved by FDA to become the first nanodrug in the clinic application in 1995 for ovarian and metastatic breast cancer. Since then, a variety of nanomedicine entered into clinical trials, some of which have been approved for clinical treatment of tumors. There are already 15 nanomaterials approved for cancer treatments by regulatory agencies such as the US FDA or the European Medicines Agency, and more than 50 nanomaterials are currently in clinical trials. At present, anti-tumor nanomaterials can be roughly divided into liposomes, polymer micelles, nanoparticles, polymer-drug couplings, and inorganic nanoparticles according to their chemical compositions. Among the clinically approved nanomaterials, there are 10 kinds of liposomes, 2 kinds of polymer micelles, 2 kinds of nanoparticles, and 1 kind of inorganic nanoparticles under investigation.

Intelligent hydrogels have been widely used in daily life and in the medical field. For example, they can be used as slow-release drug carriers, tissue engineering materials, embolic microspheres, skin wound dressings, and surgical anti-sticking agents. The raw materials for synthesis with natural products exhibited great potentials in biological application. The raw materials include sodium alginate, hyaluronic acid, chitosan, cellulose and modified starch, and small molecular monomers. The advantages of hydrogels prepared from natural materials are excellent biocompatibility and degradability, low cost, and high added value. The innovation of using modern advanced synthesis methods is applied with nanomaterials preparation technology, and the obtained hydrogel products have excellent environmental responsiveness and appropriate mechanical and thermal properties.

Overall, the clinical applications of biomedical materials include trauma biomaterials, 3D-printing biomaterials, cardiovascular biomaterials and implants, bone repairing and regeneration biomaterials, cartilage repairing and regeneration materials, epithelial replacement biomaterials, electrostatic spinning materials, bioelectronics and biosensors, tissue engineering and stem cells, etc. Recently, the clinical development of stem cells attracted great attention, and the emergence of cell engineering promotes a new generation of stem cell-based therapies, greatly expanding their therapeutic application. The next-generation stem cells are used as delivery systems to improve the delivery of drugs and oncolytic viruses to refractory tumors with the properties of angiogenesis, neurotrophic, and anti-inflammatory molecules to accelerate the repair of damaged or diseased tissues. In addition, gene therapy and gene editing techniques are being used to create stem cell derivatives that are more functional, specific, and responsive than natural cells. The approved products in the market combining intelligent biomaterials with cytokines, vaccines, tumor-directed monoclonal antibodies, and immune checkpoint blockers continue to grow. One promising strategy is to use CAR-T cells in combination with vaccines that boost the immune response to fight solid tumors. The immune enhancement helps the body produce more host T cells that can target other tumor antigens that CAR-T cells cannot kill.

12.2 Next-Generation Intelligent Biomaterials

12.2.1 Challenges

Although intelligent biomaterials are rapidly developing, only a few of them can be translated for clinical uses. Next-generation intelligent biomaterials rely on the innovation of advanced composite biomaterials, cutting-edge fabrication strategies as well as innovative applications to meet the requirements for convenient, portable, and remote disease monitoring, diagnosis, and precision treatments. Besides that, new emerging fields such as organoids, lab-on-a-chip, nerve engineering, etc. should be fully considered to expand the scope of intelligent biomaterials for multiple technical integration. Regenerated biomaterials, for example, are often difficult to use in clinical practices because of its inconvenient traits, shapes, and properties not meeting the clinical needs. There also exist challenges in establishing contact and communication between clinical and material sciences experts. Besides that, the sources of intelligent biomaterials are limited, and standards are not uniform which make it difficult to closely combine clinical applications with these intelligent materials. For example, 4D-printed shape memory biomaterials can be fabricated for varied treatment plans, but there is a huge gap between clinical practices and material manufacturing for total personalization and customization (Fig. 12.3). Therefore, clinical applications of 4D-printing face the following problems of (1) 4D-printing techniques are time-consuming and not mature enough, and biomaterial types are not biocompatible; (2) Printing accuracy is not stable, and the corresponding technology needs to be further improved to afford high-precision medical models; (3) The performance of printing biomaterial needs to be optimized since the mechanical strength and biodegradability of these materials cannot meet the requirements for medical implanted devices. For these reasons, it is necessary to develop multi-incentive responsive composite biomaterials in order that 4D-printing of intelligent biomaterials makes breakthroughs to meet the clinical requirements with excellent biological performances and mechanical features.

12.2.2 Opportunities

The development of intelligent biomaterials affords us a new vision to look into upcoming trends of modern medicine, and therefore drives us to contribute to medical practices. Next-generation intelligent biomaterials from piezoelectric biomaterials to interactive biomaterials are expanding the scope of their biomedical and clinical applications, and more importantly they are integrating brand-new technologies for advanced paradigms for human disease treatments. On one hand, artificial intelligence and innovative algorithms enable modern medicine digitalization and automation. On the other hand, the integration of biomaterials, medical devices, and physical fields accelerated the evolution and upgrading on the basis of

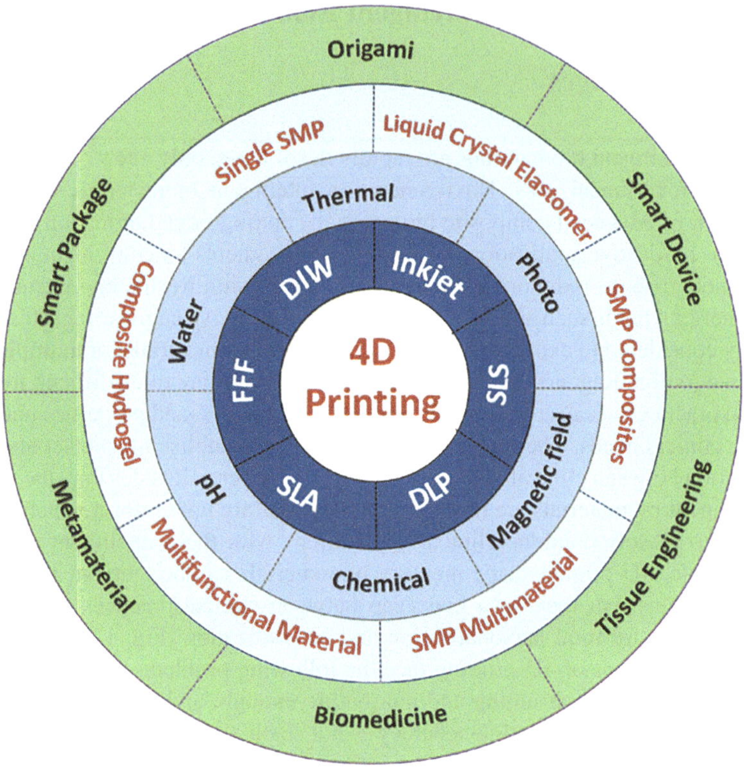

Fig. 12.3 4D-printing, referred to 3D-printing+time, enables shape or function evolution by using different biomaterials under the internal stimuli such as the self-folding of a 4D-printed flat sheet into a box. The elements and the categories involved in 4D-printing are shown. (1) 3D-printing technology: FFF, DIW, DLP, SLS, inkjet, and SLA. (2) The stimulus for 4D-printing: thermal, photo, water, pH, chemical, and magnetic field. (3) Material systems for 4D printing: single SMP, liquid crystal elastomer, composite hydrogel, SMP composites, SMP multi-material, and other multifunctional material. (4) Applications for 4D printing: origami, smart device, smart package, metamaterial, tissue engineering, and biomedicine. (Reprinted with permission from (Kuang et al. 2019))

drug-device combination (DDC). For instance, the concept of "organ chip" integrates chemistry and bioengineering techniques, and it was listed as one of the "Top 10 Emerging Technologies" by the Davos Forum in 2016. The organ chip is a multichannel 3D-microfluidic cell culture chip, which simulates the in vitro construction and functions of different tissues/organs in human body. Most of the organ chip models only culture one cell line, and mainly use patient-derived stem cells to realize the engineered construction of induced pluripotent stem cell (iPSC)-derived organ models, so as to realize individualized disease prediction, drug efficacy evaluation, toxicological evaluation, and prognostic analysis (Fig. 12.4) (Liu et al. 2019).

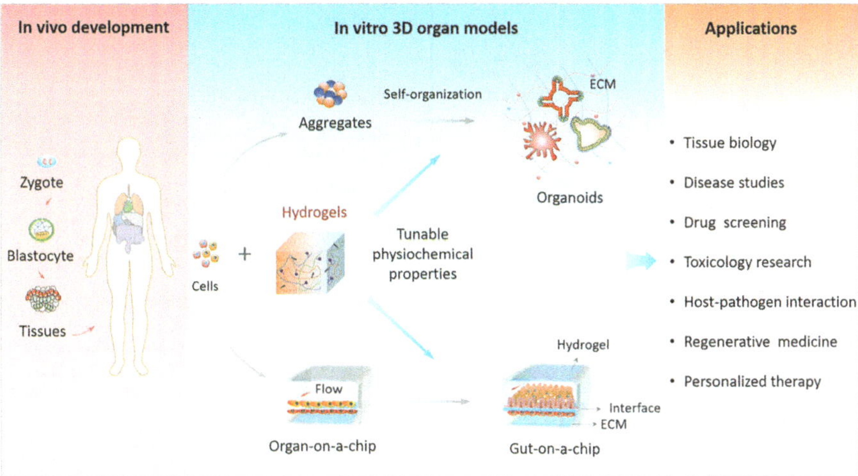

Fig. 12.4 Illustration of hydrogels in organoids and organ-on-a-chip applications. In vivo organs originate from a zygote followed by blastocyte stage, and eventually give rise to a mature organism. Organoids are 3D multicellular clusters derived from stem cells, including human ASCs (isolated from mature tissues), ESCs (isolated from blastocyte), and iPSCs (reprogrammed from adult tissues), undergoing a process of self-organization of resembled native tissues. Organ-on-a-chip is a bioengineered microfluidic cell culture device resided with diverse types of human living cells that can recapitulate the functional units of human organs. These two model systems can be used to build in vitro 3D tissue/organ models with physiological relevance. Engineered hydrogels with well-defined and tunable physiochemical properties can be incorporated into organoids and organ-on-a-chip systems for building high-fidelity 3D tissue/organ models, thus facilitating their potential applications in tissue biology, disease modeling, drug screening, toxicology research, host-pathogen interaction, regenerative medicine, and personalized therapy. (Reprinted with permission from (Liu et al. 2019))

The organ chip can be used as a vehicle for the reconstruction of human organs, not only as a predictor of drug responses, but also to reduce the time-span of animal experiments. Scientists speculate that the final form of human organ chips will have more than 10 organ types, including liver, intestine, heart, kidney, brain, lungs as well as the reproductive immune system, blood vessels, and skins (Takebe et al. 2017; Wu et al. 2023).

The regeneration of articular cartilage has always been a hot topic in the field of bone repair. However, due to its non-vascular and nearly cell-free characteristics, the regeneration ability of cartilage tissues is limited to a certain extent. As shown in Fig. 12.5, smart biomaterials were used to regulate cell behavior and provide an excellent microenvironment for tissue repair, thereby inducing joint cartilage regeneration. This process was modulated by responding to exogenous/endogenous stimuli, customizing material structure and function similar to natural cartilage, or providing physiochemical and physical signaling factors to control the release of drug/bioactive factors (Takebe et al. 2017; Wu et al. 2023; Gu et al. 2023).

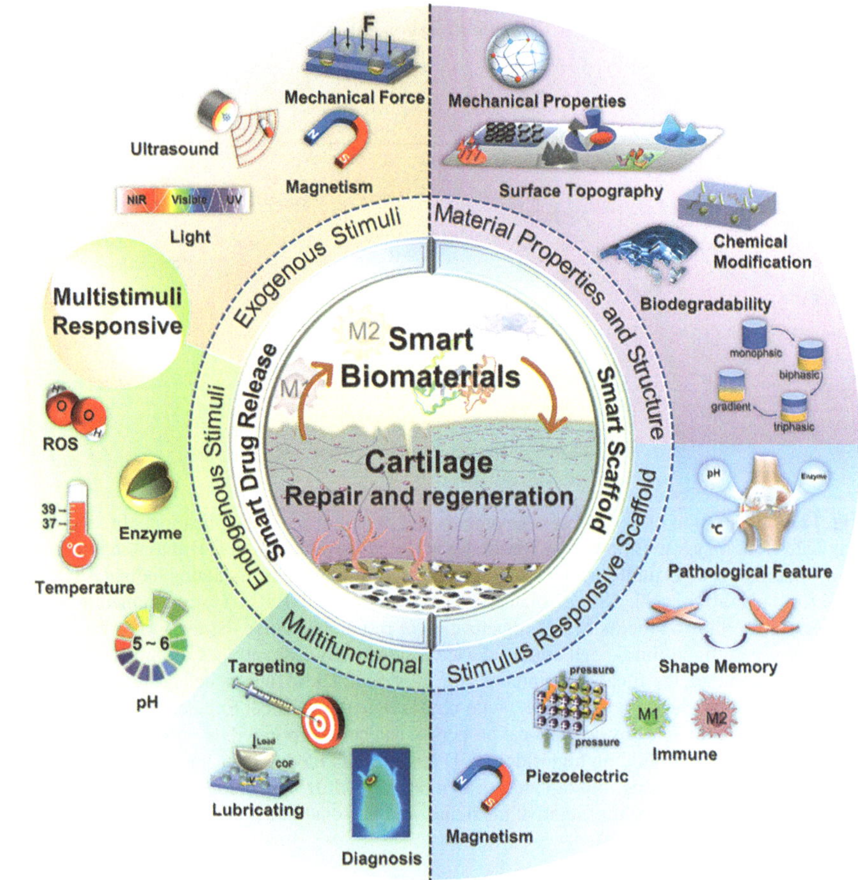

Fig. 12.5 A schematic illustration of smart biomaterials for articular cartilage repairing and regeneration. (Reprinted with permission from (Gu et al. 2023))

12.3 Conclusions and Prospect

How to apply intelligent biomaterials to clinical applications has received a lot of attention. For drug delivery, the treatment efficiency is directly related to the drug accumulation, which requires the targeted release of therapeutic drugs. For diagnosis, intelligent biomaterials enable non-invasive or minimally invasive real-time monitoring and imaging. In tissue engineering and regenerative medicine, intelligent biomaterials should be able to interact with cells and undergo the growth of engineered tissues with specific formulations, scaffolds, or devices. For example, shape memory polymer is a kind of intelligent biomaterial which transforms itself from temporary shape to initial shape under external stimulus to complete a shape memory cycle (Ze et al. 2020). At the same time, according to different shape memory mechanisms, shape memory polymers also have multi-shape memory effects including reversible shape memory effects. Shape memory polymer has the characteristics of lightweight, strong recovery performance, mild recovery conditions, biodegradability, and non-biotoxicity. They have shown great potentials and have been used in aerospace, additive manufacturing, clothing materials such as shape memory sutures, dental orthotics, aneurysm occludes, etc. However, their structure is always simply linear and hardly satisfy the needs in heart stents, bone stents, and tracheal stents. In this regard, 4D printing of shape memory polymers as a novel fabrication technique has achieved a series of remarkable results to produce com plex structures with excellent properties. Magnetic Fe_3O_4 nanoparticles can be added to the shape memory polymers to prepare shape memory composites that can be deformable by magnetic drives (Yue et al. 2021). Experiments showed that the spiral scaffold structure based on this material could be independently developed in a magnetic field within 10 seconds. The shape memory composites were capable to self-deploy vascular stents in non-contact conditions. Researchers used a 3D light curing printer and printed out shape memory scaffolds by epoxidized acrylic materials, and the printed biomaterials proved efficient to support the growth of human bone marrow mesenchymal stem cells. It also showed that the stents could be fixed into a temporary shape at −18 °C, and fully returned to its original shape at 37 °C. The stent can be implanted into the heart through a minimally invasive operation, and its mechanical properties are comparable to those conventional Nitinol stents currently used for heart valve implantations.

Osteoarthritis is a global epidemic that imposes a significant burden on individuals and healthcare systems. As a cell-free scaffold for cartilage regeneration in vivo, intelligent shape memory biomaterials have shown ultra-durable mechanical properties and controlled drug release behavior, which validated their availability for minimally invasive surgery to achieve cartilage regeneration in arthritis patients. In addition, intelligent hydrogels may be used to reduce inflammatory responses and regulate oxidative stresses, thereby establishing a microenvironment conducive to wound healing. These gels can load Kartogenin drug and deposit continuously at the joint, which promotes the migration of bone marrow mesenchymal stem cells to the hydrogel scaffolds and induces chondrocyte differentiation for full-thickness cartilage regeneration in vivo (Yang et al. 2023). Piezoelectric biomaterials are able to self-generate electricity under mechanical stresses such as joint loading or ultrasonic pressure, which may be combined to regulate the microenvironment at the joint and improve the osteogenic treatment. On this basis, an injectable and biodegradable piezoelectric hydrogel was developed and injected into the joint, and the local electrical stimulation under ultrasonic activation helped the cartilage healing without any harms to the body (Vinikoor et al. 2023).

The development of magnetically driven micro/nanorobots opens up new avenues for multifunctional biomedical applications. The magnetic spiral miniature swimmer is capable of generating spiral motion in response to rotating magnetic fields, and its self-actuated mechanism and controllable navigation make it suitable for a variety of environments and applications. In addition, the wireless control microtransporter of archimedes screw pumping mechanism is also applied to the spatio-temporal control of particles and magnetic nanospirals in microfluidic channels. Light-triggered microrobots are designed to serve as miniature containers for material transport, using light and heat to drive convection to load and release cargo (Palagi and Fischer 2018).

In addition, the medical electronics provides a forward-looking solution for human health monitoring and disease diagnosis (Wang et al. 2023). A set of wireless charging and energy storage module (zinc ion hybrid supercapacitors) was proposed as one of the flexible power systems (Sheng et al. 2023). It showed the advantages of high power density and long life of supercapacitors, and further made use of electromagnetic coupling wireless charging technology to make up for the application bottleneck of low energy density. Specifically, through the construction of heterojunction and the use of gel electrolyte ion limiting effect, the synergistic effect gives zinc ion hybrid supercapacitors excellent energy storage performance. The degradation experiments and cytotoxicity evaluation also showed that the electrode materials and degraded products had good biocompatibility and biodegradability. Thus, by integration with wireless charging module and wireless energy transmission/storage system, the rectified voltage was not only used to charge the energy storage unit, but also directly supplied power to external devices. The application of drug release based on electric field driving mechanism further confirmed its energy supply potential, providing a new solution to solve the energy supply bottleneck faced by implantable medical electronic devices (Fig. 12.6) (Guo et al. 2023).

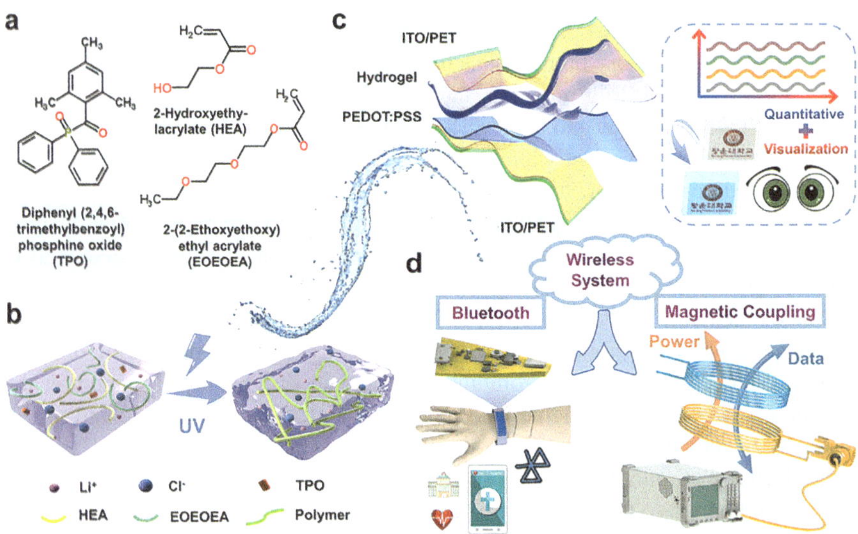

Fig. 12.6 (**a**) Chemical structure of three critical constituents of the polymer matrix. (**b**) Schematic depiction of synthesis of the photocurable ionic hydrogel. (**c**) Overall structure of the wearable device capable of quantitative and visual analysis of pressure. (**d**) Illustration of the wireless integrated systems composed of wearable devices with the assistance of Bluetooth and magnetic coupling. (Reprinted with permission from (Guo et al. 2023))

References

Bulbake U, Doppalapudi S, Kommineni N, Khan W (2017) Liposomal formulations in clinical use: an updated review. Pharmaceutics 9(2):12

Chen J, Liu F, Abdiryim T, Liu X (2024) An overview of conductive composite hydrogels for flexible electronic devices. Adv Compos Hybrid Mater 7(2):35

Dormont F, Varna M, Couvreur P (2018) Nanoplumbers: biomaterials to fight cardiovascular diseases. Mater Today 21(2):122–143

Gu Z, Wang J, Fu Y, Pan H, He H, Gan Q, Liu C (2023) Smart biomaterials for articular cartilage repair and regeneration. Adv Funct Mater 33(10):2212561

Guo Y, Yin F, Li Y, Shen G, Lee J-C (2023) Incorporating wireless strategies to wearable devices enabled by a photocurable hydrogel for monitoring pressure information. Adv Mater 35(29):2300855

Heinrich MA, Liu W, Jimenez A, Yang J, Akpek A, Liu X, Pi Q, Mu X, Hu N, Schiffelers RM, Prakash J, Xie J, Zhang YS (2019) 3D bioprinting: from benches to translational applications. Small 15(23):1805510

Jin S, Choi H, Seong D, You C-L, Kang J-S, Rho S, Lee WB, Son D, Shin M (2023) Injectable tissue prosthesis for instantaneous closed-loop rehabilitation. Nature 623(7985):58–65

Kong H, Chu Q, Fang C, Cao G, Han G, Li X (2021) Cu–Ferrocene-functionalized CaO2 nanoparticles to enable tumor-specific synergistic therapy with GSH depletion and calcium overload. Adv Sci 8(14):2100241

Kuang X, Roach DJ, Wu J, Hamel CM, Ding Z, Wang T, Dunn ML, Qi HJ (2019) Advances in 4D printing: materials and applications. Adv Funct Mater 29(2):1805290

Li T, Wei H, Zhang Y, Wan T, Cui D, Zhao S, Zhang T, Ji Y, Algadi H, Guo Z, Chu L, Cheng B (2023) Sodium alginate reinforced polyacrylamide/xanthan gum double network ionic hydrogels for stress sensing and self-powered wearable device applications. Carbohydr Polym 309:120678

Liu H, Wang Y, Cui K, Guo Y, Zhang X, Qin J (2019) Advances in hydrogels in organoids and organs-on-a-chip. Adv Mater 31(50):1902042

Murphy SV, De Coppi P, Atala A (2020) Opportunities and challenges of translational 3D bioprinting. Nature Biomed Eng 4(4):370–380

Palagi S, Fischer P (2018) Bioinspired microrobots. Nat Rev Mater 3(6):113–124

Pertici V, Pin-Barre C, Rivera C, Pellegrino C, Laurin J, Gigmes D, Trimaille T (2019) Degradable and injectable hydrogel for drug delivery in soft tissues. Biomacromolecules 20(1):149–163

Sharma N, Bietar K, Stochaj U (2022) Targeting nanoparticles to malignant tumors. Biochimica et Biophysica Acta (BBA) Rev Cancer 1877(3):188703

Sheng H, Jiang L, Wang Q, Zhang Z, Lv Y, Ma H, Bi H, Yuan J, Shao M, Li F, Li W, Xie E, Liu Y, Xie Z, Wang J, Yu C, Lan W (2023) A soft implantable energy supply system that integrates wireless charging and biodegradable Zn-ion hybrid supercapacitors. Sci Adv 9(46):eadh8083

Shodeinde AB, Murphy AC, Oldenkamp HF, Potdar AS, Ludolph CM, Peppas NA (2020) Recent advances in smart biomaterials for the detection and treatment of autoimmune diseases. Adv Funct Mater 30(37):1909556

Soto F, Wang J, Ahmed R, Demirci U (2020) Medical micro/nanorobots in precision medicine. Adv Sci 7(21):2002203

Takebe T, Zhang B, Radisic M (2017) Synergistic engineering: organoids meet organs-on-a-chip. Cell Stem Cell 21(3):297–300

Vinikoor T, Dzidotor GK, Le TT, Liu Y, Kan H-M, Barui S, Chorsi MT, Curry EJ, Reinhardt E, Wang H, Singh P, Merriman MA, D'Orio E, Park J, Xiao S, Chapman JH, Lin F, Truong C-S, Prasadh S, Chuba L, Killoh S, Lee S-W, Wu Q, Chidambaram RM, Lo KWH, Laurencin CT, Nguyen TD (2023) Injectable and biodegradable piezoelectric hydrogel for osteoarthritis treatment. Nat Commun 14(1):6257

Wang H, Zhang L-M (2024) Intelligent biobased hydrogels for diabetic wound healing: a review. Chem Eng J 484:149493

Wang Y, Li G, Yang L, Luo R, Guo G (2022a) Development of innovative biomaterials and devices for the treatment of cardiovascular diseases. Adv Mater 34(46):2201971

Wang X, He Y, Liu Y, Leng J (2022b) Advances in shape memory polymers: remote actuation, multi-stimuli control, 4D printing and prospective applications. Mater Sci Eng R Reports 151:100702

Wang Z, Ruan R, Lin G, He S, Liu Y, Gong C, Xiao P, Chen J, Lu Y, Cao Y, Lin X, Zhang J (2023) Smart healthcare sensor with dual autonomous self-healing, repeatable skin adhesion/detachment, and hydrophilic/hydrophobic Janus surface for aquatic ambulatory electrophysiological monitoring. Chem Eng J 468:143603

Waters R, Alam P, Pacelli S, Chakravarti AR, Ahmed RPH, Paul A (2018) Stem cell-inspired secretome-rich injectable hydrogel to repair injured cardiac tissue. Acta Biomater 69:95–106

Weng H, Jia W, Li M, Chen Z (2022) New injectable chitosan-hyaluronic acid based hydrogels for hemostasis and wound healing. Carbohydr Polym 294:119767

Wu L, Ai Y, Xie R, Xiong J, Wang Y, Liang Q (2023) Organoids/organs-on-a-chip: new frontiers of intestinal pathophysiological models. Lab Chip 23(5):1192–1212

Wu J, Yun Z, Song W, Yu T, Xue W, Liu Q, Sun X (2024) Highly oriented hydrogels for tissue regeneration: design strategies, cellular mechanisms, and biomedical applications. Theranostics 14(5):1982–2035

Xia Y, He Y, Zhang F, Liu Y, Leng J (2021) A review of shape memory polymers and composites: mechanisms, materials, and applications. Adv Mater 33(6):2000713

Xu X, Han C, Zhang C, Yan D, Ren C, Kong L (2021) Intelligent phototriggered nanoparticles induce a domino effect for multimodal tumor therapy. Theranostics 11(13):6477–6490

Yang Y, Zhao X, Wang S, Zhang Y, Yang A, Cheng Y, Chen X (2023) Ultra-durable cell-free bioactive hydrogel with fast shape memory and on-demand drug release for cartilage regeneration. Nat Commun 14(1):7771

Yu C, Shi M, He S, Yao M, Sun H, Yue Z, Qiu Y, Liu B, Liang L, Zhao Z, Yao F, Zhang H, Li J (2023) Chronological adhesive cardiac patch for synchronous mechanophysiological monitoring and electrocoupling therapy. Nat Commun 14(1):6226

Yue C, Li M, Liu Y, Fang Y, Song Y, Xu M, Li J (2021) Three-dimensional printing of cellulose nanofibers reinforced PHB/PCL/Fe3O4 magneto-responsive shape memory polymer composites with excellent mechanical properties. Addit Manuf 46:102146

Ze Q, Kuang X, Wu S, Wong J, Montgomery SM, Zhang R, Kovitz JM, Yang F, Qi HJ, Zhao R (2020) Shape memory polymers: magnetic shape memory polymers with integrated multifunctional shape manipulation. Adv Mater 32(4):2070025

Zhang K, Wang S, Zhou C, Cheng L, Gao X, Xie X, Sun J, Wang H, Weir MD, Reynolds MA, Zhang N, Bai Y, Xu HHK (2018) Advanced smart biomaterials and constructs for hard tissue engineering and regeneration. Bone Res 6(1):31

The manufacturer's authorised representative in the EU is Springer Nature Customer Service Centre GmbH, Europaplatz 3, 69115 Heidelberg, Germany. If you have any concerns regarding our products, please contact ProductSafety@springernature.com

Printed and bound by CPI Group (UK) Ltd, Croydon, CR0 4YY
30/09/2026
02225300-0018